Editorial Policy

§ 1. Lecture Notes aim to report new developments - quickly, informally, and at a high level. The texts should be reasonably self-contained and rounded off. Thus they may, and often will, present not only results of the author but also related work by other people. Furthermore, the manuscripts should provide sufficient motivation, examples and applications. This clearly distinguishes Lecture Notes manuscripts from journal articles which normally are very concise. Articles intended for a journal but too long to be accepted by most journals, usually do not have this "lecture notes" character. For similar reasons it is unusual for Ph. D. theses to be accepted for the Lecture Notes series.

§ 2. Manuscripts or plans for Lecture Notes volumes should be submitted (preferably in duplicate) either to one of the series editors or to Springer- Verlag, Heidelberg . These proposals are then refereed. A final decision concerning publication can only be made on the basis of the complete manuscript, but a preliminary decision can often be based on partial information: a fairly detailed outline describing the planned contents of each chapter, and an indication of the estimated length, a bibliography, and one or two sample chapters - or a first draft of the manuscript. The editors will try to make the preliminary decision as definite as they can on the basis of the available information.

§ 3. Final manuscripts should preferably be in English. They should contain at least 100 pages of scientific text and should include

- a table of contents;
- an informative introduction, perhaps with some historical remarks: it should be accessible to a reader not particularly familiar with the topic treated;
- a subject index: as a rule this is genuinely helpful for the reader.

Further remarks and relevant addresses at the back of this book.

Lecture Notes in Mathematics 1686

Editors:
A. Dold, Heidelberg
F. Takens, Groningen
B. Teissier, Paris

Subseries:
Institut de Mathématiques, Université de Strasbourg
Adviser: J.-L. Loday

Springer
Berlin
Heidelberg
New York
Barcelona
Budapest
Hong Kong
London
Milan
Paris
Santa Clara
Singapore
Tokyo

J. Azéma M. Émery
M. Ledoux M. Yor (Eds.)

Séminaire de Probabilités XXXII

Springer

Editors

Jacques Azéma
Marc Yor
Laboratoire de Probabilités
Université Pierre et Marie Curie
Tour 56, 3 ème étage
4, Place Jussieu
F-75252 Paris, France

Michel Émery
Institut de Recherche Mathématique Avancée
Université Louis Pasteur
7, rue René Descartes
F-67084 Strasbourg, France

Michel Ledoux
Laboratoire de Statistiques et Probabilités
Université Paul Sabatier
118, route de Narbonne
F-31601 Toulouse Cedex, France

Cataloging-in-Publication Data applied for

Die Deutsche Bibliothek - CIP-Einheitsaufnahme

Séminaire de probabilités .... - Berlin ; Heidelberg ; New York ; ; Barcelona ; Budapest ; Hong Kong ; London ; Milan ; Paris ; Santa Clara ; Singapore ; Tokyo : Springer
ISSN 0720-8766
32 (1998)
(Lecture notes in mathematics ; Vol. 1686)
ISBN 3-540-64376-1 (Berlin ...)

Mathematics Subject Classification (1991): 60GXX, 60HXX, 60JXX

ISSN 0075-8434
ISBN 3-540-64376-1 Springer-Verlag Berlin Heidelberg New York

Typesetting: Camera-ready TEX output by the author
SPIN: 10649814 41/3143-543210 - Printed on acid-free paper

SÉMINAIRE DE PROBABILITÉS XXXII

TABLE DES MATIÈRES

Sous-mesures symétriques sur un ensemble fini

C. Dellacherie A. Iwanik

Introduction

Écartant de notre propos les "sous-mesures pathologiques" de [4], nous entendrons ici par "sous-mesure" une fonction croissante et sous-additive d'ensembles définie sur une tribu, et égale à l'enveloppe supérieure ponctuelle des mesures qu'elle majore. De telles fonctions d'ensemble se rencontrent communément en analyse, et tout particulièrement en théorie du potentiel. Dans cette théorie, il y a une forte analogie entre les opérateurs linéaires du type "potentiel" et les non-linéaires du type "réduite", l'opération "$\vee$" jouant ici le rôle de l'opération "+" là. Aussi était-il tentant de conjecturer qu'on avait, pour un espace de base fini de cardinalité n, un analogue du théorème de Carathéodory sur les enveloppes convexes qui majorerait le nombre minimal de mesures pour engendrer une sous-mesure par une fonction de n à croissance linéaire. Nous montrons ici qu'il n'en est rien. Plus précisément, si on se limite aux sous-mesures symétriques, i.e. dont la valeur sur un ensemble ne dépend que de la cardinalité de celui-ci et qui donc ne peuvent prendre qu'au plus $n+1$ valeurs distinctes, nous montrons que, dans le pire des cas, il faut exactement 2^{n-1} mesures pour engendrer la sous-mesure. Enfin, nous déterminons, toujours dans le cas fini mais de manière analogue à [3], les génératrices extrémales du cône convexe des sous-mesures symétriques : la simplicité de cette caractérisation, fournissant exactement 2^{n-1} éléments extrémaux, contraste avec l'inextricable complexité du cas général (cf. [1, 2]) des sous-mesures (pathologiques ou non).

1 Sous-mesures symétriques

Soient $E = \{0, 1, \ldots, n-1\}$ et $\mathcal{M}_n$ l'ensemble des mesures positives finies sur E. En particulier, la probabilité uniforme λ_A sur une partie non vide A de E appartient à $\mathcal{M}_n$. Notons $\mathcal{S}_n$ l'ensemble des fonctions η, définies sur les parties de E, telles que η soit positive, croissante, sous-additive et $\eta(\emptyset) = 0$. À tout sous-ensemble non vide et borné H de $\mathcal{M}_n$ on associe la fonction

$$I_H(A) = \sup_{\mu \in H} \mu(A).$$

On a $I_H \in \mathcal{S}_n$ et on note $\mathcal{I}_n$ l'ensemble des sous-mesures I_H où H est non vide borné dans $\mathcal{M}_n$.

Nous allons considérer aussi des familles de fonctions symétriques $\eta \in \mathcal{S}_n$, c'est-à-dire celles dont les valeurs $\eta(A)$ ne dependent que du cardinal $|A|$ (cf. [4]). Autrement dit, η symétrique signifie η invariant par rapport à l'action du groupe symétrique sur E. On écrit, respectivement, ${\mathcal{S}_n}^{\mathrm{sym}}$ et ${\mathcal{I}_n}^{\mathrm{sym}}$ pour indiquer qu'il s'agit de fonctions symétriques. Il est facile de voir que toutes ces familles sont des cônes convexes fermés dans l'espace $\mathbf{R}^{2^n}$.

Le lemme suivant est évident :

LEMME — *Soient $B, A_1, \ldots, A_k \subset E$ et $m \in \mathbf{N}$ tels que*

$$\sum_{i=1}^{k} 1_{A_i}(x) = m\, 1_B(x)$$

sur E. Alors $m\mu(B) = \sum_{i=1}^{k} \mu(A_i)$ pour tout $\mu \in \mathcal{M}_n$.

Si $J \in {\mathcal{I}_n}^{\mathrm{sym}}$, on écrit $J(A) = J_k$ quand $|A| = k$. La fonction J est alors complètement caracterisée par la suite $J_1, \ldots, J_n$.

THÉORÈME 1 — *Soit $0 \le x_1 \le x_2 \le \ldots \le x_n$. Il existe $J \in {\mathcal{I}_n}^{\mathrm{sym}}$ tel que $J_k = x_k$ $(k = 1, \ldots, n)$ si et seulement si*

$$\frac{x_1}{1} \ge \frac{x_2}{2} \ge \ldots \ge \frac{x_n}{n}.$$

Démonstration. La condition est nécessaire : J étant la borne supérieure de mesures, on obtient du lemme, pour un B à $k+1$ points,

$$kJ_{k+1} = kJ(B) \le \sum_{i \in B} J(B \setminus \{i\}) = (k+1)J_k.$$

La condition est aussi suffisante. En effet, on note que $kx_l \le x_k$ pour tout $k \le l$. Posons $J = I_H$, où

$$H = \{x_{|A|}\lambda_A : \emptyset \ne A \subset E\}.$$

Pour $|B| = k > 0$ on vérifie que $J(B) = x_k$, donc $J_k = x_k$. En effet,

$$\begin{aligned} J(B) &= \sup\{\mu(B) : \mu \in H\} = \sup\{\mu(B) : \mu \in H,\ |\mathrm{supp}\,\mu| \ge k\} \\ &= \sup\{\mu(B) : \mu \in H,\ B \subset \mathrm{supp}\,\mu\} = \sup\{x_l\, k/l : l \ge k\} = x_k. \end{aligned}$$

2 Mesures engendrant une sous-mesure

Pour tout $n = 1, 2, \dots$, $E = \{0, \dots, n-1\}$ et $J \in \mathcal{I}_n$ posons $s(n, J) = \min\{|H| : J = I_H\}$ et

$$s(n) = \max_{J \in \mathcal{I}_n} s(n, J).$$

On écrit $s^{\mathrm{sym}}(n)$ si $\mathcal{I}_n$ est remplacée par $\mathcal{I}_n{}^{\mathrm{sym}}$. On constate aisément que $s^{\mathrm{sym}} \le s$, $s(1) = s^{\mathrm{sym}}(1) = 1$, $s(2) = s^{\mathrm{sym}}(2) = 2$ et que les fonctions s, s^{sym} sont croissantes. Le résultat suivant implique en particulier que $s(3) = 4$ et que $n^{-1} \log s(n) \to \log 2$.

THÉORÈME 2 — *Pour tout* $n \ge 3$ *on a* $2^{n-1} \le s(n) \le 2^n - n - 1$. *En outre,*

$$s^{\mathrm{sym}}(n) = 2^{n-1}$$

pour tout $n \ge 1$.

Démonstration.

1. Afin de démontrer la minoration choisissons d'abord des nombres γ_k tels que

$$\frac{n}{n+1} \le \gamma_1 < \gamma_2 < \dots < \gamma_{[(n+1)/2]} \le 1$$

et posons $J = I_H$, où

$$H = \{\gamma_k \lambda_A : \emptyset \ne A \subset E,\ k = [(|A|+1)/2]\}.$$

Évidemment $J \in \mathcal{I}_n{}^{\mathrm{sym}}$. Montrons que

$$J(B) = \gamma_l,$$

où $l = [(|B|+1)/2]$, pour toute partie non vide B de E. On a clairement $J(B) \ge \gamma_l \lambda_B(B) = \gamma_l$. D'autre part, $J(B) = \gamma_k \lambda_A(B)$ pour un certain A avec $|A| = 2k-1$. Lorsque $k \le l$, nous avons $J(B) \le \gamma_l$, tandis que pour $k > l$ on écrit:

$$J(B) = \gamma_k \, |B \cap A| \, / \, |A| \le \gamma_k \frac{2l}{2k-1} \le \frac{2l}{2k-1}$$

$$\le 1 - \frac{1}{2l+1} \le 1 - \frac{1}{n+1} \le \gamma_l.$$

Supposons alors que $J = I_L$ et démontrons que $|L| \ge 2^{n-1}$. Comme on peut supposer que L est fini, il existe des mesures $\mu_0, \dots, \mu_{n-1}$ dans L telles que $\mu_i(\{i\}) = \gamma_1$. Pour tout $j \ne i$ on obtient:

$$\mu_i(\{j\}) = \mu_i(\{i,j\}) - \mu_i(\{i\}) \le J(\{i,j\}) - \gamma_1 = 0,$$

donc μ_i est concentrée sur i et égale à $\gamma_1 \lambda_{\{i\}}$. De même, il existe, pour tout A de 3 points, une mesure $\mu_A \in L$ satisfaisant $\mu_A(A) = \gamma_2$. La mesure μ_A est concentrée sur A, car si $j \notin A$ alors $[(|A \cup \{j\}| + 1)/2] = 2$, donc

$$\mu_A(A \cup \{j\}) \le J(A \cup \{j\}) = \gamma_2 = \mu_A(A).$$

On constate que les mesures μ_A, où $|A| = 1, 3$, sont distinctes entre elles. De même façon on obtient pour toute partie impaire $B \subset E$ une mesure $\mu_B \in L$ telle que μ_B est concentrée sur B et $\mu_B(B) = \gamma_{(|B|+1)/2}$. Notons que μ_B n'est concentrée sur aucune partie propre C de B, car on a

$$\mu_B(C) \leq J(C) = \gamma_{[(|C|+1)/2]} < \mu_B(B).$$

Par conséquent,

$$|L| \geq \binom{n}{1} + \binom{n}{3} + \ldots = 2^{n-1}.$$

2. Pour obtenir la majoration de $s(n)$ supposons $J \in \mathcal{I}_n$. Par compacité, il existe, pour toute partie non vide A de E, une mesure $\mu_A \leq J$ telle que $J(A) = \mu_A(A)$. Donc $J = I_H$, où $H = \{\mu_A : \emptyset \neq A \subset E\}$. Ceci démontre que $s(n) \leq 2^n - 1$. Or, on peut choisir les mesures $\mu_{\{i,i'\}}$, où $i = 0, 1, \ldots, n-1$ et $i' = i + 1 \mod n$, de telle manière que $\mu_{\{i,i'\}}(\{i\}) = J(\{i\})$, $\mu_{\{i,i'\}}(\{i'\}) = J(\{i,i'\}) - J(\{i\})$ et $\mu_{\{i,i'\}}(\{k\}) = 0$ si $k \neq i, i'$. Comme J est sous-additive, on a bien $\mu_{\{i,i'\}} \leq J$ et les mesures $\mu_{\{i,i'\}}$ réalisent les valeurs de J en $\{i\}$ et en $\{i, i'\}$ à la fois. Par conséquent, $J = I_L$, où $L = \{\mu_A : |A| \geq 2\}$, d'où la majoration souhaitée.

3. La majoration de $s^{\mathrm{sym}}(n)$ exige d'autres économies sur le choix de H. Soit $J \in \mathcal{I}_n{}^{\mathrm{sym}}$. Comme $J(A)$, où $\emptyset \neq A \subset E$, ne depend que de $|A|$, on peut bien poser $J(A) = J_k$, où $k = |A| = 1, \ldots, n$.

Pour trouver un assez petit H tel que notre J soit égale à I_H on définit des mesures μ_B $(0 \in B \subset E)$ de telle manière que $\mu_B \leq J$, $\mu_B(B) = J_{j+1}$ et $\mu_B(B \setminus \{0\}) = J_j$, où $j = |B| - 1$, comme suit

$$\mu_B = J_j \lambda_{B \setminus \{0\}} + (J_{j+1} - J_j)\lambda_{\{0\}}.$$

La vérification de la relation $\mu_B \leq J$ est facile grâce au Théorème 1 (notons qu'il suffit de vérifier l'inégalité $\mu_B(C) \leq J(C)$ pour $C \subset B$; dans le cas où $0 \in C$ et $j > 0$ utilisons la relation $J_{j+1} - J_j \leq J_j/j$). Ainsi les 2^{n-1} mesures μ_B réalisent les valeurs de J sur $\{B : 0 \in B\}$ aussi bien que sur $\{A : 0 \notin A\}$, ce qui achève de démontrer le théorème.

3 Points extrémaux

Notons $\mathcal{P}_n^{\mathrm{sym}}$ l'ensemble des sous-mesures J dans $\mathcal{I}_n^{\mathrm{sym}}$ telles que $J(E) = 1$. Cet ensemble est convexe et compact dans $\mathbf{R}^{2^n}$ donc tout élément de $\mathcal{P}_n^{\mathrm{sym}}$ s'écrit comme une combinaison convexe de points extrémaux. Nous allons caractériser et compter les points extrémaux de $\mathcal{P}_n^{\mathrm{sym}}$.

Toute sous-mesure symétrique $J \in \mathcal{I}_n^{\mathrm{sym}}$ est déterminée par sa fonction de répartition $x_k = J(\{0, \ldots, k-1\})$ (cette idée a été utilisée dans [3]). On sait du théorème 1 que les fonctions de répartition sont exactement les suites croissantes de nombres positifs $(x_1, \ldots, x_n)$ telles que la suite x_k/k décroisse. Il est clair que la correspondance entre les sous-mesures dans $\mathcal{P}_n^{\mathrm{sym}}$ et leurs fonctions de répartition établit un isomorphisme affine de $\mathcal{P}_n^{\mathrm{sym}}$ et l'ensemble convexe $\mathcal{F}_n$ des suites de nombres positifs caractérisées par les trois conditions suivantes :

a) $x_n = 1$,

b) x_k est croissante,

c) les pentes des vecteurs (k, x_k) décroissent.

THÉORÈME 3 — *Soit x_k la fonction de répartition d'une sous-mesure J de $\mathcal{P}_n^{\mathrm{sym}}$. Alors J est un point extrémal de $\mathcal{P}_n^{\mathrm{sym}}$ si et seulement si*

$$x_{k-1} = x_k \quad \text{ou} \quad \frac{k-1}{k} x_k$$

pour tout $k = 2, \ldots, n$. Le nombre des points extrémaux est égal à 2^{n-1}.

Démonstration.

Soit $k \geq 2$. Il résulte des conditions b) et c) que

$$\frac{k-1}{k} x_k \leq x_{k-1} \leq x_k .$$

Si les deux inégalités sont strictes alors on s'aperçoit qu'il existe $\epsilon > 0$ tel que les suites x^+, x^-, où

$$x_t^{\pm} = (1 \pm \epsilon) x_t$$

pour $t = 0, 1, \ldots, k-1$ et $x_t^{\pm} = x_t$ ailleurs, appartiennent à $\mathcal{F}_n$. Or, $x_t = (x_t^+ + x_t^-)/2$ d'où ni x ni J ne sont extrémales. D'autre part il est évident que si une des inégalités est toujours une égalité (pour $k = n, n-1, \ldots, 2$), alors x est un point extrémal dans $\mathcal{F}_n$ donc J en est un dans $\mathcal{P}_n^{\mathrm{sym}}$.

Comme pour tout $k = n, n-1, \ldots, 2$ il y a un choix entre les deux égalités, le nombre total des choix monte à 2^{n-1}, ce qui donne le résultat souhaité.

Références

[1] M. Bąk, *On the extremal submeasures on a 3-set*, Discuss. Math. **10** (1990), 41–45

[2] J. Ball, S. Eigen, J. Lindhe, *Extreme probability submeasures on 3 points*, Real Analysis Exchange **20**(1) (1994/95), 94–101

[3] J. B. Kadane, L. Wasserman, *Symmetric coherent Choquet capacities*, prépublication

[4] F. Topsøe, *Some remarks concerning pathological submeasures*, Math. Scand. **38** (1976), 159–166

C. Dellacherie, Mathématiques, URA CNRS 1378,
Université de Rouen, 76821 Mt-St-Aignan
e-mail : Claude.Dellacherie@univ-rouen.fr

A. Iwanik, Instytut Matematyki,
Politechnika Wrocławska, 50-370 Wrocław, Pologne
e-mail : iwanik@im.pwr.wroc.pl

Sur une inégalité de Sobolev logarithmique pour une diffusion unidimensionnelle

Mireille CAPITAINE

Résumé

Par une méthode élémentaire de calcul stochastique, nous établissons, dans cet article, une inégalité de Sobolev logarithmique pour une fonction cylindrique d'une diffusion unidimensionnelle, solution d'une équation différentielle stochastique. L'idée est de considérer cette diffusion comme une fonctionnelle du mouvement Brownien intervenant dans l'équation différentielle et de lui appliquer l'inégalité de Sobolev logarithmique connue pour le mouvement Brownien sur $\mathbb{R}$. Pour cela, nous déterminons dans un premier temps, une expression "adéquate" de la dérivée au sens de Malliavin des marginales de la diffusion.

1 Introduction et notations

E. P.Hsu [8] d'une part, et S. Aida et D. Elworthy [1] d'autre part, ont démontré des inégalités de Sobolev logarithmiques pour le mouvement Brownien sur une variété Riemannienne; le premier a utilisé un procédé d'itération et, pour ce faire, le procédé de couplage de Kendall; les seconds ont plongé isométriquement la variété dans un espace euclidien $\mathbb{R}^d$ et ont appliqué l'inégalité de Sobolev logarithmique pour le mouvement Brownien sur $\mathbb{R}^d$. F.Y Wang [9] a étendu le résultat de E. P. Hsu à certaines diffusions sur les variétés. Dans cet article, nous obtenons une inégalité de Sobolev logarithmique pour une diffusion sur $\mathbb{R}$, par une méthode élémentaire de calcul stochastique, ne faisant pas appel à la géométrie différentielle.

Soit $\mathcal{C}_0([0,1];\mathbb{R})$ l'espace des fonctions continues sur $[0,1]$, à valeurs dans $\mathbb{R}$ s'annulant en zéro, et soit $\mathcal{B}$ la tribu Borélienne de $(\mathcal{C}_0([0,1];\mathbb{R}), \|\cdot\|_\infty)$ où $\|\cdot\|_\infty$ désigne la norme de la convergence uniforme sur $[0,1]$; nous notons P la mesure de Wiener définie sur $\mathcal{B}$ et $\omega = \{\omega_t, t \in [0,1]\}$ le mouvement Brownien canonique sur l'espace de Wiener $(\mathcal{C}_0([0,1];\mathbb{R}), \mathcal{B}, P)$. Soit $X = \{X_t, t \in [0,1]\}$, le processus de diffusion solution de l'équation différentielle stochastique

$$dX_t = \sigma(X_t)\, d\omega_t + b(X_t)\, dt,\ X_0 = x,\ X_t \in \mathbb{R},$$

où x appartient à $\mathbb{R}$, $\sigma : \mathbb{R} \to \mathbb{R}$ et $b : \mathbb{R} \to \mathbb{R}$ sont lipschitziennes, respectivement de classe $\mathcal{C}^2$ et $\mathcal{C}^1$, σ ne s'annulant pas sur $\mathbb{R}$. Nous regardons chaque X_t comme une fonction de ω et appliquons l'inégalité de Sobolev logarithmique pour le mouvement Brownien sur $\mathbb{R}$ (notre approche rejoint celle de S. Aida et D. Elworthy [1] sur ce point). Cette méthode fait apparaître la dérivée au sens de Malliavin de X_t; aussi, dans un premier temps, nous déterminons une expression de cette dérivée, se prêtant à la majoration désirée.

2 Inégalité de Sobolev logarithmique pour la loi de X_t

Dans ce paragraphe, nous établissons l'inégalité de Sobolev logarithmique pour la loi d'une coordonnée de la diffusion X. Cette inégalité résulte déjà des méthodes de semigroupes (opérateur Γ_2) développées par D. Bakry et M. Emery (cf. [3]). Notre approche probabiliste s'appuie déjà sur l'inégalité de Sobolev logarithmique du mouvement Brownien.

Proposition 1 *Soit t dans $[0,1]$. Notons L, le générateur infinitésimal de X, défini par $Lg(x) = \frac{1}{2}\sigma^2(x)g''(x) + b(x)g'(x)$, pour toute fonction $g : \mathbb{R} \to \mathbb{R}$ de classe C^2. Soit $c \geq 0$. Si $\frac{L\sigma - b'\sigma}{\sigma} \geq -c$, alors, pour toute fonction $f : \mathbb{R} \to \mathbb{R}$ de classe C^1 dont la dérivée est bornée,*

$$E\left[f^2(X_t)\log f^2(X_t)\right] - E\left[f^2(X_t)\right]\log E\left[f^2(X_t)\right] \leq \frac{e^{2ct}-1}{c}E\left[\sigma^2(X_t)\{f'(X_t)\}^2\right].$$

Démonstration Appliquons l'inégalité de Sobolev logarithmique pour le mouvement Brownien sur $\mathbb{R}$, en considérant X_t comme une fonction de ω. Il vient

$$E\left[f^2(X_t)\log f^2(X_t)\right] - E\left[f^2(X_t)\right]\log E\left[f^2(X_t)\right] \leq 2E\left[\int_0^1 (f'(X_t)\, D_rX_t)^2\, dr\right]$$

où $D.X_t$ désigne la dérivée au sens de Malliavin de X_t. D_rX_t vérifie une équation différentielle stochastique (cf [7]), et il est immédiat en dimension 1 d'obtenir la forme explicite:

$$\begin{aligned} D_rX_t &= \sigma(X_r)\exp\left[\int_r^t \sigma'(X_s)\,dB_s + \int_r^t \left(b'(X_s) - \frac{1}{2}\sigma'^2(X_s)\right)ds\right] \\ &\quad \textit{si } r \leq t \\ &= 0 \textit{ si } r > t. \end{aligned}$$

Nous allons déterminer une autre expression de $D.X_t$ faisant intervenir le générateur L.

Lemme 1 *Si $0 \leq r \leq t$, alors*

$$D_rX_t = \sigma(X_t)\exp\left[-\int_r^t \frac{L\sigma - b'\sigma}{\sigma}(X_s)\,ds\right].$$

Démonstration du lemme 1 Définissons, pour tout u dans $[0,1]$,

$$M_u = \exp\left[-\int_0^u \sigma'(X_s)\,dB_s - \int_0^u \left(b'(X_s) - \frac{1}{2}\sigma'^2(X_s)\right)ds\right],$$

$$A_u = \exp\left[-\int_0^u \frac{L\sigma - b'\sigma}{\sigma}(X_s)\,ds\right].$$

Par la formule d'Itô, nous obtenons, pour tous $0 \leq r \leq t$,

$$\begin{aligned}
A_t\sigma(X_t)M_t &= A_r\sigma(X_r)M_r - \int_r^t \frac{L\sigma - b'\sigma}{\sigma}(X_s)A_s\sigma(X_s)M_s ds \\
&+ \int_r^t A_s\sigma'(X_s)M_s[\sigma(X_s)dB_s + b(X_s)ds] \\
&- \int_r^t A_s\sigma(X_s)M_s\left[\sigma'(X_s)dB_s + \left(b'(X_s) - \frac{1}{2}\sigma'^2(X_s)\right)ds\right] \\
&+ \frac{1}{2}\int_r^t A_s\sigma(X_s)M_s\sigma'^2(X_s)ds + \frac{1}{2}\int_r^t A_s\sigma''(X_s)M_s\sigma^2(X_s)ds \\
&- \int_r^t A_s\sigma'(X_s)M_s\sigma(X_s)\sigma'(X_s)ds \\
&= A_r\sigma(X_r)M_r.
\end{aligned}$$

Il s'ensuit que, pour tous $0 \le r \le t$,

$$\sigma(X_r)\exp\left[\int_r^t \sigma'(X_s)dB_s + \int_r^t \left(b'(X_s) - \frac{1}{2}\sigma'^2(X_s)\right)ds\right]$$
$$= \sigma(X_t)\exp\left[-\int_r^t \frac{L\sigma - b'\sigma}{\sigma}(X_s)ds\right],$$

ce qui termine la démonstration du lemme 1.

La proposition 1 se déduit aisément du lemme 1. En effet, nous avons

$$\begin{aligned}
E\left[f^2(X_t)\log f^2(X_t)\right] &- E\left[f^2(X_t)\right]\log E\left[f^2(X_t)\right] \\
&\le 2E\left[\int_0^1 (f'(X_t)D_rX_t)^2 dr\right] \\
&= 2E\left[(f'(X_t))^2\sigma^2(X_t)\int_0^t \exp\left[-2\int_r^t \frac{L\sigma - b'\sigma}{\sigma}(X_s)ds\right]dr\right] \\
&\le \frac{e^{2ct}-1}{c}E\left[(f'(X_t))^2\sigma^2(X_t)\right],
\end{aligned}$$

ce qui correspond au résultat annoncé.
Remarque Il est en fait inutile de supposer σ non nul pour obtenir la proposition 1. En effet, par le lemme de Gronwall, nous obtenons de la même façon, sous la condition $(L\sigma - b'\sigma)\sigma \ge -c\sigma^2$, la majoration:

$$\int_0^1 (D_rX_t)^2 dr \le \frac{e^{2ct}-1}{2c}\sigma^2(X_t).$$

Il nous a semblé néanmoins intéressant de donner la forme explicite de D_rX_t.

3 Extension à toute loi marginale de X

La théorie des semigroupes ne semble pas permettre la généralisation de l'inégalité de Sobolev logarithmique à toute fonction cylindrique du processus de diffusion.

Dans ce paragraphe, nous obtenons cette extension, en gardant pour points de départ l'inégalité de Sobolev logarithmique pour le mouvement Brownien et le lemme 1.

Proposition 2 *Soient* $0 \leq t_1 < \cdots < t_n \leq 1$, $c \geq 0$ *et* $K(c) = 4(c^2+1)e^{2c}$. *Si* $\left|\frac{L\sigma - b'\sigma}{\sigma}\right| \leq c$, *alors, pour toute fonction* $f : \mathbb{R}^n \to \mathbb{R}$ *de classe* C^1 *dont les dérivées partielles sont bornées,*

$$E\left[f^2(X_{t_1},\ldots,X_{t_n})\log f^2(X_{t_1},\ldots,X_{t_n})\right] - E\left[f^2(X_{t_1},\ldots,X_{t_n})\right]\log E\left[f^2(X_{t_1},\ldots,X_{t_n})\right]$$

$$\leq K(c)\sum_{i,j=1}^{n}\min(t_i,t_j)E\left[\sigma(X_{t_i})\,\sigma\left(X_{t_j}\right)\frac{\partial f}{\partial x_i}(X_{t_1},\ldots,X_{t_n})\,\frac{\partial f}{\partial x_j}(X_{t_1},\ldots,X_{t_n})\right]$$

Démonstration Elle suit le même principe que celle de la proposition 1. Appliquons de même l'inégalité de Sobolev logarithmique connue pour le mouvement Brownien sur $\mathbb{R}$, en considérant chaque X_{t_i} comme une fonction de ω. Nous obtenons ainsi

$$E\left[f^2(X_{t_1},\ldots,X_{t_n})\log f^2(X_{t_1},\ldots,X_{t_n})\right] - E\left[f^2(X_{t_1},\ldots,X_{t_n})\right]\log E\left[f^2(X_{t_1},\ldots,X_{t_n})\right]$$

$$\leq 2E\left[\int_0^1\left(\sum_{i=1}^{n}\frac{\partial f}{\partial x_i}(X_{t_1},\ldots,X_{t_n})\,D_rX_{t_i}\right)^2 dr\right].$$

Utilisant l'expression de $D_rX_{t_i}$ fournie par le lemme 1, il s'agit donc de majorer:

$$E\left[\int_0^1\left(\sum_{i=1}^{n}\frac{\partial f}{\partial x_i}(X_{t_1},\ldots,X_{t_n})\,\sigma(X_{t_i})\exp\left[-\int_r^{t_i}\frac{L\sigma - b'\sigma}{\sigma}(X_s)\,ds\right]1_{[0,t_i]}(r)\right)^2 dr\right]$$

qui s'écrit encore (en posant $t_0 = 0$)

$$E\left[\sum_{i=1}^{n}\int_{t_{i-1}}^{t_i}\left\{\sum_{j=i}^{n}\frac{\partial f}{\partial x_j}(X_{t_1},\ldots,X_{t_n})\,\sigma\left(X_{t_j}\right)\exp\left[-\int_r^{t_j}\frac{L\sigma - b'\sigma}{\sigma}(X_s)\,ds\right]\right\}^2 dr\right].$$

Pour ce faire, nous nous inspirons de la démonstration du lemme 4.3 de [8]. Soit i dans $\{1,\ldots,n\}$ et r dans $[t_{i-1},t_i]$. Remarquons que nous pouvons écrire (avec la convention $\sum_{j=i+1}^{n} = 0$ si $i = n$)

$$\sum_{j=i}^{n}\frac{\partial f}{\partial x_j}(X_{t_1},\ldots,X_{t_n})\,\sigma\left(X_{t_j}\right)\exp\left[-\int_r^{t_j}\frac{L\sigma - b'\sigma}{\sigma}(X_s)\,ds\right]$$

$$= \left\{\sum_{j=i}^{n}\frac{\partial f}{\partial x_j}(X_{t_1},\ldots,X_{t_n})\,\sigma\left(X_{t_j}\right)\right\}\exp\left[-\int_r^{t_i}\frac{L\sigma - b'\sigma}{\sigma}(X_s)\,ds\right]$$

$$+\sum_{j=i+1}^{n}\left\{\exp\left[-\int_r^{t_j}\frac{L\sigma - b'\sigma}{\sigma}(X_s)\,ds\right] - \exp\left[-\int_r^{t_{j-1}}\frac{L\sigma - b'\sigma}{\sigma}(X_s)\,ds\right]\right\}$$

$$\times\left\{\sum_{k=j}^{n}\frac{\partial f}{\partial x_k}(X_{t_1},\ldots,X_{t_n})\,\sigma(X_{t_k})\right\}.$$

Il vient

$$\left|\sum_{j=i}^{n}\frac{\partial f}{\partial x_j}(X_{t_1},\ldots,X_{t_n})\,\sigma\left(X_{t_j}\right)\exp\left[-\int_r^{t_j}\frac{L\sigma-b'\sigma}{\sigma}(X_s)\,ds\right]\right|$$

$$\leq \exp(c)\left|\sum_{j=i}^{n}\frac{\partial f}{\partial x_j}(X_{t_1},\ldots,X_{t_n})\,\sigma\left(X_{t_j}\right)\right|$$

$$+\sum_{j=i+1}^{n}\left|\exp\left[-\int_r^{t_j}\frac{L\sigma-b'\sigma}{\sigma}(X_s)\,ds\right]-\exp\left[-\int_r^{t_{j-1}}\frac{L\sigma-b'\sigma}{\sigma}(X_s)\,ds\right]\right|$$

$$\times\left|\sum_{k=j}^{n}\frac{\partial f}{\partial x_k}(X_{t_1},\ldots,X_{t_n})\,\sigma\left(X_{t_k}\right)\right|.$$

De l'égalité

$$\exp\left[-\int_r^{t_j}\frac{L\sigma-b'\sigma}{\sigma}(X_s)\,ds\right]-\exp\left[-\int_r^{t_{j-1}}\frac{L\sigma-b'\sigma}{\sigma}(X_s)\,ds\right]$$

$$=-\int_{t_{j-1}}^{t_j}\exp\left[-\int_r^{s}\frac{L\sigma-b'\sigma}{\sigma}(X_u)\,du\right]\frac{L\sigma-b'\sigma}{\sigma}(X_s)\,ds,$$

nous pouvons déduire

$$\left|\exp\left[-\int_r^{t_j}\frac{L\sigma-b'\sigma}{\sigma}(X_s)\,ds\right]-\exp\left[-\int_r^{t_{j-1}}\frac{L\sigma-b'\sigma}{\sigma}(X_s)\,ds\right]\right|\leq ce^c\,(t_j-t_{j-1}),$$

puis

$$\left|\sum_{j=i}^{n}\frac{\partial f}{\partial x_j}(X_{t_1},\ldots,X_{t_n})\,\sigma\left(X_{t_j}\right)\exp\left[-\int_r^{t_j}\frac{L\sigma-b'\sigma}{\sigma}(X_s)\,ds\right]\right|$$

$$\leq e^c\left|\sum_{j=i}^{n}\frac{\partial f}{\partial x_j}(X_{t_1},\ldots,X_{t_n})\,\sigma\left(X_{t_j}\right)\right|$$

$$+ce^c\sum_{j=i+1}^{n}(t_j-t_{j-1})\left|\sum_{k=j}^{n}\frac{\partial f}{\partial x_k}(X_{t_1},\ldots,X_{t_n})\,\sigma\left(X_{t_k}\right)\right|.$$

Définissons une fonction g sur $[0,t_n]$ en posant $g(t_n)=0$ et, pour tout j dans $\{1,\ldots,n\}$ et tout v dans $[t_{j-1},t_j[$,

$$g(v)=\left|\sum_{k=j}^{n}\frac{\partial f}{\partial x_k}(X_{t_1},\ldots,X_{t_n})\,\sigma\left(X_{t_k}\right)\right|.$$

Remarquons alors que:

$$\sum_{j=i+1}^{n}(t_j-t_{j-1})\left|\sum_{k=j}^{n}\frac{\partial f}{\partial x_k}(X_{t_1},\ldots,X_{t_n})\,\sigma\left(X_{t_k}\right)\right|=\int_{t_i}^{t_n}g(v)dv.$$

Nous avons ainsi:

$$\left|\sum_{j=i}^{n} \frac{\partial f}{\partial x_j}(X_{t_1},\dots,X_{t_n})\,\sigma\left(X_{t_j}\right)\exp\left[-\int_r^{t_j} \frac{L\sigma - b'\sigma}{\sigma}(X_s)\,ds\right]\right|^2$$

$$\le 2e^{2c}\left|\sum_{j=i}^{n} \frac{\partial f}{\partial x_j}(X_{t_1},\dots,X_{t_n})\,\sigma\left(X_{t_j}\right)\right|^2 + 2c^2e^{2c}\left\{\int_{t_i}^{t_n} g(v)dv\right\}^2.$$

$$\le 2e^{2c}\left|\sum_{j=i}^{n} \frac{\partial f}{\partial x_j}(X_{t_1},\dots,X_{t_n})\,\sigma\left(X_{t_j}\right)\right|^2$$

$$+2c^2e^{2c}\sum_{j=i+1}^{n}(t_j - t_{j-1})\left|\sum_{k=j}^{n} \frac{\partial f}{\partial x_k}(X_{t_1},\dots,X_{t_n})\,\sigma(X_{t_k})\right|^2$$

(en utilisant $\left\{\int_{t_i}^{t_n} g(v)dv\right\}^2 \le \int_{t_i}^{t_n}\{g(v)\}^2 dv$).

Nous obtenons finalement la majoration:

$$\left|\sum_{j=i}^{n} \frac{\partial f}{\partial x_j}(X_{t_1},\dots,X_{t_n})\,\sigma\left(X_{t_j}\right)\exp\left[-\int_r^{t_j} \frac{L\sigma - b'\sigma}{\sigma}(X_s)\,ds\right]\right|^2$$

$$\le 2e^{2c}\left|\sum_{j=i}^{n} \frac{\partial f}{\partial x_j}(X_{t_1},\dots,X_{t_n})\,\sigma\left(X_{t_j}\right)\right|^2$$

$$+2c^2e^{2c}\sum_{j=1}^{n}(t_j - t_{j-1})\left|\sum_{k=j}^{n} \frac{\partial f}{\partial x_k}(X_{t_1},\dots,X_{t_n})\,\sigma(X_{t_k})\right|^2.$$

Nous en déduisons:

$$\sum_{i=1}^{n}\int_{t_{i-1}}^{t_i}\left\{\sum_{j=i}^{n} \frac{\partial f}{\partial x_j}(X_{t_1},\dots,X_{t_n})\,\sigma\left(X_{t_j}\right)\exp\left[-\int_r^{t_j} \frac{L\sigma - b'\sigma}{\sigma}(X_s)\,ds\right]\right\}^2 dr$$

$$\le 2e^{2c}\sum_{i=1}^{n}(t_i - t_{i-1})\left|\sum_{j=i}^{n} \frac{\partial f}{\partial x_j}(X_{t_1},\dots,X_{t_n})\,\sigma\left(X_{t_j}\right)\right|^2$$

$$+2c^2e^{2c}\sum_{j=1}^{n}(t_j - t_{j-1})\left|\sum_{k=j}^{n} \frac{\partial f}{\partial x_k}(X_{t_1},\dots,X_{t_n})\,\sigma(X_{t_k})\right|^2$$

$$\leq 2(c^2+1)e^{2c}\sum_{i=1}^{n}(t_i-t_{i-1})\left|\sum_{j=i}^{n}\frac{\partial f}{\partial x_j}(X_{t_1},\ldots,X_{t_n})\,\sigma\left(X_{t_j}\right)\right|^2.$$

Remarquant que

$$\sum_{i=1}^{n}(t_i-t_{i-1})\left|\sum_{j=i}^{n}\frac{\partial f}{\partial x_j}(X_{t_1},\ldots,X_{t_n})\,\sigma\left(X_{t_j}\right)\right|^2$$
$$=\sum_{i,j=1}^{n}\min(t_i,t_j)\sigma(X_{t_i})\,\sigma\left(X_{t_j}\right)\frac{\partial f}{\partial x_i}(X_{t_1},\ldots,X_{t_n})\frac{\partial f}{\partial x_j}(X_{t_1},\ldots,X_{t_n}),$$

nous obtenons la majoration attendue. La proposition est établie. Nous n'avons pas cherché à optimiser la constante dans l'inégalité de la proposition 2.

4 Remarques et conclusion

Lorsque l'on considère une diffusion solution d'une équation différentielle stochastique sur $\mathbb{R}^d$, $d \geq 2$, une étude similaire permet de conclure lorsque les champs commutent et sous la condition d'ellipticité. En dimension $d \geq 2$, notre méthode n'est plus appropriée et il est naturel de considérer la diffusion comme vivant sur la variété Riemannienne $\left(\mathbb{R}^d, (\sigma\sigma^*)^{-1}\right)$.

Signalons pour finir que la technique développée dans ce travail permet d'étendre de la même façon l'inégalité isopérimétrique de [4] à la loi d'une diffusion satisfaisant les hypothèses de la proposition 2. Ainsi qu'il est établi en [4], cette inégalité isopérimétrique renforce l'inégalité de Sobolev logarithmique correspondante.

Références

[1] S. Aida and D. Elworthy, "Logarithmic Sobolev Inequalities on Path and Loop Spaces" (Preprint).

[2] D. Bakry, "L'hypercontractivité et son utilisation en théorie des semi-groupes". Ecole d'Eté de Probabilités de Saint-Flour. Lecture Notes in Math. 1581, 1-114 (1994), Springer, Berlin Heidelberg New York.

[3] D. Bakry, " On Sobolev and logarithmic Sobolev inequalities for Markov semigroups", in the Proceedings of the Taniguchi Symposium, Warwick 1994 (à paraître).

[4] D. Bakry, M. Ledoux, "Lévy-Gromov's isoperimetric inequality for an infinite dimensional diffusion generator", *Invent. math. 123, 259-281 (1996).*

[5] Y. Hu, "Itô-Wiener Chaos expansion with exact residual and correlation, variance inequalities"(Preprint).

[6] N. Ikeda and S. Watanabe, Stochastic Differential Equations and Diffusion Processes, North-Holland, 1981.

[7] D. Nualart, The Malliavin Calculus and Related Topics, Springer 1995.

[8] E. P.Hsu, "Logarithmic Sobolev Inequalities on path spaces over Riemannian manifolds" (Preprint).

[9] F. Y. Wang, "Logarithmic Sobolev Inequalities for diffusion processes with application to path space" (Preprint).

Laboratoire de Statistique et Probabilités
Université Paul Sabatier
118, route de Narbonne
31062 Toulouse Cedex, France
capitain@cict.fr

Sur les minorations des constantes de Sobolev et de Sobolev logarithmiques pour les opérateurs de Jacobi et de Laguerre

Éric FONTENAS
Laboratoire de Modélisation et de Calcul
Domaine Universitaire BP 53
38041 Grenoble cedex 9

Résumé

Dans cet article, nous nous intéressons à la minoration des constantes dans les inégalités de Sobolev pour l'opérateur associé aux polynômes de Jacobi ainsi qu'à celle des constantes de Sobolev logarithmiques pour l'opérateur associé aux polynômes de Laguerre. Ces constantes ont pour particularité d'être différentes de la première valeur propre non nulle associée à ces opérateurs. Nous proposons ici une méthode basée sur l'existence de fonctions extrémales et sur de simples études de signe de polynômes du second degré. Elle nous permet d'obtenir une nouvelle minoration de ces constantes qui fait le lien, dans le cas de l'opérateur de Jacobi non symétrique, entre la constante de Sobolev pour l'exposant optimal due à Bakry et la constante de Sobolev logarithmique obtenue par Saloff-Coste.

Résumé

This paper concerns Sobolev constants for Jacobi operators and logarithmic Sobolev constants for Laguerre operators. These constants are different from the first non-negative eigenvalue associated to these operators. In this work, we suggest a method based on the existence of extremal functions and on the study of the sign of second degree polynomials to bound these constants. These new lower bounds allow us to recover, on one hand, the Sobolev constant due to Bakry for optimal exponent and, on the other hand, the logarithmic Sobolev constant obtained by Saloff-Coste.

1 Introduction et définitions

Considérons l'espace $(]-1,1[, d\mu)$ où μ est la mesure de probabilités :

$$d\mu = C(a,b)(1-x)^{(a-1)/2}(1+x)^{(b-1)/2}dx$$

avec a et b des réels strictement positifs et $C(a,b)$ est la constante qui fait de μ une mesure de probabilités :

$$C(a,b) = \frac{\Gamma(\frac{a+b}{2}+1)}{\Gamma(\frac{a+1}{2})\Gamma(\frac{b+1}{2})}2^{-(\frac{a+b}{2})}$$

avec $\Gamma(x) = \int_0^{+\infty} t^{x-1}e^{-t}\,dt$. Soit L l'opérateur associé aux polynômes de Jacobi agissant sur les fonctions f de classe C^2 par

$$Lf(x) = (1-x^2)f''(x) + \frac{1}{2}[a - b - (a+b+2)x]f'(x).$$

Le cas symétrique, i.e, $a = b$, a été très étudié : les meilleures constantes de Sobolev et de Sobolev logarithmiques sont connues (voir par exemple [13]). Dans le cas non symétrique, i.e $a \neq b$, on ne dispose que de deux résultats intéressants, celui de Bakry [2] concernant la minoration de la constante de Sobolev pour l'exposant optimal et celui de Saloff-Coste [15] lors de l'étude de la constante de Sobolev logarithmique. Le but de cet exposé est de montrer que ces deux constantes sont liées et que le passage de l'une à l'autre s'obtient par un simple passage à la limite de l'exposant vers 2 dans les inégalités de Sobolev. Nous donnons alors une nouvelle minoration de la constante de Sobolev pour tous les exposants dans les inégalités de Sobolev. Bakry et Saloff-Coste ont obtenu ces minorations de constantes de Sobolev et de Sobolev logarithmiques en utilisant une inégalité de courbure-dimension associée à l'opérateur L (voir [3] pour plus de détails). Ici, la méthode repose sur une équation différentielle du second degré associée aux inégalités de Sobolev et, par de simples changements de variables et d'études du signe de polynômes du second degré, on arrive à une nouvelle minoration de ces constantes. De plus, cette méthode permet de retrouver les meilleures constantes de Sobolev et de Sobolev logarithmiques pour l'opérateur de Jacobi symétrique, i.e dans le cadre $a = b$.

Voici tout d'abord quelques remarques utiles concernant les opérateurs de Jacobi : l'espace $L^2(d\mu)$ admet pour base orthogonale les polynômes de Jacobi $(J_k)_{k\geq 0}$ définis par la série génératrice suivante (voir [10]) :

$$2^{(2-a-b)/2}\sum_k t^k J_k =$$
$$(1-2xt+t^2)^{-1/2}(1-t+(1-2xt+t^2)^{1/2})^{-(b-1)/2}(1+t+(1-2xt+t^2)^{1/2})^{-(a-1)/2}.$$

Ces polynômes satisfont la relation $L(J_k) = -\lambda_k J_k$, avec

$$\lambda_k = k\left[\frac{a+b}{2} + k\right].$$

Ce qui signifie que chaque J_k est vecteur propre associé à la valeur propre λ_k de $-L$. On peut trouver la forme explicite des polynômes J_k dans [15] et toutes les formules précédentes dans [12]. L'opérateur L ainsi défini est symétrique par rapport à la mesure μ au sens suivant : pour toutes fonctions (f, g) de classe C^2,

$$\int_{-1}^{1} fLg\,d\mu = \int_{-1}^{1} gLf\,d\mu.$$

On associe à L un opérateur Γ carré du champ défini par : pour toutes fonctions f et g de classe C^1,

$$\Gamma(f,g) = (1-x^2)f'g'.$$

On vérifie aisément que pour toute fonction f de classe C^2,

$$\int_{-1}^{1} \Gamma(f,f)\,d\mu = \int_{-1}^{1} f(-Lf)\,d\mu.$$

Nous avons de plus les formules suivantes de changement de variable pour L et Γ: pour toute fonction $\phi \in C^\infty(\mathbb{R},\mathbb{R})$ et pour toute fonction $f \in C^2$,

$$L(\phi(f)) = \phi'(f)Lf + \phi''(f)\Gamma(f,f)$$

et

$$\Gamma(f,f) = \phi'^2(f)\Gamma(f,f).$$

Voici, maintenant quelques exemples d'opérateurs de Jacobi qui ne se différencient que par des valeurs bien précises des paramètres a et b:

- Polynômes de Gegenbauer: on pose $a = b = n-1$ avec $n > 0$.
- Polynômes de Legendre: on choisit $a = b = 1$.
- Espaces projectifs réels: on choisit $b = 0$ et $a > 1$ (Lors de l'étude des fonctions extrémales pour les inégalités de Sobolev, nous verrons que cet exemple est un cas tout à fait particulier).

Nous définissons maintenant la notion d'inégalité de Sobolev de dimension n:

Définition 1 *Nous dirons que L satisfait une inégalité de Sobolev de dimension $n > 2$ si il existe deux constantes A et B strictement positives telles que*

$$\forall f \in C^2, \quad \left(\int_{-1}^1 f^p\,d\mu\right)^{2/p} \le A\int_{-1}^1 f^2\,d\mu + B\int_{-1}^1 (1-x^2)f'^2\,d\mu \tag{1}$$

avec $p = 2n/(n-2)$.

En appliquant l'inégalité précédente à la fonction $f = 1$, on obtient que la constante A est supérieure ou égale à 1. Cette constante peut être prise égale à 1 quitte à modifier la constante B (cf. [1] et [2]). Dans cet article, nous nous intéressons donc aux inégalités de Sobolev avec $A = 1$. Bakry, [2], a démontré l'existence de telles inégalités pour $2 < p \le 2(a+1)/(a-1)$ où $a+1$ est la meilleure dimension possible dans l'inégalité de Sobolev (1). Nous recherchons alors des estimations précises de la constante B_p pour $2 < p \le 2(a+1)/(a-1)$ dans l'inégalité suivante:

$$\frac{B_p}{p-2}\left[\left(\int_{-1}^1 f^p\,d\mu\right)^{2/p} - \int_{-1}^1 f^2\,d\mu\right] \le \int_{-1}^1 (1-x^2)f'^2\,d\mu. \tag{2}$$

Dans le cadre où $(a = b)$, nous verrons que la méthode utilisée fournit la meilleure constante dans (2) et que celle-ci est alors égale à $a+1 = \lambda_1$ (voir [13] et [7]). Nous allons surtout étudier les constantes de Sobolev pour les opérateurs de Jacobi avec $a \ne b$.

1.1 Majoration de la constante de Sobolev

La première valeur propre λ_1 de $-L$ peut être caractérisée (voir par exemple [5]) comme le plus grand des λ strictement positifs tels que, pour toute fonction f de classe C^2, on ait:

$$\lambda\left[\int_{-1}^1 f^2\,d\mu - \left(\int_{-1}^1 f\,d\mu\right)^2\right] \le \int_{-1}^1 (1-x^2)f'^2\,d\mu.$$

Si nous remplaçons f par $1+\epsilon f$ dans l'inégalité (2), par passage à la limite de ϵ vers 0, cette inégalité s'écrit :

$$B_p\left[\int_{-1}^1 f^2\,d\mu - \left(\int_{-1}^1 f\,d\mu\right)^2\right] \leq \int_{-1}^1 (1-x^2) f'^2\,d\mu.$$

De part la définition de λ_1, il vient que, pour tout $2 < p \leq 2(a+1)/(a-1)$, $B_p \leq \lambda_1$. Nous allons voir que, si $a \neq b$, cette inégalité est stricte :

Théorème 1 *Si $a \neq b$, la constante de Sobolev B_p dans les inégalités (2) est strictement inférieure à λ_1.*

Démonstration : nous reprenons la démonstration de Rothaus [14] lors de l'étude de la constante de Sobolev logarithmique sur une variété riemannienne compacte. Considérons la fonction

$$H = Q \Big/ \left(\int_{-1}^1 Q^2\,d\mu\right)^{1/2},$$

avec $Q = (-(a+b+2)x + a - b)/2$. On constate que la fonction H est fonction propre de $-L$ associée à la première valeur propre λ_1. De plus, on vérifie aisément, dans le cas où $a \neq b$ et contrairement au cas $a = b$, que $\int_{-1}^1 H^3\,d\mu \neq 0$. Comme H est vecteur propre de $-L$, on a :

$$\int_{-1}^1 \Gamma(H,H)\,d\mu = \lambda_1 \int_{-1}^1 H^2\,d\mu = \lambda_1$$

et $\int_{-1}^1 H\,d\mu = 0$. Le résultat du théorème 1 repose sur les propriétés de cette fonction H. Considérons la fonctionnelle suivante :

$$\varphi(f) = \left(\int_{-1}^1 f^p\,d\mu\right)^{-2/p} \left\{\int_{-1}^1 f(-Lf)\,d\mu - \frac{B_p}{p-2}\left[\left(\int_{-1}^1 f^p\,d\mu\right)^{2/p} - \int_{-1}^1 f^2\,d\mu\right]\right\}.$$

Posons $f = 1+\epsilon H$. Par un développement limité, les termes de la fonction φ s'écrivent :

$$\int_{-1}^1 (1+\epsilon H)^p\,d\mu = 1 + \frac{p(p-1)}{2}\epsilon^2 + \frac{p(p-1)(p-2)}{6}\epsilon^3 \int_{-1}^1 H^3\,d\mu + \epsilon^3 g(\epsilon)$$

où $g(\epsilon)$ tend vers 0 quand ϵ tend vers 0. On a donc :

$$\left(\int_{-1}^1 (1+\epsilon H)^p\,d\mu\right)^{2/p} = 1 + (p-1)\epsilon^2 + \frac{(p-1)(p-2)}{3}\epsilon^3 \int_{-1}^1 H^3\,d\mu + \epsilon^3 g_1(\epsilon)$$

et

$$\left(\int_{-1}^1 (1+\epsilon H)^p\,d\mu\right)^{-2/p} = 1 - (p-1)\epsilon^2 - \frac{(p-1)(p-2)}{3}\epsilon^3 \int_{-1}^1 H^3\,d\mu + \epsilon^3 g_2(\epsilon)$$

où g_1 et g_2 tendent vers 0 avec ϵ. De plus, H étant fonction propre de $-L$, $\varphi(f)$ s'écrit

$$\varphi(f) = [\lambda_1 - B_p]\epsilon^2 - B_p\frac{(p-1)}{3}\epsilon^3 \int_{-1}^1 H^3\,d\mu + \epsilon^3 g_3(\epsilon)$$

avec $g_3(\epsilon)$ qui tend vers 0 avec ϵ. Nous voyons que, si $B_p = \lambda_1$ et $\int_{-1}^1 H^3\,d\mu \neq 0$, alors $\varphi(f) < 0$. On en déduit, finalement,que, si $\int_{-1}^1 H^3\,d\mu \neq 0$, $B_p < \lambda_1$. Ce qui termine la démonstration du théorème 1.

Remarque : pour les opérateurs de Jacobi tels que $a = b$, on vérifie que $\int_{-1}^1 Q^3\,d\mu = 0$. Il est démontré dans [6] que cette condition est équivalente au fait que la constante de Sobolev B_p est égale à λ_1.

1.2 Minorations des constantes de Sobolev

Nous en venons maintenant au résultat principal de cet article concernant la minoration des constantes de Sobolev pour l'opérateur de Jacobi dans le cadre $a \neq b$. Nous proposons ici une nouvelle minoration de ces constantes, qui permet de faire le lien entre des résultats déjà connus sur la constante de Sobolev pour l'exposant optimal [2] et la constante de Sobolev logarithmique [15]. Notre approche est directement inspirée des travaux de Rothaus [14] lors de l'étude de la constante de Sobolev logarithmique, méthode reprise par Ledoux [1] lors de l'étude des constantes de Sobolev pour un générateur markovien abstrait. Elle repose sur l'existence d'une fonction extrémale satisfaisant l'égalité dans les inégalités de Sobolev. Seule cette notion nous permettra de conclure à cette minoration. Nous n'utiliserons pas, contrairement à Rothaus et à Bakry, la notion d'inégalité de courbure-dimension.
Notre résultat principal est le suivant :

Théorème 2 *Si $a = b = \lambda_1 - 1$ avec $\lambda_1 > 2$ alors, la constante de Sobolev est optimale et on a, pour tout p, $1 \leq p \leq 2\lambda_1/(\lambda_1 - 2)$, $B_p = \lambda_1$.*
Si $a > b > 0$ et $a > 1$ alors, pour tout réel p, $1 \leq p \leq \frac{2(a+1)}{a-1}$, on a $B_p \geq \frac{(1-V_p)\left[-2abV_p+a+b+2\sqrt{ab(1+aV_p)(1+bV_p)}\right]}{4}$ où $V_p = \frac{2-p}{p+2}$.
Si $a > b > 0$ et $a < 1$, on a la même minoration mais pour tout $p \geq 1$.

Remarques : – Pour l'exposant optimal $p = 2(a+1)/(a-1)$, le théorème 2 fournit la minoration suivante pour la constante de Sobolev :

$$B_p \geq \frac{(a+1)(3b+a)}{4a}.$$

On retrouve ainsi un résultat de Bakry [2], résultat obtenu à partir d'une inégalité de courbure-dimension.

– Si on pose

$$F(p) = \left[\int_{-1}^{1} f^p d\mu\right]^{2/p},$$

l'inégalité de Sobolev du théorème s'écrit :

$$\frac{B_p}{p-2}[F(p) - F(2)] \leq \int_{-1}^{1} (1 - x^2) f'^2 \, d\mu. \tag{3}$$

Par passage à la limite de p vers 2 dans l'inégalité précédente, on obtient

$$B_2 F'(2) \leq \int_{-1}^{1} (1 - x^2) f'^2 \, d\mu$$

avec

$$F'(p) = \frac{2}{p}\left(\int_{-1}^{1} f^p \log f \, d\mu\right)\left(\int_{-1}^{1} f^p \, d\mu\right)^{1-2/p} - \frac{2}{p^2}\left(\int_{-1}^{1} f^p \, d\mu\right)^{2/p} \log \int_{-1}^{1} f^p \, d\mu$$

et

$$B_2 = \frac{a + b + 2\sqrt{ab}}{4}.$$

On retrouve donc “l'inégalité de Sobolev logarithmique”, introduite par Gross [8],

$$B_2\left(\int_{-1}^{1} f^2\log|f|\,d\mu-\int_{-1}^{1} f^2\,d\mu\log\left[\int_{-1}^{1} f^2\,d\mu\right]^{1/2}\right)\le\int_{-1}^{1}(1-x^2)f'^2\,d\mu$$

ainsi que la minoration obtenue par Saloff-Coste [15] lors de l'étude de la constante de Sobolev logarithmique de l'opérateur de Jacobi avec $a\neq b$. L'intérêt de cette nouvelle minoration est qu'elle permet de faire le lien entre la constante de Sobolev pour l'exposant optimal due à Bakry et cla constante de Sobolev logarithmique obtenue par Saloff-Coste.

Démonstration : nous considérons, tout d'abord, les opérateurs de Jacobi avec $a\neq b$. Toutefois, cette démonstration reste valable dans le cadre $a=b$ et nous verrons que celle-ci fournit la meilleure constante de Sobolev à savoir $B_p=\lambda_1$.
Pour alléger les notations intervenant dans cette démonstration, on posera $D=(a-b)/2$. Nous noterons, par la suite, P_X tout polynôme de second degré de la variable X et Δ_X le discriminant associé à ce polynôme. Nous considérons l'inégalité suivante

$$\frac{C}{p-2}\left[\left(\int_{-1}^{1} f^p\,d\mu\right)^{2/p}-(1+\epsilon)\int_{-1}^{1} f^2\,d\mu\right]\le\int_{-1}^{1}(1-x^2)f'^2\,d\mu \tag{4}$$

où $2<p<2(a+1)/(a-1)$. On notera par C_ϵ le plus grand des C strictement positif satisfaisant l'inégalité (4). Il a été démontré (cf. [1] et [7] pour le cas $a=b$) l'existence d'une fonction f positive non constante, bornée inférieurement et supérieurement, satisfaisant l'égalité dans (4) et vérifiant l'équation non linéaire suivante :

$$Lf(x)=\frac{C_\epsilon}{p-2}[(1+\epsilon)f(x)-f^{p-1}] \tag{5}$$

avec

$$\int_{-1}^{1} f^p\,d\mu=1$$

où $\epsilon>0$ et $C_\epsilon>B_p$. Le fait d'avoir ajouté un terme en ϵ nous assure que cette fonction ne peut être constante. L'existence d'une telle fonction va nous permettre d'obtenir la minoration annoncée. Notre méthode de calcul est basée sur des changements de variables successifs opérant sur l'équation (5). Posons $f=u^\beta$ avec $\beta>0$ dans l'équation (5). Cette équation implique le lemme suivant :

Lemme 1

$$\begin{aligned}&[C_\epsilon(1+\epsilon)-\lambda_1]\int_{-1}^{1}(1-x^2)u'^2\,d\mu \qquad (6)\\ &=\int_{-1}^{1}(1-x^2)^2\left[u''^2-2\beta(r-1)\frac{u''u'^2}{u}\right]d\mu+2\beta(r-1)\int_{-1}^{1}x(1-x^2)\frac{u'^3}{u}\,d\mu\\ &\quad+[(\beta-1)(1+\beta(r-2))+\beta(r-1)]\int_{-1}^{1}(1-x^2)^2\frac{u'^4}{u^2}\,d\mu.\end{aligned}$$

Démonstration : d'après la formule de changement de variable pour L, (5) devient :

$$\beta u^{\beta-1}Lu+\beta(\beta-1)\Gamma(u,u)u^{\beta-2}=(1+\epsilon)\frac{C_\epsilon}{p-2}u^\beta-\frac{C_\epsilon}{p-2}u^{\beta(p-1)}. \tag{7}$$

Multiplions (7) par $\Gamma(u,u)/u^\beta$, puis intégrons par rapport à la mesure μ:

$$\beta\int_{-1}^{1}\frac{Lu\Gamma(u,u)}{u}\,d\mu+\beta(\beta-1)\int_{-1}^{1}\frac{\Gamma^2(u,u)}{u^2}\,d\mu \tag{8}$$
$$=(1+\epsilon)\frac{C_\epsilon}{p-2}\int_{-1}^{1}\Gamma(u,u)\,d\mu-\frac{C_\epsilon}{p-2}\int_{-1}^{1}u^{\beta(p-2)}\Gamma(u,u)\,d\mu.$$

De plus,

$$\int_{-1}^{1}u^{\beta(p-2)}\Gamma(u,u)\,d\mu=-\frac{1}{\beta(p-2)+1}\int_{-1}^{1}u^{\beta(p-2)+1}Lu\,d\mu.$$

En multipliant l'égalité (7) par $u^{1-\beta}Lu$ puis en l'intégrant par rapport à la mesure μ, il vient:

$$\frac{C_\epsilon[\beta(p-2)+1]}{p-2}\int_{-1}^{1}u^{\beta(p-2)}\Gamma(u,u)\,d\mu$$
$$=\beta\int_{-1}^{1}(Lu)^2\,d\mu+\beta(\beta-1)\int_{-1}^{1}\frac{Lu\Gamma(u,u)}{u}\,d\mu+(1+\epsilon)\frac{C_\epsilon}{p-2}\int_{-1}^{1}\Gamma(u,u)\,d\mu.$$

En reportant ce terme dans (8), on obtient ainsi l'égalité suivante:

$$C_\epsilon(1+\epsilon)\int_{-1}^{1}\Gamma(u,u)\,d\mu \tag{9}$$
$$=(q-1)[1+q(p-2)]\int_{-1}^{1}\frac{\Gamma^2(u,u)}{u^2}\,d\mu+q(p-1)\int_{-1}^{1}\frac{\Gamma(u,u)Lu}{u}\,d\mu+\int_{-1}^{1}(Lu)^2\,d\mu.$$

De plus, on a:

$$\int_{-1}^{1}(Lu)^2\,d\mu=\int_{-1}^{1}(1-x^2)^2u''^2\,d\mu+\lambda_1\int_{-1}^{1}(1-x^2)u'^2\,d\mu^{(\alpha)}$$

et

$$\int_{-1}^{1}\frac{Lu\Gamma(u,u)}{u}\,d\mu=-2\int_{-1}^{1}(1-x^2)^2\frac{u''u'^2}{u}\,d\mu+\int_{-1}^{1}(1-x^2)^2\frac{u'^4}{u^2}\,d\mu+2\int_{-1}^{1}x(1-x^2)\frac{u'^3}{u}\,d\mu.$$

En reportant ces deux termes dans l'équation (9), on retrouve le résultat du lemme. La démonstration du lemme est terminée.

D'après le théorème 1, la constante de Sobolev pour l'opérateur de Jacobi est strictement inférieure à λ_1. Nous recherchons alors une minoration de la constante de Sobolev sous la forme:

$$B_p\geq(1-c)\lambda_1$$

où $0\leq c<1$. Pour établir cette minoration, on considère l'équation (6) après avoir ajouté un terme $c\lambda_1\int_{-1}^{1}(1-x^2)u'^2\,d\mu$. On arrive alors à l'équation suivante:

$$[C_\epsilon(1+\epsilon)-\lambda_1(1-c)]\int_{-1}^{1}(1-x^2)u'^2\,d\mu \tag{10}$$
$$=\int_{-1}^{1}(1-x^2)^2\left[u''^2-2\beta(p-1)\frac{u''u'^2}{u}\right]d\mu+2\beta(p-1)\int_{-1}^{1}x(1-x^2)\frac{u'^3}{u}\,d\mu$$
$$+[(\beta-1)(1+\beta(p-2))+\beta(p-1)]\int_{-1}^{1}(1-x^2)^2\frac{u'^4}{u^2}\,d\mu+c\lambda_1\int_{-1}^{1}(1-x^2)u'^2\,d\mu.$$

Toute notre étude est basée sur le signe du membre de droite de cette équation. Nous recherchons le plus petit c possible de manière à ce que ce terme soit positif. Nous en déduirons alors, que, pour tout $\epsilon > 0$, $C_\epsilon(1+\epsilon) \geq (1-c)\lambda_1$. Par passage à la limite de ϵ vers 0, et pour la valeur minimum de c permise, nous obtiendrons le résultat du théorème 2.

Posons $g(x) = u'u^{-\alpha}$ avec $\alpha \neq 1/3$. Après avoir remarqué que $u'' = g'u^\alpha + \alpha g^2 u^{2\alpha-1}$, l'équation (6) devient :

$$[C_\epsilon(1+\epsilon) - \lambda_1(1-c)] \int_{-1}^1 (1-x^2)u^{2\alpha}g^2\,d\mu = 2\beta(p-1)\int_{-1}^1 (1-x^2)xu^{3\alpha-1}g^3\,d\mu$$
$$+[\alpha^2 - 2\alpha\beta(p-1) + (\beta-1)(1+\beta(p-2)) + \beta(p-1)]\int_{-1}^1 (1-x^2)^2u^{4\alpha-2}g^4\,d\mu$$
$$+\int_{-1}^1 (1-x^2)^2\left\{u^{2\alpha}g'^2 + [2\alpha - 2\beta(p-1)]u^{3\alpha-1}g^2g'\right\}d\mu. \tag{11}$$

En remarquant que

$$\int_{-1}^1 (1-x^2)^2g^2g'u^{3\alpha-1}\,d\mu = \frac{2+\lambda_1}{3}\int_{-1}^1 (1-x^2)xu^{3\alpha-1}g^3\,d\mu$$
$$-\frac{3\alpha-1}{3}\int_{-1}^1 (1-x^2)^2u^{4\alpha-2}g^4\,d\mu - \frac{D}{3}\int_{-1}^1 (1-x^2)u^{3\alpha-1}g^3\,d\mu$$

et en reportant ce terme dans l'équation (11), on obtient

$$[C_\epsilon(1+\epsilon) - \lambda_1(1-c)]\int_{-1}^1 (1-x^2)u^{2\alpha}g^2\,d\mu = \int_{-1}^1 (1-x^2)^2u^{2\alpha}g'^2\,d\mu \tag{12}$$
$$+\left[-\alpha^2 + \frac{2}{3}\alpha + \beta^2(p-2) + \frac{2}{3}\beta(4-p) - 1\right]\int_{-1}^1 (1-x^2)^2u^{4\alpha-2}g^4\,d\mu$$
$$+2\left[\beta(p-1) + \frac{(\lambda_1+2)(\alpha-\beta(p-1))}{3}\right]\int_{-1}^1 (1-x^2)xu^{3\alpha-1}g^3\,d\mu$$
$$+c\lambda_1\int_{-1}^1 (1-x^2)u^{2\alpha}g^2\,d\mu - \frac{2}{3}[\alpha-\beta(p-1)]D\int_{-1}^1 (1-x^2)u^{3\alpha-1}g^3\,d\mu.$$

Notre but est de faire apparaître dans le membre de droite de cette équation sous le signe intégral un polynôme du second degré en la variable g'. Nous choisissons α de manière à ce que le coefficient de p^2g^4 soit nul :

$$\alpha = \frac{1+\sqrt{(3\beta(p-2)+4)(3\beta-2)}}{3}. \tag{13}$$

De plus, par de simples intégrations par parties, on arrive à

$$\int_{-1}^1 (1-x^2)xu^{3\alpha-1}g^3\,d\mu = -\frac{1}{2\alpha}\int_{-1}^1 u^{2\alpha}[Qx + (1-x^2)]g^2\,d\mu - \frac{1}{\alpha}\int_{-1}^1 (1-x^2)xu^{2\alpha}gg'\,d\mu$$

et

$$\int_{-1}^1 (1-x^2)u^{3\alpha-1}g^3\,d\mu = -\frac{1}{2\alpha}\int_{-1}^1 Qu^{2\alpha}g^2\,d\mu - \frac{1}{\alpha}\int_{-1}^1 (1-x^2)u^{2\alpha}gg'\,d\mu$$

où Q est la fonction utilisée au cours de la démonstration du théorème 1. En remplaçant ces deux termes dans l'équation (12) et en prenant pour α la valeur indiquée au (13), on arrive à l'équation suivante

$$[C_\epsilon(1+\epsilon) - \lambda_1(1-c)] \int_{-1}^{1} (1-x^2)u^{2\alpha} g^2 \, d\mu =$$

$$\int_{-1}^{1} u^{2\alpha} \left[(1-x^2)^2 g'^2 + 2pg'g[VD - (1+V(\lambda_1-1))x] + DVQg^2\right] d\mu$$

$$-[1+V(\lambda_1-1)] \int_{-1}^{1} (Qx + (1-x^2))u^{2\alpha} g^2 \, d\mu + c\lambda_1 \int_{-1}^{1} (1-x^2)u^{2\alpha} g^2 \, d\mu,$$

avec

$$V = \frac{1}{3}\left[1 - \frac{\beta(p-1)}{\alpha}\right].$$

Il nous reste maintenant à étudier le signe du membre de droite de cette équation. En fait, nous nous intéressons au signe du polynôme $P_{g'}$ de la variable g' défini par

$$P_{g'} = (1-x^2)^2 g'^2 + 2(1-x^2)g'g[VD - (1+V(\lambda_1-1))x]$$
$$+ g^2 \left[DVQ - (1+V(\lambda_1-1))(Qx + (1-x^2)) + c\lambda_1(1-x^2)\right].$$

Le discriminant du polynôme $P_{g'}$ de la variable g' est égal à :

$$\Delta_{g'} = 4(1-x^2)^2 g^2 P_x$$

avec

$$P_x = [(1+V(\lambda_1-1))(V(\lambda_1-1)-\lambda_1) + c\lambda_1]\, x^2 +$$
$$D\left[-2V^2(\lambda_1-1) + V(2\lambda_1-3) + 1\right] x + V^2D^2 - c\lambda_1 - D^2V + 1 + V(\lambda_1-1).$$

Nous recherchons c de manière à ce que $P_{g'}$ soit positif ou nul, ce qui revient à démontrer que le discriminant $\Delta_{g'}$ de $P_{g'}$ est négatif ou nul. Pour cela, nous calculons le discriminant Δ_x du polynôme P_x. Si $-1/a \leq V \leq 1$, on obtient :

$$\Delta_x = 4(\lambda_1)^2(c-c_1)(c-c_2),$$

où

$$c_1 = 1 - \frac{(1-V)\left[-2abV + a + b + 2\sqrt{ab(1+aV)(1+bV)}\right]}{4\lambda_1}$$

et

$$c_2 = 1 - \frac{(1-V)\left[-2abV + a + b - 2\sqrt{ab(1+aV)(1+bV)}\right]}{4\lambda_1}$$

En posant $c = c_1$, on en déduit que $\Delta_x = 0$. On vérifie que, pour cette valeur de c, le coefficient de x^2 dans P_x est négatif. Le fait que $\Delta_x = 0$ entraîne alors que $P_x \leq 0$ et donc que $\Delta'_g \leq 0$. On en déduit alors que $P_{g'}$ est positif et donc la minoration suivante pour la constante C_ϵ :

$$C_\epsilon(1+\epsilon) \geq \frac{(1-V)\left[-2abV + a + b + 2\sqrt{ab(1+aV)(1+bV)}\right]}{4}. \tag{14}$$

Il nous reste maintenant à choisir un V, $-1/a \leq V \leq 1$, optimisant cette minoration. De part la valeur de α (13) et la définition de V, il est facile de vérifier que

$$V \leq \frac{2-p}{2+p}.$$

De plus, cette majoration de V et la minoration $V \geq -1/a$ pour $a > 1$ sont compatibles dès que

$$p \leq \frac{2(a+1)}{a-1}$$

où $2(a+1)/(a-1)$ est l'exposant optimal dans ce cadre et, si $a < 1$, il y a compatibilité pour tout $p > 1$. En posant $V = (2-p)/(p+2)$ et par passage à la limite de ϵ vers 0 dans l'inégalité (14), on trouve la minoration annoncée. La démonstration du théorème 2 est terminée.

Remarques : – Pour l'exposant optimal $p = 2(a+1)/(a-1)$, notre méthode de calcul ne peut fournir une meilleure minoration étant donné que la seule valeur de V permise est $-1/a$.

– Pour des exposants p, $p < 2(a+1)/(a-1)$, le résultat du théorème 2 n'est pas optimal. En effet, si nous considérons la minoration de la constante de Sobolev comme fonction de V que nous notons $m(V)$, on vérifie que $m''(V) \leq 0$ pour $V \in [-1/a, 0]$. De plus, $m'(x)$ tend vers $+\infty$ pour x tendant vers $-1/a$ tend vers $+\infty$ avec $x > -1/a$ et $m'(0) < 0$. On en déduit que la fonction $m(V)$ admet un maximum sur l'intervalle $[-1/a, 0]$. La détermination de ce maximum nous conduit à des calculs presque inextricables et surtout peu explicites.

– Ce choix de $V = (2-p)(p+2)$ nous permet d'atteindre tous les exposants de l'intervalle $[-1/a, 0]$. V étant majoré par $(2-p)(p+2)$, en choisissant une autre valeur pour V, notre minoration ne concernerait qu'un intervalle plus restreint d'exposants p.

– Le cas $a = b$ correspond à l'opérateur associé aux polynômes de Gegenbauer. La minoration s'écrit alors :

$$B_p \geq \frac{(1-V)a}{4}.$$

Naturellement, cette minoration est meilleure pour $V = -1/a = -1/(\lambda_1 - 1)$ et on arrive au résultat suivant :

$$\forall p,\ 2 \leq p \leq \frac{2\lambda_1}{\lambda_1 - 2}, \quad B_p = \lambda_1.$$

Le théorème 2 étend à tous les p de $[2, \frac{2\lambda_1}{\lambda_1-2}]$ les conclusions d'un travail récent de Pearson [13] sur les constantes de Sobolev pour le générateur associé aux polynômes de Gegenbauer.

– Pour $b = 0$, soit $a = 2(\lambda_1 - 1)$, on a

$$B_p \geq (1-V)\frac{a}{4}.$$

La meilleure minoration que nous fournit notre méthode de calcul est atteinte pour $V = -1/a$ ou encore $V = -1/(\lambda_1 - 1)$ et on a alors

$$\forall 2 \le p \le \frac{2(2\lambda_1 - 1)}{2\lambda_1 - 3}, \quad B_p \ge \frac{2\lambda_1 - 1}{4}.$$

De plus, seulement pour cette valeur de V et sous la condition $b = 0$, le polynôme P_x est égal à 0. Nous reprendrons ce cas tout à fait particulier lors de la recherche de fonctions extrémales.

2 Constantes de Sobolev et de Sobolev logarithmiques pour l'opérateur de Laguerre

2.1 Introduction et définition

Nous étudions d'une manière analogue les constantes de Sobolev pour des exposants $1 \le p < 2$ pour l'opérateur associé aux polynômes de Laguerre. Par passage à la limite de p vers 2 dans ces inégalités, nous retrouverons la meilleure constante due à [9] dans l'inégalité de Sobolev logarithmique.

Considérons les opérateurs suivants agissant sur des fonctions de classe $C^2([0, +\infty[)$ définis par

$$Lf(x) = xf''(x) + Qf'(x)$$

où $Q = -x + D$ et $D \ge 1/2$. Les polynômes de Laguerre forment une base orthonormée de l'espace $L^2(\rho)$ où ρ désigne la mesure de probabilités :

$$d\rho(x) = (x^{D-1}e^{-x}/\Gamma(D))dx.$$

. On associe à L un opérateur gradient itéré Γ défini pour toute fonction de classe C^1 par $\Gamma(f, f) = xf'^2$. De plus, la fonction $H = Q / \int_0^{+\infty} Q^2\, d\mu$ est fonction propre de $-L$ associée à la première valeur propre $\lambda_1 = 1$. On vérifie que $\int_0^{+\infty} H^3\, d\rho$ est non nul. En reprenant la démonstration du théorème 1 appliquée à l'opérateur de Laguerre, on vérifie, alors, que la constante de Sobolev est strictement inférieure à 1. Nous en venons maintenant à la minoration de ces constantes.

2.2 Minorations des constantes de Sobolev pour l'opérateur de Laguerre

Nous étudions les constantes B_p dans les inégalités suivantes :

$$\frac{B_p}{p-2}\left[\left(\int_0^{+\infty} f^p\, d\rho\right)^{2/p} - \int_0^{+\infty} f^2\, d\rho\right] \le \int_0^{+\infty} (1 - x^2) f'^2\, d\rho$$

avec $1 \le p < 2$. Pour l'opérateur associé aux polynômes de Laguerre, il n'existe pas d'inégalités de Sobolev avec des exposants $p > 2$. Pour s'en convaincre, il suffit de considérer les fonctions suivantes $f(x) = e^{\alpha\sqrt{x} - \alpha^2/2}$. On vérifie que $\lim_{\alpha \to +\infty} \int_0^{+\infty} f^2\, d\rho$

et $\lim_{\alpha\to+\infty}\int_0^{+\infty} xf'^2\,d\rho$ sont finies, tandis que $\lim_{\alpha\to+\infty}\int_0^{+\infty} f^p\,d\rho = \infty$ dès que $p > 2$. Notre étude ne va donc concerner que les constantes de Sobolev relatives aux inégalités de Sobolev d'exposant p, $1 \le p < 2$. Notre résultat est le suivant :

Proposition 1 *Pour tout réel* $1 \le p < 2$, *on a*

$$B_p \ge \frac{(1-V)}{2}\left[2(1-2D)V + 1 + 2\sqrt{(2D-1)V[(2D-1)V+1]}\right],$$

avec

$$V = \frac{2-p}{p+2}.$$

Remarque : Si nous faisons tendre p vers 2 dans les inégalités de Sobolev précédentes, on obtient (voir remarque théorème 2) l'inégalité de Sobolev logarithmique suivante :

$$B_2\left(\int_0^{+\infty} f^2\log|f|\,d\rho - \int_0^{+\infty} f^2\,d\rho \log\left[\int_0^{+\infty} f^2\,d\rho\right]^{1/2}\right) \le \int_0^{+\infty} xf'^2\,d\rho$$

avec $B_2 = 1/2$. La meilleure constante dans l'inégalité de Sobolev logarithmique pour $D = 1/2$ a été calculée dans [9] et correspond bien au résultat obtenu.

Démonstration : nous adoptons une démarche analogue à celle effectuée au cours de la démonstration du théorème 2. On considère l'équation non linéaire suivante :

$$Lf = \frac{C_\epsilon}{p-2}[(1+\epsilon)f - f^{p-1}]$$

où L désigne l'opérateur de Laguerre. Par des transformations identiques à celles effectuées sur l'équation (5), on arrive à l'équation suivante :

$$\begin{aligned}
&[C_\epsilon(1+\epsilon) - (1-c)]\int_0^{+\infty} xu^{2\alpha}g^2\,d\mu \qquad (15)\\
&= \int_0^{+\infty} x^2\left\{u^{2\alpha}g'^2 + [2\alpha - 2\beta(p-1)]u^{3\alpha-1}g^2g'\right\}d\mu - \beta(p-1)\int_0^{+\infty} xu^{3\alpha-1}g^3\,d\mu\\
&+ [\alpha^2 - 2\alpha\beta(p-1) + (\beta-1)(1+\beta(p-2)) + \beta(p-1)]\int_0^{+\infty} x^2u^{4\alpha-2}g^4\,d\mu\\
&+ c\int_0^{+\infty} xu^{2\alpha}g^2\,d\mu
\end{aligned}$$

où $0 \le c < 1$. Remarquons de plus que

$$\int_0^{+\infty} x^2g^2g'u^{3\alpha-1}\,d\mu = -\frac{1}{3}\int_0^{+\infty} x(Q+1)u^{3\alpha-1}g^3\,d\rho - \frac{3\alpha-1}{3}\int_0^{+\infty} x^2u^{4\alpha-2}g^4\,d\rho.$$

En reportant ce terme dans l'équation précédente, on obtient

$$\begin{aligned}
&[C_\epsilon(1+\epsilon) - (1-c)]\int_0^{+\infty} xu^{2\alpha}g^2\,d\rho = \qquad (16)\\
&\int_0^{+\infty} x^2u^{2\alpha}g'^2\,d\rho - \frac{2}{3}[\alpha - \beta(p-1)]\int_0^{+\infty} xQu^{3\alpha-1}g^3\,d\rho\\
&+ \left[-\alpha^2 + \frac{2}{3}\alpha + \beta^2(p-2) + \frac{2}{3}\beta(4-p) - 1\right]\int_0^{+\infty} x^2u^{4\alpha-2}g^4\,d\rho\\
&+ c\int_0^{+\infty} xu^{2\alpha}g^2\,d\rho - \left[\beta(p-1) + \frac{2(\alpha - \beta(p-1))}{3}\right]\int_0^{+\infty} xu^{3\alpha-1}g^3\,d\rho.
\end{aligned}$$

Nous reprenons pour α la valeur calculée au (13) de manière à ce que le coefficient de x^2g^4 soit nul. Par de simples intégrations par parties, on calcule :

$$\int_0^{+\infty} xu^{3\alpha-1}g^3\,d\rho = -\frac{1}{2\alpha}\int_0^{+\infty} u^{2\alpha}[Qg^2 + 2xgg']\,d\rho$$

et

$$\int_0^{+\infty} xQu^{3\alpha-1}g^3\,d\rho = -\frac{1}{2\alpha}\int_0^{+\infty} u^{2\alpha}[Q^2xg^2 - xg^2 + 2xQgg']\,d\rho.$$

En reportant ces termes ainsi que la valeur de α (13) dans l'équation (16), il vient que :

$$[C_\epsilon(1+\epsilon) - (1-c)]\int_0^{+\infty} xu^{2\alpha}g^2\,d\rho = \int_0^{+\infty} u^{2\alpha}P_{g'}\,d\rho$$

avec

$$P_{g'} = x^2g'^2 + xg'g[2VQ + (1-V)] + \left[V(Q^2 - x) + \frac{1-V}{2}Q + cx\right]g^2$$

où

$$V = \frac{1}{3}\left[1 - \frac{\beta(p-1)}{\alpha}\right].$$

Nous cherchons c et V de manière à ce que ce polynôme soit positif. Le discriminant de $P_{g'}$ est égal à $\Delta'_g = p^2P_x$ où

$$\begin{aligned}P_x = 4V(V-1)x^2 + 2x\left[(V-1)(2V-1-4DV) + 2V - 2c\right]\\ +(V-1)[4VD^2 - 2D(2V-1) + V - 1].\end{aligned}$$

Pour que Δ'_g soit négatif, il est nécessaire que $0 < V < 1$. De plus, ce discriminant sera négatif, si celui associé au polynôme P_x est négatif. On a

$$\Delta_x = 4(2c - c_1)(2c - c_2)$$

avec

$$c_1 = 2V - (1-V)(2V-1) - 4DV(V-1) - 2(1-V)\sqrt{(2D-1)[(2D-1)V+1]}$$

et

$$c_2 = 2V - (1-V)(2V-1) - 4DV(V-1) + 2(1-V)\sqrt{(2D-1)[(2D-1)V+1]}.$$

En posant $c = c_1/2$, $\Delta_x = 0$ et donc P_x est négatif. On en déduit que $P_{g'}$ est positif pour toute fonction g. On obtient, finalement, la minoration suivante pour la constante B_p

$$B_p \geq \frac{(1-V)}{2}\left[2(1-2D)V + 1 + 2\sqrt{(2D-1)V[(2D-1)V+1]}\right].$$

De part la valeur de α choisie et la définition de V, on vérifie que $V \leq (2-p)/(p+2)$. Cette minoration et la condition $0 < V < 1$ sont compatibles dès que $1 \leq p < 2$. Finalement, en posant $V = (2-p)/(p+2)$, on trouve la minoration de la constante de Sobolev énoncée. La démonstration de la proposition 1 est terminée.

Remarques : – comme pour le cas des polynômes de Jacobi, nous ne pouvons conclure à l'optimalité de cette minoration. La valeur de V choisie nous permet seulement d'atteindre des exposants p avec $1 \leq p < 2$. Toutefois, on peut espérer trouver une meilleure minoration par un autre choix de V mais pour des valeurs de p plus restreintes.

– Ce choix de $V = (2-p)/(p-2)$ nous permet d'atteindre par passage à la limite l'exposant $p = 2$ et donc la meilleure constante dans l'inégalité de Sobolev logarithmique pour $D = 1/2$ due à [9]. De plus, dans ce cadre, notre méthode de calcul n'autorise qu'une valeur pour V à savoir $V = 0$.

– Si $D = 1/2$ et $p = 2$, on pose $V = 0$. Le polynôme P_x s'écrit alors $P_x = 2x[1-2c]$. Nous voyons que la meilleure minoration est atteinte pour $c = 1/2$, ce qui entraîne que P_x est nul. Cette remarque sera reprise lors de l'étude des fonctions extrémales relatives à ces opérateurs.

2.3 Quelques remarques sur les fonctions extrémales

Nous avons vu que dans le cadre d'opérateurs de la forme

$$Lf = (1-x^2)f'' + (-\lambda_1 x + D)f',$$

la constante de Sobolev est différente du trou spectral λ_1. L'intérêt de cette minoration est qu'elle permet de faire le lien entre la constante de Sobolev obtenue par Bakry [2] pour l'exposant optimal et la constante de Sobolev logarithmique due à Saloff-Coste [15] pour l'opérateur associé aux polynômes de Jacobi. D'après la remarque suivant le théorème 2, le cas $b = 0$ et $a \neq 0$ semble être un exemple tout à fait particulier. En effet, considérons l'équation non linéaire suivante :

$$Lf = \frac{B_p}{p-2}[f - f^{p-1}] \tag{17}$$

où $p = 2(a+1)/(a-1)$ et $B_p = (a+1)/4$. En faisant des opérations identiques à celles effectuées sur l'équation (5), on arrive à l'équation

$$\int_{-1}^{1} u^{2\alpha} P_{g'}\, d\mu = 0.$$

Pour de telles valeurs de a, b et B_p, on vérifie que le polynôme P_x est égal à 0. On en déduit que $\Delta_{g'} = 0$ et donc que $P_{g'} \geq 0$. D'après l'équation précédente, nécessairement $P_{g'} = 0$. Il est facile alors de déterminer une fonction g telle que $P_{g'} = 0$. On trouve $g(x) = (1-x)^{-1/2}$. On en déduit, que

$$f(x) = \left[d - 2\eta(1-x)^{1/2}\right]^{(a-1)/2}.$$

En reportant cette valeur de f dans l'équation (17), d et η doivent vérifier la condition suivante $d^2 = 8\eta^2 + 1$. Maintenant, en multipliant cette équation par f puis en l'intégrant par rapport à la mesure μ, il vient que :

$$\frac{a+1}{4(p-2)}\left[\int_{-1}^{1} f^p\, d\mu - \int_{-1}^{1} f^2\, d\mu\right] = \int_{-1}^{1}(1-x^2)f'^2\, d\mu \tag{18}$$

avec $p = 2(a+1)/(a-1)$. Cependant, excepté pour $\eta = 0$ c'est-à-dire $f = 1$, la fonction f trouvée n'est pas fonction extrémale pour l'inégalité de Sobolev car elle ne satisfait pas la condition de normalisation $\int_{-1}^{1} f^p\, d\mu = 1$. En fait, on vérifie que cette norme est strictement supérieure à 1. L'équation (18) implique seulement l'inégalité de Sobolev suivante :

$$\frac{a+1}{4(p-2)}\left[\left(\int_{-1}^{1} f^p\, d\mu\right)^{2/p} - \int_{-1}^{1} f^2\, d\mu\right] \leq \int_{-1}^{1} (1-x^2) f'^2\, d\mu.$$

D'une manière analogue, dans le cadre d'opérateurs de la forme

$$Lf = xf'' + (-x+D)f',$$

la proposition 1 nous donne la meilleure minoration pour la constante de Sobolev logarithmique à savoir $B_2 = 1/2$. Nous recherchons, alors, une fonction extrémale relative à cette constante. Cette valeur de la constante correspond dans la proposition 1, à $V = 0$ et donc à $\alpha = \beta = 1$. De plus, d'après la remarque de la proposition 1, si $D = 1/2$ et sous l'hypothèse $B_2 = 1/2$ soit $c = 1/2$, le polynôme P_x est égal à 0. Pour ce cas particulier, on peut s'attendre alors à trouver une fonction extrémale pour cette inégalité de Sobolev logarithmique. Mais, par notre méthode, nous arrivons seulement à une fonction non constante ne satisfaisant pas la condition de normalisation $\int_0^{+\infty} f^2\, d\rho = 1$ et vérifiant l'égalité

$$\frac{1}{2}\int_0^{+\infty} f^2 \log f\, d\rho = \int_0^{+\infty} \Gamma(f,f)\, d\rho.$$

Finalement, dans le cadre où les constantes de Sobolev ou de Sobolev logarithmiques sont différentes de $\lambda_1^{(\alpha)}$, notre méthode montrerait que seules les fonctions constantes sont extrémales pour les inégalités de Sobolev ou de Sobolev logarithmiques.

Je tiens à remercier M. Ledoux pour ses idées et ses conseils lors de l'élaboration de cet article.

Références

[1] D. Bakry, "L'Hypercontractivité et son utilisation en théorie des semi-groupes," Ecole d'été de Probabilités de Saint-Flour XXII-1992, p. 1-114, Lecture Notes in Math., Vol. 1581, Springer-Verlag, New-York/Berlin, 1994.

[2] D. Bakry, Remarques sur les semi-groupes de Jacobi, *Astérisque, Hommage à P. A. Meyer et J. Neveu*, p. 23-39, **236**, (1996).

[3] D. Bakry et M. Emery, Diffusions hypercontractives, Séminaire de Probabilités XIX, pp.175-206, Lecture Notes in Math., Vol. 1123, Springer-Verlag, New-York/Berlin, 1985.

[4] A. Bentaleb, Inégalité de Sobolev pour l'opérateur ultrasphérique, C. R. Acad. Sci. Paris, Série I, t. 317, (1993), 187-190.

[5] M. Berger, P. Gauduchon, et E. Mazet, Le spectre d'une variété riemannienne, Lecture Notes in Math., Vol. 194, Springer-Verlag, New-York/Berlin, 1971.

[6] E. Fontenas, Constantes dans les inégalités de Sobolev et fonctions extrémales, Thèse de l'Université Paul Sabatier, Toulouse III, (1995).

[7] E. Fontenas, Sur les constantes de Sobolev des variétés riemanniennes compactes et les fonctions extrémales des sphères, *Bulletin des Sciences Mathématiques*, p. 71-96, Vol. 121, 2, (1997).

[8] L. Gross, Logarithmic Sobolev inequalities, *Amer. J. Math.*, **97**, (1975), 1061-1083.

[9] A. Korzeniowski et D. Stroock, An example in the theory of hypercontractive semigroups, *Proc. A. M. S.*, **94**, No 1, (1985), 87-90.

[10] O. Mazet, Caractérisation des semi-groupes de diffusion sur un intervalle de IR associés à des familles de polynômes orthogonaux, Séminaire de Probabilités XXXI, Lecture Notes in Math, 1655, Springer-Verlag, (1997).

[11] C. Mueller et F. Weissler, Hypercontractivity for the heat semigroup for ultraspherical polynomials and on the n-sphere, *J. Funct. Anal.* **48** (1982), 252-283.

[12] A. Nikiforov, V. Ouvarov, "Fonctions spéciales de la physique mathématique," Éditions Mir, Moscou, 1983.

[13] J. M. Pearson, Best Constants in Sobolev Inequalities for Ultraspherical Polynomials, *Arch. Rational Mech. Anal.* **116** (1991), 361-374.

[14] O. S. Rothaus, Hypercontractivity and the Bakry-Emery criterion for compaet Lie groups, *J. Funct. Anal.* **65** (1986), 358-367.

[15] L. Saloff-Coste, Precise estimates on the rate at which certain diffusions tend to equilibrium, *Math. Z.* **217** (1994), 641-677.

QUAND L'INEGALITE LOG-SOBOLEV IMPLIQUE L'INEGALITE DE TROU SPECTRAL

P.Mathieu.
LATP-CMI. Rue Joliot-curie.
F-13013 Marseille.
PMathieu@gyptis.univ-mrs.fr.

Résumé: Nous montrons que, sur une variété Riemannienne de volume fini, l'inégalité de Sobolev logarithmique implique l'inégalité de trou spectral.

Soit E un espace localement compact séparable. Soit μ une mesure de Radon sur E. Nous supposons que μ est une probabilité et que son support est E. Soit $(\mathcal{E}, \mathcal{F})$ une forme de Dirichlet symétrique sur (E, μ) au sens de Fukushima 1980. Nous supposerons que $1 \in \mathcal{F}$ (1 est la fonction constante égale à 1) et que $\mathcal{E}(u, 1) = 0$ pour tout $u \in \mathcal{F}$. On pose $\mathcal{E}(u) = \mathcal{E}(u, u)$. Nous utilisons aussi la notation $\|u\|_p$ pour la norme de la fonction u dans $L_p = L_p(E, \mu)$, $1 \leq p \leq +\infty$. Pour une fonction positive $u \in L_2$, on notera

$$E_2(u) = \int u^2 \log u^2 d\mu - \int u^2 d\mu \log \int u^2 d\mu$$

Soient $m \in I\!R_+$ et $\Lambda > 0$. Nous dirons que la forme de Dirichlet $(\mathcal{E}, \mathcal{F})$ satisfait l'inégalité log-Sobolev de constantes Λ et m si, pour toute fonction positive $u \in \mathcal{F}$, on a

$$E_2(u) \leq \frac{1}{\Lambda}(\mathcal{E}(u) + m \int u^2 d\mu) \tag{1}$$

On appelle "inégalité log-Sobolev tendue" l'inégalité (1) si $m = 0$. Il est connu que l'inégalité (1) est équivalente à certaines propriétés de contraction du semi-groupe Markovien associé à la forme de Dirichlet $(\mathcal{E}, \mathcal{F})$. Par exemple on a, pour tout $t > 0$, $\|P_t u\|_q \leq e^{m'} \|u\|_2$, où $q = 1 + \exp(4\Lambda t)$ et $m' = (1/2 - 1/q)m/\Lambda$. Remarquer que si $m = 0$ - i.e. si l'inégalité est tendue - alors $m' = 0$ et P_t est en fait une contraction de L_2 dans L_q. Nous renvoyons à Bakry 1992 pour d'autres résultats.
Soit maintenant

$$\Lambda(2) = \inf_{u \in \mathcal{F}, \int u d\mu = 0} \frac{\mathcal{E}(u)}{\int u^2 d\mu} \tag{2}$$

le trou spectral du générateur de $(\mathcal{E}, \mathcal{F})$. Nous dirons que la forme de Dirichlet satisfait l'inégalité de trou spectral si $\Lambda(2) > 0$.
D'après les propositions 3.7 et 3.9 de Bakry 1992, $(\mathcal{E}, \mathcal{F})$ satisfait une inégalité log-Sobolev tendue si et seulement si elle satisfait une inégalité log-Sobolev et l'inégalité de trou spectral. Des estimées du trou spectral en fonction de constantes dans des inégalités de Sobolev, log-Sobolev... ont également été obtenues par S.Aida 1997.
L'objet de cette note est de montrer que, sous certaines conditions d'ergodicité plus faibles que l'inégalité de trou spectral, on conserve l'implication: "log-Sobolev" implique "log-Sobolev tendue".
Pour cela nous introduisons la propriété suivante:

(P) Pour toute suite de fonctions $u_n \in \mathcal{F}$ telles que $\int u_n d\mu = 0$, $\|u_n\|_\infty \le 1$ et $\mathcal{E}(u_n) \to 0$, alors u_n converge vers 0 en μ mesure.

THEOREME

Si $(\mathcal{E}, \mathcal{F})$ satisfait la propriété (P) et une inégalité log-Sobolev, alors $(\mathcal{E}, \mathcal{F})$ satisfait l'inégalité de trou spectral.

Avant de donner la preuve du théorème, voici un exemple:

COROLLAIRE 1

Soit E une variété Riemannienne connexe complète. On choisit pour μ la mesure de volume Riemannienne et nous supposerons donc que E est de volume égal à 1. Soit $\mathcal{E}(u) = \int_E |du|^2 d\mu$, la forme de Dirichlet associée à l'opérateur de Laplace Beltrami sur E. $\mathcal{F}$ est l'espace de Sobolev H_1. Alors $(\mathcal{E}, \mathcal{F})$ satisfait la propriété (P). D'après le théorème, cela implique que, si $(\mathcal{E}, \mathcal{F})$ satisfait l'inégalité log-Sobolev alors elle satisfait aussi l'inégalité de trou spectral.

Preuve

Nous recopions la preuve donnée dans Mathieu 1996-2. Soit K un sous-ensemble compact connexe régulier de E. Il existe une constante C telle que, pour toute fonction $u \in C^\infty$, on ait

$$\mu(K) \int_K \left(u - \frac{\int_K u d\mu}{\mu(K)}\right)^2 d\mu \le \frac{1}{C} \int_K |du|^2 d\mu \tag{3}$$

(En d'autres termes, nous avons une inégalité de Poincaré sur K. Nous donnons une preuve de l'inégalité (3) en annexe).

Soit u_n une suite de fonctions de $\mathcal{F}$ telles que $\int u_n d\mu = 0$, $\|u_n\|_\infty \le 1$ et $\mathcal{E}(u_n) \to 0$. Alors

$$\begin{aligned}
&\int u_n^2 d\mu \\
&= \int_K u_n^2 d\mu + \int_{E-K} u_n^2 d\mu \\
&\le \frac{(\int_K u_n d\mu)^2}{\mu(K)^2} + \frac{1}{C\mu(K)} \int_K |du_n|^2 d\mu + \int_{E-K} u_n^2 d\mu \\
&\le \frac{\mu(E-K)^2}{\mu(K)^2} + \frac{1}{C\mu(K)} \mathcal{E}(u_n) + \mu(E-K)
\end{aligned}$$

Donc

$$\limsup \int u_n^2 d\mu \le \frac{\mu(E-K)^2}{\mu(K)^2} + \mu(E-K)$$

Il suffit maintenant de choisir K assez gros pour montrer que u_n converge vers 0 dans L_2 donc en μ mesure.

Remarques

1-Pour une forme de Dirichlet symétrique (comme c'est le cas ici), la propriété (P) est équivalente à la propriété suivante:

$$\sup_{\|u\|_\infty \leq 1} \|P_t u - \int u d\mu\|_1 \to 0$$

quand t tend vers $+\infty$.
2-La propriété (P) permet également d'obtenir des estimées pour les temps d'atteinte pour le processus de Markov associé à $(\mathcal{E}, \mathcal{F})$: soit $(X_t, t \geq 0)$ le processus de Markov associé à $(\mathcal{E}, \mathcal{F})$. Pour un ensemble mesurable A, on note $\tau_A = \inf\{t > 0 \ t.q. \ X_t \in A\}$ le temps d'atteinte de A par X. Alors (P) a lieu si et seulement si $\lim_{T\to\infty} \sup_A \mu(A) P_\mu[\tau_A \geq T] = 0$. (P_μ désigne la loi du processus X quand la loi de X_0 est μ). Voir Mathieu 1996-2 et 1996-3.
3-Tous les processus de Markov symétriques ergodiques ne satisfont pas (P). Par exemple le processus d'exclusion simple symétrique en dimension 1 ne satisfait pas (P). (Voir Mathieu 1996-2)

COROLLAIRE 2

Comme dans le corollaire 1, nous choisissons pour $(\mathcal{E}, \mathcal{F})$ la forme de Dirichlet associée à l'opérateur de Laplace-Beltrami sur une variété connexe complète de volume 1. Soit $n > 2$. Supposons que $(\mathcal{E}, \mathcal{F})$ satisfait l'inégalité de Sobolev de dimension n:

$$\|u\|_p \leq \frac{1}{\Lambda}(\mathcal{E}(u) + m \int u^2 d\mu)$$

où $p = 2n/(n-2)$.
Alors la variété E est compacte.

Preuve
L'inégalité de Sobolev implique l'inégalité de Sobolev faible et donc l'inégalité log-Sobolev. (Voir Bakry 1992, chapitre 4). D'après le théorème, on a donc une inégalité de trou spectral. $(\mathcal{E}, \mathcal{F})$ satisfait donc une inégalité de Sobolev et l'inégalité de trou spectral. D'après Bakry 1992, Lemme 4.1, cela implique que $(\mathcal{E}, \mathcal{F})$ satisfait en fait l'inégalité de Sobolev tendue i.e. avec $m = \Lambda$.
D'après la remarque 1 suivant le théorème 5.4 de Bakry 1992, le diamètre de E est fini. Donc E est compacte.

Preuve du théorème
Supposons l'inégalité (1) satisfaite ainsi que (P).
1- Reformulons (P): soit $u_n \in \mathcal{F}$, une suite de fonctions telle que $\int u_n d\mu = 0$, $\int u_n^2 d\mu \leq 1$ et $\mathcal{E}(u_n) \to 0$. Soit $k > 0$ et posons $v_n = u_n \wedge k \vee (-k)$ et $w_n = v_n - \int v_n d\mu$. Alors

$$\frac{k}{2}\mu(u_n > \frac{k}{2}) \leq \int_{u_n > \frac{k}{2}} u_n d\mu = -\int_{u_n \leq \frac{k}{2}} u_n d\mu \leq \sqrt{\mu(u_n \leq \frac{k}{2})}$$

car $\int u_n^2 d\mu \leq 1$. Donc

$$\int (k - v_n) d\mu \geq \frac{k}{2}\mu(v_n \leq \frac{k}{2}) = \frac{k}{2}\mu(u_n \leq \frac{k}{2})$$
$$\geq \frac{k^3}{8}\mu(u_n > \frac{k}{2})^2 \geq \frac{k^3}{8}\mu(u_n \geq k)^2$$

Sur l'ensemble où $u_n \geq k$, on a $v_n = k$, donc

$$\int |w_n| d\mu \geq \mu(u_n \geq k) \int (k - v_n) d\mu \geq \frac{k^3}{8}\mu(u_n \geq k)^3$$

Comme nous avons supposé que $\mathcal{E}(1, u) = 0$ et comme v_n est une contraction normale de u_n, on a $\mathcal{E}(w_n) = \mathcal{E}(v_n) \leq \mathcal{E}(u_n)$. De plus $\|w_n\|_\infty \leq 2k$. La suite w_n satisfait donc les hypothèses de (P), donc w_n converge vers 0 en μ mesure et donc dans L_1. D'après les inégalités ci-dessus, cela implique que $\mu(u_n \geq k) \to 0$. De même on montrerait que $\mu(u_n \leq -k) \to 0$. On a ainsi prouvé que u_n converge vers 0 en μ mesure. Comme la suite u_n est bornée dans L_2 elle converge en fait vers 0 dans L_1.
Nous avons donc montré que (P) implique
(P'): pour toute suite de fonctions $u_n \in \mathcal{F}$ telles que $\int u_n d\mu = 0$, $\int u_n^2 d\mu \leq 1$ et $\mathcal{E}(u_n) \to 0$, alors u_n converge vers 0 dans L_1.
2- Nous allons aussi ré-écrire l'inégalité log-Sobolev sous une forme un peu différente. Soient $p < 2 < p'$ et $q < 0 < q'$ tels que $1/p + 1/q = 1/p' + 1/q' = 1/2$. L'inégalité de Holder implique que

$$q \log \frac{\|u\|_p}{\|u\|_2} \leq q' \log \frac{\|u\|_{p'}}{\|u\|_2}$$

En faisant tendre p' vers 2 et en choisissant $p = 1$, on obtient

$$-2 \log \frac{\|u\|_1}{\|u\|_2} \leq \frac{E_2(u)}{\|u\|_2^2} - \tag{4}$$

(1) et (4) impliquent que

$$2 \log \frac{\|u\|_2}{\|u\|_1} \leq \frac{1}{\Lambda}\left(\frac{\mathcal{E}(u)}{\|u\|_2^2} + m\right) \tag{5}$$

pour toute fonction positive $u \in \mathcal{F}$.
3- Montrons que $\Lambda(2) \neq 0$. Il suffit de prouver que, pour toute suite $u_n \in \mathcal{F}$ telle que $\int u_n d\mu = 0$, $\int u_n^2 d\mu \leq 1$ et $\mathcal{E}(u_n) \to 0$, on a $\int u_n^2 d\mu \to 0$ Or d'après (P'), u_n converge vers 0 dans L_1. D'après (5), u_n converge donc vers 0 dans L_2.

Annexe

Nous donnons ici une démonstration de l'inégalité (3): on notera n la dimension de la variété E, d la distance sur K et B la boule unité de $\mathbb{R}^n$.
Soit $\varepsilon_1 > 0$. Soit $(x_i, i = 1...N)$ une famille de points de l'intérieur de K tels que $K \subset \cup_i B(x_i, \varepsilon_1)$, où $B(x, r)$ désigne la boule géodésique de centre x et rayon r.
Soient $k, l \in \{1, ...N\}$. Soit $(l(t), t \in [0, 1])$ un chemin dans l'intérieur de K tel que $l(0) = x_k$ et $l(1) = x_l$. Soit $\varepsilon_2 = \inf_{k,l} \inf_t d(l(t), \partial K)$. Soit $\varepsilon_3 < \varepsilon_1, \varepsilon_3 < \varepsilon_2/3$. Soient $x \in B(x_k, \varepsilon_1) \cap K$ et $y \in B(x_l, \varepsilon_1) \cap K$. Soit $(l_{x,y}(t), t \in [-1/4, 5/4])$

un chemin de x à y tel que $l_{x,y}(t) \in K$, $d(l_{x,y}(0), x_k) \leq \varepsilon_3$, $d(l_{x,y}(1), x_l) \leq \varepsilon_3$, $\sup_{t\in[0,1]} d(l_{x,y}(t), l(t)) < \varepsilon_2$. Nous supposons de plus que $\sup_{x,y} \sup_t |\frac{d}{dt} l_{x,y}(t)| < \infty$ et que

(*) $\int_{K\cap B(x_k,\varepsilon_1)} d\mu(x) \int_{K\cap B(x_l,\varepsilon_1)} d\mu(y) \Phi(l_{x,y}(t)) \leq C \int_K \Phi(z) d\mu(z)$

pour toute fonction positive régulière Φ. Ici C est une constante indépendante de Φ et t. Pour construire $(l_{x,y}(t), -1/4 \leq t \leq 0)$, on utilise un difféomorphisme entre $B(x_k, \varepsilon_1)$ et B. Rappelons que, par hypothèse, K est régulier, c'est-à-dire qu' il est possible de choisir ε_1 et les x_i tels que soit $B(x_i, \varepsilon_1) \subset K$, soit il existe un difféomorphisme F de $B(x_i, \varepsilon_1)$ vers B tel que l'image de $B(x_i, \varepsilon_1) \cap \partial K$ soit $B \cap [\mathrm{x}_n = 0]$, et l'image de $B(x_i, \varepsilon_1) \cap Int(K)$ soit $B \cap [\mathrm{x}_n < 0]$. On procède de même pour $t \in [1, 5/4]$. Soient maintenant $e_0 \in T_{x_k} K$ et $e_1 \in T_{x_l} K$ tels que $l_{x,y}(0) = \exp_{x_k}(e_0)$ et $l_{x,y}(1) = \exp_{x_l}(e_1)$. On identifie les espaces tangents $(T_{l(t)} K, t \in [0,1])$ par transport parallèle le long de la courbe l. Pour $t \in [0,1]$, posons $e_t = e_0 + t(e_1 - e_0) \in T_{l(t)} K$ et $l_{x,y}(t) = \exp_{l(t)}(e_t)$. Le chemin ainsi construit satisfait nos hypothèses.
On a alors

$$
\begin{aligned}
&2\mu(K) \int_K (u(x) - \frac{\int_K u(y) d\mu(y)}{\mu(K)})^2 d\mu(x) \\
&= \int_K d\mu(x) \int_K d\mu(y) (u(x) - u(y))^2 \\
&\leq \Sigma_{k,l} \int_{K\cap B(x_k,\varepsilon_1)} d\mu(x) \int_{K\cap B(x_l,\varepsilon_1)} d\mu(y) (u(x) - u(y))^2
\end{aligned}
$$

Et

$$
\begin{aligned}
&\int_{K\cap B(x_k,\varepsilon_1)} d\mu(x) \int_{K\cap B(x_l,\varepsilon_1)} d\mu(y) (u(x) - u(y))^2 \\
&= \int_{K\cap B(x_k,\varepsilon_1)} d\mu(x) \int_{K\cap B(x_l,\varepsilon_1)} d\mu(y) (\int_{-1/4}^{5/4} dt\ du(l_{x,y}(t)). \frac{d}{dt} l_{x,y}(t))^2 \\
&\leq C' \int_{K\cap B(x_k,\varepsilon_1)} d\mu(x) \int_{K\cap B(x_l,\varepsilon_1)} d\mu(y) \int_{-1/4}^{5/4} dt |du(l_{x,y}(t))|^2 \\
&\leq C' \sup_t \int_{K\cap B(x_k,\varepsilon_1)} d\mu(x) \int_{K\cap B(x_l,\varepsilon_1)} d\mu(y) |du(l_{x,y}(t))|^2 \\
&\leq C' \int_K |du(x)|^2 d\mu(x) \text{ d'après (*)} \\
&= C' \mathcal{E}(u)
\end{aligned}
$$

où la constante C' ne dépend ni de k, ni de l, ni de u.

L'auteur remercie D.Bakry et M.Ledoux pour leurs suggestions qui ont permis de simplifier la preuve du théorème.

Références:

Aida.S. (1997): Uniform positivity improving property, Sobolev inequality and spectral gaps. Preprint.

Bakry.D. (1992): L'hypercontractivité et son utilisation en théorie des semi-groupes. Ecole d'été de St Flour 1992. LNM 1581.

Fukushima.M. (1980): Dirichlet forms and Markov processes. (North Holland).

Mathieu.P. (1996-1): Hitting times and spectral gap inequalities. A paraître dans Ann. IHP.

Mathieu.P (1996-2): On the law of the hitting time of a small set by a Markov process. Preprint

Mathieu.P (1996-3): Uniform estimates of the hitting times of big sets by a Markov process. Preprint

Trous spectraux à basse température : un contre-exemple à un comportement asymptotique escompté

Laurent Miclo
CNRS, UMR C5583
E-mail : miclo@cict.fr

Summary : On a general separable state space, let P be a Markovian kernel, reversible with respect to a probability μ, such that there is a spectral gap for the operator $\mathrm{Id} - P$ in $\mathbb{L}^2(\mu)$. Let P_β be the Metropolis kernel associated to P and to a potential βU, where U is a measurable bounded function and $\beta \geq 0$ is the inverse temperature. We are interested in its spectral gap $\lambda(\beta)$ and specially in the behaviour for large β of the quantity $-\beta^{-1}\ln(\lambda(\beta))$. By analogy with some classical cases, we have conjectured that it should converge to c, the largest secondary well exit height, but as we will see on a counter-example, this is not true, since $-\beta^{-1}\ln(\lambda(\beta))$ can asymptotically oscillate, and even if it converges, the limit can be different from c. We also get similar results for the asymptotical behaviour at small temperature of the isoperimetric constants $I(\beta)$, but for them the study is a little more satisfactory, as we can conclude to the convergence of $-\beta^{-1}\ln(I(\beta))$ to c, if it is equal to $\operatorname{ess\,sup}_\mu U - \operatorname{ess\,inf}_\mu U$.

Résumé : Sur un espace mesurable et séparable quelconque, soit P un noyau markovien réversible par rapport à une probabilité μ, tel que $\mathrm{Id} - P$ admette un trou spectral dans $\mathbb{L}^2(\mu)$. Considérons le noyau de Métropolis P_β associé à P et au potentiel βU, où U est une fonction mesurable bornée et où $\beta \geq 0$ est l'inverse de la température, et notons $\lambda(\beta)$ son trou spectral. On s'intéresse au comportement pour β grand de la quantité $-\beta^{-1}\ln(\lambda(\beta))$, car par analogie avec des cas classiques connus, on aurait voulu montrer qu'elle converge vers c, la plus grande hauteur de sortie de puits secondaires associée à (P, μ, U). Mais on verra sur un contre-exemple que ceci est faux, $-\beta^{-1}\ln(\lambda(\beta))$ pouvant asymptotiquement toujours osciller, et même si elle converge, la limite n'est pas nécessairement c. On obtiendra également des résultats similaires pour le comportement asymptotique à basse température des constantes isopérimétriques $I(\beta)$, cependant pour elles l'étude sera un peu plus satisfaisante, car on pourra conclure à la convergence de $-\beta^{-1}\ln(I(\beta))$ vers c, si cette constante vaut $\operatorname{ess\,sup}_\mu U - \operatorname{ess\,inf}_\mu U$.

1 Une conjecture fausse

Soit P un noyau markovien défini sur un espace mesurable et séparable $(S, \mathcal{S})$. Supposons que P admette une probabilité réversible μ, i.e. qui satisfait pour toutes fonctions mesurables et bornées f, g définies sur S,

$$\int_{S\times S} f(x)g(y)\,\mu(dx)P(x,dy) = \int_{S\times S} g(x)f(y)\,\mu(dx)P(x,dy)$$

De manière usuelle, on peut alors considérer P comme un opérateur auto-adjoint sur l'espace de Hilbert $\mathbb{L}^2(\mu)$, positif et de norme égale à 1. Notons I l'opérateur identité et faisons l'hypothèse supplémentaire (de type ergodicité forte pour le processus de Markov homogène qui admet P pour noyau d'intensités de transitions et qui part d'une probabilité initiale absolument continue par rapport à μ) que $I-P$ admet un trou spectral, c'est-à-dire que la quantité suivante est strictement positive :

$$\lambda(I-P) \stackrel{\text{déf.}}{=} \inf_{f\in\mathbb{L}^2(\mu),\,\mu(f)=0} \frac{\mathcal{E}_{I-P}(f,f)}{\mathcal{E}_{I-E_\mu}(f,f)}$$

où pour tout opérateur borné Q, on a noté

$$\forall\, f\in\mathbb{L}^2(\mu),\qquad \mathcal{E}_Q(f,f)=\int fQf\,d\mu$$

et où E_μ est l'opérateur d'espérance par rapport à μ

$$\forall\, f\in\mathbb{L}^2(\mu),\qquad E_\mu f=\mu(f)\mathbb{I}$$

$\mathbb{I}$ désignant la fonction prenant toujours la valeur 1.

Rappelons comment on peut approcher P (puis surtout $\lambda(I-P)$) par des noyaux markoviens définis sur des ensembles finis : si $\mathcal{A}$ est une sous-tribu de $\mathcal{S}$, notons $Q_\mathcal{A}$ la projection orthogonale sur $\mathbb{L}^2(S,\mathcal{A},\mu_\mathcal{A})$ (qui n'est autre que l'espérance conditionnelle sachant $\mathcal{A}$), où $\mu_\mathcal{A}$ est la restriction de μ à $\mathcal{A}$, et $P_\mathcal{A}$ l'opérateur sur $\mathbb{L}^2(S,\mathcal{A},\mu_\mathcal{A})$ défini par $P_\mathcal{A}=Q_\mathcal{A}P$. Il est clair que $P_\mathcal{A}$ est encore auto-adjoint (ainsi que positif et de norme égale à 1), i.e. $\mu_\mathcal{A}$ est réversible pour le noyau de Markov généralisé $P_\mathcal{A}$ sur $(S,\mathcal{A})$. De plus le trou spectral de $P_\mathcal{A}$ est alors plus grand que celui de P. En effet,

$$\begin{aligned}
\lambda(I-P_\mathcal{A}) &= 1-\sup_{f\in\mathbb{L}^2(\mathcal{A},\mu_\mathcal{A}),\,\mu_\mathcal{A}(f)=0,\,\mu_\mathcal{A}(f^2)=1}\int fQ_\mathcal{A}Pf\,d\mu_\mathcal{A}\\
&= 1-\sup_{f\in\mathbb{L}^2(\mathcal{A},\mu_\mathcal{A}),\,\mu(f)=0,\,\mu(f^2)=1}\int fPf\,d\mu\\
&\geq \lambda(I-P)
\end{aligned}$$

Si la sous-tribu $\mathcal{A}$ est finie, disons si elle est engendrée par une partition $S=\sqcup_{1\leq i\leq N}A_i$, indexée de telle sorte que pour un $1\leq N_0\leq N$, on a pour tout $1\leq i\leq N_0$, $\mu(A_i)>0$ et pour tout $i>N_0$, $\mu(A_i)=0$, on peut interpréter $P_\mathcal{A}$ comme un noyau de probabilités de transitions sur l'ensemble fini $\{A_1,\cdots,A_{N_0}\}$ donné par

$$\forall\, 1\leq i,j\leq N_0,\qquad P_\mathcal{A}(A_i,A_j)=\frac{1}{\mu(A_i)}\int\mu(dx)P(x,dy)\mathbb{I}_{A_i}(x)\mathbb{I}_{A_j}(y)$$

qui admet pour probabilité réversible celle donnant le poids $\mu(A_i)$ à A_i, pour $1\leq i\leq N_0$.

Du fait que $\mathcal{S}$ est séparable, il existe au moins une suite croissante $(\mathcal{A}_n)_{n\in\mathbb{N}}$ de sous-tribus finies telle que $\mathcal{S}=\sigma(\mathcal{A}_n\,;\,n\in\mathbb{N})$ (on notera désormais Υ l'ensemble des telles suites $(\mathcal{A}_n)_{n\in\mathbb{N}}$). Par la procédure précédente, on en déduit alors une suite $(P_{\mathcal{A}_n})_{n\in\mathbb{N}}$ de noyaux markoviens finis telle que $(\lambda(I-P_{\mathcal{A}_n}))_{n\in\mathbb{N}}$ soit une suite décroissante dont la limite est au moins supérieure à $\lambda(I-P)$. Vérifions qu'en fait

$$\lim_{n\to\infty}\lambda(I-P_{\mathcal{A}_n})=\lambda(I-P)$$

Pour ceci soit $\epsilon > 0$ et $f \in \mathbb{L}^2(\mu)$ tels que $\mu(f) = 0$ et

$$1 - \frac{\mu(fPf)}{\mu(f^2)} \leq \lambda(I - P) + \epsilon$$

il suffit de considérer pour $n \in \mathbb{N}$, f_n la projection orthogonale de f sur $\mathbb{L}^2(S, \mathcal{A}_n, \mu_{\mathcal{A}_n})$ et d'utiliser le fait que f_n converge pour n grand vers f dans $\mathbb{L}^2(\mu)$ (théorème de convergence des martingales $\mathbb{L}^2$) pour se rendre compte qu'à la limite $\lambda(I - P) + \epsilon \geq \lim_{n\to\infty} \lambda(I - P_{\mathcal{A}_n})$, puis obtenir le résultat annoncé.

Soit maintenant U un potentiel, qui sera simplement une fonction mesurable et bornée sur S, et $\beta \geq 0$ qui représentera l'inverse de la température. On peut leur associer un nouveau noyau markovien P_β défini par

$$\forall\, x \in S, \forall\, A \in \mathcal{S}, \quad P_\beta(x, A) = \int \exp(-\beta(U(y) - U(x))_+)\mathbb{I}_A(y)P(x, dy) + d_\beta(x)\delta_x(A)$$

où $d_\beta(x) = 1 - \int \exp(-\beta(U(y) - U(x))_+)P(x, dy)$.

On vérifie immédiatement que la probabilité μ_β qui admet la densité

$$S \ni x \mapsto \frac{\exp(-\beta U(x))}{Z_\beta}$$

par rapport à μ, avec $Z_\beta = \mu(\exp(-\beta U))$, est réversible pour P_β. On conviendra de poser

$$\lambda(\beta) \stackrel{\text{déf.}}{=} \lambda(I - P_\beta)$$

En remarquant que pour $f \in \mathbb{L}^2(\mu_\beta)$ (qui n'est autre que $\mathbb{L}^2(\mu)$, car pour $\beta \geq 0$ fixé, les normes $\|\cdot\|_{\mathbb{L}^2(\mu_\beta)}$ et $\|\cdot\|_{\mathbb{L}^2(\mu)}$ sont équivalentes), on peut écrire

$$\begin{aligned}
\mathcal{E}_{I-P_\beta}(f, f) &= Z_\beta^{-1} \int \mu(dx)P(x, dy)\exp(-\beta(U(x) \vee U(y)))(f(y) - f(x))^2/2 \\
\mathcal{E}_{I-\mathbb{E}_{\mu_\beta}}(f, f) &= Z_\beta^{-2} \int \mu(dx)\mu(dy)\exp(-\beta(U(x) + U(y)))(f(y) - f(x))^2/2
\end{aligned}$$

il apparaît que l'on a l'encadrement pour tout $\beta \geq 0$,

$$\lambda(0)\exp(-2\beta(\operatorname*{ess\,sup}_\mu U - \operatorname*{ess\,inf}_\mu U)) \leq \lambda(\beta) \leq 2 \vee [\lambda(0)\exp(2\beta(\operatorname*{ess\,sup}_\mu U - \operatorname*{ess\,inf}_\mu U))]$$

et on peut se demander si la limite suivante

$$\lim_{\beta\to+\infty} \beta^{-1}\ln(\lambda(\beta))$$

existe?

Plus précisément, soit $\mathcal{A}$ une sous-tribu finie de $\mathcal{S}$, que l'on suppose à nouveau engendrée par une partition $S = \sqcup_{1\leq i\leq N} A_i$, avec pour tout $1 \leq i \leq N$, $\mu(A_i) > 0$. On peut considérer comme précédemment le noyau de probabilités de transitions $P_{\beta,\mathcal{A}}$ sur l'ensemble fini $\{A_1, \cdots, A_N\}$ associé à P_β. On a clairement

$$\begin{aligned}
\forall\, 1 \leq i \leq N, \quad & \lim_{\beta\to+\infty} -\beta^{-1}\ln(\mu_\beta(A_i)) &= \operatorname*{ess\,inf}_{\mathbb{I}_{A_i}\mu} U \\
\forall\, 1 \leq i, j \leq N, \quad & \lim_{\beta\to+\infty} -\beta^{-1}\ln(P_{\beta,\mathcal{A}}(A_i, A_j)) &= \operatorname*{ess\,inf}_{\mathbb{I}_{A_i\times A_j}m} W - \operatorname*{ess\,inf}_{\mathbb{I}_{A_i}\mu} U
\end{aligned}$$

où $\mathbb{I}_{A_i}\mu$ désigne la mesure admettant $\mathbb{I}_{A_i}$ pour densité par rapport à μ, où $m(dx,dy)$ est la probabilité sur $S\times S$ donnée par $\mu(x)P(x,dy)$ et où W est défini sur $S\times S$ par $W(x,y)=U(y)\vee U(x)$.

Remarquons que la dernière limite, que l'on appelera désormais $V(A_i,A_j)$ et qui est à valeurs dans $\overline{\mathbb{R}_+}$, est infinie si et seulement si $m(A_i\times A_j)=0$. On notera aussi pour $1\leq i\leq N$, $U(A_i)=\operatorname{ess\,inf}_{\mathbb{I}_{A_i}\mu}U$. Des résultats classiques (voir Holley et Stroock [3] et la généralisation donnée dans [5], car ici on n'a pas nécessairement $V(A_i,A_j)=(U(A_j)-U(A_i))_+$ pour $1\leq j\neq i\leq N$, que l'on peut encore étendre pour traiter notre situation actuelle, grâce à des résultats de comparaisons comme par exemple celui décrit dans le lemme 2.2.12 de [8]) permettent alors de déduire des convergences précédentes que la limite suivante existe

$$c(\mathcal{A})\stackrel{\text{déf.}}{=}-\lim_{\beta\to+\infty}\beta^{-1}\ln(\lambda(I-P_{\beta,\mathcal{A}}))\geq 0 \tag{1}$$

et on peut décrire cette quantité uniquement en termes des $(V(A_i,A_j))_{1\leq i\neq j\leq N}$.

Il est clair que l'application $c(\cdot)$ est croissante sur l'ensemble des sous-tribus finies de $\mathcal{S}$, ce qui nous conduit ensuite à poser

$$c\stackrel{\text{déf.}}{=}\sup_{\mathcal{A}\text{ sous-tribu finie de }\mathcal{S}}c(\mathcal{A})$$

Si on se donne une suite croissante $(\mathcal{A}_n)_{n\in\mathbb{N}}$ de sous-tribus finies convergeante vers $\mathcal{S}$, la suite croissante des constantes associées $c(\mathcal{A}_n)$ ne satisfait pas nécessairement

$$\lim_{n\to\infty}\uparrow c(\mathcal{A}_n)=c$$

comme le montre le contre-exemple suivant :

L'espace des états S sera ici le cercle $\mathbb{R}/\mathbb{Z}$ que l'on munit de sa tribu borélienne $\mathcal{S}$. Soit $0<h\leq 1/2$, on considère le noyau markovien P donné par

$$\forall\, x\in S,\qquad P(x,dy)=\mathbb{I}_{]x-h,x+h[}(y)\frac{dy}{2h}$$

Il est clair que P admet pour unique probabilité invariante la mesure de Lebesgue μ qui de plus est bien réversible. Pour vérifier que $I-P$ admet un trou spectral, on utilise l'analyse de Fourier : soit $f\in\mathbb{L}^2(\mu)$ telle que $\mu(f)=0$, on peut l'écrire dans $\mathbb{L}^2(\mu)$ sous la forme

$$f=\sum_{n\in\mathbb{Z}^*}a_ne_n$$

où pour tout $n\in\mathbb{Z}^*$, e_n est la fonction trigonométrique $S\ni x\mapsto\exp(i2\pi nx)$ et où $(a_n)_{n\in\mathbb{Z}^*}$ est une suite de complexes satisfaisant pour tout $n\in\mathbb{N}^*$, $a_{-n}=\overline{a_n}$ (car on suppose f à valeurs réelles).

On vérifie immédiatement que Pf est alors donné par

$$Pf=\sum_{n\in\mathbb{Z}^*}\widetilde{a}_ne_n$$

où pour tout $n\in\mathbb{Z}^*$,

$$\widetilde{a}_n=a_n\frac{\sin(2\pi nh)}{2\pi nh}$$

En utilisant le fait que

$$\begin{aligned}|2\sin(2\pi nh)| &\leq |\exp(i2\pi nh) - \exp(-i2\pi nh)| \\ &\leq \sum_{0\leq j\leq n-1} |\exp(i2\pi(n-2j)h) - \exp(-i2\pi(n-2j-2)h)| \\ &= n\,|\exp(i4\pi h) - 1|\end{aligned}$$

il apparaît que

$$\begin{aligned}\int fPf\,d\mu &= \sum_{n\in\mathbb{Z}^*} a_n\tilde{a}_{-n} \\ &\leq \frac{|\exp(i4\pi h)-1|}{4\pi h}\sum_{n\in\mathbb{Z}^*}|a_n|^2 \\ &= \frac{|\exp(i4\pi h)-1|}{4\pi h}\int f^2\,d\mu\end{aligned}$$

ce qui entraîne que

$$\lambda(I-P) \geq 1 - \frac{|\exp(i4\pi h)-1|}{4\pi h} > 0$$

Introduisons ensuite le potentiel U donné par

$$\forall\, x \in S, \qquad U(x) = -\cos(4\pi x)$$

dont les minima globaux sont 0 et 1/2. Il n'est pas très difficile de se convaincre que $c = (1+\cos(4\pi h))_+$, qui est donc strictement positif si $h < 1/4$, or on peut construire une suite $(\mathcal{A}_n)_{n\in\mathbb{N}} \in \Upsilon$ satisfaisant de plus

$$\forall\, n \in \mathbb{N}, \qquad c(\mathcal{A}_n) = 0$$

de la manière suivante :

Pour $n \geq 1$ et $0 \leq k < 2^n$, on note $A_{n,k}$ l'ensemble $]k2^{-n} - 2^{-2n}, k2^{-n} + 2^{-2n}]$ et $\mathcal{A}_n$ est alors la tribu engendrée par les ensembles $\{0\}$, $(A_{n,0} \cup A_{n,2^n-1}) \setminus \{0\}$ et $A_{n,k}$, pour $0 < k < 2^{n-1}$ et $2^{n-1} < k < 2^n$.

De manière plus générale, on aurait pu penser qu'il suffit que μ soit diffuse pour qu'il existe toujours une suite $(\mathcal{A}_n)_{n\in\mathbb{N}} \in \Upsilon$ telle que pour tout $n \in \mathbb{N}$, $c(\mathcal{A}_n) = 0$, mais ceci n'est pas vrai. En effet, considérons $S = [0,1] \times \{0,1,2\}$ que l'on munit de sa tribu naturelle $\mathcal{B}([0,1]) \otimes \mathcal{P}(\{0,1,2\})$. Pour $i \in \{0,1,2\}$, soit la probabilité $\mu_i = m \otimes \delta_i$ sur S, où m est la mesure de Lebesgue. On définit alors un noyau markovien en posant

$$\forall\, x \in S, \qquad P(x,\cdot) = \begin{cases} \mu_1(\cdot) & \text{, si } x \in [0,1]\times\{0\} \\ [\mu_0(\cdot) + \mu_2(\cdot)]/2 & \text{, si } x \in [0,1]\times\{1\} \\ \mu_1(\cdot) & \text{, si } x \in [0,1]\times\{2\} \end{cases}$$

Il est clair que P est réversible par rapport à la probabilité diffuse $\mu = (\mu_0 + \mu_1 + \mu_2)/3$ et que $I - P$ admet un trou spectral : en effet, soit $\mathcal{A}$ la tribu finie engendrée par la partition $S = \sqcup_{i=0}^{3} [0,1] \times \{i\}$. Pour tout $f \in \mathbb{L}^2(\mu)$, Pf est $\mathcal{A}$-mesurable, ainsi

$$\mu(fPf) \;=\; \mu(Q_\mathcal{A}(f)Pf) = \mu(P(Q_\mathcal{A}(f))f) = \mu(Q_\mathcal{A}(f)P(Q_\mathcal{A}(f)))$$

ce qui montre que

$$\begin{aligned}\lambda(I-P) &= 1-\sup_{f\in\mathbb{L}^2(\mu),\mu(f)=0}\frac{\mu(fPf)}{\mu(f^2)}\\ &= 1-\sup_{f\in\mathbb{L}^2(\mu),\mu(f)=0}\frac{\mu(Q_{\mathcal{A}}(f)P(Q_{\mathcal{A}}(f)))}{\mu(f^2)}\\ &\geq 1-\sup_{f\in\mathbb{L}^2(\mu),\mu(f)=0}\frac{\mu(Q_{\mathcal{A}}(f)P(Q_{\mathcal{A}}(f)))}{\mu((Q_{\mathcal{A}}(f))^2)}\\ &= \lambda(I-P_{\mathcal{A}})\end{aligned}$$

d'où en fin de compte $\lambda(I-P)=\lambda(I-P_{\mathcal{A}})=1$.

Introduisons ensuite le potentiel U donné par

$$\forall\, x\in S,\qquad U(x)=\begin{cases}0 & \text{, si } x\in[0,1]\times\{0\}\\ 1 & \text{, si } x\in[0,1]\times\{1\}\\ 0 & \text{, si } x\in[0,1]\times\{2\}\end{cases}$$

de sorte que dès que la sous-tribu $\mathcal{B}$ est non μ-p.s. triviale, $c(\mathcal{B})=1$.

Cependant il est évidemment toujours possible de trouver une suite $(\mathcal{A}_n)_{n\in\mathbb{N}}\in\Upsilon$ telle que la suite croissante des constantes associées $c(\mathcal{A}_n)$ admette c pour limite, et de telles suites seraient donc d'honnêtes candidats pour fournir de relativement « bonnes » approximations de P_β à basse température. On remarque également que

$$(2)\qquad\begin{cases}\liminf_{\beta\to+\infty}-\beta^{-1}\ln(\lambda(\beta)) \geq c\\ \limsup_{\beta\to+\infty}-\beta^{-1}\ln(\lambda(\beta)) \leq 2(\operatorname{ess\,sup}_\mu U-\operatorname{ess\,inf}_\mu U)\end{cases}$$

Tout naturellement, on est donc amené à se demander si le résultat suivant est juste :

Conjecture (fausse) 1

| Pour β grand, la quantité $-\beta^{-1}\ln(\lambda(\beta))$ converge et sa limite vaut c.

Le fait que même la seule convergence ci-dessus n'est pas toujours vérifiée implique en quelque sorte qu'il n'y a pas de « comportements canoniques agréables » pour $\mathbb{R}_+\ni\beta\mapsto\beta^{-1}\ln(\lambda(\beta))$, ou pour des quantités proches, qui en assurerait la convergence en l'infini par des arguments généraux (de type sous-additivité). Ainsi par exemple on ne peut pas avoir pour tout noyau irréductible et réversible fini P et pour tout potentiel U,

$$\lim_{\beta\to+\infty}\beta^{-1}\ln(\lambda(\beta)/\lambda(0))=\inf_{\beta\geq 0}\beta^{-1}\ln(\lambda(\beta)/\lambda(0))$$

mais ce dernier point peut se voir directement sur un ensemble à deux points, où c'est d'ailleurs le comportement contraire qui apparaît :

Soit $0<a,b<1$ et pour $\beta\geq 0$ posons (ce qui correspond au cas générique)

$$P_\beta=\begin{pmatrix}1-a\exp(-\beta) & a\exp(-\beta)\\ b & 1-b\end{pmatrix}$$

on a alors par la formule de la trace $\lambda(\beta)=a\exp(-\beta)+b$. Une application classique de l'inégalité de Hölder permet de voir sans effort que $]0,+\infty[\ni t\mapsto t\ln(\lambda(1/t)/\lambda(0))$

est convexe, et comme elle admet une limite en $+\infty$ (car λ est dérivable en 0), elle est nécessairement décroissante, d'où la croissance de $\mathbb{R}_+ \ni \beta \mapsto \beta^{-1}\ln(\lambda(\beta)/\lambda(0))$.

Cependant dès que l'espace à plus de trois points, même cette croissance (qui si elle s'était généralisée aurait été utile pour l'existence d'une limite, mais en l'occurence aurait été insuffisante pour son identification avec c, contrairement à une décroissance) n'est plus assurée.

Il reste toutefois à se demander s'il n'existerait pas des critères naturels (voir la dernière remarque de la section suivante) qui assurent que la conjecture précédente soit satisfaite. La motivation provient de l'étude du recuit simulé sur des espaces généraux, où il est très important de bien connaître $\limsup_{\beta\to+\infty} -\beta^{-1}\ln(\lambda(\beta))$ (voir par exemple le corollaire 14 de [6]), notamment on voudrait savoir si l'on peut l'exprimer simplement en termes des données (P, μ, U) ou du moins comment l'approximer par des quantités plus maniables pour en avoir des estimées moins grossières que la majoration par $2(\text{ess sup}_\mu U - \text{ess inf}_\mu U)$.

On peut considérer des questions similaires pour la constante isopérimétrique $I(P,\mu)$ relative au couple (P,μ), qui est définie de la manière suivante : à tout $A \in \mathcal{S}$, on associe la quantité

$$Q_P(A) = \int \mu(dx)P(x,dy)\,\mathbb{I}_A(x)\mathbb{I}_{S\setminus A}(y)$$

ce qui permet de poser

$$I(P,\mu) = \inf_{A\in\mathcal{S},\,\mu(A)\leq 1/2} \frac{Q_P(A)}{\mu(A)}$$

(avec la convention que le quotient ci-dessus vaut 1 si $\mu(A) = 0$), qui est une constante appartenant a priori à $[0,1]$.

Notons que l'on a la même propriété d'approximation par des sous-tribus finies que pour le trou spectral. En effet, si $\mathcal{A}$ est une sous-tribu de $\mathcal{S}$, on vérifie immédiatement que

$$I(P_{\mathcal{A}},\mu_{\mathcal{A}}) = \inf_{A\in\mathcal{A},\,\mu(A)\leq 1/2} \frac{Q_P(A)}{\mu(A)}$$

ce qui montre que cette quantité est décroissante sur l'ensemble des sous-tribus de $\mathcal{S}$, et comme précédemment on montre en fait que si $(\mathcal{A}_n)_{n\in\mathbb{N}} \in \Upsilon$, alors

$$\lim_{n\to\infty} I(P_{\mathcal{A}_n},\mu_{\mathcal{A}_n}) = I(P,\mu)$$

En effet, soit $A \in \mathcal{S}$ avec $\mu(A) \leq 1/2$, il suffit de se rappeler que par un théorème de classes monotones, il existe une suite $(A_n)_{n\in\mathbb{N}}$ d'éléments de $\cup_{n\in\mathbb{N}}\mathcal{A}_n$ telle que $\mathbb{I}_{A_n}$ converge vers $\mathbb{I}_A$ dans $\mathbb{L}^1(\mu)$ et donc aussi dans $\mathbb{L}^2(\mu)$, et de noter que par réversibilité, $Q_P(A_n) = Q_P(S\setminus A_n)$, ce qui montre qu'en posant

$$\forall\, n\in\mathbb{N},\qquad B_n = \begin{cases} A_n & ,\ \text{si } \mu(A_n)\leq 1/2 \\ S\setminus A_n & ,\ \text{sinon} \end{cases}$$

on obtient une suite $(B_n)_{n\in\mathbb{N}}$ d'éléments de $\{B \in \cup_{n\in\mathbb{N}}\mathcal{A}_n \,/\, \mu(B) \leq 1/2\}$ telle que $\lim_{n\to\infty} Q_P(B_n)/\mu(B_n) = Q_P(A)/\mu(A)$, d'où découle le résultat annoncé.

Ceci permet d'ailleurs d'étendre l'inégalité de Cheeger (originellement prouvée dans un cadre de géométrie riemannienne [1]) pour les noyaux markoviens finis (voir

par exemple [2] ou [8] et les références qui y sont données), qui dit que pour toute tribu finie $\mathcal{A}$,

$$\frac{I(P_{\mathcal{A}},\mu_{\mathcal{A}})^2}{2} \leq \lambda(I-P_{\mathcal{A}}) \leq 2I(P_{\mathcal{A}},\mu_{\mathcal{A}})$$

(dans cet encadrement seule la première inégalité est non-triviale) au cadre général (mais la preuve de [8] peut aussi s'adapter directement sans difficulté, voir également [4]) :

$$\frac{I(P,\mu)^2}{2} \leq \lambda(I-P) \leq 2I(P,\mu)$$

Notamment il apparaît que l'hypothèse d'existence d'un trou spectral est équivalente au fait que la constante isopérimétrique est strictement positive.

Le potentiel U étant toujours donné, on reconsidère pour tout $\beta \geq 0$ les noyaux markoviens P_β, et on se rend compte que l'on a l'encadrement

$$I(0)\exp(-\beta(\operatorname*{ess\,sup}_{\mu} U - \operatorname*{ess\,inf}_{\mu} U)) \leq I(\beta) \leq 1 \vee [I(0)\exp(\beta(\operatorname*{ess\,sup}_{\mu} U - \operatorname*{ess\,inf}_{\mu} U))]$$

où on a convenu de noter $I(\beta) = I(P_\beta, \mu(\beta))$

Par ailleurs, soit $\mathcal{A}$ une sous-tribu de $\mathcal{S}$ finie, en utilisant d'une part l'inégalité facile

$$\lambda(I-P_{\beta,\mathcal{A}}) \leq 2I(P_{\beta,\mathcal{A}},\mu_{\beta,\mathcal{A}})$$

(plus simplement, on peut montrer directement que $\limsup_{\beta\to+\infty} -\beta^{-1}\ln(I(P_{\beta,\mathcal{A}},\mu_{\beta,\mathcal{A}})) \leq c(\mathcal{A})$ en se servant du théorème 3.3.6 de [8], qui est une version isopérimétrique de l'inégalité de Poincaré que l'on utilise pour la minoration de $\lambda(I-P_{\beta,\mathcal{A}})$ intervenant dans la preuve de (1)) et d'autre part le fait que pour la majoration de $\lambda(I-P_{\beta,\mathcal{A}})$ dans la preuve de (1), on fait seulement intervenir la fonction indicatrice d'un bon « cycle » (voir la démonstration donnée dans [3], qui se généralise facilement aux cas décrits dans [5]), qui est donc aussi utilisable pour une majoration de $I(P_{\beta,\mathcal{A}},\mu_{\beta,\mathcal{A}})$, car on peut s'arranger pour en prendre un de masse plus petite que $1/2$ pour $\mu_{\beta,\mathcal{A}}$ (du moins pour β assez grand), on prouve facilement que l'on a aussi

$$c(\mathcal{A}) = -\lim_{\beta\to+\infty}\beta^{-1}\ln(I(P_{\beta,\mathcal{A}},\mu_{\beta,\mathcal{A}}))$$

Ce résultat et la remarque qui précède permettent en fait de voir que

$$c = \sup_{A\in\mathcal{S}} c(\mathcal{A}_A)$$

où $\mathcal{A}_A$ est la tribu $\{S, \emptyset, A, S\setminus A\}$. Or la quantité $c(\mathcal{A}_A)$ est facile à décrire, car elle vaut

$$\min(V(A,S\setminus A); V(S\setminus A, A)) = \operatorname*{ess\,inf}_{\mathbb{1}_{A\times(S\setminus A)}m} W - \max(\operatorname*{ess\,inf}_{\mathbb{1}_A\mu} U; \operatorname*{ess\,inf}_{\mathbb{1}_{S\setminus A}\mu} U)$$

et on retrouve donc bien l'interprétation usuelle de c en terme de « hauteurs de puits secondaires ».

Si S est dénombrable, on pourrait croire que dans le sup précédent on peut également imposer que A soit fini, mais ceci est faux, comme on le voit sur l'exemple

suivant : on prend $S = \{0\} \sqcup (\mathbb{N} \times \{0,1,2\})$ muni de sa tribu naturelle $\mathcal{S}$. Soit $(q_i)_{i\in\mathbb{N}}$ une probabilité sur $\mathbb{N}$ qui en charge tous les points avec $q_0 = 1/2$, et posons

$$\forall\, x \in S, \qquad \mu(x) = \begin{cases} 2/3 & \text{, si } x = 0 \\ q_i/9 & \text{, si } x = (i,j) \in \mathbb{N} \times \{0,1,2\} \end{cases}$$

On définit ensuite pour tout $x \in S$,

$$\forall\, i \in \mathbb{N}^*, \forall\, j \in \{0,2\},$$

$$\begin{aligned} P((i,j),x) &= \begin{cases} 1/3 & \text{, si } x = 0 \text{ ou } x = (0,j) \text{ ou } x = (i,1) \\ 0 & \text{, sinon} \end{cases} \\ P((0,j),x) &= \begin{cases} 2q_i/3 & \text{, si } x = (i,j) \\ 1/3 & \text{, si } x = 0 \text{ ou } x = (0,1) \\ 0 & \text{, sinon} \end{cases} \end{aligned}$$

$$\forall\, i \in \mathbb{N},$$

$$\begin{aligned} P((i,1),x) &= \begin{cases} 1/3 & \text{, si } x = 0 \text{ ou } x = (i,0) \text{ ou } x = (i,2) \\ 0 & \text{, sinon} \end{cases} \\ P(0,x) &= \begin{cases} q_i/18 & \text{, si } x = (i,0) \text{ ou } x = (i,1) \text{ ou } x = (i,2) \\ 5/6 & \text{, si } x = 0 \end{cases} \end{aligned}$$

et il est clair que ce noyau est réversible par rapport à μ.

De plus, puisque $\mu(0) > 1/2$ et que pour tout $x \in S$, $P(x,0) = 1/3$, il apparaît que $I(P,\mu) = 1/3$ et toutes nos hypothèses sont donc bien satisfaites. Soit U le potentiel défini par

$$\forall\, x \in S, \qquad U(x) = \begin{cases} 0 & \text{, si } x = (i,0) \text{ ou } x = (i,2) \text{, avec } i \in \mathbb{N} \\ 1 & \text{, sinon} \end{cases}$$

on vérifie que $c = 1$, mais que si $A \subset S$ est fini, alors $c(\mathcal{A}_A) = 0$. Ceci permet aussi de voir que même si S est dénombrable, on n'a pas nécessairement

$$\forall\, (\mathcal{A}_n)_{n\in\mathbb{N}} \in \Upsilon, \qquad \lim_{n\to\infty} \uparrow c(\mathcal{A}_n) = c$$

car ceci implique notamment que $c = \sup_{A\in\mathcal{S},\,\mathrm{card}(A)<\infty} c(\mathcal{A}_A)$ (par contre la réciproque est fausse, car il est possible de construire un exemple vérifiant nos hypothèses sur $\mathbb{N} \cup \{\infty\}$ tel que tout élément de $\mathbb{N}$ soit relié à ses quatre plus proches voisins, tel que ∞ soit relié seulement aux entiers pairs et tel que U prenne la valeur 1 sur $2\mathbb{N}$ et 0 ailleurs).

On peut se demander si l'analogue de la conjecture précédente est satisfaite pour les constantes isopérimétriques :

Conjecture (fausse) 2

Pour β grand, la quantité $-\beta^{-1}\ln(I(\beta))$ converge et sa limite vaut c.

Posons $b = \operatorname{ess\,sup}_\mu U - \operatorname{ess\,inf}_\mu U$, et remarquons que du moins on a toujours clairement

$$\text{(3)} \qquad \begin{cases} \liminf_{\beta\to+\infty} -\beta^{-1}\ln(I(\beta)) & \geq & c \\ \limsup_{\beta\to+\infty} -\beta^{-1}\ln(I(\beta)) & \leq & b \end{cases}$$

Or on peut avoir égalité entre b et c (si on reprend l'exemple précédent sur $\mathbb{R}/\mathbb{Z}$, ceci est satisfait pour le potentiel défini par $\forall\, x \in S$, $U(x) = \mathbb{1}_{\cos(4\pi x) \geq 0}$, si $0 < h \leq 1/4$, et on a aussi $b = c = 1$ pour l'exemple ci-dessus sur $\{0\} \cup \mathbb{N} \times \{0,1,2\}$) et (3) permet alors de conclure que dans ce cas la conjecture 2 est vérifiée, par contre, on ne peut jamais utiliser (2) (sauf dans le cas sans intérêt où U est μ-p.s. constant) pour en déduire la validité de la conjecture 1, c'est d'ailleurs ce qui a motivé notre présentation des inégalités isopérimétriques (voir également la remarque (a) de la fin de la section suivante).

Notre but dans la section suivante est de décrire un exemple relativement simple avec $b > c$ et pour lequel les deux inégalités dans (3) sont en fait des égalités. En nous servant de la majoration du trou spectral par deux fois la constante isopérimétrique, ceci nous fournira déjà un contre-exemple à la première conjecture, mais on verra aussi directement que la quantité $\beta^{-1}\ln(\lambda(\beta))$ ne converge pas non plus nécessairement pour β grand.

2 La preuve

Nous allons donner ici deux exemples construits suivant un même principe, pour le premier $-\beta^{-1}\ln(I(\beta))$ va converger pour β grand vers $b > c$, et pour le second cette quantité oscillera asymptotiquement entre b et c.

Dans un premier temps l'espace des états S sera l'ensemble dénombrable $\{0\} \sqcup (\mathbb{N} \times \{0,1,2\})$, que l'on munit tout naturellement de la tribu $\mathcal{S}$ formée de toutes ses sous-parties. On identifiera donc un noyau markovien P sur $(S, \mathcal{S})$ avec une matrice infinie $P = (P(x,y))_{x,y \in S}$. Avant de décrire celui que l'on va considérer, on va présenter la probabilité $\mu = (\mu(x))_{x \in S}$. Soit $q = (q_i)_{i \in \mathbb{N}}$ une probabilité sur $\mathbb{N}$ qui en charge tous les points et soit $(a_i)_{i \in \mathbb{N}}$ une suite d'éléments de $]0,1[$.

On prend

$$\forall\, x \in S, \qquad \mu(x) = \begin{cases} 2/3 & \text{, si } x = 0 \\ q_i/6 & \text{, si } x = (i,0) \text{, pour un } i \in \mathbb{N} \\ a_i q_i/6 & \text{, si } x = (i,1) \text{, pour un } i \in \mathbb{N} \\ (1-a_i)q_i/6 & \text{, si } x = (i,2) \text{, pour un } i \in \mathbb{N} \end{cases}$$

Les seules transitions permises pour P seront celles qui lient (dans les deux sens) 0 à des éléments de la forme $(i,1)$ ou $(i,2)$, pour $i \in \mathbb{N}$, et $(i,0)$ à $(i,1)$ ou $(i,2)$, plus éventuellement des boucles d'un élément de S vers lui même (le graphe associé est donc relativement « trapu » puisque son diamètre vaut 4, qui est aussi la plus grande longueur de ses chemins ne se recoupant pas, on aurait ainsi pu espérer être proche du cas fini). Le noyau P est alors uniquement déterminé par l'imposition que

$$\forall\, (i,j) \in \mathbb{N} \times \{1,2\}, \qquad P((i,j),0) = \frac{1}{2} = P((i,j),(i,0))$$

du fait de la réversibilité demandée par rapport à μ.

Ainsi on obtient que pour tout $(i,j) \in \mathbb{N} \times \{1,2\}$,

$$P(0,(i,j)) \quad = \quad \frac{3}{4}\mu((i,j))$$

$$
\begin{aligned}
P((i,0),(i,j)) &= \begin{cases} a_i/2 & \text{, si } j=1 \\ (1-a_i)/2 & \text{, si } j=2 \end{cases} \\
P((i,j),(i,j)) &= 0
\end{aligned}
$$

puis les boucles, pour $i \in \mathbb{N}$,

$$
\begin{aligned}
P(0,0) &= 7/8 \\
P((i,0),(i,0)) &= 1/2
\end{aligned}
$$

toutes les probabilités de transitions non décrites étant nulles.

Nous allons dans un premier temps montrer que la constante isopérimétrique $I(0) \stackrel{\text{déf.}}{=} I(P,\mu)$ se calcule facilement et vaut en fait 1/4. Par le biais de l'inégalité de Cheeger $\lambda(I-P) \geq I(0)^2/8$, on sera alors rassuré de savoir que toutes nos conditions sont bien remplies.

Pour calculer $I(0)$, commençons par remarquer que $\mu(A) \leq 1/2$ équivaut au fait que $0 \notin A$. Ensuite considérons d'abord le cas d'un ensemble A inclus dans $\{i\} \times \{0,1,2\}$, pour un $i \in \mathbb{N}$ fixé, et notons $J = \{j \in \{1,2\} \,/\, (i,j) \in A\}$ et $J' = \{1,2\} \setminus J$.

• Si $(i,0) \in A$, on a

$$
\begin{aligned}
Q_P(A) &= \sum_{j\in J'} \mu((i,0))P((i,0),(i,j)) + \sum_{j\in J} \mu((i,j))P((i,j),0) \\
&= \frac{1}{2}\sum_{j\in J'} \mu((i,j)) + \frac{1}{2}\sum_{j\in J} \mu((i,j)) \\
&= \frac{1}{2}\mu((i,0))
\end{aligned}
$$

et

$$
\mu(A) = \mu((i,0)) + \sum_{j\in J} \mu((i,j)) \leq 2\mu((i,0))
$$

ce qui fait apparaître que

$$
\inf_{(i,0)\in A\subset\{i\}\times\{0,1,2\}} \frac{Q_P(A)}{\mu(A)} = \frac{1}{4}
$$

• Si $(i,0) \notin A$, on a

$$
\begin{aligned}
Q_P(A) &= \sum_{j\in J} [\mu((i,j))P((i,j),0) + \mu((i,j))P((i,j),(i,0))] \\
&= \sum_{j\in J} \mu((i,j)) \\
&= \mu(A)
\end{aligned}
$$

Il resort de ces considérations que

$$
\inf_{A\subset\{i\}\times\{0,1,2\}} \frac{Q_P(A)}{\mu(A)} = \frac{1}{4}
$$

puis en fin de compte que

$$
I(0) = \frac{1}{4}
$$

Introduisons maintenant le potentiel U défini très simplement par

$$\forall\, x \in S, \qquad U(x) = \begin{cases} 0 & \text{, si } x = 0 \text{ ou si } x = (i,0) \text{ pour un } i \in \mathbb{N} \\ 1 & \text{, si } x = (i,1) \text{ pour un } i \in \mathbb{N} \\ 2 & \text{, si } x = (i,2) \text{ pour un } i \in \mathbb{N} \end{cases}$$

Soit $\mathcal{A}$ une sous-tribu finie de S, disons engendrée par une partition $S = \sqcup_{k=1}^{N} A_k$, on vérifie immédiatement que

$$c(\mathcal{A}) = \begin{cases} 0 & \text{, s'il existe } 1 \le k \le N \text{ tel que } A_k \supset \{0\} \cup \{(i,0)\,/\, i \in \mathbb{N}\} \\ 1 & \text{, sinon} \end{cases}$$

Ainsi $c = 1$, et ici pour tout $(\mathcal{A}_n)_{n\in\mathbb{N}} \in \Upsilon$, on a $\lim_{n\to\infty} c(\mathcal{A}_n) = 1$.

Par ailleurs, on peut reprendre les calculs ci-dessus (notons par exemple que l'on a pour tout $\beta \ge 0$, $\mu_\beta(0) = 2/(3Z_\beta) \ge 2/3$) pour montrer que

$$I(\beta) = \frac{1}{2} \inf_{i\in\mathbb{N}} \frac{a_i e^{-\beta} + (1-a_i)e^{-2\beta}}{1 + a_i e^{-\beta} + (1-a_i)e^{-2\beta}} = \frac{1}{2} \frac{e^{-2\beta} + (e^{-\beta} - e^{-2\beta}) \inf_{i\in\mathbb{N}} a_i}{1 + e^{-2\beta} + (e^{-\beta} - e^{-2\beta}) \inf_{i\in\mathbb{N}} a_i}$$

d'où si on choisit une suite $(a_i)_{i\in\mathbb{N}}$ telle que $\inf_{i\in\mathbb{N}} a_i = 0$,

$$I(\beta) = \frac{1}{2} \frac{e^{-2\beta}}{1 + e^{-2\beta}}$$

puis

$$\lim_{\beta\to+\infty} -\beta^{-1} \ln(I(\beta)) = 2$$

L'exemple précédent montre donc que l'on peut avoir $\limsup_{\beta\to+\infty} -\beta^{-1}\ln(I(\beta)) > c$, mais pour celui-ci la quantité $-\beta^{-1}\ln(I(\beta))$ admet une limite quand β devient grand (qui est $b = \max_S U - \min_S U$) et est en fait décroissante en $\beta \in \mathbb{R}_+$. Pour voir qu'en général on n'est même pas assuré de la convergence, on va compliquer un peu cet exemple, pour obtenir comme annoncé dans la section précédente

$$\begin{aligned} \liminf_{\beta\to+\infty} -\beta^{-1}\ln(I(\beta)) &= c \\ \limsup_{\beta\to+\infty} -\beta^{-1}\ln(I(\beta)) &= b \end{aligned}$$

On prend maintenant $S = \{0\} \sqcup (\mathbb{N} \times \{0,1,2,3\})$ que l'on munit de la probabilité μ définie de la manière suivante, à partir de la donnée d'une probabilité $(q_i)_{i\in\mathbb{N}}$ sur $\mathbb{N}$ en chargeant tous les points et de deux suites $(a_i)_{i\in\mathbb{N}}$ et $(b_i)_{i\in\mathbb{N}}$ d'éléments de $]0,1/3]$:

$$\forall\, x \in S, \qquad \mu(x) = \begin{cases} 2/3 & \text{, si } x = 0 \\ q_i/6 & \text{, si } x = (i,0), \text{ pour un } i \in \mathbb{N} \\ a_i q_i/6 & \text{, si } x = (i,1), \text{ pour un } i \in \mathbb{N} \\ (1 - a_i - b_i)q_i/6 & \text{, si } x = (i,2), \text{ pour un } i \in \mathbb{N} \\ b_i q_i/6 & \text{, si } x = (i,3), \text{ pour un } i \in \mathbb{N} \end{cases}$$

On impose ensuite que pour tout $i \in \mathbb{N}$, $1/2$ est la valeur de $P((i,1),0)$, $P((i,2),0)$, $P((i,1),(i,0))$, $P((i,2),(i,0))$ et de $P((i,3),(i,0))$, ce qui par réversibilité implique que

$$\begin{aligned} P(0,(i,j)) &= \begin{cases} a_i q_i/8 & \text{, si } j = 1 \\ (1 - a_i - b_i)q_i/8 & \text{, si } j = 2 \end{cases} \\ P((i,0),(i,j)) &= \begin{cases} a_i/2 & \text{, si } j = 1 \\ (1 - a_i - b_i)/2 & \text{, si } j = 2 \\ b_i/2 & \text{, si } j = 3 \end{cases} \end{aligned}$$

et on suppose que toutes les transitions non décrites (outre les boucles) sont nulles.

Enfin on considère le potentiel U défini par

$$\forall\, x \in S, \qquad U(x) = \begin{cases} 0 & \text{, si } x = 0 \text{ ou si } x = (i,3) \text{ pour un } i \in \mathbb{N} \\ 1 & \text{, si } x = (i,0) \text{ pour un } i \in \mathbb{N} \\ 2 & \text{, si } x = (i,1) \text{ pour un } i \in \mathbb{N} \\ 3 & \text{, si } x = (i,2) \text{ pour un } i \in \mathbb{N} \end{cases}$$

Soit $\beta \geq 0$, comme précédemment le point crucial pour évaluer $I(\beta)$ est de calculer pour $i \in \mathbb{N}$ fixé la quantité

$$\min_{A \subset \{i\} \times \{0,1,2,3\}} \frac{Q_{P_\beta}(A)}{\mu_\beta(A)}$$

et pour cela on considère à nouveau plusieurs cas, en notant toujours $J = \{j \in \{1,2\} \,/\, (i,j) \in A\}$ et $J' = \{1,2\} \setminus J$.

• Si $(i,0) \in A$ et $(i,3) \in A$, on calcule que

$$\begin{aligned} Q_{P_\beta}(A) &= \sum_{j \in J'} \mu_\beta((i,0)) P_\beta((i,0),(i,j)) + \sum_{j \in J} \mu_\beta((i,j)) P_\beta((i,j),0) \\ &= \frac{1}{2} \sum_{j \in J'} \mu_\beta((i,j)) + \frac{1}{2} \sum_{j \in J} \mu_\beta((i,j)) \\ &= \frac{1}{2} \sum_{j=1,2} \mu_\beta((i,j)) \end{aligned}$$

qui est une quantité ne dépendant pas de J, ainsi dans cette situation le cas le pire sera celui où $J = \{1,2\}$, i.e. $A = A_1 \stackrel{\text{déf.}}{=} \{(i,0),(i,1),(i,2),(i,3)\}$ et

$$\frac{Q_{P_\beta}(A_1)}{\mu_\beta(A_1)} = \frac{1}{2} \frac{a_i e^{-2\beta} + (1 - a_i - b_i) e^{-3\beta}}{b_i + e^{-\beta} + a_i e^{-2\beta} + (1 - a_i - b_i) e^{-3\beta}}$$

• De manière analogue, si $(i,0) \in A$ et $(i,3) \notin A$, la situation la pire est celle qui correspond à $J = \{1,2\}$, i.e. $A = A_2 \stackrel{\text{déf.}}{=} \{(i,0),(i,1),(i,2)\}$, auquel cas

$$\frac{Q_{P_\beta}(A_2)}{\mu_\beta(A_2)} = \frac{1}{2} \frac{a_i e^{-2\beta} + (1 - a_i - b_i) e^{-3\beta} + b_i}{e^{-\beta} + a_i e^{-2\beta} + (1 - a_i - b_i) e^{-3\beta}}$$

qui est manifestement minoré par $Q_{P_\beta}(A_1)/\mu_\beta(A_1)$.

• Si $(i,0) \notin A$ et $(i,3) \in A$, on a

$$\begin{aligned} Q_{P_\beta}(A) &= \sum_{j \in J} \mu_\beta((i,j)) [P_\beta((i,j),0) + P_\beta((i,j),(i,0))] + \mu_\beta((i,3)) P_\beta((i,3),(i,0)) \\ &= \sum_{j \in J} \mu_\beta((i,j)) + \frac{e^{-\beta}}{2} \mu_\beta((i,3)) \\ \mu_\beta(A) &= \sum_{j \in J} \mu_\beta((i,j)) + \mu_\beta((i,3)) \end{aligned}$$

et clairement la situation la pire correspond à $J = \emptyset$, i.e. $A = A_3 \stackrel{\text{déf.}}{=} \{(i,3)\}$, et alors

$$\frac{Q_{P_\beta}(A_3)}{\mu_\beta(A_3)} = \frac{e^{-\beta}}{2}$$

• **Enfin si $(i,0) \notin A$ et $(i,3) \notin A$, on s'aperçoit que l'on a toujours $Q_{P_\beta}(A)/\mu_\beta(A) = 1$.**

En résumé, puisque pour tout $\beta \geq 0$,

$$\frac{Q_{P_\beta}(A_1)}{\mu_\beta(A_1)} < \frac{Q_{P_\beta}(A_3)}{\mu_\beta(A_3)} < 1$$

on a obtenu

$$\min_{A \subset \{i\} \times \{0,1,2,3\}} \frac{Q_{P_\beta}(A)}{\mu_\beta(A)} = \frac{1}{2} \frac{a_i e^{-2\beta} + (1 - a_i - b_i)e^{-3\beta}}{b_i + e^{-\beta} + a_i e^{-2\beta} + (1 - a_i - b_i)e^{-3\beta}}$$

puis pour tout $\beta \geq 0$,

$$I(\beta) = \frac{1}{2} \inf_{i \in \mathbb{N}} \frac{a_i e^{-2\beta} + (1 - a_i - b_i)e^{-3\beta}}{b_i + e^{-\beta} + a_i e^{-2\beta} + (1 - a_i - b_i)e^{-3\beta}}$$

et notamment pour $\beta = 0$,

$$I(0) = \frac{1}{2} \inf_{i \in \mathbb{N}} \frac{1 - b_i}{2} \geq \frac{1}{6}$$

car on a imposé que pour tout $i \in \mathbb{N}$, $b_i \leq 1/3$, ainsi le modèle initial $(S, \mathcal{S}, P, \mu)$ admet bien une constante isopérimétrique strictement positive.

Notons que l'on a l'encadrement suivant pour tous $i \in \mathbb{N}$ et $\beta \geq 0$,

$$\frac{1}{6} e^{-2\beta} \frac{a_i + e^{-\beta}}{b_i + e^{-\beta}} \leq \frac{a_i e^{-2\beta} + (1 - a_i - b_i)e^{-3\beta}}{b_i + e^{-\beta} + a_i e^{-2\beta} + (1 - a_i - b_i)e^{-3\beta}} \leq e^{-2\beta} \frac{a_i + e^{-\beta}}{b_i + e^{-\beta}}$$

de sorte que pour étudier le comportement en β grand de $\beta^{-1}\ln((I(\beta))$, il suffit de s'intéresser à la quantité

$$g(\beta) \stackrel{\text{déf.}}{=} \beta^{-1} \ln\left(\inf_{i \in \mathbb{N}} \frac{a_i + e^{-\beta}}{b_i + e^{-\beta}}\right)$$

On aura déjà réalisé qu'ici $c = 2$ et $b = 3$, ainsi prouver le résultat annoncé revient à voir que l'on peut trouver des suites $(a_i)_{i \in \mathbb{N}}$ et $(b_i)_{i \in \mathbb{N}}$ telles que

$$(4) \qquad \begin{cases} \liminf_{\beta \to +\infty} g(\beta) = -1 \\ \limsup_{\beta \to +\infty} g(\beta) = 0 \end{cases}$$

La construction qui suit est due au referee: soit $(\chi_i)_{i \in \mathbb{N}}$ une suite d'éléments de $]0,1[$ telle que $\chi_{2i} > \chi_{2j+1}$ pour tous $i,j \in \mathbb{N}$ et telle que $\lim_{i \to \infty} \chi_{2i} = 1$ et $\lim_{i \to \infty} \chi_{2i+1} = 0$. Il suffit de trouver trois suites décroissantes, $(a_i)_{i \in \mathbb{N}}$, $(b_i)_{i \in \mathbb{N}}$ et $(\epsilon_i)_{i \in \mathbb{N}}$, d'éléments de $]0,1/3]$, satisfaisant les conditions suivantes pour tout $i \in \mathbb{N}$:

$$(5) \qquad \frac{a_i}{b_i} < 1$$

$$(6) \qquad \frac{a_{i+1} + \epsilon_i}{b_{i+1} + \epsilon_i} = \frac{a_i + \epsilon_i}{b_i + \epsilon_i}$$

$$(7) \qquad \frac{a_i + \epsilon_i}{b_i + \epsilon_i} = \epsilon_i^{\chi_i}$$

$$(8) \qquad \lim_{i \to \infty} \epsilon_i = 0$$

En effet, (6) implique que $(a_i+\epsilon_i)/(b_i+\epsilon_i)$ est la pente qui permet d'aller des points (b_{i+1}, a_{i+1}) à (b_i, a_i) dans $\mathbb{R}^2$. Or du fait de (5), l'application $\mathbb{R}_+ \ni \epsilon \mapsto (a_i+\epsilon)/(b_i+\epsilon)$ est strictement croissante, d'où

$$\frac{a_i + \epsilon_i}{b_i + \epsilon_i} \geq \frac{a_{i+1} + \epsilon_{i+1}}{b_{i+1} + \epsilon_{i+1}}$$

ce qui montre que les points (b_i, a_i), $i \in \mathbb{N}$, forment les sommets d'un graphe polygonal convexe de $\mathbb{R}^2$.

Ainsi, puisque $\inf_{i\in\mathbb{N}}(a_i + \epsilon)/(b_i + \epsilon)$ est la plus petite pente de (0,0) à l'un des sommets $(b_i + \epsilon, a_i + \epsilon)$ du translaté par (ϵ, ϵ) de ce graphe, il apparaît par convexité (faire un dessin) que pour $0 < \epsilon \leq \epsilon_0$, cet infimum est atteint et vaut $(a_j + \epsilon)/(b_j + \epsilon)$ où j est tel que $\epsilon \in [\epsilon_j, \epsilon_{j-1}]$.

En effectuant la conversion $\beta = -\ln(\epsilon)$ et en nous servant de (7) et (8), on obtient alors (4).

Il reste donc à construire par récurrence les trois suites :

• Pour l'initialisation, posons (par exemple) $b_0 = 1/3$, et choisissons a_0 et ϵ_0 tels que $0 < a_0 < 1/3$, $0 < \epsilon_0 < 1/3$ et

$$\frac{a_0 + \epsilon_0}{b_0 + \epsilon_0} = \epsilon_0^{\chi_0}$$

ce qui est toujours possible, car pour $\epsilon > 0$ suffisament petit, $0 < \epsilon^{\chi_0}(b_0+\epsilon)-\epsilon < 1/3$.

• Passage de $2i$ à $2i+1$, $i \in \mathbb{N}$: il suffit de poser $a_{2i+1} = a_{2i}$ et $b_{2i+1} = b_{2i}$, car du fait que l'expression $(a_{2i+1} + \epsilon)/(b_{2i+1} + \epsilon) - \epsilon^{\chi_{2i+1}}$ est strictement négative en $\epsilon = \epsilon_{2i}$ (rappelons que $\chi_{2i+1} < \chi_{2i}$) et strictement positive en 0, elle s'annule au moins en un point appelé $\epsilon_{2i+1} \in]0, \epsilon_{2i}[$.

• Passage de $2i-1$ à $2i$, $i \in \mathbb{N}^*$: considérons la relation (6) comme la définition d'une fonction b_{2i} de la variable $a_{2i} \in]0, a_{2i-1}[$, fonction qui est croissante et qui admet une limite strictement positive en 0_+ (et b_{2i-1} en $a_{2i-1}-$). Choisissons alors $a_{2i} < \epsilon_{2i-1}$ suffisament petit de sorte que

$$\frac{a_{2i} + a_{2i}}{b_{2i} + a_{2i}} < a_{2i}^{\chi_{2i}}$$

(en utilisant que $\chi_{2i} < 1$), et considérons l'application

$$]0, \epsilon_{2i-1}] \ni \epsilon \mapsto \frac{a_{2i} + \epsilon}{b_{2i} + \epsilon} - \epsilon^{\chi_{2i}}$$

Celle-ci est strictement négative (resp. strictement positive) en a_{2i} (resp. ϵ_{2i-1}, car $\chi_{2i} > \chi_{2i-1}$) et par continuité s'annule donc en un point $\epsilon_{2i} \in]a_{2i}, \epsilon_{2i-1}[$ qui convient.

Notons que par construction les trois suites sont bien décroissantes, et admettent donc respectivement des limites a_∞, b_∞ et ϵ_∞. Si ϵ_∞ devait être strictement positive, en passant à la limite dans les équations (7), séparément pour les indices pairs et impairs, on aboutirait à la contradiction $\epsilon_\infty = 1$, d'où $\epsilon_\infty = 0$.

Ainsi la quantité $\beta^{-1}\ln(I(\beta))$ peut asymptotiquement osciller. Pour voir qu'il peut en être de même pour $\beta^{-1}\ln(\lambda(\beta))$, on reprend l'exemple ci-dessus sur $\{0\} \sqcup (\mathbb{N} \times \{0, 1, 2, 3\})$.

Notons que pour tout $\beta \geq 0$, on a

$$\lambda(\beta) = \inf_{f\in\mathbb{L}^2(\mu_\beta)\setminus\mathrm{Vect}(\mathbb{I})} \frac{\sum_{x,y\in S}(f(y) - f(x))^2\alpha_\beta(x, y)}{\sum_{x,y\in S}(f(y) - f(x))^2\mu_\beta(x)\mu_\beta(y)}$$

avec pour tous $x, y \in S$, $\alpha_\beta(x,y) = \mu_\beta(x)P_\beta(x,y)$.

Pour $i \in \mathbb{N}$, posons $S_i = \{0\} \sqcup (\{i\} \times \{0,1,2,3\})$ et remarquons qu'en utilisant le fait que $\sum_{x\neq 0} \mu_\beta(x) \leq \mu_\beta(0)$, on a

$$\begin{aligned}
&\sum_{i\neq j\in \mathbb{N}} \sum_{x\in S_i\setminus\{0\},\, y\in S_j\setminus\{0\}} (f(y)-f(x))^2\mu_\beta(x)\mu_\beta(y) \\
&\leq \sum_{i\neq j\in \mathbb{N}} \sum_{x\in S_i\setminus\{0\},\, y\in S_j\setminus\{0\}} 2[(f(y)-f(0))^2 + (f(x)-f(0))^2]\mu_\beta(x)\mu_\beta(y) \\
&\leq 4\sum_{i\in \mathbb{N}} \sum_{x\in S_i} (f(x)-f(0))^2\mu_\beta(x)\mu_\beta(0) \\
&\leq 4\sum_{i\in \mathbb{N}} \sum_{x\in S_i,\, y\in S_i} (f(y)-f(x))^2\mu_\beta(x)\mu_\beta(y)
\end{aligned}$$

ce qui fait resortir l'encadrement

$$\begin{aligned}
\sum_{i\in \mathbb{N}} \sum_{x,y\in S_i} (f(y)-f(x))^2\mu_\beta(x)\mu_\beta(y) &\leq \sum_{x,y\in S} (f(y)-f(x))^2\mu_\beta(x)\mu_\beta(y) \\
&\leq 5\sum_{i\in \mathbb{N}} \sum_{x,y\in S_i} (f(y)-f(x))^2\mu_\beta(x)\mu_\beta(y)
\end{aligned}$$

D'autre part on a aussi

$$\sum_{x,y\in S} (f(y)-f(x))^2\alpha_\beta(x,y) = \sum_{i\in \mathbb{N}} \sum_{x,y\in S_i} (f(y)-f(x))^2\alpha_\beta(x,y)$$

et on est donc amené à poser pour $i \in \mathbb{N}$

$$\lambda_i(\beta) = \inf_{f\in F(S_i)\setminus \mathrm{Vect}(\mathbb{1})} \frac{\sum_{x,y\in S_i}(f(y)-f(x))^2\alpha_\beta(x,y)}{\sum_{x,y\in S_i}(f(y)-f(x))^2\mu_\beta(x)\mu_\beta(y)}$$

(où $F(S_i)$ est l'ensemble des fonctions définies sur S_i), car les considérations précédentes montrent que

$$\inf_{i\in \mathbb{N}} \lambda_i(\beta) \leq \lambda(\beta) \leq 5 \inf_{i\in \mathbb{N}} \lambda_i(\beta)$$

Intéressons-nous donc à $\lambda_i(\beta)$. Théoriquement, il est possible d'expliciter cette quantité, car cela revient à trouver les valeurs propres d'une matrice 5×5, dont on sait a priori que l'une des valeurs propres est 0. Mais en pratique de tels calculs sont déjà trop compliqués et on va plutôt se ramener au cas de trois points. Pour simplifier les notations, on laisse tomber un moment l'indice $i \in \mathbb{N}$, supposé fixé, et on pose $1 = (i,0)$, $2 = (i,1)$, $3 = (i,2)$ et $4 = (i,3)$. Notons également $S^{(1)} = \{0,1,4\}$ et $S^{(2)} = \{0,2,3\}$, sur lesquels ensembles on considère les « matrices hors diagonale »

$$\alpha_\beta^{(1)} = \begin{pmatrix} * & \alpha_\beta^{(1)}(0,1) & 0 \\ \alpha_\beta^{(1)}(0,1) & * & \alpha_\beta(1,4) \\ 0 & \alpha_\beta(1,4) & * \end{pmatrix}$$

avec $\alpha_\beta^{(1)}(0,1) = (\alpha_\beta(0,3) \wedge \alpha_\beta(3,1)) + (\alpha_\beta(0,2) \wedge \alpha_\beta(2,1)) = \alpha_\beta(0,3) + \alpha_\beta(0,2)$, car on aura noté que pour tout $\beta \geq 0$, $\alpha_\beta(0,3) = \alpha_\beta(3,1)$ et $\alpha_\beta(0,2) = \alpha_\beta(2,1)$, et

$$\alpha_\beta^{(2)} = \begin{pmatrix} * & \alpha_\beta(0,2) & \alpha_\beta(0,3) \\ \alpha_\beta(0,2) & * & 0 \\ \alpha_\beta(0,3) & 0 & * \end{pmatrix}$$

Pour $j = 1, 2$, posons alors

$$\lambda^{(j)}(\beta) = \inf_{f \in F(S^{(j)}) \setminus \mathrm{Vect}(\mathbb{1})} \frac{\sum_{x \neq y \in S^{(j)}} (f(y) - f(x))^2 \alpha_\beta^{(j)}(x,y)}{\sum_{x \neq y \in S^{(j)}} (f(y) - f(x))^2 \mu_\beta(x) \mu_\beta(y)}$$

En utilisant des inégalités du type

$$\forall f \in F(S), \qquad (f(1) - f(3))^2 + (f(3) - f(0))^2 \geq \frac{1}{2}(f(0) - f(3))^2 + \frac{1}{4}(f(1) - f(0))^2$$

on voit que pour tout $f \in F(S)$,

$$\begin{aligned} &\sum_{x \neq y \in S} (f(y) - f(x))^2 \alpha_\beta(x,y) \\ &\geq \frac{1}{4} \left(\sum_{x \neq y \in S^{(1)}} (f(y) - f(x))^2 \alpha_\beta^{(1)}(x,y) + \sum_{x \neq y \in S^{(2)}} (f(y) - f(x))^2 \alpha_\beta^{(2)}(x,y) \right) \end{aligned}$$

or comme précédemment, on vérifie que pour tout $f \in F(S)$, on a aussi

$$\begin{aligned} &\sum_{x \neq y \in S} (f(y) - f(x))^2 \mu_\beta(x) \mu_\beta(y) \\ &\leq 3 \left(\sum_{x \neq y \in S^{(1)}} (f(y) - f(x))^2 \mu_\beta(x) \mu_\beta(y) + \sum_{x \neq y \in S^{(2)}} (f(y) - f(x))^2 \mu_\beta(x) \mu_\beta(y) \right) \end{aligned}$$

ainsi il resort que

$$\lambda(\beta) \geq \frac{1}{12}(\lambda^{(1)}(\beta) \wedge \lambda^{(2)}(\beta))$$

D'autre part, en considérant les fonctions dont les valeurs sur $\{0, 1, 3\}$ sont égales, il est clair que

$$\lambda^{(1)}(\beta) \geq \lambda(\beta)$$

Ceci nous conduit donc à estimer $\lambda^{(1)}(\beta)$: manifestement il s'agit là de la plus petite valeur propre non nulle de la matrice

$$(\mu_\beta(S^{(1)}))^{-1} \begin{pmatrix} \alpha_\beta^{(1)}(0,1)/\mu_\beta(0) & -\alpha_\beta^{(1)}(0,1)/\mu_\beta(0) & 0 \\ -\alpha_\beta^{(1)}(0,1)/\mu_\beta(1) & (\alpha_\beta^{(1)}(0,1) + \alpha_\beta(1,4))/\mu_\beta(1) & -\alpha_\beta(1,4)/\mu_\beta(1) \\ 0 & -\alpha_\beta(1,4)/\mu_\beta(4) & \alpha_\beta(1,4)/\mu_\beta(4) \end{pmatrix}$$

Toujours à partir de la formulation variationnelle de cette valeur propre, on s'aperçoit qu'à un multiple près, qui est minoré et majoré par des constantes strictement positives universelles, il suffit de considérer la plus petite valeur propre non nulle $\widetilde{\lambda}^{(1)}(\beta)$ associée à la matrice

$$\widetilde{\alpha}_\beta^{(1)} = \begin{pmatrix} q(ae^{-2\beta} + e^{-3\beta}) & -q(ae^{-2\beta} + e^{-3\beta}) & 0 \\ -(ae^{-\beta} + e^{-2\beta}) & ae^{-\beta} + e^{-2\beta} + b & -b \\ 0 & -e^{-\beta} & e^{-\beta} \end{pmatrix}$$

Ceci nous amène à une équation du second degré que l'on résoud pour obtenir

$$\widetilde{\lambda}^{(1)}(\beta) = \frac{1}{2}(A_\beta - \sqrt{A_\beta^2 - 4B_\beta})$$

avec

$$\begin{aligned} A_\beta &= qe^{-3\beta} + (1+aq)e^{-2\beta} + (a+1)e^{-\beta} + b \\ B_\beta &= e^{-2\beta}(a+e^{-\beta})(1+q(e^{-\beta}+b)) \end{aligned}$$

Remarquons que l'on a, pour $\beta \geq 0$,

$$\begin{aligned} 0 \leq 4\frac{B_\beta}{A_\beta^2} &\leq \frac{4e^{-2\beta}(a+e^{-\beta})(1+(1+1/3))}{(b+e^{-\beta})^2} \\ &\leq \frac{28}{3}\frac{a+e^{-\beta}}{b+e^{-\beta}}e^{-\beta} \\ &\leq \frac{28}{3}e^{-\beta} \end{aligned}$$

car par notre choix des a_i et des b_i, pour $i \in \mathbb{N}$, on a toujours $a \leq b$, d'où $(a+e^{-\beta})/(b+e^{-\beta}) \leq 1$.

Ainsi on voit qu'il existe deux constantes universelles $C, D > 0$ (désormais C et D désigneront de telles constantes, dont les valeurs changeront de lignes en lignes, l'important étant qu'elles ne dépendent pas de l'indice $i \in \mathbb{N}$ caché) telles que pour tout $\beta \geq 3$, $1 - CB_\beta/A_\beta^2 \leq \sqrt{1-4B_\beta/A_\beta^2} \leq 1 - DB_\beta/A_\beta^2$, ce qui implique que

$$CB_\beta/A_\beta^2 \leq \widetilde{\lambda}^{(1)}(\beta) \leq DB_\beta/A_\beta^2$$

puis en utilisant que $0 \leq q, a, b, e^{-\beta} \leq 1$, on fait plus précisément apparaître que

$$C\frac{a+e^{-\beta}}{b+e^{-\beta}}e^{-2\beta} \leq \widetilde{\lambda}^{(1)}(\beta) \leq D\frac{a+e^{-\beta}}{b+e^{-\beta}}e^{-2\beta}$$

D'une manière similaire (et même plus simplement), on montre un encadrement

$$C \leq \lambda^{(2)}(\beta) \leq D$$

En fin de compte, en quittant S_i (que l'on notait S !) pour revenir à S, on a donc

$$C \inf_{i\in \mathbb{N}} \frac{a_i+e^{-\beta}}{b_i+e^{-\beta}}e^{-2\beta} \leq \lambda(\beta) \leq D \inf_{i\in \mathbb{N}} \frac{a_i+e^{-\beta}}{b_i+e^{-\beta}}e^{-2\beta}$$

ce qui fait apparaître que pour étudier le comportement asymptotique de $\beta^{-1}\ln(\lambda(\beta)) - 2$, il suffit de s'intéresser à celui de

$$\beta^{-1}\ln\left(\inf_{i\in \mathbb{N}} \frac{a_i+e^{-\beta}}{b_i+e^{-\beta}}\right)$$

ce qui a déjà été fait. On obtient donc finalement

$$\begin{aligned} \liminf_{\beta\to+\infty} -\beta^{-1}\ln(\lambda(\beta)) &= c = 2 \\ \limsup_{\beta\to+\infty} -\beta^{-1}\ln(\lambda(\beta)) &= b = 3 \end{aligned}$$

Remarques :

a) Notons que pour les exemples présentés dans cette section on a toujours

$$(9) \qquad \lim_{\beta\to+\infty} \beta^{-1} \ln\left(\frac{\lambda(\beta)}{I(\beta)}\right) = 0$$

et il serait intéressant de savoir dans quelle mesure il s'agit là d'un résultat général (peut-être cela n'est-il dû qu'au fait que les longueurs des chemins ne se recoupant pas sont bornées pour les graphes considérés).

b) Si l'on cherche à trouver des critères assurant la validité des conjectures 1 ou 2, la première idée est d'introduire une topologie et des hypothèses de compacité et de continuité.

Ainsi par exemple on pourrait supposer de plus que S est un espace métrique compact, que $\mathcal{S}$ est sa tribu borélienne, que P est fellerien (i.e. l'image d'une fonction continue bornée est elle-même continue) et que U est continu. Mais ceci est insuffisant, car on peut transformer l'exemple précédent en un contre-exemple sur $S = \{0\} \sqcup ([0,1] \times \{0,1,2,3\})$ que l'on munit de la distance définie par

$\forall\, x, y \in S,$

$$d(x,y) = \begin{cases} |t-s| & \text{, si } x = (s,i) \text{ et } y = (t,i) \text{ avec } 0 \le s,t \le 1 \text{ et } i \in \{0,1,2,3\} \\ 0 & \text{, si } x = y \\ 1 & \text{, sinon} \end{cases}$$

Ainsi S est compact et $\mathcal{S}$ désignera sa tribu borélienne.

Par analogie avec ce qui précède, donnons nous une probabilité q sur $[0,1]$ dont le support est $[0,1]$ et deux fonctions continues $a, b : [0,1] \to [0,1/3]$ dont le seul point d'annulation éventuelle soit 1. Il existe alors une unique probabilité μ sur $\mathcal{S}$ telle que pour tout borélien A de $[0,1]$, on ait

$$\mu(A \times \{i\}) = \begin{cases} q(A)/6 & \text{, si } i = 0 \\ \int_A a_t\, q(dt)/6 & \text{, si } i = 1 \\ \int_A (1 - a_t - b_t)\, q(dt)/6 & \text{, si } i = 2 \\ \int_A b_t\, q(dt)/6 & \text{, si } i = 3 \end{cases}$$

On considère ensuite le noyau markovien P fellerien donné d'une part par

$\forall\, 0 \le s \le 1,$

$$\begin{aligned} P((s,0),\cdot) &= \frac{1}{2}(a_t\delta_{(s,1)}(\cdot) + (1 - a_t - b_t)\delta_{(s,2)}(\cdot) + b_t\delta_{(s,3)}(\cdot) + \delta_{(s,0)}(\cdot)) \\ P((s,1),\cdot) &= \frac{1}{2}(\delta_{(s,0)}(\cdot) + \delta_0(\cdot)) \\ P((s,2),\cdot) &= \frac{1}{2}(\delta_{(s,0)}(\cdot) + \delta_0(\cdot)) \\ P((s,3),\cdot) &= \frac{1}{2}(\delta_{(s,0)}(\cdot) + \delta_{(s,3)}(\cdot)) \end{aligned}$$

et d'autre part en posant pour tout $A \in \mathcal{S}$,

$$P(0,A) = \frac{7 + \int b_s\, q(ds)}{8}\delta_0(A) + \frac{1}{8}\int \mathbb{I}_A(s,1)\, a_s q(ds) + \frac{1}{8}\int \mathbb{I}_A(s,2)\,(1 - a_s - b_s) q(ds)$$

Puis on définit le potentiel U en posant

$$\forall\, x \in S, \qquad U(x) = \begin{cases} 0 & \text{, si } x = 0 \text{ ou si } x = (s,3) \text{ pour un } s \in [0,1] \\ 1 & \text{, si } x = (s,0) \text{ pour un } s \in [0,1] \\ 2 & \text{, si } x = (s,1) \text{ pour un } s \in [0,1] \\ 3 & \text{, si } x = (s,2) \text{ pour un } s \in [0,1] \end{cases}$$

Il est alors possible de reprendre les preuves précédentes pour obtenir les mêmes résultats de divergence (faire décrire à $(b_t, a_t)_{t\geq 0}$ le graphe convexe polygonal dont les sommets sont les points $(b_i, a_i)_{i\in\mathbb{N}}$ construits précédemment, et voir aussi la fin de l'introduction de [7], où l'on vient de vérifier directement que (9) est généralement satisfait pour le type d'exemples considérés ici).

L'avantage de cet exemple est de suggérer que ce qui manque encore à P est peut-être une « propriété d'ellipticité », au sens par exemple où les probabilités $P(x, \cdot)$ seraient uniformément (en $x \in S$) absolument continues par rapport à $\mu(\cdot)$.

Remerciements :

Je tiens tout particulièrement à exprimer ma gratitude envers le referee qui m'a fourni la preuve de (4) présentée ici, la démonstration initiale était beaucoup moins élégante.

Références

[1] J. Cheeger. A lower bound for the smallest eigenvalue of the Laplacien. In R. C. Gunning, editor, *Problems in Analysis: A Symposium in Honor of S. Bochner*, pages 195–199. Princeton University Press, 1970.

[2] P. Diaconis and D. Stroock. Geometric bounds for eigenvalues of Markov chains. *The Annals of Applied Probability*, 1(1):36–61, 1991.

[3] R. Holley and D. Stroock. Simulated annealing via Sobolev inequalities. *Communications in Mathematical Physics*, 115:553–569, 1988.

[4] G. Lawler and A. Sokal. Bounds on the L^2 spectrum for Markov chains and Markov processes: a generalization of Cheeger's inequality. *Transactions of the American mathematical society*, 309(2):557–580, October 1988.

[5] L. Miclo. Recuit simulé sans potentiel sur un ensemble fini. In J. Azéma, P.A. Meyer, and M. Yor, editors, *Séminaire de Probabilités XXVI*, Lecture Notes in Mathematics 1526, pages 47–60. Springer-Verlag, 1992.

[6] L. Miclo. Convergence sous-exponentielle de l'entropie des chaînes de Markov à trou spectral. Préprint, 1996.

[7] L. Miclo. Une variante de l'inégalité de Cheeger pour les chaînes de Markov finies. Préprint, 1997.

[8] L. Saloff-Coste. Lectures on finite Markov chains. Cours à l'Ecole d'Eté de Probabilités de Saint-Flour XXVI-1996, provisional draft, 1996.

SOME REMARKS ON THE OPTIONAL DECOMPOSITION THEOREM

C. Stricker and J.A. Yan

Summary : Let S be a vector-valued semimartingale and $\mathcal{Z}(S)$ the set of all strictly positive local martingales Z with $Z_0 = 1$ such that ZS is a local martingale. Assume V (resp. U) is a nonnegative process such that for each $Z \in \mathcal{Z}(S)$ ZV is a supermartingale (resp. ZU is a local submartingale with $\sup_{Z\in\mathcal{Z}(S),\tau\in\mathcal{T}^f} E(Z_\tau U_\tau) < +\infty$ where $\mathcal{T}^f$ denotes the set of all finite stopping times). Then V (resp. U) admits a decomposition $V = V_0 + \phi\cdot S - C$ (resp. $U = U_0 + \psi \cdot S + A$) where C and A are adapted increasing processes with $C_0 = A_0 = 0$. The first result is a slight generalization of the optional decomposition theorem (see [2,4,7]) and the second one is new. As an application to mathematical finance, if S is interpreted as the discounted price process of the stocks, we show $\mathcal{Z}(S)$ contains exactly one element iff the market is complete.

1. Introduction and motivations.

Let $(\Omega, \mathcal{F}, (\mathcal{F}_t), P)$ be a usual stochastic basis. For simplicity we assume $\mathcal{F}_0$ is *trivial*. Consider a model of a security market which consists of $d+1$ assets : one bond and d stocks. We choose the bond as a numéraire and denote by $S = (S^1, \dots, S^d)$ the discounted price process of the stocks. We fix a time horizon $[0,T]$. Let B be a discounted European contingent claim (i.e. a nonnegative $\mathcal{F}_T$-measurable r.v.). Assume that there exists at least one probability measure Q, equivalent to P such that S is a local martingale under Q. We denote by $\mathcal{P}(S)$ the set of all such probability measures Q. Put

$$(1.1) \qquad V_t := \text{ess sup}_{\,Q\in\mathcal{P}(S)} E^Q[B|\mathcal{F}_t], \qquad 0 \le t \le T.$$

If $\sup_{Q\in\mathcal{P}(S)} E^Q[B] < +\infty$, then (V_t) is a supermartingale under each $Q \in \mathcal{P}(S)$. When S is a diffusion process, El Karoui and Quenez [2] proved that V admits a decomposition of the form

$$(1.2) \qquad V_t = V_0 + (\phi\cdot S)_t - C_t, \qquad 0 \le t \le T,$$

where ϕ is a vector-valued predictable process, integrable w.r.t. S and C is an adapted increasing process with $C_0 = 0$. Notice that since $V_0 + \phi \cdot S$ is nonnegative, $\phi \cdot S$ is also a local martingale under $Q \in \mathcal{P}(S)$ (see Emery [3] and Ansel/Stricker [1]). The process V is called the value process associated to the problem of hedging the contingent claim B. Here the financial meaning of the processes ϕ and C is clear : If the "option-writer"(i.e. the seller of the contingent claim B) invests the initial capital V_0 in the market and uses the hedging strategy ϕ, he can obtain a cumulative profit called the cumulated consumptions C during the time interval $[0,T]$ and replicate the contigent claim B at time T. V_0 is called the "selling price" of the contingent claim B, because with this price as initial capital one can hedge the contingent claim without risk. Inspired by El Karoui/Quenez [2], Kramkov [7] showed that a decomposition of the form (1.2) is valid in a more general situation, i.e. (S_t) is a locally bounded process and (V_t) is a nonnegative process such that for each $Q \in \mathcal{P}(S)$, (V_t) is a Q-supermartingale. This unpleasant "local boundedness" condition has been removed in the recent paper of Föllmer/Kabanov [4] by means of the "Lagrange Multipliers" method. The following theorem is the main result of their paper :

Theorem 1.1. *Let S be a vector-valued process defined on $(\Omega, (\mathcal{F}_t), \mathcal{F}, P)$. Denote by $\mathcal{P}(S)$ the set of all probability measures $Q \sim P$ such that S is a Q-local martingale. Suppose that $\mathcal{P}(S) \neq \emptyset$ and V is a nonnegative process. Then the following statements are equivalent :*
i) V is a Q-supermartingale for each $Q \in \mathcal{P}(S)$.
ii) There exist a predictable process ϕ, integrable w.r.t. S and an adapted increasing process C with $C_0 = 0$ such that $V = V_0 + \phi \cdot S - C$ and $E^Q(C_\infty) < +\infty$ for each $Q \in \mathcal{P}(S)$.

This theorem is an important contribution both to the theory of semimartingales and to mathematical finance. The first aim of this paper is to weaken the assumption $\mathcal{P}(S) \neq \emptyset$ in Theorem 1.1 . It turns out that if there exists at least one strictly positive local martingale Z with $Z_0 = 1$ such that ZS is a local martingale (Z is called *a strict martingale density for S*), then any nonnegative process V satisfying the property that ZV is a supermartingale for each strict martingale density Z for S admits a decomposition of the form (1.2)(see Theorem 2.1 below). When no strict martingale density for S exists, we give a characterization of those stopping times τ such that S^τ has a strict martingale density (see Theorem 2.2 below). The second aim of our paper is to investigate the submartingale case. From the option-buyer's point of view the following process

$$U_t := \operatorname{ess\,inf}_{Q \in \mathcal{P}(S)} E^Q[B|\mathcal{F}_t], \qquad 0 \le t \le T$$

should be the value process associated to the problem of hedging the contingent claim B. Observe that (U_t) is a submartingale under each $Q \in \mathcal{P}(S)$. A very natural question is the following : does U admit a decomposition of the form

$$(1.3) \qquad U = U_0 + (\psi \cdot S)_t + A_t, \qquad 0 \le t \le T,$$

where (A_t) is an adapted increasing process with $A_0 = 0$. It turns out that in the general setting of Theorem 1.1 a necessary and sufficient condition for a process U to admit a decomposition of form (1.3) is available. We shall prove this result in the setting of Theorem 1.1 as well as in our generalized setting (see Theorem 2.3 and 2.4 below). The decomposition (1.3) has the following financial meaning. In contrast to the selling price V_0, U_0 is called the purchase price of the contingent claim B, because with this price as the initial capital one should save the amount A during the time interval $[0,T]$ in order to hedge the contingent claim B at time T. So this price U_0 is favourable to the option buyer.

As an application of the above two results we give a characterization of the replicability of a contingent claim which generalizes a previous result of Ansel/Stricker [1] and Jacka [5] .

2. Main results.

Throughout this paper we consider a stochastic basis $(\Omega, \mathcal{F}, (\mathcal{F}_t), P)$ which satisfies the usual conditions. We denote by $\mathcal{M}_{loc}$ the set of all local martingales. A strictly positive local martingale Z with $Z_0 = 1$ is called a *strict martingale density* for a vector-valued process S if $ZS \in \mathcal{M}_{loc}$. We denote by $\mathcal{Z}(S)$ the set of all strict martingale densities for S, and by $\mathcal{P}(S)$ the set of all laws $Q \sim P$ such that S is a local martingale under Q. In the sequel we also denote by $\mathcal{P}(S)$ the set of density processes $M_t := E(\frac{dQ}{dP}|\mathcal{F}_t)$ associated to $Q \in \mathcal{P}(S)$. For a semimartingale X we denote by $\mathcal{E}(X)$ the Doléans-Dade exponential of X, i.e. the unique solution Y of the stochastic differential equation $dY = Y_- dX$ and $Y_0 = 1$. The following three lemmas that are probably known, are the key for the proof of our main results. For the sake of completeness we shall give the proofs.

Lemma 2.1. *Let V be an optional process, Y a local martingale and T a stopping time. If $Y^T V^T$ is a local supermartingale, then YV^T is also a local supermartingale.*

Proof. Notice that

$$YV^T = (Y - Y^T)V^T + Y^T V^T = (V_T 1_{]T,\infty[}) \cdot Y + Y^T V^T.$$

Since Y is a local martingale and $V_T 1_{]T,\infty[}$ is a locally bounded predictable process, $(V_T 1_{]T,\infty[}) \cdot Y$ is a local martingale. Hence YV^T is a local supermartingale.

Lemma 2.2. *Let Z and M be two strictly positive uniformly integrable martingales and T be a stopping time. Assume $Z = Z^T$ and $M_T > 0$. Put $Z' := ZM(M^T)^{-1}$. Then Z' is a strictly positive uniformly integrable martingale.*

Proof. First of all Z' is a local martingale because $Z = Z^T$. Since Z' is positive, it is a supermartingale. Thus it remains to prove that $E(Z'_\infty) = E(Z'_0)$. We have $M_T = E[M_\infty|\mathcal{F}_T] = E[M_T M_\infty M_T^{-1}|\mathcal{F}_T] = M_T E[M_\infty M_T^{-1}|\mathcal{F}_T]$. Thus $E[M_\infty M_T^{-1}|\mathcal{F}_T] = 1$. Now $E[Z'_\infty] = E(Z_T E[M_\infty M_T^{-1}|\mathcal{F}_T]) = E(Z_T) = E(Z_0) = E(Z'_0)$.

Lemma 2.3. *Let S be a vector-valued process and T, T_1, T_2 be stopping times.*
i) We have $\mathcal{Z}(S) \subset \mathcal{Z}(S^T)$.
ii) If $Z' \in \mathcal{Z}(S^{T_1})$ and $Z'' \in \mathcal{Z}(S^{T_2})$, then $Z := (Z')^{T_1}(Z'')^{T_2}[(Z'')^{T_1 \wedge T_2}]^{-1}$ belongs to $\mathcal{Z}(S^{T_1 \vee T_2})$.
iii) If $\mathcal{P}(S^{T_1}) \neq \emptyset$ and $\mathcal{P}(S^{T_2}) \neq \emptyset$, then $\mathcal{P}(S^{T_1 \vee T_2}) \neq \emptyset$.

Proof. We only need to prove the lemma for the case when S is a real-valued process. Since $ZS^T = (Z - Z^T)S^T + (ZS)^T = (S_T 1_{]T,,+\infty[}) \cdot Z + (ZS)^T$, we conclude that $\mathcal{Z}(S) \subset \mathcal{Z}(S^T)$. Now we are going to prove the second assertion. We have $Z(S^{T_1 \vee T_2}) = (ZS)^{T_1 \vee T_2} = (ZS)^{T_2} - (ZS)^{T_1 \vee T_2} + (ZS)^{T_1} = \frac{Z'_{T_1}}{Z''_{T_1}} 1_{T_1 < +\infty}((Z''S)^{T_2} -$

$(Z''S)^{T_1 \vee T_2}) + (Z'S)^{T_1} = \left(\frac{Z'_{T_1 \wedge T_2}}{Z''_{T_1 \wedge T_2}} 1_{]T_1 \wedge T_2, T_2]}\right) \cdot (Z''S)^{T_2} + (Z'S)^{T_1}$. The last assertion follows from ii) and Lemma 2.2 applied to $(Z')^{T_1}$ and $(Z'')^{T_2}$ with $T = T_1$.

Definition 2.1. *A subset $\mathcal{H}$ of $\mathcal{Z}(S)$ is called dense in $\mathcal{Z}(S)$ if for each $Z \in \mathcal{Z}(S)$ there exists a sequence $Z^n \in \mathcal{H}$ such that $\forall t \geq 0$ $Z^n_t \longrightarrow Z_t$ a.e.*

If $\mathcal{P}(S) \neq \emptyset$, then $\mathcal{P}(S)$ is dense in $\mathcal{Z}(S)$. Indeed, let $Z \in \mathcal{Z}(S)$ and (T_n) be an increasing sequence of finite stopping times converging to $+\infty$ such that Z^{T_n} is a uniformly integrable martingale. Let $Q \in \mathcal{P}(S)$, $M := E(\frac{dQ}{dP}|\mathcal{F}_t)$ and $Z^{(n)} := Z^{T_n} M (M^{T_n})^{-1}$. Then by Lemma 2.2 and 2.3 $Z^{(n)}$ is a uniformly integrable martingale which belongs to $\mathcal{P}(S)$. Moreover $\forall t \geq 0$ $Z^{(n)}_t$ converges to Z_t. Thus the next theorem is a generalization of Theorem 1.1.

Theorem 2.1. *Let S be a vector-valued semimartingale such that $\mathcal{Z}(S) \neq \emptyset$, $\mathcal{H}$ a dense subset of $\mathcal{Z}(S)$, $\mathcal{T}$ the set of all stopping times and V a nonnegative process. The following statements are equivalent :*
i) For each $Z \in \mathcal{H}$, ZV is a supermartingale.
ii) For each $Z \in \mathcal{Z}(S)$, ZV is a supermartingale.
iii) V admits a decomposition of the form :

$$V = V_0 + \phi \cdot S - C, \tag{2.1}$$

such that ϕ is a predictable process integrable w.r.t. S, $Z(\phi \cdot S)$ a local martingale for each $Z \in \mathcal{Z}(S)$, C an adapted increasing process, $C_0 = 0$ and $\forall Z \in \mathcal{Z}(S)$, $\forall \tau \in \mathcal{T}$ $E(Z_\tau C_\tau) < +\infty$.
Moreover, if i), ii) or iii) holds, then $\sup\limits_{Z \in \mathcal{Z}(S), \tau \in \mathcal{T}} E(Z_\tau C_\tau) \leq V_0$.

Proof. We first prove i) $\Longrightarrow$ii). Let $Z \in \mathcal{Z}(S)$. Then there exists a sequence $Z^n \in \mathcal{H}$ such that $\forall t \geq 0$ $Z^n_t \longrightarrow Z_t$ a.e.. Since $Z^n V$ is a supermartingale, for $s \leq t$ we have $E(Z^n_t V_t | \mathcal{F}_t) \leq Z^n_s V_s$. According to Fatou's lemma $E(Z_t V_t | \mathcal{F}_t) \leq Z_s V_s$, i.e. ZV is a supermartingale.
Now we prove ii) $\Longrightarrow$ iii). We take an arbitrary but fixed element $M \in \mathcal{Z}(S)$ and take an increasing sequence (T_n) of finite stopping times with $T_n \longrightarrow +\infty$ such that each M^{T_n} is a uniformly integrable martingale. We shall show that each pair (S^{T_n}, V^{T_n}) satisfies the assumption of Theorem 1.1. First of all, $\mathcal{P}(S^{T_n}) \neq \emptyset$, because $M^{T_n} \in \mathcal{P}(S^{T_n})$. Now fix an n and let $Q \in \mathcal{P}(S^{T_n})$. Put $Y_t := E(\frac{dQ}{dP}|\mathcal{F}_t)$. Then $Y \in \mathcal{Z}(S^{T_n})$. Set $Y' := Y^{T_n} M (M^{T_n})^{-1}$. By Lemma 2.3 $Y' \in \mathcal{Z}(S)$. Thus $Y'V$ is a supermartingale under P. Since $(Y')^{T_n} = Y^{T_n}$, we know that $(YV)^{T_n}$ is also a supermartingale under P. According to Lemma 2.1 and taking in account the fact that YV^{T_n} is nonnegative with $Y_0 V_0^{T_n} = V_0$ being integrable, YV^{T_n} is a supermartingale under P. This is equivalent to saying that V^{T_n} is a supermartingale under Q. Therefore we can apply Theorem 1.1 to get the following decomposition of V :

$$V^{T_n} = V_0 + \phi^{(n)} \cdot S^{T_n} - C^{(n)},$$

where $\phi^{(n)}$ is a vector-valued predictable process which is integrable w.r.t. S^{T_n} and $C^{(n)}$ is an adapted increasing process with $(C^{(n)})^{T_n} = C^{(n)}$. Obviously we can assume

$\phi^{(n)}1_{]T_n,+\infty[} = 0$. Put

$$\phi := \phi^{(1)} + \sum_{k=1}^{\infty}(\phi^{(k+1)} - \phi^{(k)}) \text{ and } C := C^{(1)} + \sum_{k=1}^{\infty}(C^{(k+1)} - C^{(k)}).$$

Then we obtain a decomposition of the form (2.1). Remark that for $n \geq k$, we have $\phi^{(k)} \cdot S^{T_n} = \phi^{(k)} \cdot S^{T_k}$. So it is easy to see that we have

$$(\phi \cdot S)^{T_n} = \phi^{(n)} \cdot S^{T_n}.$$

Next we show that for each $Z \in \mathcal{Z}(S)$, $Z(\phi \cdot S)$ is a local martingale. Let $Z \in \mathcal{Z}(S)$. Take an increasing sequence T_n of finite stopping times with $T_n \longrightarrow +\infty$ such that each Z^{T_n} is a uniformly integrable martingale. Now S^{T_n} is a local martingale under the law $Q^n := Z_{T_n}P$, $V_0 + \phi \cdot S \geq 0$ and $(\phi \cdot S)^{T_n} = \phi \cdot S^{T_n}$. Therefore $(\phi \cdot S)^{T_n}$ is a Q^n local martingale (see Ansel/Stricker [1] Corollaire 3.5). Hence $(Z(\phi \cdot S))^{T_n}$ is a P-local martingale for each T_n and therefore $Z(\phi \cdot S)$ is a P-local martingale. Moreover since $V_0 + \phi \cdot S$ is nonnegative, $Z(\phi \cdot S + V_0)$ is also a P-supermartingale. So $\sup_{Z \in \mathcal{Z}(S), \tau \in CT} E(Z_\tau C_\tau) \leq V_0$.

It remains to prove iii) $\Longrightarrow$ i). Let $Z \in \mathcal{Z}(S)$. Since for each stopping time τ $E(Z_\tau C_\tau) < +\infty$, the process ZC is locally integrable and $ZC = C_- \cdot Z + Z \cdot C$ is a submartingale. Therefore the assumption that $Z(\phi \cdot S)$ is a P local martingale implies ZV is a P-supermartingale.

Next we give a straightforward application of the previous theorem to mathematical finance.

Corollary 2.1. *Let S be a vector-valued process such that $\mathcal{Z}(S) \neq \emptyset$. Then $\mathcal{Z}(S)$ contains only one element iff there exists $Z \in \mathcal{Z}(S)$ such that for any finite stopping time R, each bounded $\mathcal{F}_R$-measurable random variable ξ admits a representation of the form*

$$\xi = x + \int_0^R \phi_s dS_s \tag{2.2}$$

where ϕ is a vector-valued predictable process which is integrable w.r.t S and $Z(x + \phi \cdot S)$ is a uniformly integrable martingale.

Proof. Assume $\mathcal{Z}(S)$ contains only one element Z. We can and will assume ξ is nonnegative. Put

$$V_t := Z_t^{-1} E(Z_R \xi | \mathcal{F}_t),\ t \geq 0.$$

Then ZV is a nonnegative martingale. By Theorem 2.1 V admits a decomposition of the form (2.1) with $C = 0$:

$$V_t = V_0 + (\phi \cdot S)_t.$$

Since $V_R = \xi$ we get (2.2).

The converse is straightforward. Let Z' be another strict martingale density for S and τ a stopping time such that Z^τ and $(Z')^\tau$ are uniformly integrable martingales. Then it is well-known (see for instance Corollary 11.4 page 340 of [6]) that (2.2) implies $Z^\tau = (Z')^\tau$. Therefore we get $Z = Z'$ and the proof is completed.

Now, given a RCLL adapted vector-valued process S, we will investigate those stopping times R for which $\mathcal{Z}(S^R) \neq \emptyset$. Recall that a subset B of $\mathbb{R}^+ \times \Omega$ is said of type $[\![0, \cdot\|$ if there exists a nonnegative r.v. R such that each section $B(\omega)$ is not empty and equal to $[0, R(\omega)]$ or $[0, R(\omega))$. According to Lemma 5.2 of Jacod [6], a set B of type $[\![0, \cdot\|$ is predictable iff there exists an increasing sequence (T_n) of stopping times such that $T_n \to R$ and $B = \cup_{n=1}^{\infty} [\![0, T_n]\!]$.

Theorem 2.2. *Let S be a vector-valued RCLL adapted process. Then there exists a unique (up to an evanescent set) predictable set B of type $[\![0, \cdot\|$ such that for each stopping time τ, $[\![0, \tau]\!] \subset B$ iff $\mathcal{Z}(S^\tau) \neq \emptyset$. Moreover there exists an increasing sequence of stopping times (T_n) such that $\mathcal{Z}(S^{T_n}) \neq \emptyset$ and $B = \cup_n [\![0, T_n]\!]$.*

Proof. Denote by $\mathcal{T}$ the set of all stopping times. Put $\mathcal{C} := \{\tau \in \mathcal{T} : \mathcal{Z}(S^\tau) \neq \emptyset\}$. Then $\tau \equiv 0$ belongs to $\mathcal{C}$. Set $R :=$ ess sup $\mathcal{C}$ and $\mathcal{C}_1 := \{\tau \in \mathcal{C} : \tau \equiv 0 \text{ or } P(\tau = R) > 0\}$. By lemma 2.3, if T_1 and T_2 belong to $\mathcal{C}$, then $T_1 \vee T_2 \in \mathcal{C}$. Consequently, there exists an increasing sequence (R^n) of elements of $\mathcal{C}$ such that $R^n \uparrow R$. If T_1 and T_2 belong to $\mathcal{C}_1$, it is easy to check that $T_1 \vee T_2 \in \mathcal{C}_1$. Hence there exists an increasing sequence (U_n) of elements of $\mathcal{C}_1$ such that $U^n \uparrow$ ess sup $\mathcal{C}_1$. We put

$$T_n := R^n \vee U^n \quad \text{and} \quad B := \cup_n [\![0, T^n]\!].$$

Then B is predictable. It is obvious that

$$\tau \in \mathcal{C} \Longrightarrow [\![0, \tau]\!] \subset B.$$

Now we are going to show the converse. Assume τ is a stopping time such that $[\![0, \tau]\!] \subset B$. Put

$$R_n := T^n_{[T^n < \tau]}.$$

It is easy to see that $R_n \uparrow +\infty$ a.s. and $R_n \wedge \tau = T^n \wedge \tau$. Let $V^{(n)} \in \mathcal{Z}(S^{T_n})$. We can assume $(V^{(n)})^{T_n} = V^{(n)}$. Now if we successively apply Lemma 2.3 to $Z^{(n)}$ and $V^{(n+1)}$, we can construct a sequence $Z^{(n)} \in \mathcal{Z}(S^{T_n})$ such that $(Z^{(n+1)})^{T_n} = Z^{(n)}$. Put $Z_t := Z_t^{(n)}$ for $t \in [0, T_n]$. Since $(Z^\tau)^{R_n} = Z^{T_n \wedge \tau} = (Z^{(n)})^\tau$ and $((ZS)^\tau)^{R_n} = (Z^{(n)} S^{T_n})^\tau$ are local martingales, Z^τ and $(ZS)^\tau$ are local martingales too. Hence $\mathcal{Z}(S^\tau) \neq \emptyset$ and the proof of Theorem 2.2 is complete.

The next theorem is a counterpart of Theorem 1.1 for the local submartingale case.

Theorem 2.3. *Let S be a vector-valued semimartingale such that $\mathcal{P}(S) \neq \emptyset$ and U be a nonnegative process. The following statements are equivalent :*
i) U admits a decomposition of the form

$$U = U_0 + \psi \cdot S + A, \tag{2.3}$$

where ψ is a predictable process, integrable w.r.t. S, such that $\psi \cdot S$ is a local martingale under each $Q \in \mathcal{P}(S)$, and A is an adapted increasing process with $A_0 = 0$ and $\sup_{Q \in \mathcal{P}(S)} E^Q(A_\infty)) < +\infty$.
ii) U is a local submartingale under each law $Q \in \mathcal{P}(S)$ and $\sup_{Q \in \mathcal{P}(S),\, \tau \in \mathcal{T}^f} E^Q(U_\tau) < +\infty$ where $\mathcal{T}^f$ is the set of all finite stopping times.

Proof. We first prove i) $\Longrightarrow$ ii). Since $\psi \cdot S \geq -A$ and $E^Q(A_\infty)) < +\infty$, $\psi \cdot S$ is a supermartingale under each $Q \in \mathcal{P}(S)$ and $\sup_{Q\in\mathcal{P}(S),\ \tau\in\mathcal{T}^f} E^Q((\psi \cdot S)_\tau) < +\infty$. Therefore we get $\sup_{Q\in\mathcal{P}(S),\ \tau\in\mathcal{T}^f} E^Q(U_\tau) < +\infty$.

Now we prove ii) $\Longrightarrow$ i). Denote by $\mathcal{T}_t^f$ the set of all stopping times taking values in $[t, +\infty)$. Put

$$V_t := \text{ess sup}_{\,Q\in\mathcal{P}(S),\tau\in\mathcal{T}_t^f} E^Q(U_\tau|\mathcal{F}_t).$$

Kramkov showed in [7] (Proposition 4.3) that (V_t) is a Q-supermartingale for each $Q \in \mathcal{P}(S)$. According to Theorem 1.1 V admits a decomposition of the form (2.1)

$$V_t = V_0 + \phi^{(1)} \cdot S - C^{(1)}.$$

Put $W := V_0 + \phi^{(1)} \cdot S - U = V - U + C^{(1)}$. Then under each $Q \in \mathcal{P}(S)$ W is a nonnegative supermartingale. Thus, according to Theorem 1.1 , W admits a decomposition of the form (2.1)

$$W = W_0 + \phi^{(2)} \cdot S - C^{(2)}.$$

Put

$$\psi := \phi^{(1)} - \phi^{(2)},\ A := C^{(2)}.$$

We get a decomposition of the form (2.3). Since $A_\infty = C^{(2)}$, $\sup_{Q\in\mathcal{P}(S)} E^Q(A_\infty)) < +\infty$.

The following lemma is a slight generalization of a result due to Kramkov (see [7] Proposition 4.3).

Lemma 2.4. *Let S be a vector-valued process with $\mathcal{Z}(S) \neq \emptyset$. Let $f := (f_t)_{t\geq 0}$ be a nonnegative adapted RCLL process such that*

$$(*) \qquad a := \sup_{Z\in\mathcal{Z}(S),\tau\in\mathcal{T}^f} E(Z_\tau f_\tau) < +\infty.$$

Then there exists an adapted RCLL process V such that V dominates f and for each $Z \in \mathcal{Z}(S)$ ZV is a supermartingale.

Proof. First of all, we are going to show that for any stopping time T such that $\mathcal{P}(S^T) \neq \emptyset$ we have

$$\sup_{Q\in\mathcal{P}(S^T),\tau\in\mathcal{T}^f} E^Q(f_{T\wedge\tau}) < +\infty.$$

In fact, let $Q \in \mathcal{P}(S^T)$ and M be the associated density process w.r.t. P. For $\tau \in \mathcal{T}^f$ $M^{T\wedge\tau} \in \mathcal{Z}(S^{T\wedge\tau})$. By Lemma 2.3 there exists a $Z \in \mathcal{Z}(S)$ such that $Z^{T\wedge\tau} = M^{T\wedge\tau}$. Thus we have

$$E^Q(f_{T\wedge\tau}) = E(M_\infty f_{T\wedge\tau}) = E(M_{T\wedge\tau} f_{T\wedge\tau}) = E(Z_{T\wedge\tau} f_{T\wedge\tau}) \leq a.$$

Now let $T_n \in \mathcal{T}^f$ be such that $T_n \uparrow +\infty$ and for each n $\mathcal{P}(S^{T_n}) \neq \emptyset$. By Proposition 4.3 in [7], for each n there exists an adapted RCLL process $V^{(n)}$ such that

$$V_t^{(n)} = \text{ess sup}_{\,Q\in\mathcal{P}(S^{T_n}),\tau\in\mathcal{T}_t^f} E^Q(f_{T_n\wedge\tau}|\mathcal{F}_t).$$

$V^{(n)}$ is a Q supermartingale for each $Q \in \mathcal{P}(S^{T_n})$. We claim that this latter property implies that for each $Z \in \mathcal{Z}(S^{T_n})$ $ZV^{(n)}$ is a supermartingale. In fact, take an increasing sequence (τ_m) of finite stopping times such that each Z^{τ_m} is a uniformly integrable martingale. Now fix m. According to Lemma 2.2, we can construct a uniformly integrable martingale M such that $M \in \mathcal{Z}(S^{T_n})$ and $M^{\tau_m} = Z^{\tau_m}$. Since $MV^{(n)}$ is a supermartingale, $(ZV^{(n)})^{\tau_m} = (MV^{(n)})^{\tau_m}$ is also a supermartingale. Consequently, $ZV^{(n)}$ is a supermartingale. In particular for each $Z \in \mathcal{Z}(S)$ $ZV^{(n)}$ is a supermartingale because $\mathcal{Z}(S) \subset \mathcal{Z}(S^{T_n})$. Moreover, we have for each $\tau \in \mathcal{T}^f$

$$E(Z_\tau V_\tau^{(n)}) \leq V_0^{(n)} = \sup_{Q \in \mathcal{P}(S^{T_n}), \tau \in \mathcal{T}^f} E^Q(f_{T_n \wedge \tau}) \leq a.$$

Put $V_t := \sup_n V_t^{(n)}$. By Lemma 2.2 and 2.3, for any $Q \in \mathcal{P}(S^{T_n})$ we can construct a $Q' \in \mathcal{P}(S^{T_{n+1}})$ such that $Q'_{|\mathcal{F}_{T_n}} = Q_{|\mathcal{F}_{T_n}}$. Therefore $V_t^{(n)} \uparrow V_t$ and $(Z_t V_t)$ is a supermartingale. Now we are going to prove that V is a RCLL process. Let $\tau_n \in \mathcal{T}^f$, $\tau_n \downarrow \tau$. We have $\lim_{n\to\infty} E(Z_{\tau_n} V_{\tau_n}) = \lim_{n\to\infty} \lim_{m\to\infty} E(Z_{\tau_n} V_{\tau_n}^{(m)}) = \lim_{m\to\infty} \lim_{n\to\infty} E(Z_{\tau_n} V_{\tau_n}^{(m)})$ $= \lim_{m\to\infty} E(Z_\tau V_\tau^{(m)}) = E(Z_\tau V_\tau)$. Here $E(Z_{\tau_n} V_{\tau_n}^{(m)})$ increases both in m and in n, so the interchange of the limits is allowed. Since ZV is an optional process, we conclude that ZV is a RCLL supermartingale. Thus V is an adapted RCLL process. Obviously V dominates f.

Remark 1. If $\mathcal{P}(S) \neq \emptyset$ and $b := \sup_{Q \in \mathcal{P}(S), \tau \in \mathcal{T}^f} E^Q(f_\tau) < +\infty$, then (*) holds. In fact, let $Z \in \mathcal{Z}(S)$ and $\tau \in \mathcal{T}^f$. Take an increasing sequence of finite stopping times T_n converging to $+\infty$ such that Z^{T_n} is a uniformly integrable martingale. By Lemma 2.2 we can construct a uniformly integrable martingale M such that $M \in \mathcal{P}(S)$ and $M^{T_n \wedge \tau} = Z^{T_n \wedge \tau}$. Thus we have

$$E(Z_{T_n \wedge \tau} f_{T_n \wedge \tau}) = E(M_{T_n \wedge \tau} f_{T_n \wedge \tau}) = E(M_\infty f_{T_n \wedge \tau}) \leq \sup_{Q \in \mathcal{P}(S), \tau \in \mathcal{T}_0^f} E^Q(f_\tau).$$

By Fatou's lemma, we get $E(Z_\tau f_\tau) \leq b$. Thus, Lemma 2.4 extends Proposition 4.3 of [7].

Remark 2. It is easy to prove that the process V constructed in the proof of Lemma 2.4 is the smallest process dominating f such that ZV is a supermartingale for each $Z \in \mathcal{Z}(S)$ and, since $\cup_n \mathcal{P}(S^{T_n})$ is dense in $\mathcal{Z}(S)$, we have

$$V_t = \operatorname{ess\,sup}_{Z \in \mathcal{Z}(S), \tau \in \mathcal{T}_t^f} (E(Z_\tau f_\tau | \mathcal{F}_t) Z_t^{-1}).$$

The next theorem is a counterpart of Theorem 2.1 for the local submartingale case.

Theorem 2.4. *Let S be a vector-valued process such that $\mathcal{Z}(S) \neq \emptyset$ and U be a nonnegative process. The following statements are equivalent :*
i) U admits a decomposition of the form :

$$(2.4), \qquad U = U_0 + \psi \cdot S + A,$$

where ψ is a predictable process, integrable w.r.t. S, such that $Z(\psi \cdot S)$ is a local martingale for each $Z \in \mathcal{Z}(S)$, A is an adapted increasing process with $A_0 = 0$ and $\sup_{Z \in \mathcal{Z}(S), \tau \in \mathcal{T}^f} E(Z_\tau A_\tau) < +\infty$.

ii) For each $Z \in \mathcal{Z}(S)$, ZU is a local submartingale and $\sup_{Z\in\mathcal{Z}(S),\tau\in\mathcal{T}^f} E(Z_\tau U_\tau) < +\infty$.
iii) For each $Z \in \mathcal{Z}(S)$, ZU is a local submartingale and there exists a process V dominating U such that ZV is a supermartingale for each $Z \in \mathcal{Z}(S)$.

Proof. First we are going to prove i) $\Longrightarrow$ ii). Let $Z \in \mathcal{Z}(S)$, $\tau \in \mathcal{T}^f$ and (τ_n) be an increasing sequence of finite stopping times converging to $+\infty$ such that Z^{T_n} and $[Z(\psi \cdot S)]^{T_n}$ are uniformly integrable martingales. By (2.4) we have

$$E(Z_{\tau\wedge T_n} U_{\tau\wedge T_n}) = U_0 + E(Z_{\tau\wedge T_n} A_{\tau\wedge T_n}) \leq U_0 + \sup_{Z\in\mathcal{Z}(S),\tau\in\mathcal{T}^f} E(Z_\tau A_\tau).$$

Let $n \longrightarrow +\infty$ we get $E(Z_\tau U_\tau) \leq U_0 + \sup_{Z\in\mathcal{Z}(S),\tau\in\mathcal{T}^f} E(Z_\tau A_\tau) < +\infty$.
ii) $\Longrightarrow$ iii) is a consequence of Lemma 2.4.
Next we are going to prove iii) $\Longrightarrow$ i). Assume iii) holds. By Theorem 2.1 V admits a decomposition of the form

$$V = V_0 + \phi_0^{(1)} \cdot S - C^{(1)}.$$

Put $W := V_0 + \phi_0^{(1)} \cdot S - U = V - U + C^{(1)}$. Then ZW is a nonnegative supermartingale. Again by Theorem 2.1 W admits a decomposition of the form

$$W = U_0 + \phi^{(2)} \cdot S - C^{(2)}.$$

By putting $\psi := \phi^{(1)} - \phi^{(2)}, A := C^{(2)}$, we get (2.4). It remains to show that $\sup_{Z\in\mathcal{Z}(S),\tau\in\mathcal{T}^f} E(Z_\tau A_\tau) < +\infty$. Since V dominates U, we have

$$\sup_{Z\in\mathcal{Z}(S),\tau\in\mathcal{T}^f} E(Z_\tau U_\tau) \leq \sup_{Z\in\mathcal{Z}(S),\tau\in\mathcal{T}^f} E(Z_\tau U_\tau) \leq V_0 < +\infty$$

By a same argument as in the proof of i) $\Longrightarrow$ ii) we conclude that

$$\sup_{Z\in\mathcal{Z}(S),\tau\in\mathcal{T}^f} E(Z_\tau A_\tau) \leq V_0 - U_0.$$

3. Application to Mathematical Finance.

Now we fix a time horizon T. By stopping at T, all results of section 2 are applicable to the present case. As mentioned in section 1, the vector-valued semimartingale S can be interpreted as the discounted price process of stocks in a security market. We suppose that $\mathcal{Z}(S) \neq \emptyset$, but we don't assume $\mathcal{P}(S) \neq \emptyset$. In general, the market is incomplete. So there exist contingent claims which are not replicable. Here, by a contingent claim, we mean a nonnegative $\mathcal{F}_T$-random variable. The contingent claim B is said to be *replicable* if there exist $x \geq 0$, a vector-valued predictable process ϕ integrable w.r.t. S and $Z \in \mathcal{Z}(S)$ such that $Z(x + \phi \cdot S)$ is a martingale and $B = x + (\phi \cdot S)_T$. In that case $\phi \cdot S$ is uniquely defined. In the sequel we consider only those contingent claims B which satisfy the following condition :

$$\sup_{Z\in\mathcal{Z}(S)} E(Z_T B) < \infty.$$

Such a contingent claim will be called *tradable.* Of course, each bounded contingent claim is tradable. If B is a tradable claim, we put :

(3.1) $$V_t := \text{ess sup}_{Z\in\mathcal{Z}(S)}(E[Z_TB|\mathcal{F}_t]Z_t^{-1}),\ 0\le t\le T.$$

(3.2) $$U_t := \text{ess inf}_{Z\in\mathcal{Z}(S)}(E[Z_TB|\mathcal{F}_t]Z_t^{-1}),\ 0\le t\le T.$$

The following theorem, an immediate consequence of Theorem 2.1 and 2.4, shows that (V_t) and (U_t) can be interpreted as the value processes associated to the problem of hedging the contingent claim B for option-writer and option-buyer respectively.

Theorem 3.1. *Assume $\mathcal{Z}(S)\neq\emptyset$. Then (V_t) and (U_t) admit the following decompositions :*

(3.3) $$V_t = V_0 + (\phi\cdot S)_t - C_t,\ 0\le t\le T,$$

(3.4) $$U_t = U_0 + (\psi\cdot S)_t + A_t,\ 0\le t\le T,$$

where ϕ and ψ are predictable vector-valued processes, integrable w.r.t. S and C and A are adapted, increasing processes with $C_0 = A_0 = 0$, such that for each $Z\in\mathcal{Z}(S)$, $Z(\phi\cdot S)$ and $Z(\psi\cdot S)$ are local martingales.

If $\mathcal{P}(S)\neq\emptyset$, it is easy to see that we have

(3.3)′ $$V_t := \text{ess sup}_{Q\in\mathcal{P}(S)}E^Q[B|\mathcal{F}_t],\ 0\le t\le T,$$

(3.4)′ $$U_t := \text{ess inf}_{Q\in\mathcal{P}(S)}E^Q[B|\mathcal{F}_t],\ 0\le t\le T.$$

In this case, Ansel/Stricker [1] proved that B is replicable iff there exists a $Q\in\mathcal{P}(S)$ such that $V_0 = E^Q(B)$. The following theorem extends this result to the general case.

Theorem 3.2. *Assume $\mathcal{Z}(S)\neq\emptyset$. Then the following statements are equivalent :*
i) B is replicable.
ii) $\exists Z'\in\mathcal{Z}(S)$ such that $E(Z'_TB) = \sup_{Z\in\mathcal{Z}(S)}E(Z_TB)$.
iii) $\exists Z'\in\mathcal{Z}(S)$ such that $Z'V$ is a martingale.

Proof. i) $\Longrightarrow$ ii). Asume $B = x + (\phi\cdot S)_T$ and there is a $Z'\in\mathcal{Z}(S)$ such that $Z'(x+\phi\cdot S)$ is a martingale. Since $x+\phi\cdot S$ is nonnegative, by Corollary 3.5 in Ansel/Stricker [1], $\forall Z\in\mathcal{Z}(S)$ $Z(x+\phi\cdot S)$ is a nonnegative local martingale. Thus, it is a supermartingale and we have $E(Z_TB)\le E(Z_0x) = x = E(Z'_TB)$.
ii) $\Longrightarrow$ iii). Assume ii) holds. Since $Z'V$ is a supermartingale and $E(Z'_TV_T) = EZ'_TB) = \sup_{Z\in\mathcal{Z}(S)}E(Z_TB) = V_0 = E(Z'_TV_T)$, $Z'V$ is a martingale.
iii) $\Longrightarrow$ i). Assume $Z'V$ is a martigale. By (3.3) we must have $C_t = C_0 = 0$. So i) holds.

Acknowledgements. The authors would like to thank T. Choulli and Yu. M. Kabanov for fruitful discussions and the referee for pointing out an error in the first version of this paper. The second author would like to thank the first author for kind invitation and hospitality during his visit to Université de Franche-Comté. The financial support from Université Franche-Comté and the NSF of China is acknowledged by the second author.

References.

[1] J.P. Ansel and C. Stricker, Couverture des actifs contingents et prix maximum, Ann. Inst. Henri Poincaré, vol. 30. n^0 2, p. 303-315,1994.

[2] N. El Karoui and M.C. Quenez, Dynamic programming and pricing of contingent claims in an incomplete market, SIAM Journal on Control and Optimization, 33 (1), p. 27-66, 1995.

[3] M. Émery, Compensation de processus à variation finie non localement intégrables, Séminaire Prob. XIV, LN in Math. 784, p. 152-160, Springer 1980.

[4] H. Föllmer and Y. Kabanov, On the optional decomposition theorem and the Lagrange multipliers, to appear in Finance and Stochastics, 1996.

[5] S.D. Jacka, A Martingale Representation Result and an Appplication to Incomplete Financial Markets, Mathematical Finance 2, p. 239-250, 1992.

[6] J. Jacod, Calcul stochastique et problèmes de martingales, LN in Math. 714, Springer 1979.

[7] D.O. Kramkov, Optional decomposition of supermartingales and hedging contingent claims in incomplete security markets. To appear in Prob. Theory and Related Fields, 1996.

C. Stricker, Équipe de Mathématiques, URA 741, Université de Franche-Comté, 16 route de Gray, 25030 Besançon Cedex France.

J.A. Yan, Institute of Applied Mathematics, Academia Sinica, Beijing 100080, China.

Note added in proofs : When this work was completed, we learned that Föllmer and Kabanov modified their proof of Theorem 1.1 in order to remove the assumption that V is nonnegative. So Theorem 1.1 can also be stated for submartingales. Recently Delbaen and Schachermayer provided the proof of this general result using completely different ideas.

SÉPARATION D'UNE SUR- ET D'UNE SOUSMARTINGALE PAR UNE MARTINGALE

Tahir Choulli et Christophe Stricker

Équipe de Mathématiques, URA CNRS 741
Université de Franche-Comté Route de Gray,
25030 Besançon cedex FRANCE

0. Introduction.

Il est désormais bien connu que l'absence d'opportunités d'arbitrage dans un marché financier sans coûts de transactions est essentiellement équivalente à l'existence d'une loi de martingale pour le processus des prix actualisés des actifs de base, c'est-à-dire qu'il existe une loi Q équivalente à la loi initiale P telle que les processus des prix actualisés soient des martingales sous Q.
Lorsque le marché présente des coûts de transactions, ce qui est évidemment plus réaliste, Jouini et Kallal ont montré que le marché était viable si et seulement s'il existait une loi Q équivalente à P telle que les processus des prix actualisés d'achat et de vente puissent être séparés par une Q-martingale.
Dans ce papier nous allons préciser un peu le résultat de Jouini et Kallal en montrant que si X est une surmartingale et Y une sousmartingale vérifiant $\inf_{t\geq 0}(Y_t - X_t) > 0$, alors il existe un processus prévisible λ à variation finie à valeurs dans [0, 1] tel que le processus $Z := X + \lambda(Y - X)$ soit une martingale.
Nous remercions vivement C. Dellacherie pour des discussions fructueuses sur ce sujet, en particulier dans le cas discret.

1. Notations et préliminaires.

Soit $(\Omega, \mathcal{F}, (\mathcal{F}_t)_{0\leq t\leq T}, P)$ un espace probabilisé vérifiant les conditions habituelles. On note $\mathcal{V}$ (resp. $\mathcal{V}^+$) l'ensemble des processus càdlàg adaptés à variation finie (resp. croissants), $\mathcal{P}$ la tribu prévisible et $\mathcal{M}$ (resp. $\mathcal{M}_{loc}$) l'ensemble des martingales (resp. martingales locales).
Si W est une semimartingale, $\mathcal{E}(W)$ désigne l'exponentielle stochastique de W, c'est-à-dire $\mathcal{E}(W)$ est la solution de l'équation différentielle

$$dU = U_- dW, \quad \text{avec } U_0 = 1.$$

Lorsque W est à variation finie, cette exponentielle s'écrit simplement:

$$\mathcal{E}(W) = e^{W'_t} \prod_{0<s\leq t} (1 + \Delta W_s)$$

où W' est la partie continue de W.

Enfin pour tout processus càdlàg H on pose : $H_t^* = \sup_{0\leq s\leq t}|H_s|$. Pour plus de détails et pour les notations non expliquées nous renvoyons le lecteur intéressé à Dellacherie/Meyer (1980).
Dans toute la suite de cet article nous désignerons par X une surmartingale et par Y une sousmartingale réelles telles que $X \leq Y$.

2. Cas à variation finie prévisible.
Dans cette partie nous étudions l'existence et l'unicité d'un processus prévisible λ à variation finie et à valeurs dans [0,1] tel que $Z := X + \lambda(Y - X)$ soit une martingale. Nous commençons par un résultat presque trivial mais qui sera utile pour la suite.

Lemme 2.1. Toute martingale locale Z comprise entre X et Y est une martingale.

Démonstration. Comme X est une surmartingale et Y est une sousmartingale, nous avons les inégalités suivantes :

$$E(X_T \mid \mathcal{F}_t) \leq X_t \leq Z_t \leq Y_t \leq E(Y_T \mid \mathcal{F}_t).$$

Ainsi $\{Z_U \ : \ U \leq T$ et U temps d'arrêt $\}$ est uniformément intégrable, si bien que Z est une martingale.

Les décompositions de Doob-Meyer de X et de Y s'écrivent :

$$\begin{cases} X = M - A \\ Y = N + B \end{cases}$$

où $M, N \in \mathcal{M}_{loc}(P)$ et $A, B \in \mathcal{P} \cap \mathcal{V}^+$.
Posons :

$$\begin{cases} S = Y - X \\ C = A + B. \end{cases}$$

Lemme 2.2. Soit λ un processus prévisible à variation finie. Alors $Z = X + \lambda S \in \mathcal{M}_{loc}(P)$ si et seulement si λ est solution de l'équation différentielle

$$(E) \qquad S_- d\lambda = -\lambda dC + dA.$$

Démonstration. D'après la formule d'intégration par parties (voir Dellacherie/Meyer (1980) page 343) on a :

$$dZ = dM + \lambda d(N - M) + S_- d\lambda + \lambda dC - dA,$$

Ainsi Z est une martingale locale si et seulement si l'équation (E) est vérifiée.

Voici le résultat principal de ce paragraphe.

Proposition 2.3. *Supposons que* $\inf_{t\in[0,T]} S_t > 0$. *Alors pour toute v.a.* λ_0 $\mathcal{F}_0$ *mesurable et comprise entre 0 et 1, l'équation* (E) *admet une solution unique* λ *dans* $\mathcal{V} \bigcap \mathcal{P}$ *et à valeurs dans* $[0,1]$. *De plus elle est donnée par :*

$$\lambda = \mathcal{E}(-\frac{1}{{}^p(S)} \cdot C)(\lambda_0 + \frac{1}{\mathcal{E}(-\frac{1}{{}^p(S)} \cdot C)\,{}^p(S)} \cdot A).$$

Par conséquent pour toute v.a. λ_0 $\mathcal{F}_0$ mesurable et comprise entre 0 et 1 il existe une martingale Z et une seule telle que le processus $\lambda := (Z - X)S^{-1}$ appartienne à $\mathcal{V} \cap \mathcal{P}$ et soit à valeurs dans $[0,1]$.

Démonstration. Rappelons que la projection prévisible de $S := N - M + C$ est égale à ${}^p(S) := (N-M)_- + C = S_- + \Delta C$ si bien que

$$
{}^p(S)d\lambda + \lambda_- dC = S_- d\lambda + \Delta C d\lambda + \lambda_- dC = S_- d\lambda + \lambda dC.
$$

Ainsi l'équation (E) est équivalente à

$$
(E') \qquad {}^p(S)d\lambda = -\lambda_- dC + dA.
$$

Puisque ${}^p(S) = \Delta C + S_- \geq S_- \geq \inf_{t\in[0,T]} S_t > 0$, (E') est aussi équivalente à

$$
(E'') \qquad d\lambda = -\frac{\lambda_-}{{}^p(S)} dC + \frac{1}{{}^p(S)} dA.
$$

Bien que cette équation soit classique, nous allons détailler sa résolution. Comme $1 - \frac{\Delta C}{{}^p(S)} = \frac{S_-}{{}^p(S)} > 0$, nous en déduisons que $\mathcal{E}(-\frac{1}{{}^p(S)} \cdot C) > 0$ et par suite son inverse est un processus à variation finie.
Nous procédons par la méthode de la variation de la constante en considérant le processus $\lambda := \mathcal{E}(-\frac{1}{{}^p(S)} \cdot C)V$ et en explicitant $V \in \mathcal{P} \cap \mathcal{V}$ pour que λ satisfasse (E').
Par la formule de changement de variable on a

$$
d\lambda = -\lambda_- \frac{1}{{}^p(S)} dC + \mathcal{E}(-\frac{1}{{}^p(S)} \cdot C) dV = -\lambda_- \frac{1}{{}^p(S)} dC + \frac{1}{{}^p(S)} dA,
$$

donc

$$
V = \lambda_0 + \frac{1}{\mathcal{E}(-\frac{1}{{}^p(S)} \cdot C)\,{}^p(S)} \cdot A.
$$

Nous obtenons alors

$$
\lambda = \mathcal{E}(-\frac{1}{{}^p(S)} \cdot C)(\lambda_0 + \frac{1}{\mathcal{E}(-\frac{1}{{}^p(S)} \cdot C)\,{}^p(S)} \cdot A).
$$

Et donc $\lambda \in \mathcal{V} \cap \mathcal{P}$ et $\lambda \geq 0$.
Le processus $\lambda' := 1 - \lambda$ vérifie l'équation ${}^p(S)d\lambda' = -\lambda'_- dC + dB$. Cette équation est du même type que l'équation (E') et $\lambda'_0 \geq 0$, si bien que $\lambda' \geq 0$ et $\lambda \leq 1$.

Remarque 2.4. Lorsque $X = 0$, on retrouve la décomposition multiplicative d'une sousmartingale positive Z telle que Z et Z_- ne s'annulent jamais (voir théorème 2 de Meyer/Yoeurp (1976)).

Nous allons montrer grâce à un contre-exemple que si $\inf_{t\in[0,T]} S_t = 0$, l'équation (E) n'a pas de solution en général, si bien qu'il n'existe pas de processus prévisible λ à variation finie à valeurs dans [0,1] tel que le processus $Z := X + \lambda(Y - X)$ soit une martingale.

Exemple 2.5. Soit $(W_t)_{0\le t\le 1}$ un mouvement brownien unidimensionnel issu de 1, $(\mathcal{F}_t)_{0\le t\le 1}$ sa filtration naturelle et $S := |W|$. La formule de Tanaka nous dit que S s'écrit :

$$S_t = 1 + (\text{sign}(W)\cdot W)_t + L_t^0 = \beta + L_t^0$$

où L^0 est le temps local de W en 0 et β est un $(\mathcal{F}_t)$-mouvement brownien. Puisque $P(L_1^0 > 0) > 0$, il existe $\alpha > 0$ telle que $P(L_1^0 > \alpha) > 0$. Soient

$$T_0 = 0, \quad T_{n+1} = \inf\{1 \ge t > T_n \mid L_t - L_{T_n} > \alpha 2^{-n}\}$$

une suite de temps d'arrêt et

$$f = \sum \frac{1}{\sqrt{n+1}} 1_{[T_{2n},T_{2n+1}[}$$

un processus prévisible, positif et borné par 1. Posons $X := \beta - f\cdot L^0$ et $Y = 2\beta + (1-f)\cdot L^0$. Ainsi $C = L^0$, $A = f\cdot C$ et l'équation (E) s'écrit :

$$|W|d\lambda = -\lambda dL^0 + f dL^0.$$

En intégrant $\mathbf{1}_{\{W=0\}}$ par rapport aux deux membres de cette équation on obtient que

$$(f-\lambda)\mathbf{1}_{\{W=0\}}dL^0 = 0.$$

Comme L^0 est un processus continu, on a $L_{T_{n+1}}^0 - L_{T_n}^0 > 0$ sur $\{L_1^0 > \alpha\}$ et la variation de λ ne peut pas être finie sur $\{L_1^0 > \alpha\}$, ce qui est absurde.
Cet exemple montre même plus : il n'existe pas de semimartingale λ avec $0 \le \lambda \le 1$ telle que $Z := X + \lambda S$ soit une martingale. En effet $dZ = dX + \lambda dS + S_- d\lambda$ et la martingale Z qui est une intégrale stochastique par rapport à W en vertu de la propriété de représentation prévisible de W, ne charge pas $\{W = 0\}$, si bien que

$$0 = \mathbf{1}_{\{W=0\}}dZ = (\lambda - f)dL^0.$$

Puisque f n'est pas à variation quadratique bornée, λ ne peut pas être une semimartingale.
Toutefois Jouini et Kallal (1995) ont montré que X et Y peuvent être séparées par une martingale. La proposition suivante montre qu'on peut même choisir une combinaison convexe prévisible de X et Y.

Proposition 2.6. *Soit X une surmartingale, Y une sousmartingale et $X \le Y$. Alors il existe un processus prévisible λ à valeurs dans $[0,1]$ tel que $Z := X + \lambda(Y - X)$ soit une martingale.*

Démonstration. La proposition 2.3 nous dit que pour tout $n \in \mathbb{N}^*$ il existe un processus prévisible λ^n tel que $Z^n := X + \lambda^n(Y - X + \frac{1}{n})$ est une martingale. Comme $X_T \le Z_T^n \le Y_T - X_T + \frac{1}{n}$, la suite (Z_T^n) est uniformément intégrable, si bien qu'il existe une sous-suite qui converge faiblement dans L^1 vers une variable aléatoire Z_T. Désignons par $\mathcal{C}(Z_T^n, Z_T^{n+1}, \ldots)$ l'enveloppe convexe de $\{Z_T^n, Z_T^{n+1}, \ldots\}$. Il est bien connu qu'on peut choisir une suite (W_T^n) de $\mathcal{C}(Z_T^n, Z_T^{n+1}, \ldots)$ telle que (W_T^n) converge dans L^1 vers Z_T. On en déduit qu'il existe des processus prévisibles μ^n et ϵ^n tels que $0 \leq \mu^n \leq 1$, $0 \leq \epsilon^n \leq \frac{1}{n}$ et pour tout $t \in [0,T]$

$$W_t^n := E(W_T^n|\mathcal{F}_t) = X_t + \mu^n(Y_t - X_t) + \epsilon_t^n.$$

Soit $W_t := E[Z_T|\mathcal{F}_t]$. Le lemme maximal de Doob nous dit que la suite $(W - W^n)_T^*$ converge vers 0 en probabilité. Quitte à prendre une nouvelle sous-suite on peut supposer qu'elle converge p.s. vers 0. On note A l'ensemble prévisible où la limite de la suite (μ^n) existe et on pose $\mu := \lim \mu^n 1_A$. Dans ce cas le processus $X + \mu(Y - X) = W$ est une martingale et la proposition est démontrée.

3. Cas général.

Dans ce paragraphe nous étudions l'existence et l'unicité d'une semimartingale λ à valeurs dans [0,1] telle que $Z := X + \lambda S$ soit une martingale. Soient

$$\begin{cases} \lambda' = \mathcal{E}(-\frac{1}{P(S)} \cdot C)(\frac{1}{\mathcal{E}(-\frac{1}{P(S)} \cdot C)P(S)} \cdot A) \\ \lambda'' = \mathcal{E}(-\frac{1}{P(S)} \cdot C)(1 + \frac{1}{\mathcal{E}(-\frac{1}{P(S)} \cdot C)P(S)} \cdot A) \end{cases}$$

les solutions de l'équation (E) correspondant aux valeurs initiales $\lambda_0' = 0$ et $\lambda_0'' = 1$.

Proposition 3.1. *Supposons que* $\inf_{t \in [0,T]} S_t > 0$. *Pour toute v.a.* λ_T *vérifiant les inégalités* $\lambda_T' \leq \lambda_T \leq \lambda_T''$, *il existe une et une seule semimartingale* λ *à valeurs dans* $[0,1]$ *telle que* $Z := X + \lambda(Y - X)$. *De plus* λ *est prévisible à variation finie si et seulement s'il existe deux v.a. positives* a *et* b, $\mathcal{F}_0$*-mesurables, telles que* $\lambda_T = a\lambda_T' + b\lambda_T''$ *et* $a + b = 1$.

Démonstration. Le processus $Z := X + \lambda S$ est une martingale si et seulement si $\lambda_t = \dfrac{E(X_T + \lambda_T S_T \mid \mathcal{F}_t) - X_t}{S_t}$. Ainsi λ est unique. Pour l'existence de λ on pose $\lambda_t := \dfrac{E(X_T + \lambda_T S_T \mid \mathcal{F}_t) - X_t}{S_t}$. Ce processus est certainement une semimartingale et de plus les inégalités $\lambda_T' \leq \lambda_T \leq \lambda_T''$ entraînent que $0 \leq \lambda_t' \leq \lambda_t \leq \lambda_t'' \leq 1$ $\forall t \in [0,T]$. Enfin la dernière assertion de la proposition 3.1 est une conséquence immédiate de la proposition 2.3.

Remarque 3.2. Lorsque $\lambda_T < \lambda_T'$ p.s. (resp. $\lambda_T > \lambda_T''$ p.s.), il n'existe pas de semimartingale λ à valeurs dans $[0, 1]$ telle que $Z := X + \lambda(Y - X)$ soit une martingale. En effet si $\lambda_T < \lambda_T'$ p.s. (resp. $\lambda_T > \lambda_T''$ p.s.) , la démonstration de la proposition 3.1 nous dit que $\lambda_0' < \lambda_0 = 0$ (resp. $\lambda_0'' > \lambda_0 = 1$) , ce qui est absurde.

Références.

C. Dellacherie et P.A. Meyer (1980) "Probabilités et Potentiel", chapitre V à VIII, Hermann.

J. Jacod (1979) "Calcul Stochastique et Problèmes de Martingales", LN 714, Springer Verlag.

E. Jouini et H. Kallal (1995) "Martingales and Arbitrage in Securities Markets with Transaction Costs", Journal of Econ. Theory, 54, p. 259-304.

P.A. Meyer et C. Yoeurp (1976) " Sur la décomposition multiplicative de sousmartingales positives", Séminaire de Probabilités X, LN 511, p. 501-504, Springer Verlag.

Closedness of some spaces of stochastic integrals

Peter Grandits*
Institut für Statistik
Universität Wien
Brünnerstraße 72,A-1210 Wien
Austria

Leszek Krawczyk
Department of Mathematics
Warsaw University
Banacha 2, 02-097 Warsaw
Poland

1991 Mathematics Subject Classification : 60H05,90A09

Abstract

We consider an R^d-valued continous semimartingale $(X_t)_{t\in[0,T]}$, the space of processes $\mathcal{G}^p = \{\theta \cdot X \mid \theta \cdot X \text{ is a semimartingale in } S^p\}$ and the space of their terminal values $\mathcal{G}^p_T$. We give necessary and sufficient conditions for completeness of $\mathcal{G}^p$ in the norm $||(\theta \cdot X)^*||_p$ and closedness of $\mathcal{G}^p_T$ in L^p. These results are related to some problems in mathematical finance and have been given for $p = 2$ in [DMSSS].

1 Introduction

As the results of our paper are given in [DMSSS] for the case $p = 2$, we have tried to be as short as possible and refer for a motivating section on the financial interpretation to [DMSSS] (see also [DM] especially for the discontinuous case).

By construction, the stochastic integral with respect to a square integrable martingale M is an isometry. The space

$$\left\{ \int_0^T \theta dM \,\Big|\, \int_0^\bullet \theta dM \text{ is a square integrable martingale} \right\}$$

*Supported by "Fonds zur Förderung der wissenschaftlichen Forschung in Österreich",Project Nr. P11544

is therefore closed in L^2. Quite recently a characterization of the closedness of the space of stochastic integrals with respect to a continuous *semimartingale* has been given in an L^2-setting [DMSSS]. The aim of this paper is to generalize some of these results to the case L^p $(p \neq 2)$.

After some definitions in section 2 we consider in section 3 the space of processes $\mathcal{G}^p = \{(\theta \cdot X)_\bullet | \theta \in \Theta^p\}$ for a fixed continuous semimartingale X and an appropriately chosen space of integrands Θ^p. We give necessary and sufficient conditions for the closedness of this space with respect to the norm $||H||_{\mathcal{R}^p} = ||H^*||_{L^p}$. The idea of the proof of Theorem 3.1 comes from [DMSSS]. We have simplified parts $(i) \Rightarrow (ii)$ and $(i) \Rightarrow (iv)$. It is remarkable that if this space is closed for some $p > 1$, then it is closed for all $p > 1$.

In section 4 we consider $\mathcal{G}^p_T = \{(\theta \cdot X)_T | \theta \in \Theta^p\}$, the terminal values of the space of processes of section 3. Again we give necessary and sufficient conditions for the closedness in L^p. In [DMSSS] the variance optimal martingale measure plays a prominent rôle in the characterization of the closedness of $\mathcal{G}^p_T$. Here the q-optimal measure, i.e. the martingale measure for X, which has minimal L^q norm (q conjugate to p) is an important tool for the main result of section 4.

2 Definitions

Let $(\Omega, \mathcal{F}, P, (\mathcal{F}_t)_{0 \le t \le T})$ be a stochastic basis satisfying the usual conditions. Let $1 < p < \infty$ be fixed in the whole paper.

Definition 2.1 *For a real valued adapted continuous process H we define*

$$||H||_{\mathcal{R}^p} = ||H^*_t||_{L^p(P)},$$

where $H^*_t = \sup_{0 \le s \le t} |H_s|$,
and $\mathcal{R}^p = \{H \,|\, ||H||_{\mathcal{R}^p} < \infty\}$ *is a Banach space.*

Definition 2.2 *For a continuous martingale Y we denote*

$$||Y||_{H^p} = (E[Y]_T^{\frac{p}{2}})^{\frac{1}{p}}.$$

Definition 2.3 *We say that a (not necessarily continuous) martingale N is in bmo_p (see [P]), if there is a constant C, such that for any t*

$$E(|N_T - N_t|^p | \mathcal{F}_t) \le C,$$

or equivalently, if there is a constant C such that for any t

$$E(([N]_T - [N]_t)^{\frac{p}{2}} | \mathcal{F}_t) \le C.$$

If N is continuous then above conditions are equivalent for all p. The class of such processes is called BMO (see [K]).

In the sequel $(X_t)_{t \in [0,T]}$ denotes always a fixed R^d-valued continuous semimartingale with decomposition $X = M + A$ and $X_0 = 0$, where M is a continuous local martingale and A is a continuous finite variation process.

Definition 2.4 *For an R^d-valued predictable process θ we define*

$$||\theta||_{L^p(M)} = (E(\int_0^T \theta_t' d[M]_t \theta_t)^{\frac{p}{2}})^{\frac{1}{p}}$$

and $L^p(M) = \{\theta : ||\theta||_{L^p(M)} < +\infty\}$.

Definition 2.5 *For an R^d-valued predictable process θ we define*

$$||\theta||_{L^p(A)} = (E(\int_0^T |\theta_t dA_t|)^p)^{1/p}$$

and $L^p(A) = \{\theta : ||\theta||_{L^p(A)} < +\infty\}$.

Definition 2.6 *We define*

$$\Theta^p = L^p(M) \cap L^p(A),$$

equipped with the seminorm

$$||\theta||_{\Theta^p} = ||\theta||_{L^p(M)} + ||\theta||_{L^p(A)} .$$

This is the space of θ, for which $\theta \cdot X$ is in the space S^p of semimartingales. We denote by $\tilde{\Theta}^p$ the quotient Banach space obtained from Θ in the canonical way (i.e by identification of zero-seminorm processes with zero).

Definition 2.7 *We define a mapping $i : \Theta^p \mapsto \mathcal{R}^p$ by*

$$i(\theta) = \theta \cdot X .$$

Since by the Burkholder-Davis-Gundy inequality

$$||(\theta \cdot X)^*||_p \leq ||(\theta \cdot M)^*||_p + ||(\theta \cdot A)^*||_p \leq C||\theta||_{L^p(M)} + ||\theta||_{L^p(A)},$$

i is continuous. Moreover, the above inequality yields, that i induces a continuous mapping $\tilde{i} : \tilde{\Theta}^p \mapsto \mathcal{R}^p$. This mapping is one-to-one. Indeed, if $\theta \cdot X$ is the zero process, then so are its finite variation part $\theta \cdot A$ and its local martingale part $\theta \cdot M$. Thus $||\theta||_{\Theta^p} = 0$.
The image of the mapping i (or $\tilde{i}$) will be denoted by $\mathcal{G}^p$.
We define $\mathcal{G}_T^p$ as the space of terminal values of processes in $\mathcal{G}^p$, i.e.

$$\mathcal{G}_T^p = \{Y_T | Y \in \mathcal{G}^p\} .$$

We treat it as a subspace of $L^p(P)$.

The notation $\mathcal{G}_S^p = \{Y_S | Y \in \mathcal{G}^p\}$, ${}_S\mathcal{G}^p = \{Y_T - Y_S | Y \in \mathcal{G}^p\} \subset L^p(\Omega, \mathcal{F}, P)$, where S is a stopping time is also used.

Definition 2.8 *$M^s(X)$ denotes the set of signed martingale measures for the process X, i.e. measures $\mu << P$, which fulfill $\mu(\Omega) = 1$, $\frac{d\mu}{dP} \in L^q(P)$ and $E((\theta \cdot X)_T \frac{d\mu}{dP}) = 0$ for all $\theta \in \Theta^p$. q is conjugate to the fixed p, which we use in our paper. $M^e(X) \subset M^s(X)$ denotes the set of equivalent martingale measures for X. We always identify a martingale measure with its density $\frac{d\mu}{dP}$.*

Definition 2.9 *For $1 < q < \infty$, we call the solution of the minimum problem*

$$\int SZdP = \langle S, Z\rangle = 0 \qquad \forall S \in \mathcal{G}_T^p \subset L^p(\Omega, P)$$

$$\int ZdP = \langle 1, Z\rangle = 1$$

$$\|Z\|_{L^q}^q = \int |Z|^q dP \to min$$

$Z^{(q)} = \frac{dQ^{(q)}}{dP}$ q-optimal martingale measure for the process X. Noting that $M^s(X)$ is closed in $L^q(P)$ and that L^q for $1 < q < \infty$ is uniformly convex, we can conclude that there is always a unique solution of the minimum problem above, if $M^s(X) \neq \emptyset$.

Definition 2.10 *We say that X satisfies (SC) (structure condition) iff there exists a predictable process λ such that $A = \int_0^T d[M]\lambda$.*

Proposition 2.1 *The following conditions are equivalent*
(i) $\exists C \ \ \forall \theta \in L^p(M) \ \ \|\theta\|_{L^p(A)} \leq C\|\theta\|_{L^p(M)}$,
(ii) $\exists C \ \ \forall \theta \in L^p(M) \ \ \|\theta\|_{\Theta^p} \leq C\|\theta\|_{L^p(M)}$,
(iii) $\exists C \ \ \forall \theta \in \Theta^p \ \ \|\theta\|_{\Theta^p} \leq C\|\theta\|_{L^p(M)}$

Proof. Equivalence of (i) and (ii), and implication (ii) $\Rightarrow$ (iii) are obvious. To prove *(iii)* $\Rightarrow$ *(ii)*, it is enough to consider the case $\theta \in L^p(M)$, $\|\theta\|_{L^p(A)} = \infty$. Define stopping times $\tau_n = \inf\{t : \int_0^t |\theta dA| = n\} \wedge T$. Processes $\theta^n = \theta I_{\{t \leq \tau_n\}}$ are in $L^p(A)$, so also in Θ^p. By (iii) $\|\theta^n\|_{\Theta^p} \leq C\|\theta^n\|_{L^p(M)} \leq C\|\theta\|_{L^p(M)}$. Taking the limit $n \to \infty$ we get a contradiction, that proves (ii).

Definition 2.11 *If one of the conditions (i)-(iii) is fulfilled, then we say that the D_p inequality holds.*

Definition 2.12 *If L is a uniformly integrable martingale such that $L_0 = 1$ and $L_T > 0$ P-a.s, then we say that L satisfies the reverse Hölder inequality under P, denoted by $R_p(P)$, where $1 < p < +\infty$, if and only if there is a constant K such that for every t, we have $E((\frac{L_T}{L_t})^p \mid \mathcal{F}_t) \leq K$.*

Definition 2.13 *Let Z be a positive process. Z satisfies condition (S), if there exists a constant $C > 0$ such that $\frac{1}{C}Z_- \leq Z \leq CZ_-$.*

Definition 2.14 *Let F be a Banach space. Then two vectors $x \in F, x^* \in F^*$ are aligned if $< x, x^* >= \|x\|\|x^*\|$ holds. For L^p-spaces this means equality in the Hölder inequality.*

3 On the closedness of $\mathcal{G}^p$ in $\mathcal{R}^p$.

Throughout this section C denotes a constant, which may vary at each occurrence.
The aim of this section is to give necessary and sufficient conditions for closedness of $\mathcal{G}^p$ in $\mathcal{R}^p$. We shall prove

Theorem 3.1 *Let $1 \le p < \infty$ and let X be a continuous semimartingale. Then the following conditions are equivalent*
(i) D_p holds,
(ii) X satisfies (SC) and $\lambda \cdot M$ is in BMO,
(iii) $\tilde{i}$ is an isomorphism, i.e there exist a constant C such that for any $\theta \in \Theta^p$

$$||\theta||_{\Theta^p} \le C||\theta \cdot X||_{\mathcal{R}^p}$$

(iv) $\mathcal{G}^p$ is closed in $\mathcal{R}^p$.

Remark 1.
Since (ii) does not depend on p, (iv) is valid for all $p > 1$ if it is valid for some $p > 1$.

Remark 2.
Let $K_t = \int_0^t \lambda' d[M]\lambda = [\lambda \cdot M]_t$. This process is called the mean-variance trade-off process. If (ii) is fulfilled then $(\lambda \cdot M)_T$ possesses all moments and hence so does K_T.

For the proof we will need the following generalization of Fefferman's inequality.

Lemma 3.1 *If $p \ge 1$, $Y \in H^p$, $N \in BMO$ are continuous martingales then*

$$|| \int_0^T |d[Y,N]_T| \; ||_p \le C_p ||N||_{BMO} ||Y||_{H^p}.$$

The proof of the lemma is given in [Y], Corollaire 1.1, p.116.

Proof of Theorem 3.1.

(i)$\Rightarrow$ (ii)
Since by D_p condition $\theta \cdot M = 0$ implies $\theta \cdot A = 0$, the process λ in (SC) exists by the predictable Radon-Nikodym theorem ([DS1]). Let $0 \le u < T$ and $B \in \mathcal{F}_u$. We define a predictable set $D_n = \{|\lambda| \le n, \; ||[M]|| \le n, \; t > u, \; \omega \in B\}$. By (SC) and the D_p inequality we get

$$E(\int I_{D_n} \lambda' d[M]\lambda)^p = E(\int I_{D_n} \lambda' d[A])^p \le C^p E(\int I_{D_n} \lambda' d[M]\lambda)^{\frac{p}{2}} =$$

$$= C^p E I_B (\int I_{D_n} d[\lambda . M])^{\frac{p}{2}} \le C^p P(B)^{\frac{1}{2}} (E(\int I_{D_n} d[\lambda \cdot M])^p)^{\frac{1}{2}}$$

by Schwarz inequality. Since the last expression is finite, the above estimation yields

$$EI_B(\int I_{D_n} d[\lambda\cdot M])^p = E(\int I_{D_n} d[\lambda\cdot M])^p \le C^{2p}P(B)$$

and taking the limit $n\to\infty$

$$E(\int_u^T I_B(\omega)\lambda' d[M]\lambda)^p \le C^{2p}P(B).$$

Since $B\in\mathcal{F}_u$ was arbitrary, we get

$$E([\lambda\cdot M]_T - [\lambda\cdot M]_u|\mathcal{F}_u) \le C^2$$

i.e $\lambda\cdot M$ is in BMO.

(ii)⇒ (i)
By Lemma 3.1 for $Y=\theta\cdot M$, $N=\lambda\cdot M$ we have

$$||\int|\theta dA|\,||_p = ||\int|\theta\lambda d[M]|\,||_p = ||\int|d[Y,N]|\,||_p \le C||\theta\cdot M||_{H^p},$$

i.e the R_p inequality.

(i),(ii)⇒ (iii) First, we will prove that for $\theta\in\Theta^p$

$$||\theta||^2_{L^p(M)} = (E[\theta\cdot X]^{\frac{p}{2}})^{\frac{2}{p}} \le C||\theta\cdot X||^2_{\mathcal{R}^p} \tag{1}$$

Denote $Y=\theta\cdot M$, $L=\theta\cdot X$. We have $[Y]=[L]$. By Ito's formula $L^2 = 2L\cdot L+[L]$, so for $p\ge 2$ we have the estimation

$$([E[L]^{\frac{p}{2}})^{\frac{2}{p}} = ||L^2-2L\cdot L||_{\frac{p}{2}} \le ||L^{*2}||_{\frac{p}{2}} + 2||L\cdot Y||_{\frac{p}{2}} + 2||L\cdot(\theta\cdot A)||_{\frac{p}{2}}.$$

The first term is equal to $||L^*||_p^2$. The second one can be estimated using Burkholder-Davis-Gundy inequality by

$$C(E[L\cdot Y]^{\frac{p}{4}})^{\frac{2}{p}} = C(E(L^2\cdot[Y])^{\frac{p}{4}})^{\frac{2}{p}} \le CE(L^*)^{\frac{p}{2}}[Y]^{\frac{p}{4}}))^{\frac{2}{p}} \le$$

$$\le C((EL^{*p})^{\frac{1}{2}}(E[Y]^{\frac{p}{2}})^{\frac{1}{2}})^{\frac{2}{p}} = C(EL^{*p})^{\frac{1}{p}}(E[L]^{\frac{p}{2}})^{\frac{1}{p}}.$$

The third term can be estimated using Lemma 3.1 (with $a=\frac{p}{2}$) by

$$||L\cdot(\theta\cdot A)||_{\frac{p}{2}} \le ||L\cdot(\theta'\cdot[M]\cdot\lambda)||_{\frac{p}{2}} = ||L\cdot[\theta\cdot M,\lambda\cdot M]||_{\frac{p}{2}} =$$

$$= ||[L\cdot Y,\lambda\cdot M]||_{\frac{p}{2}} \le C(E[L\cdot Y]^{\frac{p}{4}})^{\frac{2}{p}},$$

which can be estimated as the second term. Putting these estimations together we get a second degree inequality

$$(E[L]^{\frac{p}{2}})^{\frac{2}{p}} \le ||L^*||_p^2 + C||L^*||_p(E[Y]^{\frac{p}{2}})^{\frac{1}{p}},$$

from which we get (1) for $p\ge 2$.
The case $p<2$ can be derived from the case $p=2$ using

Lemma 3.2 (Lenglart's domination) *If A and B are positive continuous adapted processes, A is increasing and for any bounded stopping time S*

$$EB_S \leq CEA_S\,,$$

then for any $0 < k < 1$

$$EB_\infty^{*k} \leq C_k EA_\infty^k\,.$$

The proof of this lemma can be found in [RY], Proposition 4.7 in chapter IV.

In order to prove (1) for $p < 2$ it is enough to take $A = (\theta \cdot X)^{*2}$, $B = [\theta \cdot X]$. The assumption of Lemma 3.2 is satisfied by (1) for $p = 2$ and the stopped process X^S.
Combining (1) and D_p inequality we get (iii).

(iii)⇒ (i) We know, that there exists a constant C such that for all $\theta \in \Theta^p$

$$||\theta||_{\Theta^p} \leq C||\theta \cdot X||_{\mathcal{R}^p}.$$

Fix $\delta > 0$ and take an arbitrary $\theta \in \Theta^p$. Since A is a continuous finite variation process, there exist a predictable process ϵ, taking values in $\{1, -1\}$, such that

$$\sup_t |(\epsilon \cdot \theta \cdot A)_t| \leq \delta \tag{2}$$

(Lemma 3.8 in [DMSSS]). Thus

$$||\theta||_{\Theta^p} = ||\epsilon\theta||_{\Theta^p} \leq C||(\epsilon\theta) \cdot X)||_{\mathcal{R}^p} \leq C||\epsilon\theta \cdot M||_{\mathcal{R}^p} + C||\epsilon\theta \cdot A||_{\mathcal{R}^p} \leq$$
$$\leq C||\epsilon\theta||_{L^p(M)} + C\delta \leq C||\theta||_{L^p(M)} + C\delta$$

by Burkholder-Davis-Gundy inequality and (2), where C does not depend on δ. Taking the limit $\delta \to 0$ we get inequality D_p.
(iii)⇔ (iv) $\bar{i}$ is a continuous linear, one-to-one mapping between two Banach spaces. By Banach's closed graph theorem its image is closed if and only if this mapping is an isomorphism, completing the proof of Theorem 3.1.□

4 The closedness of $\mathcal{G}_T^p$ in $L^p(P)$

In this section we investigate the closedness of $\mathcal{G}_T^p$ in $L^p(P)$ for a fixed continuous semimartingale X. Our main theorem is analogous to Theorem 4.1 of [DMSSS] for the case $p \neq 2$.

Theorem 4.1 *Let X denote a continuous semimartingale, let $1 < p < \infty$ and q conjugate to p. Then the following are equivalent*

(1) There is a martingale measure Q in $M^e(X)$ and $\mathcal{G}_T^p$ is closed in $L^p(P)$.
(2) There is a martingale measure Q in $M^e(X)$ that satisfies the $R_q(P)$ condition.

(3) The q-optimal martingale measure $Q^{(q)}$ is in $M^e(X)$ and satisfies $R_q(P)$.
(4) $\exists C$ such that for all $\theta \in \Theta^p$ we have

$$\|\theta \cdot X\|_{\mathcal{R}^p} = \|(\theta \cdot X)^*_T\|_{L^p(P)} \leq C\|(\theta \cdot X)_T\|_{L^p(P)}.$$

(5) $\exists C$ such that for all $\theta \in \Theta^p$ and all $\lambda \geq 0$ we have

$$\lambda P[(\theta \cdot X)^*_T > \lambda]^{1/p} \leq C\|(\theta \cdot X)_T\|_{L^p(P)}.$$

(6) $\exists C > 0$ such that for every stopping time S, every $A \in \mathcal{F}_S$ and every $\theta \in \Theta^p$ with $\theta = \theta 1_{]S,T]}$ we have $\|1_A - (\theta \cdot X)_T\|_{L^p(P)} \geq CP[A]^{1/p}$.

In order to prove the theorem we shall need some auxiliary results.

Lemma 4.1 *The density of the q-optimal martingale measure $Z^{(q)}$ is aligned to $(1-f)$, i.e.*

$$Z^{(q)} = \gamma sgn(1-f)|1-f|^{\frac{p}{q}},$$

where $\gamma = (E(sgn(1-f)|1-f|^{\frac{p}{q}}))^{-1} > 0$ holds. f is the solution of the minimum problem

$$\min_{g \in \overline{\mathcal{G}^p_T}} \|1-g\|_p,$$

p is conjugate to q, and the closure is understood here and in the sequel with respect to the norm of $L^p(P)$.

Proof. The fact that $Z^{(q)}$ is aligned to $1-f$ for some $f \in \overline{\mathcal{G}^p_T}$ is standard in the theory of minimum norm problems, c.f. [Lb] Theorem 5.8.1. What remains to be proved is that f is the solution of $\min_{g \in \overline{\mathcal{G}^p_T}} \|1-g\|_p$. The following inequality holds for all $g \in \overline{\mathcal{G}^p_T}$

$$1 = \langle Z^{(q)}, 1-g \rangle \leq \|Z^{(q)}\|_q \|1-g\|_p,$$

but equality holds only, if $1-g$ is aligned to $Z^{(q)}$. By the equality

$$1 = \langle Z^{(q)}, 1-f \rangle = \gamma \langle |1-f|, |1-f|^{\frac{p}{q}} \rangle$$

we finally get $\gamma > 0$. □

The next lemma gives a special feature of the q-optimal density.

Lemma 4.2 *If the q-optimal measure $Q^{(q)} \in M^e(X)$ exists and the cadlag martingale L defined as*

$$L_t = E(\frac{dQ^{(q)}}{dP}|\mathcal{F}_t)$$

satisfies $R_q(P)$, then L satisfies (S).

Proof. Define for each $f_T \in \overline{\mathcal{G}_T^p}$ the $Q^{(q)}$-martingale $f_t := E_{Q^{(q)}}(f_T|\ \mathcal{F}_t)$. Let (f_T^n) be a sequence in $\mathcal{G}_T^p$ converging to f_T with respect to the $L^p(P)$-norm, then the sequence (f_t^n) converges uniformly in t with respect to the norm of $L^1(Q^{(q)})$ and hence in probability to (f_t). As each (f_t^n) is a continuous martingale, the $Q^{(q)}$-martingale (f_t) is continuous whenever $f_T \in \overline{\mathcal{G}_T^p}$. From Lemma 4.1 we know that $L_T^{q-1} = \alpha(1-f)$ holds for some $\alpha > 0$ and an $f \in \overline{\mathcal{G}_T^p}$. W_t defined by

$$W_t = E_{Q^{(q)}}(L_T^{q-1}|\mathcal{F}_t) = \frac{E_P(L_T^q|\mathcal{F}_t)}{L_t}$$

is therefore continuous. By assumption L satisfies

$$1 \leq \frac{E_P(L_T^q|\mathcal{F}_t)}{L_t^q} \leq C,$$

and we conclude that $L_t^{q-1} \leq W_t \leq CL_t^{q-1}$ holds. Since W is continuous, L satisfies the condition (S). □

The proof of the next lemma can be found in [DS2] (Lemma 3.4) with the only difference that we have to use once Hölder's inequality instead of Cauchy-Schwarz.

Lemma 4.3 *If $U = (U_t)_{0\leq t\leq T}$ is a non-negative $L^q(P)$-martingale ($1 < q < \infty$), if $U_0 > 0$, if the stopping time $\tau = \inf\{t|U_t = 0\}$ is predictable and announced by a sequence of stopping times $(\tau_n)_{n\geq 1}$, then*

$$E(\frac{U_\tau^q}{U_{\tau_n}^q}|\mathcal{F}_{\tau_n}) \to +\infty$$

on the $\mathcal{F}_{\tau-}$-measurable set $\{U_\tau = 0\}$.

Our next lemma shows roughly that we can give an upper bound for the L^q-norm of a martingale measure for X, if we have a lower bound for the L^p-distance between 1 and $\overline{\mathcal{G}_T^p}$.

Lemma 4.4 *If there is a constant $C > 0$ such that for every stopping time S, every $A \in \mathcal{F}_S$ and every $U \in {}_S\mathcal{G}^p$*

$$||1_A - U||_{L^p} \geq CP[A]^{1/p},$$

then for each stopping time S there is an element $g \in L_+^q(P)$ such that $E(g|\mathcal{F}_S) = 1$, $E(g^q|\mathcal{F}_S) \leq C^{-q}$ and $E(gU) = 0$ for each $U \in {}_S\mathcal{G}^p$.

Proof. From Lemma 4.1 applied for the space ${}_S\mathcal{G}^p$ instead of $\mathcal{G}_T^p$ we get a $\hat{Z}^{(q)} = \gamma sgn(1-f)|1-f|^{\frac{p}{q}}$ with $f \in {}_S\mathcal{G}^p$ and $\gamma > 0$. Since $\langle \hat{Z}^{(q)}, 1_A f\rangle = 0$,

$$\langle \hat{Z}^{(q)}, 1_A\rangle = \langle \hat{Z}^{(q)}, 1_A(1-f)\rangle = \langle \gamma sgn(1-f)|1-f|^{\frac{p}{q}}, 1_A(1-f)\rangle =$$
$$\gamma \int_A |1-f|^{\frac{p}{q}+1} dP = \gamma \int_A |1-f|^p dP \geq \gamma C^p P(A)$$

holds, and because A was arbitrary in $\mathcal{F}_S$, we get the estimate

$$E(\hat{Z}^{(q)}|\mathcal{F}_S) = \gamma E(|1-f|^p|\mathcal{F}_S) \geq \gamma C^p.$$

Defining now $g = \frac{\hat{Z}^{(q)}}{E(\hat{Z}^{(q)}|\mathcal{F}_S)}$, yields

$$E(|g|^q|\mathcal{F}_S) = \frac{E(|\hat{Z}^{(q)}|^q|\mathcal{F}_S)}{(E(\hat{Z}^{(q)}|\mathcal{F}_S))^q} = \frac{E(\gamma^q|1-f|^p|\mathcal{F}_S)}{\gamma^q(E(|1-f|^p|\mathcal{F}_S))^q} =$$
$$\frac{1}{(E(|1-f|^p|\mathcal{F}_S))^{q-1}} \leq \frac{1}{C^q}.$$

By construction $E(gU) = 0$ holds for all $U \in {}_S\mathcal{G}^p$.

The positivity of g is shown exactly as in Theorem 3.1 of [DS2], if we bear in mind Lemma 4.1, which tells us that f in the formula for $Z^{(q)}$ is the element of $\overline{\mathcal{G}_T^p}$ with minimal L^p-distance from 1.□

We also need the following characterization of closedness of $\mathcal{G}_T^p$ in $L^p(P)$.

Lemma 4.5 *If there is an equivalent local martingale measure Q for X with density in $L^q(P)$, then $\mathcal{G}_T^p$ is closed in $L^p(P)$ if and only if there is a constant $C > 0$ such that*

$$\forall \theta \in \Theta^p, \|\theta\|_{\Theta^p} \leq C\|(\theta \cdot X)_T\|_{L^p(P)}.$$

Proof. Using Doob's L^q inequality instead of the L^2-version and exploiting the duality of L^q and L^p, we can prove the lemma in the same way as Proposition 3.6 of [DMSSS]. □

Finally we need the subsequent technical lemma.

Lemma 4.6 *If L is a uniformly integrable positive martingale, that satisfies the R_q inequality, then $L = \mathcal{E}(N)$, where N is in bmo_q.*

Proof.
Since L is a positive martingale, $\frac{1}{L_-}$ is locally bounded, and hence its stochastic logarithm $N = \frac{1}{L_-}.L$ is a well-defined local martingale. Fix $s \geq 0$ and as in [DMSSS] define the sequence of stopping times T_n by

$$T_0 = s, \;\; T_n = \inf\{t > T_{n-1} | L_t \leq \frac{1}{2} L_{T_{n-1}}\} \wedge T.$$

We have

$$1 = E(\frac{L_{T_n}}{L_{T_{n-1}}}|\mathcal{F}_{T_{n-1}}) = E(\frac{L_{T_n}}{L_{T_{n-1}}} I_{\{T_n<T\}}|\mathcal{F}_{T_{n-1}}) + E(\frac{L_{T_n}}{L_{T_{n-1}}} I_{\{T_n=T\}}|\mathcal{F}_{T_{n-1}}) \leq$$
$$\leq \frac{1}{2} P(T_n < T|\mathcal{F}_{T_{n-1}}) + C(1 - P(T_n < T|\mathcal{F}_{T_{n-1}}))^{\frac{1}{p}},$$

by Hölder's and R_q inequalities. The obtained inequality implies, that there exists a $\gamma < 1$ such that $P(T_n < T|\mathcal{F}_{T_{n-1}}) \le \gamma$. By induction it implies that $P(T_n < T) \le \gamma^n$. From the conditional Burkholder-Davis-Gundy inequality and the definition of T_n we get

$$E(|N_{T_n} - N_{T_{n-1}}|^q|\mathcal{F}_{T_{n-1}}) \le CE(([N]_{T_n} - [N]_{T_{n-1}})^{\frac{q}{2}}|\mathcal{F}_{T_{n-1}}) =$$

$$= CE([\int_{T_{n-1}}^{T_n} \frac{1}{L_{t-}} dL_t]^{\frac{q}{2}}|\mathcal{F}_{T_{n-1}}) = CE((\int_{T_{n-1}}^{T_n} \frac{1}{L_{t-}^2} . [L])^{\frac{q}{2}}|\mathcal{F}_{T_{n-1}}) \le$$

$$\le 2^q CE((\frac{1}{L_{T_{n-1}}^2}[L]_{T_n})^{\frac{q}{2}}|\mathcal{F}_{T_{n-1}}) \le CE(\frac{L_{T_n}^q}{L_{T_{n-1}}^q}|\mathcal{F}_{T_{n-1}}) \le C$$

by the R_q inequality. Finally we have

$$E(|N_T - N_s|^q|\mathcal{F}_s)^{\frac{1}{q}} \le \sum_n (E|N_{T_n} - N_{T_{n-1}}|^q|\mathcal{F}_s)^{\frac{1}{q}} =$$

$$= \sum E(E(|N_{T_n} - N_{T_{n-1}}|^q|\mathcal{F}_{T_{n-1}})|\mathcal{F}_s)^{\frac{1}{q}} \le$$

$$\le \sum E(I_{\{T_{n-1}<T\}} E(|N_{T_n} - N_{T_{n-1}}|^q|\mathcal{F}_{T_{n-1}})|\mathcal{F}_s) \le$$

$$\le C\sum P(T_{n-1} < T) \le C\sum \gamma^k .$$

Since γ does not depend on s, the above inequality completes the proof of the lemma.

Remark.
If X, Y are strongly orthogonal (i.e [X,Y] is a local martingale) then $[X, Y] = 0$ (indeed, [X,Y] is a continuous local martingale of finite variation).

Corollary.
If $X + Y = N \in bmo_q$, where X continuous and X, Y strongly orthogonal, then $X \in bmo_q$ (so from continuity in BMO).

After these preparatory results we prove now the main result of this section
Proof of Theorem 4.1 First we prove the equivalence of (2)-(6). Obviously (3) implies (2). By Theorem 2.16 of [DMSSS] and Lemma 4.2, (3) implies (4) and (2) implies (5). The strong inequality (4) certainly implies the weak inequality (5). The proof of the equivalence of (5) and (6) works with the same reflection argument as for the case $p = 2$ in Theorem 2.18 of [DMSSS].

We prove now that (6) together with (5) implies (3).
By Lemma 4.1 and the proof of Lemma 4.4 (positivity of the q-optimal measure) we have

$$(\frac{dQ^{(q)}}{dP})^{q-1} = \alpha(1 - f),$$

where $f \in \overline{\mathcal{G}^p_T}$, $f \leq 1$ and $\alpha > 0$ holds. Therefore we can find a sequence $Y^n \in \mathcal{G}^p$ obeying $\|Y^n_T - Y^{n+1}_T\|_{L^p} \leq 3^{-n}$ and $Y^n_T \to f$ in $L^p(P)$. The weak inequality (5) yields

$$\sum_{n\geq 1} P(\sup_{0\leq t\leq T} |Y^n_t - Y^{n+1}_t| > 2^{-n}) < +\infty,$$

and we can conclude that Y^n_t converges uniformly in t a.s to a continuous process denoted by f_t $(f_T = f)$. We define now $|\widetilde{Z}_t|^{q-1} sgn(\widetilde{Z}_t) = \alpha(1 - f_t)$ and write L_t for the density process of the q-optimal measure. Because

$$L_t Y_t = E_P(L_T Y_T | \mathcal{F}_t)$$

holds for all $Y \in \mathcal{G}^p$, $L_T, L_t \in L^q(P)$ and Y^n_T tends to f with respect to the norm of $L^p(P)$, we infer that

$$L_t |\widetilde{Z}_t|^{q-1} sgn(\widetilde{Z}_t) = E_P(L_T |\widetilde{Z}_T|^{q-1} sgn(\widetilde{Z}_T)|\mathcal{F}_t) = E_P[|L_T|^q|\mathcal{F}_t)$$

holds. Defining the stopping time $\tau = \inf\{t \mid L_t\widetilde{Z}_t = 0\}$, we have

$$0 = \int_{\tau<T} |L_T|^q$$

yielding $L_T = 0$ on $\{\tau < T\}$ and hence $L_\tau = 0$ on $\{\tau < T\}$. Using the continuity of $\widetilde{Z}$, Lemma 4.3 and Lemma 4.4, we can finish the proof in completely the same way as the proof of Theorem 2.18 in [DMSSS], if we replace 2 by q at the appropriate places.
Since we have now proved the equivalence of (2)-(6), it is still to be shown that (1) is equivalent to (2)-(6). Assuming (1) yields by the continuity of the map i (see Definition 2.7) and Lemma 4.5 the validity of (4).
Conversely assuming (2)-(6) there is an equivalent martingale measure Q satisfying $R_q(P)$. The density process of Q denoted by L_t is necessarily of the form $L = \mathcal{E}(-\lambda \cdot M + U)$, where U is a local martingale strongly orthogonal to M (see [AS]).). Lemma 4.6 and its corollary show that $-\lambda \cdot M + U$ as well as $-\lambda \cdot M$ are in bmo_q. By Theorem 3.1, our hypothesis (4) and the Lemma 4.5 we conclude that $\mathcal{G}^p_T$ is closed. □

References

[AS] Ansel J.P. and Stricker C, 'Lois de martingale, densités et décomposition de Föllmer-Schweizer', Annales de l'Institut Henri Poincaré 28, 375-392(1992).
[DM] Doleans-Dade C. and Meyer P.A., 'Inégalités de normes avec poids', Séminaire de Probabilités XIII, Lecture Notes In Mathematics 721, 313-331 (1977).
[DMSSS] Delbaen F., Monat P., Schachermayer W., Schweizer M., Stricker C., 'Weighted Norm Inequalities and Closedness of a Space of Stochastic Integrals', to appear in Finance and Stochastics.

[DS1] Delbaen F., Schachermayer W., 'The Existence of Absolutely Continuous Local Martingale Measures', Annals of Applied Probability Vol.5,No.4,926-945(1995).

[DS2] Delbaen F.,Schachermayer W., 'The Variance Optimal Martingale Measure for Continuous Processes', Bernoulli 2(1),81-105(1996).

[K] Kazamaki N., 'Exponential Continuous martingales and BMO', Lecture Notes in Mathematics 1579, Springer-Verlag 1994.

[Lb] Luenberger D. G., 'Optimization by Vector Space Methods', Wiley and Sons 1969.

[P] Pratelli M., 'Sur certains espaces de martingales localement de carré intégrable', Séminaire de Probabilités X, Lecture Notes In Mathematics 511, 401-413 (1975).

[RY] Revuz D.,Yor M., 'Continuous Martingales and Brownian Motion', Springer-Verlag Berlin 1991.

[Y] Yor M., 'Inegalités de martingales continues arrêtées à un temps quelconque', Lecture Notes In Mathematics 1118, Springer-Verlag 1985.

Homogeneous diffusions on the Sierpinski gasket

Matthias K. Heck

FB. 9 - Mathematik, Universität des Saarlandes, Postfach 151150, Saarbrücken, Germany

ABSTRACT: We prove that certain diffusions on the Sierpinski gasket may be characterized, up to a multiplicative constant in the time scale, by a parameter $\alpha \in [0, \frac{1}{4}]$. The diffusions considered have the Feller property and certain natural symmetry properties, but they are not necessarily scale invariant. The case $\alpha = 0$ corresponds roughly speaking to one-dimensional Brownian motion and the case $\alpha = \frac{1}{4}$ corresponds to Brownian motion on the Sierpinski gasket.

1 Introduction

In the last few years diffusions on fractals have emerged as an area of probability theory in which an intensive research activity has taken place. Diffusions on the Sierpinski gasket in particular have attracted great attention, perhaps because of their remarkable accessibility.

The construction of Brownian motion on the Sierpinski gasket is due to Goldstein [3], Kusuoka [8] and Barlow and Perkins [1] (in the following referred to as B.-P.). These authors used different methods to obtain their respective results.

Possibly one of the most remarkable properties of Brownian motion on the Sierpinski gasket is the invariance under natural symmetries of the gasket, to be precise under bijective isometries of open subsets of the gasket. Here the distance between two points in an open subset is the length of a shortest path in the set which connects the two points, or ∞ if there is no connecting path. This property sets Brownian motion apart from all other diffusions on the Sierpinski gasket. It also implies another important property of Brownian motion, the so called scale invariance.

Kumagai [7] succeeded in constructing diffusions on the Sierpinski gasket which fulfill less stringent symmetry requirements than Brownian motion, but which are still scale invariant. Hattori, K., Hattori, T. and Watanabe [4] on the other hand constructed interesting diffusions on the Sierpinski gasket which lack both types of invariance.

Apart from the metric mentioned, there is another natural metric, namely the one inherited from $\mathbb{R}^2$. This metric is fairly natural if one is interested in forces, such as gravitational forces, which act through the space surrounding the gasket rather than through the gasket itself.

If $\phi : U \to V$ is a bijective isometry with respect to the second metric, then for every path in U the image of the path under ϕ is a path in V of the same length. Hence ϕ is also an isometry with respect to the first metric.

Moreover, we shall see that the bijective isometries related to the second metric form a proper subset of the bijective isometries related to the first metric.

In the present paper we shall construct a class of diffusions, which, though generally not scale invariant, are invariant under isometries relative to the metric on the gasket inherited from $\mathbb{R}^2$.

Our construction follows the pattern of the construction given by B.-P. That construction is based on the convergence of certain random walks with scaled time. In that case the choice of the time scaling factors is rather natural because of an underlying spatial scale invariance. In our construction without spatial scale invariance, however, establishing the existence of time scaling factors presents a major problem. Our way to overcome this problem is to apply a perturbation result on matrix powers (see [5]). This perturbation result seems interesting in itself. We shall use it here to obtain convergence results for fairly general multi-type branching processes with varying environment – convergence results which play the key role in the construction of our diffusions.

To explain our main result, we need some notations and definitions:

1. We recall the definition of the Sierpinski gasket G as a subset of $\mathbb{R}^2$. To this end, let $F_0 := \{(0,0),(1,0),(\frac{1}{2},\frac{\sqrt{3}}{2})\}$, $F_{n+1} := F_n \cup [(2^n,0)+F_n] \cup [(2^{n-1},2^{n-1}\sqrt{3})+F_n]$, $\hat{G}^{(0)} := \bigcup_{n=0}^{\infty} F_n$. We denote by $G^{(0)}$ the union of $\hat{G}^{(0)}$ and $\hat{G}^{(0)}$, reflected at the y-axis. The sequence $G^{(n)} := 2^{-n}G^{(0)}, n \in \mathbb{N}_0$ is increasing, and if we let $G^{(\infty)} := \bigcup_{n=0}^{\infty} G^{(n)}$, then the Sierpinski gasket is $G := \text{closure}(G^{(\infty)})$. The Sierpinski gasket is endowed with the metric and the topology inherited from $\mathbb{R}^2$.

For the reader's convenience, we include an illustration of a section of the Sierpinski gasket (Figure 1).

2. By a symmetry of the Sierpinski gasket G we understand a bijective isometry, say $\psi : U \to V$, between two open subsets of G. We note that obviously ψ can be extended uniquely to a bijective isometry (also called ψ) between the closures of U and V.

If we denote by $\mu : \mathbb{R}^2 \to \mathbb{R}^2$ the reflection at the y-axis and let $\nu : \mathbb{R}^2 \to \mathbb{R}^2$ be the affine transformation of $\mathbb{R}^2$, with $\nu(0,0) := (\frac{3}{4},\frac{\sqrt{3}}{4})$, $\nu(-\frac{1}{2},0) := (1,0)$ and $\nu(-\frac{1}{4},\frac{\sqrt{3}}{4}) := (\frac{1}{2},0)$, then the restrictions of ν and $\nu \circ \mu$ to sufficiently small open neighborhoods of $(0,0)$ are symmetries of the Sierpinski gasket (see figure 1).

In the same way one can easily obtain for all $x, y \in G^{(\infty)}$ two affine transformations of $\mathbb{R}^2$, say $\phi_{x,y}$ and $\lambda_{x,y}$, with $\phi_{x,y}(x) = \lambda_{x,y}(x) = y$ such that $\phi_{x,y}$ preserves the orientation and $\lambda_{x,y}$ reverses the orientation and the restrictions of $\phi_{x,y}$ and $\lambda_{x,y}$ to sufficiently small open neighborhoods of x are symmetries. It is not very hard to see that conversely every symmetry $\phi : U \to V$ of the Sierpinski gasket is the restriction to U of such an affine transformation $\phi_{x,y}$ or $\lambda_{x,y}$.

3. A diffusion on G is a family of probability measures $(P_x;\, x \in G)$ on $C([0,\infty),G)$ which are connected by the strong Markov property and satisfy

$P_x\{\omega(0) = x\} = 1$.

4. For every set $A \subset G$ and a continuous path ω in G we denote by $T_A(\omega)$ the first time ω leaves A.

5. We call a diffusion $(P_x; x \in G)$ homogeneous, if for every symmetry $\psi : U \to V$ and every $x \in U$ the P_x-distribution of $\psi \circ \omega(\cdot \wedge T_U(\omega))$ equals the $P_{\psi(x)}$-distribution of $\omega(\cdot \wedge T_V(\omega))$. A diffusion on G is thus, roughly speaking, homogeneous, if it is invariant under the symmetries of G.

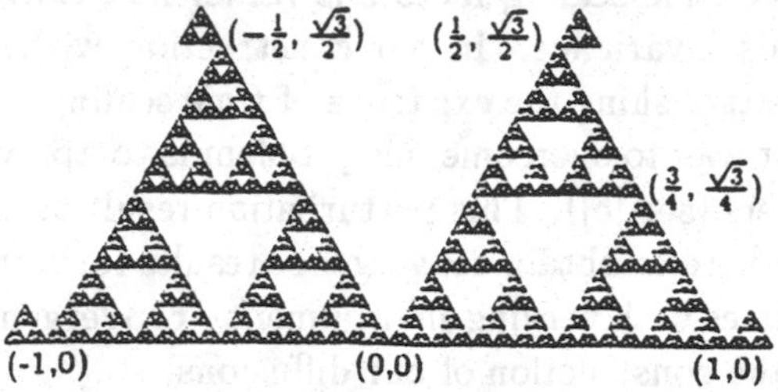

Figure 1

The purpose of the present paper is the description of homogeneous diffusions on G in terms of one real parameter, which may be interpreted as a certain exit probability. We shall explain briefly how this description is achieved.

If we denote by N the interior in G of the set of all points in G, which lie in one of the two closed triangles $\Delta((0,0),(-1,0),(-\frac{1}{2},\frac{\sqrt{3}}{2}))$ and $\Delta((0,0),(1,0),(\frac{1}{2},\frac{\sqrt{3}}{2}))$, then the open set N contains the point $(0,0)$, and its boundary in G consists of the two collinear boundary points $(-1,0), (1,0)$ and the two non-collinear boundary points $(-\frac{1}{2}, \frac{\sqrt{3}}{2})$, $(\frac{1}{2}, \frac{\sqrt{3}}{2})$. A continuous path starting at $(0,0)$ can leave N only through one of these four points. Excluding the trivial case $P_x\{\omega(t) = x, \text{ for all } t \geq 0\} = 1$ for all $x \in G$, we shall see, that for a homogeneous diffusion on G with Feller property, the exit time from N, i.e. T_N, is finite $P_{(0,0)}$-almost everywhere. In this case the $P_{(0,0)}$-exit-distribution coincides for the two non-collinear boundary points of N as well as for the two collinear boundary points of N, as the reflection ν of $\mathbb{R}^2$ at the y-axis defines a symmetry $\nu|_N : N \to N$ with $\nu((0,0)) = (0,0)$. In particular, the $P_{(0,0)}$-exit-distribution on the boundary of N is uniquely determined by the probability $\alpha = P_{(0,0)}\{\omega \text{ leaves } N \text{ through } (\frac{1}{2}, \frac{\sqrt{3}}{2})\}$.

The following theorem states that this parameter α characterizes completely the homogeneous diffusions with Feller property and that the parameter set is the interval $[0, \frac{1}{4}]$.

1.1 Theorem

a) *For every non trivial homogeneous diffusion on G, say $(P_x; x \in G)$, with Feller property, the probability $\alpha = P_{(0,0)}\{\omega$ leaves N through $(\frac{1}{2}, \frac{\sqrt{3}}{2})\}$ is a number in $[0, \frac{1}{4}]$.*

b) *Conversely, for each $\alpha \in [0, \frac{1}{4}]$ there exists a homogeneous diffusion $(P_x; x \in G)$ with Feller property, such that $P_{(0,0)}\{\omega$ leaves N through $(\frac{1}{2}, \frac{\sqrt{3}}{2})\} = \alpha$. This diffusion is uniquely determined up to a linear scaling of time.*

Remark *a) For $\alpha = \frac{1}{4}$ the diffusion of Theorem 1.1 is Brownian motion as constructed by B.-P. and for $\alpha = 0$ it is a modification of the ordinary one-dimensional Brownian motion. With the exception of these two values of α, the diffusions of Theorem 1.1 are not scale-invariant.*
b) It should be noted that a diffusion on G is already homogeneous, if it is invariant under symmetries of G of a certain local nature.
c) If $(P_x;\, x \in G)$ is a non trivial diffusion on G with Feller property which not only is homogeneous, but also is invariant under bijective isometries related to the metric of the length of a shortest path, then Theorem 1.1 implies that $(P_x;\, x \in G)$ is Brownian motion, as constructed by B.-P.

Indeed, if we denote by $r : \mathbf{R}^2 \to \mathbf{R}^2$ the reflection at the axis through $(0,0)$ and $(\frac{3}{4}, \frac{\sqrt{3}}{4})$ and define $\phi : G \to G$ by

$$\phi(x) := \begin{cases} x & : \text{if the first coordinate of } x \text{ is negative} \\ r(x) & : \text{otherwise} \end{cases},$$

then obviously ϕ is not a symmetry, but it is an isometry with respect to the metric of the length of a shortest path. Now the invariance under ϕ and under the reflection at the y-axis implies that the $P_{(0.0)}$-exit-distribution for the open set N coincides for all four boundary points and hence $\alpha = \frac{1}{4}$, i.e. $(P_x;\, x \in G)$ is Brownian motion. The corresponding uniqueness result for Brownian motion in B.-P. is obviously stronger since Barlow and Perkins do not assume that the diffusion has the Feller property.

2 Multi-type branching processes with varying environment

We start with limit results for a class of supercritical multi-type branching processes with varying environment, which we will present as Theorem 2.1 below.

For $d \in \mathbf{N}$ we consider a d-type branching process with varying environment which we shall denote by $(Z^{(n)};\, n \in \mathbf{N}_0)$. This process is completely described by the distribution of the initial generation $Z^{(0)}$ and the generating functions $(F^{(n)})_{n \in \mathbf{N}_0}$, which determine the branching mechanism for each generation $Z^{(n)}$, $n \in \mathbf{N}_0$.

We shall assume that

(A) the distribution of $Z^{(0)}$ has finite second moments,

and that the generating functions $F^{(n)}$, $n \in \mathbf{N}_0$ fulfill the following conditions:

(B) $M_1 := \sup\limits_{1 \le i,j \le d,\, n \in \mathbf{N}_0} \dfrac{\partial^2 F_i^{(n)}}{\partial x_j{}^2}(1,\ldots,1) < \infty.$

(C) $a_{i,j}^{(n)} := \dfrac{\partial F_i^{(n)}}{\partial x_j}(1,\ldots,1) > 0$ for all $1 \le i,j \le d$ and $n \in \mathbf{N}_0$.

(D) $a_{i,j} := \lim_{n\to\infty} a_{i,j}^{(n)}$ exists for all $1 \le i,j \le d$ and the matrix $A := (a_{i,j})_{i,j=1,\ldots,d}$ is invertible.

(E) There exists a real eigenvalue λ_A of A such that:

- (α) $\lambda_A > 1$.
- (β) $|\mu| < \lambda_A$ for all eigenvalues μ of A, different from λ_A.
- (γ) λ_A has multiplicity one.
- (δ) The entries of every non zero eigenvector of A corresponding to λ_A are either all strictly positive or all strictly negative.

If there exists $n \in \mathbb{N}$ such that all entries of A^n are strictly positive, then by the Perron–Frobenius Theorem assumption (E) reduces to:

(E′) There exists an eigenvalue ν of A with $|\nu| > 1$.

The probabilistic meaning of $a_{i,j}^{(n)}$ is the expected number of j-type descendants of a i-type individual of the n'th generation and according to (D) this expected number stabilizes if n tends to ∞. In the case of branching processes with non varying environment, i.e. if all $F^{(n)}$ coincide, our assumption (E·α) just describes the supercritical case.

In order to formulate Theorem 2.1 we introduce the following notations: For a matrix A satisfying (E) we denote by c_A the eigenvector of A corresponding to λ_A which has strictly positive entries and Euclidean norm 1. If in addition there exists $n \in \mathbb{N}$ such that all entries of A^n are strictly positive, then A^t, the transpose of A, also satisfies (E′) and hence (E) and we let $\tilde{c}_A := \frac{1}{c_A^t c_{A^t}} c_{A^t}$. Here the row vector c_A^t is the transpose of the column vector c_A.

2.1 Theorem *Let $(Z^{(n)}; n \in \mathbb{N}_0)$ be a d-type branching process with varying environment which satisfies conditions (A) to (E).*
Then there exists a sequence $(a_n)_{n\in\mathbb{N}_0}$ of strictly positive numbers converging to 0 such that:

a) $W^{(\infty)} := \lim_{n\to\infty} a_n\, c_A^t Z^{(n)}$ *exists a.e. and in* L^2.

b) $0 < E(W^{(\infty)}) < \infty$.

c) For a.a. ω: $W^{(\infty)}(\omega) = 0$ *iff* $Z^{(n)}(\omega) = (0,\ldots,0)^t$ *for all sufficiently large* n.

d) If there exists $n \in \mathbb{N}$ such that all entries of A^n are strictly positive, then

$$\lim_{n\to\infty} a_n Z^{(n)} = W^{(\infty)} \tilde{c}_A \quad \text{a.e.}$$

Remark *The sequence $(a_n)_{n\in\mathbb{N}_0}$ in the preceding Theorem 2.1 is obviously determined up to asymptotic equivalence by a) and b) in Theorem 2.1.*

Moreover, any sequence $(a_n)_{n\in\mathbb{N}_0}$ of Theorem 2.1 satisfies

$$\lim_{n\to\infty} a_n^{\frac{1}{n}} = \frac{1}{\lambda_A}.$$

For the proof of Theorem 2.1 we need two Lemmas.
These two Lemmas are results on convergent sequences of matrices, which apply in particular to the sequence $\left((a_{i,j}^{(n)})_{i,j=1,\ldots,d}\right)_{n\in\mathbb{N}_0}$ introduced in (C) above.

We let $\mathbb{R}_+ := \{x \in \mathbb{R};\, x > 0\}$ and denote by $\|\ \ \|$ the Euclidean norm on $\mathbb{R}^d$.

2.2 Lemma *Let $A, A^{(n)}, n \in \mathbb{N}_0$ be invertible $d \times d$–matrices, such that $A = \lim_{n\to\infty} A^{(n)}$ and there exists an eigenvalue ν of A with multiplicity 1 such that $|\mu| < |\nu|$ for all eigenvalues μ of A different from ν.*
Then there exists a sequence $(c_n)_{n\in\mathbb{N}_0}$ of vectors in $\mathbb{R}^d$, satisfying:

a) $c^{(n)} = A^{(n)} c^{(n+1)}$ for all $n \in \mathbb{N}_0$.

b) All limit points of $(\frac{c^{(n)}}{\|c^{(n)}\|})_{n\in\mathbb{N}_0}$ are contained in the eigenspace of A corresponding to ν.

The sequence $(c^{(n)})_{n\in\mathbb{N}_0}$ is unique up to a multiplicative constant.

Lemma 2.2 is an immediate consequence of Theorem 1.1 in [5]. □

2.3 Lemma *Let $A^{(n)} := (a_{i,j}^{(n)})_{i,j=1,\ldots,d}$, $n \in \mathbb{N}_0$ be matrices with strictly positive entries, satisfying (D) and (E). Then there exists a sequence $(c^{(n)})_{n\in\mathbb{N}_0}$ of vectors in $\mathbb{R}_+^d$ satisfying:*

$$c^{(n)} = A^{(n)} \cdot c^{(n+1)} \text{ for all } n \in \mathbb{N}_0. \tag{2.1a}$$

$\lim_{n\to\infty} \frac{c^{(n)}}{\|c^{(n)}\|} = c_A$, *where c_A is the eigenvector of the limit matrix A introduced previously.* (2.1b)

The sequence $(c^{(n)})_{n\in\mathbb{N}_0}$ is unique up to a strictly positive multiplicative constant.

Remark *It is easy to see that under the assumptions of Lemma 2.3 we obtain*

$$\lim_{n\to\infty} \| c^{(n)} \|^{\frac{1}{n}} = \frac{1}{\lambda_A} \tag{2.1c}$$

from (2.1a) and (2.1b). Moreover, Theorem 1.1 in [5] implies that any sequence $(c^{(n)})_{n\in\mathbb{N}_0}$ of vectors in $\mathbb{R}_+^d$ satisfying (2.1a) and (2.1c) also satisfies (2.1b).

Proof of Lemma 2.3: By (D) there exists $n_0 \in \mathbb{N}_0$, such that $A^{(n)}$ is invertible for all $n \geq n_0$. Using (D) and (E), Lemma 2.2 implies that there exists $c^{(n)}; n \geq n_0$ which satisfy the equation in (2.1a) for all $n \geq n_0$ and for which c_A and $-c_A$ are the only possible limit points of $(\frac{c^{(n)}}{\|c^{(n)}\|})_{n \geq n_0}$. Lemma 2.2 also implies, that this sequence is unique up to a multiplicative constant. It follows that c_A is a limit point of $(\frac{c^{(n)}}{\|c^{(n)}\|})_{n \geq n_0}$ or of $(\frac{-c^{(n)}}{\|-c^{(n)}\|})_{n \geq n_0}$. Without loss of generality we may assume that the former is true.

If all entries of $c^{(k)}$ are strictly positive for a $k \geq n_0$ then all entries of $c^{(n)}$ are strictly positive for $n_0 \leq n \leq k$ by (2.1a) and the fact that all entries of $A^{(n)}$ are strictly positive. Since c_A is a limit point of $(\frac{c^{(n)}}{\|c^{(n)}\|})_{n \geq n_0}$ and c_A has strictly positive entries, we conclude that all entries of $c^{(n)}$ are strictly positive for infinitely many $n \geq n_0$ and hence for all $n \geq n_0$. This implies (2.1b), since c_A and $-c_A$ are the only possible limit points of $(\frac{c^{(n)}}{\|c^{(n)}\|})_{n \geq n_0}$. For $n < n_0$ we define $c^{(n)}$ by (2.1a). □

Proof of Theorem 2.1: Let $(c^{(n)})_{n \in \mathbb{N}_0}$ be a sequence of vectors in $\mathbb{R}_+^d$ which satisfies (2.1a) and (2.1b) and let $p \in (\frac{1}{\lambda_A}, 1)$. Furthermore, let $W^{(n)} := c^{(n)t} Z^{(n)}$ for $n \in \mathbb{N}_0$.

Obviously (2.1a) implies for $n, l \in \mathbb{N}$

$$\frac{\| c^{(n+l)} \|}{\| c^{(n)} \|} = \prod_{k=1}^{l} \frac{\| c^{(n+k)} \|}{\| c^{(n+k-1)} \|} = \prod_{k=1}^{l} \left\| A^{(n+k-1)} \frac{c^{(n+k)}}{\| c^{(n+k)} \|} \right\|^{-1}.$$

This together with (D) and (2.1b) implies

$$\lim_{n \to \infty} \frac{\| c^{(n+l)} \|}{\| c^{(n)} \|} = \lambda_A^{-l} \quad \text{for all } l \in \mathbb{N} \tag{2.2}$$

and for sufficiently large $n \in \mathbb{N}$: $\quad \frac{\|c^{(n+l)}\|}{\|c^{(n)}\|} \leq p^l \quad$ for all $l \in \mathbb{N}$.

Observing that $\inf_{n \in \mathbb{N}_0, 1 \leq i \leq d} \frac{c_i^{(n)}}{\|c^{(n)}\|} > 0$ by (2.1b), we conclude that there exists $M_2 > 0$, such that

$$\sup_{n \in \mathbb{N}_0, 1 \leq i,j \leq d} \frac{c_i^{(n+l)}}{c_j^{(n)}} \leq M_2 p^l \quad \text{for all } l \in \mathbb{N}_0. \tag{2.3}$$

Obviously $(W^{(n)}; n \in \mathbb{N}_0)$ is a positive martingale. A simple but lengthy computation using (B), (C), (2.1a) and (2.3) shows L^2 boundedness of this martingale, to be more precise that

$$E[(W^{(n)})^2] \leq d^2\,(M_1 + 1)\, M_2^2 \,\| c^{(0)} \|\, E(W^{(0)})\, \frac{p^2}{1-p} + E[(W^{(0)})^2].$$

Hence the martingale convergence theorem implies:

$$W^{(\infty)} := \lim_{n \to \infty} W^{(n)} \quad \text{exists a.e. and in } L^2. \tag{2.4a}$$

$$E(W^{\langle\infty\rangle}) = E(W^{\langle 0\rangle}). \tag{2.4b}$$

$$E[(W^{\langle\infty\rangle})^2] \le d^2\,(M_1+1)\,M_2^2\,\|c^{\langle 0\rangle}\|\,E(W^{\langle 0\rangle})\,\frac{p^2}{1-p} + E[(W^{\langle 0\rangle})^2]. \tag{2.4c}$$

Letting $a_n := \|c^{\langle n\rangle}\|$, we obtain by (2.1b) $|a_n c_A^t Z^{\langle n\rangle} - W^{\langle n\rangle}| \le \varepsilon W^{\langle n\rangle}$ for $\varepsilon > 0$ and n sufficiently large. Thus part a) and b) of Theorem 2.1 are easy consequences of (2.4a) to (2.4c).

For the proof of part c) of Theorem 2.1 let $(X^{\langle n\rangle}(i,j,r);\ n \in \mathbf{N}_0)$, $1 \le j \le d, i, r \in \mathbf{N}$ be independent branching processes with generating functions $(F^{\langle n+r\rangle})_{n\in\mathbf{N}_0}$ such that $X^{\langle 0\rangle}(i,j,r)_m$ equals 1 for $m = j$ and 0 otherwise.

Letting $W^{\langle n\rangle}(i,j,r) := \frac{1}{c_j^{\langle r\rangle}} c^{\langle n+r\rangle t} X^{\langle n\rangle}(i,j,r)$, we conclude from equations (2.4a) to (2.4c) that $W^{\langle\infty\rangle}(i,j,r) := \lim_{n\to\infty} W^{\langle n\rangle}(i,j,r)$ exists a.e. and that $E(W^{\langle\infty\rangle}(i,j,r)) = 1$ and $E[W^{\langle\infty\rangle}(i,j,r)^2] \le d^3(M_1+1)M_2^3\frac{p^2}{1-p} + 1 =: \rho$. By Hölder's inequality we conclude that

$$\begin{aligned} 1 &= (E[W^{\langle\infty\rangle}(i,j,r)])^2 \le E[W^{\langle\infty\rangle}(i,j,r)^2]P\{W^{\langle\infty\rangle}(i,j,r) > 0\} \\ &\le \rho P\{W^{\langle\infty\rangle}(i,j,r) > 0\} \end{aligned}$$

and that hence $P\{W^{\langle\infty\rangle}(i,j,r) = 0\} \le 1 - \frac{1}{\rho}$ for $1 \le j \le d, i, r \in \mathbf{N}$.

Since $(Z^{\langle n\rangle};\ n \in \mathbf{N}_0)$ is a branching process, we have

$$P\{W^{\langle\infty\rangle} = 0 | Z^{\langle r\rangle}, \ldots, Z^{\langle 0\rangle}\}(\tilde\omega) = P\left\{\sum_{j=1}^{d}\sum_{i=1}^{Z_j^{\langle r\rangle}(\tilde\omega)} W^{\langle\infty\rangle}(i,j,r) = 0\right\}$$

$$= \prod_{j=1}^{d}\prod_{i=1}^{Z_j^{\langle r\rangle}(\tilde\omega)} P\{W^{\langle\infty\rangle}(i,j,r) = 0\} \le (1 - \frac{1}{\rho})^{\sum_{j=1}^d Z_j^{\langle r\rangle}(\tilde\omega)} \quad \text{a.e.}$$

Since $1_{\{W^{\langle\infty\rangle}=0\}}$ is measurable with respect to $\sigma(Z^{\langle n\rangle}, n \in \mathbf{N}_0)$, the well known martingale convergence theorem implies

$$\lim_{r\to\infty} P\{W^{\langle\infty\rangle} = 0 | Z^{\langle r\rangle}, \ldots, Z^{\langle 0\rangle}\} = P\{W^{\langle\infty\rangle} = 0 | Z^{\langle n\rangle}, n \in \mathbf{N}_0\} = 1_{\{W^{\langle\infty\rangle}=0\}}.$$

Hence $1_{\{W^{\langle\infty\rangle}=0\}} \le (1 - \frac{1}{\rho})^{\limsup_{r\to\infty}\sum_{j=1}^d Z_j^{\langle r\rangle}}$ a.e.

Thus, for a.a. ω: $\quad W^{\langle\infty\rangle}(\omega) = 0$ implies $\limsup_{n\to\infty}\sum_{j=1}^d Z_j^{\langle n\rangle}(\omega) = 0$.

The converse implication is obvious. Since $\sum_{j=1}^d Z_j^{\langle n\rangle} \in \mathbf{N}_0$, we conclude part c).

The proof of part d) of Theorem 2.1 is similar to that of Lemma 8.2 in [10]. For the reader's convenience we shall sketch the proof.

Let $\lambda := \lambda_A$, $c := c_A$ and $\tilde c := \tilde c_A$. If we define $B := A - \lambda c\tilde c^t$, then $A^k = \lambda^k c\tilde c^t + B^k$ for $k \in \mathbf{N}$. Moreover, it can be shown that for some $q \in (0, \lambda)$ and $M_3 > 0$

$$\max_{1\le i,j\le d} |(B^k)_{i,j}| < M_3\,q^k \quad \text{for } k \in \mathbf{N}. \tag{2.5}$$

For details – under somewhat weaker assumptions – we refer to Karlin [6], Appendix 2.

Let $1 \le j \le d$. For $\varepsilon > 0$ let $l \in \mathbf{N}$ such that $M_3 \left(\frac{q}{\lambda}\right)^l \le \varepsilon$. Considering $\| c^{\langle n \rangle} \| Z^{\langle n \rangle} - W^{\langle \infty \rangle} \tilde{c}$ for $n > l$, we obtain for its j'th entry the estimate

$$\begin{aligned} & \left| \| c^{\langle n \rangle} \| Z_j^{\langle n \rangle} - W^{\langle \infty \rangle} \tilde{c}_j \right| \\ \le & \left| \| c^{\langle n \rangle} \| (Z^{\langle n-l \rangle t} A^l)_j - W^{\langle \infty \rangle} \tilde{c}_j \right| + \left| \| c^{\langle n \rangle} \| \left(Z_j^{\langle n \rangle} - (Z^{\langle n-l \rangle t} A^l)_j \right) \right|. \end{aligned} \tag{2.6}$$

Substituting $A^l = \lambda^l c \tilde{c}^t + B^l$ we have

$$\begin{aligned} & \| c^{\langle n \rangle} \| (Z^{\langle n-l \rangle t} A^l)_j \\ = & \left(\| c^{\langle n-l \rangle} \| Z^{\langle n-l \rangle t} c \right) \tilde{c}_j \left(\lambda^l \frac{\| c^{\langle n \rangle} \|}{\| c^{\langle n-l \rangle} \|} \right) + \left(\| c^{\langle n \rangle} \| (Z^{\langle n-l \rangle t} B^l)_j \right). \end{aligned} \tag{2.7}$$

The first summand on the right sight of (2.7) tends to $W^{\langle \infty \rangle} \tilde{c}_j$ a.e. by part a) of Theorem 2.1 and (2.2). For the second summand on the right sight of (2.7) we obtain the asymptotic estimate

$$\begin{aligned} & \limsup_{n \to \infty} \left| \| c^{\langle n \rangle} \| (Z^{\langle n-l \rangle t} B^l)_j \right| \\ = & \limsup_{n \to \infty} \left| \left((\| c^{\langle n-l \rangle} \| Z^{\langle n-l \rangle})^t \left(\frac{\| c^{\langle n \rangle} \|}{\| c^{\langle n-l \rangle} \|} B^l \right) \right)_j \right| \le \max_{1 \le i \le d} \left(\frac{1}{c_i} \right) W^{\langle \infty \rangle} \varepsilon \text{ a.e.}, \end{aligned}$$

by (2.2), (2.5) and part a) of Theorem 2.1. We therefore obtain from (2.7) for the first summand on the right sight of (2.6) the estimate

$$\limsup_{n \to \infty} \left| \| c^{\langle n \rangle} \| (Z^{\langle n-l \rangle t} A^l)_j - W^{\langle \infty \rangle} \tilde{c}_j \right| \le \max_{1 \le i \le d} \left(\frac{1}{c_i} \right) W^{\langle \infty \rangle} \varepsilon \text{ a.e.}$$

Thus it remains to prove that the second summand of the right sight of (2.6) tends to zero. Indeed, using (D) and part a) of Theorem 2.1 we see that

$$\begin{aligned} & \limsup_{n \to \infty} \left| \| c^{\langle n \rangle} \| \left(Z_j^{\langle n \rangle} - (Z^{\langle n-l \rangle t} A^l)_j \right) \right| \\ \le \quad & \limsup_{n \to \infty} \left| \| c^{\langle n \rangle} \| \left(Z_j^{\langle n \rangle} - (Z^{\langle n-l \rangle t} A^{\langle n-l \rangle} \cdots A^{\langle n-1 \rangle})_j \right) \right| \\ & + \limsup_{n \to \infty} \| c^{\langle n \rangle} \| \sum_{k=1}^{d} Z_k^{\langle n-l \rangle} \max_{1 \le k,j \le d} |(A^{\langle n-l \rangle} \cdots A^{\langle n-1 \rangle})_{k,j} - (A^l)_{k,j}| \\ = \quad & \limsup_{n \to \infty} \left| \| c^{\langle n \rangle} \| \left(Z_j^{\langle n \rangle} - (Z^{\langle n-l \rangle t} A^{\langle n-l \rangle} \cdots A^{\langle n-1 \rangle})_j \right) \right| \qquad a.e. \end{aligned}$$

We obtain for $n > l$

$$\begin{aligned} & E \left[\left(Z_j^{\langle n \rangle} - (Z^{\langle n-l \rangle t} A^{\langle n-l \rangle} \cdots A^{\langle n-1 \rangle})_j \right)^2 \right] \| c^{\langle n \rangle} \|^2 \\ \le \; & E \left[\left(\sum_{r=0}^{l-1} (Z^{\langle n-r \rangle t} A^{\langle n-r \rangle} \cdots A^{\langle n-1 \rangle} - Z^{\langle n-r-1 \rangle t} A^{\langle n-r-1 \rangle} \cdots A^{\langle n-1 \rangle})_j \right)^2 \right] \| c^{\langle n \rangle} \|^2 \end{aligned}$$

$$\leq l \sum_{r=0}^{l-1} E\left[\left(\left[Z^{(n-r)^t} - Z^{(n-r-1)^t} A^{(n-r-1)}\right] A^{(n-r)} \cdots A^{(n-1)}\right)_j^2\right] \|c^{(n)}\|^2$$

$$\leq l(1 + \sup_{\substack{r \in \{0,\ldots,l\} \\ k \in \mathbb{N}_0, 1 \leq i,j \leq d}} (A^{(k)} \cdots A^{(k+r)})_{i,j}^2)$$

$$\times \sum_{r=0}^{l-1} E\left[\left(\sum_{k=1}^{d}\left[Z_k^{(n-r)} - \sum_{m=1}^{d} Z_m^{(n-r-1)} a_{m,k}^{(n-r-1)}\right]\right)^2\right] \|c^{(n)}\|^2$$

Here we used the convention $(A^{(i)} \cdots A^{(j)}) =: Id$ for $i > j$.

Keeping in mind that $(Z^{(n)}; n \in \mathbb{N}_0)$ is a branching process for which the matrices $A^{(n)}$ are the matrices of the first moments, a lengthy but standard computation using (B) and (2.3) shows that the last term is majorized by

$$l(1 + \sup_{\substack{r \in \{0,\ldots,l\} \\ k \in \mathbb{N}_0, 1 \leq i,j \leq d}} (A^{(k)} \cdots A^{(k+r)})_{i,j}^2) d^3 (M_1 + 1) M_2^2 \| c^{(0)} \| E(W^{(0)}) \frac{p^n}{1-p}.$$

Since the supremum in the last majorant is finite by (D) it follows that the last majorant is summable. The fact that the second summand on the right sight of (2.6) tends to 0 a.e. now follows by Chebyshev's inequality and the Borel-Cantelli Lemma. □

With a_n and $W^{(n)}$ as in the preceding proof, we conclude from (2.1b) that $|a_n c_A^t Z^{(n)} - W^{(n)}| \leq \varepsilon a_n c_A^t Z^{(n)}$ for $\varepsilon > 0$ and n sufficiently large. Hence, Theorem 2.1 implies the following corollary.

2.4 Corollary *Under the assumptions of Theorem 2.1, a) to c) in Theorem 2.1 hold with $a_n c_A^t Z^{(n)}$ replaced by $c^{(n)t} Z^{(n)}$, where $(c^{(n)})_{n \in \mathbb{N}_0}$ is any sequence in $\mathbb{R}_+^d$ satisfying (2.1a) and (2.1b).*

3 Consistent random walks

For every $k \in \mathbb{N}_0$ and $x \in G^{(k)}$ there are exactly four points in $G^{(k)}$ which have distance 2^{-k} from x. We call the two points, whose connecting line contains x collinear k-neighbors of x and the remaining two points non-collinear k-neighbors of x.

For $k \in \mathbb{N}_0$ let

$$\Omega^{(k)} := \{\omega : \mathbb{N}_0 \to G^{(k)}; \text{ with } \|\omega_{i+1} - \omega_i\| \leq 2^{-k} \text{ for all } i \in \mathbb{N}_0\}.$$

Given $\alpha \in [0, \frac{1}{2}]$, a Markov process $(P_x^{(k)}; x \in G^{(k)})$ on $\Omega^{(k)}$ is an (α, k)-random walk if

$$P_x^{(k)}\{\omega_1 = y\} = \begin{cases} \alpha & ; \text{if } y \text{ is an non-collinear } k\text{-neighbor of } x \\ \frac{1}{2} - \alpha & ; \text{if } y \text{ is a collinear } k\text{-neighbor of } x \\ 0 & ; \text{otherwise} \end{cases} .$$

In the following let $k, l \in \mathbb{N}_0$ such that $k \geq l$.

For $\omega \in \Omega^{\langle k\rangle}$ let $T_0^{\langle l,k\rangle}(\omega)$ be the first time ω hits $G^{\langle l\rangle}$ and let $T^{\langle l,k\rangle}(\omega)$ be the first time ω hits $G^{\langle l\rangle}\backslash\{\omega_0\}$. Starting with the stopping time $T_0^{\langle l,k\rangle}(\omega)$ we define inductively for $i \in \mathbf{N}_0$ the stopping times $T_{i+1}^{\langle l,k\rangle}(\omega)$

$$T_{i+1}^{\langle l,k\rangle}(\omega) := \begin{cases} T^{\langle l,k\rangle}(\omega_{T_i^{\langle l,k\rangle}+\cdot}) + T_i^{\langle l,k\rangle}(\omega) & ;\text{ if } T_i^{\langle l,k\rangle}(\omega) < \infty \\ \infty & ;\text{ otherwise} \end{cases} .$$

For $\omega \in \Omega^{\langle k\rangle}$ and $i \in \mathbf{N}_0$ we define $\Psi^{\langle l,k\rangle}(\omega)_i$ by

$$\Psi^{\langle l,k\rangle}(\omega)_i := \begin{cases} \omega_{T_i^{\langle l,k\rangle}} & ;\text{ if } T_i^{\langle l,k\rangle}(\omega) < \infty \\ \Psi^{\langle l,k\rangle}(\omega)_{i-1} & ;\text{ if } T_i^{\langle l,k\rangle}(\omega) = \infty \end{cases} .$$

Here we use the convention $\Psi^{\langle l,k\rangle}(\omega)_{-1} := \mathcal{O} := (0,0)$.
Obviously $\|\Psi^{\langle l,k\rangle}(\omega)_i - \Psi^{\langle l,k\rangle}(\omega)_{i-1}\| \leq 2^{-l}$ for $i \in \mathbf{N}$, so that $\Psi^{\langle l,k\rangle}$ maps $\Omega^{\langle k\rangle}$ into $\Omega^{\langle l\rangle}$.
For $\omega \in C([0,\infty), G)$ we obtain in an analogous way stopping times $T^{\langle l\rangle}$ and $T_i^{\langle l\rangle}$ and functions $\Psi^{\langle l\rangle}$.
Let $(\alpha^{\langle n\rangle})_{n\in\mathbf{N}_0}$ be a sequence of real numbers in $[0,\frac{1}{2}]$ and for each $n \in \mathbf{N}_0$ let $(P_x^{\langle n\rangle};\, x \in G^{\langle n\rangle})$ be an $(\alpha^{\langle n\rangle}, n)$-random walk.
We call the sequence $(P_x^{\langle n\rangle};\, x \in G^{\langle n\rangle})_{n\in\mathbf{N}_0}$ consistent, if

$$P_x^{\langle l\rangle} = P_x^{\langle k\rangle} \circ \Psi^{\langle l,k\rangle -1} \text{ for all } x \in G^{\langle l\rangle},\ k,l \in \mathbf{N}_0 \text{ with } k \geq l, \tag{3.8}$$

where $P_x^{\langle k\rangle} \circ \Psi^{\langle l,k\rangle -1}$ denotes the image of $P_x^{\langle k\rangle}$ under $\Psi^{\langle l,k\rangle}$.
We shall remark, that this condition is very natural in the following sense:

Remark *We shall see later, that for a non trivial homogeneous diffusion $(P_x;\ x \in G)$ with Feller property, we obtain random walks $(P_x \circ \Psi^{\langle n\rangle -1};\ x \in G^{\langle n\rangle})$ for all $n \in \mathbf{N}_0$. Since*

$$\Psi^{\langle l,k\rangle} \circ \Psi^{\langle k\rangle} = \Psi^{\langle l\rangle} \tag{3.9}$$

the sequence $(P_x \circ \Psi^{\langle n\rangle -1};\ x \in G^{\langle n\rangle})_{n\in\mathbf{N}_0}$ is consistent.

In the following we shall use the indicator functions χ_1 and χ_2 on $G^{\langle\infty\rangle} \times G^{\langle\infty\rangle}$

$$\chi_i(x,y) := \begin{cases} 1 & ;\text{ if } i=1 \text{ and } y \text{ is a non-collinear } k\text{-neighbor of } x \text{ for some } k \\ & \ \ \text{or if } i=2 \text{ and } y \text{ is a collinear } k\text{-neighbor of } x \text{ for some } k\,. \\ 0 & ;\text{ otherwise} \end{cases}$$

For $m \in \{1,2\}$, $i \in \mathbf{N}_0$ and $\omega \in \Omega^{\langle k\rangle}$ let $R_i^{\langle l,k\rangle}(m,\omega) := \sum_{j=1}^{T_i^{\langle l,k\rangle}(\omega)} \chi_m(\omega_{j-1},\omega_j)$ denote the number of 'steps to non-collinear neighbors' ($m=1$), respectively the number of 'steps to collinear neighbors' ($m=2$), up to the time $T_i^{\langle l,k\rangle}(\omega)$.
For $k \in \mathbf{N}$, $x \in G^{\langle k-1\rangle}$, $i \in \{1,2\}$, $\alpha \in [0,\frac{1}{2}]$ and $a,b \geq 0$ we denote by

$$g_{\alpha,i}^{\langle k\rangle,x}(a,b) := E_x^{\langle k\rangle}\left[a^{R_1^{\langle k-1,k\rangle}(1)}\, b^{R_1^{\langle k-1,k\rangle}(2)};\, \chi_i(x, \Psi^{\langle k-1,k\rangle}(\cdot)_1) = 1, T_1^{\langle k-1,k\rangle} < \infty\right]$$

where $E_x^{\langle k\rangle}$ denotes the expectation relative to $P_x^{\langle k\rangle}$ for the (α,k)-random walk $(P_x^{\langle k\rangle};\, x \in G^{\langle k\rangle})$.

3.1 Lemma *For all $a, b \in [0,1)$ we have*

$$g_{\alpha,1}^{\langle k\rangle,x}(a,b) = g_{\alpha,1}(a,b) := \frac{2b\alpha a(1-2\alpha)[(1-2\alpha)b+1]}{2(1-2\alpha a)(1+\alpha a) - b(1-2\alpha)[2\alpha a + (1-2\alpha)b]},$$

$$g_{\alpha,2}^{\langle k\rangle,x}(a,b) = g_{\alpha,2}(a,b) := \frac{b(1-2\alpha)[(1-2\alpha)b+4\alpha^2 a^2]}{2(1-2\alpha a)(1+\alpha a) - b(1-2\alpha)[2\alpha a + (1-2\alpha)b]}.$$

In the case $\alpha \neq \frac{1}{2}$, the equations are valid for all $a, b \in [0,1]$.

The proof is similar to the proof of Lemma 2.2 in B.-P. □

Since $R_i^{\langle k-1,k\rangle} - R_{i-1}^{\langle k-1,k\rangle}$, $i \in \mathbb{N}$ have the same distribution by Lemma 3.1 and the strong Markov property, Lemma 3.1 implies for $\alpha < \frac{1}{2}$, $k, i \in \mathbb{N}$ and $x \in G^{\langle k-1\rangle}$ that the random variables

$$R_i^{\langle k-1,k\rangle} \text{ and hence } T_i^{\langle k-1,k\rangle} \text{ are } P_x^{\langle k\rangle}\text{-square integrable.} \tag{3.10}$$

Let $f(\alpha) := \frac{2\alpha(1-\alpha)}{1+2\alpha}$. By the strong Markov property and Lemma 3.1 we obtain:

3.2 Lemma *For $k \in \mathbb{N}$ let $(P_x^{\langle k\rangle}\ ;\ x \in G^{\langle k\rangle})$ be an (α, k)-random walk and let $l < k$.*

a) For $\alpha \neq \frac{1}{2}$: $\quad (P_x^{\langle k\rangle} \circ \Psi^{\langle l,k\rangle^{-1}};\ x \in G^{\langle l\rangle})$ *is a* $(\underbrace{f \circ \cdots \circ f}_{k-l \text{ times}}(\alpha), l)$*-random walk.*

b) For $\alpha = \frac{1}{2}$: $\quad \Psi^{\langle l,k\rangle}(\omega) \equiv \omega_0 \quad P_x^{\langle k\rangle}$*-a.s. for all $x \in G^{\langle l\rangle}$.*

□

3.3 Proposition *For $k \in \mathbb{N}_0$ let $\alpha^{\langle k\rangle} \in [0, \frac{1}{2}]$ and let $(P_x^{\langle k\rangle};\ x \in G^{\langle k\rangle})$ be an $(\alpha^{\langle k\rangle}, k)$-random walk. The sequence $(P_x^{\langle k\rangle};\ x \in G^{\langle k\rangle})_{k\in\mathbb{N}_0}$ is consistent, iff $\alpha^{\langle 0\rangle} \in [0, \frac{1}{4}]$ and*

$$\alpha^{\langle k+1\rangle} = \delta(\alpha^{\langle k\rangle})\alpha^{\langle k\rangle} \text{ for all } k \in \mathbb{N}_0 \tag{3.11}$$

with $\delta(\alpha) := (1 - \alpha + \sqrt{(1-\alpha)^2 - 2\alpha})^{-1}$. In this case

$$\begin{aligned} &\alpha^{\langle\cdot\rangle} \equiv \alpha^{\langle 0\rangle} && ;\ \text{if } \alpha^{\langle 0\rangle} = 0 \text{ or } \tfrac{1}{4}. \\ &2^{-k}\alpha^{\langle 0\rangle} \leq \alpha^{\langle k\rangle} \leq \delta(\alpha^{\langle 0\rangle})^k \alpha^{\langle 0\rangle} \text{ with } \delta(\alpha^{\langle 0\rangle}) < 1 && ;\ \text{if } \alpha^{\langle 0\rangle} \in (0, \tfrac{1}{4}). \end{aligned}$$

Remark *We remark that for $\alpha^{\langle 0\rangle} \in [0, \frac{1}{4}]$ equation (3.11) defines a sequence of real numbers in $[0, \frac{1}{4}]$.*

Proof: Let $(P_x^{\langle k\rangle};\ x \in G^{\langle k\rangle})_{k\in\mathbb{N}_0}$ be consistent. By Lemma 3.2 b) $\alpha^{\langle k\rangle} \neq \frac{1}{2}$ for all $k \in \mathbb{N}$. By Lemma 3.2 a) consistency is then equivalent to $f(\alpha^{\langle k+1\rangle}) = \alpha^{\langle k\rangle}$ for $k \in \mathbb{N}_0$, i.e. $\alpha^{\langle k+1\rangle} \in f^{-1}\{\alpha^{\langle k\rangle}\}$. The set on the right sight consists of the (possibly complex) points $\frac{1}{2}\left(1 - \alpha^{\langle k\rangle} + \sqrt{(1-\alpha^{\langle k\rangle})^2 - 2\alpha^{\langle k\rangle}}\right)$ and $\frac{1}{2}\left(1 - \alpha^{\langle k\rangle} - \sqrt{(1-\alpha^{\langle k\rangle})^2 - 2\alpha^{\langle k\rangle}}\right)$.

We shall prove that consistency is equivalent to

$$\alpha^{(k+1)} = \frac{1}{2}\left(1 - \alpha^{(k)} - \sqrt{(1-\alpha^{(k)})^2 - 2\alpha^{(k)}}\right) \quad \text{for } k \in \mathbb{N}_0.$$

Since $\alpha^{(k+1)}$ has to be real, consistency is equivalent to $\alpha^{(k+1)} \in f^{-1}\{\alpha^{(k)}\}$ and $\alpha^{(k)} \in [0, 2-\sqrt{3}]$ for all $k \in \mathbb{N}_0$.
But for $\alpha^{(k)} \in [0, 2-\sqrt{3}]$, $\frac{1}{2}\left(1-\alpha^{(k)} + \sqrt{(1-\alpha^{(k)})^2 - 2\alpha^{(k)}}\right) > 2-\sqrt{3}$. Hence consistency is equivalent to
$\alpha^{(k+1)} = \frac{1}{2}\left(1 - \alpha^{(k)} - \sqrt{(1-\alpha^{(k)})^2 - 2\alpha^{(k)}}\right) = \delta(\alpha^{(k)})\alpha^{(k)}$ for $k \in \mathbb{N}_0$.

Using strict monotony of $\delta|_{[0,2-\sqrt{3})}$ and $\delta(0) = \frac{1}{2}, \delta(\frac{1}{4}) = 1$ the remaining part of the Proposition follows now easily. □

4 Consistent velocities

For the following we fix a consistent sequence $(P_x^{(k)}; x \in G^{(k)})_{k\in\mathbb{N}_0}$ of $(\alpha^{(k)}, k)$-random walks on $G^{(k)}$, where $(\alpha^{(k)})_{k\in\mathbb{N}_0}$ is a sequence of real numbers in $[0, \frac{1}{4}]$ which satisfies equation (3.11).

In order to embed the sample spaces $\Omega^{(k)}$ of the random walks into $C([0,\infty), G)$ we fix velocities $c^{(k)} \in \mathbb{R}_+^2$ and define for $\omega \in \Omega^{(k)}$ the function $Y_{c^{(k)}}^{(k)}(\omega) \in C([0,\infty), G)$ by $Y_{c^{(k)}}^{(k)}(\omega)_t = \omega_i$ for $t = c_1^{(k)} R_i^{(k,k)}(1,\omega) + c_2^{(k)} R_i^{(k,k)}(2,\omega)$, $i \in \mathbb{N}_0$, and by linear interpolation for all other times t.

For $k \in \mathbb{N}_0$, $x \in G^{(k)}$ let the measure $\tilde{P}_x^{(k)}$ on $C([0,\infty), G)$ be the image measure of $P_x^{(k)}$ on $\Omega^{(k)}$ under the embedding $Y_{c^{(k)}}^{(k)}$.

In turns of this measures $(\tilde{P}_x^{(k)}; x \in G^{(k)})_{k\in\mathbb{N}_0}$ the previously discussed consistency of the random walks $(P_x^{(k)}; x \in G^{(k)})_{k\in\mathbb{N}_0}$ translates into

$$\tilde{P}_x^{(k)}\{X_{T^{(k-1)}} \in \cdot\} = \tilde{P}_x^{(k-1)}\{X_{T^{(k-1)}} \in \cdot\}$$

for $k \in \mathbb{N}$ and $x \in G^{(k-1)}$, where $X_t(\omega) = \omega_t$.

The sequence of velocities $(c^{(k)})_{k\in\mathbb{N}_0}$ underlying the measures $\tilde{P}_x^{(k)}$, $x \in G^{(k)}$, $k \in \mathbb{N}_0$ is called consistent if

$$\tilde{E}_x^{(k)}(T^{(k-1)} \mid X_{T^{(k-1)}}) = \tilde{E}_x^{(k-1)}(T^{(k-1)} \mid X_{T^{(k-1)}}) \quad \text{for } k \in \mathbb{N} \text{ and } x \in G^{(k-1)},$$

where $\tilde{E}_x^{(k)}$ denotes integration with respect to $\tilde{P}_x^{(k)}$.

4.1 Proposition *For every consistent sequence of random walks $(P_x^{(k)}; x \in G^{(k)})_{k\in\mathbb{N}_0}$ there exists a consistent sequence of velocities $(c^{(k)})_{k\in\mathbb{N}_0}$. For two consistent sequences of velocities $(c^{(k)})_{k\in\mathbb{N}_0}$ and $(c^{(k)\prime})_{k\in\mathbb{N}_0}$ the resulting measures $\tilde{P}_x^{(k)}$ and $\tilde{P}_x^{(k)\prime}$ are related in a simple way: There exists $c > 0$ such that $\tilde{P}_x^{(k)}\{\omega(\cdot) \in \star\} = \tilde{P}_x^{(k)\prime}\{\omega(c\,\cdot) \in \star\}$ for $k \in \mathbb{N}_0$, $x \in G^{(k)}$.*

For the proof of the Proposition we shall use Lemma 4.2 below. This Lemma will again be used later on and is therefore formulated slightly more general than would be necessary in the present context. To formulate this Lemma we introduce the variables $V_i^{(l,k)} := R_i^{(l,k)} - R_{i-1}^{(l,k)}$, $i \in \mathbb{N}$ and $k, l \in \mathbb{N}_0, k \geq l$.

4.2 Lemma *For $x \in G^{\langle k-1\rangle}$, $k, i \in \mathbf{N}$*

$$P_x^{\langle k\rangle}\{V_i^{\langle k-1,k\rangle} \in \star | X^{\langle k\rangle}_{\cdot\wedge T_{i-1}^{\langle k-1,k\rangle}}, X^{\langle k\rangle}_{T_i^{\langle k-1,k\rangle}+\cdot}\} = P_x^{\langle k\rangle}\{V_i^{\langle k-1,k\rangle} \in \star | V_i^{\langle k-1,k-1\rangle} \circ \Psi^{\langle k-1,k\rangle}\}$$

where $X_i^{\langle k\rangle}(\omega) = \omega_i$ for $\omega \in \Omega^{\langle k\rangle}$. Furthermore,

$$E_x^{\langle k\rangle}(a^{V_i^{\langle k-1,k\rangle}(1)} b^{V_i^{\langle k-1,k\rangle}(2)} | V_i^{\langle k-1,k-1\rangle} \circ \Psi^{\langle k-1,k\rangle}) = \frac{g_{\alpha^{\langle k\rangle},1}(a,b)}{g_{\alpha^{\langle k\rangle},1}(1,1)} V_i^{\langle k-1,k-1\rangle}(1) \circ \Psi^{\langle k-1,k\rangle} + \frac{g_{\alpha^{\langle k\rangle},2}(a,b)}{g_{\alpha^{\langle k\rangle},2}(1,1)} V_i^{\langle k-1,k-1\rangle}(2) \circ \Psi^{\langle k-1,k\rangle}$$

for $a, b \in [0,1]$, where $\frac{g_{0,1}(a,b)}{g_{0,1}(1,1)} := 1$.

As the proof of Lemma 4.2 is similar to the proof of Lemma 2.5 in B.-P. we omit the proof. Notation: $G_\alpha(a,b) := \left(\frac{g_{\alpha,i}(a,b)}{g_{\alpha,i}(1,1)}\right)_{i=1,2}$.

Proof of Proposition 4.1: We observe, that $T^{\langle k-1\rangle}(Y^{\langle k\rangle}_{c^{\langle k\rangle}}(\omega)) = c^{\langle k\rangle t} V_1^{\langle k-1\rangle}(\omega)$ for $\omega \in \Omega^{\langle k\rangle}$ with $\omega_0 \in G^{\langle k-1\rangle}$.

Hence, if $\alpha^{\langle 0\rangle} = 0$, i.e. $\alpha^{\langle\cdot\rangle} \equiv 0$, Lemma 4.2 implies that consistency of the sequence of velocities $(c^{\langle k\rangle})_{k\in\mathbf{N}_0}$ is equivalent to

$$c_2^{\langle k\rangle} = \frac{\partial G_{0,2}}{\partial x_2}(1,1)\, c_2^{\langle k+1\rangle} = 4c_2^{\langle k+1\rangle}, \tag{4.12}$$

which in turn is equivalent to $c_2^{\langle k\rangle} = (\frac{1}{4})^k c_2^{\langle 0\rangle}$, for all $k \in \mathbf{N}_0$. This implies in particular for two consistent sequences of velocities $(c^{\langle k\rangle})_{k\in\mathbf{N}_0}$ and $(c^{\langle k\rangle\prime})_{k\in\mathbf{N}_0}$ that there exists $c > 0$ such that $c\,c_2^{\langle k\rangle} = c_2^{\langle k\rangle\prime}$ for all $k \in \mathbf{N}_0$.

Moreover, since the assumption $\alpha^{\langle\cdot\rangle} \equiv 0$ implies that $R_i^{\langle k,k\rangle}(1) = 0$ $P_x^{\langle k\rangle}$-a.s., we conclude that

$$Y^{\langle k\rangle}_{c^{\langle k\rangle}}(\omega)_t = Y^{\langle k\rangle}_{c^{\langle k\rangle\prime}}(\omega)_{ct} \qquad \text{for } P_x^{\langle k\rangle}\text{-a.a. } \omega$$

which implies the statement in the Proposition regarding the measures $\tilde{P}_x^{\langle k\rangle}$ and $\tilde{P}_x^{\langle k\rangle\prime}$.

In the case $\alpha^{\langle 0\rangle} \in (0, \frac{1}{4}]$ Lemma 4.2 implies that consistency of the sequence of velocities $(c^{\langle k\rangle})_{k\in\mathbf{N}_0}$ is equivalent to

$$c^{\langle k\rangle} = \left(\frac{\partial G_{\alpha^{\langle k+1\rangle},i}}{\partial x_j}(1,1)\right)_{i,j=1,2} \cdot c^{\langle k+1\rangle}. \tag{4.13}$$

Since the conditions (D) and (E) hold for $a_{i,j}^{(n)} := \frac{\partial G_{\alpha^{\langle n+1\rangle},i}}{\partial x_j}(1,1) > 0$, Lemma 2.3 implies the existence of a sequence $(c^{\langle k\rangle})_{k\in\mathbf{N}_0}$ which satisfies (2.1a) and (2.1b) and hence (4.13).

For the remaining part of Proposition 4.1 it suffices to prove, that the sequence $(c^{\langle k\rangle})_{k\in\mathbf{N}_0}$ is uniquely determined by (4.13) up to a multiplicative constant $c > 0$. In the case that $\alpha^{\langle 0\rangle} = \frac{1}{4}$ this is easily seen by Corollary 1.1 in [5] and in the case that $\alpha^{(0)} \in (0, \frac{1}{4})$ it follows from Lemma 2.3 and Lemma 4.3 below. □

4.3 Lemma *For $\alpha^{(0)} \in (0, \frac{1}{4})$ define $(\alpha^{(k)})_{k\in\mathbb{N}_0}$ by (3.11) and let $c^{(k)}$, $k \in \mathbb{N}_0$ be vectors in $\mathbb{R}^2_+$, satisfying (4.13) for all $k \in \mathbb{N}_0$. Then the sequence $(c^{(k)})_{k\in\mathbb{N}_0}$ satisfies (2.1b) with $A^{(n)} := (\frac{\partial G_{\alpha^{(n+1)},i}}{\partial x_j}(1,1))_{i,j=1,2}$.*

Proof: $\left(\frac{\partial G_{\alpha,i}}{\partial x_j}(1,1)\right)_{i,j=1,2}$ is invertible for $\alpha \in (0, \frac{1}{4})$ and

$$\left(\frac{\partial G_{\alpha,i}}{\partial x_j}(1,1)\right)^{-1}_{i,j=1,2} = \begin{pmatrix} \frac{2-3\alpha+4\alpha^2}{2(1+\alpha)(1-2\alpha-2\alpha^2)} & -\frac{(7-6\alpha-4\alpha^2)(1+2\alpha+4\alpha^2)}{8(1+\alpha)(1-2\alpha-2\alpha^2)(1-\alpha)} \\ -\frac{\alpha(1-\alpha)(1+2\alpha+4\alpha^2)}{2(1+\alpha)(1-2\alpha-2\alpha^2)(1-2\alpha)} & \frac{(1+4\alpha)(1-2\alpha+4\alpha^2)}{4(1+\alpha)(1-2\alpha-2\alpha^2)(1-2\alpha)} \end{pmatrix}.$$

Let $(b^{(k)}_{i,j})_{i,j=1,2} := \left(\frac{\partial G_{\alpha^{(k)},i}}{\partial x_j}(1,1)\right)^{-1}_{i,j=1,2}$.
Since $\alpha^{(k)} \to 0$ by Proposition 3.3, there exists $k_0 \in \mathbb{N}$ such that for all $k \geq k_0$

$$b^{(k)}_{11} + b^{(k)}_{12}\frac{1}{10} > \frac{9}{10}, \quad \frac{\alpha^{(k)}}{20} < -b^{(k)}_{21} \text{ and } \frac{1}{5} < b^{(k)}_{22} < \frac{1}{3}. \tag{4.14}$$

Another way of saying that the sequence $(c^{(k)})_{k\in\mathbb{N}_0}$ satisfies (4.13) is to say that it satisfies (2.1a).

We assume now, that the sequence did not satisfy (2.1b). We know by Proposition 2.3 in [5] that $\frac{c^{(n)}}{\|c^{(n)}\|} \to (1,0)^t$ or equivalently $\gamma^{(n)} := \frac{c_2^{(n)}}{c_1^{(n)}} \to 0$. Obviously $c^{(n)} \in \mathbb{R}^2_+$ implies $\gamma^{(n)} > 0$.

If $k \geq k_0$ and $\gamma^{(k)} < \frac{1}{10}$ then by (4.14)

$$\gamma^{(k+1)} = \frac{b^{(k+1)}_{21} + b^{(k+1)}_{22}\gamma^{(k)}}{b^{(k+1)}_{11} + b^{(k+1)}_{12}\gamma^{(k)}} \leq \frac{b^{(k+1)}_{22}}{b^{(k+1)}_{11} + b^{(k+1)}_{12}\gamma^{(k)}}\gamma^{(k)} \leq \frac{10}{27}\gamma^{(k)}. \tag{4.15}$$

Inductively we obtain for $l \in \mathbb{N}$ and some $k_1 \geq k_0$

$$\gamma^{(k_1+l)} \leq \frac{1}{10}\left(\frac{10}{27}\right)^l. \tag{4.16}$$

As by (4.14) and Proposition 3.3 $b^{(k)}_{21} < -\frac{\alpha^{(0)}}{20}2^{-k}$ for $k \geq k_0$ we conclude from (4.14) and (4.16) that $b^{(k_1+l+1)}_{21} + b^{(k_1+l+1)}_{22}\gamma^{(k_1+l)} < 0$ for l sufficiently large. Using this inequality in combination with (4.14) and (4.15) we obtain that $\gamma^{(k_1+l)} < 0$ for l sufficiently large. This contradicts $c^{(n)} \in \mathbb{R}^2_+$ for all $n \in \mathbb{N}_0$. □

5 Proof of Theorem 1.1

In this section we shall conclude the proof of Theorem 1.1, of which part b) is by far more delicate than part a). The essential tools for the proof of part b) have already been developed in sections 2, 3 and 4 and for the conclusion of the proof of part b) we may follow arguments used in B.-P. Thus we shall only sketch this part of the proof.

We remind the reader that we shall refer to isometries of open subsets of G, relative to the Euclidean metric, simply as symmetries. We introduce first a special class of symmetries. For $k \in \mathbf{N}_0$ and $x \in G^{\langle k \rangle}$ consider the two closed equilateral triangles, which have as vertices x, one collinear and one non-collinear k-neighbor of x. We denote by $N^{\langle k \rangle}(x)$ the interior (relative to the topology of G) of the set of all points in G, which lie in one of those two triangles. It is not hard to see that $\{N^{\langle k \rangle}(x); x \in G^{\langle k \rangle}, k \in \mathbf{N}_0\}$ is a base for the topology of G. For $x, y \in G^{\langle k \rangle}$ there exist exactly two isometries of $\mathbf{R}^2$, which map x onto y and $N^{\langle k \rangle}(x)$ onto $N^{\langle k \rangle}(y)$. One of these isometries, say $\phi_{x,y}$, preserves the orientation the other isometry, say $\lambda_{x,y}$, reverses the orientation. The restrictions of $\phi_{x,y}$ and $\lambda_{x,y}$ to $N^{\langle k \rangle}(x)$ are symmetries of the gasket and shall be denoted by $\phi_{x,y}^{\langle k \rangle}$ and $\lambda_{x,y}^{\langle k \rangle}$.

For $A \subset G$ let T_A be the first exit time from A. It is not hard to see, that for $\omega \in C([0,\infty), G)$ with $\omega_0 \in G^{\langle k \rangle}$ we have $T_1^{\langle k \rangle}(\omega) = T^{\langle k \rangle}(\omega) = T_{N^{\langle k \rangle}(\omega_0)}(\omega)$.

In the following two lemmas $(Q_x; x \in G)$ will denote a non trivial, homogeneous diffusion on G with Feller property.

5.1 Lemma *There exists $k_0 \in \mathbf{N}_0$, such that*

a) $\sup_{x \in G^{\langle k_0 \rangle}} \sup_{y \in N^{\langle k_0 \rangle}(x)} Q_y\{T_{N^{\langle k_0 \rangle}(x)} > 1\} < 1,$

b) $T_{N^{\langle k \rangle}(x)}$ *is Q_x-square integrable for all $x \in G^{\langle k \rangle}, k \geq k_0$.*

Proof: Homogeneity of the diffusion $(Q_x; x \in G)$ implies for all $x \in G^{\langle k \rangle}, k \in \mathbf{N}_0$ and $y \in N^{\langle k \rangle}(x)$

$$Q_y\{\phi_{x,\mathcal{O}}(\omega(\cdot \wedge T_{N^{\langle k \rangle}(x)})) \in \star\} = Q_{\phi_{x,\mathcal{O}}(y)}\{\omega(\cdot \wedge T_{N^{\langle k \rangle}(\mathcal{O})}) \in \star\}. \tag{5.17}$$

The invariance identity (5.17) and the fact that the diffusion is non trivial imply that $Q_\mathcal{O}$ is non trivial, i.e. $Q_\mathcal{O}\{\omega(t) = \mathcal{O}$ for all $t \geq 0\} < 1$. Indeed, if $Q_\mathcal{O}$ were trivial the invariance identity (5.17) would imply that Q_x is trivial for all $x \in G^{\langle \infty \rangle}$, and since $G \backslash G^{\langle \infty \rangle}$ is totally disconnected this would imply that the diffusion $(Q_x; x \in G)$ is trivial.

The invariance identity (5.17) also implies that for all $y \in N^{\langle k \rangle}(x)$, $x \in G^{\langle k \rangle}, k \in \mathbf{N}_0$:

$$Q_y\{T_{N^{\langle k \rangle}(x)} \in \cdot\} = Q_{\phi_{x,\mathcal{O}}^{\langle k \rangle}(y)}\{T_{N^{\langle k \rangle}(\mathcal{O})} \in \cdot\}. \tag{5.18}$$

As $Q_\mathcal{O}$ is non trivial and $\{N^{\langle k \rangle}(\mathcal{O}); k \in \mathbf{N}_0\}$ is a base for the neighborhood system of $\mathcal{O}$, the strong Markov property implies, that there exists $k_1 \in \mathbf{N}_0$ with

$$Q_\mathcal{O}\{T_{N^{\langle k_1 \rangle}(\mathcal{O})} > 1\} < 1. \tag{5.19}$$

Let $\varphi : G \to [0,1]$ be a continuous function which vanishes on the closed set $\overline{N^{\langle k_1+1 \rangle}(\mathcal{O})}$ and is strictly positive on its complement. By the Feller property the function $H : G \to [0,1]$ defined by

$$H(\star) = \int \int_0^1 \varphi(\omega_t)dt\, dQ_\star$$

is continuous. Moreover, $H(\mathcal{O}) > 0$ by (5.19).

If assertion a) did not hold, (5.18) would imply, that for all $k \in \mathbf{N}$ there exists $y_k \in N^{\langle k\rangle}(\mathcal{O})$ such that $\lim_{k\to\infty} Q_{y_k}\{T_{N^{\langle k\rangle}(\mathcal{O})} > 1\} = 1$. Hence we would have $\lim_{k\to\infty} H(y_k) = 0$.

On the other hand $y_k \in N^{\langle k\rangle}(\mathcal{O})$ implies $\|y_k\| \le 2^{-k}$ and by the continuity of H $\lim_{k\to\infty} H(y_k) = H(\mathcal{O}) > 0$.

As for assertion b) we observe first that $T_{N^{\langle k\rangle}(x)}$ decreases in k. By a standard argument involving the strong Markov property we conclude from assertion a) that

$$\sup_{x\in G^{\langle k_0\rangle}} Q_x\{T_{N^{\langle k_0\rangle}(x)} > k\} \le \Big(\sup_{x\in G^{\langle k_0\rangle}} \sup_{y\in N^{\langle k_0\rangle}(x)} Q_y\{T_{N^{\langle k_0\rangle}(x)} > 1\}\Big)^k.$$

This implies immediately assertion b). □

5.2 Lemma

a) *For each $k \in \mathbf{N}_0$ $(Q_x \circ \Psi^{\langle k\rangle -1};\, x \in G^{\langle k\rangle})$ is a $(\beta^{\langle k\rangle}, k)$-random walk on $G^{\langle k\rangle}$, where $\beta^{\langle k\rangle} := \frac{1}{2}Q_{\mathcal{O}}\{\chi_1(\mathcal{O}, \omega_{T^{\langle k\rangle}}) = 1\}$.*
Moreover, $(Q_x\circ\Psi^{\langle k\rangle -1};\, x \in G^{\langle k\rangle})_{k\in\mathbf{N}_0}$ is a consistent sequence of random walks.

b) *For all $k \in \mathbf{N}_0$ we have that*

$$T_i^{\langle k\rangle} \text{ is } Q_x\text{-square integrable for all } x \in G^{\langle k\rangle} \text{ and } i \in \mathbf{N}. \tag{5.20}$$

Proof: Fix $k_0 \in \mathbf{N}_0$ according to Lemma 5.1. Then the invariance identity (5.17), Lemma 5.1 part b) and the strong Markov property imply that $(Q_x\circ\Psi^{\langle k\rangle -1};\, x \in G^{\langle k\rangle})$ is a $(\beta^{\langle k\rangle}, k)$-random walk on $G^{\langle k\rangle}$, for indices $k \ge k_0$. For indices $k < k_0$ this follows from (3.9), Lemma 3.2 and the fact that $\beta^{\langle k_0\rangle} = f(\beta^{\langle k_0+1\rangle}) \in [0, \frac{1}{2})$.

In order to conclude part a) we observe that consistency of the sequence of random walks is an easy consequence of (3.9).

As for part b) we observe that for indices $k \ge k_0$ assertion (5.20) follows by (5.17), Lemma 5.1 and the strong Markov property.

It remains to verify (5.20) for indices $k < k_0$. To this end we shall prove that if (5.20) holds for some $k \in \mathbf{N}$, it also holds for $k-1$.

Arguments similar to those in the proof of B.-P. Lemma 8.2 a) show that for $i \in \mathbf{N}$ and $x \in G^{\langle k-1\rangle}$

$$\begin{aligned} &Q_x\{T_i^{\langle k\rangle} - T_{i-1}^{\langle k\rangle} \in \star \mid X_{\cdot\wedge T_{i-1}^{\langle k\rangle}}, X_{T_i^{\langle k\rangle}+\cdot}\} \\ =\ &Q_x\{T_i^{\langle k\rangle} - T_{i-1}^{\langle k\rangle} \in \star \mid \chi_j(\Psi^{\langle k\rangle}(\cdot)_{i-1}, \Psi^{\langle k\rangle}(\cdot)_i), j = 1,2\} \end{aligned} \tag{5.21}$$

and that for $j \in \{1,2\}$

$$\begin{aligned} & Q_x\{T_i^{(k)} - T_{i-1}^{(k)} \in \star \mid \chi_j(\Psi^{(k)}(\cdot)_{i-1}, \Psi^{(k)}(\cdot)_i) = 1\} \\ = \; & Q_O\{T_1^{(k)} - T_0^{(k)} \in \star \mid \chi_j(\Psi^{(k)}(\cdot)_0, \Psi^{(k)}(\cdot)_1) = 1\}. \end{aligned} \tag{5.22}$$

Moreover, for $\omega \in C([0,\infty), G)$ with $\omega_0 \in G^{(k)}$

$$T_i^{(k-1)}(\omega) = \sum_{j=1}^{T_i^{(k-1,k)}(\Psi^{(k)}(\omega))} T_j^{(k)}(\omega) - T_{j-1}^{(k)}(\omega). \tag{5.23}$$

Now Q_x-square integrability of $T_i^{(k)}, i \in \mathbb{N}_0$ implies Q_x-square integrability of $T_i^{(k-1)}, i \in \mathbb{N}_0$ by (3.10), (5.21) and (5.22). □

Remark *Lemma 5.2 part a) and Proposition 3.3 imply $\beta^{(k)} \in [0, \frac{1}{4}]$. Moreover, as $\beta^{(0)} = Q_O\{\omega$ leaves N through $(\frac{1}{2}, \frac{\sqrt{3}}{2})\}$ we have proved part a) of Theorem 1.1.*

For the proof of part b) of Theorem 1.1 fix $\alpha \in [0, \frac{1}{4}]$ and define the sequence $(\alpha^{(k)})_{k \in \mathbb{N}_0}$ by (3.11) with $\alpha^{(0)} = \alpha$. Let $(P_x^{(k)};\, x \in G^{(k)})_{k \in \mathbb{N}_0}$ be the sequence of $(\alpha^{(k)}, k)$-random walks. By Proposition 3.3 this sequence is consistent. By Bochner's version of Kolomogorov's Extension Theorem (Theorem 5.1.1. in [2]) it is possible to show that there exist a probability space $(\Omega, \mathfrak{F}, P)$ and random functions $\tilde{X}(k,x) : \Omega \to \Omega^{(k)}$, $k \in \mathbb{N}_0$, $x \in G^{(k)}$, such that for all $k, l \in \mathbb{N}_0, l \geq k$ and $x \in G^{(k)}$:

a) $P \circ \tilde{X}(k,x)^{-1} = P_x^{(k)}$.

b) $\tilde{X}(k,x) = \Psi^{(k,l)} \circ \tilde{X}(l,x)$.

c) The σ-algebras $\mathfrak{F}_y := \sigma\{\tilde{X}(k,y); k \in \mathbb{N}_0$ with $y \in G^{(k)}\}$, $y \in G^{(\infty)}$ are independent.

The consistency of our $(\alpha^{(k)}, k)$-random walks is just the consistency condition in Bochner's Version of Kolomogorov's Theorem.

Starting from the functions $\tilde{X}(n,x)$ we construct inductively for each $k \in \mathbb{N}_0$ random functions $X(k+l, x) : \Omega \to \Omega^{(k+l)}$, $l \in \mathbb{N}_0$ and $x \in G^{(k)}$:
For $x \in G^{(0)}$ let $X(l,x) := \tilde{X}(l,x)$ for all $l \in \mathbb{N}_0$.
If $X(k,x), X(k+1,x), \ldots$ are defined for all $x \in G^{(k)}$, we define for $i \in \mathbb{N}_0$, $x \in G^{(k+1)} \setminus G^{(k)}$ and $l \geq k+1$

$$X(l,x)_i := \begin{cases} \tilde{X}(l,x)_i & ; \text{if } i < T^{(k,l)}(\tilde{X}(l,x)) \\ X(l, \tilde{X}(l,x)_{T^{(k,l)}})_{i - T^{(k,l)}(\tilde{X}(l,x))} & ; \text{if } i \geq T^{(k,l)}(\tilde{X}(l,x)) \end{cases}.$$

We remark that the strong Markov property implies that $X(k,x)$ is $P_x^{(k)}$-distributed.

Let $(c^{(k)})_{k\in\mathbb{N}_0}$ be a consistent sequence of velocities for the consistent sequence of random walks under consideration, whose existence is assured by Proposition 4.1, and define $Y(k,x) := Y^{(k)}_{c^{(k)}}(X(k,x))$. Furthermore, we denote by $L^0(C([0,\infty),G))$ the space of $C([0,\infty),G)$-valued measurable functions on Ω endowed with the topology of convergence in probability.

5.3 Proposition

a) $Y(x) := \lim_{k\to\infty} Y(k,x)$ exists P-a.e. in $C([0,\infty),G)$ for all $x \in G^{(\infty)}$.

b) The function $Y : G^{(\infty)} \to L^0(C([0,\infty),G))$ is uniformly continuous on bounded subsets of $G^{(\infty)}$, and hence there exists a unique continuous extension of Y to all of G (also called Y).

c) $(P\circ Y(x)^{-1};\ x\in G)$ is a diffusion with Feller property and $P\circ Y(\mathcal{O})^{-1}\{\omega$ leaves N through $(\frac{1}{2},\frac{\sqrt{3}}{2})\} = \alpha^{(0)} = \alpha$.

Proposition 5.3 corresponds to Theorem 2.8, Proposition 2.13 and Theorem 2.15 in B.-P. and the proofs can easily be adapted, with one exception.

For the proof of part a) we have to show that $\lim_{k\to\infty} T_i^{(l)}(Y(k,x))$ exists and that this limit is strictly increasing as a function of i P-a.e. for $l\in\mathbb{N}_0$ and $x\in G^{(l)}$.

In B.-P. the corresponding result is derived from the classical limit theorem for single-type branching processes with constant environment. In the present case we may apply our limit result on multi-type branching processes with varying environment. Indeed, using Lemma 4.2 it is not hard to see, that $\{V_i^{(l,k+l)}(X(l+k,x))\,;\,k\in\mathbb{N}_0\}$ is a two-type branching process with varying environment with generating functions $F^{(k)} = G_{\alpha^{(k+l+1)}}$ and that this process does not die out a.s. Since $T_i^{(l)}(Y(l+k,x)) - T_{i-1}^{(l)}(Y(l+k,x)) = c^{(l+k)t}V_i^{(l,l+k)}(X(l+k,x))$ and $T_0^{(k)}(Y(l+k,x)) = 0$ P-a.e. we may conclude that $\lim_{k\to\infty} T_i^{(l)}(Y(k,x))$ exists and is strictly increasing as a function of i a.e., by Corollary 2.4.

Next we will show that the diffusion of Proposition 5.3 part c) is homogeneous. In the following we shall use the notation $P_x = P\circ Y(x)^{-1}$.

We remark first that the operator Θ_A mapping the path ω into the stopped path $\Theta_A(\omega) := \omega_{\cdot\wedge T_A}$ is continuous in P_x-a.a. paths $\omega\in C([0,\infty),G)$, for $A\subset G$ open with $\partial A\subset G^{(\infty)}$ finite. This follows from an argument which is essentially the same as that for Lemma 2.16 in B.-P. and shall be omitted here.

Applying this result to the sets $A = N^{(k)}(x)$ we conclude by Proposition 5.3, part a) and the corresponding properties of $(P_x^{(l)};\ x\in G^{(l)})$ that for $l\geq k$

$$P_x\{\phi_{x,y}\circ\Theta_{N^{(k)}(x)}(\omega)\in\cdot\} = P_x\{\lambda_{x,y}\circ\Theta_{N^{(k)}(x)}(\omega)\in\cdot\} = P_y\{\Theta_{N^{(k)}(y)}(\omega)\in\cdot\} \tag{5.24}$$

for $x,y\in G^{(k)}$, $k\in\mathbb{N}_0$. If $\eta: U\to V$ is a symmetry with domain

$$U = \cup_{i=1}^n N^{(k)}(x_i) \tag{5.25}$$

for some $k \in \mathbb{N}_0$, $n \in \mathbb{N}$ and $x_i \in G^{\langle k \rangle}$, then $\eta|_{N^{\langle k \rangle}(x_i)}$ equals $\phi^{(k)}_{x_i,\eta(x_i)}$ or $\lambda^{(k)}_{x_i,\eta(x_i)}$ and thus by the strong Markov property and (5.24) we obtain

$$P_x\{\eta \circ \Theta_U(\omega) \in \cdot\} = P_{\eta(x)}\{\Theta_V(\omega) \in \cdot\} \tag{5.26}$$

for $x \in \{x_1, \ldots, x_n\}$.

Now for U of the form (5.25) we have $U = \cup_{y \in G^{\langle l \rangle} \cap U} N^{\langle l \rangle}(y)$ and $G^{\langle l \rangle} \cap U$ is finite for all $l \geq k$. Therefore (5.26) holds for all $x \in G^{\langle l \rangle} \cap U$, $l \geq k$ and hence all $x \in G^{\langle \infty \rangle} \cap U$. By the continuity of Θ_U Proposition 5.3 b) implies (5.26) for all $x \in U$.

Finally let $\eta : U \to V$ be a symmetry with an arbitrary open set $U \subseteq G$ as domain. There exists a sequence $U_n \subset G$, increasing to U, such that each U_n is of the form (5.25). Since (5.26) holds for $\eta|_{U_n} : U_n \to \eta(U_n)$ and $x \in U_n$ it follows by a continuity argument that (5.26) also holds for $\eta : U \to V$ and $x \in U$. This proves that the diffusion $(P_x\, ;\, x \in G)$ is homogeneous.

In order to verify uniqueness of the diffusion we assume that $(Q_x\, ;\, x \in G)$ is another homogeneous diffusion with Feller property, for which

$$\alpha = Q_O\{\omega \text{ leaves } N \text{ through } (\frac{1}{2}, \frac{\sqrt{3}}{2})\}.$$

Lemma 5.2 a) and Proposition 3.3 imply

$$Q_x \circ \Psi^{\langle k \rangle -1} = P_x^{\langle k \rangle} \text{ for all } x \in G^{\langle k \rangle}, k \in \mathbb{N}_0. \tag{5.27}$$

If we denote by E^{Q_x} integration with respect to Q_x, we obtain by Lemma 5.2 b) that $E^{Q_O}(T_1^{\langle k \rangle}) < \infty$ for $k \in \mathbb{N}_0$. Therefore

$$d_i^{\langle k \rangle} := E^{Q_O}[T_1^{\langle k \rangle} | \chi_i(O, X_{T_1^{\langle k \rangle}}) = 1] \quad i = 1, 2,$$

are finite. In the case $\alpha = 0$ we have $Q_O\{\chi_1(O, X_{T_1^{\langle k \rangle}}) = 1\} = 0$ and we let $d_1^{\langle k \rangle} = 1$.

We now conclude from (5.21), (5.22) and (5.23) that for $x \in G^{\langle k-1 \rangle}$, $k \in \mathbb{N}$

$$\begin{aligned}
&E^{Q_x}(T_1^{\langle k-1 \rangle}(Y^{\langle k \rangle}_{d^{\langle k \rangle}}(\Psi^{\langle k \rangle})) \mid \Psi^{\langle k-1 \rangle}(.)_1) \\
= \;& E^{Q_x}(\sum_{i=1}^{2} R_1^{\langle k-1,k \rangle}(i, \Psi^{\langle k \rangle}) d_i^{\langle k \rangle} \mid \Psi^{\langle k-1 \rangle}(.)_1) \\
= \;& E^{Q_x}(\sum_{i=1}^{2} \sum_{j=1}^{T_1^{\langle k-1,k \rangle}(\Psi^{\langle k \rangle})} 1_{\{\chi_i(\omega_{T_j^{\langle k \rangle}}, \omega_{T_{j-1}^{\langle k \rangle}}) = 1\}} E^{Q_x}(T_j^{\langle k \rangle} - T_{j-1}^{\langle k \rangle} | \Psi^{\langle k \rangle}) \mid \Psi^{\langle k-1 \rangle}(.)_1) \\
= \;& E^{Q_x}(\sum_{j=1}^{T_1^{\langle k-1,k \rangle}(\Psi^{\langle k \rangle})} T_j^{\langle k \rangle} - T_{j-1}^{\langle k \rangle} \mid \Psi^{\langle k-1 \rangle}(.)_1) \\
= \;& E^{Q_x}(T^{\langle k-1 \rangle} \mid \Psi^{\langle k-1 \rangle}(.)_1).
\end{aligned}$$

In view of (5.27) this implies that $(d^{(k)})_{k\in\mathbb{N}_0}$ is a consistent sequence of velocities for the consistent sequence of $(\alpha^{(k)}, k)$-random walks $(P_x^{(k)}; x \in G^{(k)})$. By Proposition 4.1, Proposition 5.3, a) and (5.27) there exists $c > 0$ such that

$$\lim_{k\to\infty} Q_x \circ \Psi^{(k)-1} \circ Y_{d^{(k)}}^{(k)\;-1} = P_x \circ X_{c\cdot}^{\;-1} \quad \text{for all } x \in G^{(\infty)}. \tag{5.28}$$

Moreover,

$$\lim_{k\to\infty} Y_{d^{(k)}}^{(k)}(\Psi^{(k)}(\omega)) = \omega \quad Q_x\text{-a.e.} \quad \text{for all } x \in G^{(\infty)}. \tag{5.29}$$

The proof of (5.29) uses Lemma 5.1 and 5.2 and is similar to the proof of Theorem 8.1 in B.-P. Details shall be omitted.

Now, (5.28) and (5.29) imply that $Q_x = P_x \circ X_{c\cdot}^{\;-1}$ for all $x \in G^{(\infty)}$. Hence $Q_x = P_x \circ X_{c\cdot}^{\;-1}$ for all $x \in G$ by the Feller property. □.

Acknowledgment: The results of this paper are partially contained in the author's 1992 Research Report at Saarbücken University.

The author would like to thank Prof. Brosamler for drawing his attention to diffusions on fractals, for helpful comments and many useful discussions.

The author would like to thank also the referee, for suggesting a simplification in the proof of Theorem 2.1.

References

1. Barlow, M.T. and Perkins, E.A. (1988). Brownian motion on the Sierpinski gasket. *Probab. Theory Related Fields* **79**, 543-623.

2. Bochner, S. (1955). *Harmonic analysis and the theory of probability*, Univ. of California Press, Berkeley.

3. Goldstein, S. (1987). Random walks and diffusions on fractals. In: Kesten, H. (ed.) *Percolation theory and ergodic theory of infinite particle systems.* (IMA Math. Appl., vol.8.) Springer, New York, p. 121-129.

4. Hattori, K., Hattori, T. and Watanabe, H. (1994). Asymptotically one-dimensional diffusions on the Sierpinski gasket and the abc-gaskets. *Probab. Theory Related Fields* **100**, 85-116.

5. Heck, M. (1996). A perturbation result for the asymptotic behavior of matrix powers. *J. Theoret. Probability* **9**, 647-658.

6. Karlin, S. (1966). *A first course in stochastic processes.* Academic Press, New York.

7. Kumagai, T. Construction and some properties of a class of non-symmetric diffusion processes on the Sierpinski gasket. In: Elworthy, K.D. and Ikeda, N. *Asymptotic Problems in Probability Theory*, Pitman.

8. Kusuoka, S. (1987). A diffusion process on a fractal. In: Ito, K. and Ikeda, N. (eds.) *Symposium on Probabilistic Methods in Mathematical Physics.* Proceedings Taniguchi Symposium, Katata 1985. Academic Press, Amsterdam, p. 251-274.

9. Lindstrøm, T. (1990). Brownian motion on nested fractals. *Mem. Amer. Math. Soc.* **420**.

10. Mode, C.J. (1971). *Multi type branching processes,* American Elsevier Publishing Company, New York.

Almost Sure Path Properties of Branching Diffusion Processes

Y. Git, School of Mathematical Sciences, Bath University, Bath BA2 7AY, UK
ygl@maths.bath.ac.uk

Abstract

We consider a one-dimensional Branching Brownian Motion. We present a large deviations result concerning the almost sure number of particles along any given path. We then observe the implications of this result by studying Branching Integrated Brownian Motion.
Key Words: Strassen Law, Large Deviations, Branching Diffusion Processes, Reaction Diffusion Equations

1 Introduction

Branching Diffusion Processes (BDPs) have been studied extensively over the last decades. The behaviour in expectation is well understood, but to study the almost sure behaviour of a BDP, one must study the associated Reaction-Diffusion equation using martingales theory. We refer the reader to [Neveu] for an excellent exposition on the subject.
We analyse the almost-sure behaviour using large deviations techniques while concentrating on the study of a dyadic Branching Brownian Motion (BBM). We formulate a large deviations principle for the *almost sure* rate of growth of particles along any given path.

This work is divided into two sections.

- The derivation of the almost sure rate of growth function, measuring the number of BBM-particles along any path.
- An example of how this derivation can be utilised.

We begin by studying the rate of growth of the *expected* number of particles along any path of a BBM. The result follows directly from work by [Schilder] who first described the large-deviation principle associated with the paths of a single particle. We combine his result with a simple many-to-one picture to deduce the rate function for each BBM-path. This provides the upper bound for the *almost sure* rate function. We then pull together results by [Uchiyama] and [Chauvin] to prove the lower bound.

In the second section, we consider a system of breeding particles, whose velocity is given by a Brownian Motion. We arrive at a two dimensional point-process on the plane $\left(B_i(t), \int_0^t B_i(s)ds\right)$. We study the phase plane to discover that the behaviour almost surely and in expectation is extremely different.

The remainder of this introduction sets up the background with some definitions.

1.1 Constructing The Branching Brownian Motion Model

Let each Brownian particle wait an exponential time of rate 1 before dying while giving birth to $1 + C$ offspring. We excluded the possibility of death so that $1 + C$ is a $\mathcal{Z}^+$–valued random variable. We also impose that $E(C \log C) < \infty$. At birth, the parent particle and its offspring share the same spatial position, but from then on, each offspring follows an independent Brownian path. We follow Neveu's construction. A finite sequence i of numbers will label each particle, starting with the first particle labelled $\emptyset$. Each particle i has $1 + C_i$ descendants $i0, i1 \ldots iC_i$. Let $I = \cup_{n \in \mathcal{N}} \mathcal{Z}^{+n}$ be the space of labels. Let τ_i be the lifetime of particle i ($i \in I$). Particle i will thus be born at time

$$T_i = \sum_{k=0}^{k=n-1} \tau_{j_1 \ldots j_k} \quad \text{if } i = j_1 \ldots j_n.$$

The τ_i are assumed to be strictly positive random variables satisfying the non-explosion condition: $\{i : T_i \leq t\}$ is finite for all t. The trajectories of particles are continuous maps B_i of the time intervals $[T_i, T_i + \tau_i]$ into $\mathcal{R}$ such that $B_{ic}(T_{ic}) = B_i(T_i + \tau_i)$ for every $i \in I$ and $c \leq C_i$.
A point $\omega \in \Omega$ is a collection $\{\tau_i, B_i, C_i : i \in I\}$ satisfying the above conditions. Let $N_t(\omega) = \{i : T_i \leq t < T_i + \tau_i\}$ be the set of particles alive at time t. The filtration $\{\mathcal{F}_t : t \in \mathcal{R}^+\}$ on Ω is generated by $\{N_t, (B_i(t), i \in N_t)\}_{t \in \mathcal{R}^+}$. There exists a unique probability measure P on $(\Omega, \mathcal{F}_\infty)$ such that $\{B_i : i \in I\}$ is an independent family of Brownian-motion processes with each B_i started at $B_i(T_i)$, stopped after an exponential time τ_i of mean 1, and giving birth to C_i offspring at its time of death.
Although C need only satisfy $E(C \log C) < \infty$, we will restrict ourselves to the simple dyadic Branching Brownian Motion where $C = 2$ almost surely. Our results fail if $E(C \log C) = \infty$ because result 3 is no longer valid.

1.2 Scaling The Branching Brownian Motion

At time T (now a fixed time), let us scale the BBM by a factor of T in both the space and time coordinates. We get a branching process on the time-parameter set $[0, 1]$. Specifically, for every $i \in N_T$, let $x_i^T \in C^0([0, 1], \mathcal{R})$ – the space of continuous functions from $[0, 1]$ to $\mathcal{R}$ – be the T-scaled path of particle i, defined as

$$x_i^T(t) := \tfrac{1}{T} B_{a_{tT}(i)}(tT).$$

Here, $a_{tT}(i)$ denotes the unique ancestor of particle i at time tT. Clearly all T-scaled paths satisfy $x_i^T(0) = 0$. We let C_0 be the space of continuous paths started at 0 with the supremum norm $||x - z|| = \sup_{0 \leq t \leq 1} |x(t) - z(t)|$. Let C_1 be the space of paths which are also absolutely continuous with $\int_0^1 \dot{x}^2 dt < \infty$. If $D \subset C_0$, then let $M_D(T)$ denote the set of particles at time T whose T-scaled path is in D. Also if

$D|_\theta := \{x \in \mathcal{C}^0([0,\theta],\mathcal{R}) : \exists z \in D, \quad x(t) = z(t) \quad \forall t \in [0,\theta]\}$, let $M_D(T,\theta)$ denote set of particles whose T-scaled path is in $D|_\theta$ up to time $\theta \leq 1$.

$$M_D(T) := \{i \in N_T : x_i^T \in D\},$$
$$M_D(T,\theta) := \{i \in N_{\theta T} : x_i^T|_\theta \in D|_\theta\}.$$

The function $x|_\theta \in \mathcal{C}^0([0,\theta],\mathcal{R})$ is x truncated at time θ.

2 Rate of Growth in Expectation

We denote the law of a standard Brownian Motion run until time 1 as P_1. We also denote the law of an individual T-scaled path x_i^T path by $P_{1/T}$ which is the same in law as P_ϵ defined in [Varadhan, Section 5]. The large deviation principle associated with $\{P_{1/T} : T \in \mathcal{R}\}$ was first proved by [Schilder] with a rate function

$$I(x) := \begin{cases} \frac{1}{2}\int_0^1 \dot{x}^2 dt & \text{if } x \in C_1, \\ \infty & \text{otherwise.} \end{cases}$$

Of course, nothing stops us running the process only until time θT where $\theta \in [0,1]$. We get a slightly modified rate function $I(x,\theta) = \frac{1}{2}\int_0^\theta \dot{x}^2 dt$.

Let D be a subset of C_0. By conditioning on the first birth, we arrive at a many-to-one picture:

$$E(|M_D(T,\theta)|) = E(|N_{\theta T}|)P(x^T|_\theta \in D|_\theta),$$

whence

$$\begin{aligned} T^{-1}\log E(|M_D(T,\theta)|) &= T^{-1}\log\left\{E(|N_{\theta T}|)P(x^T|_\theta \in D|_\theta)\right\}, \\ &= \theta + T^{-1}\log P(x^T|_\theta \in D|_\theta), \\ &\approx \theta - I(x,\theta). \end{aligned}$$

In fact, combining [Varadhan] and the many-to-one particle picture, the following result is immediate:

Result 1. *Let $J(x,\theta) := \theta - I(x,\theta)$. If A and D are an open subset and a closed subset of $C_0|_\theta$ respectively, then*

$$\liminf_{T\to\infty} T^{-1}\log E(|M_A(T,\theta)|) \geq \sup_{x\in A} J(x,\theta),$$
$$\limsup_{T\to\infty} T^{-1}\log E(|M_D(T,\theta)|) \leq \sup_{x\in D} J(x,\theta).$$

As a matter of convenience, for all θ and for all sets B, we let

$$I(B,\theta) := \inf_{x\in B} I(x,\theta),$$
$$J(B,\theta) := \sup_{x\in B} J(x,\theta).$$

We note that I is lower-semicontinuous while J is upper-semicontinuous in the sense that $\lim_{z\to x} I(z) \geq I(x)$ and $\lim_{z\to x} J(z) \leq J(x)$.

3 Rate of Growth Almost-Surely

We wish to transform the result in probability to an almost sure result, so that for some function $K(x)$ to be determined later (which we might hope looks like $J(x)$), we have almost surely;

$$\lim_{T\to\infty} T^{-1}\log|M_A(T)| \geq \sup_{x\in A} K(x),$$
$$\lim_{T\to\infty} T^{-1}\log|M_D(T)| \leq \sup_{x\in D} K(x).$$

We certainly expect $K(x) \leq J(x)$ for all $x \in C_1$. We can improve this upper bound by considering the following: Suppose that for some $\theta \in [0,1]$ we have $J(D,\theta) < 0$. Then, using result 1 and Chebychev inequality, we deduce that as T tends to infinity,

$$P(|M_D(T,\theta)| > 0) \leq \exp\{TJ(D,\theta)\} \to 0.$$

Intuitively, this implies that $\lim_{T\to\infty}|M_D(T,\theta)| = 0$, and consequently also $\lim_{T\to\infty}|M_D(T)| = 0$ almost surely. This is a better indication as to how J "controls" K. It turns out that this upper bound is actually tight and distinguishes exactly between the different rates of growth. We now begin the rigorous study.

3.1 Upper Bound

Lemma 1. *Let D be a closed subset of C_0. Then for every $\theta \in [0,1]$, we have almost surely:*

$$\limsup_{T\to\infty} T^{-1}\log|M_D(T,\theta)| \leq J(D,\theta).$$

Proof. Suppose that the result is false. Then there exists a θ and an event W with $P(W) > 0$ such that, for every $\omega \in W$, $\limsup_{T\to\infty} T^{-1}\log|M_D(T,\theta)| > J(D,\theta)$. Hence if

$$W_n := \{\omega \in \Omega : \limsup_{T\to\infty} T^{-1}\log|M_D(T,\theta)| > J(D,\theta) + n^{-1}\},$$

then $P(W_n) > 0$ for some n. It is now clear that

$$\limsup T^{-1} \log E(|M_D(T,\theta)|) \geq J(D,\theta) + n^{-1}$$

contradicting result 1.

□

In particular, we see that if for some $\theta \leq 1$ we have $J(D,\theta) < 0$, then, almost surely, $\lim_{T\to\infty} |M_D(T,\theta)| = 0$. Since $x_i^T \in D$ implies that $x_i^T|_\theta \in D|_\theta$ we must also have that $\lim_{T\to\infty} |M_D(T)| = 0$ almost surely. This leads us to the following definition and the upper bound result:

Definition (The Almost Sure Rate Function). *Let $\theta_0 \in [0,1] \cup \{\infty\}$ be the last time at which $J(x,\theta)$ is non-negative, $\theta_0 := \inf\{\theta \in [0,1] : J(x,\theta) < 0\}$. Define $K(x,\theta)$ as:*

$$K(x,\theta) := \begin{cases} J(x,\theta) & \text{if } \theta \leq \theta_0, \\ -\infty & \text{otherwise} \end{cases}$$

Result 2. *Let $\theta \in [0,1]$ and let $D \subset C_0$ be closed. Then*

$$\limsup_{T\to\infty} T^{-1} \log |M_D(T)| \leq \sup_{x\in D} K(x).$$

3.2 Lower Bound

We shall prove the lower bound in stages. We first consider open sets around linear functions, then open sets around piecewise-linear functions, and finally arbitrary open sets. We use the following definition of an open ϵ-neighbourhood:

$$A(x,\epsilon) := \{z \in C_0 : ||z - x|| < \epsilon\} = \{z \in C_0 : \sup_t |x(t) - z(t)| < \epsilon\}.$$

Lemma 2. *Let $x(t) = \lambda t$ be a linear function with $0 \leq \lambda < \sqrt{2}$. For every $\epsilon > 0$, we have almost surely,*

$$\liminf_{T\to\infty} T^{-1} \log |M_{A(x,\epsilon)}(T)| \geq 1 - \tfrac{1}{2}\lambda^2.$$

Our proof relies on work by [Uchiyama] with slight modifications using change of measure by [Warren]. Their result involves the convergence of expressions of the the form $e^{-t}\sqrt{t}\sum_{i\in N_t} g(B_i(t) - \lambda t)e^{\lambda B_i(t) - \frac{1}{2}\lambda^2}$. We consider a special case of their result, when $g(x) = 1_{[0,1]}e^{-\lambda x}$ to deduce the following:

Result 3. *Let $|\lambda| < \sqrt{2}$. Let $N_{[\lambda T, \lambda T+1]}(T) = \{i \in N_T : \lambda T \leq B_i(T) \leq \lambda T + 1\}$ represent the particles at time T with spacial position between λT and $\lambda T + 1$. Then, almost surely,*

$$\lim_{T\to\infty} \sqrt{T} e^{-(1-\frac{1}{2}\lambda^2)T} |N_{[\lambda T,\lambda T+1]}(T)| \to \text{constant} \times Z_\lambda(\infty),$$

where $Z_\lambda(\infty)$ is a strictly positive random variable.

Before we prove lemma 2, please note that this result is true for an arbitrary birth process C as long as $E(C \log C) < \infty$. It is for this reason that we imposed this condition on the birth process.

Proof of Lemma 2. Consider the total number of particles at time $T > 1/\delta$ in the interval $[\lambda T, (\lambda+\delta)T]$. Result 3 clearly implies that almost surely

$$\lim_{T\to\infty} T^{-1} \log |N_{[\lambda T,(\lambda+\delta)T]}| \geq 1 - \tfrac{1}{2}\lambda^2.$$

Next, define the closed sets

$$D_\delta := \{z \in C_0 \setminus A(x,\epsilon) : z(0) = 0, z(1) \in [\lambda, \lambda+\delta]\}.$$

If $\delta = 0$ then $I(D_0)$ is minimised by the piecewise-linear path $z(0) = 0, z(\frac{1}{2}) = \frac{1}{2}\lambda+\epsilon, z(1) = \lambda$ and so $I(D) = \frac{1}{2}\lambda^2+2\epsilon^2$. By definition, a lower semicontinuous function satisfies $\lim_{z\to x} I(z) \geq I(x)$. Since I is lower-semicontinuous $\lim_{\delta\downarrow 0} I(D_\delta) \geq I(D_0)$, so that for a sufficiently small δ, we can ensure that $I(D_\delta) > \frac{1}{2}\lambda^2$.

We now use the upper-bound (result 2) with the knowledge that D_δ is closed to deduce that, almost surely,

$$\limsup_{T\to\infty} T^{-1} \log |M_{D_\delta}(T)| < 1 - \tfrac{1}{2}\lambda^2.$$

Since $N_{[\lambda T,(\lambda+\delta)T]}(T) \subset M_{A(x,\epsilon)}(T) \cup M_{D_\delta}(T)$ the lemma is complete.

□

We now wish to glue together several linear functions.

Definition. *Let x be a piece-wise linear function. We say x satisfies the lower bound condition until time $\theta_1 > 0$, if for all $\epsilon > 0$, almost surely*

$$\liminf_{T\to\infty} T^{-1} \log |M_{A(x,\epsilon)}(T,\theta_1)| \geq J(x,\theta_1).$$

Suppose x satisfies the lower bound condition until θ_1. If from θ_1 until θ_2, x is a linear function satisfying $\dot{x} = \lambda$, then we wish to show that x satisfies the lower bound condition until θ_2. We first assume $|\lambda| < \sqrt{2}$. We will run the process until time $\theta_1 T$, arriving at $M_{A(x,\epsilon)}(T,\theta_1)$ particles. We will then run $M_{A(x,\epsilon)}(T,\theta_1)$ independent copies from time $\theta_1 T$ to time $\theta_2 T$, and add them all together. We require the following two definitions. The first simply introduces the change in the rate function over the interval $[\theta_1, \theta_2]$. The second defines a random variable, very much like $M_{A(x,\epsilon)}(T)$, for each $i \in N(\theta_1 T)$ which simply counts the offspring of i whose T-scaled paths follow x closely over the interval $[\theta_1, \theta_2]$. Formally,

$$J(x,\theta_1,\theta_2) := \theta_2 - \theta_1 - \tfrac{1}{2}\int_{\theta_1}^{\theta_2} \dot{x}^2(t)\,dt,$$

$$M^i_{A(x,\epsilon)}(T,\theta_1,\theta_2) :=$$
$$\{j \in N(\theta_2 T) : a(j) = i, \left|(x_j^T(t) - x_j^T(\theta_1)) - (x(t) - x(\theta_1))\right| < \epsilon \text{ for all } t \in [\theta_1, \theta_2]\}.$$

It is a simple matter to verify that since all particles are independent, M^i all share the same law and that

$$M_{A(x,\epsilon)}(T,\theta_2) \supseteq \sum_{i \in M_{A(x,\frac{1}{2}\epsilon)}(T,\theta_1)} M^i_{A(x,\frac{1}{2}\epsilon)}(T,\theta_1,\theta_2).$$

Also apparent is the additivity of the rate function:

$$J(x,\theta_2) = J(x,\theta_1) + J(x,\theta_1,\theta_2),$$

Lemma 3. *Let $x \in C_1$ be piecewise-linear satisfying the lower bound condition up until time θ_1. Let $\dot{x} = \lambda$ on $[\theta_1,\theta_2]$, with $|\lambda| < \sqrt{2}$. Then, for every $\epsilon > 0$,*

$$\liminf_{T\to\infty} T^{-1} \log |M_{A(x,\epsilon)}(T,\theta_2)| \geq J(x,\theta_2).$$

Proof. Let $\delta > 0$ be arbitrary. We use the Strong Law of Large Numbers. From previous discussion, we have

$$e^{-(J(x,\theta_2)-2\delta)T}|M_{A(x,\epsilon)}(T,\theta_2)| \geq e^{-(J(x,\theta_2)-2\delta)T} \sum_{i \in M_{A(x,\frac{1}{2}\epsilon)}(T,\theta_1)} |M^i_{A(x,\frac{1}{2}\epsilon)}(T,\theta_1,\theta_2)|$$

$$\geq e^{-(J(x,\theta_1)-\delta)T}|M_{A(x,\frac{1}{2}\epsilon)}(T,\theta_1)| \ldots$$
$$\times \frac{1}{|M_{A(x,\frac{1}{2}\epsilon)}(T,\theta_1)|} \sum_{i \in M_{A(x,\frac{1}{2}\epsilon)}(T,\theta_1)} e^{-(J(x,\theta_1,\theta_2)-\delta)T}|M^i_{A(x,\frac{1}{2}\epsilon)}(T,\theta_1,\theta_2)|$$

The i.i.d. random variables $e^{-(J(x,\theta_1,\theta_2)-\delta)T}|M^i_{A(x,\frac{1}{2}\epsilon)}(T,\theta_1,\theta_2)|$ inside the summation tend almost surely to infinity as T tends to infinity (using lemma 2). We average over *an independent* random number $|M_{A(x,\frac{1}{2}\epsilon)}(T,\theta_1)|$ of particles, but this random variable tends to infinity almost surely as T tends to infinity, so that the SLLN still holds. By the induction hypothesis, x satisfies the lower bound condition until time θ_1, and thus the random variable $e^{-(J(x,\theta_1)-\delta)T}M_{A(x,\frac{1}{2}\epsilon)}(T,\theta_1)$ also tends almost surely to infinity. We conclude that the RHS (and hence the LHS) tends to infinity almost surely as T tends to infinity, and hence, almost surely,

$$\liminf_{T\to\infty} T^{-1} \log |M_{A(x,\epsilon)}(T,\theta_2)| \geq J(x,\theta_2) - 2\delta.$$

Letting $\delta \downarrow 0$ concludes the proof.

□

We turn our attention to the case where x satisfies the lower bound condition until time θ_1 , while between θ_1 and θ_2, the gradient $\dot{x} = \lambda > \sqrt{2}$. We of course insist on $J(x,\theta_2) > 0$.
Heuristics: The following proof is in principle the same as that of lemma 3 above. We run the process until time θ_1 arriving (using lemma 2) at an almost sure $M_{A(x,\epsilon)}(T,\theta_1)$

particles. We then run independent copies on $[\theta_1 T, \theta_2 T]$.

We replace the almost sure number of particles $M^i_{A(x,\epsilon)}(T,\theta_1,\theta_2)$ produced by an independent copy, with $P^i_{A(x,\epsilon)}(T,\theta_1,\theta_2)$, the probability of an independent copy with $x^T_i(\theta_1) = x(\theta_1)$, still remaining close to x by time $\theta_2 T$. Formally, we define

$$P^i_{A(x,\epsilon)}(T,\theta_1,\theta_2) := P\left(|M^i_{A(x,\epsilon)}(T,\theta_1,\theta_2)| > 0\right)$$

These probabilities are identical for all i, and are equal to the probability of finding a particle started at 0, at an ϵ-neighbourhood of $y = \lambda t$ at time $(\theta_2 - \theta_1)T$. [Chauvin] showed that the probability of the right-most particle starting at 0 ascending to level λT at time T decays at the rate $1 - \frac{1}{2}\lambda^2$. We will need to modify her result slightly to prove that $\liminf_{T\to\infty} P^i_{A(x,\epsilon)}(T,\theta_1,\theta_2) \geq J(x,\theta_1,\theta_2)$. This will be done by a method analogous to the one used in lemma 2.

We think of $e^{J(x,\theta_1)T}$ copies, each performing an independent trial, with probability of success $P^i \approx e^{J(x,\theta_1,\theta_2)T}$. We see that since $J(x,\theta_1) + J(x,\theta_1,\theta_2) = J(x,\theta_2) > 0$, the expected number of particles succeeding, increases exponentially. Using an estimate on the Binomial distribution, we show that the probability that the growth rate is less than $J(x,\theta_2) - \delta$, decays exponentially for all $\delta > 0$. Finally, this result is true only *in probability*. To get an almost sure result, we have to use some sort of Borel-Cantelli Lemma. Basically, we show that if we had a particle inside $A(x,r)$ at time t, for some $r < \epsilon$. Then the particle was inside $A(x,\epsilon)$ for some interval before t. This allows us to divide time into countably many intervals, and use BCL.

We state and prove the three supporting lemmas.

Lemma 4. *Let $x \in C_1$ be a piece-wise linear function. We claim that for every $\epsilon > 0$, there exists $r > 0$, such that, for all sufficiently large T, if $x^T \in A(x,r)$, then $x^\tau \in A(x,\epsilon)$ for all $\tau \in [T-1,T]$.*

Proof. We define the look-back transformation for all $\tau < T$:

$$L^T_\tau z(t) := \tfrac{T}{\tau} z(t\tfrac{\tau}{T}).$$

$L^T_\tau : A(x,r) \to A(L^T_\tau x, \frac{T}{\tau} r)$ and $\lim_{T\to\infty} \sup_{T-1<\tau<T} ||L^T_\tau x - x|| = 0$. We let $r = \frac{1}{4}\epsilon$. Pick T sufficiently large such that $\sup_{T-1<\tau<T} ||L^T_\tau y - y|| < r$ and $\frac{T}{T-1} < 2$. We deduce that for such T sufficiently large,

$$L^T_\tau A(x,r) \subseteq A(x,3r) \subset A(x,\epsilon) \quad \text{for all } \tau \in [T-1,T].$$

□

Result 4 (Right-Most Particle At The Subcritical Region - Chauvin). *Let $\lambda > \sqrt{2}$ and let R_T be the position of the right-most particle of a dyadic Branching Brownian Motion at time T. Then*

$$\liminf_{T\to\infty} T^{-1} \log P(R_T > \lambda T) = 1 - \tfrac{1}{2}\lambda^2$$

Corollary 1. *Let $y(t) = \lambda t$ where $\lambda > \sqrt{2}$. Then for all $r > 0$*

$$\liminf_{T\to\infty} T^{-1} \log P\left(|M_{A(y,r)}(T)| > 0\right) \geq 1 - \tfrac{1}{2}\lambda^2.$$

It follows that if $\dot{x} = \lambda$ on $[\theta_1, \theta_2]$, then for all $r > 0$

$$\liminf_{T\to\infty} T^{-1} \log P^i_{A(x,r)}(T, \theta_1, \theta_2) \geq J(x, \theta_1, \theta_2).$$

Proof. Define the closed set $D := \{z \in C_0 \setminus A(y,r) : z(1) \geq \lambda_2\}$. It is easy to show that $J(D) < 1 - \frac{1}{2}\lambda^2$ and hence

$$\limsup_{T\to\infty} T^{-1} \log P(|M_D(T)| > 0) < 1 - \tfrac{1}{2}\lambda^2.$$

Since $P(R_T \geq \lambda T) \leq P(|M_D(T)| > 0) + P(|M_{A(y,r)}(T)| > 0)$, the result follows. □

Finally, an estimate on the binomial distribution $\mathcal{B}(n,p)$.

Lemma 5. *Let $\alpha < 1$. Then $P\left(\mathcal{B}(n,p) < pn\alpha\right) < e^{-np\alpha}$.*

Proof. For $x \in [0,1]$ we know that $E(x^{\mathcal{B}(n,p)}) = (q + px)^n$. Since $\mathcal{B}(n,p) < np\alpha$ if and only if $x^{\mathcal{B}(n,p)} > x^{np\alpha}$, we deduce that

$$P\left(\mathcal{B}(n,p) < pn\alpha\right) < x^{-np\alpha}(q + px)^n$$

Picking $x = \alpha(1-p)/(1-p\alpha)$ which minimises the above expression we deduce that

$$\begin{aligned}
\log P\left(\mathcal{B}(n,p) < pn\alpha\right) &\leq -n\left\{p\alpha \log\alpha + (1-p\alpha)\log(1-p\alpha) - (1-p\alpha)\log(1-p)\right\}, \\
&\sim -n\left\{p\alpha\log\alpha - (1-p\alpha)p\alpha + (1-p\alpha)p\right\}, \\
&\sim -np\left\{\alpha\log\alpha + (1-\alpha)(1-p\alpha)\right\}, \\
&\sim -np, \\
&< -np\alpha.
\end{aligned}$$

□

Let us now state and prove the main lemma.

Lemma 6. *Let $x \in C_1$ be piece-wise linear satisfying the lower bound condition until time θ_1. Let $\dot{x} = \lambda$ on $[\theta_1, \theta_2]$, with $\lambda > \sqrt{2}$, but with $J(x, \theta_2) > 0$. Then, for every $\epsilon > 0$,*

$$\liminf_{T\to\infty} T^{-1} \log |M_{A(x,\epsilon)}(T, \theta_2)| \geq J(x, \theta_2).$$

Proof. Pick $r < \epsilon$ as in lemma 4. At integer times $T_m := m$ define the following events:

$$U_m := \{\omega \in \Omega : |M_{A(x,\frac{1}{2}r)}(T_m, \theta_1)| < e^{(J(x,\theta_1)-\delta)T_m}\},$$
$$V_m := \{\omega \in \Omega \setminus U_m : |M_{A(x,r)}(T_m, \theta_2)| < e^{(J(x,\theta_2)-3\delta)T_m}\}.$$

Since x satisfies the lower bound condition until time θ_1, we know that almost surely, U_m will not occur. To work out the probability of V_m we use lemma 5 with the values $n \geq e^{(J(x,\theta_1)-\delta)T_m}$, $p \geq e^{(J(x,\theta_1,\theta_2)-\delta)T_m}$ and $\alpha = e^{-\delta T_m}$. We take n to represent the number of particles which stayed within $A(x, \frac{1}{2}r)$ up to time $\theta_1 T_m$. Since we are not in U_m we know that n is large (i.e. $n > e^{(J(x,\theta_1)-\delta)T}$). We take p to represent the probability for each of these particles that we could find a descendent in $A(x,r)$ by time $\theta_2 T_m$. This probability is greater than $P^i_{A(x,\frac{1}{2}r)}(T, \theta_1, \theta_2)$ which was evaluated in corollary 1.
We deduce that $P(V_m)$ decays exponentially, and using BCL, V_m does not occur almost surely. Thus, almost surely,

$$\liminf_{m\to\infty} T_m^{-1} \log |M_{A(x,r)}(T_m)| \geq J(x, \theta_2) - 3\delta.$$

We now use lemma 4 to deduce that for all m and for all $\tau \in [T_{m-1}, T_m]$

$$\liminf_{\tau\to\infty} \tau^{-1} \log |M_{A(x,r)}(\tau)| \geq J(x, \theta_2) - 3\delta.$$

□

Corollary 2. *Let $x \in C_1$ be a piecewise-linear function such that $K(x) > 0$. Then, for every $\epsilon > 0$,*

$$\liminf_{T\to\infty} T^{-1} \log |M_{A(x,\epsilon)}(T)| \geq K(x).$$

Proof. Clearly, $\dot{x}(0) \leq \sqrt{2}$. Since $A(x,\epsilon)$ is open, there is no problem of finding a piecewise-linear function in $A(x,\epsilon)$ with $\dot{z}(0) < \sqrt{2}$. Now, proceed to piece together each linear segment of z using the previous lemmas. Please note that we avoided the case where $K(x) = 0$.

□

We now have the lower bound result for the almost sure rate function. We ignored the case where $K(x) = 0$ because if A is any open set, and $x \in A$ satisfies $K(x) = 0$, then for some $0 < \alpha < 1$ we have $\alpha x \in A$ and $K(\alpha x) > 0$.

Theorem 1. *Let A be an open subset in C_0. Then, almost surely,*

$$\liminf_{T\to\infty} T^{-1} \log |M_A(T)| \geq \sup_{x\in A} K(x).$$

Proof. Since the piecewise-linear functions are dense in C_0 and $I(x)$ is lower-semicontinuous in the supremum topology, the result follows directly from the above corollary. □

To conclude, we state and sketch-prove a more general result. Consider a BBM with a birth process C satisfying $E(C) = \mu$. We assume that $E(C \log C) < \infty$. Let each particle die at an exponential rate $r(B_i(t)/t)$. The breeding rate $r \geq 0$ is assumed to be a continuous function. For every $x \in C_1$, the adjusted expectation rate function is defined as:

$$\bar{J}(x,\theta) := \int_0^\theta \mu r(x) - \tfrac{1}{2}\dot{x}^2 dt.$$

As before, let $\theta_0 := \inf\{\theta : \bar{J}(x,\theta) < 0\}$. Also let the almost sure rate function $\bar{K}$ be defined as

$$\bar{K}(x,\theta) := \begin{cases} \bar{J}(x,\theta) & \text{if } \theta \leq \theta_0, \\ -\infty & \text{otherwise.} \end{cases}$$

Theorem 2. *Let A, D be open and closed sets in C_0. Then*

$$\limsup_{T\to\infty} T^{-1} \log E(|M_D(T)|) \leq \bar{J}(D),$$
$$\liminf_{T\to\infty} T^{-1} \log E(|M_A(T)|) \geq \bar{J}(A).$$

Also, almost surely,

$$\limsup_{T\to\infty} T^{-1} \log |M_D(T)| \leq \bar{K}(D),$$
$$\liminf_{T\to\infty} T^{-1} \log |M_A(T)| \geq \bar{K}(A).$$

Proof: Almost-sure lower bound. We prove the lower bound for an open neighbourhood of a piece-wise linear function x. Take $\mathcal{D}$, a partition of $[0,1]$. $\mathcal{D} := \{0 = t_0 < t_1 \ldots < t_n = 1\}$. Over the interval $[t_i, t_{i+1}]$, along the path $\{x(t) : t_i < t < t_{i+1}\}$ the process "observes" breeding at a rate greater or equal to $\inf\{r(x(t)) : t_i \leq t \leq t_{i+1}\}$. Thus, using the lower bound lemmas 2, 3 and 5, almost surely,

$$\liminf_{T\to\infty} T^{-1} \log |M_A(T)| \geq \mu \sum_{i<n} (t_{i+1} - t_i) \inf_{t_i \leq t \leq t_{i+1}} r(x(t)) - \int_0^1 \tfrac{1}{2}\dot{x}^2 dt.$$

Since x is piece-wise continuous, by taking the supremum over all partitions we get the result. We cheated slightly, as we are only allowed to consider partitions which satisfy for all $j < n$,

$$\mu \sum_{i\leq j} (t_{i+1} - t_i) \inf_{t_i \leq t \leq t_{i+1}} r(x(t)) > \int_0^{t_{j+1}} \tfrac{1}{2}\dot{x}^2 dt.$$

Since $\bar{K}(x,\theta) > 0$ for all θ, it can be shown that this constraint does not matter.

□

4 Application: Branching Integrated Brownian Motion

4.1 An Integrated Brownian Motion

We take a Brownian path and use it to describe the velocity of a particle. The position is then defined as

$$Y(t) = Y(0) + \int_0^t B(s)ds.$$

We assume $Y(0) = 0$ for simplicity. $Y(t)$ is an integral with respect to a continuous path and is therefore a differentiable finite-variation process. It is also Gaussian and its variance is given by

$$2E\left(\int_0^t B_r dr \int_r^t B_s ds\right) = 2\int_0^t \int_r^t r ds dr = \frac{1}{3}t^3$$

We will proceed to show using current methods that there *must* be a difference between the behaviour in expectation and almost surely. We will do so by considering the wavefront speeds. We will then give the full phase-plane picture using the new techniques which we have developed.

4.2 The Expectation Wavefront

If we let $N_{[x,\infty)}(t) = \{i \in N_t : Y_i(t) \geq x\}$ by conditioning on the first birth we find the following many-to-one picture holds:

$$E(|N_{[u_t,\infty)}(t)|) = E(|N_t|)P(Y_t > u_t) \approx e^t \exp(-\frac{3}{2t^3}u_t^2),$$

and we deduce the expectation wavefront travels at the speed $t^2\sqrt{2/3}$ in the sense that

$$\lim_{t\to\infty} \frac{u_t}{t^2} = \sqrt{2/3}. \tag{1}$$

4.3 The Almost-Sure Wavefront

[Neveu] observed that almost surely R_t, the rightmost particle of a branching Brownian Motion satisfies $\limsup_{t\to\infty} R_t - t\sqrt{2} = -\infty$. Because $\sup_{i\in N_t} Y_i(t) \leq \int_0^t R_s ds$ by integrating the bound on R_t we get an instant upper bound on v_t, the almost-sure wavefront speed.

$$\limsup_{t\to\infty} \frac{v_t}{t^2} \leq 1/\sqrt{2}. \tag{2}$$

Before proving that equality holds in equation 2 we want to point out that the almost-sure wavefront speed is below the expectation wavefront speed already in the first order of magnitude! A more comprehensive explanation of this phenomenon will be offered when we study the phase plane.

Theorem 3. *Let v_t denote the rightmost particle's position of a branching Integrated Brownian Motion. Then, almost surely,*

$$\lim_{t\to\infty} \frac{v_t}{t^2} = 1/\sqrt{2}.$$

Proof (Lower Bound). We look at the Branching Brownian Motion. We follow [Neveu] and define Z_s^λ to be the number of particles which first among their ancestors crossed the line $x = s - \lambda t$.

$$Z_s^\lambda = |\{i \in I : \exists t \in [T_i, T_i + \tau_i] \quad B_i(t) > s - \lambda t, \quad \forall t < T_i \quad B_{a(i)}(t) < s - \lambda t\}|.$$

Definition (Infinitesimal Generator Function of a Galton-Watson Process). *Consider a Markovian birth-death process $Z : \mathcal{R}^+ \to \mathcal{N}$ representing the number of particles alive. Each particle lives for an exponential time of rate α and gives particle to $n \in \mathcal{Z}^+{}_\infty$ particles with probability a_n. The infinitesimal generator function is then defined as*

$$a(x) = \alpha\{\sum_{\mathcal{Z}^+{}_\infty} a_n x^n - x\}. \tag{3}$$

Note that $a(1) = 0$ while $\lim_{x\uparrow 1} a(x) = -\alpha a_\infty$. On $0 < x < 1$, $a(x)$ is convex and has a unique root $a(\sigma) = 0$ with $a(x) < 0$ on $(\sigma, 1)$. If $a_0 = 0$ then $\sigma = 0$. If a is also continuous at 1, then the solution of the equation

$$a = \psi' \circ \psi^{-1} \quad \text{on } (0,1)$$

has a unique (modulo translation) monotone decreasing solution $\psi : \mathcal{R} \to (0,1)$.

Result 5 (Neveu, Proposition 3). *For each $\lambda \geq \sqrt{2}$ the integer valued process $(Z_s^\lambda, s > 0)$ is a Galton-Walton process without extinction whose infinitesimal generating function a is given by*

$$a = \psi' \circ \psi^{-1} \quad \textit{on } (0,1)$$

where $\psi : \mathcal{R} \to (0,1)$ is the solution of Kolmogorov's equation

$$\tfrac{1}{2}\psi'' - \lambda\psi' = \psi - \psi^2. \tag{4}$$

We now consider what happens if $0 < \lambda < \sqrt{2}$. Z_s^λ can still be defined as a birth-death process. Since a Brownian Motion almost surely hits the downward sloping line $x(t) = s - \lambda t$ we see that Z_s^λ is without extinction. Reproducing [Neveu]'s proof we arrive at Kolmogorov's equation. From differential equations theory we know that Kolmogorov's equation 4 does not have a monotone solution on $(0,1)$. Looking at definition 3 this implies that a possesses a discontinuity at 1 which means $a_\infty > 0$ so that the process explodes almost surely. We let $T(\omega)$ denote the explosion time. Spatially this corresponds to there being, at all times, a particle below the line $x = T(\omega) - \lambda t$ whose all ancestors have also been below that line. (If after some time τ, there is no such particle, then $Z_T^\lambda \leq N_\tau < \infty$). Integrating the Brownian Path of this particle and its unique ancestors we deduce that almost surely $v_t(\omega) > \frac{1}{2}\lambda t^2 - T(\omega)t$ and hence, almost surely,

$$\liminf_{t\to\infty} \frac{v_t}{t^2} \geq \lambda/2.$$

We now let $\lambda \uparrow \sqrt{2}$ to complete the proof. □

The Phase Plane Picture

We use the projection from the BBM to the Branching point process, to project the space of paths of BBM to the phase plane. We deduce a large deviations principle for the phase plane, both in expectation and almost surely. We find the two rate functions to be different, and the difference explains the different wavefront speeds we observed earlier.

4.4 Scaling The Process

As before, at a fixed time T, let us scale the branching Brownian Motion by a factor of T in both the space and time coordinates. We get a branching process on $[0, 1]$. For every $i \in N_T$ let $x_i^T \in \mathcal{C}^0([0,1], \mathcal{R})$ and $y_i^T \in \mathcal{C}^0([0,1], \mathcal{R})$ be defined as

$$x_i^T(t) := \frac{1}{T} B_{a(i)}(tT),$$
$$y_i^T(t) := \int_0^t x_i^T(s)ds = \frac{1}{T^2} Y_{a(i)}(tT).$$

We define the projection map $\Pi : \mathcal{C}^0 \to \mathcal{R}^2$ as

$$\Pi(z) := (z(1), \int_0^1 z(t)dt)$$

which is clearly continuous in the $||z||_\infty$ norm. For every $D \subset \mathcal{C}^0$ we let $M_D(T)$ denote the particles at time T whose path is in D. For every $D \subset \mathcal{R}^2$ we define let $\Pi M_D(T)$ to be $M_{\Pi^{-1}D}(T)$.

$$M_D(T) := \{i \in N_T : z_i^T \in D\},$$
$$\Pi M_D(T) := \{i \in N_T : \Pi z_i^T \in D\}.$$

We must apologise to the reader for the slight change of notations which is about to occur. From now on, we will use $z \in C_1$ to denote a path of a BBM. x and y will now represent the coordinates in the phase plane.

4.5 The Expectation Picture

The Expectation large deviations result tells us that if $D \subset \mathcal{R}^2$ is closed and $A \subset \mathcal{R}^2$ is open then

$$\limsup_{T\to\infty} \tfrac{1}{T} \log E|\Pi M_D(T)| \leq \sup_{z\in\Pi^{-1}D} J(z) = \sup_{(x,y)\in D} \sup_{\Pi z=(x,y)} J(z),$$
$$\liminf_{T\to\infty} \tfrac{1}{T} \log E|\Pi M_A(T)| \geq \sup_{z\in\Pi^{-1}A} J(z) = \sup_{(x,y)\in A} \sup_{\Pi z=(x,y)} J(z).$$

We recall that $J(z) = J(z,1)$ is the expectation rate function for a Branching Brownian Motion

$$J(z,\theta) := \theta - \tfrac{1}{2} \int_0^\theta \dot{z}^2(t)dt.$$

For every $(x,y) \in \mathcal{R}^2$ we define $\Pi J(x,y) := \sup\{J(z) : \Pi z = (x,y)\}$ and use Calculus of Variation with Lagrange Multiplier optimising procedure (see section 4.7) to find that there is a unique z maximising $\Pi J(x,y)$:

$$z(t) = 3(x-2y)t^2 + 2(3y-x)t.$$

Accordingly, $\Pi J(x,y) = J(z) = 1 - \frac{1}{2}x^2 - 6(y - \frac{1}{2}x)^2$. We immediately have the following result.

Result 6. *Let $D \subset \mathcal{R}^2$ be closed and let $A \subset \mathcal{R}^2$ be open. Then*

$$\limsup_{T\to\infty} \tfrac{1}{T} \log E|\Pi M_D(T)| \leq \sup_{(x,y)\in D} \Pi J(x,y),$$
$$\liminf_{T\to\infty} \tfrac{1}{T} \log E|\Pi M_A(T)| \geq \sup_{(x,y)\in A} \Pi J(x,y).$$

Before carrying on please take time to consider how natural this formula is. For particles whose Brownian position at time T is xT we know that their rate function is given by $1 - \frac{1}{2}x^2$. Conditioning on their final position we know that the particles have the law of a Brownian Motion with drift x so that most of these particles arrive in a straight line $x_i^T(t) = xt$ yielding $y_i^T(1) = \frac{1}{2}x$. Some of them will deviate from that path and are penalised by the amount $6(y - \frac{1}{2}x)^2$.

Corollary 3. *Let u_t be the expectation wavefront speed. Then*

$$\lim_{t\to\infty} \frac{u_t}{t^2} = \sqrt{2/3}.$$

Proof. The boundary of the region $\{(x,y) : \Pi J(x,y) \geq 0\}$ defines the expectation wavefront. In particular, maximising y subject to $\Pi J(x,y) \geq 0$, we find $x = \sqrt{3/2}$ and $y = \sqrt{2/3}$. Since $y = \frac{1}{T^2} Y_i(T)$ the result follows.

□

4.6 The Almost-Sure Picture

We know from the large deviation result that almost surely

$$\limsup_{T\to\infty} \tfrac{1}{T}\log|\Pi M_D(T)| \le \sup_{z\in\Pi^{-1}D} K(z) = \sup_{(x,y)\in D}\ \sup_{\Pi z=(x,y)} K(z),$$
$$\liminf_{T\to\infty} \tfrac{1}{T}\log|\Pi M_A(T)| \ge \sup_{z\in\Pi^{-1}A} K(z) = \sup_{(x,y)\in A}\ \sup_{\Pi z=(x,y)} K(z).$$

Here $K(z) = K(z,1)$, is the almost-sure rate function for a Branching Brownian Motion. If $\theta_0 := \inf\{\theta \in [0,1] : J(z,\theta) < 0\}$ the K is defined as

$$K(z,\theta) := \begin{cases} J(z,\theta) & 0 < \theta \le \theta_0 \\ -\infty & \theta_0 \le \theta \le 1 \end{cases}$$

Finding $\Pi K(x,y) := \sup\{K(z) : \Pi z = (x,y)\}$ is more involved as in addition to $\Pi z = (x,y)$, we also impose that $J(z,\theta) \ge 0$ for all $\theta \in [0,1]$, but see section 4.8. We find that for the half-plane $\{y \ge \frac{1}{2}x\}$ the following holds.
Let $\alpha, \beta, \gamma \in \mathcal{R}^2$ be defined as

$$\alpha = (\sqrt{2}, 1/\sqrt{2}),\quad \beta = (-1/\sqrt{2}, 0),\quad \gamma = (-\sqrt{2}, -1/\sqrt{2}).$$

Let f, g, h be the functions defined as

$$f(x) = \tfrac{1}{3}x + \tfrac{1}{3\sqrt{2}},\quad g(x) = \tfrac{1}{2}x + \tfrac{1}{6}\sqrt{6-3x^2},\quad h(x) = \tfrac{1}{2}x.$$

Note that f links α to β, the function g links β to γ while clearly h links γ and α. So that they form a region D_1 (also see diagram)

$$D_1 = \left\{(x,y) : x \in [-\sqrt{2}, \sqrt{2}], y \in [h(x), f(x) \wedge g(x)]\right\}.$$

In addition let $l(x) = \frac{1}{\sqrt{2}} - \frac{\sqrt{2}}{9}(\sqrt{2}-x)^2$ be another function linking α and β and let D_2 be the region enclosed by $l(x)$ and $f(x)$.

$$D_2 = \left\{(x,y) : x \in [-1/\sqrt{2}, \sqrt{2}], y \in [f(x), l(x)]\right\}.$$

We find that in D_1 (and by symmetry in $-D_1$ too) the almost-sure behaviour and the behaviour in expectation agree so that the z which maximises the expectation satisfies $K(z) = J(z)$ and so $\Pi K = \Pi J$. Inside D_2 we find that the z maximising is given by

$$\dot{z}(t) = \begin{cases} \sqrt{2} & 0 \le t \le \theta \\ \sqrt{2} - \mu(t-\theta) & \theta \le t \le 1 \end{cases}$$

with

$$\theta = 1 - 3(\tfrac{1}{\sqrt{2}} - y)/(\sqrt{2} - x),$$
$$\mu = \frac{2}{9}(\sqrt{2} - x)^3/(\tfrac{1}{\sqrt{2}} - y)^2.$$

We conclude that for $(x, y) \in D_2$

$$\Pi K(x, y) = \sqrt{2}(\sqrt{2} - x)\left(1 - \frac{2}{9}(\sqrt{2} - x)^2/(1 - \sqrt{2}y)\right).$$

Otherwise we find $\Pi K(x, y) = -\infty$. We now do the same analysis for the other half plane $\{y \leq \frac{1}{2}x\}$ and get the almost-sure result.

Result 7. *Let $D \subset \mathcal{R}^2$ be closed and let $A \subset \mathcal{R}^2$ be open. Then*

$$\limsup_{T\to\infty} \tfrac{1}{T} \log |\Pi M_D(T)| \leq \sup_{(x,y)\in D} \Pi K(x, y),$$
$$\liminf_{T\to\infty} \tfrac{1}{T} \log |\Pi M_A(T)| \geq \sup_{(x,y)\in A} \Pi K(x, y).$$

Corollary 4. *Let v_t be the almost-sure wavefront speed, then*

$$\lim_{t\to\infty} \frac{v_t}{t^2} = 1/\sqrt{2}.$$

Proof. The boundary of the region $D_1 \cup D_2 = \{(x, y) : \Pi K(x, y) \geq 0\}$ defines the almost-sure wavefront. In particular, maximising y subject to $\Pi K(x, y) \geq 0$, we find that $\frac{dl}{dx}|_{x=\sqrt{2}} = 0$ and the supremum is attained at $x = \sqrt{2}$ and $y = 1/\sqrt{2}$. Since $y = \frac{1}{T^2} Y_i(T)$ the result follows.

□

Optimisation of The Rate Functions

In this section we explain briefly how the optimisations for $\Pi J(x, y)$ and $\Pi K(x, y)$ were derived.

4.7 The Expectation Rate Function

If $(x, y) \in \mathcal{R}^2$ we wish to maximise $\{J(z) : \Pi(z) = (x, y)\}$. Alternatively, we minimise $\frac{1}{2}\int_0^1 \dot{z}^2$ subject to the constraint $z(0) = 0, z(1) = x, \int_0^1 z(t)dt = y$. Using Lagrange multiplier we get the unconstrained problem of minimising F.

$$F(z, \dot{z}, t) = \int_0^1 \tfrac{1}{2}\dot{z}^2 - \lambda(y - z)dt.$$

From Calculus of Variations we have $F_z - \dot{F_{\dot z}} = 0$ from which we get that $\ddot{z} =$ constant and so $z(t) = (x-\alpha)t^2 + \alpha t$. Substituting $\int_0^1 z dt = y$ we arrive at the optimal path in expectations

$$z = 3(x-2y)t^2 + 2(3y-x)t,$$

from which we deduce that

$$J(z) = 1 - \tfrac{1}{2}\int_0^1 \dot{z}^2 dt = 1 - \tfrac{1}{2}x^2 - 6(y - \tfrac{1}{2}x)^2.$$

4.8 The Almost-Sure Rate Function

Throughout, we assume that $y \geq \frac{1}{2}x$. When $y < \frac{1}{2}x$ we use the symmetry $\Pi K(-x,-y) = \Pi K(x,y)$. Clearly if the z which optimises $\{J(z) : \Pi(z) = (x,y)\}$ also has $K(z) = J(z)$ we are done. This amounts to ensuring $\dot{z}(0) \leq \sqrt{2}$ and we find that if $(x,y) \in D_1$, this is indeed the case.

Outside D_1, although we can not follow the same optimising procedure, some points are clear. Keeping x fixed, as y increases ΠJ and ΠK are decreasing in y. To maximise y while keeping J constant, we must have $\dot{z}$ as a non increasing function. From the in-expectation optimisation procedure, another way of maximising y while keeping J constant is by ensuring $\dot{z}$ is piece-wise linear. Conditioning on the first time θ when $\dot{z} < \sqrt{2}$ we find that on $[\theta, 1]$, the in-expectation optimisation is also valid almost surely so that z must be of the form

$$\dot{z}(t) = \begin{cases} \sqrt{2} & 0 \leq t \leq \theta \\ \sqrt{2} - \mu(t-\theta) & \theta \leq t \leq 1 \end{cases}$$

which we integrate to get

$$z(t) = \begin{cases} \sqrt{2}t & 0 \leq t \leq \theta \\ \sqrt{2}t - \frac{1}{2}\mu(t-\theta)^2 & \theta \leq t \leq 1 \end{cases}$$

and finally we deduce that

$$\int_0^1 z dt = \tfrac{1}{\sqrt{2}} - \tfrac{1}{6}\mu(1-\theta)^3.$$

We substitute boundary conditions $z(1) = x$ and $\int z = y$ to complete the analysis.

4.9 The Phase Plane Diagram

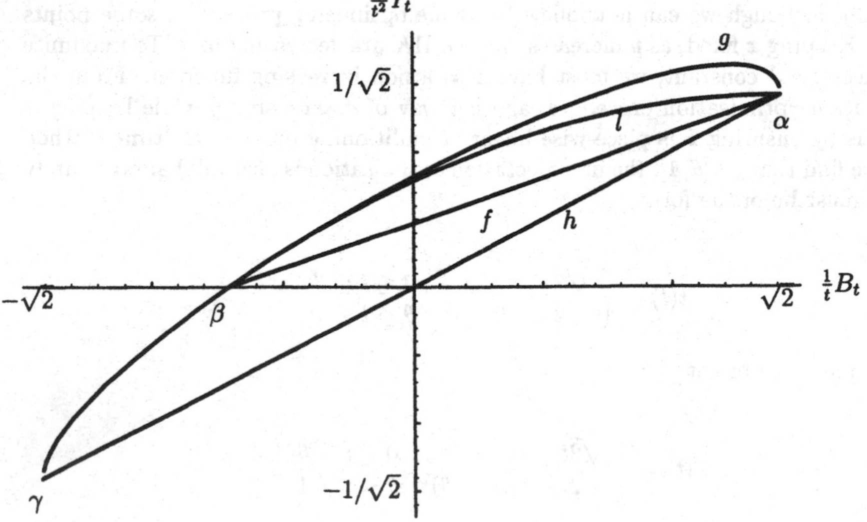

References

[1] Chauvin, B. & Rouault, A. *KPP equation and branching Brownian Motion in the subcritical speed-area. Application to spatial trees.*
Prob. Th. and Rel. Fields 80, 1988. p299-314.

[2] Neveu, J. *Multiplicative martingales for spatial branching process.*
Sem. Stochastic Processes Princeton. p223-242.

[3] Schilder, M. *Some Asymptotic formulae for Wiener integrals.*
Trans. Amer. Math. Soc. 125, 1966. p63-85

[4] Strassen, V. *An invariance principle for the law of the iterated logarithm.*
Z. Wahrsch. Verw. Gebiete 3, 1964. p227-246.

[5] Uchiyama, K. *Spatial growth of a branching process of particles living in $\mathcal{R}^d$.*
Annals of Probability 10, 1982. p896-918.

[6] Varadhan, S.R.S. *Large Deviations and Applications.*
CBMS-NSF Regional Conference Series in Applied Mathematics, 1984.

[7] Warren, J. *Some Aspects of Branching Processes.*
Ph.D. dissertation, University of Bath, 1995

Acknowledgements

This work was inspired by Professor David Williams, guided by the probability group in Bath University and supported financially by EPSRC. To all, I am deeply indebted.

A Final Note

Since submitting this manuscript, we discovered a book by P. Revesz (Random walks of infinitely many particles, 1994). He considered a split-at-integer-times Branching Brownian Motion and showed the space of paths to be the closure of $\{f : J(f) \geq 0\}$ without counting the actual growth rate along each path. His result is similar in nature although the methods he used are different.

Criteria of regularity at the end of a tree

S. Amghibech

Université de Rouen
U.F.R des sciences, mathématique, URA CNRS 1378
76821 Mont Saint Aignan Cedex

Abstract

For a random walk on a tree, we give analogues of Wiener's test relatively to Dirichlet's problem for the endpoints of the tree.

Résumé

Étant donnée une marche aléatoire sur un arbre, nous établissons pour les points de la frontière des critères de régularité analogues à des critères classiques relatifs au problème de Dirichlet pour le mouvement brownien dans $\mathbb{R}^n$, dont celui de Wiener pour $n = 2$.

Keywords Dirichlet problem, resistance, regularity criteria.

1 Introduction

Let $\mathcal{A} = (A, \mathcal{U}, 0)$ be a non oriented infinite tree with a root : A is the set of vertices x, y, α etc., $\mathcal{U}$ the set of edges (x, y) or $[x, y]$, and 0 a fixed point in A. We denote by $x \sim y$ the symmetric relation $(x, y) \in \mathcal{U}$ and $d(x)$ the cardinality of $\{y \in A : x \sim y\}$. We suppose

$$2 \leq \inf_{x \in A} d(x) \leq \sup_{x \in A} d(x) < \infty$$

verified; in particular A is countably infinite. A *geodesic ray* (starting at 0) of $\mathcal{A}$ is any one to one sequence $\eta = (x_n)$ of vertices such that $x_0 = 0$ and $x_n \sim x_{n+1}$ for all $n \in \mathbb{N}$, and the *end* of $\mathcal{A}$ is the set of all geodesic rays.

We consider a *resistance* R on $\mathcal{A}$, i.e. a function from $\mathcal{U}$ to $\mathbb{R}_+$ such that $R[x, y] = R[y, x]$ for all $[x, y] \in \mathcal{U}$ and we associate to R a random walk $X = (X_n)_{n \geq 0}$ with transition $P(X_{n+1} = x / X_n = y) = p_{xy} = \dfrac{R[x, y]^{-1}}{\sum_{\{z: y \sim z\}} R[y, z]^{-1}}$ if $x \sim y$ and $= 0$ otherwise, where $p_{xy} = 1/d(y)$ if $x \sim y$ in the simple random walk ($R \equiv 1$). We denote by P_x the law of $X_0 = x$, $T_y = \inf\{n \geq 0 : X_n = y\}$ the first hitting time of $y \in A$, and $S_B = \inf\{n > 0 : X_n \in B\}$ the first return time to the subset B of A. *We assume in all this article* that X is *transient*, i.e. $P_x[S_{\{x\}} = \infty] > 0$ for all $x \in A$.

Following [1] we say that a geodesic ray $\eta = (x_n)_{n \in \mathbb{N}}$ is *regular* for the Dirichlet problem if $\lim_{n \to \infty} P_{x_n}[T_0 < \infty] = 0$; this is analogous to classical definition of a regular point of a Dirichlet problem. In [4] and [11], Wiener's test in the continuous case is presented. In [12], [6] the description of the Dirichlet problem on graph and conditions to obtain a regular problem are given. In [2] another description is given.

In §2 we establish a criterion of regularity, for geodesic ray for random walk on a tree, analogue in the simple case to Wiener's test [11], [4] for the brownian motion in $\mathbb{R}^2$, and we give the analogue of Frostman criterion.

In §3 we give a characterization of the regularity of a geodesic ray, analogous in the simple case to Wiener's test which we find in [5] for brownian motion in $\mathbb{R}^n$, $n \geq 2$. This characterization is based on the behaviour of the potential kernel in the neighbourhood of geodesic ray.

2 Electrical network and Wiener's test

To each $\alpha \in A$ we associate a partial order (orientation) $<_\alpha$ on A as : for $x \neq y$ we have $x <_\alpha y$ if and only if x belongs to a geodesic ray between y and α. We call a *flow started at* α any function I^α from $\mathcal{U}$ to $\mathbb{R}$ such that

1. $\sum_{y:\alpha\sim y} I^\alpha([\alpha, y]) = 1$ and $\sum_{y:\beta\sim y} I^\alpha([\beta, y]) = 0$ for all $\beta \neq \alpha$;
2. $I^\alpha([x,y]) = -I^\alpha([y,x])$ for all $[x,y] \in \mathcal{U}$ and $I^\alpha([x,y]) \geq 0$ if $x <_\alpha y$.

The *energy* of the flow I^α is the number $E(I^\alpha) = \frac{1}{2}\sum_{x\sim y} R[x,y] I^\alpha([x,y])^2$. Since the random walk X is transient, there exists a flow $\tilde{I}^\alpha$ starting at α with finite minimal energy (see [8] and [10]) $\tilde{E}^\alpha$ which we call the *resistance* of $\mathcal{A}$ *at* α and we denote it by $R_{\mathcal{A}}(\alpha)$. We think of $R_{\mathcal{A}}(\alpha)$ as the inverse of the ordinary capacity. If $\mathcal{B}$ is a subtree of $\mathcal{A}$ rooted at α, we define in the same way the resistance $R_{\mathcal{B}}(\alpha)$ of $\mathcal{B}$ at α if $\mathcal{B}$ is transient, and we put $R_{\mathcal{B}}(\alpha) = \infty$ if $\mathcal{B}$ is recurrent. Finally, if $\eta = (x_n)_{n\in\mathbb{N}}$ is a geodesic ray we denote by $\mathcal{A}_\eta(x_n)$ the subtree of $\mathcal{A}$ which has

$$\{x_n\} \cup \{x \in A,\ x_n <_{x_{n+1}} x,\ x_n <_0 x\},$$

as vertices and we denote by $R_\eta(k)$ the resistance $R_{\mathcal{A}_\eta(x_k)}(x_k)$ of $\mathcal{A}_\eta(x_k)$ at x_k.

We now give the analogue of Wiener's test for the tree [11]

Theorem 1 *Suppose that $R[x,y] \geq 1$ for all $[x,y] \in \mathcal{U}$. Then a geodesic ray $\eta = (x_n)$ is non regular if and only if*

$$\sum_{n=1}^{\infty} \frac{1}{R_\eta(n)} \sum_{k=1}^{n} R[x_k, x_{k-1}] < \infty;$$

in particular, if we have $R[x_k, x_{k+1}] = 1$ for all $k \in \mathbb{N}$, the geodesic ray $\eta = (x_n)_n$ is non regular if and only if

$$\sum_{n=1}^{\infty} \frac{n}{R_\eta(n)} < \infty.$$

We give an example before proving our theorem.

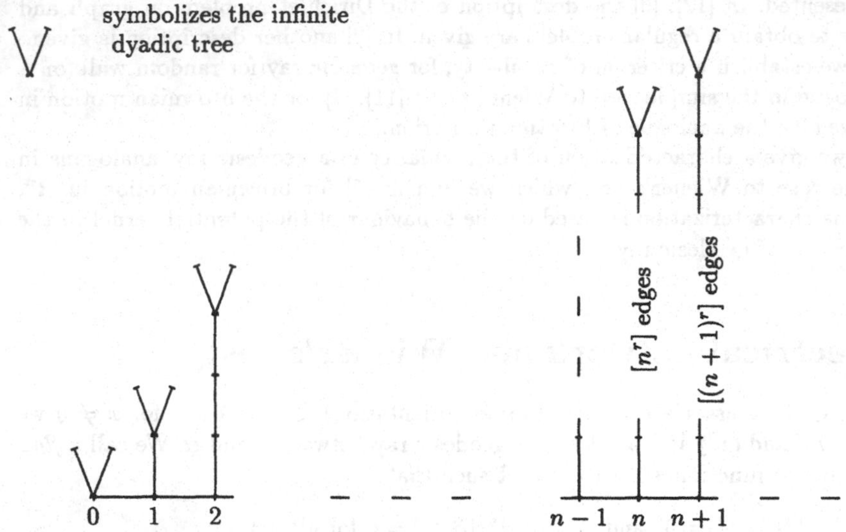

figure 1

Let us consider the simple random walk X in the tree $\mathcal{A}$ depending on the parameter r. Using the symmetry of $\mathcal{A}_\eta(n)$ we obtain

$$R_\eta(n) = [n^r] + 1.$$

According to Theorem 1 $\eta = (k)_{k\in N}$ is non regular if and only if $\sum_{n\in\mathbb{N}} n/R_\eta(n) < \infty$, which is equivalent to $r > 2$. Furthermore if $\alpha \leq 2$ then $\lim_{n\to\infty} P_n[T_0 < \infty] = 0$ but if $r > 2$ then $\lim_{n\to\infty} P_n[T_0 < \infty] > 0$.

We prove Theorem 1 in several steps.

First, if $R_\eta(k)$ is infinite for all k then the ray η is non regular because $R \geq 1$. Hence we can suppose that there exists k such that $R_\eta(k)$ is finite and then we can change x_k to 0. For simplification we suppose $R_\eta(0)$ is finite, which implies $\tilde{I}^{x_n}([x_k, x_{k-1}]) > 0$ for all $n \geq k \geq 1$.

Proposition 1 *Let $\eta = (x_n)_{n\in\mathbb{N}}$ be a geodesic ray.*

1) For all $k > 0$ the quantity

$$c_k = \frac{\tilde{I}^{x_n}([x_k, x_{k-1}])}{\tilde{I}^{x_n}([x_{k+1}, x_k])}$$

is independent of $n > k$.

2) A geodesic ray η is non regular if and only if $\prod_{n\in\mathbb{N}} c_n > 0$.

Proof of proposition 1 Part 1) is trivial. To prove 2) note that the flow starting at α defined by

$$I^\alpha[x, y] = \sum_{k\in\mathbb{N}} P_\alpha[X_k = x, X_{k+1} = y]\, P_y[T_x = \infty] \text{ if } x <_\alpha y,$$

is the flow of minimal energy $\tilde{E}^{x_n}$. This means that

$$\tilde{I}^{x_n}([x_1, x_0]) \leq P_{x_n}[T_0 < \infty].$$

Combining this inequality and the transience of X we easily deduce 2).

Proof of Theorem 1 Suppose $R_\eta(k) < \infty$ and $R_\eta(k+1) < \infty$. Using the minimality of the energy $\tilde{I}^{x_{k+2}}$, we obtain the equation of equilibrium

$$\begin{aligned} R[x_{k+1}, x_k]\tilde{I}^{x_{k+2}}([x_k, x_{k+1}]) \; &+ \; R_\eta(k)\{\tilde{I}^{x_{k+2}}([x_{k+1}, x_k]) - \tilde{I}^{x_{k+2}}([x_k, x_{k-1}])\} \\ &= \; R_\eta(k+1)\{\tilde{I}^{x_{k+2}}([x_{k+2}, x_{k+1}]) - \tilde{I}^{x_{k+2}}([x_{k+1}, x_k])\}. \end{aligned}$$

Dividing each term by $\tilde{I}^{x_{k+2}}([x_{k+2}, x_{k+1}])$ we obtain

$$(1 - c_{k+1})R_\eta(k+1) = c_{k+1}R[x_{k+1}, x_k] + c_{k+1}(1 - c_k)R_\eta(k).$$

Multiplying by $\prod_{i=k+2}^n c_i$ for $n \geq k+1$, we obtain

$$R_\eta(k+1)(1 - c_{k+1}) \prod_{i=k+2}^{n} c_i = R[x_{k+1}, x_k] \prod_{i=k+1}^{n} c_i + R_\eta(k)(1 - c_k) \prod_{i=k+1}^{n} c_i$$

and finally

$$(1 - c_n)R_\eta(n) = \sum_{k=0}^{n-1} R[x_{k+1}, x_k] \prod_{i=k+1}^{n} c_i + R_\eta(0) \prod_{i=1}^{n} c_i, \tag{1}$$

if $R_\eta(k) < \infty$ for $k = 0, \ldots, n$. We show easily that (1) is true if we suppose only $R_\eta(n)$ is finite and $R_\eta(k)$, for $k = 1, \ldots n-1$ are finite or infinite. This implies the inequality

$$1 - c_n \geq \frac{1}{R_\eta(n)} \sum_{k=1}^{n} (R[x_k, x_{k-1}] \prod_{i=k}^{n} c_i)$$

and therefore

$$\sum_{n=1}^{\infty} (1 - c_n) \geq \sum_{n=1}^{\infty} \frac{1}{R_\eta(n)} \sum_{k=1}^{n} (R[x_k, x_{k-1}] \prod_{i=k}^{n} c_i). \tag{2}$$

If $\eta = (x_n)$ is irregular, by Proposition 1 $\prod_{n=1}^{\infty} c_n$ is finite, and so the series

$$\sum_{n=1}^{\infty} \frac{1}{R_\eta(n)} \sum_{k=1}^{n} R[x_k, x_{k-1}] \prod_{i=k}^{n} c_i$$

converges, and by inequality (2), we have

$$\sum_{n=1}^{\infty} \frac{1}{R_\eta(n)} \sum_{k=1}^{n} R[x_k, x_{k-1}] < \infty.$$

If η is regular, by Proposition 1, $R[x_1, x_0] \geq 1$ and (1), we have

$$(1 - c_n) \frac{1}{R_\eta(0) + 1} \leq \frac{1}{R_\eta(n)} \sum_{k=1}^{n} (R[x_k, x_{k-1}] \prod_{i=k}^{n} c_i) \tag{3}$$

for all n. Since $c_i \in]0,1]$ and $\lim_{n\to\infty}\prod_{i=1}^n c_i = 0$, $\sum_{n\in\mathbb{N}}(1-c_n)$ diverges, and by using inequality (3) we obtain

$$\sum_1^\infty \frac{1}{R_\eta(n)} \sum_{k=1}^n (R[x_k, x_{k-1}] \prod_{i=k}^n c_i) = \infty.$$

This completes the proof because $c_i \in]0,1]$.

The consequence following Theorem 1 has an interesting physical interpretation. Let us denote by $\tilde{I}^0$ and $\tilde{E}^0$ the flow and the energy at equilibrium starting at 0 i.e. the flow with minimal energy starting at 0.

Proposition 2 *Let $\eta = (x_n)_{n\in\mathbb{N}}$ be a geodesic ray and define the equilibrium potential $\tilde{V}^0(\eta)$ at η by*

$$\tilde{V}^0(\eta) = \sum_{k=0}^\infty R[x_k, x_{k+1}]\tilde{I}^0[x_k, x_{k+1}].$$

The ray η is irregular if and only if $\tilde{V}^0(\eta) < \tilde{E}^0$.

Proof First let us define, for $n > 0$, the variation potential at equilibrium $\tilde{V}_\eta^0(n)$ in $\mathcal{A}_\eta(x_n)$ by

$$\tilde{V}_\eta^0(n) = R_\eta(n)\{\tilde{I}^0([x_{n-1}, x_n]) - \tilde{I}^0([x_n, x_{n+1}])\}.$$

Suppose η is irregular. Using the equilibrium for $n < m$ at $R_\eta(m)$ and $R_\eta(n)$ we obtain

$$\tilde{V}_\eta^0(n) = \sum_{k=n}^{m-1} R[x_k, x_{k+1}]\tilde{I}^0[x_k, x_{k+1}] + \tilde{V}_\eta^0(m),$$

if $R_\eta(m) < \infty$ and $R_\eta(n) < \infty$. Since $\tilde{I}^0[x_k, x_{k+1}] \to 0$ if $k \to \infty$ we obtain

$$\tilde{I}_\eta^0[x_{k+1}, x_k] = \sum_{i=k+1}^\infty \frac{\tilde{V}_\eta^0(i)}{R_\eta(i)}.$$

For simplification, we put

$$\tilde{V}_\eta^0(p_n + k) = \tilde{V}_\eta^0(p_{n+1}) \text{ for } k = 1, \cdots, p_{n+1} - p_n$$

where $R_\eta(p_n) < \infty$, $R_\eta(p_n + 1) = \infty, \cdots, R_\eta(p_{n+1} - 1) = \infty$, $R_\eta(p_{n+1}) < \infty$. Thus we have

$$\tilde{V}_\eta^0(p_n) = \tilde{V}_\eta^0(p_{n+1}) + \sum_{k=p_n}^{p_{n+1}-1} R[x_k, x_{k+1}] \sum_{i=k+1}^\infty \frac{\tilde{V}_\eta^0(i)}{R_\eta(i)}; \tag{4}$$

since η is irregular the series of general term

$$R[x_k, x_{k+1}] \sum_{i=k+1}^\infty \frac{1}{R_\eta(i)}$$

is convergent by Theorem 1. Therefore we have

$$\prod_{n\geq 1}\left(1 + \sum_{k=p_n}^{p_{n+1}-1} R[x_k, x_{k+1}] \sum_{i=k+1}^\infty \frac{1}{R_\eta(i)}\right) < \infty;$$

by applying the inequality (4) and the nonincrease of $V_\eta^0(n)_n$ we obtain $\lim_{n\to\infty} \tilde{V}_\eta^0(n) > 0$, which proves the first implication.

Conversely, $\tilde{V}^0(\eta) < \tilde{E}$ implies $\inf_n \tilde{V}_\eta^0(n) \geq \tilde{E} - \tilde{V}^0(\eta) > 0$ and equation (3) gives

$$\tilde{V}_\eta^0(p_n) \geq \inf_k \tilde{V}_\eta^0(k) \left(1 + \sum_{k=p_n}^{p_{n+1}-1} R[x_k, x_{k+1}] \sum_{i=k+1}^{\infty} \frac{1}{R_\eta(i)}\right)$$

and therefore the result follows.

3 A third criterion of irregularity

In this section we assume that

$$0 < \inf_{y\in A} \sum_{\{z:y\sim z\}} R[y,z]^{-1} \leq \sup_{y\in A} \sum_{\{z:y\sim z\}} R[y,z]^{-1} < \infty$$

and we denote by G the potential kernel of the transient random walk X. Let $\eta = (x_n)$ be a geodesic ray.

Since $n \mapsto P_{x_n}[T_x < \infty]$ is nonincreasing for large value of n, and since $G(x_n, x) = P_{x_n}[T_x < \infty]\, G(x,x)$ then $\lim_{n\to\infty} G(x_n, x)$ exists. We denote it by $G(\eta, x)$.

For a subset B of A we define its *capacity* $\mathrm{Cap}(B)$ as in [9] by

$$\mathrm{Cap}(B) = \sum_{x\in B} C(x) P_x[S_B = \infty],$$

which is equivalent to other classical definitions.

Theorem 2 *Let $\eta = (x_n)_{n\in\mathbb{N}}$ be a geodesic ray and put for all $k \in \mathbb{N}$*

$$A_k = \{x \in A : 2^k \leq \lim_{n\to\infty} G(x_n, x) \leq 2^{k+1}\}.$$

If η is irregular we have

$$\limsup_{n\to\infty} 2^n \mathrm{Cap}(A_n) > 0.$$

If η is regular we have $G(\eta, x) = 0$ for all $x \in A$.

We begin the proof with two lemmas.

Lemma 1 *If $\lim_{n\to\infty} P_{x_n}[T_0 < \infty] > 0$ then $\lim_{n\to\infty} P_0[T_{x_n} < \infty] = 0$.*

Proof of lemma 1 Suppose the result is not true, i.e. $\lim_{n\to\infty} P_0[T_{x_n} < \infty] \neq 0$. By proposition 2.6 of [3] (strong Markov property) we have

$$P_{x_n}[T_0 < \infty] = P_{x_n}[T_{x_{n-1}} < \infty]\, P_{x_{n-1}}[T_0 < \infty]$$

and

$$P_0[T_{x_n} < \infty] = P_0[T_{x_{n-1}} < \infty]\, P_{x_{n-1}}[T_{x_n} < \infty].$$

By this equality we have

$$\lim_{n\to\infty} P_{x_{n-1}}[T_{x_n} < \infty] = P_{x_n}[T_{x_{n-1}} < \infty] = 1$$

which gives $\lim_{n\to\infty} P_{x_n}[S_{x_n} < \infty] = 1$ and so $\lim_{n\to\infty} G(x_n, x_n) = \infty$. By [3] we have

$$C(x_n)P_{x_n}[T_0 < \infty]\, G(0,0) = C(0)P_0[T_{x_n} < \infty]\, G(x_n, x_n) \tag{5}$$

because

$$C(x_n)\, G(x_n, 0) = C(0)\, G(0, x_n)$$

and for all x, y in A we have

$$G(x,y) = P_x[T_y < \infty] G(y,y).$$

Since the graph is bounded, this gives $\lim\limits_{n\to\infty} P_0[T_{x_n} < \infty] = 0$, which contradicts the hypotheses. Hence the lemma is proven.

Lemma 2 *If η is irregular, then, for all $\epsilon > 0$, the subsets $\{x \in A : P_x[T_0 < \infty] \geq \epsilon\}$ and $\{x \in A : \lim_{n\to\infty} G(x_n, x) \geq \epsilon\}$ are non recurrent.*

Proof of lemma 2 Let us denote by $\mathcal{A}_{\eta,\epsilon}$ the tree induced by $\{x \in A : P_x[T_0 < \infty] \geq \epsilon\}$. If $\mathcal{A}_{\eta,\epsilon}$ is finite, $\{x \in A : P_x[T_0 < \infty] \geq \epsilon\}$ is non recurrent; if $\mathcal{A}_{\eta,\epsilon}$ is infinite, every geodesic ray of $\mathcal{A}_{\eta,\epsilon}$ is irregular, and hence applying Proposition 4.3 of [8], we obtain that $\mathcal{A}_{\eta,\epsilon}$ is recurrent, and so $\{x \in A : P_x[T_0 < \infty] \geq \epsilon\}$ is non recurrent.

To prove the non recurrence of the second subset in lemma 1, we apply the first part to each element of the decomposition of $\{x \in A : \lim_{n\to\infty} G(x_n, x) \geq \epsilon\}$ in the subgraph $\mathcal{A}_\eta(x_k)$, $k \in \mathbb{N}$, and so we easily obtain the conclusion.

Proof of theorem 2 Suppose the geodesic ray is irregular; by Lemma 1 and equality (6) we have

$$\lim_{n\to\infty} G(x_n, x_n) = \infty.$$

Let $n \in \mathbb{N}^*$ such that, in A_n we have a vertex x_k of η and i_n the largest $k \in \mathbb{N}$ such that $x_k \in A_n$. Let u_n be the equilibrium measure of A_n, i.e. the non negative function u_n such that $Gu_n = 1$ in A_n and vanishes in the complement of A_n. In fact A_n is non recurrent by Lemma 1, and we have $u_n(x) = P_x[S_{A_n} = \infty]$ for $x \in A_n$. By definition we have

$$2^{n+1}\text{Cap}\,(A_n) = 2^{n+1} \sum_{x\in A_n} u_n(x)C(x);$$

since $G(\eta, x) \in [2^n, 2^{n+1}]$ for $x \in A_n$, we have

$$2^{n+1}\text{Cap}\,(A_n) \geq \sum_{x\in A_n} \lim_{k\to\infty} G(x_k, x)u_n(x)C(x),$$

and

$$\sum_{x\in A_n} \lim_{k\to\infty} G(x_k, x)u_n(x)C(x) \geq 2^n\text{Cap}\,(A_n).$$

On the other hand, since x_{i_n} is in the geodesic ray between x_k and x for large values of k, we have

$$G(x_k, x) = P_{x_k}[T_{x_{i_n}} < \infty]\, G(x_{i_n}, x).$$

This implies

$$2^{n+1}\text{Cap}\,(A_n) \geq \sum_{x\in A_n} \lim_{k\to\infty} P_{x_k}[T_{x_{i_n}} < \infty]\, G(x_{i_n}, x)\, u_n(x)\, C(x)$$

and

$$\sum_{x\in A_n} \lim_{k\to\infty} P_{x_k}[T_{x_{i_n}} < \infty]\, G(x_{i_n}, x)\, u_n(x)\, C(x) \geq 2^n \mathrm{Cap}\ (A_n)$$

so

$$2^{n+1}\mathrm{Cap}\ (A_n) \geq \lim_{k\to\infty} P_{x_k}[T_{x_{i_n}} < \infty] \sum_{x\in A_n} G(x_{i_n}, x)\, u_n(x)\, C(x)$$

therefore

$$2^{n+1}\mathrm{Cap}\ (A_n) \geq \lim_{k\to\infty} P_{x_k}[T_{x_{i_n}} < \infty].$$

With the same argument we have

$$\lim_{k\to\infty} P_{x_k}[T_{x_{i_n}} < \infty] \geq 2^n \mathrm{Cap}(A_n)/[\max_{x\in A} C(x)].$$

which finishes the result and the theorem.

Remark Here we use a geodesic ray for the determination of A_n. We have an analogous result if we replace the geodesic by a vertex : we obtain in the case of a tree Wiener's test for Markov chains in [7].

4 Appendix

We give an example in which there are infinitely many non countable irregular points which are in the support of the harmonic measure starting at the root.

Let Γ be the dyadic tree which has vertices, root and edges denoted respectively by $(x_n)_{n\in\mathbb{N}}$, x_0 and (i,j) if $x_i \sim x_j$. Let (n_p) be a increasing sequence of $\mathbb{N}^*$ such that the series $\sum_{p\geq 1} n_p/n_{p+1}$ is convergent. We construct the tree Λ (see figure 2) as follows. We introduce $n_p - 1$ vertices, in each edge (i,j) such $d(x_0, x_i) = p$ and $d(x_0, x_j) = p+1$ and we attach at each vertex x_i of Γ a tree $\Gamma_{2,p}$ where $\Gamma_{2,p}$ is the tree obtained by attaching at the root of a dyadic tree a geodesic ray formed with $n_p - 1$ edges.

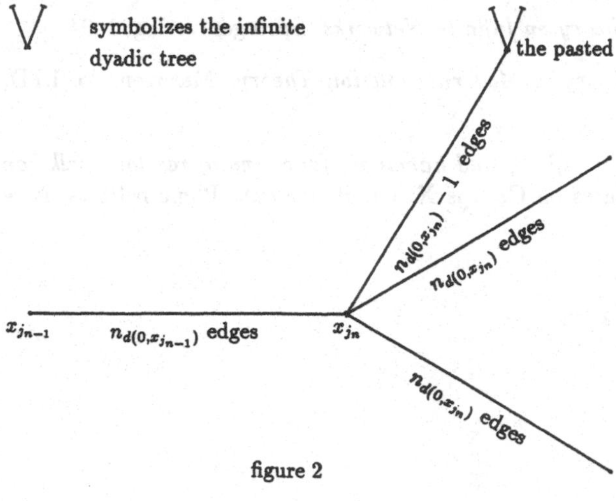

figure 2

We show easily by applying theorem 1 that a geodesic ray of Λ which contains an infinite mumber of x_i's and passes through x_0, is irregular and is in the support of the harmonic measure $\mu(\cdot) = P_{x_0}(\cdot)$. Then we have an uncountable set of irregular points in the support of the harmonic measure

Acknowledgments. The author thanks C. DELLACHERIE, V. KAIMANOVICH, A. BENASSI, R. LYONS and Y. PERES for useful suggestions.

References

[1] BENJAMINI, I., AND PERES, Y. Random Walk on Tree and Capacity in the Interval. *Ann. Inst. H. Poincaré sect B.* *28*, 4 (1992), 557–592.

[2] BENJAMINI, I, R. PEMANTLE, AND Y. PERES. Martin capacity for Markov chains. *Ann. Probability.* *23*, 3 (1995), 1332–1346.

[3] CARTIER, P. Fonctions harmoniques sur un arbre. *Symposia. Math. Acadi 3* (1972), 203–270.

[4] CONWAY, J. *Fonctions of One Complex Variable II.* Springer-Verlag, 1995.

[5] DOOB, J. L. *Classical Potential Theory.* Springer-Verlag, 1984.

[6] KAIMANOVICH, V., AND WOESS, W. The Dirichlet problem at infinity for random walks on graphs with a strong isoperimetric inequality. *Probab. Theory Relat. Fields* *91*, 3-4 (1992), 445–466.

[7] LAMPERTI, J. Wiener's Test and Markov Chains. *J. Math. Anal. Appl.* *6* (1963), 58–66.

[8] LYONS, R. Random Walk and Percolation on Trees. *Ann. Probability.* *18*, 3 (1990), 931–958.

[9] REVUZ, D. *Markov Chains.* North Holland, 1975.

[10] SOARDI, P. *Potential Theory on Infinite Networks.* Springer-Verlag, 1994.

[11] TSUJI, M. *Potential Theory in Modern Function Theory.* Maruzen Co. LTD, Tokyo, 1959.

[12] WOESS, W. *Behaviour at infinity and harmonic functions of random walks on graphs.* Probability Mesures on Groups X. ed. H. HEYER. Plenum Press, New York, 1991.

Normalized Stochastic Integrals in Topological Vector Spaces

R. Mikulevicius and B. L. Rozovskii*
Institute of Mathematics and Informatics
2600 Vilnius
Lithuania
and
Department of Mathematics and
Center for Applied Mathematical Sciences
University of Southern California
Los Angeles, CA 90089-1113

1 Introduction

Stochastic integration in infinite dimensional spaces is a mature area. Several important classes of stochastic integrals were introduced and studied in depth by Kunita [12], Métivier and Pistone [15], Meyer [14], Métivier and Pellaumail [16], Gyöngi and Krylov [7], Grigelionis and Mikulevicius [5], Walsh [22], Korezlioglu [9], Kunita [11], etc. Not surprisingly, the approaches to infinite dimensional stochastic integration proposed in these works have some similarities but also some distinct features. The latter are mainly related to the specifics of the spaces and processes involved. For example, the integral with respect to a stochastic flow (see Kunita [11], and also Gihman, Skorohod [4]) and the integrals with respect to orthogonal martingale measures (see Gyöngy, Krylov [7], [6], Walsh [22]) seem to have very little in common. In fact, the relation between these two integrals as well as others mentioned above is stronger than it might appear. More specifically, it will be shown below that all these integrals and some others are particular cases of one stochastic integral with respect to a locally square integrable cylindrical martingale in a topological vector space.

Let E be a quasicomplete locally convex topological vector space with weakly separable dual space E', i.e. E is a locally convex topological vector space so that all its bounded closed subsets are complete. Let $(Q_s)_{s\geq 0}$ be a predictable family of symmetric non-negative linear forms from E' into $\bar{E}$ and λ_s be a predictable increasing process. By a locally square integrable cylindrical martingale in E

*This paper was partially supported by ONR Grant N00014-91-J-1526

(with covariance operator function Q_s and quadratic variation $\int_0^t Q_s d\lambda_s$) we understand a family of real valued locally square integrable martingales $M_t(y)$, $y \in E'$, such that

$$\langle M(y), M(y')\rangle_t = \int_0^t \langle Q_s y, y'\rangle_{E,E'}\, d\lambda_s .$$

The stochastic integral is constructed in three steps. To begin with, we define an Ito integral for integrands from the set $\tilde{L}^2_{loc}(Q)$ consisting of E'-valued predictable functions f_s such that

$$\int_0^1 \langle Q_s f_s,\, f_s\rangle_{E,\,E'}\, d\lambda_s < \infty \qquad \mathbf{P}\text{-a.e.}$$

(see Proposition 9). Below this integral is denoted $\int_0^t f_s dM_s$ or $I_t(f)$. In our approach, the set S_b of simple (elementary) functions consists of all finite linear combinations $\sum_1^N f_s^k y_k$, $y_k \in E'$, of real valued predictable functions so that

$$\int_0^1 \sum_{k,j=1}^{N} f_s^k f_s^j \langle Q_s y_k,\, y_j\rangle_{E,\,E'}\, d\lambda_s < \infty \qquad \mathbf{P}\text{-a.e.}$$

The choice of the set of simple functions is almost the only nonstandard feature of the first part of our construction.

Unfortunately, the above integral is not quite satisfactory in that the space of integrands, $\tilde{L}^2_{loc}(Q)$, is not complete. So the next important step is to find a natural completion of this space. To address this problem we rely on the L. Schwartz theory of reproducing kernels [21]. The results in [21] allow to construct a family of Hilbert subspaces $H_s \subset E$ naturally associated with the covariance operator function Q_s; below these spaces are referred to as *covariance spaces*. The covariance space H_s is defined as the completion of $Q_s E'$ with respect to the inner product

$$(Q_s y,\, Q_s y')_{H_s} := \langle Q_s y,\, y'\rangle_{E,\,E'} \tag{1}$$

Using these results we demonstrate (Proposition 10) that the closure of $\tilde{L}^2_{loc}(Q)$ is isometric to the space $L^2_{loc}(Q) := \{$predictable E-valued g : $\int_0^1 |g_s|^2_{H_s} d\lambda_s < \infty$ $\mathbf{P}$-a.s.$\}$.

The third and final step of our construction is to extend the stochastic integral from $\tilde{L}^2_{loc}(Q)$ onto $L^2_{loc}(Q)$. To achieve this goal, we introduce a *normalized* stochastic integral for E-valued integrands. We denote this integral $\int_0^t g_s * dM_s$ or $\mathcal{R}_t(g)$. Loosely speaking, the integral is defined by the equality

$$\mathcal{R}_t(g) = \int_0^t g_s * dM_s := \int_0^t g_s d(M_s/Q_s)\,. \tag{2}$$

Of course, this "definition" is formal; it explains the origins of the term "normalized" rather than defines the integral. However, if $g_s = Q_s f_s$ and $f \in \tilde{L}^2_{loc}(Q)$,

(2) can be made meaningful by setting

$$\int_0^t g_s * dM_s := \int_0^t Q_s f_s d(M_s/Q_s) := \int_0^t f_s dM_s,$$

(see section 4.1 as well).

Since $Q_s f_s \in L^2_{loc}(Q)$, the idea now is to extend the Ito stochastic integral $\int_0^t f_s dM_s$ by extending the normalized integral $\int_0^t Q_s f_s * dM_s$ to all integrands belonging to $L^2_{loc}(Q)$. We prove that this is indeed possible (Proposition 11), and for every $g \in L^2_{loc}(Q)$, $\mathcal{R}_t(g)$ is a local square integrable martingale such that

$$\langle \mathcal{R}(g) \rangle_t = \int_0^t |g_s|^2_{H_s} d\lambda_s \,.$$

In addition, we show (Proposition 11) that the range of Ito stochastic integrals, $R(\mathcal{I}(f), f \in \tilde{L}^2_{loc}(Q))$ is a dense subset of the range of normalized stochastic integrals, $R(\mathcal{R}.(f), f \in L^2_{loc}(Q))$ in the topology generated by uniform in t convergence in probability.

The linkage between the normalized integral and other extensions of Ito stochastic integral is considered in detail in Sections 3.3 and 4.1.

The normalized integrals arise naturally in many problems of stochastic analysis. Indeed, their utility is quite evident in the characterization of measures that are absolutely continuous with respect to the measure generated by a given martingale. For example,consider the pair of 1-dimensional processes:

$$\begin{cases} dX_t = a_t dt + \sigma_t dW_t \\ dM_t = \sigma_t dW_t \end{cases}$$

Then

$$\begin{aligned} dP_X/dP_M &= \exp\{\textstyle\int_0^t a_s \sigma_s^{-2} dM_s - \frac{1}{2}\int_0^t a_s^2 \sigma_s^{-2} ds\} \\ &= \exp\{\textstyle\int_0^t a_s * dM_s - \frac{1}{2}\int_0^t |a_s|^2_{H_s} ds\}, \\ &\text{where } |f_s|_{H_s} := |\sigma_s^{-2} f_s|. \end{aligned}$$

In the forthcoming paper [19] we prove that all absolutely continuous shifts of a local square integrable cylindrical martingale M_t introduced above are of the form $\int_0^t g_s d\lambda_s, g \in L^2_{loc}(Q)$, and the corresponding Radon-Nikodym derivative is given by $\exp\{\int_0^t g_s * dM_s - \frac{1}{2}\int_0^t |g_s|^2_{H_s} d\lambda_s\}$.

Another interesting example arises in the characterization of the stable subspaces of local martingales. It is well known that this problem is of central importance for the representation theorem in martingale problems (see e.g. [8]). In Section 3 (Proposition 12) we prove that the stable space of a locally square integrable continuous cylindrical martingale $\{M_t(y'), y' \in E'\}$ coincides with the set of normalized integrals

$$\mathcal{L}^1(M) = \{\mathcal{R}(f) : E[(\int_0^1 |f_s|^2_{H_s} d\lambda_s)^{1/2}] < \infty \,.$$

In Section 4 of the paper we discuss various particular cases of the normalized integral. These include Hilbert-valued stochastic integrals, stochastic integrals with respect to orthogonal martingale measures, stochastic integrals with respect to stochastic flows, etc.

In Section 5 we apply the same ideas for integration of vector valued functions with respect to martingale measures.

Our construction obviously does not cover the more difficult case of Banach space valued integrands with respect to one-dimensional Brownian motion where the geometry of Banach space is involved (see [1], [2], etc.). Also, we leave aside the complicated problem of the existence of the factorization $Q_s d\lambda_{s_x}$ in the most general case. In many particular cases this factorization is known. It was established in [15], [14] in the Hilbert space setting, in [9] for nuclear space valued square integrable martingales, in [22], [6] and [7] for orthogonal martingale measures, etc. The stochastic integral for Banach space valued square integrable martingales constructed by Métivier-Pellaumail [16] is based on an a priori estimate of simple integrals. In the Appendix we show that this estimate actually implies the existence of the factorization $Qd\lambda$.

2 Ito Stochastic Integrals

Suppose we have a probability space $(\Omega, \mathcal{F}, \mathbf{P})$ with the right-continuous filtration of σ-algebras $\mathbb{F} = (\mathcal{F}_t)_{0\le t\le 1}$. Let $\mathcal{P}(\mathbb{F})$ be the $\mathbb{F}$-predictable σ-algebra. Let E be a quasi-complete locally convex topological vector space, i.e. E is locally convex topological vector space so that all its bounded closed subsets are complete. Let E' be its topological dual. Denote by $\langle x, y\rangle$ $(x \in E',\ \ y \in E)$ the canonical bilinear form. We suppose that there exists a countable weakly dense subset of E'. Let $\mathcal{L}^+(E)$ be the space of symmetric non-negative definite forms Q from E' to E, i.e:

$$\langle y', Qy''\rangle = \langle y'', Qy'\rangle, \langle y', Qy'\rangle \ge 0 \quad \forall y', y'' \in E'$$

Definition 1. We say that a family of real valued random processes $M_t = (M_t(y'))_{y'\in E'}$ is a locally square integrable cylindrical martingale in E with covariance operator function Q_s and quadratic variation $\int_0^{\cdot} Q_s d\lambda_s$ if for each $y' \in E'$ $M_t(y') \in \mathcal{M}^2_{loc}(\mathbb{F}, \mathbf{P})$ and

$$M_t(y')M_t(y'') - \int_0^t \langle y', Q_s y''\rangle\, d\lambda_s \in \mathcal{M}_{loc}(\mathbb{F}, \mathbf{P}), \tag{3}$$

where $Q : [0,1] \times \Omega \to \mathcal{L}^+(E)$ is a $\mathcal{P}(\mathbb{F})$-measurable function (i.e. $\forall y', y'' \in E'$, $\langle y', Q_s y''\rangle$ is $\mathcal{P}(\mathbb{F})$-measurable), and λ_t is an increasing $\mathcal{P}(\mathbb{F})$-measurable process.

Here and below $\mathcal{M}_{loc}(\mathbb{F}, \mathbf{P})$ is the space of real-valued local $(\mathbb{F}, \mathbf{P})$-martingales and $\mathcal{M}^2_{loc}(\mathbb{F}, \mathbf{P})$ is the space of locally square integrable real valued $(\mathbb{F}, \mathbf{P})$- martingales.

Our next step is to construct a $\mathcal{P}(\mathbb{F})$-measurable family of Hilbert subspaces of E generated by the covariance operator of the cylindrical martingale M.

According to [21], for any $K \in \mathcal{L}^+(E)$, one can define an inner product in KE' by the formula $(Ky', Ky'')_K = \langle y', Ky'' \rangle \quad \forall y', y'' \in E'$.

The following statements hold true (see Appendix for the proofs).

Proposition 2. *(See Proposition 10 in [21]). There exists a completion H_K of KE' with respect to the inner product $(\cdot,\cdot)_K$ such that $H_K \subset E$ and the natural imbedding is continuous.*

Corollary 3. *(cf. Corollary to Proposition 7 in [21]). Let T be a countable weakly dense subset of E'. Then KT is strongly dense in H_K, i.e., H_K is a separable Hilbert space.*

Denoting $H_s = H_{Q_s}$, we can rewrite (3) as

$$M_t(y')M_t(y'') - \int_0^t (Q_s y', Q_s y'')_{H_s}\, d\lambda_s \in \mathcal{M}_{loc}(\mathbb{F}, \mathbf{P}), \tag{4}$$

Definition 4. We say that $(H_s) = (H_{Q_s})$ is the family of covariance spaces of M.

Let $L(Q)$ be the set of all vector fields $f = f_s = f(s,\omega)$ such that $f_s \in H_s$ and $(f_s, Q_s y')_{H_s}$ are $\mathcal{P}(\mathbb{F})$-measurable for each $y' \in E'$. Denote $L^2_{loc}(Q) = \{f \in L(Q) : \int_0^1 |f|^2_{H_s}\, d\lambda_s < \infty \quad \mathbf{P}\text{-a.s.}\} = L^2_{loc}(Q, \mathbf{P})$.

Let $T = \{e'_1, \ldots\}$ be a countable weakly dense subset of E'. We define a sequence of E-valued $\mathcal{P}(\mathbb{F})$-measurable functions:

$$\begin{aligned} e^1_s &= \begin{cases} Q_s e'_1/|Q_s e'_1|_{H_s}, & \text{if } Q_s e'_1 \neq 0 \\ 0, & \text{if } Q_s e'_1 = 0 \end{cases}, \\ &\ldots \\ e^{k+1}_s &= \begin{cases} \left(Q_s e'_{k+1} - \sum_{i=1}^k (Q_s e'_{k+1}, e^i_s) e^i_s\right)/d^{k+1}_s, & \text{if } d^{k+1}_s \neq 0 \\ 0, & \text{if } d^{k+1}_s = 0 \end{cases}, \\ &\ldots \end{aligned} \tag{5}$$

where $d^{k+1}_s = |Q_s e'_{k+1} - \sum_{i=1}^k (Q_s e'_{k+1}, e^i_s)_{H_s} e^i_s|_{H_s}$. It follows from the definition of the sequence (e^n_s) that for each n, there exists an E'-valued $\mathcal{P}(\mathbb{F})$-measurable function $\tilde{e}^n_s$ such that

$$e^n_s = Q_s \tilde{e}^n_s\,. \tag{6}$$

According to Corollary 3, $Q_s T$ is a dense subset of H_s, then (5) is the Hilbert-Schmidt orthogonalization procedure. This yields that for each s, the vectors (e^k_s) form a basis in H_s. Thus we arrive at the following statement.

Corollary 5. *Let $f \in L(Q)$. Then for each s, we have the expansion in $H_{Q_s} = H_s$*

$$f_s = \sum_k (f_s, e^k_s)_{H_s} e^k_s, \quad \textit{and}\ |f_s|^2_{H_s} = \sum_k (f_s, e^k_s)^2_{H_s}, \tag{7}$$

In particular, this expansion implies that $|f_s|_{H_s}$ is a predictable function.

Remark 6. Assume that **P**-a.s. for each $y \in E'$,

$$\int_0^1 \langle Q_s y, y \rangle d\lambda_s < \infty .$$

Then it follows that **P**-a.s. for each $y \in E'$, $\int_0^1 \langle Q_s y, \cdot \rangle d\lambda_s \in E'^*$, where E'^* is the algebraic dual of E'.

Remark 7. Assume that **P**-a.s. $\int_0^1 \langle Q_s y, \cdot \rangle d\lambda_s \in E$ for each $y \in E'$. Let $f \in L^2_{loc}(Q)$. Then $\int_0^t f_s d\lambda_s \in E$ **P**-a.s. for each t.

Indeed, for each $y \in E'$,

$$|\int_0^t \langle f_s, y \rangle d\lambda_s| \le (\int_0^t |f_s|^2_{H_s} d\lambda_s)^{1/2} (\int_0^1 \langle Q_s y, y \rangle d\lambda_s)^{1/2}$$

and the statement follows.

Remark 8. We remark that for a predictable increasing process A_t, $t \in [0, 1]$, (we assume $A_0 = 0$) the condition $A_1 < \infty$ **P**-a.s. is equivalent to the existence of a sequence of stopping times (τ_n) such that $\mathbf{P}(\tau_n < 1) \to 0$, and $\mathbf{E} A_{\tau_n} < \infty$ for each n (see Lemma 1.37 in [8]).

Now we can construct the Ito stochastic integral for the class $\tilde{L}^2_{loc}(Q)$ of all $\mathcal{P}(\mathbb{F})$-measurable E'-valued functions f such that

$$\int_0^1 \langle Q_s f_s, f_s \rangle \, d\lambda_s < \infty \qquad \mathbf{P}\text{-a.s.}$$

We start with the set of simple functions $S_b = \{ f \in \tilde{L}^2_{loc}(Q) : f = \sum_1^n f_s^k h_k,\ f^k$ are $\mathcal{P}(\mathbb{F})$-measurable bounded scalar functions, $h_k \in E', k = 1, \ldots, n, \quad n \ge 1 \}$. For $f = \sum_1^n f_s^k h_k \in S_b$, we define the Ito integral by

$$\mathcal{I}_t(f) = \int_0^t f_s dM_s = \sum_1^n \int_0^t f_s^k dM_s(h_k) .$$

We see immediately that the map $f \rightsquigarrow \mathcal{I}(f)$ defined on S_b (with values in $\mathcal{M}^2_{loc}(\mathbb{F}, \mathbf{P})$) is linear up to evanescence and for each $f \in S_b$,

$$\langle \mathcal{I}(f) \rangle_t = \int_0^t \langle Q_s f_s, f_s \rangle \, d\lambda_s . \tag{8}$$

Proposition 9. *The map $f \rightsquigarrow \mathcal{I}(f)$ defined on S_b has a further extension to the set $\tilde{L}^2_{loc}(Q)$ (still denoted $f \rightsquigarrow \mathcal{I}_t(f) = \int_0^t f_s dM_s$) such that:*

1. $\mathcal{I}(f) \in \mathcal{M}^2_{loc}(\mathbb{F}, \mathbf{P})$ and (8) holds;

2. $f \rightsquigarrow \mathcal{I}(f)$ is linear up to evanescence;

3. If $f^n, f \in \tilde{L}^2_{loc}(Q)$ and $\int_0^1 \langle Q_s(f_s^n - f_s), f_s^n - f_s \rangle d\lambda_s \to 0$ in probability, then $\sup_{s \le 1} |\mathcal{I}_s(f^n) - \mathcal{I}_s(f)| \to 0$ in probability, as $n \to \infty$.

Moreover, this extension is unique (up to evanescence).

Proof. 1^0. Firstly, we extend the Ito integral to

$$\tilde{L}^2(Q) = \left\{ f \in \tilde{L}^2_{\text{loc}} : E \int_0^1 \langle Q_s f_s, f_s \rangle d\lambda_s < \infty \right\}.$$

Let $S = \left\{ f \in \tilde{L}^2(Q) : f = \sum_1^n f_s^k h_k \,,\, f^k \text{ are real valued } \mathcal{P}(\mathbb{F})\text{-measurable functions}, h_k \in E',\, k = 1, \dots, n, n \geq 1 \right\}$. Fix $f_s = \sum_1^p f_s^k h_k \in S$ and define

$$g_s^n = f_s 1_{\{\max_{1\leq k\leq p} |f_s^k| \leq n\}}.$$

Obviously $g^n \in S_b \cap \tilde{L}^2(Q)$ and by the Lebesgue dominated convergence theorem

$$\mathbf{E} \int_0^1 \langle Q_s (f_s - g_s^n), f_s - g_s^n \rangle d\lambda_s \to 0, \quad \text{as } n \to \infty.$$

Now it follows from (8) that

$$\mathbf{E} \sup_t |\mathcal{I}_t(g^n) - \mathcal{I}_t(g^m)|^2 \to 0, \quad \text{as } n, m \to \infty.$$

Thus we can extend $\mathcal{I}$ to S linearly so that for each $f \in S$, $\mathcal{I}_t(f) \in \mathcal{M}^2(\mathbb{F}, P)$ and (8) holds.

Now fix $f \in \tilde{L}^2(Q)$. Then $Q_s f_s \in L^2_{\text{loc}}(Q)$ and by Corollary 5 (see (7)) $Q_s f_s = \sum_k (Q_s f_s, e_s^k)_{H_s} e_s^k$. Let $g_s^N = \sum_1^N (Q_s f_s, e_s^k)_{H_s} e_s^k$. By the Lebesgue dominated convergence theorem

$$\mathbf{E} \int_0^1 |Q_s f_s - g_s^N|^2_{H_s} d\lambda_s \to 0 \tag{9}$$

as $N \to \infty$. By the definition of g_s^N and e_s^k (see (5), (6)) it follows that there exists $f^N \in S$ such that $g_s^N = Q_s f_s^N$. Therefore we can write (9) as

$$\mathbf{E} \int_0^1 |Q_s f_s - Q_s f_s^N|^2_{H_s} d\lambda_s = \mathbf{E} \int_0^1 \langle Q_s (f_s - f_s^N), f_s - f_s^N \rangle d\lambda_s \to 0, \text{ as } N \to \infty.$$

Thus $\mathcal{I}_t(f^N)$ is a Cauchy sequence and we can find $\mathcal{I}(f) \in M^2(\mathbb{F}, \mathbf{P})$ such that (8) holds and

$$\mathbf{E} \sup_t |\mathcal{I}_t(f) - \mathcal{I}_t(f^N)|^2 \to 0, \text{ as } N \to \infty.$$

Obviously this extension is linear (up to evanescence) and unique by the Property 3.

2°. In order to extend $\mathcal{I}$ to $\tilde{L}^2_{\text{loc}}(Q)$ we apply the standard localization procedure. Fix $f \in \tilde{L}^2_{\text{loc}}(Q)$, then there exists a sequence of stopping times $\tau_m \uparrow 1$ such that $f_s 1_{\{s \leq \tau_m\}} \in \tilde{L}^2(Q)$ for each m and we can find $\mathcal{I}(f) \in \mathcal{M}^2_{\text{loc}}(\mathbb{F}, \mathbf{P})$ such that $\mathcal{I}_{t\wedge\tau_m}(f) = \mathcal{I}_t\left(f 1_{\{\cdot \leq \tau_m\}}\right)$. Properties 1,2 of the extension are obvious.

3°. Finally, we prove that property 3 holds for $\mathcal{I}(f), f \in \tilde{L}^2_{\text{loc}}(Q)$. Let $\int_0^1 \langle Q_s(f^n_s - f_s), f^n_s - f_s \rangle d\lambda_s \to 0$ in probability as $n \to \infty$. There exists a sequence (τ_n) of stopping times such that

$$\mathbf{E} \int_0^{\tau_n} \langle Q_s(f^n_s - f_s), f^n_s - f_s \rangle d\lambda_s + P(\tau_n < 1) \to 0$$

as $n \to \infty$, and for each n $\mathbf{E} \int_0^{\tau_n} \langle Q_s f_s, f_s \rangle d\lambda_s < \infty$. Thus $\mathbf{E} \sup_t |\mathcal{I}_{t \wedge \tau_n}(f_n) - \mathcal{I}_{t \wedge \tau_n}(f)|^2 \to 0$, as $n \to \infty$. Since $P(\tau_n < 1) \overset{n \to \infty}{\to} 0$, we derive easily that $\sup_t |\mathcal{I}_t(f_n) - \mathcal{I}_t(f)| \to 0$ in probability as $n \to \infty$. Then the statement follows.

3 Normalized and Ito Stochastic Integrals

3.1 Normalized stochastic integrals

If $f^n \in \tilde{L}^2_{loc}(Q)$ and $\int_0^1 \langle Q_s(f^n_s - f^m_s,), f^m_s - f^n_s \rangle \, d\lambda_s \to 0$ in probability as $n, m \to \infty$, then there exists $\mathcal{I}_t \in \mathcal{M}^2_{loc}(\mathbb{F}, P)$ such that $\sup_t |\mathcal{I}_t - \mathcal{I}_t(f_n)| \to 0$ in probability, as $n \to \infty$. In order to describe $\mathcal{I}_t$ we need to complete $\tilde{L}^2_{loc}(Q)$.

Let $\tilde{\mathcal{O}} = \left\{ f \in \tilde{L}^2_{loc}(Q) : \int_0^1 \langle Q_s f_s, f_s \rangle \, d\lambda_s = 0 \quad P\text{-a.s.} \right\}$ and $\tilde{\mathcal{L}}^2_{loc}(Q) = \tilde{L}^2_{loc}(Q)/\tilde{\mathcal{O}}$. For $f \in \tilde{L}^2_{loc}(Q)$ we denote $\hat{f} = f + \tilde{\mathcal{O}}$ and define the distance of the convergence in probability.

$$\tilde{d}(\hat{f}, \hat{g}) = E[\int_0^1 \langle Q_s(f_s - g_s), f_s - g_s \rangle \, d\lambda_s^{1/2} \wedge 1] .$$

Let $\mathcal{O} = \left\{ f \in L^2_{loc}(Q) : \int_0^1 |f_s|^2_{H_s} \, d\lambda_s = 0 \quad P\text{-a.s.} \right\}$, $\mathcal{L}^2_{loc}(Q) = L^2_{loc}(Q)/\mathcal{O}$. For $f \in L^2_{loc}(Q)$ we denote $\hat{f} = f + \mathcal{O}$ and define the distance

$$d(\hat{f}, \hat{g}) = E[\int_0^1 |f_s - g_s|^2_{H_s} d\lambda_s^{1/2} \wedge 1], f, g \in L^2_{loc}(Q) .$$

It is easy to see that these definitions do not depend on the particular representative of the equivalence class and $\mathcal{L}_{loc}(Q)$ is a complete metric space.

Proposition 10. *(see [18, 19]) The map $\hat{\mathcal{G}} : \hat{f}_s \rightsquigarrow \widehat{Q_s f_s}$ is an isometric imbedding of $\tilde{\mathcal{L}}^2_{loc}(Q)$ into $\mathcal{L}^2_{loc}(Q)$ and $\hat{\mathcal{G}}(\tilde{\mathcal{L}}^2_{loc}(Q))$ is a dense subset of $\mathcal{L}^2_{loc}(Q)$, i.e., $\mathcal{L}^2_{loc}(Q)$ is the completion of $\tilde{\mathcal{L}}^2_{loc}(Q)$.*

Proof. For each $y \in E'$, $\langle Q_s y, y \rangle = 0$ if and only if $Q_s y = 0$ and the first part of the statement follows from the definitions.

Let $f_s \in L^2_{loc}(Q)$, $f^N_s = \sum_1^N (f_s, e^k_s) e^k_s$. Then by Corollary 5

$$\int_0^1 |f_s - f^N_s|^2_{H_s} d\lambda_s \to 0, \qquad \text{as } N \to \infty, \qquad \mathbf{P}\text{-a.s.} \tag{10}$$

From the definition of (e_s^k) (see (5), (6)) it follows that there exists a sequence $\tilde{f}^N \in \tilde{L}^2_{loc}(Q)$ such that $f_s^N = Q_s \tilde{f}_s^N$. We can rewrite (10) as

$$\int_0^1 |f_s - Q_s \tilde{f}_s^N|^2_{H_s} d\lambda_s \to 0, \qquad \text{as } N \to \infty, \qquad \mathbf{P}\text{-a.s.}$$

Now the second part of the statement follows.

Let $\mathcal{G}$ be the map $f_s \rightsquigarrow Q_s f_s$ from $\tilde{L}^2_{loc}(Q)$ to $L^2_{loc}(Q)$. If $f, g \in \tilde{L}^2_{loc}(Q)$ and $f - g \in \tilde{\mathcal{O}}$, we have $\mathcal{T}(f) = \mathcal{T}(g)$. Thus according to Propositions 9 and 10, we can define the stochastic integral on $\mathcal{G}(\tilde{L}^2_{loc}(Q)) \subset L^2_{loc}(Q)$ by

$$\mathcal{R}_t(\hat{g}) = \mathcal{R}_t(g) = \int_0^t g_s * dM_s = \int_0^t f_s dM_s = \mathcal{I}_t(f) = \mathcal{I}_t(\hat{f}), \tag{11}$$

where $g_s = Q_s f_s$, $f_s \in \tilde{L}^2_{loc}(Q)$. Obviously $\langle \mathcal{R}(g) \rangle_t = \int_0^t |g_s|^2_{H_s} d\lambda_s$.

Proposition 11. *(see [18, 19])The map $f \rightsquigarrow \mathcal{R}(f)$ defined on $\mathcal{G}(\tilde{L}^2_{loc}(Q))$ has a unique extension to the set $L^2_{loc}(Q)$, still denoted $f \rightsquigarrow \mathcal{R}_t(f) = \int_0^t f_s * dM_s$, with these properties:*

1. $\mathcal{R}(f) \in \mathcal{M}^2_{loc}(\mathbb{F}, \mathbf{P})$, $\langle \mathcal{R}(f) \rangle_t = \int_0^t |f_s|^2_{H_s} d\lambda_s$;

2. $\mathcal{R}(f)$ is linear up to evanescence;

3. If $f_n, f \in L^2_{loc}(Q)$ and $\int_0^1 |f_s^n - f_s|^2_{H_s} d\lambda_s \to \infty$ in probability, as $n \to \infty$, then $\sup_{s \le 1} |\mathcal{R}_s(f_n) - \mathcal{R}_s(f)| \to 0$ in probability, as $n \to \infty$.

Proof. 1°. Let $f \in L^2_{loc}(Q)$, $f^N = \sum_1^N (f_s, e_s^k)_{H_s} e_s^k$. By the definition of (e_s^k) (see (5), (6)), it follows that $f^N \in \mathcal{G}(\tilde{L}^2_{loc}(Q))$. By Corollary 5, we have P-a.e.

$$\int_0^1 |f_s - f_s^N|^2_{H_s} d\lambda_s \to 0, \quad \text{as } N \to \infty.$$

Thus there exists increasing sequences of stopping times $(\tau_{N,p}), (\tau_p)$ such that $\tau_{N,p} \le \tau_p \le 1$ for each N, p, and

$$\mathbf{P}(\tau_p < 1) \overset{p\to\infty}{\longrightarrow} 0, \quad \mathbf{P}(\tau_{N,p} < \tau_p) \overset{n\to\infty}{\longrightarrow} 0,$$

$$\mathbf{E} \int_0^{\tau_p} |f_s|^2_{H_0} d\lambda_s < \infty, \quad \int_0^{\tau_{N,p}} |f_s^N|^2_{H_s} d\lambda_s \le \int_0^{\tau_p} |f_s|^2 d\lambda_s + 1.$$

Then for each p,

$$\mathbf{E} \sup_t |\mathcal{T}_{t \wedge \tau_{N,p}}(f^N) - \mathcal{T}_{t \wedge \tau_{M,p}}(f^M)|^2 \to 0,$$

as $N, M \to \infty$. Thus the existence of an extension satisfying 1,2 follows immediately.

2°. Now we prove Property 3 of the extension. Let $f^n, f \in L^2_{loc}(Q)$ and

$$\int_0^1 |f_s^n - f_s|^2_{H_s} d\lambda_s \to 0, \quad \text{as} \quad n \to \infty.$$

Then there exists a sequence of stopping times (τ_n) such that $\mathbf{P}(\tau_n < 1) + \mathbf{E}\int_0^{\tau_n} |f_s^n - f_s|_{H_s}^2 d\lambda_s \overset{n\to\infty}{\longrightarrow} 0$.

Hence,

$$\mathbf{E}\sup_t |\mathcal{R}_{t\wedge\tau_n}(f^n) - \mathcal{R}_{t\wedge\tau_n}(f)|^2 \leq$$
$$\leq C\mathbf{E}\int_0^{\tau_n} |f_s^n - f_s|_{H_s}^2 d\lambda_s \to 0, \text{ in probability, as } n \to \infty .$$

Thus the property 3 holds for the extension which is obviously unique.

For the martingale representation theorem, it is important to describe the stable subspace generated by $M(y)$, $y \in E'$. According to the definition (see [8]), this is the smallest subspace, $\mathcal{L}^1(M)$, of the closure of $H^1 = \{M \in \mathcal{M}_{loc}(\mathbb{F}, P) : |M|_1 = \mathbf{E}\sup_t |M_t| < \infty\}$ with respect to the norm $|\cdot|_1$ that contains all the integrals $\int_0^t h_s\, dM_s(y)$ where h_s is a real valued predictable function such that

$$\mathbf{E}[\int_0^1 h_s^2\, d < M(y) >_s]^{1/2} < \infty, y \in E'.$$

Proposition 12. *Let $M(y) \in \mathcal{M}_{loc}^c(\mathbb{F}, \mathbf{P})$ for each $y \in E$. Then the stable subspace of H^1generated by $M(y)$ is*

$$\mathcal{L}^1(M) = \Big\{\mathcal{R}_t(f) = \int_0^t f_s * dM_s : \quad f \in L_{loc}^2(Q)$$
$$\text{and} \quad \mathbf{E}\left[\left(\int_0^1 |f_s|_{H_s}^2 d\lambda_s\right)^{1/2}\right] < \infty\Big\} .$$

Proof. From Burkholder's inequality (see [8]) it follows that $\mathcal{L}^1(M)$ is a closed subspace of H^1. Now the statement follows by the definition of the basis (e^i) and the normalized integrals $\mathcal{R}(f)$.

3.2 Linear transformations of integrands and covariance spaces

Let F be a quasicomplete locally convex topological vector space and F' its topological dual. Denote $\mathcal{L}_w(E, F)$ the set of weakly continuous linear forms from E to F. Let $L_{loc}^2(Q, \mathcal{L}_w(E, F))$ be the set of all predictable $\mathcal{L}_w(E, F)$-valued functions u_s such that for each $f' \in F$,

$$\int_0^1 \langle u_s' f', Q_s u_s' f'\rangle d\lambda_s < \infty \quad \mathbf{P}\text{-a.s.}$$

where $u_s' : F' \to E'$ is an adjoint linear form.

Definition 13. We define the stochastic integral $\int_0^t u_s' dM_s$ as the cylindrical locally square integrable martingale $\bar{M}_t = (\bar{M}_t(f'))$ in F such that

$$\bar{M}_t(f') = \int_0^t u_s' f' dM_s, f' \in F' .$$

Remark 14. It follows immediately by the definition that the covariance operator function of $\bar{M}$

$$\bar{Q}_s = u_s Q_s u'_s .$$

Obviously, $f'_s \in \tilde{L}^2_{loc}(\bar{Q})$ if and only if $u'_s f'_s \in \tilde{L}^2_{loc}(Q)$.

Let E_1 be a quasicomplete locally convex topological vector space and E'_1 be its topological dual with weakly dense countable subset. Let u be a weakly continuous linear form from E_1 to E, (i.e., $u \in \mathcal{L}_w(E_1, E)$), $K \in \mathcal{L}^+(E_1)$, $H = H_K \subset E_1$. We define a Hilbert structure on $u(H) \subset E$ by (see [21])

$$|f|_G = \inf_{u(y)=f} |y|_H .$$

We shall need the following statement from [21] (see Appendix for the proof).

Proposition 15. *(see [21], Proposition 21).*
1. The set $Ku'E'$ is a dense subset of the orthogonal complement $\mathcal{K}$ to $\mathcal{N} = u^{-1}(0) = Ker\ u$ in H and u is an isometry between $\mathcal{K}$ and $u(H)$;
2. $u(H) = H_{\bar{K}}$, where $\bar{K} = uKu' \in \mathcal{L}^+(E)$.

Corollary 16. *For each $y \in H$, $|u(y)|_{u(H)} \leq |y|_H$.*

Proof. The statement follows obviously from part 1) of Proposition 15.

These statements can be generalized a little. Consider a finite number of quasicomplete locally convex topological vector spaces E_i, $(i = 1, \ldots, N)$ with topological duals E'_i having weakly dense countable subsets. Let $K^i \in \mathcal{L}^+(E_i)$, $H^i = H_{K^i} \subset E_i$, $u_i \in \mathcal{L}_w(E_i, E)$. We define a Hilbert structure on $G = \sum_1^N u_i(H^i)$ by (see [21])

$$|f|^2_G = \inf_{f=\sum_1^N u_i(y_i)} \sum_1^N |y_i|^2_{H^i} .$$

This setting can be reduced to the previous one by setting $E = E_1 \oplus \ldots \oplus E_N$, $H = H^1 \oplus \ldots \oplus H^N$, and $u(y_1 \oplus \ldots \oplus y_N) = u_1(y_1) + \ldots + u_N(y_N)$.
Hence we obtained the following result.

Corollary 17. *1. $G = H_{\bar{K}}$ where $\bar{K} = \sum_1^N u^i K^i u^{i\prime}$.*
2. $|u_1(y_1) + \ldots + u_N(y_N)|^2_G \leq |y_1|^2_{H^1} + \ldots + |y_N|^2_{H^N}$.

Let Q^1 be a predictable $\mathcal{L}^+(E_1)$-valued function and u_s be a predictable $\mathcal{L}_w(E_1, E)$-valued function. Denote $H^1_s = H_{Q^1_s}$, $H_s = H_{Q_s}$.

Proposition 18. *Let $Q_s = u_s Q^1_s u'_s$. Then*
a) $H_s = u_s(H^1_s)$;
b) $f_s \in L^2(Q)$ if and only if there exists $g_s \in L^2(Q^1)$ such that $f_s = u_s(g_s)$.

Proof. Part a) follows immediately from part 2) of Proposition 15. Since by Corollary 5 $|f_s|_{H_s} \le |g_s|$, one of the implications in b) is obvious. Assume now that $f_s \in L^2(Q)$. Let $f_s^n = \sum_1^n (f_s, e_s^k)_{H_s} e_s^k$. Then $\int_0^1 |f_s - f_s^n|^2_{H_s} d\lambda_s \overset{n\to\infty}{\to} 0$. By the definition of e_s^k, we see that $f_s^n = Q_s \bar{f}_s^n$ for some predictable E'-valued function $\bar{f}^n$. Thus $\int_0^1 |Q_s \bar{f}_s^n - f_s|^2_{H_s} d\lambda_s \overset{n\to\infty}{\longrightarrow} 0$. Let $g_s^n := Q_s^1 u_s' \bar{f}_s^n$. Then $Q_s \bar{f}_s^n = u_s(g_s^n)$. By Proposition 15, g_s^n takes values in the orthogonal complement of $u_s^{-1}\{0\}$, and

$$|Q_s \bar{f}_s^n|^2_{H_s} = |g_s^n|^2_{H_s^1}, \; |Q_s \bar{f}_s^n - Q_s \bar{f}_s^m|^2_{H_s} = |g_s^n - g_s^m|^2_{H_s^1} \overset{m,n\to\infty}{\longrightarrow} 0 .$$

This completes the proof.

It is readily checked that Corollary 17 and Proposition 15 yield the following statement.

Corollary 19. *Let $E_1, \ldots, E_n$ be quasicomplete topological vector spaces with topological duals $E'_1, \ldots E'_N$, respectively, having weakly dense countable subsets. Let $Q_s = \sum_1^N u_s^i Q_s^i u_s^{i\prime}$ for some predictable $\mathcal{L}^+(E_i)$-valued functions Q_s^i and some predictable $\mathcal{L}_w(E_i, E)$- valued functions u^i. Then*

a) $H_s = \sum_1^N u_s^i(H_s^i)$ $(H_s = H_{Q_s}, H_s^i = H_{Q_s^i})$,

b) $f_s \in L^2(Q)$ if and only if there exists $g_s^i \in L^2(Q^i)$ such that $f_s = \sum_1^N u_s^i(g_s^i)$.

Remark 20. By Corollary 17,

$$|f_s|^2_{H_s} \le \sum_1^N |g_s^i|^2_{H_s^i}, \text{ if } f_s = u_s^i(g_s^i) .$$

Let $M^i (i = 1, \ldots, N)$ be cylindrical locally square integrable martingales in E_i with covariance operator functions Q_s^i and quadratic variations $\int_0^\cdot Q_s^i d\lambda_s$. Assume that for each $y_i' \in E_i', y_j' \in E_j'$,

$$\langle M^i(y_i'), M^j(y_j') \rangle = 0, \text{ if } i \ne j .$$

Proposition 21. *Let $u^i \in L^2_{loc}(Q^i, \mathcal{L}_w(E_i, E))$. Then $M_t = \sum_1^N \int_0^t u_s^{i\prime} dM_s^i$ is a cylindrical locally square integrable martingale in E with covariance operator function $Q_s = \sum_1^N u_s^i Q_s^i u_s^{i\prime}$ and quadratic variation $\int_0^\cdot Q_s d\lambda_s$.*

Proof. By the definition for each $f' \in E'$ $M_t(f') = \sum_1^N \int_0^t u_s^{i\prime} f' dM_s^i$. By our assumption

$$\langle \int_0^1 u_s^{i\prime} f' dM_s^i, \int_0^i u_s^{j\prime} g' dM_s^j \rangle = 0,$$

if $i \ne j$, for each $f', g' \in E'$. Now the statement follows.

3.3 Linkage between normalized and Ito integrals

Now we shall discuss the relation of the normalized and Ito integrals. Let $T = \{e'_1, e'_2, \ldots\}$ be a countable weakly dense subset of E'. For any $K \in \mathcal{L}^+(E)$, using the Hilbert- Schmidt orthogonalization procedure, we obtain an orthogonal basis (e^n) in H_K :

$$e^1 = \begin{cases} Ke'_1/|Ke'_1|_{H_K}, \text{ if } |Ke'_1|_{H_K} \neq 0, \\ 0, \text{ otherwise,} \end{cases}$$

$$\ldots$$

$$e^{k+1} = \begin{cases} (Ke'_{k+1} - \sum_{i=1}^{k}(Ke'_{k+1}, e^i)_{H_K} e^i)/d^{k+1}, \text{ if } d^{k+1} \neq 0, \\ 0, \text{ otherwise,} \end{cases}$$

$$\ldots$$

where $d^{k+1} = |Ke'_{k+1} - \sum_{i=1}^{k}(Ke'_{k+1}, e^i)_{H_K} e^i|_{H_K}$.

Remark 22. From the definition of (e^k), it follows that for each n there exists $\tilde{e}^k \in E'$ such that

$$e^n = K\tilde{e}^n, \; n = 1, 2, \ldots \tag{12}$$

Lemma 23. *a) If $h \in H_K$, then there exists a unique $F \in H'_K$ such that $h = F \circ K$ (here $F \circ K(e') = F(Ke')$, $e' \in E'$);*

b) Let $F \in H'_K$. Then $F \circ K \in H_K$ and

$$|F \circ K|^2_{H_K} = \sup_{e'} |F \circ K(e')|^2 / \langle Ke', e' \rangle = \textstyle\sum_n F \circ K(\tilde{e}^n)^2 = |F|^2_{H'_K},$$
$$(F \circ K, Ke')_{H_K} = F \circ K(e') \text{ for all } e' \in E'.$$

Proof. a) If $h \in H_K$, then for all $e' \in E'$,

$$\langle h, e' \rangle = (h, Ke')_{H_K}. \tag{13}$$

Define $F(Ke') = \langle h, e' \rangle$. Since (13) holds, $F \in (KE')^*$ and $h = F \circ K$.

b) Let $F \in H'_K$. Then by Riesz theorem, there exists $h \in H_K$ such that for all $e' \in E'$,

$$F(Ke') = F \circ K(e') = (h, Ke')_{H_K} = \langle h, e' \rangle .$$

Thus $h = F \circ K$ and

$$|h|^2_{H_K} = \sup_{e'} |F \circ K(e')|^2 / \langle Ke', e' \rangle = \sum_n F \circ K(\tilde{e}^n)^2 < +\infty.$$

The statement is proved.

Remark 24. Since the imbedding of H_K into E is continuous, a continuous form on E is continuous on H_K, i.e. $E' \subset H'_K$.

Corollary 25. *a) If $f' \in E'$, we have $f' \circ K = Kf'$ and $|f'|_{H'_K} = |Kf'|_{H_K}$. Also,*

$$f' \circ K(e') = \langle f', Ke' \rangle_{E',E} = \langle e', Kf' \rangle_{E',E} = (Ke', Kf')_{H_K} \qquad (14)$$
$$= \langle f', Ke' \rangle_{H'_K, H_K}.$$

b) the sequence $(\tilde{e}^n)$ defined by (12) is an orthogonal basis of H'_K, i.e. for $F \in H'_K$ we have an expansion $F = \sum_n F \circ K(\tilde{e}^n)\tilde{e}^n$ in H'_K, and E' is a dense subset of H'_K.

c) the kernel K on E' can be continuously extended from E' to H'_K, and it defines a canonical isometry from H'_K onto H_K. Also, for all $F \in H'_K$, $e' \in E'$, and $h \in H_K$,

$$\langle e', KF \rangle_{E',E} = \langle F, Ke' \rangle_{H'_K, H_K} = (KF, Ke')_{H_K} = F \circ K(e'), \qquad (15)$$
$$(KF, h)_{H_K} = \langle F, h \rangle_{H'_K, H_K}, \langle F, KF \rangle_{H'_K, H_K} = |KF|^2_{H_K} = |F|^2_{H'_K}.$$

Proof. a) Indeed, for each $e' \in E'$ we have

$$f' \circ K(e') = \langle f', Ke' \rangle = \langle e', Kf' \rangle.$$

Thus $f' \circ K = Kf'$ and obviously $|f'|_{H'_K} = |Kf'|_{H_K}$.

b) By Lemma 23 for each $e' \in E'$, we have $(F \circ K, Ke')_{H_K} = F \circ K(e')$. Therefore

$$F \circ K = \sum_n (F \circ K, K\tilde{e}^n)_{H_K} K\tilde{e}^n = \sum_n F \circ K(\tilde{e}^n)\, \tilde{e}^n \circ K,$$

i.e. $F = \sum_n F \circ K(\tilde{e}^n)\tilde{e}^n$ in H'_K. Since $\tilde{e}^n \in E'$, it follows obviously that E' is dense in H'_K.

c) By (14),

$$\langle e', Kf' \rangle_{E',E} = (Ke', Kf')_{H_K} = \langle f', Ke' \rangle_{H'_K, H_K}.$$

By a), a sequence $f'_n \in E'$ is a Cauchy sequence in H'_K if and only if Kf'_n is Cauchy in H_K. If $f'_n \to F$ in H'_K then (Kf'_n) converges to an element $g \in H_K$. We denote $g = KF$ and obtain (15) by continuity.

Remark 26. We note that the set of restrictions $E'|_{H_K} = \{e'|_{H_K} : e' \in E'\}$ is isomorphic to $E'/\text{Ker } K$ and $K : E'|_{H_K} \to KE'$ is an algebraic isomorphism. It follows from the Corollary 25 that K can be continuously extended to a canonical isomorphism from H'_K onto H_K.

Let $\hat{L}(Q)$ be the set of all functions $F_s = F_s(\omega)$ such that $F_s \in H'_{Q_s}$ and $F_s \circ Q_s(e') = \langle e', Q_s F_s \rangle$ is $\mathcal{P}(\mathbb{F})$-measurable for all $e' \in E'$. Define

$$\hat{L}^2_{loc}(Q) = \{F \in \hat{L}(Q) : \int_0^1 |F_s|^2_{H'_s}\, d\lambda_s < \infty \ \mathbf{P}\text{-a.s.}\}.$$

Lemma 27. *The inclusion $\tilde{L}^2_{loc}(Q) \subset \hat{L}^2_{loc}(Q)$ holds, and for each $F \in \hat{L}^2_{loc}(Q)$, there exists a sequence $F^n \in \tilde{L}^2_{loc}(Q)$ such that*

$$\int_0^1 |F^n_s - F_s|^2_{H'_s} \, d\lambda_s \to 0$$

in probability, as $n \to \infty$.

Proof. Obviously, $\tilde{L}^2_{loc}(Q) \subset \hat{L}^2_{loc}(Q)$ by Corollary 25. Let (e^n_s) and $(\tilde{e}^n_s)$ be sequences defined by (5) and (6), i.e. (e^n_s) is a basis in $H_s = H_{Q_s}$, and $(\tilde{e}^n_s)$ is a sequence of E'-valued predictable processes such that $Q_s \tilde{e}^n_s = e^n_s$. Let

$$F^n_s = \sum_{k=1}^n F \circ Q_s(\tilde{e}^k_s)\, \tilde{e}^k_s \, .$$

Then the statement follows immediately by Corollary 25.

Now we show that there exists a natural extension of the Ito integral related to the normalized integral.

Proposition 28. *The map $F \rightsquigarrow \mathcal{I}_t(F)$ defined on $\tilde{L}^2_{loc}(Q)$ has a further extension to the set $\hat{L}^2_{loc}(Q)$ (still denoted $F \rightsquigarrow \mathcal{I}_t(F) = \int_0^t F_s dM_s$) with these properties:*

1. $\mathcal{I}(F) \in \mathcal{M}^2_{loc}(\mathbb{F}, \mathbf{P})$ and $\langle \mathcal{I}(F) \rangle_t = \int_0^t |F_s|^2_{H'_s} \, d\lambda_s$ holds;

2. $F \rightsquigarrow \mathcal{I}_t(F)$ is linear up to evanescence;

3. If $F^n, F \in \hat{L}^2_{loc}(Q)$ and $\int_0^1 |F^n_s - F_s|^2_{H'_s} \, d\lambda_s \to 0$ in probability, then $\sup_{s \le 1} |\mathcal{I}_s(F^n) - \mathcal{I}_s(F)| \to 0$ in probability, as $n \to \infty$.

Moreover, this extension is unique (up to evanescence), and for all $F \in \hat{L}^2_{loc}(Q)$, $g \in L^2_{loc}(Q)$,

$$\int_0^t F_s dM_s = \int_0^t (F_s \circ Q_s) * dM_s = \int_0^t Q_s F_s * dM_s ,$$

$$\int_0^t g_s * dMs = \int_0^t G_s dM_s ,$$

where $G_s \circ Q_s = g_s$ (recall that if $F \in \tilde{L}^2_{loc}(Q)$, then $F_s \circ Q_s = Q_s F_s$ and the first equality is simply (11)).

Proof. We claim that the extension with the properties specified above is given by

$$\mathcal{I}_t(F) = \int_0^t F_s \circ Q_s * dM_s .$$

This fact is an obvious consequence of the isometry of H'_s and H_s induced by Q_s (see Corollary 25, (c), definition of the normalized integral and Lemma 27).

Example 29. Let Y be a separable Hilbert space. Consider a Y-valued locally square integrable martingale with a deterministic covariance operator Q and $\lambda_t = t$. In this case there exists a CONS (e_k) of eigenvectors of Q, and

$$Q = \sum_k \lambda_k (e_k, \cdot)_Y \, e_k \text{ (see [14]).}$$

Then $H_Q = Q^{1/2}H$ is the set of all $h = \sum_k \lambda_k^{1/2} h_k e_k$ such that $\sum_k h_k^2 < \infty$. The dual space H'_Q is the set of all $g = \sum_{\lambda_k>0} \lambda_k^{-1/2} g_k \, e_k$ such that $\sum_k g_k^2 < \infty$. We have an obvious duality here:

$$g(h) = \sum_{\lambda_k>0} h_k g_k$$

where $h = \sum_k \lambda_k^{1/2} h_k e_k \in H_Q$, $g = \sum_{\lambda_k>0} \lambda_k^{-1/2} g_k e_k \in H'_Q$.

Let $g^n = \sum_{\lambda_k>0,\, k\le n} \lambda_k^{-1/2} g_k e_k$. Then

$$\langle Q(g^n - g^m), g^n - g^m \rangle = \sum_{\lambda_k>0,\, m<k\le n} g_k^2 \to 0,$$

as $n, m \to \infty$. It was explained above that in this case $g \circ Q = Qg = \sum_k \lambda_k^{1/2} g_k \in H_Q$.

Remark 30. A stochastic integral with respect to a Hilbert space-valued martingale with $\lambda_t = t$, and $Q = \sum_k 2^{-k} (e_k, \cdot)_Y \, e_k$ was constructed in [16] (p. 171).It was shown in there that the unbounded deterministic linear operator $g = \sum_k k e_k$ is an admissible integrand.

Obviously the aforementioned integral fits into the setting of Example 29 In particular, it follows from Example 29 that in the case of [16] (p. 171) the set of all deterministic integrands coincides with

$$H'_Q = \{g = \sum_k 2^{k/2} g_k e_k : \sum_k g_k^2 < \infty\}.$$

4 Some Particular Cases

4.1 Case when E' is a subspace of E.

In this subsection it is assumed that we are given a symmetric injection $I : E' \to E$. So we identify E' by I to a subspace of E and call I an identity. Obviously I is weakly continuous. Since I is injective, then $I' = I$ has a dense image, i.e. in this case E' is necessarily a dense subset of E . We think here most frequently about the example $E = \mathcal{D}'(X)$, the space of distributions on an open set $X \subset \mathbf{R}^n$, $E' = \mathcal{D}(X)$, the space of test functions on X.

Remark 31. If E is an arbitrary quasicomplete locally convex space, $K \in \mathcal{L}^+(E)$, and $H = H_K$ is a dense Hilbert subspace of E, then K is an injection and can play the role of I.

Also, in this case $f' \circ K = Kf'$ $(\langle Kf', e'\rangle = \langle Ke', f'\rangle)$.

We introduce now an important class of Hilbert subspaces of E.

Definition 32. We say that a Hilbert subspace $H \subset E$ with weakly continuous injection $H \to E$ is *normal* if E' is a dense subspace of H.

Remark 33. If $H \subset E$ is a normal Hilbert subspace, then H' is a normal Hilbert subspace of E.

Proof. Indeed, if H is normal we have weakly continuous dense injections $E' \xrightarrow{i} H \xrightarrow{j} E$. Passing to the adjoints we get weakly continuous dense injections $E' \xrightarrow{j'} H' \xrightarrow{i'} E$. Thus we identify H' with a subspace of E which is normal as well: for each e', $f' \in E'$.

$$\langle e', f'\rangle_{H,H'} = \langle e', f'\rangle_{E,E'} \,.$$

One can say here (see [13]) that H' is identified with the subspace of elements $e \in E$ such that the linear form $e' \to \langle e, e'\rangle_{E,E'}$ is continuous on E' with respect to the topology induced by H.

Example 34. Let $X \subset \mathbf{R}^n$ be a bounded open subset. Let H^s be a completion of $E' = \mathcal{D}(X)$ with respect to the norm

$$|\varphi|_s^2 = \int_X (-\Delta)^s \varphi(x)\varphi(x)dx, s \geq 0 \,.$$

Define $H^{-s} = (H^s)', s \geq 0$. All the $H^s, s \in \mathbf{R}$, are normal subspaces of $E = \mathcal{D}'(X)$. Obviously $H^{-s} = H_{(-\Delta)^{-s}}$ and $(-\Delta)^{-s} : (H^s)' = H^{-s} \to H^s$ is a canonical isomorphism between $(H^s)'$ and H^s.

The following statement is true (Propositions 28 bis, 29 in [21], see Appendix for the proof).

Proposition 35. *For $K \in \mathcal{L}^+(E)$, assume that $H = H_K$ is a normal Hilbert subspace and $H' = H_{\hat{K}}$ is its dual. Then*

a) we can extend the E'- restrictions of K and $\hat{K}$ to the continuous linear forms $K : H' \to H$, $\hat{K} : H \to H'$ which are canonical isomorphisms between H and H', i.e., $\hat{K} = K^{-1}$, $KH' = H$, $K^{-1}H = H'$;

b) for each e', $f' \in E'$

$$(e', f')_{H'} = (Ke', Kf')_H = \langle Ke', f'\rangle_{E,E'} = \langle Kf', e'\rangle_{E,E'}$$
$$(e', f')_H = (\hat{K}e', \hat{K}f')_{H'} = \langle \hat{K}e', f'\rangle_{E,E'} = \langle \hat{K}f', e'\rangle_{E,E'} .$$

Remark 36. Let $H = H_K$ be a normal Hilbert subspace. If $F \in H'_K$, then $F \circ K = KF \in H_K$; if $h \in H_K$, then $K^{-1}h \in H'_K$.

Denote $\mathcal{L}_n^+(E) = \{K \in \mathcal{L}^+(E) : H_K \text{ is a normal subspace}\}$. By Proposition 28 and the previous remark we get the following statement.

Proposition 37. *Let Q be a predictable $\mathcal{L}_n^+(E)$-valued function. In this case $F_s \in \hat{L}^2_{loc}(Q)$ if and only if $Q_s F_s \in L^2_{loc}(Q)$. Also, $h_s \in L^2_{loc}(Q)$ if and only if $Q_s^{-1} h_s \in \hat{L}^2_{loc}(Q)$,and for each $f \in \hat{L}^2_{loc}(Q)$, $g \in L^2_{loc}(Q)$,*

$$\int_0^t f_s dM_s = \int_0^t Q_s f_s * dM_s, \quad \int_0^t g_s * dMs = \int_0^t Q_s^{-1} g_s dM_s.$$

Now we shall discuss briefly some criterions of normality.

Proposition 38. *(see [21], p. 215). Assume that we can extend $K \in \mathcal{L}^+(E)$ to a weakly continuous linear form $K : E \to E$ and $E' \subset H \subset H_K$. Then H is normal.*

Proof. Since $K = K'$, we have $KE' \subset E' \subset H = H_K$ and HE' is dense in H. Then E' is dense in H_K and the statement follows.

Corollary 39. *Let $E_i (i = 1, \ldots, N)$ be quasi-complete locally convex topological vector spaces, let $u^i : E_i \to E$ be weakly continuous linear forms, $K^i \in \mathcal{L}_n^+(E_i)$ and $K = \sum_1^N u^i k^i u^{i\prime}$. Assume that $u^{i\prime}$ is extendable to a weakly continuous linear form $u^{i\prime} : E \to H^i = H_{K_i}$, and there exists i such that $E' \subset u^i(H^i)$. Then $K \in \mathcal{L}_n^+(E)$.*

Proof. By the assumptions K is extendable to a weakly continuous linear form $K : E \to E$ and $E' \subset \sum_1^N u^i(H^i) = H_K$. The statement follows by Proposition 38.

4.2 Orthogonal martingale measures.

Let $(U, \mathcal{B}(U))$ be a countable measurable space (i.e., $\mathcal{B}(U)$ is generated by a countable subset of $\mathcal{B}(U)$). Let $\mathcal{A}$ be an algebra generating $\mathcal{B}(U)$ and for each $A \in \mathcal{A}$ we have $M_t(A) \in \mathcal{M}^2_{loc}(\mathbb{F}, P)$. Suppose that there exists an increasing $\mathcal{P}(\mathbb{F})$-measurable process λ and $\mathcal{P}(\mathbb{F})$-measurable family $q_s(dx)$of non-negative measures on $(U, \mathcal{B}(U))$ such that

$$M_t(A)M_t(B) - \int_0^t \int_{A\cap B} q_s(dx)\, d\lambda_s \in \mathcal{M}_{loc}(\mathbb{F}, \mathbf{P}) \quad \forall A, B \in \mathcal{A}.$$

Suppose that there exists an increasing sequence of $\mathcal{B}(U)$-measurable sets (U_n) such that $U_n \uparrow U$ and

$$\int_0^t \int_{U_n} q_s(dx)\, d\lambda_s < \infty \quad \mathbf{P}\text{-a.s.} \quad \forall n\ .$$

Let E' be the set of bounded $\mathcal{B}(U)$-measurable functions f such that supp $f \subset U_n$ for some n. Let E be the set of measures μ on $(U, \mathcal{B}(U))$ such that $\mu|_{U_n}$ is bounded for each n with the weak topology $\sigma(E, E')$. Now $Q_s f = f(x) q_s(dx)$ for each $f \in E'$, i.e., it is a measure on $(U, \mathcal{B}(U))$ from E.

Proposition 40. *(see [18]) For each s, H_s is the set of all measures on $(U, \mathcal{B}(U))$ of the form $f(x) q_s(dx)$ such that $\int_U (f(x))^2 q_s(dx) < \infty$, and H_s' is the set of all measurable functions f on U such that $\int_U (f(x))^2 q_s(dx) < \infty$.*

$L^2_{loc}(Q)$ is the set of all $\mathcal{P}(\mathbb{F})$-measurable measures on $(U, \mathcal{B}(U))$ of the form $f(s, x) q_s(dx)$ (f is a $\mathcal{P}(\mathbb{F}) \otimes \mathcal{B}(U)$-measurable function) such that.

$$\int_0^t \int_U |f(s,x)|^2 q_s(dx)\, d\lambda_s < \infty \qquad \mathbf{P}\text{-}a.s., \forall t,$$

$$\text{and} \qquad |f_s q_s|^2_{H_s} = \int_U |f(s,\, x)|^2 q_s(dx).$$

$\hat{L}^2_{loc}(Q)$ is the set of all $\mathcal{P}(\mathbb{F}) \otimes \mathcal{B}(U)$-measurable functions f such that

$$\int_0^t \int_U |f(s,x)|^2 q_s(dx)\, d\lambda_s < \infty \ \mathbf{P}\text{-}a.s., \forall t,$$

$$|f_s|^2_{H_s'} = \int_U |f(s,\, x)|^2 q_s(dx).$$

Proof. The set $Q_s E'$ consists of all measures of the form $f(u) q_s(du)$, $f \in E'$. We will treat it as the space of classes of equivalent measures, two measures $f(u) q_s(du)$ and $g(u) q_s(du)$ are equivalent if $f = g$ q-a.s. This vector space endowed with the norm

$$|f(u) q_s(du)|^2_{H_s} = \int_\Omega f^2(u) q_s(du) = \langle Q_s f,\, f \rangle$$

becomes a Banach space isometrically isomorphic to $L_2(U, \mathcal{B}(U), q_s) = H_s'$, and hence complete. Now our statement follows simply from the definitions and Theorem IV.Q.4 in [3].

4.3 Integrals with respect to stochastic flows.

Here we generalize some results of H. Kunita in [11] concerning the integrals with respect to stochastic flows.

Let X be a locally compact metrisable space (there exists a countable dense subset of X). For each $x \in X$ we are given $M_t(x) \in \mathcal{M}^2_{loc}(\mathbb{F}, \mathbf{P})$. Suppose that there exists a $\mathcal{P}(\mathbb{F}) \otimes \mathcal{B}(X) \otimes \mathcal{B}(X)$ measurable function $Q_s(x, y)$ and an increasing $\mathcal{P}(\mathbb{F})$-measurable process λ_t such that

$$M_t(x) M_t(y) - \int_0^t Q_s(x,\, y)\, d\lambda_s \in \mathcal{M}_{loc}(\mathbb{F}, \mathbf{P}) \tag{16}$$

We assume that Q_s is symmetric and non-negative definite, i.e. $\forall(\xi_n)$, $\forall(x_j)$,

$$\sum_{n,j} \xi_n \xi_j Q_s(x_n, x_j) \geq 0 .$$

Let E be the set of all real valued functions with the topology of simple convergence and E' its dual space, i.e., the space of finite combinations of Dirac measures. If $\mu \in E'$, we have $\mu = \sum_x c_x \delta_x$ and only a finite number of $c_x \neq 0$. For $\mu = \sum_x c_x \delta_x$ we can define

$$M_t(\mu) = \sum_x c_x M_t(\delta_x) = \sum_x c_x M_t(x) .$$

Obviously, (16) yields

$$M_t(\mu) M_t(\mu') - \int_0^t \sum_{x,y} c_x c'_y Q_s(x, y) \, d\lambda_s \in \mathcal{M}_{loc}(\mathbb{F}, \mathbf{P}) ,$$

where

$$\mu = \sum_x c_x \delta_x , \quad \mu' = \sum_x c'_x \delta_x .$$

For $\mu = \sum_x c_x \delta_x$, we define $\bar{Q}_s \mu(x) = \sum_y c_y Q_s(y, x)$, i.e., $\bar{Q}_s \in \mathcal{L}^+(E)$ and $\forall \mu$, $\mu' \in E'$

$$M_t(\mu) M_t(\mu') - \int_0^t \langle \mu', \bar{Q}_s \mu \rangle_{E',E} \, d\lambda_s \in \mathcal{M}_{loc}(\mathbb{F}, \mathbf{P}).$$

In this case the corresponding Hilbert subspaces $H_s = H_{\bar{Q}_s}$ can be described using their reproducing kernels $Q_s(x, y)$ (see [21]). We consider two particular cases.

(a.) Let Q_s be separately continuous on $X \times X$, continuous on the diagonal and locally bounded.

Denote $\mathcal{E}^0(X)$ the space of continuous functions on X with the topology of uniform convergence on compact sets. Let $\mathcal{E}^0(X)'$ be its dual space which is the space of Radon measures with compact support. We can extend the kernel $\bar{Q}_s$ to $\mathcal{E}^0(X)'$. If $\nu \in \mathcal{E}^0(X)'$, we set

$$\bar{Q}_s \nu(x) = \int_X Q_s(x, y) \nu(dy) .$$

Lemma 41. *For each $\nu \in \mathcal{E}^0(X)'$, there exists a sequence (μ_n) from E' and $M_t(\nu) \in \mathcal{M}^2_{\rm loc}(\mathbb{F}, \mathbf{P})$ such that*

$$\sup_t |M_t(\mu_n) - M_t(\nu)| + \int_0^1 \langle \bar{Q}_s(\mu_n - \nu), \mu_n - \nu \rangle \, d\lambda_s \overset{n \to \infty}{\longrightarrow} 0$$

in probability. Moreover, for each $\mu, \nu \in \mathcal{E}^0(X)'$,

$$M_t(\mu) M_t(\nu) - \int_0^t \langle \bar{Q}_s(\mu - \nu), (\mu - \nu) \rangle d\lambda s \in \mathcal{M}_{\rm loc}(\mathbb{F}, \mathbf{P}) .$$

Proof. For each n there exists a finite measurable partition (A^n_k) of $\operatorname{supp}\nu$ such that $\operatorname{diam}(A^n_k) \le 1/n$. We choose arbitrary $x_k \in A^n_k$ and define $\mu_n = \sum_k \nu(A^n_k)\delta_{x_k}$. Obviously, we have inequality $|\mu^K_n| \le |\nu|$ for their variations and for each continuous function f on X $\mu_n(f) \to \nu(f)$. Thus by our assumptions $\int_0^1 \langle \bar{Q}_s(\mu_n - \nu), \mu_n - \nu\rangle d\lambda_s \overset{n\to\infty}{\longrightarrow} 0$ in probability. Therefore there exists increasing sequences of stopping times $(\tau_{n,p})$, (τ_p) such that $\tau_{n,p} \le \tau_p \le 1$ and

$$\mathbf{P}(\tau_p < 1) \overset{p\to\infty}{\longrightarrow} 0,\ \mathbf{P}(\tau_{n,p} < \tau_p) \overset{n\to\infty}{\longrightarrow} 0,$$

$$\mathbf{E}\int_0^{\tau_p} \langle Q_s\nu, \nu\rangle d\lambda_s < \infty,\ \int_0^{\tau_{n,p}} \langle Q_s\mu_n, \mu_n\rangle d\lambda_s \le$$

$$\le \int_0^{\tau_p} \langle Q_s\nu, \nu\rangle d\lambda_s + 1\,.$$

Then for each p

$$\mathbf{E}\sup_t |M_{t\wedge\tau_{n,p}}(\mu_n) - M_{t\wedge\tau_{n',p}}(\mu_{n'})|^2 \to 0\,,$$

as $n, n' \to \infty$. Thus the existence of a limit with required properties follows immediately.

Proposition 42. *Suppose that for each s, $\bar{Q}_s \in \mathcal{L}^+(\mathcal{E}^0(X))$, and $\int_0^1 \bar{Q}_s\, d\lambda_s \in \mathcal{L}^+(\mathcal{E}^0(X))$. If ν_s is a $\mathcal{E}^0(X)'$-valued $\mathcal{P}(\mathbb{F})$-measurable function such that*

$$\int_0^1 \int_X \nu_s(dx) \int_X Q_s(x,y)\nu_s(dy)\, d\lambda_s < \infty \quad \mathbf{P}\text{-}a.s.,$$

we can define the Ito integral $\int_0^t \nu_s dM_s = \mathcal{I}_t(\nu) \in \mathcal{M}^2_{loc}(\mathbb{F}, \mathbf{P})$ such that $\langle \mathcal{I}(\nu)\rangle_t = \int_0^t \int_X \nu_s(dx) \int_X Q_s(x,y)\nu_s(dy) d\lambda_s$.

Proof. The statement follows from Lemma 41, the definitions and Proposition 9.

Remark 43. In [11] the case $\nu_s = \delta_{f_s}$ was considered, where f_s is an X-valued $\mathcal{P}(\mathbb{F})$-measurable function.

(b.) In addition to the assumptions made in (a) let us suppose that X is an open subset of $\mathbf{R}^d$. Assume that Q_s has all derivatives up to order m and for $|p| \le m$, $|q| \le m$, $D^p_x D^q_y Q_s(x,y)$ are separately continuous, locally bounded and continuous on the diagonal. Let $\mathcal{E}^m(X)$ be the space of m times continuously differentiable functions with the topology of uniform convergence on compact sets of all derivatives up to order m. The dual space $\mathcal{E}^m(X)'$ will be the space of the generalized functions of order $\le m$ with compact support. We can easily extend $\bar{Q}_s$ to $\mathcal{E}^m(X)'$ by $\bar{Q}_s T(x) = \int_X Q_s(x,y)T(y)dy$, where $T \in \mathcal{E}^m(X)'$ (obviously $\bar{Q}_s T \in \mathcal{E}^m(X)$).

Lemma 44. *For each $T \in \mathcal{E}^m(X)'$ there exist a sequence $\mu_n \in E'$ and a process $M_t(T) \in \mathcal{M}^2_{\text{loc}}(\mathbb{F}, P)$ such that*

$$\int_0^1 \int_X \bar{Q}_s(T - \mu_n)(x)(T(x) - \mu_n(x))dx\, d\lambda_s +$$
$$\sup_t |M_t(T) - M_t(\mu_n)| \stackrel{n\to\infty}{\longrightarrow} 0 \quad \textit{in probability.}$$

Proof. Fix $T \in \mathcal{E}^m(X)'$. Then there is a family of Radon measures $\{\nu_p\}$ ($|p| \le m$, $p \in \mathbf{N}^d$) with compact supports such that

$$T = \sum_{|p|\le m} \left(\frac{\partial}{\partial x}\right)^p \nu_p \,.$$

It is enough, obviously, to prove the statement for $T = \left(\frac{\partial}{\partial x}\right)^p \nu_p$. As in the proof of Lemma 41, for each n we take a measurable partition (A^n_k) of the support of ν_p such that $\operatorname{diam}(A^n_k) \le 1/n$. Then we choose an arbitrary $x_k \in A^n_k$ and define $\bar{\mu}_n = \sum_k \nu_p(A^n_k)\delta_{x_k}$. Obviously, the variations satisfy the inequality $|\bar{\mu}_n| \le |\nu_p|$, and $(\bar{\mu}_n)$ converges weakly to ν_p. Thus,

$$\int_0^1 \langle \bar{Q}_s \left(\left(\frac{\partial}{\partial x}\right)^p \bar{\mu}_m - T\right), \quad \left(\frac{\partial}{\partial x}\right)^p \bar{\mu}_n - T\rangle d\lambda_s \stackrel{n\to\infty}{\longrightarrow} 0$$

in probability. Let $(e_k)_{1\le k\le d}$ be a canonical basis in $\mathbf{R}^d$, $p = (p_1, \dots, p_d)$. For sufficiently small $h > 0$ and $x \in X$, define $\Delta^k_h \delta_x = \frac{1}{h}(\delta_{x+he_k} - \delta_x)$. Consider $\bar{\mu}^h_n = (\Delta^1_h)^{p_1} \dots (\Delta^d_h)^{p_d} \bar{\mu}_n = \sum_k \nu_p(A^n_k)(\Delta^1_h)^{p_n} \dots (\Delta^d_h)^{p_d} \bar{\delta}_{x_k}$. It is well defined for small $h > 0$ and is an element of E'. By our assumptions, for each n,

$$\int_0^1 \langle \bar{Q}_s \left(\mu^h_n - \left(\frac{\partial}{\partial x}\right)^p \bar{\mu}_n\right), \bar{\mu}^h_n - \left(\frac{\partial}{\partial x}\right)^p \bar{\mu}_n\rangle \, d\lambda_s \stackrel{h\downarrow 0}{\longrightarrow} 0 \,.$$

Thus we can find a sequence (h_n) such that

$$\int_0^1 \langle \bar{Q}\left(\bar{\mu}^{h_n}_n - T\right), \left(\bar{\mu}^{h_n}_n - T\right)\rangle \, d\lambda s \stackrel{n\to\infty}{\longrightarrow} 0$$

in probability, and $\mu_n = \bar{\mu}^{h_n}_n \in E'$. We complete the proof as in the case of Lemma 41. There exist sequences of stopping times $(\tau_{n,p}), (\tau_p)$ such that $\tau_{n,p} \le \tau_p \le 1$ and

$$\mathbf{P}(\tau_p < 1) \stackrel{p\to\infty}{\longrightarrow} 0, \quad \mathbf{P}(\tau_{n,p} < \tau_p) \stackrel{n\to\infty}{\longrightarrow} 0,$$

$$\mathbf{E} \int_0^{\tau_p} \langle \bar{Q}_s, T, T\rangle d\lambda_s < \infty, \quad \int_0^{\tau_{n,p}} \langle \bar{Q}_s \mu_n, \mu_n\rangle d\lambda_s \le$$

$$\le \int_0^{\tau_p} \langle \bar{Q}_s T, T\rangle d\lambda_s + 1$$

Then for each $p \sup_t |M_{t\wedge\tau_{n,p}}(\mu_n) - M_{t\wedge\tau_{n',p}}(\mu_{n'})|^2 \to 0$, as $n, n' \to \infty$. The existence of a limit with required properties follows immediately.

Proposition 45. *For each s, $\bar{Q}_s \in \mathcal{L}^+(\mathcal{E}^m(X))$, $\int_0^1 \bar{Q}_s\, d\lambda_s \in \mathcal{L}^+(\mathcal{E}^m(X))$. If T_s is an $\mathcal{E}^m(X)'$-valued $\mathcal{P}(\mathbb{F})$-measurable function such that*

$$\int_0^1 \int_X T_s(x)dx \int_X Q_s(x,y)T_s(y)\, dy\, d\lambda_s < \infty \ \mathbf{P}\text{-a.s.},$$

we can define the integral $\mathcal{I}_t(T) = \int_0^t T_s dM_s \in \mathcal{M}^2_{loc}(\mathbb{F}, \mathbf{P})$ such that

$$\langle \mathcal{I}(T) \rangle_t = \int_0^t \int_X T_s(x)dx \int_X Q_s(x,y)T_s(y)dy\, d\lambda_s\,.$$

Proof. The statement follows from Lemma 44, the definitions and Proposition 9.

Remark 46. Let $T_s = \sum_{|\alpha|\le m} c_s^\alpha \partial^\alpha \mu_s^\alpha$ be such that

$$\int_0^1 \int_X \int_X \sum_{\alpha,\alpha'} (-1)^{|\alpha|+|\alpha'|} c_s^\alpha c_s^{\alpha'} \partial_y^{\alpha'} \partial_x^\alpha Q_s(x,y) \mu_s^\alpha(dx) \mu_s^{\alpha'}(dy)\, ds < \infty\,,$$

where c_s^α are one dimensional predictable functions and μ_s^α are predictable functions with values in the space $\mathcal{E}^0(X)$ of Radon measures on X with compact support. Then the function T is M-integrable.

4.4 Integrals with respect to purely discontinuous martingales

In this Section we apply the ideas discussed above to integration of infinite-dimensional vector functions with respect to purely discontinuous martingales.

Let $(U, \mathcal{U})$ be a measurable space and $p(dt, du)$ be a non-negative point measure on $([0,1] \times U, \mathcal{B}([0,1]) \otimes \mathcal{U})$. Assume that there exists a $\mathcal{P}(\mathbb{F})$-measurable family of measures $\pi_s(dx)$ and an increasing continuous process λ_t such that $q(dt, du) = p(dt, du) - \pi_t(du)d\lambda_t$ is a martingale measure. Let (U_n) be an increasing sequence of measurable subsets of U such that $\cup_n U_n = U$ and for every n,

$$\int_0^1 \int_{U_n} \pi_s(du)d\lambda_s < \infty \qquad \mathbf{P}\text{-a.s.}$$

Let $f(s,u)$ be a $\mathcal{P}(\mathbb{F})$-measurable E-valued function such that

$$\int_0^1 \int \langle f(s,u), y\rangle^2 \wedge |\langle f(s,u), y\rangle| \pi_s(du)d\lambda_s < \infty$$

for each $y \in E'$, P-a.e.

For each $y \in E'$, define

$$M_t(y) = \int_0^t \int_u \langle f(s,u), y\rangle q(ds, du) \in \mathcal{M}_{loc}(\mathbb{F}, \mathbf{P})\,.$$

The function $f(s,u)$ defines a $\mathcal{P}(\mathbb{F})\otimes\mathcal{U}$-measurable family $\bar{Q}_{s,u}$ of kernels from $\mathcal{L}^+(E)$ such that

$$\langle \bar{Q}_{s,u}y, y'\rangle = \langle f(s,u), y\rangle\langle f(s,u), y'\rangle \text{ for all } y, y' \in E$$

Obviously, the corresponding Hilbert space $\bar{H}_{s,u}$ is one dimensional:

$$\bar{H}_{s,u} = \begin{cases} \mathbf{R}f(s,u), & \text{if } f(s,u) \neq 0 \\ 0, & \text{if } f(s,u) = 0\,. \end{cases}$$

In this case we can integrate functions of the variables (ω, s, u). Let $\tilde{D}^1 = \tilde{D}^1(\mathbf{P})$ be the set of all E'-valued $\mathcal{P}(\mathbb{F})\otimes\mathcal{U}$-measurable functions g such that

$$\mathbf{E}[\int_0^1\int_U \langle g(s,u), f(s,u)\rangle^2 p(ds,du)^{1/2}] < \infty\,.$$

Define $\bar{D}^1 = \bar{D}^1(\mathbf{P})$ as the set of all $\mathcal{P}(\mathbb{F})\otimes\mathcal{U}$-measurable scalar functions ρ such that

$$\rho = \rho 1_{\{f\neq 0\}} \quad \text{and} \quad \mathbf{E}[\int_0^1\int_U \rho(s,u)^2 p(ds,du)^{1/2}] < \infty\,.$$

Let $D^1 = D^1(\mathbf{P}) = \{g = \rho f : \rho \in \bar{D}^1\}$. Since $\bar{H}_{s,u}$ is one dimensional and $q(ds,du)$ is a scalar measure, the integration is elementary. Now let $\{y_1, y_2, \ldots\}$ be a weakly dense subset of E'. Write

$$\begin{aligned} \bar{N} = \bar{n}(s,u) &= \begin{cases} \inf(n : \langle f(s,u), y_n\rangle \neq 0, & \text{if } f(s,u) \neq 0 \\ 1, & \text{if } f(s,u) = 0 \end{cases} \\ \bar{e}'(s,u) &= \begin{cases} y_{\bar{n}(s,u)}\langle f(s,u), y_{\bar{n}(s,u)}\rangle^{-2}, & \text{if } f(s,u) \neq 0 \\ 0, & \text{if } f(s,u) = 0\,. \end{cases} \end{aligned} \tag{17}$$

Then $\langle f(s,u), \bar{e}'(s,u)\rangle f(s,u)$ is a measurable basis in $\bar{H}_{s,u}$. We define the integrals

$$\begin{aligned} \mathcal{T}_t(\tilde{g}) &= \textstyle\int_0^t \tilde{g}_s dM_s = \int_0^t \int_U \langle f(s,u), \tilde{g}(s,u)\rangle q(ds,du), \tilde{g} \in \tilde{D}^1\,, \\ \mathcal{R}_t(g) &= \textstyle\int_0^t g_s * dM_s = \int_0^t \int_U \rho(s,u) q(ds,du), g = \rho f \in D^1, \end{aligned}$$

(see [8] for the definition of the right sides). Now we write

$$\begin{aligned} \tilde{\mathcal{O}} &= \left\{g \in \tilde{D}^1 : \langle f, g\rangle = 0 \quad \pi_s(du)d\lambda_s d\mathbf{P}\text{-a.s}\right\} \\ \mathcal{O} &= \left\{g = \rho f \in D^1 : \rho = 0 \quad \pi_s(du)d\lambda_s d\mathbf{P}\text{-a.s.}\right\}, \\ \bar{\mathcal{O}} &= \left\{\rho \in \bar{D}^1 : \rho = 0 \quad \pi_s(du)d\lambda_s d\mathbf{P}\text{-a.s.}\right\}, \\ \tilde{\mathcal{D}}^1 &= \tilde{D}^1/\tilde{\mathcal{O}}, \quad \mathcal{D}^1 = D^1/\mathcal{O}, \bar{\mathcal{D}}^1 = \bar{D}^1/\bar{\mathcal{O}}\,. \end{aligned}$$

For $g \in \tilde{D}^1$, we denote $\hat{g} = g + \tilde{\mathcal{O}}$ and define the distance

$$\tilde{d}(\hat{g}, \hat{g'}) = \mathbf{E}\left[\int_0^1 \langle f(s,u), g(s,u) - g'(s,u)\rangle^2 p(ds,du)^{1/2}\right]\,.$$

For $g \in D^1$, we denote $\hat{g} = g + \mathcal{O}$ and define the distance ($g = \rho f, g' = \rho' f \in D^1$) $d(\hat{g}, \hat{g}') = \mathbf{E}\left[\int_0^1 \int_U (\rho(s,u) - \rho'(s,u))^2 p(ds, du)^{1/2}\right]$. For $\rho \in \bar{D}^1$, we write $\hat{\rho} = \rho + \hat{\mathcal{O}}$ and define the distance

$$\bar{d}(\hat{\rho}, \hat{\rho}') = \mathbf{E}[\int_0^1 \int_U (\rho(s,u) - \rho'(s,u))^2 p(ds, du)^{1/2}] .$$

Proposition 47. *The maps* $\mathcal{G}_1 : \hat{g}_s \rightsquigarrow \widehat{\langle f_s, g_s \rangle f_s}$, *and* $\mathcal{G}_2 : \widehat{\rho_s f_s} \rightsquigarrow \hat{\rho}_s$ *are isometries from* $\tilde{\mathcal{D}}^1$ *to* $\bar{\mathcal{D}}^1$ *and from* $\mathcal{D}^1$ *to* $\bar{\mathcal{D}}^1$, *respectively, i.e., all the spaces are complete. Moreover,*

$$\mathcal{T}_t(g) = \mathcal{R}_t(\langle f, g \rangle f) = \int_0^t \int_U \langle f, g \rangle q(ds, du),$$
$$\mathcal{R}_t(\rho f) = \int_0^t \int_U \rho q(ds, du), g \in \tilde{D}^1, \rho f \in D^1 .$$

Proof. Let $g = \rho f \in D^1$ and $g' = \rho(s,u)\bar{e}'(s,u)$ where $\bar{e}'(s,u)$ is defined by (17). Then $\langle f, g' \rangle = \rho$, i.e., $g' \in \tilde{D}^1$ and the statement follows from the definitions.

Remark 48. We can localize the definitions and integrate the functions from D^1_{loc}, $\tilde{D}^1_{loc}$ and $\bar{D}^1_{loc}$ (a function $g \in \mathcal{A}_{loc}$, if there exists a sequence of stopping times (τ_n) such that $\tau_n \uparrow 1$ and $g 1_{[0,\tau_n]} \in \mathcal{A}$ where $\mathcal{A} = D^1, \tilde{D}^1, \bar{D}^1$).

Let $G^1_{loc} = G^1_{loc}(\pi d\lambda, \mathbf{P})$ be the set of all $\mathcal{P}(\mathbb{F}) \otimes U$-measurable scalar functions ρ such that $\mathbf{P}$-a.e.

$$\int_0^1 \int_U |\rho(s,u)|^2 \wedge |\rho(s,u)| \pi_s(du) d\lambda_s < \infty .$$

Remark 49. By Proposition (3.71) in [8] we have the following predictable characterizations of the classes D^1_{loc}, $\bar{D}^1_{loc}$, $\tilde{D}^1_{loc}$:

a) $\rho \in \bar{D}^1_{loc} \Leftrightarrow \rho \in G^1_{loc}$,

b) $g \in \tilde{D}^1_{loc} \Leftrightarrow \langle g, f \rangle \in G^1_{loc}$,

c) $g = \rho f \in D^1_{loc} \Leftrightarrow \rho \in G^1_{loc} \Leftrightarrow \rho \in \bar{D}^1_{loc}$.

5 Appendix

Proof of Proposition 2. Let $\hat{\mathcal{H}}_0$ be a completion of $\mathcal{H}_0 = KE$. Since the natural embedding $j : \mathcal{H}_o \to E$ is continuous we can extend it continuously to a linear continuous mapping $\hat{j} : \hat{\mathcal{H}}_0 \to \hat{E}$ where $\hat{E}$ is the completion of E and $\hat{j}|_{\mathcal{H}_0} = j$. Let $x_n \in \mathcal{H}_0$ and $x_n \to x$ in $\hat{\mathcal{H}}_0$. Then $\hat{j}(x_n) = j(x_n) \to \hat{j}(x)$ in $\hat{E}$. Therefore $\hat{j}(x_n) = j(x_n)$ is a bounded Cauchy sequence in E. From the quasi-completeness of E it follows that $\hat{j}(x) \in E$, i.e. $\hat{j}$ is a continuous linear map from $\hat{\mathcal{H}}_0$ to E. We shall prove that $\hat{j}$ is an injection. Let $k \in \hat{\mathcal{H}}_0$ and $\hat{j}(k) = 0$. Then for each $y \in E'$,

$$0 = \langle \hat{j}(k), y \rangle = (k, Ky)_{\hat{\mathcal{H}}_0} .$$

This equality is obvious for $k \in \mathcal{H}_0$ and the general case follows by continuity of both sides. Thus $k \in \hat{\mathcal{H}}_0$ is orthogonal to $\mathcal{H}_0$, i.e., $k = 0$. Thus $\hat{\jmath} : \hat{\mathcal{H}}_0 \to E$ is a continuous injection. Let $H_K = \hat{\jmath}(\hat{\mathcal{H}}_0) \subset E$ and define $(x, y)_K = (\hat{\jmath}^{-1}(x), \hat{\jmath}^{-1}(y))_{\hat{\mathcal{H}}_0}$ for every $x, y \in H_K$. Then $H_K \in E$ is the completion of $\mathcal{H}_0$ and the natural imbedding $H_K \to E$ is continuous by the continuity of $\hat{\jmath}$. The statement is proved.

Proof of Corollary 3. Let $\tilde{L}$ be a linear subspace generated by T. It is weakly dense in E'. Denote by L the closure of $K\tilde{L}$ in KE. If $L \neq KE$, we can find $x_0 = Ky_0 \in KE \setminus L$ and a continuous linear form $l : KE \to \mathbf{R}$ such that $l(x_0) = l(Ky_0) > 0$ and $l|_L = 0$. By Riesz theorem there exists $h \in H_K$ such that $l(Ky) = (h, Ky)_K$ for each $y \in E'$. On the other hand $(h, Ky)_K = \langle h, y \rangle$. Indeed, this equality is obvious for $h \in KE$ and follows from continuity of both sides in the general case. Thus $\langle h, y_0 \rangle > 0$ and $\langle h, y \rangle = 0$ for each $y \in \tilde{L}$, and we get a contradiction. It means that $L = KE$. Since KE is obviously strongly dense in H_K, our statement is proved.

Proof of Proposition 15. For each $f' \in E'$ and $h \in H = H_K \subset E_1$,

$$(h, Ku'f')_H = \langle h, u'f' \rangle_{E_1, E_1'} = \langle uh, f' \rangle_{E, E'} .$$

So, h is orthogonal to $Ku'E'$ if and only if $uh = 0$, i.e., $h \in u^{-1}\{0\}$ =Ker u. Then the orthogonal complement to $\mathcal{N}$ in H is the closure in H of the set $Ku'E'$. Then the scalar product $(h, Ku'f')_H$ is always equal to the scalar product of the corresponding images in $u(H)$:

$$\langle uh, f' \rangle_{E,E'} = (h, Ku'f')_H = (uh, uKu'f')_{u(H)}.$$

This proves that uKu' is a reproducing kernel of $u(H)$, i.e., $u(H) = H_{uKu'}$.

Proof of Proposition 35. If $H = H_K$, then $K \in \mathcal{L}^+(E)$ is the composition of linear forms: $K : E' \xrightarrow{j'} H' \xrightarrow{\theta} H \xrightarrow{j} E$, where j is the injection from H to E, j' is the injection from E' to H' and θ is the canonical isomorphism between Hilbert space H and its dual H'. Similarly with the identifications discussed in the proof of Remark 33, $\hat{K}$ is the composition of the linear forms j', θ^{-1}, and j:

$$\hat{K} : E' \xrightarrow{j'} H \xrightarrow{\theta^{-1}} H' \xrightarrow{j} E .$$

Now we shall discuss briefly the stochastic integral for Banach space valued martingales (see [16]). The Métivier-Pellaumail construction is based on an a priori estimate for simple integrals. It will be shown below that this estimate guarantees the existence of the factorization $Qd\lambda$. For the sake of simplicity, we consider a particular situation. Let L be a separable Banach-space with its dual L'. Let X be a square integrable L-valued martingale and $\tilde{\mathcal{P}}$ be a boolean ring generated by the sets of the form $]s,t] \times F$, $F \in \mathcal{F}_s$. $\mathcal{E}(L')$ will denote the vector space of the L'-valued and $\tilde{\mathcal{P}}$-simple processes, i.e., the processes Y such that $Y = \sum_i a_i 1_{A_i}$, where (a_i) is a finite family of L' elements and (A_i) is a finite family of $\tilde{\mathcal{P}}$ elements. If $Y \in \mathcal{E}(L')$ the definition of the stochastic integral $\int_0^t Y dX$ is obvious (see [16]). In [16], p. 20, the following assumption is made .

[i] there exists a finite positive measure α on predictable sets, vanishing on evanescent sets and such that for every L-valued $\tilde{\mathcal{P}}$-simple process Y,

$$\mathbf{E}(\int_0^1 Y_s dX_s)^2 \leq \int_{\Omega'} |Y_s|^2_{L'} d\alpha < \infty ,$$

where $\Omega' = [0,1] \times \Omega$.

Note that this assumption is always satisfied if L is a Hilbert space.

Let $\mathcal{L}(L', L'')$ be the space of continuous linear operators from L' to its dual L'' and

$$\mathcal{L}^+(L', L'') = \{A \in \mathcal{L}(L', L'') : (Ay)y \geq 0, (Ay)z = (Az)y \; \forall y, z \in L'\} .$$

Proposition 50. *Let L' be separable and the assumption (i) be satisfied. Then there exists an increasing $\mathcal{P}(\mathbb{F})$-measurable process λ and $\mathcal{L}^+(L', L'')$-valued $\mathcal{P}(\mathbb{F})$-measurable function Q such that for each $z, y \in L'$*

$$z(X_t)y(X_t) - \int_0^t (Q_s y) z d\lambda_s \in \mathcal{M}_{loc}(\mathbb{F}, \mathbf{P}) .$$

Proof. Let $\{y_1, y_2, \ldots\}$ be a countable dense subset in L'. Denote $\lambda^i_t = \langle y_i(X)\rangle_t$. Choose a sequence (c_i), $c_i > 0$ such that $E \sum_i c_i \lambda^i_1 < \infty$ and define $\lambda_t = \sum_i c_i \lambda^i_t$. Then by assumption (i), we have that for each $y \in L'$,

$$d\langle y(X)\rangle_s d\mathbf{P} \ll d\lambda_s d\mathbf{P} \qquad \text{on } \mathcal{P}(\mathbb{F}).$$

Thus for each $y, z \in L'$, there exists a $\mathcal{P}(\mathbb{F})$-measurable function $C_s(y, z)$ such that

$$y(X_t)z(X_t) - \int_0^t C_s(y, z) d\lambda_s \in \mathcal{M}_{loc}(\mathbb{F}, \mathbf{P}) .$$

Obviously, for each $y, z, u \in L'$ and $a, b \in \mathbf{R}$

$$C_s(y,z) = C_s(z,y), C_s(y,y) \geq 0, C_s(ay+bz,u) = aC_s(y,u) + bC_s(z,u)$$
$d\lambda_s d\mathbf{P}$-a.s.

Let $\mathcal{J}$ be the vector space generated by $\{y_1, y_2, \ldots\}$. Then it is easy to find a bilinear form $\tilde{Q}_s$ on $\mathcal{J} \times \mathcal{J}$ such that $d\lambda_s d\mathbf{P}$-a.e. for each $y, z \in \mathcal{J}$, $\tilde{Q}(y,z) = \tilde{Q}_s(z,y)$, $\tilde{Q}_s(y,y) \geq 0$ and, moreover, $\tilde{Q}_s(y,z) = C_s(y,z)$.

By the Lebesgue theorem, there exists a $\mathcal{P}(\mathbb{F})$-measurable function $f \geq 0$ and a finite measure $\bar{\alpha}$ on $\mathcal{P}(\mathbb{F})$ orthogonal to $d\lambda d\mathbf{P}$ such that

$$d\alpha = f d\lambda d\mathbf{P} + d\bar{\alpha} .$$

Let $l \in \mathcal{J}$, $A \in \mathcal{P}(\mathbb{F})$, $y = l 1_A$. Then

$$\mathbf{E}(\int_0^1 l 1_A dX_s)^2 = \mathbf{E}\int_0^1 1_A \tilde{Q}(l,l) d\lambda_s \leq \mathbf{E}\int_0^1 1_A |l|^2_{L'} f s d\lambda_s .$$

Since A is arbitrary, $\tilde{Q}_s(l,l) \leq f_s |l|^2_{L'}$ $d\lambda_s d\mathbf{P}$-a.e. Thus we can find a $d\lambda d\mathbf{P}$-modification of $\tilde{Q}_s$ such that

$$0 \leq \tilde{Q}_s(y,y) \leq f_s |y|^2_{L'}, \forall y \in \mathcal{J} ,$$

everywhere and we can extend $\tilde{Q}_s$ continuously as a bilinear form on the whole $L' \times L'$. So there exists $Q_s : L' \to L''$ such that $\tilde{Q}_s(y,z) = (Q_s y)z$ for every $y, z \in L'$ and we are done.

References

1. Daletskii, Yu. L., "Infinite dimensional elliptic operators and parabolic equations connected with them," *Uspekhi Mat. Nauk, Translations, Russ. Math. Surveys*, **22**, 1967, pp. 1–53.
2. Detweiler, E., "Banach space valued processes with independent increments and stochastic integration," In: *Probability in Banach Spaces IV, Proceedings*, Oberwolfach 1982, *Lecture Notes in Math.* **990**, Berlin–Heidelberg–New York, Springer, pp. 54–83.
3. Dunford, N. and Schwartz, J.T., *Linear Operators*, **1**, Interscience Publishers, New York, 1958.
4. Gihman, I.I. and Skorohod, A.V., *Stochastic Differential Equations*, Springer Verlag, 1972.
5. Grigelionis, B. and Mikulevicius, R., "Stochastic evolution equations and derivatives of the conditional distributions," *Lecture Notes in Control and Inf. Sci.*, **49**, 1983, pp. 43–86.
6. Gyöngi, I. and Krylov, N.V. "On stochastic equations with respect to semimartingales I," *Stochastics*, **4(1)**, 1980, pp. 1–21.
7. Gyöngi, I. and Krylov, N.V., "On stochastic equations with respect to semimartingles II.," *Stochastics*, **6(3-4)**, 1982, pp. 153–173.
8. Jacod, J., "Calcul Stochastique et Problèmes de Martingales," *Lecture Notes in Math.*, **714**, Springer-Verlag, Berlin, 1979.
9. Korezlioglu, H., "Stochastic integration for operator valued processes on Hilbert spaces and on nuclear spaces," *Stochastics*, **24**, 1988, pp. 171–213.
10. Krylov, N.V. and Rozovskii, B.L., "Stochastic evolution systems," *J. Soviet Math.*, **16**, 1981, pp. 1233–1276.
11. Kunita, H., '*Stochastic Flows and Stochastic Differential Equations*, Cambridge University Press, Cambridge, 1990.
12. Kunita, H., "Stochastic integrals based on martingales taking values in Hilbert space," *Nagoya Math. J.*, **38**, 1970, pp. 41–52.

13. Liptser, R.S. and Shiryaev, A.N., *Statistics of Random Processes I*, Springer-Verlag, Berlin, 1977.
14. Meyer, P.A., "Notes sur les intégrales stochastiques I. Intégrales hilbertiennes," Séminaire de Proba XI, *Lecture Notes in Math*, **581**, 1977, pp. 446–463.
15. Métivier, M. and Pistone, G. "Une formule d'isométrie pour l'intégrale stochastique hilbertienne et équations d'évolution linéaires stochastiques," *Zeitschrift Wahrscheinlichkeitstheorie*, **33**, 1975, pp. 1–18.
16. Métivier, M. and Pellaumail, J., *Stochastic Integration*, Academic Press, 1980.
17. Métivier, M., *Stochastic Partial Differential Equations in Infinite Dimensional Spaces*, Scuola Normale Superiore, Pisa, 1988.
18. Mikulevicius, R. and Rozovskii, B.L., "Uniqueness and absolute continuity of weak solutions for parabolic SPDE's," Acta Appl. Math., **35**, 1994, pp. 179-192.
19. Mikulevicius, R. and Rozovskii, B.L., "Martingale Problems for Stochastic PDE's", In: "Stochastic Partial Differential Equations. Six Perspectives " R. Carmona and B.L. Rozovskii Ed., AMS, Mathematical Surveys and Monographs series, 1998 (to appear).
20. Rozovskii, B., *Stochastic Evolution Systems*, Kluwer Academic Publishers, Dordrecht, 1990.
21. Schwartz, L., "Sous-espaces hilbertiens d'espaces vectoriels topologiques et noyaux associés (noyaux reproduisants)," Journal d'Analyse Math., **13**, 1964, pp. 115–256.
22. Walsh, J.B., "An introduction to partial differential equations," Ecole d'Eté de Probabilités de Saint-Flour XIV, 1984, *Lecture Notes in Math.*, **1180**, 1986, pp. 265–439.

PATHWISE UNIQUENESS AND APPROXIMATION OF SOLUTIONS OF STOCHASTIC DIFFERENTIAL EQUATIONS

Khaled BAHLALI
Dépt de maths, Université de Toulon et du Var B.P132
83 957 La Garde Cedex FRANCE
Brahim MEZERDI
Institut d'hydraulique, Centre universitaire M[ed] Khider
B.P 145 Biskra ALGERIE
Youssef OUKNINE
Dépt de maths, Fac. des sciences Semlalia
Université Cadi Ayyad Marrakech MAROC

Abstract

We consider stochastic differential equations for which pathwise uniqueness holds. By using Skorokhod's selection theorem we establish various strong stability results under perturbation of the initial conditions, coefficients and driving processes. Applications to the convergence of successive approximations and to stochastic control of diffusion processes are also given. Finally, we show that in the sense of Baire, almost all stochastic differential equations with continuous and bounded coefficients have unique strong solutions.

1 INTRODUCTION

We consider the following stochastic differential equation:

$$\begin{cases} dX_t = \sigma(t, X_t)\, dB_t + b(t, X_t)\, dt \\ X_0 = x \end{cases} \tag{1}$$

where $\sigma : \mathbf{R}_+ \times \mathbf{R}^d \longrightarrow \mathbf{R}^d \otimes \mathbf{R}^r$ and $b : \mathbf{R}_+ \times \mathbf{R}^d \longrightarrow \mathbf{R}^d$ are measurable functions, B is a given r-dimensional Brownian motion defined on a probability space $(\Omega, \mathcal{F}, P)$ with a filtration $\mathcal{F}_t$ satisfying the usual conditions. Throughout this paper we assume that equation (1) has a unique strong solution $X_t(x)$ for each initial value $x \in \mathbf{R}^d$.

It is a well known fact that if the coefficients are Lipschitz continuous, then equation (1) has a unique strong solution $X_t(x)$, which is continuous with respect to the initial condition and coefficients. Moreover, the solution may be constructed by means of various numerical schemes.

Our purpose in this paper, is to study strong stability properties of the solution of (1) under pathwise uniqueness of solutions and a minimal assumption on the coefficients. Such minimal assumption ensures the existence of a weak solution, and is either the continuity of b, σ in the state variable [19], or the uniform ellipticity of the diffusion coefficient [15]. According to Yamada-Watanabe's theorem [22], existence of a weak solution and pathwise uniqueness imply existence of a unique strong solution.

The paper is organized as follows. In the second section we study the variation of the solution with respect to initial data and parameters. Extension of the above result to the Hölder space is the subject of section 3. The fourth section is devoted to the study of successive approximations. We know from the theory of ordinary differential equations with bounded continuous coefficients that, the uniqueness of a solution is not sufficient for successive approximations to converge. Under pathwise uniqueness, we give a necessary and sufficient condition for this convergence. Moreover, we introduce a class of moduli of continuity for which the method of successive approximations converges, covering the results of many authors. In the fifth section, we study the stability of solutions of stochastic differential equations driven by continuous semi-martingales, with respect to the driving processes. Note that we don't suppose pathwise uniqueness for the approximating equations, as it is usually done in the literature.

In section 6, we give an application to optimal control of diffusions. Namely we prove that under pathwise uniqueness, the trajectories associated to relaxed controls are approximated in L^2- sense by trajectories associated to ordinary controls. This result extends a theorem of S. Méléard [17] which is proved under Lipschitz condition. Extension of some of the previous results to the case where the coefficients are merely measurable, with uniformly elliptic diffusion matrix is the subject of section 7.

At the end of this work, we prove that in the sense of Baire, almost all stochastic differential equations with bounded continuous coefficients have unique strong solutions.

The main tool used in the proofs is the Skorokhod selection theorem given by the following

Lemma 1.1 ([11] page 9) *Let (S,ρ) be a complete separable metric space, $P_n, n = 1,2,...$ and P be probability measures on $(S,\mathcal{B}(S))$ such that $P_n \underset{n\to+\infty}{\longrightarrow} P$. Then, on a probability space $\left(\widehat{\Omega},\widehat{\mathcal{F}},\widehat{P}\right)$, we can construct S-valued random variables X_n, $n = 1,2,...,$ and X such that:*

(i) $P_n = \widehat{P}^{X_n}, n = 1,2,...,$ and $P = \widehat{P}^X$.

(ii) X_n converges to X, $\widehat{P}$ almost surely.

We'll make use of the following result, which gives a criterion for tightness of sequences of laws associated to continuous processes.

Lemma 1.2 ([11] page 18) *Let $(X_n(t))$, $n = 1, 2, \ldots$, be a sequence of d-dimensional continuous processes satisfying the following two conditions:*

(i) There exist positive constants M and γ such that $E[|X_n(0)|^\gamma] \leq M$ for every $n = 1, 2, \ldots$.

(ii) There exist positive constants α, β, M_k, $k = 1, 2, \ldots$, such that: $E[|X_n(t) - X_n(s)|^\alpha] \leq M_k |t - s|^{1+\beta}$ for every n and $t, s \in [0, k], (k = 1, 2, \ldots)$.

Then there exist a subsequence (n_k), a probability space $\left(\widehat{\Omega}, \widehat{\mathcal{F}}, \widehat{P}\right)$ and d-dimensional continuous processes $\widehat{X}_{n_k}$, $k = 1, 2, \ldots$, and $\widehat{X}$ defined on it such that

1) The laws of $\widehat{X}_{n_k}$ and X_{n_k} coincide.

2) $\widehat{X}_{n_k}(t)$ converges to $\widehat{X}(t)$ uniformly on every finite time interval $\widehat{P}$ almost surely.

2 VARIATION OF SOLUTIONS WITH RESPECT TO INITIAL CONDITIONS AND PARAMETERS

Definition 2.3 *We say that pathwise uniqueness holds for equation (1) if whenever $(X, B, (\Omega, \mathcal{F}, P), \mathcal{F}_t)$ and $(X', B, (\Omega, \mathcal{F}, P), \mathcal{F}'_t)$ are weak solutions of equation (1) with common probability space and Brownian motion B (relative to possibly different filtrations) such that $P[X_0 = X'_0] = 1$, then X and X' are indistinguishable.*

In the theory of ordinary differential equations with continuous coefficients, uniqueness of solutions is sufficient for continuous dependence of the solution with respect to the initial condition [3]. The following theorem gives the analogue of the above result to the stochastic case.

Theorem 2.4 *Let $\sigma(t, x)$ and $b(t, x)$ be continuous functions satisfying the linear growth condition: for each $T \geq 0$, there exists M such that:*

$$|\sigma(t, x)| + |b(t, x)| \leq M \ (1 + |x|) \text{ for every } t \in [0, T] \tag{2}$$

Then, if pathwise uniqueness holds for equation (1), we get:

$$\lim_{x \to x_0} E\left[\sup_{t \leq T} |X_t(x) - X_t(x_0)|^2\right] = 0, \text{ for every } T \geq 0 .$$

Proof Suppose that the conclusion of our theorem is false, then there exist a positive number δ and a sequence (x_n) converging to x such that:

$$\inf_{n\in\mathbf{N}} E\left[\sup_{t\le T}\left|X_t(x_n)-X_t(x)\right|^2\right]\ge\delta. \tag{3}$$

Let us denote by X^n (resp. X) the solution of (1) corresponding to the initial data x_n (resp. x).

By standard arguments from the theory of stochastic differential equations ([13], page 289) we can show that the sequence (X^n, X, B) satisfies conditions i) and ii) of lemma 1.2 with $\alpha=4$ and $\beta=1$. Then there exist a probability space $\left(\widehat{\Omega},\widehat{\mathcal{F}},\widehat{P}\right)$ and a sequence $\left(\widehat{X}^n_t,\widehat{Y}^n_t,\widehat{B}^n_t\right)$ of stochastic processes defined on it such that:

α) The laws of (X^n, X, B) and $\left(\widehat{X}^n,\widehat{Y}^n,\widehat{B}^n\right)$ coincide for every $n\in\mathbf{N}$.

β) There exists a subsequence $\left(\widehat{X}^{n_k},\widehat{Y}^{n_k},\widehat{B}^{n_k}\right)$ converging to $\left(\widehat{X},\widehat{Y},\widehat{B}\right)$ uniformly on every finite time interval $\widehat{P}$-a.s.

If we denote by $\widehat{\mathcal{F}}^n_t=\sigma\left(\widehat{X}^n_s,\widehat{Y}^n_s,\widehat{B}^n_s;s\le t\right)$ and $\widehat{\mathcal{F}}_t=\sigma\left(\widehat{X}_s,\widehat{Y}_s,\widehat{B}_s;s\le t\right)$, then $\left(\widehat{B}^n_t,\widehat{\mathcal{F}}^n_t\right)$ and $\left(\widehat{B}_t,\widehat{\mathcal{F}}_t\right)$ are Brownian motions.

According to property α) and the fact that X^n_t and X_t satisfy (1) with initial data x_n and x, it can be proved ([15] page 89) that $\forall n\in\mathbf{N}$, $\forall t\ge 0$

$$E\left|\widehat{X}^n_t-x_n-\int_0^t\sigma\left(s,\widehat{X}^n_s\right)d\widehat{B}^n_s-\int_0^t b\left(s,\widehat{X}^n_s\right)ds\right|^2=0.$$

In other words, $\widehat{X}^n$ satisfies the stochastic differential equation:

$$\widehat{X}^n_t=x_n+\int_0^t\sigma\left(s,\widehat{X}^n_s\right)d\widehat{B}^n_s+\int_0^t b\left(s,\widehat{X}^n_s\right)ds.$$

Writing similar relations, we obtain:

$$\widehat{Y}^n_t=x+\int_0^t\sigma\left(s,\widehat{Y}^n_s\right)d\widehat{B}^n_s+\int_0^t b\left(s,\widehat{Y}^n_s\right)ds.$$

By using property (β) and a limit theorem of Skorokhod [19] page 32, it holds that

$$\int_0^t\sigma\left(s,\widehat{X}^{n_k}_s\right)d\widehat{B}^{n_k}_s\underset{k\to+\infty}{\longrightarrow}\int_0^t\sigma\left(s,\widehat{X}_s\right)d\widehat{B}_s$$

and $\displaystyle\int_0^t b\left(s,\widehat{X}^{n_k}_s\right)ds\underset{k\to+\infty}{\longrightarrow}\int_0^t b\left(s,\widehat{X}_s\right)ds$ in probability.

Therefore $\widehat{X}$ and $\widehat{Y}$ satisfy the same stochastic differential equation (1), on $\left(\widehat{\Omega}, \widehat{\mathcal{F}}, \widehat{P}\right)$, with the same Brownian motion $\widehat{B}_t$ and initial condition x. Then, by pathwise uniqueness, we conclude that $\widehat{X}_t = \widehat{Y}_t$, $\forall t \quad \widehat{P}$ a.s.

By uniform integrability, it holds that:

$$\delta \leq \liminf_{n \in \mathbb{N}} E\left[\sup_{t \leq T} \left|X_t(x_n) - X_t(x)\right|^2\right] \leq \liminf_{k \in \mathbb{N}} \widehat{E}\left[\sup_{t \leq T} \left|\widehat{X}_t^{n_k} - \widehat{Y}_t^{n_k}\right|^2\right] =$$

$$= \widehat{E}\left[\sup_{t \leq T} \left|\widehat{X}_t - \widehat{Y}_t\right|^2\right]$$

which contradicts (3). ■

We shall next state a variant of the first theorem. Let us consider a family of functions depending on a parameter λ, and consider the stochastic differential equation:

$$\begin{cases} dX_t^\lambda = \sigma\left(\lambda, t, X_t^\lambda\right) dB_t + b\left(\lambda, t, X_t^\lambda\right) dt \\ X_0^\lambda = \varphi(\lambda). \end{cases} \tag{4}$$

Theorem 2.5 *Suppose that $\sigma(\lambda, t, x)$ and $b(\lambda, t, x)$ are continuous. Further suppose that for each $T > 0$, and each compact set K there exists $L > 0$ such that*

i) $\sup_{t \leq T} \left(|\sigma(\lambda, t, x)| + |b(\lambda, t, x)|\right) \leq L\ (1 + |x|)$ uniformly in λ,

ii) $\lim_{\lambda \to \lambda_0} \sup_{x \in K} \sup_{t \leq T} \left(|\sigma(\lambda, t, x) - \sigma(\lambda_0, t, x)| + |b(\lambda, t, x) - b(\lambda_0, t, x)|\right) = 0$,

iii) $\varphi(\lambda)$ is continuous at $\lambda = \lambda_0$.

If pathwise uniqueness holds for equation (4) at λ_0, we have:

$$\lim_{\lambda \to \lambda_0} E\left[\sup_{t \leq T} \left|X_t^\lambda - X_t^{\lambda_0}\right|^2\right] = 0\ ,\ \textit{for every } T \geq 0\ .$$

Proof Similar to the proof of theorem 2.4. ■

Remark 2.6 *Though (4) need not have a pathwise unique solution for $\lambda \neq \lambda_0$, nevertheless its solutions are continuous in the parameter λ at λ_0.*

The same method may be applied to show the convergence of many approximation schemes such as Euler scheme, approximation by stochastic delay equations [6], the splitting up method [7] and polygonal approximation [12].

3 EXTENSION OF THE RESULTS TO THE HÖLDER SPACE

Let $\alpha > 0$ and denote by $C^{\alpha}\left([0,1]\,;R^{d}\right)$ the set of α-Hölder continuous functions equiped with the norm defined by:

$$\|f\|_{\alpha} = \sup_{0\le t\le 1} |f(t)| + \sup_{0\le s\le t\le 1} \frac{|f(t)-f(s)|}{|t-s|^{\alpha}}$$

It follows from [13], page 53 that the solutions of (1) are α-Hölder continuous for any $\alpha \in \left[0, \frac{1}{2}\right[$. Let X (resp. X^{n}) denote the solution of (1) corresponding to the intial condition x (resp. x_n) and $Y_n = X - X_n$.

Lemma 3.7 *For any $p > 1$, $\delta > 0$ and any $\gamma < \dfrac{p-1}{2p}$ the following estimates hold:*

(i) $\sup_n E\,|Y_n(t) - Y_n(s)|^{2p} \le c(p)\,|t-s|^{p}$;

(ii) $\sup_n P\left(\sup_{s<t} \frac{|Y_n(t)-Y_n(s)|}{|t-s|^{\gamma}} > \delta\right) \le c(p,\gamma)\,\delta^{-2p}$.

Proof Part (i) is a consequence of Ito's formula. (ii) is a simple consequence of the Garcia-Rodemich-Rumsey lemma ([20] page 49). ■

Proposition 3.8 *Under the hypothesis of theorem 2.4 and for any $\alpha \in \left[0; \frac{1}{2}\right[$, $\varepsilon > 0$ we have:*

$$\lim_{n\to+\infty} P\left(\|X_n - X\|_{\alpha} > \varepsilon\right) = 0$$

Proof $P\left(\|X_n - X\|_{\alpha} > \varepsilon\right) \le P\left(\sup_{0\le t\le 1} |X_n(t) - X(t)| > \frac{\varepsilon}{2}\right) +$

$$+P\left(\sup_{s<t} \frac{|Y_n(t)-Y_n(s)|}{|t-s|^{\alpha}} > \frac{\varepsilon}{2}\right).$$

According to theorem 2.4, the first term in the right hand side goes to 0 as n tends to $+\infty$.

Let $\eta > 0$ such that $\alpha + \eta < \dfrac{p-1}{2p}$ and let $\mu > 0$, we have

$$P\left(\sup_{s<t} \frac{|Y_n(t)-Y_n(s)|}{|t-s|^{\alpha}} > \frac{\varepsilon}{2}\right) = P\left(\sup_{s<t} \frac{|Y_n(t)-Y_n(s)|}{|t-s|^{\alpha}} > \frac{\varepsilon}{2}; |t-s| < \mu\right) +$$

$$+P\left(\sup_{s<t} \frac{|Y_n(t)-Y_n(s)|}{|t-s|^{\alpha}} > \frac{\varepsilon}{2}; |t-s| > \mu\right)$$

$$\le P\left(\sup_{s<t} \frac{|Y_n(t)-Y_n(s)|}{|t-s|^{\alpha+\eta}} > \frac{\varepsilon}{2}\mu^{-\eta}\right) + 2P\left(\sup_{0\le t\le 1} |X_n(t) - X(t)| > \frac{\varepsilon}{4}\mu^{\alpha}\right)$$

$$\le c\left(\frac{\varepsilon}{2}\mu^{-\eta}\right)^{-2p} + 2P\left(\sup_{0\le t\le 1} |X_n(t) - X(t)| > \frac{\varepsilon}{4}\mu^{\alpha}\right).$$

By taking μ small enough and using theorem 2.4 we get the desired result. ■

4 PATHWISE UNIQUENESS AND SUCCESSIVE APPROXIMATIONS

Let σ and b as in theorem 2.4 and consider the stochastic differential equation (1). The sequence of successive approximations associated to (1) is defined as follows

$$\begin{cases} X_t^{n+1} = \int_0^t \sigma(s, X_s^n)\, dB_s + \int_0^t b(s, X_s^n)\, ds \\ X^0 = x. \end{cases} \tag{5}$$

If we assume that the coefficients are Lipschitz continuous, then the sequence (X^n) converges in quadratic mean and gives an effective way for the construction of the unique solution X of equation (1) (see [11]). Now if we drop the Lipshitz condition and assume only that equation (1) admits a unique strong solution, does the sequence (X^n) converge to X? The answer is negative even in the deterministic case, see ([4] pp.114-124).

The aim of the following theorem is to establish an additional necessary and sufficient condition which ensures the convergence of successive approximations.

Theorem 4.9 *Let σ and b as in theorem 2.4. Under pathwise uniqueness for s.d.e (1), (X^n) converges in quadratic mean to the unique solution of (1) if and only if $X^{n+1} - X^n$ converges to 0.*

Lemma 4.10 *Let (X^n) be defined by (5), then:*

1) For every $p > 1$, $\sup_n E\left[\sup_{t \le T} |X_t^n|^{2p}\right] < +\infty$.

2) For every $T > 0$ and $p > 1$, there exists a constant C independant of n such that for every $s < t$ in $[0,T]$, $E\left[|X_t^n - X_s^n|^{2p}\right] \le C\,|t-s|^p$.

Proof 1) For all $t > 0$ and $n > 1$, we have

$$|X_t^n|^{2p} \le C_1 \left[|x|^{2p} + \left| \int_0^t b(s, X_s^{n-1})\, ds \right|^{2p} + \left| \int_0^t \sigma(s, X_s^{n-1})\, dB_s \right|^{2p} \right].$$

By applying Hölder's inequality, it holds that

$$\left| \int_0^t b(s, X_s^{n-1})\, ds \right|^{2p} \le \left[\sum_{i=1}^d \left(\int_0^t b_i(s, X_s^{n-1})\, ds \right)^2 \right]^p$$

$$\le t^p \left[\int_0^t |b(s, X_s^{n-1})|^2\, ds \right]^p \le t^{2p-1} . \int_0^t |b(s, X_s^{n-1})|^{2p}\, ds.$$

Burkholder Davis Gundy and Hölder inequalities provide the following estimate

$$E\left[\sup_{0 \le t \le T} \left| \int_0^t \sigma(s, X_s^{n-1})\, dB_s \right|^{2p} \right] \le C_2 E\left[\left(\int_0^T |\sigma(s, X_s^{n-1})|^2\, ds \right)^p \right]$$

$$\le C_2 T^{p-1} E\left[\int_0^T |\sigma(s, X_s^{n-1})|^{2p}\, ds \right].$$

Taking expectations, we obtain

$$E\left[\sup_{t\le T}|X_t^n|^{2p}\right]\le C_3\left[|x|^{2p}+C_4E\int_0^T\left(|\sigma|^{2p}+|b|^{2p}\right)(s,X_s^{n-1})\,ds\right].$$

By using the linear growth condition we get

$$E\left[\sup_{t\le T}|X_t^n|^{2p}\right]\le C_5\left(1+|x|^{2p}\right)+C_5\int_0^T E\left[\sup_{s\le t}|X_s^{n-1}|^{2p}\right]dt$$

where the various constants C_k depend only on T,m,d.

Iteration of the last inequality gives

$$E\left[\sup_{t\le T}|X_t^n|^{2p}\right]\le C_5\left(1+|x|^{2p}\right)\left[1+CT+\frac{(CT)^2}{2!}+...+\frac{(CT)^n}{n!}\right].$$

Then $\sup_n E\left[\sup_{t\le T}|X_t^n|^{2p}\right]\le C_5\left(1+|x|^{2p}\right)\exp(CT)$.

2) If we fix $s<t$ in $[0,T]$, we may proceed as before to obtain

$$E\left[|X_t^n-X_s^n|^{2p}\right]\le C_6\,|t-s|^{p-1}\int_s^t\left(1+E\left[\sup_{v\le u}|X_v^{n-1}|^{2p}\right]du\right).$$

Then by using the previous result, we get

$E\left[|X_t^n-X_s^n|^{2p}\right]\le C_7\,|t-s|^p$, where C_7 depends on x,p,d,T. ■

Proof of theorem 4.9. Suppose that $X^{n+1}-X^n$ converges to 0 and there is some $\delta>0$ such that

$$\inf_n E\left[\max_{0\le t\le T}|X_t^n-X_t|^2\right]\ge\delta$$

According to lemma 4.10, the family (X^n,X^{n+1},X,B) satisfies conditions i) and ii) of lemma 1.2. Then by Skorokhod's selection theorem, there exists some probability space $\left(\widehat{\Omega},\widehat{\mathcal{F}},\widehat{P}\right)$ carrying a sequence of stochastic processes $\left(\widehat{X}^n,\widehat{Z}^n,\widehat{Y}^n,\widehat{B}^n\right)$ with the following properties:

i) the laws of $\left(\widehat{X}^n,\widehat{Z}^n,\widehat{Y}^n,\widehat{B}^n\right)$ and (X^n,X^{n+1},X,B) coincide for each $n\in\mathbf{N}$,

ii) there exists a subsequence $\{n_k\}$ such that $\left(\widehat{X}^{n_k},\widehat{Z}^{n_k},\widehat{Y}^{n_k},\widehat{B}^{n_k}\right)$ converges to $\left(\widehat{X},\widehat{Z},\widehat{Y},\widehat{B}\right)$ uniformly on every finite time interval $\widehat{P}$ a.s.

But we know that $X^{n+1}-X^n$ converges to 0, then we can show easily that $\widehat{X}=\widehat{Z}$, $\widehat{P}$ a.s.

Proceeding as in the proof of theorem 2.4, we can show that

$$\widehat{Z}^{n_k}=x+\int_0^t\sigma\left(s,\widehat{X}^{n_k}\right)d\widehat{B}_s^{n_k}+\int_0^t b\left(s,\widehat{X}^{n_k}\right)ds;$$

$$\widehat{Y}^{n_k}=x+\int_0^t\sigma\left(s,\widehat{Y}_s^{n_k}\right)d\widehat{B}_s^{n_k}+\int_0^t b\left(s,\widehat{Y}_s^{n_k}\right)ds.$$

Taking the limit as k goes to $+\infty$, it holds that

$$\widehat{X}_t = x + \int_0^t \sigma\left(s, \widehat{X}_s\right) d\widehat{B}_s + \int_0^t b\left(s, \widehat{X}_s\right) ds;$$
$$\widehat{Y}_t = x + \int_0^t \sigma\left(s, \widehat{Y}_s\right) d\widehat{B}_s + \int_0^t b\left(s, \widehat{Y}_s\right) ds.$$

In other words, $\widehat{X}$ and $\widehat{Y}$ solve equation (1). Then by pathwise uniqueness we have $\widehat{X} = \widehat{Y}$, $\widehat{P}$ a.s.

Using uniform integrability, we obtain:

$$\begin{aligned} \delta &\le \liminf_k E\left[\max_{0\le t\le T} \left|X_t^{n_k} - X_t\right|^2\right] = \liminf_k \widehat{E}\left[\max_{0\le t\le T} \left|\widehat{X}_t^{n_k} - \widehat{Y}_t^{n_k}\right|^2\right] \\ &= \widehat{E}\left[\max_{0\le t\le T} \left|\widehat{X}_t - \widehat{Y}_t\right|^2\right] \end{aligned}$$

which is a contradiction. ■

Remark 4.11 *Roughly speaking, under pathwise uniqueness, the series* $\sum \left(X^{n+1} - X^n\right)$ *converges if and only if* $\left(X^{n+1} - X^n\right)$ *converges to* 0.

As a generalization of the condition which S. Kawabata [14] has already considered, we'll assume the following:

Condition A 1) There exist measurable functions m and ρ such that:

$$|\sigma(t,x) - \sigma(t,y)|^2 + |b(t,x) - b(t,y)|^2 \le m(t)\,\rho\left(|x-y|^2\right)$$

where m is in L^1_{loc}.

2) ρ is a continuous, non decreasing and concave function defined on $\mathbf{R}_+$ such that:

$$\int_{0+} \frac{du}{\rho(u)} = +\infty.$$

Theorem 4.12 *Assume that σ and b satisfy condition A. Then the successive approximations converge in quadratic mean to the unique solution of (1).*

Proof Under condition A, it is well known that pathwise uniqueness holds [22]. Now it is sufficient to prove that $X^{n+1} - X^n$ converges to 0 in quadratic mean. Let

$$\varphi_n(t) = E\left[\max_{0\le s\le t} \left|X_s^{n+1} - X_s^n\right|^2\right].$$

Using Doob and Schwarz inequalities, we have:

$$\begin{aligned} \varphi_{n+1}(t) \le (8+2T)\, E\Bigg[&\int_0^t \left|\sigma\left(t, X_s^{n+1}\right) - \sigma\left(t, X_s^n\right)\right|^2 ds + \\ &+ \int_0^t \left|b\left(t, X_s^{n+1}\right) - b\left(t, X_s^n\right)\right|^2 ds\Bigg]. \end{aligned}$$

Since ρ is increasing and concave, then:

$$\varphi_{n+1}(t) \leq (8+2T)\int_0^t \rho(\varphi_n(s)).m(s)\,ds.$$

Let ψ_n the sequence defined by:

$$\psi_0(t) = u(t)$$
$$\psi_{n+1}(t) = (8+2T)\int_0^t \rho(\psi_n(s)).m(s)\,ds \text{ for every } t \in [0,T].$$

By using a lemma in [4] page 114-124, see also [14], it is possible to choose a function u such that:

1) $\forall t \in [0,T]$, $u(t) \geq \varphi_0(t)$.

2) $u(t) \geq (8+2T)\int_0^t \rho(u(s)).m(s)\,ds$

Hence by induction we have: $\varphi_n(s) \leq \psi_n(s)$ $\forall n \in \mathbf{N}$ and the sequence ψ_n is decreasing.

Let $\psi(t) = \lim_{n\to+\infty} \psi_n(t)$, note that this convergence is uniform and ψ is continuous and satisfies the equation:

$$\psi(t) = (8+2T)\int_0^t \rho(\psi(s)).m(s)\,ds \text{ for every } t \in [0,T].$$

Condition A 2) implies that $\psi = 0$ and hence $\lim_{n\to+\infty} \varphi_n(t) = 0$. ■

Some examples of functions which are not Lipschitz but satisfy condition A2) are given by:

$$\rho(u) = u\,|\log u|^{\alpha} \ (0 < \alpha < 1)$$

or

$$\rho(u) = u\,|\log u|\,|\log|\log u||^{\alpha} \ (0 < \alpha < 1).$$

Let us introduce a different class of moduli of continuity $g(t,x)$ which are not necessarily written in the form $l(t).m(x)$, covering the classes considered in [21], [8].

Let Ξ be the set of functions $g :]0,T] \times \mathbf{R}_+ \longrightarrow \mathbf{R}_+$ satisfying:

i) g is continuous, non decreasing and concave with respect to the second variable.

ii) $\lim_{t\to 0} g(t,0) = 0$.

iii) If $F : [0,T] \longrightarrow \mathbf{R}_+$ is continuous such that $F(0) = 0$ and $F(t) \leq \int_0^t g(s, F(s))\,ds$, then $F = 0$ on $[0,T]$.

Theorem 4.13 *Let σ and b be continuous functions satisfying the linear growth condition (2). Moreover suppose that there exists $g \in \Xi$ such that:*

$$|\sigma(t,x) - \sigma(t,y)|^2 + |b(t,x) - b(t,y)|^2 \leq g\left(t, |x-y|^2\right)$$

Then pathwise uniqueness holds, and the sequence X^n converges in quadratic mean to the unique solution of (1).

Proof i) Pathwise uniqueness is an immediate consequence of the properties of the function g.

Let us prove the convergence of successive approximations. According to theorem 4.9, it is sufficient to prove that $\lim_{n\to+\infty} E\left[\max_{0\le s\le t}\left|X_s^{n+1}-X_s^n\right|^2\right]=0.$

Let

$$\varphi_n(t)=E\left[\max_{0\le s\le t}\left|X_s^{n+1}-X_s^n\right|^2\right]$$

and

$$A_T=\left\{t\in[0,T]\,;\ \lim_{n\to+\infty}\varphi_n(t)=0\right\}.$$

Since A_T is a non empty subset ($0\in A_T$), it is enough to establish that A_T is open and closed in $[0,T]$.

1) First step: A_T is closed.

Let $0<t_1\in\overline{A}_T$, $\varepsilon>0$ and $\delta\le\min(t_1,\varepsilon)$. Hence there exists $t_0\in A_T$ such that $t_1-t_0\le\delta$.

Since A_T is an interval which contains 0 (because $t\to\varphi_n(t)$ is an increasing function), it suffices to prove that:

$$\lim_{n\to+\infty} E\left[\max_{t_1-\delta\le s\le t_1}\left|X_s^{n+1}-X_s^n\right|^2\right]=0.$$

By Doob inequality and the non explosion condition, there exists $n_0\in\mathbb{N}$ such that for any $n\ge n_0$:

$$E\left[\max_{t_1-\delta\le s\le t_1}\left|X_s^{n+1}-X_s^n\right|^2\right]\le\varepsilon+12\,(8+2T)\,K_T\,(1+H)\,\delta$$

where $H=\sup_n E\left[\sup_{t\le T}\left|X_t^n\right|^2\right]<+\infty$ (see lemma 4.10). This proves that $t_1\in A_T$.

2) Second step: A_T is open.

Let $t_0\in A_T$; $t_0\neq0$ and $t_0\neq T$.

We'll prove that $\exists r>0$ such that $t_0+r\in A_T$, which means that $\exists r>0$ such that $\lim_{n\to+\infty}\varphi_n(t)=0$ on $[0,t_0+r[$.

Since A_T is an interval (and $t_0\in A_T$), it is sufficient to show that:

$$\lim_{n\to+\infty} E\left[\max_{t_0\le s\le t}\left|X_s^{n+1}-X_s^n\right|^2\right]=0.$$

It is easy to see that there exists a sequence (ε_n) of positive real numbers decreasing to 0 as n goes to $+\infty$, such that:

$$E\left[\max_{t_0\le s\le t_0+r}\left|X_s^{n+1}-X_s^n\right|^2\right]\le 3\varepsilon_n+ct \qquad ,\forall t\le T-t_0$$

where c is a constant which depends only on T and x.

Let

$$M=\sup\{g(t,v):(t,v)\in[t_0,T]\times[0,3\varepsilon_0+2H]\}$$
$$r=\min\left(T-t_0,\frac{2H}{3(8+2T)\sup(c,M)}\right)$$

We consider the following ordinary differential equation:

$$(*)\begin{cases} u'(t)=g_1(t,u(t)) & t\in[t_0,t_0+r]\\ u(t_0)=0\end{cases}$$

with $g_1(t,u(t))=3(8+2T)\,g(t,u(t))$.

Define the successive approximation for $(*)$ by:

$$\begin{cases} u_0(\tau)=3\varepsilon_0+3(8+2T)\sup(c,M)(\tau-t_0)\\ u_{n+1}(t)=3\varepsilon_{n+1}+\displaystyle\int_{t_0}^{\tau}g_1(t,u_n(t))\,dt.\end{cases}$$

It is obvious to see, by induction on $n\in\mathbf{N}$, that $(u_n(\tau))_{n\in\mathbf{N}}$ is a positive decreasing sequence. By using the monotone convergence theorem and the continuity of g_1, we obtain:

$$u(\tau)=\lim_{n\to+\infty}u_n(\tau)=\lim_{n\to+\infty}\varepsilon_n+\lim_{n\to+\infty}\int_{t_0}^{\tau}g_1(t,u_{n-1}(t))\,dt$$
$$=\int_{t_0}^{\tau}g_1(t,u(t))\,dt$$

Since $g_1\in\Xi$, then: $u(\tau)=0 \quad \forall\tau\in[t_0,t_0+r]$.

Let

$$\psi_n(t)=E\left[\max_{t_0\le s\le t}\left|X_s^{n+1}-X_s^n\right|^2\right].$$

We remark that $\psi_n(t)$ is majorized by $u_n(t)$ on $[t_0,t_0+r]$. Therefore $\lim_{n\to+\infty}\psi_n(t)=0$, which implies that $\lim_{n\to+\infty}\varphi_n(t)=0$ on $[t_0,t_0+r]$.

This achieves the proof. ■

We recall a condition given by S. Nakao, which guarantees pathwise uniqueness, but under which the problem of convergence of successive approximations is not solved.

σ and b are $\mathbf{R}$-valued measurable, bounded functions and σ is of bounded variation such that $\sigma\ge\varepsilon$ for some $\varepsilon>0$.

The problem of convergence of successive approximations is still open for an important class of stochastic differential equations involving the local time of the unknown process.

5 STABILITY OF STOCHASTIC EQUATIONS DRIVEN BY CONTINUOUS SEMI MARTINGALES

In this section, we consider stochastic differential equations driven by continuous semi-martingales. We establish a continuity result with respect to the driving processes when pathwise uniqueness of solutions holds.

Let $b : [0,1]\times\mathbf{R}^d \longrightarrow \mathbf{R}^d$ and $\sigma : [0,1]\times\mathbf{R}^d \longrightarrow \mathbf{R}^{d\times r}$ be bounded continuous functions.

We consider the stochastic differential equation:

$$\begin{cases} dX_t = \sigma(t,X_t)\,dM_t + b(t,X_t)\,dA_t \\ X_0 = x \end{cases} \tag{6}$$

where A_t is an adapted continuous process of bounded variation and M_t is a continuous local martingale.

Definition 5.14 *Pathwise uniqueness property holds for equation (6) if whenever $(X,M,A,(\Omega,\mathcal{F},P),\mathcal{F}_t)$ and $(X',M',A',(\Omega,\mathcal{F},P),\mathcal{F}_t')$ are two weak solutions such that $(M,A)=(M',A')$ P a.s, then $X=X'$ P a.s.*

Let (M^n) be a sequence of continuous $(\mathcal{F}_t,P)$ −local martingales and (A^n) be a sequence of $\mathcal{F}_t$-adapted continuous processes with bounded variation.

We consider the following equations:

$$\begin{cases} dX_t^n = \sigma(t,X_t^n)\,dM_t^n + b(t,X_t^n)\,dA_t^n \\ X_0^n = x. \end{cases} \tag{7}$$

Let us suppose that (A,A^n,M,M^n) satisfy the following conditions:
(H_1) The family (A,A^n,M,M^n) is bounded in probability in $C([0,1])^4$.
(H_2) $M^n - M \longrightarrow 0$ in probability in $C([0,1])$.
(H_3) $Var(A^n - A) \longrightarrow 0$ in probability.
(Var means the total variation).

Theorem 5.15 *If conditions H_1,H_2,H_3 are satisfied and if pathwise uniqueness holds for equation (6) then:*

$$\text{For any } \varepsilon>0, \lim_{n\to+\infty} P\left[\sup_{t\le 1}|X_t^n - X_t| > \varepsilon\right] = 0.$$

We need the following lemmas given in [9].

Lemma 5.16 *Let $\{f_n(t), f(t) : t\in[0,1]\}$ be a family of continuous processes and let $\{C_n(t), C(t) : t\in[0,1]\}$ be a family of continuous processes of bounded variation. Assume that:*
$\lim_{n\to+\infty} f_n = f$ *in probability in $C([0,1])$.*

$\lim_{n\to+\infty} C_n = C$ *in probability in* $C([0,1])$.
$\{Var(C_n)\ ;\ n \in \mathbf{N}\}$ *is bounded in probability.*
Then the following result holds:

$$\forall \varepsilon > 0\,, \qquad \lim_{n\to+\infty} P\left[\sup_{t\le 1}\left|\int_0^1 f_n dC_n - \int_0^1 f dC\right| > \varepsilon\right] = 0.$$

Lemma 5.17 *Consider a family of filtrations* $(F_t^n), (F_t)$ *satisfying the usual conditions. Let* $\{f_n(t), f(t) : t \in [0,1]\}$ *be a sequence of continuous adapted processes and let* $\{N_n(t), N(t) : t \in [0,1]\}$ *be a sequence of continuous local martingales with respect to* $(F_t^n), (F_t)$ *respectively. Suppose that*
$\lim_{n\to+\infty} f_n = f$ *in probability in* $C([0,1])$.
$\lim_{n\to+\infty} N_n = N$ *in probability in* $C([0,1])$.
Then

$$\forall \varepsilon > 0\,, \qquad \lim_{n\to+\infty} P\left[\sup_{t\le 1}\left|\int_0^1 f_n dN_n - \int_0^1 f dN\right| > \varepsilon\right] = 0.$$

Proof of theorem 5.15. Suppose that the conclusion of our theorem is false. Then there exists $\varepsilon > 0$ such that

$$\inf_n P\left[\|X^n - X\|_\infty > \varepsilon\right] \ge \varepsilon\,.$$

It is clear that the family $Z^n = (X^n, X, A^n, A, M^n, M)$ is tight in $\left[C\left([0,T];\mathbf{R}^d\right)\right]^6$. Then by Skorokhod's theorem, there exist a probability space $(\Omega', \mathcal{F}', P')$ and $Z'^n = \left(X'^n, \widetilde{X}^n, A'^n, \widetilde{A}^n, M'^n, \widetilde{M}^n\right)$ which satisfy

i) law $(Z^n) =$ law (Z'^n)

ii) There exists a subsequence (Z'^{n_k}) denoted also by (Z'^n) which converges P' a.s in $\left[C\left([0,T];\mathbf{R}^d\right)\right]^6$ to $Z' = \left(X', \widetilde{X}, A', \widetilde{A}, M', \widetilde{M}\right)$.

Let G_t^n denotes the completion of the σ-algebra generated by $Z_t'^n$ $(t \in [0,1])$ $\mathcal{F}_t'^n = \bigcap_{s>t} G_s^n$.

In an analogous manner we define the σ-algebra $(\mathcal{F}_t'; t \in [0,1])$ for the limiting process Z'. Then $(\Omega', \mathcal{F}', \mathcal{F}_t'^n, P')$ (resp. $(\Omega', \mathcal{F}', \mathcal{F}_t', P')$) are stochastic bases and $M_t'^n, \widetilde{M}^n$ (resp. $M_t', \widetilde{M}_t$) are $\mathcal{F}_t'^n$ (resp. $\mathcal{F}_t'$) continuous local martingales. The processes X'^n and $\widetilde{X}^n$ satisfy the following s.d.e :

$$\begin{cases} dX_t'^n = \sigma\left(t, X_t'^n\right) dM_t'^n + b\left(t, X_t'^n\right) dA_t'^n \\ X_0'^n = x \end{cases} \tag{8}$$

$$\begin{cases} d\widetilde{X}_t^n = \sigma\left(t, \widetilde{X}_t^n\right) d\widetilde{M}_t^n + b\left(t, \widetilde{X}_t^n\right) d\widetilde{A}_t^n \\ \widetilde{X}_0^n = x \end{cases} \tag{9}$$

on $(\Omega', \mathcal{F}', \mathcal{F}_t'^n, P')$.

By using lemmas 5.16 and 5.17, we see that the limiting processes satisfy the following equations:

$$\begin{cases} dX_t' = \sigma(t, X_t')\, dM_t' + b(t, X_t')\, dA_t' \\ X_0' = x \end{cases}$$

$$\begin{cases} d\widetilde{X}_t = \sigma\left(t, \widetilde{X}_t\right) d\widetilde{M}_t + b\left(t, \widetilde{X}_t\right) d\widetilde{A}_t \\ \widetilde{X}_0 = x. \end{cases}$$

By using hypothesis (H_2) and (H_3), it is easy to see that $M' = \widetilde{M}$ and $A' = \widetilde{A}$, P' *a.s.*

Hence by pathwise uniqueness, X' and $\widetilde{X}$ are indistinguishable. This contradicts our assumption. Therefore X^n converges to the unique solution X. ■

6 AN APPROXIMATION RESULT IN STOCHASTIC CONTROL

In this section, we use the ideas developped in section 2, to establish an L^2-approximation result for relaxed control problems, where the controlled process evolves according to the Ito stochastic differential equation

$$\begin{cases} dX_t^u = \sigma(t, X_t^u)\, dB_t + b(t, X_t^u, u_t)\, dt \\ X_0^u = x \end{cases} \tag{10}$$

where u is a predictable process with values in a compact Polish space E.

The cost to be minimized over the class $\mathbf{U}$ of E-valued predictable processes is defined by:

$$J(u) = E\left[\int_0^1 l(t, X_t^u, u_t)dt + g(X_1^u)\right].$$

An optimal control u^* is a process belonging to $\mathbf{U}$, such that:

$$J(u^*) = \min\{J(u) : u \in \mathbf{U}\}.$$

Usualy an optimal control in the class $\mathbf{U}$ does not exist, unless some convexity assumptions are imposed ([5]). Thus, we transform the initial problem by embedding the class $\mathbf{U}$ into the class $\mathcal{R}$ of relaxed controls which has good compactness properties.

Let $\mathbf{V}$ be the set of probability measures on $[0,1] \times E$ whose projections on $[0,1]$ coincide with the Lebesgue measure dt. $\mathbf{V}$ is equiped with the topology of weak convergence of probability measures.

$\mathbf{V}$ is a compact metrisable set.

Definition 6.18 *A relaxed control q is a random variable with values in the set* $\mathbf{V}$.

Remark 6.19 *1) Every relaxed control q can be desintegrated as* $q(\omega, dt, da) = dt.q(\omega, t, da)$, *where* $q(\omega, t, da)$ *is a predictable process with values in the space of probability measures on* E.

2) The set $\mathbf{U}$ *of ordinary controls is embedded into the set* $\mathcal{R}$ *of relaxed controls by the application* $\Psi : \mathbf{U} \longrightarrow \mathcal{R}$, $u \longrightarrow \Psi(u)(dt, da) = dt.\delta_{u(t)}(da)$ *where* δ_a *is the Dirac measure at* a.

For a full treatment of relaxed controls see [5].

Lemma 6.20 *(Chattering lemma) Let q be a relaxed control, then there exists a sequence of predictable processes* u^n *with values in* E *such that the sequence* $dt.\delta_{u^n(t)}(da)$ *converges to* $dt.q(\omega, t, da)$ P *a.s.*

Proof. See [5]. ■

Let us now define the dynamic and the cost associated with a relaxed control $q \in \mathcal{R}$. For $q \in \mathcal{R}$, we denote by X^q the solution of:

$$\begin{cases} dX_t^q = \sigma(t, X_t^q)\, dB_t + \int_E b(t, X_t^q, a)\, q(t, da)\, dt \\ X_0^q = x. \end{cases} \tag{11}$$

The cost associated to (q, X^q) is given by:

$$J(q) = E\left[\int_0^1 \int_E l(t, X_t^q, a) q(t, da)\, dt + g(X_1^q)\right].$$

Because of the compactness of the space $\mathbf{V}$, it is proved in [5] that an optimal control exists in the class $\mathcal{R}$ of relaxed controls (even when the control enters in the diffusion coefficient σ). Moreover under uniqueness in law, it is established that the family of laws of $\left(dt.\delta_{u(t)}(da), X^u\right)$ is dense in the set of laws of $(dt.q(t, da), X^q)$ on $\mathcal{R} \times C\left(\mathbf{R}_+, \mathbf{R}^d\right)$ and:

$$\inf\{J(u) : u \in \mathbf{U}\} = \inf\{J(q) : q \in \mathcal{R}\}.$$

We give now our approximation result, extending a theorem proved in [17] (where the coefficients b, σ are supposed to be Lipschitz continuous in the space variable). The novelty of our result is that the approximation procedure remains valid under any conditions on σ and b ensuring pathwise uniqueness.

Assume the following conditions:

$b : \mathbf{R}_+ \times \mathbf{R}^d \times E \longrightarrow \mathbf{R}^d$

$\sigma : \mathbf{R}_+ \times \mathbf{R}^d \times E \longrightarrow \mathbf{R}^d \otimes \mathbf{R}^r$

are continuous functions such that:

$\sup_{t \le 1} (|\sigma(t, x)| + |b(t, x, a)|) \le L(1 + |x|), \forall a \in E.$

Theorem 6.21 *Let q be a relaxed control and* X^q *be the corresponding solution of* (11). *Then if pathwise uniqueness holds for equation (11), there exists a sequence*$(u^n)_{n \in \mathbf{N}}$ *of* E*−valued predictable processes such that:*

1) $dt.\delta_{u^n(t)}(da)$ converges to $dt.q(t,da)$ P a.s in $\mathbf{V}$.

2) $\lim_{n\to+\infty} E\left[\sup_{t\le 1}|X_t^{u_n}-X_t^q|^2\right]=0.$

Proof. 1) Let $q\in\mathcal{R}$, by lemma 6.20, there exists a sequence $(u^n)\subset\mathbf{U}$ such that $q^n=dt.\delta_{u^n(t)}(da)$ converges to $dt.q(t,da)$ P a.s in $\mathbf{V}$.

2) Let $X_t^{u_n}, X_t^q$ the solutions of (10) and (11) associated with u^n and q. Suppose that 2) is false, then there exists $\delta>0$ such that:

(H) $\inf_n E\left[\sup_{t\le 1}|X_t^{u_n}-X_t^q|^2\right]\ge\delta.$

From lemma 1.2 and compactness of $\mathbf{V}$, the family of processes $\gamma^n=(q^n,q,X^{u_n},X^q,B)$ is tight in the space $\mathbf{V}^2\times C^3$, where C denotes the space of continuous functions from $[0,1]$ into $\mathbf{R}^d$ endowed with the topology of uniform convergence. By Skorokhod's theorem, there exist a probability space $(\Omega',\mathcal{F}',P')$ carrying a sequence $\gamma'^n=(q'^n,v'^n,X'^n,Y'^n,B'^n)$ such that

i) For each $n\in\mathbf{N}$, the laws of γ^n and γ'^n coincide.

ii) There exists a subsequence γ'^{n_k} which converges to γ' P' a.s on the space $\mathbf{V}^2\times C^3$, where $\gamma'=(q',v',X',Y',B')$.

We assume without loss of generality that ii) holds for the whole sequence (γ'^n). By uniform integrability we have:

$$\delta\le\liminf_{n\in\mathbf{N}} E\left[\sup_{t\le 1}|X_t^{u_n}-X_t^q|^2\right]=\liminf_{n\in\mathbf{N}} E'\left[\sup_{t\le 1}|X_t'^n-Y_t'^n|^2\right]$$

$=E'\left[\sup_{t\le 1}|X_t'-Y_t'|^2\right]$ where E' is the expectation with respect to P'.

By property i) we see that X'^n and Y'^n satisfy the following equations

$$\begin{cases} dX_t'^n=\sigma(t,X_t'^n)\,dB_t'^n+\int_E b(t,X_t'^n,a)\,q'^n(t,da)\,dt \\ X_0'^n=x \end{cases}\tag{12}$$

$$\begin{cases} dX_t'^n=\sigma(t,Y_t'^n)\,dB_t'^n+\int_E b(t,Y_t'^n,a)\,v'^n(t,da)\,dt \\ Y_0'^n=x. \end{cases}\tag{13}$$

By letting n going to infinity and using Skorokhod's limit theorem [19] page 32, we see that the processes X' and Y' satisfy equations (12) and (13) respectively, without the index n.

We know by 1) that $q^n\to q$ in $\mathbf{V}$, P a.s, then the sequence (q^n,q) converges to (q,q) in $\mathbf{V}^2$. Moreover law$(q^n,q)=$ law(q'^n,v'^n) and $(q'^n,v'^n)\longrightarrow(q',v')$, P' a.s in $\mathbf{V}^2$. Therefore law$(q',v')=$ law(q,q) which is supported by the diagonal of $\mathbf{V}^2$. Then $q'=v'$ $\quad P'$ a.s.

It follows that X' and Y' are solutions of the same stochastic differential equation driven by Brownian motion B'. Hence by pathwise uniqueness we have $X'=Y'$, P' a.s, which contradicts (H). ■

7 CASE WHERE THE COEFFICIENTS ARE NOT CONTINUOUS

In this section we drop the continuity assumption on the coefficients, nevertheless we suppose that $d = r$ and σ, b satisfy the following conditions.

a) σ and b are Borel bounded functions.

b) $\exists \lambda > 0$, such that $\forall (t, x, \xi) \in \mathbf{R}_+ \times \mathbf{R}^d \times \mathbf{R}^d : \xi^* \sigma(t,x) \xi \geq \lambda |\xi|^2$.

Theorem 7.22 *Suppose that $\sigma(t,x)$ and $b(t,x)$ satisfy conditions a) and b). If pathwise uniqueness holds for equation (1), then the conclusion of theorem 2.4 remains valid without the continuity assumption.*

Proof. The proof goes as in theorem 2.4, the only difficulty (due to the lack of continuity of b and σ) is to show that

$$\int_0^t \sigma\left(s, \widehat{X}_s^{n_k}\right) d\widehat{B}_s^{n_k} \longrightarrow \int_0^t \sigma\left(s, \widehat{X}_s\right) d\widehat{B}_s \text{ in probability,}$$

$$\int_0^t b\left(s, \widehat{X}_s^{n_k}\right) ds \longrightarrow \int_0^t b\left(s, \widehat{X}_s\right) ds \text{ in probability.}$$

For $\varepsilon > 0$, we have

$$\begin{aligned}
&\lim_{k\to+\infty} P\left[\left|\int_0^t \sigma\left(s, \widehat{X}_s^{n_k}\right) d\widehat{B}_s^{n_k} - \int_0^t \sigma\left(s, \widehat{X}_s\right) d\widehat{B}_s\right| > \varepsilon\right] \\
&\leq \limsup_{k\to+\infty} P\left[\left|\int_0^t \left(\sigma\left(s, \widehat{X}_s^{n_k}\right) - \sigma\left(s, \widehat{X}_s\right)\right) d\widehat{B}_s^{n_k}\right| > \frac{\varepsilon}{2}\right] \\
&+ \limsup_{k\to+\infty} P\left[\left|\int_0^t \sigma\left(s, \widehat{X}_s\right) d\widehat{B}_s^{n_k} - \int_0^t \sigma\left(s, \widehat{X}_s\right) d\widehat{B}_s\right| > \frac{\varepsilon}{2}\right].
\end{aligned}$$

It follows according to the Skorokhod limit theorem [19] page 32, or lemma 3, chapter 2 in [15] that the second term in the right hand side is equal to 0.

Let $\sigma^\delta(t,x) = \delta^{-d}\varphi(x/\delta) * \sigma(t,x)$ where $*$ denotes convolution on $\mathbf{R}^d$ and φ an infinitely differentiable function with support in the unit ball such that $\int \varphi(x)\, dx = 1$.

Applying Chebyshev and Doob inequalities, we obtain

$$\begin{aligned}
&P\left[\left|\int_0^t \left(\sigma\left(s, \widehat{X}_s^{n_k}\right) - \sigma\left(s, \widehat{X}_s\right)\right) d\widehat{B}_s^{n_k}\right| > \frac{\varepsilon}{2}\right] \\
&\leq \frac{4}{\varepsilon^2} E\left[\int_0^t \left|\sigma\left(s, \widehat{X}_s^{n_k}\right) - \sigma\left(s, \widehat{X}_s\right)\right|^2 ds\right] \\
&\leq \frac{16}{\varepsilon^2}\left\{E\left[\int_0^t \left|\sigma\left(s, \widehat{X}_s^{n_k}\right) - \sigma^\delta\left(s, \widehat{X}_s^{n_k}\right)\right|^2 ds\right]\right. \\
&+ E\left[\int_0^t \left|\sigma^\delta\left(s, \widehat{X}_s^{n_k}\right) - \sigma^\delta\left(s, \widehat{X}_s\right)\right|^2 ds\right] \\
&\left.+ E\left[\int_0^t \left|\sigma^\delta\left(s, \widehat{X}_s\right) - \sigma\left(s, \widehat{X}_s\right)\right|^2 ds\right]\right\} = \frac{16}{\varepsilon^2}(I_1 + I_2 + I_3).
\end{aligned}$$

It follows from the continuity of σ^δ in x and the convergence of $\widehat{X}_s^{n_k}$ to $\widehat{X}_s$ uniformly $\widehat{P}$ a.s, that I_2 goes to 0 as $k \to +\infty$ for every $\delta > 0$.

On the other hand we know that for each $p>1$, $\sup_k E\left[\sup_{t\leq T}|X_t^{n_k}|^p\right]<+\infty$ then $\lim_{M\to+\infty} P\left[\sup_{t\leq T}|X_t^{n_k}|>M\right]=0$. Therefore, without loss of generality we may suppose that σ^δ, σ have compact support in $[0,T]\times B(0,M)$.

Applying Krylov's inequality ([15], chapter 2, theorem 3.4), we obtain $I_1+I_3\leq N.\left\|\sigma^\delta-\sigma\right\|_{d+1,M}$, where N does not depend on δ, k and $\|.\|_{d+1,M}$ denotes the norm in $L^{d+1}([0,T]\times B(0,M))$.

By letting $\delta\to 0$, we obtain the desired result.

A similar claim holds for the integrals involving the drift terms. This achieves the proof. ■

By using similar techniques, one can show the following

Theorem 7.23 *Suppose that $\sigma(t,x)$ and $b(t,x,a)$ satisfy conditions a) and b). Moreover suppose that $a\longrightarrow b(t,x,a)$ is continuous. If pathwise uniqueness holds for equation* (11), *then theorem 6.21 remains valid.*

8 GENERICITY OF EXISTENCE AND UNIQUENESS

As we have seen in previous sections, pathwise uniqueness plays a key role in the proof of many stability results. It is then quite natural to raise the question whether the set of all nice functions (σ,b) for which pathwise uniqueness holds for stochastic differential equation $e(x,\sigma,b)$ is larger than its complement, in a sense to be specified. To make the question meaningful let us recall what we mean by generic property.

A property P is said to be generic for a class of stochastic differential equations F, if P is satisfied by each equation in $F-A$, where A is a set of first category (in the sense of Baire) in F. Results on generic properties for ordinary differential equations seem to go back to an old paper of Orlicz [18], see also [16]. The investigation of such questions for stochastic differential equations is carried out in [1], [10]. In this section, we show that the subset of continuous and bounded coefficients for which pathwise uniqueness holds for equation $e(x,\sigma,b)$ is a residual set. The proof is based essentially on theorem 2.5. Moreover it does not use the oscillation function introduced by Lasota & Yorke in ordinary differential equations and used in stochastic differential equations (see[10]).

Let us introduce some notations.

$e(x,\sigma,b)$ stands for equation (1) corresponding to coefficients σ, b and initial data x.

$\mathbf{M}^2=\left\{\xi:\mathbf{R}_+\times\Omega\longrightarrow\mathbf{R}^d\text{, continuous and }\forall T>0,\ E\left[\sup_{t\leq T}|\xi_t|^2\right]<+\infty\right\}$

Define a metric on $\mathbf{M}^2$ by:

$$d(\xi_1,\xi_2)=\sum_{n=1}^{+\infty}2^{-n}\frac{\left(E\sup\limits_{0\le t\le n}|\xi_t^1-\xi_t^2|^2\right)^{\frac{1}{2}}}{1+\left(E\sup\limits_{0\le t\le n}|\xi_t^1-\xi_t^2|^2\right)^{\frac{1}{2}}}$$

By using Borel-Cantelli lemma, it is easy to see that $(\mathbf{M}^2,d)$ is a complete metric space.

Let C_1be the set of functions $b:\mathbf{R}_+\times\mathbf{R}^d\longrightarrow\mathbf{R}^d$ which are continuous and bounded. Define the metric ρ_1 on C_1 as follows:

$$\rho_1(b_1,b_2)=\sum_{n=1}^{+\infty}2^{-n}\frac{\|b_1-b_2\|_{\infty,n}}{1+\|b_1-b_2\|_{\infty,n}}$$

where $\|h\|_{\infty,n}=\sup\limits_{|x|\le n,|t|\le n}|h(x)|$.

Note that the metric ρ_1 is compatible with the topology of uniform convergence on compact subsets of $\mathbf{R}_+\times\mathbf{R}^d$.

Let C_2 be the set of continuous bounded functions $\sigma:\mathbf{R}_+\times\mathbf{R}^d\longrightarrow\mathbf{R}^d\otimes\mathbf{R}^r$ with the corresponding metric ρ_2.

It is clear that the space $\mathfrak{R}=C_1\times C_2$ endowed with the product metric is a complete metric space.

Let L be the subset of $\mathfrak{R}$ consisting of functions $h(t,x)$ which are Lipschitz in both their arguments.

Proposition 8.24 *L is a dense subset in $\mathfrak{R}$.*

Proof. By truncation and regularisation arguments. ■

The main result of this section is the following

Theorem 8.25 *The subset $\mathcal{U}$ of $\mathfrak{R}$ consisting of those (σ,b) for which pathwise uniqueness holds for $e(x,\sigma,b)$ is a residual set.*

Lemma 8.26 *For each$(\sigma,b)\in L$ and $\varepsilon>0$, there exists $\delta(\varepsilon)>0$ such that for every $(\sigma',b')\in B((\sigma,b),\delta)$ and every pair of solutions X,Y of $e(x,\sigma',b')$ (defined on the same probability space and Brownian motion), we have $d(X,Y)<\varepsilon$.*

Proof Let Z be the unique strong solution of $e(x,\sigma,b)$ defined on the same probability space and Brownian motion B.

$d(X,Y)\le d(X,Z)+d(Z,Y)$, the result follows from the continuity of Z with respect to the coefficients (see theorem 2.5). ■

Proof of theorem 8.25 We put $\mathfrak{F}=\bigcap\limits_{k\ge 1}\bigcup\limits_{(\sigma,b)\in L}B\left((\sigma,b),\delta\left(\frac{1}{k}\right)\right)$.

$\mathfrak{F}$ is a G_δ dense subset in the Baire space $(\mathfrak{R},\rho)$ and for every $(\sigma,b)\in\mathfrak{F}$ pathwise uniqueness holds for $e(x,\sigma,b)$. It follows that $\mathcal{U}$ is a residual subset in $\mathfrak{R}$. ■

Remark 8.26 M.T. Barlow [2] has shown that $\mathfrak{R}-\mathfrak{F}$ is not empty.

References

[1] K. Bahlali, B. Mezerdi, Y. Ouknine: Some generic properties of stochastic differential equations. *Stochastics & stoch. reports, vol. 57, pp. 235-245 (1996).*

[2] M.T. Barlow: One dimensional stochastic differential equations with no strong solution. *J. London Math. Soc.(2) 26; 335-347.*

[3] E. Coddington, N. Levinson: Theory of ordinary differential equations. *McGraw-Hill New-york (1955).*

[4] J. Dieudonné: Choix d'oeuvres mathématiques. Tome 1, *Hermann Paris (1987).*

[5] N. El Karoui, D. Huu Nguyen, M. Jeanblanc Piqué: Compactification methods in the control of degenerate diffusions: existence of an optimal control. *Stochastics vol .20, pp.169-219 (1987).*

[6] M. Erraoui, Y. Ouknine: Approximation des équations différentielles stochastiques par des équations à retard. *Stochastics & stoch. reports vol. 46, pp. 53-63 (1994).*

[7] M. Erraoui, Y. Ouknine: Sur la convergence de la formule de Lie-Trotter pour les équations différentielles stochastiques. *Annales de Clermont II, série probabilités* (to appear).

[8] T.C. Gard: A general uniqueness theorem for solutions of stochastic differential equations. *SIAM jour. control & optim., vol.14,* 3, *pp.445-457.*

[9] I. Gyöngy: The stability of stochastic partial differential equations and applications. *Stochastics & stoch. reports, vol. 27, pp.129-150 (1989).*

[10] A.J. Heunis: On the prevalence of stochastic differential equations with unique strong solutions. *The Annals of proba., vol. 14, 2, pp 653-662 (1986).*

[11] N. Ikeda, S. Watanabe: Stochastic differential equations and diffusion processes. *North-Holland, Amsterdam (Kodansha Ltd, Tokyo)* (1981).

[12] H. Kaneko, S. Nakao: A note on approximation of stochastic differential equations. *Séminaire de proba. XXII, lect. notes in math. 1321, pp. 155-162. Springer Verlag (1988).*

[13] I. Karatzas, S.E. Shreve: Brownian motion and stochastic calculus. *Springer Verlag, New-York Berlin Heidelberg (1988).*

[14] S. Kawabata: On the successive approximation of solutions of stochastic differential equations. *Stochastics & stoch.reports, vol. 30, pp. 69-84 (1990).*

[15] N.V Krylov: Controlled diffusion processes. *Springer Verlag, New-York Berlin Heidelberg (1980).*

[16] A. Lasota, J.A. Yorke: The generic property of existence of solutions of differential equations in Banach space. *J. Diff. Equat. 13 (1973), pp. 1-12.*

[17] S.Méléard: Martingale measure approximation, application to the control of diffusions. *Prépublication du labo. de proba. , univ. Paris VI (1992).*

[18] W. Orlicz: Zur theorie der Differentialgleichung $y' = f(x, y)$. *Bull. Acad. Polon. Sci. Ser. A (1932), pp. 221-228.*

[19] A.V. Skorokhod: Studies in the theory of random processes. *Addison Wesley (1965), originally published in Kiev in (1961).*

[20] D.W. Strook, S.R.S. Varadhan: Mutidimensional diffusion processes. *Springer Verlag Berlin (1979).*

[21] T. Yamada: On the successive approximation of solutions of stochastic differential equations. *Jour. Math. Kyoto Univ. 21 (3), pp. 501-511 (1981).*

[22] T. Yamada, S. Watanabe: On the uniqueness of solutions of stochastic differential equations. *Jour. Math. Kyoto Univ. 11 n^0 1, pp. 155-167 (1971).*

Stability of stochastic differential equations in manifolds

Marc Arnaudon
Institut de Recherche Mathématique Avancée
Université Louis Pasteur et CNRS
7, rue René Descartes
F–67084 Strasbourg Cedex
France
arnaudon@math.u-strasbg.fr

Anton Thalmaier
Naturwissenschaftliche Fakultät I – Mathematik
Universität Regensburg
D–93040 Regensburg
Germany
anton.thalmaier@mathematik.uni-regensburg.de

Abstract. — We extend the so-called topology of semimartingales to continuous semimartingales with values in a manifold and with lifetime, and prove that if the manifold is endowed with a connection ∇ then this topology and the topology of compact convergence in probability coincide on the set of continuous ∇-martingales. For the topology of manifold-valued semimartingales, we give results on differentiation with respect to a parameter for second order, Stratonovich and Itô stochastic differential equations and identify the equation solved by the derivative processes. In particular, we prove that both Stratonovich and Itô equations differentiate like equations involving smooth paths (for the Itô equation the tangent bundles must be endowed with the complete lifts of the connections on the manifolds). As applications, we prove that differentiation and antidevelopment of C^1 families of semimartingales commute, and that a semimartingale with values in a tangent bundle is a martingale for the complete lift of a connection if and only if it is the derivative of a family of martingales in the manifold.

1. Introduction

Let $(\Omega, (\mathscr{F}_t)_{0\le t<\infty}, \mathbb{P})$ denote a filtered probability space, M a smooth connected manifold endowed with a connection ∇. Then the tangent bundle TM inherits a connection ∇' (usually denoted by ∇^c), the complete lift of ∇ (see [Y-I] for details). Let X be a continuous semimartingale with values in M. The antidevelopment of X

in $T_{X_0}M$ is the semimartingale Z solving the Stratonovich equation

$$p(\delta Z) = U_0 U^{-1} \delta X, \quad Z_0 = 0, \tag{1.1}$$

where U is a horizontal lift of X taking values in the frame bundle on M and p is the canonical projection in TM of a vertical vector of TTM. The map $\mathcal{A}$ will denote the antidevelopment with respect to ∇ and $\mathcal{A}'$ the antidevelopment with respect to ∇'.

The initial motivation of this paper was to answer the following question: For some open interval I in $\mathbb{R}$, consider a family $(X_t(a))_{a\in I,\, t\in[0,\xi(a)[}$ of continuous martingales $X(a)$ in M, each with lifetime $\xi(a)$, differentiable in a for the topology of compact convergence in probability. Is then also $\big(X(a), \mathcal{A}(X(a))\big)$ differentiable in a, and if the answer is positive, do we have the relation $s\big(\partial_a \mathcal{A}(X(a))\big) = \mathcal{A}'\,(\partial_a X(a))$ (where ∂_a denotes differentiation with respect to a and s is the map $TTM \to TTM$ defined by $s(\partial_a \partial_t x(t,a)) = \partial_t \partial_a x(t,a)$, if $(t,a) \mapsto x(t,a)$ is smooth and takes its values in M)?

A positive answer will be given to this question, and this result will be obtained as a particular case of general theorems on stability of stochastic differential equations.

In this paper equations of the general type

$$\mathcal{D}Z(a) = f\big(X(a), Z(a)\big)\, \mathcal{D}X(a) \tag{1.2}$$

between two manifolds M and N are studied, where $\mathcal{D}X(a)$ denotes the (formal) differential of order 2 of $X(a)$, and f is a Schwartz morphism between the second order bundles τM and τN. The topology of semimartingales, defined in [E1] for $\mathbb{R}$-valued processes, will be adapted to manifold-valued semimartingales with lifetime. In particular, it will be shown that the map $(X, f, Z_0) \mapsto (X, Z)$ is continuous, where Z is the maximal solution starting from Z_0 to $\mathcal{D}Z = f(X,Z)\mathcal{D}X$, with appropriate topologies on both sides.

When applied to a certain family of semimartingales and an appropriate Schwartz morphism, this result will tell us that if $a \mapsto X(a)$ is C^1 in the topology of semimartingales, and further if f is C^1 with locally Lipschitz derivative, $Z(a)$ the maximal solution to (1.2) with $(Z_0(a))_{a\in I}$ C^1 in probability, then $a \mapsto (X(a), Z(a))$ is C^1 in the topology of semimartingales and the derivative $\partial_a Z(a)$ is the maximal solution to

$$\mathcal{D}\partial_a Z(a) = f'\big(\partial_a X(a), \partial_a Z(a)\big)\, \mathcal{D}\partial_a X(a) \tag{1.3}$$

where f' is a Schwartz morphism between the second order bundles τTM and τTN.

As a corollary, we obtain results on differentiability of solutions to Stratonovich and Itô equations. It will be shown that they can be differentiated in the same way as solutions to ordinary differential equations (for the Itô case, the Itô differentials of the derivative process have to be defined with the complete lifts of the connections).

If M is endowed with a connection ∇, then it will be shown that, as in the flat case, the topology of semimartingales and the topology of uniform convergence in probability on compact sets coincide on the set of martingales. Using these results it will be possible to prove commutativity of antidevelopment and differentiation.

Acknowledgement. — We would like to thank Michel Émery for his comments and suggestions to improve this paper.

2. Topologies of semimartingales and of uniform convergence in probability on compact sets

2.1. $\mathbb{R}^d$-valued processes

In this section we define topologies of uniform convergence in probability and of semimartingales for processes with lifetime. We investigate their main properties.

Let $(\Omega, \mathscr{F}, (\mathscr{F}_t)_{t\geq 0}, \mathbb{P})$ be a filtered probability space satisfying the usual conditions. If ξ is a predictable stopping time, we denote by $D_c(\mathbb{R}^d;\xi)$ the set of continuous adapted $\mathbb{R}^d$-valued processes with lifetime ξ, and by $\mathscr{S}(\mathbb{R}^d;\xi)$ the set of $\mathbb{R}^d$-valued continuous semimartingales with lifetime ξ. These sets are described as follows: an element of $D_c(\mathbb{R}^d;\xi)$ (resp. $\mathscr{S}(\mathbb{R}^d;\xi)$) is the image under an isomorphic time change A: $[0,\xi[\to \{\xi>0\}\times[0,\infty)$ of an $\mathbb{R}^d$-valued continuous adapted process (resp. semimartingale) defined on the probability space $\left(\{\xi>0\}, (\mathscr{F}_{A_s^{-1}})_{s\geq 0}, \mathbb{P}(\cdot\,|\xi>0)\right)$. They can be endowed with a complete metric space structure, as in the case $\xi=\infty$, which gives respectively the topology of compact convergence in probability and the topology of semimartingales (see [E1]). Let $\mathscr{T}$ denote the set of predictable stopping times and let

$$\hat{D}_c(\mathbb{R}^d) = \bigcup_{\xi\in\mathscr{T}} D_c(\mathbb{R}^d;\xi), \quad \hat{\mathscr{S}}(\mathbb{R}^d) = \bigcup_{\xi\in\mathscr{T}} \mathscr{S}(\mathbb{R}^d;\xi).$$

The sum $X+Y$, difference $X-Y$, product (X,Y) of two processes with lifetime is a process with lifetime the infimum of the lifetimes of the two processes. The lifetime of a process X will be denoted by ξ_X.

If T is a predictable stopping time, we can define the operations of stopping at T and killing at T on the sets $\hat{D}_c(\mathbb{R}^d)$ and $\hat{\mathscr{S}}(\mathbb{R}^d)$: let X be an element of $\hat{D}_c(\mathbb{R}^d)$ or $\hat{\mathscr{S}}(\mathbb{R}^d)$. Then the process X^T stopped at time T is the continuous process with lifetime $+\infty\, 1_{\{T<\xi_X\}} + \xi_X\, 1_{\{T\geq\xi_X\}}$ which coincides with X on $[0, T\wedge\xi_X[$ and is constant on $[T,\infty[\cap\{T<\xi_X\}$; the process X^{T-} killed at time T is the continuous process which has lifetime $T\wedge\xi_X$ and coincides with X on $[0,T\wedge\xi_X[$. If ξ is any predictable stopping time, then by $T<\xi$ we will mean $T<\xi$ on $\{\xi>0\}$ and $T=0$ on $\{\xi=0\}$.

Let us define a topology on the sets $\hat{D}_c(\mathbb{R}^d)$ and $\hat{\mathscr{S}}(\mathbb{R}^d)$. If $X\in\hat{D}_c(\mathbb{R}^d)$ with lifetime ξ_X, T a predictable stopping time such that $T<\xi_X$ and $\varepsilon>0$, one defines neighbourhoods of X with the sets

$$V_{cp}(X,T,\varepsilon) = \left\{Y\in\hat{D}_c(\mathbb{R}^d),\ \mathbb{E}\Big[1\wedge \sup_{0<t\leq T}\|Y_t - X_t\|\Big] < \varepsilon\right\}$$

(with the convention that $\sup\emptyset = 0$ and $\|Z_t\| = +\infty$ if $t\geq\xi_Z$) and

$$W_{cp}(X,\varepsilon) = \left\{Y\in\hat{D}_c(\mathbb{R}^d),\ \mathbb{P}\Big(\{\xi_Y > \xi_X+\varepsilon\}\cap\{\lim_{t\to\xi_X} X_t \text{ exists}\}\Big) < \varepsilon\right\}$$

(the second condition will insure that the topology is separated).

Analogously, one defines neighbourhoods of $X\in\hat{\mathscr{S}}(\mathbb{R}^d)$ by setting

$$V(X,T,\varepsilon) = \left\{Y\in\hat{\mathscr{S}}(\mathbb{R}^d),\ \mathbb{E}[1\wedge v(Y-X)_T] < \varepsilon\right\}$$

where if $Z = Z_0 + M + A$ is the canonical decomposition of $Z \in \hat{\mathscr{S}}(\mathbb{R}^d)$,

$$v(Z)_t = \sum_{i=1}^{d} \left(|Z_0^i| + <M^i, M^i>_t^{1/2} + \int_0^t |dA^i| \right)$$

(with the convention that $v(Z)_t = +\infty$ if $t \geq \xi_Z$) and

$$W(X,\varepsilon) = \left\{ Y \in \hat{\mathscr{S}}(\mathbb{R}^d),\ \mathbb{P}\Big(\{\xi_Y > \xi_X + \varepsilon\} \cap \{\lim_{t \to \xi_X} X_t \text{ exists}\} \Big) < \varepsilon \right\}.$$

PROPOSITION AND DEFINITION 2.1. — *The basis of neighbourhoods*

$$\left.\begin{array}{ll} V_{\mathrm{cp}}(X,T,\varepsilon) \cap W_{\mathrm{cp}}(X,\varepsilon'), & X \in \hat{D}_{\mathrm{c}}(\mathbb{R}^d) \\ (\textit{resp. } V(X,T,\varepsilon) \cap W(X,\varepsilon'), & X \in \hat{\mathscr{S}}(\mathbb{R}^d)) \end{array}\right\} \begin{array}{l} \varepsilon, \varepsilon' > 0,\ T \textit{ predictable} \\ \textit{stopping time such that } T < \xi_X, \end{array}$$

defines a separated topology on $\hat{D}_{\mathrm{c}}(\mathbb{R}^d)$ (resp. $\hat{\mathscr{S}}(\mathbb{R}^d)$) such that every point has a countable basis of neighbourhoods. This topology will be called the topology of compact convergence in probability (resp. the topology of semimartingales).

REMARKS. — 1) If for the topology in $\hat{D}_{\mathrm{c}}(\mathbb{R}^d)$ (resp. $\hat{\mathscr{S}}(\mathbb{R}^d)$) $(X^n)_{n \in \mathbb{N}}$ converges to X, then $\xi_{X^n} \wedge \xi_X$ converges in probability to ξ_X and ξ_{X^n} converges to ξ_X in probability on the set $\{\lim_{t \to \xi_X} X_t \text{ exists}\}$.

2) Let $\xi \in \mathscr{T}$. The topology of the complete metric space $(D_{\mathrm{c}}(\mathbb{R}^d, \xi), d_{\mathrm{cp}})$ (resp. $(\mathscr{S}(\mathbb{R}^d, \xi), d_{\mathrm{sm}})$) defined in [E1] is exactly the topology induced by $\hat{D}_{\mathrm{c}}(\mathbb{R}^d)$ (resp. $\hat{\mathscr{S}}(\mathbb{R}^d)$) on $D_{\mathrm{c}}(\mathbb{R}^d, \xi)$ (resp. $\mathscr{S}(\mathbb{R}^d, \xi)$).

Proof of Proposition 2.1. — We are going to prove this for $\hat{D}_{\mathrm{c}}(\mathbb{R}^d)$. To see that every point has a countable basis of neighbourhoods, one shows that is is sufficient to consider an increasing sequence of predictable stopping times $(T_m)_{m \in \mathbb{N}}$ converging to ξ_X and such that $T_m < \xi_X$ for all m.

Let us show that the topology is separated. If $X \neq Y$, then two situations can occur. Either there exists $\varepsilon > 0$ and a predictable stopping time T with $T < \xi_X \wedge \xi_Y$ and $\mathbb{E}\Big[1 \wedge \sup_{0 < t \leq T} \|Y_t - X_t\| \Big] > 2\varepsilon$ in which case $V_{\mathrm{cp}}(X,T,\varepsilon) \cap V_{\mathrm{cp}}(Y,T,\varepsilon) = \emptyset$, or $Y^{\xi_X -} = X^{\xi_Y -}$ with $\mathbb{P}(\xi_Y < \xi_X) > 0$ and there exists $\varepsilon > 0$ and a predictable stopping time T satisfying $T < \xi_X$ such that $\mathbb{P}\Big(\big\{ \xi_Y + 2\varepsilon < T,\ \lim_{t \to \xi_Y} Y_t \text{ exists} \big\} \Big) > 2\varepsilon$; in this case, one verifies that $V_{\mathrm{cp}}(X,T,\varepsilon) \cap W_{\mathrm{cp}}(Y,\varepsilon) = \emptyset$. □

REMARK. — Convergence for the topology of semimartingales implies compact convergence in probability.

For $1 \leq p \leq \infty$ and $\xi \in \mathscr{T}$, let $S^p(\mathbb{R}^d, \xi)$ denote the Banach space of processes $X \in D_{\mathrm{c}}(\mathbb{R}^d, \xi)$ such that $\|X\|_{S^p(\mathbb{R}^d, \xi)} = \|X_\xi^*\|_{L^p} < \infty$, where $X_t^* = \sup_{s<t} \|X_s\|$ on $0 \leq t \leq \xi$ and $\sup \emptyset = 0$. Let $\hat{S}^p(\mathbb{R}^d) = \cup_{\xi \in \mathscr{T}} S^p(\mathbb{R}^d, \xi)$.

DEFINITION 2.2. — *We say that a sequence $(X^n)_{n\in\mathbb{N}}$ in $\hat{D}_c(\mathbb{R}^d)$ converges to $X \in \hat{D}_c(\mathbb{R}^d)$ locally in $\hat{D}_c(\mathbb{R}^d)$ (resp. $\hat{S}^p(\mathbb{R}^d)$) if the following two conditions are satisfied:*

(i) *There exists an increasing sequence of stopping times $(T_m)_{m\in\mathbb{N}}$ converging to ξ_X such that for any m, $T_m < \xi_X$, $(X^n)^{T_m-}$ belongs to $D_c(\mathbb{R}^d, T_m)$ (resp. $S^p(\mathbb{R}^d, T_m)$) for n sufficiently large and converges in $D_c(\mathbb{R}^d, T_m)$ (resp. $S^p(\mathbb{R}^d, T_m)$) to X^{T_m-}.*

(ii) *The lifetimes ξ_{X^n} converge in probability to the lifetime ξ_X on the set $\{\lim_{t\to\xi_X} X_t$ exists$\}$.*

For $1 \le p \le \infty$ and $\xi \in \mathcal{T}$, let $H^p(\mathbb{R}^d, \xi)$ be the space of processes $X \in \mathcal{S}(\mathbb{R}^d, \xi)$ such that $\|X\|_{H^p(\mathbb{R}^d,\xi)} = \|v(X)_\xi\|_{L^p} < \infty$. Let $\hat{H}^p(\mathbb{R}^d) = \cup_{\xi\in\mathcal{T}} H^p(\mathbb{R}^d, \xi)$.

DEFINITION 2.3. — *We say that a sequence $(X^n)_{n\in\mathbb{N}}$ in $\hat{\mathcal{S}}(\mathbb{R}^d)$ converges to $X \in \hat{\mathcal{S}}(\mathbb{R}^d)$ locally in $\hat{\mathcal{S}}(\mathbb{R}^d)$ (resp. $\hat{H}^p(\mathbb{R}^d)$) if the following two conditions are satisfied:*

(i) *There exists an increasing sequence of stopping times $(T_m)_{m\in\mathbb{N}}$ converging to ξ_X such that for any m, $T_m < \xi_X$, $(X^n)^{T_m-}$ belongs to $\mathcal{S}(\mathbb{R}^d, T_m)$ (resp. $H^p(\mathbb{R}^d, T_m)$) for n sufficiently large and converges in $\mathcal{S}(\mathbb{R}^d, T_m)$ (resp. $H^p(\mathbb{R}^d, T_m)$) to X^{T_m-}.*

(ii) *The lifetimes ξ_{X^n} converge in probability to the lifetime ξ_X on the set $\{\lim_{t\to\xi_X} X_t$ exists$\}$.*

Note that local convergences are not derived from topologies. Their relation to topologies is described in the following proposition which is the analogue for processes with lifetime of [E1] Proposition 1 and Theorem 2.

PROPOSITION 2.4. — *Let $p \in [1, \infty]$ and let $E \subset \hat{D}_c(\mathbb{R}^d)$ (resp. $E \subset \hat{\mathcal{S}}(\mathbb{R}^d)$). Let F be the sequential closure of E for local convergence in $\hat{D}_c(\mathbb{R}^d)$ (resp. $\hat{\mathcal{S}}(\mathbb{R}^d)$), let G be the closure of E for the topology of compact convergence in probability (resp. for the topology of semimartingales), and let K_p be the sequential closure of E for local convergence in $\hat{S}^p(\mathbb{R}^d)$ (resp. $\hat{H}^p(\mathbb{R}^d)$).*

Then $F = G = K_p$.

REMARK. — Proposition 2.4 can be rewritten as follows: let $(X^n)_{n\in\mathbb{N}}$ be a sequence of elements of $\hat{D}_c(\mathbb{R}^d)$ (resp. $\hat{\mathcal{S}}(\mathbb{R}^d)$). Then the following three conditions are equivalent:

(i) for every subsequence $(Y^n)_{n\in\mathbb{N}}$, there exists a subsubsequence $(Z^n)_{n\in\mathbb{N}}$ which converges to X^0 locally in $\hat{D}_c(\mathbb{R}^d)$ (resp. $\hat{\mathcal{S}}(\mathbb{R}^d)$),

(ii) $(X^n)_{n\in\mathbb{N}}$ converges to X^0 in the topology of compact convergence in probability (resp. in the topology of semimartingales),

(iii) for every subsequence $(Y^n)_{n\in\mathbb{N}}$, there exists a subsubsequence $(Z^n)_{n\in\mathbb{N}}$ which converges to X^0 locally in $\hat{S}^p(\mathbb{R}^d)$ (resp. $\hat{H}^p(\mathbb{R}^d)$).

Proof of Proposition 2.4. — 1) Second equality: We will give the proof for compact convergence in probability. The proof for semimartingale convergence is similar.

To prove $K_1 \subset G$, it is sufficient to verify that if X^n converges to X locally in $\hat{S}^1(\mathbb{R}^d)$, then X^n converges to X for the topology of compact convergence in probability, and this is almost evident.

We are left to prove that $G \subset K_\infty$, i.e. that for every sequence X^n converging to X for the topology of compact convergence in probability, there exists a subsequence which converges to X locally in $\hat{S}^\infty(\mathbb{R}^d)$. One easily shows that condition (ii) of local convergence is satisfied, without extracting a subsequence. By extracting a subsequence to obtain an a.s. convergence of $\xi_{X^n} \wedge \xi_X$ and by stopping at a time smaller than ξ_X but close to ξ_X in probability, one may assume that all the terms of the sequence belong to $D_c(\mathbb{R}^d, \infty)$. One can also assume that $X = 0$. It is then sufficient to show that we can find a stopping time T as big as we want for the topology of convergence in probability and a subsequence $(X^{n_k})_{k\in\mathbb{N}}$ such that $(X^{n_k})^T$ converges to 0 in $S^\infty(\mathbb{R}^d, \infty)$ (a sequence of stopping time increasing to ∞ and a diagonal subsequence give then the result). But for every $M \in \mathbb{N}^*$, $(X^n)^*_M$ converges in probability to 0. By extracting a subsequence one can assume that the convergence is almost sure. The end of the proof is similar to the proof of Egoroff's theorem: let $\varepsilon > 0$, $T^n_m = M \wedge \inf\{t > 0,\ \|(X^n)^*_t\| \geq 1/m\}$, $S^n_m = \inf_{k\geq n} T^k_m$, $n(m)$ such that $\mathbb{P}(S^{n(m)}_m < M - 1) < \dfrac{\varepsilon}{2^m}$, and $R = \inf_{m\in\mathbb{N}^*} S^{n(m)}_m$. Then R is as close to ∞ as we want and $(X^n)^R$ converges a.s. uniformly to 0.

2) The proof of the first equality is identical as the one for infinite times. □

As a corollary, using the demonstration of Theorem 2 in [E1], one can show that a sequence $(X^n)_{n\in\mathbb{N}}$ of elements of $\hat{\mathscr{S}}(\mathbb{R}^d)$ converges to $X \in \hat{\mathscr{S}}(\mathbb{R}^d)$ if and only if it converges in $\hat{D}_c(\mathbb{R}^d)$ and for all bounded predictable process H with values in $\mathbb{R}^d$, $\left(\int_0^\cdot H dX^n\right)^{\xi_X-}$ converges in $\hat{D}_c(\mathbb{R})$ to $\int_0^\cdot H dX$ (compare with the definition of the topology of semimartingales in [E1]).

Definition 2.5. — *Let $E, F = \hat{D}_c(\mathbb{R}^d)$ or $\hat{\mathscr{S}}(\mathbb{R}^d)$, and let $\phi\colon E \to F$ be a map. We will say that ϕ is lower semicontinuous if for every sequence $(X^n)_{n\in\mathbb{N}}$ of elements in E converging to $X \in E$, the sequence $\left((\phi(X^n))^{\xi_{\phi(X)}-}\right)_{n\in\mathbb{N}}$ converges to $\phi(X)$.*

An important example of a lower semicontinuous map is $X \mapsto p(\phi(X)) \in \hat{\mathscr{S}}(\mathbb{R}^d)$ if $X \mapsto \phi(X) \in \hat{\mathscr{S}}(\mathbb{R}^{d+d'})$ is continuous and $p\colon \mathbb{R}^{d+d'} \to \mathbb{R}^d$ the canonical projection.

Note also that if $X \mapsto \phi(X)$ is lower semicontinuous, and if both X and $\phi(X)$ are in $\hat{D}_c$ (or $\hat{\mathscr{S}}$) and the lifetime of $\phi(X)$ is greater or equal to the lifetime of X, then $X \mapsto (X, \phi(X))$ is continuous.

With Proposition 2.4, one can investigate continuity properties for operations on the sets of continuous adapted processes and of semimartingales. For $m \in \mathbb{N}$, let $C^m(\mathbb{R}^d)$ denote the set of real-valued C^m functions on $\mathbb{R}^d$, endowed with the topology of uniform convergence on compact sets of the derivatives up to order m.

Proposition 2.6. — 1) *The map*

$$C^0(\mathbb{R}^d) \times \hat{D}_c(\mathbb{R}^d) \longrightarrow \hat{D}_c(\mathbb{R})$$
$$(h, X) \longmapsto h(X)$$

is lower semicontinuous.

2) *The maps*

$$C^2(\mathbb{R}^d) \times \hat{\mathscr{S}}(\mathbb{R}^d) \longrightarrow \hat{\mathscr{S}}(\mathbb{R})$$
$$(h, X) \longmapsto h(X)$$

and

$$\hat{\mathscr{S}}(\mathbb{R}^d) \longrightarrow \hat{\mathscr{S}}(\mathbb{R})$$
$$X \longmapsto M^i,\ A^i,\ <M^i, M^j>$$

are lower semicontinuous, where $X = X_0 + M + A$ *is the decomposition of* X *into the value at* 0, *a local martingale and a process with finite variation.*

3) *Let* T *be a predictable stopping time. The map*

$$\hat{D}_c(\mathbb{R}^d)\ (resp.\ \hat{\mathscr{S}}(\mathbb{R}^d)) \longrightarrow \hat{D}_c(\mathbb{R}^d)\ (resp.\ \hat{\mathscr{S}}(\mathbb{R}^d))$$
$$X \longmapsto X^{T-}$$

is continuous, and

$$\hat{D}_c(\mathbb{R}^d)\ (resp.\ \hat{\mathscr{S}}(\mathbb{R}^d)) \longrightarrow \hat{D}_c(\mathbb{R}^d)\ (resp.\ \hat{\mathscr{S}}(\mathbb{R}^d))$$
$$X \longmapsto X^{T}$$

is lower semicontinuous and continuous at the points X *with lifetime* ξ_X *such that* $\mathbb{P}(\xi_X = T) = 0$.

4) *Let* U *be an open subset of* $\mathbb{R}^d$. *If* X *belongs to* $\hat{D}_c(\mathbb{R}^d)$, *let* $T_U(X)$ *denote the exit time of* X *from* U, *i.e.,* $T_U(X) = \inf\{t > 0,\ X_t \notin U\}$ *(with* $\inf \emptyset = +\infty$*). Then*

$$\hat{D}_c(\mathbb{R}^d)\ (\hat{\mathscr{S}}(\mathbb{R}^d)) \longrightarrow \hat{D}_c(\mathbb{R}^d)\ (\hat{\mathscr{S}}(\mathbb{R}^d))$$
$$X \longmapsto X^{T_U(X)-}$$

is lower semicontinuous, and

$$\hat{D}_c(\mathbb{R}^d)\ (\hat{\mathscr{S}}(\mathbb{R}^d)) \longrightarrow \hat{D}_c(U)\ (\hat{\mathscr{S}}(U))$$
$$X \longmapsto X^{T_U(X)-}$$

is continuous.

In part 4), $\hat{D}_c(U)$ $(\hat{\mathscr{S}}(U))$ is the set of elements of $\hat{D}_c(\mathbb{R}^d)$ $(\hat{\mathscr{S}}(\mathbb{R}^d))$ which take their values in U, endowed with a topology defined in the same manner.

Proof. — 1) By Proposition 2.4, is sufficient to show that for every sequence (h^n, X^n) converging to (h, X), there exists a subsequence (h^{n_k}, X^{n_k}) such that $h^{n_k}(X^{n_k})$ satisfies condition (i) of local convergence to $h(X)$ in $\hat{S}^\infty(\mathbb{R}^d)$. But using again Proposition 2.4, by extracting a subsequence, we can assume that the X^n are locally bounded and converge locally a.s. uniformly to X. We conclude using the fact that h is uniformly continuous on compact sets and h^n converges to h uniformly on compact sets.

2) The proof is analogous to 1) using the equality

$$v(h(X)) = |h(X_0)| + \left(\int_0 D_i h(X) D_j h(X)\, d\langle M^i, M^j\rangle\right)^{1/2} + \int_0 \left| \frac{1}{2} D_{ij} h(X)\, d\langle M^i, M^j\rangle + D_i h(X)\, dA^i \right|$$

and condition (i) of local convergence in $\hat{H}^\infty(\mathbb{R}^d)$.

3) The proof is left to the reader.

4) We only give a sketch of the proof for the second assertion. It is sufficient to prove that for every T satisfying $T < T_U(X) \wedge \xi_X$, $T_U(X^n) \wedge T$ converges in probability to T, and that $T_U(X^n)$ converges in probability to $T_U(X)$ on the event $\left\{ \lim_{t \to \xi_X \wedge T_U(X)} X_t \text{ exists in } U \right\}$. But this is a consequence of the existence for every subsequence $(X^{n_k})_{k \in \mathbb{N}}$ of a subsubsequence which converges locally a.s. uniformly. □

A consequence of 1) is that if F is a closed subspace of $\mathbb{R}^d$, then taking $h(x) = \mathrm{dist}(x, F)$ shows that the subset of $\hat{D}_c(\mathbb{R}^d)$ ($\hat{\mathscr{S}}(\mathbb{R}^d)$) consisting of F-valued processes is closed. This topological subspace will be denoted by $\hat{D}_c(F)$ ($\hat{\mathscr{S}}(F)$).

Property 4) is very useful for the study of manifold-valued processes and stochastic differential equations. It removes problems in connection with the exit time from domains of definition. It allows localization in time.

We are now interested in differentiability properties.

DEFINITION 2.7. — *Let $a \mapsto X(a) \in \hat{\mathscr{S}}(\mathbb{R}^d)$ be defined on some interval I in $\mathbb{R}$.*

1) *The map $a \mapsto X(a)$ is differentiable in $\hat{\mathscr{S}}(\mathbb{R}^d)$ at $a_0 \in I$ if it is continuous at a_0 and if there exists $Y \in \hat{\mathscr{S}}(\mathbb{R}^d)$ such that $\dfrac{X(a) - X(a_0)}{a - a_0}$ converges in $\hat{\mathscr{S}}(\mathbb{R}^d)$ to Y as $a \to a_0$. Then $(X(a_0), Y)$ is called the derivative of X at a_0.*

2) *The map $a \mapsto X(a)$ is C^1 in $\hat{\mathscr{S}}(\mathbb{R}^d)$ if for all $a_0 \in I$, $a \mapsto X(a)$ is differentiable in $\hat{\mathscr{S}}(\mathbb{R}^d)$ at a_0, and if the derivative $a \mapsto Y(a)$ is continuous in $\hat{\mathscr{S}}(\mathbb{R}^{2d})$. The semimartingale $Y(a)$ is denoted by $\partial_a X(a)$.*

3) *For $k \geq 1$, the map $a \mapsto X(a)$ is C^{k+1} in $\hat{\mathscr{S}}(\mathbb{R}^d)$ if $a \mapsto X(a)$ is C^1 in $\hat{\mathscr{S}}(\mathbb{R}^d)$ and $\partial_a X(a)$ is C^k in $\hat{\mathscr{S}}(\mathbb{R}^{2d})$.*

REMARKS. — 1) In the first part of the definition, one asks $a \mapsto X(a)$ to be continuous at a_0 only to guarantee that $\xi_{X(a)}$ converges in probability to $\xi_{X(a_0)}$ on the set $\left\{ \lim_{t \to \xi(a_0)} X_t(a_0) \text{ exists} \right\}$.

2) In the same manner, replacing $\hat{\mathscr{S}}(\mathbb{R}^d)$ by $\hat{D}_c(\mathbb{R}^d)$ in Definition 2.7, the notion of a map $a \mapsto X(a) \in \hat{D}_c(\mathbb{R}^d)$ being C^k in $\hat{D}_c(\mathbb{R}^d)$ can be defined.

The following proposition says that regularity of paths implies regularity in $\hat{D}_c(\mathbb{R}^d)$.

PROPOSITION 2.8. — *Let $k \geq 0$. Suppose $a \mapsto X(a) \in \hat{D}_c(\mathbb{R}^d)$, with lifetime $\xi(a)$, is defined on an open interval I in $\mathbb{R}$. Assume that ω-almost surely, $a \mapsto \xi(a)(\omega)$ is lower semicontinuous and continuous at a_0 if $\lim_{t\to\xi(a_0)(\omega)} X_t(a_0)$ exists, $a \mapsto X_t(a)(\omega)$ is of class C^k on its domain for all t, and that the map $(t,a) \mapsto \partial_a^k(X_t(a)(\omega))$, defined on $\{(t,a) \in \mathbb{R}_+ \times I,\ 0 \leq t < \xi(a)(\omega)\}$, is continuous.*

Then $a \mapsto X(a)$ is C^k in $\hat{D}_c(\mathbb{R}^d)$.

Proof. — Let us first consider the case $k = 0$. Let $(a_\ell)_{\ell\in\mathbb{N}^*}$ be a sequence of elements of I converging to $a_0 \in I$. Then $\xi(a_\ell)$ converges almost surely to $\xi(a_0)$ on the set $\left\{\lim_{t\to\xi(a_0)(\omega)} X_t(a_0) \text{ exists}\right\}$, hence for $\varepsilon > 0$, $X(a_\ell) \in W_{\text{cp}}(X(a_0),\varepsilon)$ for ℓ sufficiently large. Since $\xi(a_0) \wedge \xi(a_\ell)$ converges almost surely to $\xi(a_0)$, the stopping times $T'_m = \inf_{\ell\geq m} \xi(a_0) \wedge \xi(a_\ell)$ are predictable, increasing in m, and converge still almost surely to $\xi(a_0)$. Thus there exists a sequence of predictable stopping times $(T_m)_{m\in\mathbb{N}^*}$ increasing almost surely to $\xi(a_0)$, such that almost surely, for all m, $T_m < T'_m$ on $\{T'_m > 0\}$.

By the second part of Proposition 2.4, it is sufficient to show that $X(a_\ell)^{T_m-}$ converges in $\hat{D}_c(\mathbb{R}^d)$ to $X(a_0)^{T_m-}$ as ℓ tends to ∞. But on $\{T_m > 0\}$, almost surely, there exists $\varepsilon(\omega) > 0$ such that the map

$$\begin{aligned}[0,T_m(\omega)] \times [a_0 - \varepsilon(\omega), a_0 + \varepsilon(\omega)] &\longrightarrow \mathbb{R}^d \\ (t,a) &\longmapsto X_t(a)(\omega)\end{aligned}$$

is well-defined and uniformly continuous. Thus $\lim_{\ell\to\infty} \sup_{0\leq t\leq T_m} \|X_t(a_\ell) - X_t(a_0)\| = 0$ almost surely on $\{T_m > 0\}$, and this gives the convergence of $X(a_\ell)^{T_m-}$ to $X(a_0)^{T_m-}$ in $\hat{D}_c(\mathbb{R}^d)$. Hence we have the result.

If $k = 1$, let a_0, $(a_\ell)_{\ell\in\mathbb{N}^*}$, $(T_m)_{m\in\mathbb{N}^*}$ be as above. It is sufficient to prove that for every m,

$$\frac{X^{T_m-}(a_\ell) - X^{T_m-}(a_0)}{a_\ell - a_0}$$

converges to $\partial_a X^{T_m-}(a_0)$ in $\hat{D}_c(\mathbb{R}^d)$, as ℓ tends to ∞. Almost surely on $\{T_m > 0\}$, there exists $\varepsilon(\omega) > 0$ such that the map

$$\begin{aligned}[0,T_m(\omega)] \times [a_0 - \varepsilon(\omega), a_0 + \varepsilon(\omega)] &\longrightarrow \mathbb{R}^d \\ (t,a) &\longmapsto \partial_a X_t(a)(\omega)\end{aligned}$$

is defined and uniformly continuous. But, for such ω, t, a, we have

$$\left\|\frac{X_t(a) - X_t(a_0)}{a - a_0} - \partial_a X_t(a_0)\right\| \leq \sup_{\|b-a_0\|\leq\|a-a_0\|} \|\partial_a X_t(b) - \partial_a X_t(a_0)\|,$$

hence

$$\begin{aligned}\sup_{0\leq t\leq T_m}&\left\|\frac{X_t(a_\ell) - X_t(a_0)}{a_\ell - a_0} - \partial_a X_t(a_0)\right\| \\ &\leq \sup_{\|b-a_0\|\leq\|a_\ell-a_0\|}\ \sup_{0\leq t\leq T_m} \|\partial_a X_t(b) - \partial_a X_t(a_0)\|\end{aligned}$$

and the left-hand side converges almost surely to 0 as ℓ tends to ∞. It implies that $\dfrac{X^{T_m^-}(a_\ell) - X^{T_m^-}(a_0)}{a_\ell - a_0}$ converges to $\partial_a X^{T_m^-}(a_0)$ in $\hat{D}_c(\mathbb{R}^d)$, as ℓ tends to ∞.

If $k \geq 2$, one can prove in the same way by induction that for $\ell \leq k$, $a \mapsto X(a)$ is C^ℓ in $\hat{D}_c(\mathbb{R}^d)$, and almost surely, for all t, $(\partial_a^\ell X)_t = \partial_a^\ell (X_t)$. □

REMARK. — Proposition 2.8 is false with $\hat{\mathscr{S}}(\mathbb{R}^d)$.

2.2. *Manifold-valued processes*

Let M be a connected smooth manifold endowed with a connection ∇. With respect to some fixed filtered probability space $(\Omega, \mathscr{F}, (\mathscr{F}_t)_{t\geq 0}, \mathbb{P})$, for every predictable stopping time ξ, let let $D_c(M, \xi)$ denote the set of M-valued adapted continuous processes with lifetime ξ, and $\mathscr{S}(M, \xi)$ the set of M-valued continuous semimartingales with lifetime ξ. The spaces $D_c(F; \xi)$, $\mathscr{S}(F; \xi)$, $\hat{D}_c(F)$, $\hat{\mathscr{S}}(F)$, where F is a closed subset of M are defined by analogy with the previous definitions.

Let ϕ: $M \to \mathbb{R}^d$ be a smooth proper embedding. Then $\phi(M)$ is a closed subset of $\mathbb{R}^d$. As a consequence, $(\hat{D}_c(\phi(M)), d_{\text{cp}})$, resp. $(\hat{\mathscr{S}}(\phi(M)), d_{\text{sm}})$, is a topological subspace of $\hat{D}_c(\mathbb{R}^d)$, resp. $\hat{\mathscr{S}}(\mathbb{R}^d)$. By means of the diffeomorphism ϕ: $M \to \phi(M)$, we obtain complete topological space structures on $\hat{D}_c(M)$ and $\hat{\mathscr{S}}(M)$.

DEFINITION 2.9. — *Let ϕ: $M \to \mathbb{R}^d$ be a smooth proper embedding.*

1) *The topology of compact convergence in probability on $\hat{D}_c(M)$ is the topology induced by the diffeomorphism ϕ: $M \to \phi(M)$ and the topological space $\hat{D}_c(\phi(M))$.*

2) *The topology of semimartingales on $\hat{\mathscr{S}}(M)$ is the topology induced by the diffeomorphism ϕ: $M \to \phi(M)$ and the topological space $\hat{\mathscr{S}}(\phi(M))$.*

Since every smooth function on M is of the form $g \circ \phi$ for some smooth g: $\mathbb{R}^d \to \mathbb{R}$, it is easy to see that the induced structures are independent of the choice of the proper embedding ϕ.

Independent of the proper embedding ϕ are also the notions of local convergence in $\hat{S}^\infty(\phi(M))$ and of local convergence both in $\hat{S}^\infty(\phi(M))$ and in $\hat{H}^\infty(\phi(M))$. This is of great importance in the sequel.

With a proper embedding ϕ, we can also define differentiability for families of processes in $\hat{D}_c(M)$ (resp. $\hat{\mathscr{S}}(M)$). In this case, if $a \mapsto \phi(X(a))$ is differentiable at a_0 and Z is the derivative of $\phi(X(a))$ at a_0, then it is easy to verify that Z takes its values in $T\phi(TM)$ and the derivative of $X(a)$ at a_0 is the process $\partial_a X(a_0) = (T\phi)^{-1}(Z)$ with values in $\hat{D}_c(TM)$ (resp. $\hat{\mathscr{S}}(TM)$).

Let $\hat{\mathscr{M}}_\nabla(M)$ be the set of continuous martingales with lifetime in $\hat{D}_c(M)$. By [E4 4.43], $\hat{\mathscr{M}}_\nabla(M)$ is closed in $\hat{D}_c(M)$. This implies that it is also closed in $\hat{\mathscr{S}}(M)$.

PROPOSITION 2.10. — *On $\hat{\mathscr{M}}_\nabla(M)$, the topology of compact convergence in probability and the topology of semimartingales coincide.*

To establish this result, we need some lemmas.

Lemma 2.11. — *Every point x of M has a compact neighbourhood V, contained in the domain of a chart h, together with a smooth convex function ψ: $V \times V \to \mathbb{R}_+$ which satisfies the following conditions:*

1) *For all $x, y \in V$, $\psi(x, y) = 0$ if and only if $x = y$,*

2) *There exists a constant $c > 0$ such that for all $(X, Y) \in T_xM \times T_yM$, $x, y \in V$, with coordinates $\bar{X} = dh(X)$, $\bar{Y} = dh(Y) \in \mathbb{R}^d$,*

$$(\nabla \otimes \nabla) d\psi(x, y)\big((X, Y), (X, Y)\big) \geq c \, \|\bar{Y} - \bar{X}\|^2,$$

3) *For every Riemannian metric δ on V there exists a constant $A > 0$ such that $\psi \leq A\,\delta^2$.*

It is proven in [K] that convex geometry (the existence of a convex function ψ satisfying 1)) implies that every V-valued martingale has almost surely a limit at infinity.

Proof. — We show that the function ψ defined in [E4 4.59] has the desired properties. For $x_0 \in M$, take an exponential chart (h, V) centered at x_0, and define

$$\psi(x, y) = \frac{1}{2}\left(\varepsilon^2 + \|h(x) + h(y)\|^2\right) \|h(x) - h(y)\|^2.$$

Note that ψ satisfies 1) and 3). It is proven in [E4 4.59] that, if V is sufficiently small, one can choose $\varepsilon > 0$ and $0 < \beta < 1$ such that if $U = (U_1, U_2) \in TV \oplus TV$ is a tangent vector with coordinates $(\bar{X}, \bar{Y}) \in \mathbb{R}^d \oplus \mathbb{R}^d$ where $\bar{X} = dh(U_1)$, $\bar{Y} = dh(U_2)$, then

$$\begin{aligned}(\nabla \otimes \nabla) d\psi(U, U) &\geq (1 - \beta)\left(\varepsilon^2 \|\bar{X} - \bar{Y}\|^2 + \|h(x) - h(y)\|^2 \|\bar{X} + \bar{Y}\|^2\right) \\ &\geq (1 - \beta)\, \varepsilon^2 \, \|\bar{X} - \bar{Y}\|^2.\end{aligned}$$

This gives 2). □

Lemma 2.12. — *Let V, δ be as in Lemma 2.11. There exists a constant $C > 0$ such that if Y and Z are V-valued martingales, $h(Y) = (Y^1, \ldots, Y^d)$ and $h(Z) = (Z^1, \ldots, Z^d)$ in coordinates, then*

$$\mathbb{E}\left[\sum_{i=1}^{d} \langle Y^i - Z^i, Y^i - Z^i \rangle_\infty\right] \leq C\, \mathbb{E}\left[\delta^2(Y_\infty, Z_\infty)\right].$$

Remark. — In particular, applying this result with a constant Z, we deduce that the expectation of the quadratic Riemannian variation of Y is bounded by a constant independent of Y.

Proof of Lemma 2.12. — Let ψ be as in Lemma 2.11. The Itô formula applied to ψ and (Y, Z) gives

$$\begin{aligned}\psi(Y_\infty, Z_\infty) = \psi(Y_0, Z_0) &+ \int_0^\infty \langle d\psi, d^{\nabla \otimes \nabla}(Y, Z) \rangle \\ &+ \frac{1}{2} \int_0^\infty (\nabla \otimes \nabla) d\psi(Y, Z)\big(d(Y, Z) \otimes d(Y, Z)\big)\end{aligned}$$

where $d^{\nabla\otimes\nabla}$ denotes the Itô differential with respect to the product connection in $M \times M$. Using the fact that (Y,Z) is a martingale, we obtain

$$\mathbb{E}[\psi(Y_\infty, Z_\infty)] = \mathbb{E}[\psi(Y_0, Z_0)] + \frac{1}{2}\mathbb{E}\left[\int_0^\infty (\nabla\otimes\nabla)d\psi(Y,Z)\big(d(Y,Z)\otimes d(Y,Z)\big)\right],$$

hence by 2) and 3) of Lemma 2.11, we have

$$A\,\mathbb{E}\left[\delta^2(Y_\infty, Z_\infty)\right] \geq \mathbb{E}[\psi(Y_\infty, Z_\infty)] \geq \frac{c}{2}\,\mathbb{E}\left[\sum_{i=1}^d <Y^i - Z^i, Y^i - Z^i>_\infty\right].$$

This gives the result, with $C = 2A/c$. □

Proof of Proposition 2.10. — We may assume that M is a closed subset of $\mathbb{R}^d$, and have to show that every sequence $(X^n)_{n\in\mathbb{N}}$ of ∇-martingales converging in $\hat{D}_c(M)$ to a ∇-martingale X converges in $\hat{\mathscr{S}}(M)$ to the same limit. By means of the second equality of Proposition 2.4 with $p = 1$, it is sufficient to prove the existence of a subsequence which converges to X locally in $H^1(\mathbb{R}^d, \infty)$. Since we are allowed to extract subsequences and since we have to prove only local convergence, by using the second equality of Proposition 2.4 with $p = \infty$, we may assume that $(X^n)_{n\in\mathbb{N}}$ converges to X in $S^\infty(\mathbb{R}^d, \infty)$. Still using the fact that it is sufficient to prove local convergence, we may further assume the existence of a finite increasing sequence of stopping times such that if S and T are two consecutive times in this sequence, then on $[S, T[$ all the $(X^n)_{n\in\mathbb{N}}$ and X take values in a compact set V as considered in Lemma 2.11. Finally, since the sequence of stopping times is finite, it is sufficient to prove convergence on one of the intervals $[S, T[$. Hence we assume that $(X^n)_{n\in\mathbb{N}}$ is a sequence of V-valued ∇-martingales converging to X in $S^\infty(\mathbb{R}^d, \infty)$, and it is sufficient to prove its convergence to X locally in $H^1(\mathbb{R}^d, \infty)$.

Since we are dealing with martingales, the finite variation parts of the coordinates satisfy

$$d\widetilde{(X^n)}^i = -\frac{1}{2}\sum_{j,k=1}^d \Gamma^i_{jk}(X^n)\, d<(X^n)^j, (X^n)^k>,$$

$$d\widetilde{X}^i = -\frac{1}{2}\sum_{j,k=1}^d \Gamma^i_{jk}(X)\, d<X^j, X^k>$$

where Γ^i_{jk} are the Christoffel symbols of the connection. This gives the bound

$$\begin{aligned}\|X^n - X\|_{H^1(\mathbb{R}^d,\infty)} \leq \mathbb{E}\Bigg[&\sum_{i=1}^d |(X^n)^i_0 - X^i_0| \sum_{i=1}^d <(X^n)^i - X^i, (X^n)^i - X^i>^{1/2}_\infty \\ &+ \sum_{i,j,k=1}^d \int_0^\infty \bigg(|\Gamma^i_{jk}(X^n) - \Gamma^i_{jk}(X)|\, |d<X^j, X^k>| \\ &+ |\Gamma^i_{jk}(X^n)|\left(|d<(X^n)^j - X^j, (X^n)^k>| + |d<X^j, (X^n)^k - X^k>|\right)\bigg)\Bigg].\end{aligned}$$

The Christoffel symbols are Lipschitz on V, hence by dominated convergence, $\sum_{i,j,k=1}^{d}\int_0^{\infty}\left|\Gamma_{jk}^i(X^n)-\Gamma_{jk}^i(X)\right|\left|d{<}X^j,X^k{>}\right|$ converges to 0 almost surely, and still by dominated convergence and the remark after Lemma 2.12, its expectation converges to 0. Since the Christoffel symbols are bounded on V, the last terms can be bounded by

$$C\,\mathbb{E}\left[\sum_{i,j=1}^{d} {<}(X^n)^i - X^i,(X^n)^i - X^i{>}_{\infty}^{1/2}\left({<}(X^n)^j,(X^n)^j{>}_{\infty}^{1/2} + {<}X^j,X^j{>}_{\infty}^{1/2}\right)\right]$$

with a constant $C > 0$. Using Hölder's inequality and uniform boundedness of the expectations of the quadratic variations of V-valued martingales, we are led to show that $\mathbb{E}\left[\sum_{i=1}^{d}{<}(X^n)^i - X^i,(X^n)^i - X^i{>}_{\infty}\right]$ converges to 0. But, by means of Lemma 2.12,

$$\mathbb{E}\left[\sum_{i=1}^{d}{<}(X^n)^i - X^i,(X^n)^i - X^i{>}_{\infty}\right] \le C\,\mathbb{E}\left[\delta^2(X_{\infty}^n,X_{\infty})\right]$$

with a constant $C > 0$, and this gives the result. □

3. Regularity of solutions of stochastic differential equations

Let M and N be connected smooth manifolds. In this section, we will study stability of second order stochastic differential equations of the type

$$\mathcal{D}Z = f(X,Z)\,\mathcal{D}X \tag{3.1}$$

where $f \in \Gamma(\tau(M)^* \otimes \tau(N))$ is a Schwartz morphism, X belongs to $\hat{\mathcal{S}}(M)$ and Z to $\hat{\mathcal{S}}(N)$.

REMARK. — If P is a submanifold of $M \times N$ such that the canonical projection $P \to M$ is a surjective submersion, and if f is only defined on P and constrained to P (see [E3]), then one can extend f in a smooth way to $M \times N$, and one knows that a solution (X,Z) of (3.1) with $(X_0,Z_0) \in P$ will stay on P.

PROPOSITION 3.1. — *Let $(X^n)_{n\in\mathbb{N}}$ be a sequence of elements in $\hat{\mathcal{S}}(M)$ converging to X in $\hat{\mathcal{S}}(M)$, let $(Z_0^n)_{n\in\mathbb{N}}$ be a sequence of N-valued random variables converging to Z_0 in probability, and let $(f^n)_{n\in\mathbb{N}}$ be a sequence of locally Lipschitz Schwartz morphisms in $\Gamma(\tau(M)^*\otimes\tau(N))$ with uniform Lipschitz constant on compact sets, converging to a Schwartz morphism $f \in \Gamma(\tau(M)^*\otimes\tau(N))$. If Z^n is the maximal solution starting from Z_0^n to $\mathcal{D}Z^n = f^n(X^n,Z^n)\,\mathcal{D}X^n$, then (X^n,Z^n) converges in $\hat{\mathcal{S}}(M\times N)$ to (X,Z) where Z is the solution to $\mathcal{D}Z = f(X,Z)\,\mathcal{D}X$ starting from Z_0. Moreover, if ξ_{X^n} converges in probability to ξ_X then Z^n converges to Z in $\hat{\mathcal{S}}(N)$.*

Proof. — Let ξ_Z be the lifetime of Z. We will show that $(Z^n)^{\xi_Z-}$ converges to Z and that $\lim_{t\to\xi_Z} Z_t$ does not exist on $\{\xi_Z < \xi_X\}$, which is stronger than the results of Proposition 3.1. The second point is known, let us prove the first one. We have to show that there exists a stopping time T as close to ξ_Z as we want and a subsequence Z^{n_k} converging to Z. Hence we can assume that X^n, X take their values in a compact subset K_M, Z_0^n in a compact subset K_N and that X^n converge in $H^\infty(K_M,\infty)$ and in $S^\infty(K_M,\infty)$ to X. We can also assume that Z takes its values in K_N and has lifetime ∞. Consider Schwartz morphisms f_K^n, f_K satisfying the same convergence assumptions as f^n and f, with compact support K containing a neighbourhood of the product $K_M \times K_N$. Using the continuity results of Proposition 2.6 and [E2] theorem 0, we obtain that the solution Z_K^n of $\mathcal{D}Z_K^n = f_K^n(X^n, Z_K^n)\,\mathcal{D}X^n$ with $(Z_K^n)_0 = Z_0^n$ converge in $\mathscr{S}(N,\infty)$ to the solution Z_K of $\mathcal{D}Z_K = f_K(X, Z_K)\,\mathcal{D}X$ with $(Z_K)_0 = Z_0$. This implies that a subsequence converges locally in $\hat{H}^\infty(N)$ and in $\hat{S}^\infty(N)$, but then locally, for indices sufficiently large, the solutions to the truncated equation coincide with the solutions to the original equation. This gives the claim. □

Immediate consequences of Proposition 3.1 are the following results.

COROLLARY 3.2. — 1) *Let $\Gamma^1(\tau(M)^*\otimes\tau(N))$ be the set of C^1 Schwartz morphisms endowed with the topology of uniform convergence on compact sets of the maps and their derivatives, and let $L^0(N)$ be the set of N-valued random variables endowed with the topology of convergence in probability. Then the map*

$$\hat{\mathscr{S}}(M) \times \Gamma^1(\tau(M)^* \otimes \tau(N)) \times L^0(N) \longrightarrow \hat{\mathscr{S}}(M \times N),$$

defined by $(X, f, Z_0) \mapsto (X, Z)$ with Z the maximal solution of $\mathcal{D}Z = f(X,Z)\,\mathcal{D}X$, is continuous.

2) *Let $\Gamma^1(\tau(M)^*)$ be the set of C^1 forms of order 2 endowed with the topology of uniform convergence on compact sets of the maps and their derivatives. Then the map*

$$\begin{aligned}\Gamma^1(\tau(M)^*) \times \hat{\mathscr{S}}(M) &\longrightarrow \hat{\mathscr{S}}(\mathbb{R})\\ (\theta, X) &\longmapsto \int_0 \langle \theta(X), \mathcal{D}X\rangle\end{aligned}$$

is lower semicontinuous.

EXAMPLE. — Here we give an example of a sequence of deterministic paths converging uniformly to a constant path, but such that parallel transports above the elements of this sequence do not converge. This shows in particular that in 1) we cannot replace the topology of semimartingales in $\hat{\mathscr{S}}(M)$ by the topology of compact convergence in probability, unless we restrict for instance to the sets of martingales with respect to a given connection.

Let M be a simply connected surface endowed with a rotationally invariant metric about $o \in M$, represented in polar coordinates as $ds^2 = dr^2 + g^2(r)\,d\vartheta^2$ for some smooth function g. Let $t \mapsto x(t) \in M$ be a path in M, defined on the unit interval $[0,1]$, with polar coordinates $r(t) \equiv \varepsilon$ and $\vartheta(t) = \alpha t$ for some $\alpha > 0$. A straightforward calculation shows that the rotational speed of a parallel transport above x in polar coordinates is $-\alpha g'(\varepsilon)$. Hence the rotational speed in an exponential chart with centre o which realizes an isometry at o is $\alpha(1 - g'(\varepsilon))$ (note that this gives 0 if the metric is flat).

In the following, M is taken to be an open subset of the sphere S^2. Thus we have $g(r) = \sin r$, and $\alpha(1-g'(\varepsilon)) = \alpha(1-\cos\varepsilon)$. Consider the sequence of paths $(x^n)_{n\in\mathbb{N}}$ defined in polar coordinates by $\vartheta^n(t) = 2\pi nt$ and $r^n(t) \equiv \varepsilon_n = \arccos(1-\frac{1}{2n})$ (hence $2\pi n(1 - \cos\varepsilon_n) = \pi$). Since $\varepsilon_n \to 0$, we get uniform convergence of $(x^n)_{n\in\mathbb{N}}$ to the constant path o. But for all n, the rotation at time 1 of a parallel transport above x^n is π. Hence parallel transports above x^n do not converge to a parallel transport above o.

In the sequel we are seeking differentiability results. This requires some geometric preliminaries. We will use the maps $\phi\colon M \times N \times M \to N$ defined by Cohen [C1] and [C2] to describe stochastic differential equations in manifolds with càdlàg semimartingales.

DEFINITION 3.3. — *Let $k \in \mathbb{N}$. A Schwartz morphism $f \in \Gamma(\tau(M)^* \otimes \tau(N))$ (resp. a section $e \in \Gamma(TM^* \otimes TN)$) is said to be of class C^k_{Lip} if f (resp. e) is C^k with locally Lipschitz derivatives of order k.*

We say that a measurable map $\phi\colon M \times N \times M \to N$ is of class $C^{k,\infty}_{\mathrm{Lip}}$ if there exists a neighbourhood of the submanifold $\{(x,z,x),\ (x,z) \in M \times N\}$ on which ϕ is C^∞ with respect to the third variable and all the derivatives with respect to this variable are C^k with locally Lipschitz derivatives of order k (with respect to the three variables).

LEMMA (AND DEFINITION) 3.4. — *Let $k \in \mathbb{N}$. For every Schwartz morphism $f \in \Gamma(\tau(M)^* \otimes \tau(N))$ of class C^k_{Lip}, there exists a map $\phi\colon M \times N \times M \to N$ of class $C^{k,\infty}_{\mathrm{Lip}}$ such that for all $(x,z) \in M \times N$*

$$f(x,z) = \tau_3\,\phi(x,z,x)$$

where $\tau_3\phi$ denotes the second order derivative of ϕ with respect to the third variable. Such a map ϕ will be called a Cohen map associated to f.

In particular, a Cohen map satisfies $\phi(x,z,x) = z$ for all $(x,z) \in M \times N$.

Proof. — First, we remark that it is sufficient to construct ϕ in a neighbourhood of the submanifold $\{(x,z,x),\ (x,z) \in M \times N\}$ and to extend it then in a measurable way to $M \times N \times M$.

Let ∇^M (resp. ∇^N) be a connection on M (resp. N). There exists a neighbourhood of the diagonal of $M \times M$ on which the maps $(x,z) \mapsto v(x,z) = \dot\gamma(0)$ and $(x,z) \mapsto u(x,z) = \ddot\gamma(0)$ are smooth, where γ is the geodesic such that $\gamma(0) = x$ and $\gamma(1) = z$. There exists a neighbourhood of the null section in TN on which the exponential map, denoted by $\exp^N$, is smooth. If $u \in \tau N$ is a second order vector, denote by $F(u) \in TN$ its first order part with respect to the connection ∇^N (see [E4] for the definition).

Thus there exists a neighbourhood V of $\{(x,y,x),\ (x,y)\in M\times N\}$ such that the map

$$\begin{aligned}\phi\colon V &\to N\\ (x,y,z) &\mapsto \exp^N\Big(f(x,y)\,v(x,z)+\frac{1}{2}F\big(f(x,y)\,u(x,z)\big)\Big)\end{aligned}$$

is defined and satisfies the regularity assumptions. We have to verify the equation $\tau_3\phi(x,y,x)=f(x,y)$. For this, it is sufficient to check that these maps coincide on elements of $\tau_x M$ of the form $\dot\gamma(0)$ and $\ddot\gamma(0)$ where γ is a geodesic with $\gamma(0)=x$ and $\gamma(1)=z$. A change of time gives

$$\phi(x,y,\gamma(t))=\exp^N\Big(t\,f(x,y)\,v(x,z)+\frac{t^2}{2}F\big(f(x,y)\,u(x,z)\big)\Big)$$

Taking successively first and second order derivatives with respect to t at time 0 gives the result. □

THEOREM 3.5. — *Let $a\mapsto X(a)$ be C^1 from I to $\mathscr{S}(M)$, let $f\in\Gamma(\tau(M)^*\otimes\tau(N))$ be a Schwartz morphism of class C^1_{Lip}, and $a\mapsto Z(a)$ the maximal solution of*

$$\mathcal{D}Z(a)=f\big(X(a),Z(a)\big)\,\mathcal{D}X(a) \tag{3.2}$$

where $a\mapsto Z_0(a)$ is C^1 in probability. Then the map $a\mapsto(X(a),Z(a))$ defined on I and with values in $\mathscr{S}(M\times N)$ is C^1, and the process $\partial_a Z(a)$ is the maximal solution of

$$\mathcal{D}\partial_a Z(a)=f'\big(\partial_a X(a),\partial_a Z(a)\big)\,\mathcal{D}\partial_a X(a) \tag{3.3}$$

with initial condition $\partial_a Z_0(a)$ where f' is the Schwartz morphism of class C^0_{Lip} defined as follows: if $f(x,z)=\tau_3\,\phi(x,z,x)$ with a $C^{1,\infty}_{\mathrm{Lip}}$ Cohen map ϕ associated to f, then $f'(u,v)=\tau_3\,T\phi(u,v,u)$ for $(u,v)\in TM\times TN$, i.e., $T\phi$ is a $C^{0,\infty}_{\mathrm{Lip}}$ Cohen map associated to f'. If moreover $a\mapsto\xi_{X(a)}$ is continuous in probability, then $a\mapsto Z(a)$ is C^1 in $\mathscr{S}(N)$.

REMARK. — If P is a submanifold of $M\times N$ such that the canonical projection $P\to M$ is a surjective submersion, and if f is only defined on P and is constrained to P, then one can show that f' is constrained to TP. As a consequence, by the remark at the beginning of this section, if $\big(\partial_a X_0(a),\partial_a Z_0(a)\big)$ belongs to TP, then $\big(\partial_a X(a),\partial_a Z(a)\big)$ takes its values in TP.

LEMMA 3.6. — *Let P, Q, R, S be manifolds, $\varphi\colon Q\to P$ and $\psi\colon R\to S$ maps, and let $\phi\colon Q\times R\times Q\to R$ and $\phi'\colon P\times S\times P\to S$ be Cohen maps such that $\phi'\circ(\varphi,\psi,\varphi)=\psi\circ\phi$. Then, for all $(x,y)\in Q\times R$, we have*

$$\tau_3\phi'\big(\varphi(x),\psi(y),\varphi(x)\big)\circ\tau\varphi(x)=\tau\psi(y)\circ\tau_3\phi(x,y,x).$$

If semimartingales X, Z take values in Q, resp. R, and satisfy the equation $\mathcal{D}Z=\tau_3\,\phi(X,Z,X)\,\mathcal{D}X$, then $U=\varphi(X)$ and $V=\psi(Z)$ satisfy

$$\mathcal{D}V=\tau_3\,\phi'(U,V,U)\,\mathcal{D}U.$$

Proof. — It is sufficient and easy to prove the first equality with second order derivatives of curves. The second equality is a consequence of the first one. □

Proof of Theorem 3.5. — Assume that $0 \in I$. Using Proposition 3.1, it is sufficient to prove that $a \mapsto (X(a), Z(a))$ is differentiable at $a = 0$ and that the derivative of $a \mapsto Z(a)$ is the maximal solution of (3.3).

Let ∇^M be a connection on M. There exists an open neighbourhood Δ_1^M of the diagonal Δ^M in $M \times M$ such that for $a \neq 0$ the function

$$\begin{aligned} \varphi_a^M : \Delta_1^M &\longrightarrow U_a^M := \varphi_a^M(\Delta_1) \\ (x, y) &\longmapsto \frac{1}{a}(\exp_x^N)^{-1} y \end{aligned}$$

is well-defined and a diffeomorphism. The same objects on N are denoted with the superscript N. Let ϕ be a $C^{1,\infty}_{\mathrm{Lip}}$ Cohen map associated to f. It is easy to see that

$$\mathcal{D}(Z(0), Z(a)) = \tau_3(\phi, \phi)\big((X(0), X(a)), (Z(0), Z(a)), (X(0), X(a))\big)\mathcal{D}(X(0), X(a)).$$

Let $T^M(a)$ be the exit time of $(X(0), X(a))$ of Δ_1^M, $\widetilde{X}(a) = (X(0), X(a))^{T^M(a)-}$ and $\widetilde{Z}(a)$ be the maximal solution to

$$\mathcal{D}\widetilde{Z}(a) = \tau_3(\phi, \phi)\big(\widetilde{X}(a), \widetilde{Z}(a), \widetilde{X}(a)\big)\,\mathcal{D}\widetilde{X}(a)$$

with initial condition $(Z_0(0), Z_0(a))$. Let then $T^N(a)$ be the exit time of $\widetilde{Z}(a)$ of Δ_1^N. Using Proposition 2.6, it is easy to see that $T^M(a) \wedge \xi_{X(0)}$ converges in probability to $\xi_{X(0)}$ as a tends to 0, and then that $T^N(a) \wedge \xi_{Z(0)}$ converges in probability to $\xi_{Z(0)}$.

By Lemma 3.6, defining $\widetilde{V}(a) = \varphi_a^N(\widetilde{Z}(a)^{T^N(a)-})$ and $\widetilde{Y}(a) = \varphi_a^M(\widetilde{X}(a)^{T^N(a)-})$ for $a \neq 0$, we have that $\widetilde{V}(a)$ is the maximal solution in $\hat{\mathscr{S}}(TM)$ of

$$\mathcal{D}\widetilde{V}(a) = \tau_3\big(\varphi_a^N \circ (\phi, \phi) \circ ((\varphi_a^M)^{-1}, (\varphi_a^N)^{-1}, (\varphi_a^M)^{-1})\big)\big(\widetilde{Y}(a), \widetilde{V}(a), \widetilde{Y}(a)\big)\,\mathcal{D}\widetilde{Y}(a)$$

with initial condition $\widetilde{V}_0(a) = \varphi_a^N(Z_0(0), Z_0(a))$ on $\{(Z_0(0), Z_0(a)) \in \Delta_1^N\}$. For $u \in U_a^M$ (resp. $u \in U_a^N$), denote by $\ell_a^M(u)$ (resp. $\ell_a^N(u)$) the second coordinate of $(\varphi_a^M)^{-1}(u)$ (resp. $(\varphi_a^N)^{-1}(u)$). Then the mapping

$$(a, u, v, w) \mapsto \begin{cases} \varphi_a^N\big(\phi(\pi(u), \pi(v), \pi(w)), \phi(\ell_a^M(u), \ell_a^N(v), \ell_a^M(w))\big) & \text{if } a \neq 0, \\ T\phi(u, v, w) & \text{if } a = 0, \end{cases}$$

defined on an open subset of $(-1,1) \times TM \times TN \times TM$ containing the elements of the form $(0, u, v, u)$ with $(u, v) \in TM \times TN$, depends C^∞ on the last variable and its derivatives with respect to this variable are locally Lipschitz (as functions of all four variables). This implies the convergence of $\tau_3\big(\varphi_a^N \circ (\phi, \phi) \circ ((\varphi_a^M)^{-1}, (\varphi_a^N)^{-1}, (\varphi_a^M)^{-1})\big)$ to $\tau_3 T\phi$ as $a \to 0$, and the existence of uniform Lipschitz constants on compact sets. Since $a \mapsto X(a)$ is differentiable at $a = 0$, $T^M(a) \wedge \xi_{X(0)}$ converges in probability to $\xi_{X(0)}$ and $T^N(a) \wedge \xi_{Z(0)}$ converges in probability to $\xi_{Z(0)}$, we have that $\widetilde{Y}(a)$ converges to $Y(0)^{\xi_{Z(0)}}$ with $Y(0) := \partial_a X(0)$; on the other side, $\widetilde{V}_0(a)$ converges in probability to $\partial_a Z_0(0) = V_0(0)$ on $\{\xi_Z(0) > 0\}$; hence we get by Proposition 3.1 that $(\widetilde{Y}(a), \widetilde{V}(a))$ converges to $(Y(0), V(0))$ where $V(0)$ is the maximal solution of

$$\mathcal{D}V(0) = \tau_3 T\phi\big(Y(0), V(0), Y(0)\big)\,\mathcal{D}Y(0)$$

with initial condition $V_0(0) = \partial_a Z_0(0)$. This implies that $a \mapsto (X(a), Z(a))$ is differentiable at $a = 0$ and that its derivative is $(Y(0), V(0))$. □

We now want to investigate Stratonovich and Itô equations. In the following, if $(t,a) \mapsto x(t,a)$ is a map defined on an open subset of $\mathbb{R}^2$ and with values in a manifold M, $\dot{x}(t,a)$ will denote its derivative with respect to t, and $\ddot{x}(t,a)$ will denote the second order tangent vector such that for all smooth function g on M, $\ddot{x}(t,a)(g) = \partial_t^2(g \circ x)(t,a)$. For a smooth function g on M, d^2g will denote the second order form defined by $\langle d^2g, \ddot{x}(t,a)\rangle = \ddot{x}(t,a)(g)$ (see [E4]).

LEMMA 3.7. — *Let J, I be two intervals in $\mathbb{R}$. Suppose that $(t,a) \mapsto x(t,a) \in M$ and $(t,a) \mapsto z(t,a) \in N$ are $C^{2,1}$ maps defined on $J \times I$, and satisfy for each a*

$$\ddot{z}(0,a) = \tau_3\, \phi\big(x(0,a), z(0,a), x(0,a)\big)\, \ddot{x}(0,a) \tag{3.4}$$

where ϕ: $M \times N \times M \to N$ is a $C^{1,\infty}_{\mathrm{Lip}}$ Cohen map. Then

$$(\partial_a z)^{\cdot\cdot}(0,a) = \tau_3\, T\phi\big(\partial_a x(0,a), \partial_a z(0,a), \partial_a x(0,a)\big)(\partial_a x)^{\cdot\cdot}(0,a).$$

Proof. — It is sufficient to prove

$$\big\langle d^2\ell, (\partial_a z)^{\cdot\cdot}(0,a)\big\rangle = \big\langle d^2\ell, \tau_3\, T\phi\big(\partial_a x(0,a), \partial_a z(0,a), \partial_a x(0,a)\big)(\partial_a x)^{\cdot\cdot}(0,a)\big\rangle \tag{3.5}$$

and

$$\big\langle d\ell, (\partial_a z)^{\cdot}(0,a)\big\rangle^2 = \big\langle d\ell, T_3\, T\phi\big(\partial_a x(0,a), \partial_a z(0,a), \partial_a x(0,a)\big)(\partial_a x)^{\cdot}(0,a)\big\rangle^2 \tag{3.6}$$

for $\ell = g \circ \pi$: $TN \to \mathbb{R}$ and $\ell = dg$: $TN \to \mathbb{R}$ where g: $N \to \mathbb{R}$ is smooth. Equations (3.5) and (3.6) for $\ell = g \circ \pi$: $TN \to \mathbb{R}$, $g \in C^\infty(N,\mathbb{R})$ are direct consequences of assumption (3.4). To establish (3.5) for $\ell = dg$: $TN \to \mathbb{R}$, we define $z'(t,a) = \phi\big(x(0,a), z(0,a), x(t,a)\big)$. Then

$$\begin{aligned}
\langle d^2\ell, (\partial_a z)^{\cdot\cdot}(0,a)\rangle &= \partial_t^2\partial_a(g \circ z)(0,a) = \partial_a\partial_t^2(g \circ z)(0,a) = \partial_a\big\langle d^2g, \ddot{z}(0,a)\big\rangle \\
&= \partial_a\big\langle d^2g, \tau_3\, \phi(x(0,a), z(0,a), x(0,a))\ddot{x}(0,a)\big\rangle \\
&= \partial_a\big\langle d^2g, (z')^{\cdot\cdot}(0,a)\big\rangle = \partial_a\partial_t^2(g \circ z')(0,a) = \partial_t^2\partial_a(g \circ z')(0,a) \\
&= \partial_t^2 dg \circ T\phi\big(\partial_a x(0,a), \partial_a z(0,a), \partial_a x(t,a)\big) \\
&= \big\langle d^2\ell, \tau_3\, T\phi\big(\partial_a x(0,a), \partial_a z(0,a), \partial_a x(0,a)\big)\, (\partial_a x)^{\cdot\cdot}(0,a)\big\rangle.
\end{aligned}$$

Finally, to verify (3.6) for $\ell = dg$: $TN \to \mathbb{R}$, we have to prove that

$$\big(\partial_t\partial_a(g \circ z)(0,a)\big)^2 = \big(\partial_t\partial_a(g \circ z')(0,a)\big)^2.$$

We first derive from (3.4) that

$$\big(\partial_t(g \circ z)(0,a)\big)^2 = \big(\partial_t(g \circ z')(0,a)\big)^2,$$

and by taking the square of the derivative with respect to a,

$$\big(\partial_t(g \circ z)(0,a)\big)^2\big(\partial_t\partial_a(g \circ z)(0,a)\big)^2 = \big(\partial_t(g \circ z')(0,a)\big)^2\big(\partial_t\partial_a(g \circ z')(0,a)\big)^2.$$

Let $a_0 \in I$. If $\big(\partial_t(g\circ z)(0,a_0)\big)^2 \neq 0$, equality (3.6) is satisfied for $a = a_0$. Now consider the case $\big(\partial_t(g\circ z)(0,a_0)\big)^2 = 0$. If $\big(\partial_t\partial_a(g\circ z)(0,a_0)\big)^2 \neq 0$ or $\big(\partial_t\partial_a(g\circ z')(0,a_0)\big)^2 \neq 0$, then we have $\big(\partial_t(g \circ z)(0,a)\big)^2 \neq 0$ in a neighbourhood of a_0 (a_0 excepted) and (3.6) is satisfied for $a = a_0$ by continuity. □

Definition 3.8. — *A Cohen map $\phi\colon M \times N \times M \to N$ is said to be a Cohen map of Stratonovich type if in addition it has the following property: if a C^2 curve (γ, α) in $M \times N$ satisfies $\dot\alpha(t) = T_3\phi(\gamma(t), \alpha(t), \gamma(t))\,\dot\gamma(t)$ then $\ddot\alpha(t) = \tau_3\,\phi(\gamma(t), \alpha(t), \gamma(t))\,\ddot\gamma(t)$.*

Proposition 3.9. — *Let $k \geq 1$ and e be a C^k_{Lip} section of the vector bundle $T^*M \times TN$ over $M \times N$. Then there exists a $C^{k-1,\infty}_{\mathrm{Lip}}$ Cohen map ϕ of Stratonovich type such that $e(x,z) = T_3\phi(x,z,x)$ for all $(x,z) \in M \times N$. If ϕ is a $C^{k,\infty}_{\mathrm{Lip}}$ Cohen map of Stratonovich type, then $T\phi\colon TM \times TN \times TM \to TN$ is a $C^{k-1,\infty}_{\mathrm{Lip}}$ Cohen map of Stratonovich type.*

Proof. — The existence of ϕ of class $C^{k-1,\infty}_{\mathrm{Lip}}$ is a consequence of [E3 Theorem 8], which gives the existence of a unique Schwartz morphism of Stratonovich type f of class C^{k-1}_{Lip} associated to e, together with Lemma 3.4.

Let ϕ be a $C^{k,\infty}_{\mathrm{Lip}}$ Cohen map of Stratonovich type; we want to show that $T\phi$ is also a Cohen map of Stratonovich type. Let $t \mapsto \beta(t)$ be a smooth curve with values in TN and $t \mapsto \delta(t)$ a smooth curve with values in TM such that

$$\dot\beta(t) = T_3 T\phi(\delta(t), \beta(t), \delta(t))\,\dot\delta(t). \tag{3.7}$$

We have to prove that

$$\ddot\beta(t) = \tau_3\, T\phi(\delta(t), \beta(t), \delta(t))\,\ddot\delta(t).$$

This will be done by means of Lemma 3.7. More precisely, let $(t,a) \mapsto x(t,a)$ satisfy $\partial_a x(t,0) = \delta(t)$, and let $(t,a) \mapsto z(t,a) \in M$ be a solution of

$$\dot z(t,a) = T_3\phi(x(t,a), z(t,a), x(t,a))\,\dot x(t,a) \tag{3.8}$$

with the property $\partial_a z(0,0) = \beta(0)$. It is easy to verify that $\beta(t) = \partial_a z(t,0)$ then already for all t, by exploiting uniqueness of solutions to (3.7) with given initial conditions and by calculating $\langle dh, (\partial_a z)^{\cdot}(t,0)\rangle$ for $h = dg$ and $h = g \circ \pi$ where $g\colon N \to \mathbb{R}$ is smooth. Since ϕ is a Cohen map of Stratonovich type, together with equation (3.8), we get from Lemma 3.7

$$(\partial_a z)^{\cdot\cdot}(t,a) = \tau_3\, T\phi(\partial_a x(t,a), \partial_a z(t,a), \partial_a x(t,a))\,(\partial_a x)^{\cdot\cdot}(t,a)$$

which can be rewritten for $a = 0$ as

$$\ddot\beta(t) = \tau_3\, T\phi(\delta(t), \beta(t), \delta(t))\,\ddot\delta(t).$$

This proves that $T\phi$ is indeed a Cohen map of Stratonovich type. □

Rephrased in terms of Cohen maps of Stratonovich type, the following result is a consequence of [E3 Theorem 8].

PROPOSITION 3.10. — *Let $k \geq 1$ and let e be a C^k_{Lip} section of $T^*M \times TN$ over $M \times N$. Let ϕ be a $C^{k-1,\infty}_{\mathrm{Lip}}$ Cohen map satisfying $e(x,z) = T_3\phi(x,z,x)$. The equations $\delta Z = T_3\phi(X,Z,X)\,\delta X$ and $\mathcal{D}Z = \tau_3\,\phi(X,Z,X)\,\mathcal{D}X$ are equivalent if and only if ϕ is a Cohen map of Stratonovich type.*

For the rest of this section we assume that both M and N are endowed with connections ∇^M and ∇^N. On the tangent bundles TM and TN we consider the corresponding complete lifts $(\nabla^M)'$ and $(\nabla^N)'$ of these connections (see [Y-I] for a definition).

We will say that a Schwartz morphism $f \in \Gamma(\tau(M)^* \otimes \tau(N))$ is semi-affine if for every ∇^M-geodesic γ with values in M and defined at time 0, for every $y \in N$, $f(\gamma(0), y)\,\ddot{\gamma}(0)$ is the second derivative of a ∇^N-geodesic in N (see [E3] for details). In fact $f(\gamma(0), y)\,\ddot{\gamma}(0)$ is the second order derivative $\ddot{\alpha}(0)$ of the geodesic α which satisfies $\alpha(0) = y$ and $\dot{\alpha}(0) = f(\gamma(0), \alpha(0))\,\dot{\gamma}(0)$.

DEFINITION 3.11. — *We say that a Cohen map ϕ is a Cohen map of Itô type (with respect to the connections ∇^M and ∇^N) if $\tau_3\,\phi(x,z,x)\colon \tau_x M \to \tau_z N$ is a semi-affine morphism.*

PROPOSITION 3.12. — *Let $k \geq 0$ and let e be a C^k_{Lip} section of $T^*M \times TN$ over $M \times N$. There exists a $C^{k,\infty}_{\mathrm{Lip}}$ Cohen map ϕ of Itô type such that $e(x,z) = T_3\phi(x,z,x)$ for all $(x,z) \in M \times N$. If $k \geq 1$ and ϕ is a $C^{k,\infty}_{\mathrm{Lip}}$ Cohen map of Itô type, then $T\phi$ is a $C^{k-1,\infty}_{\mathrm{Lip}}$ Cohen map of Itô type (with respect to the connections $(\nabla^M)'$ and $(\nabla^N)'$).*

Proof. — The existence of ϕ is a consequence of [E3 Lemma 11] which gives the existence of a unique Schwartz morphism of Itô type associated to e, together with Lemma 3.4.

Let ϕ be a Cohen map of Itô type; we want to show that $T\phi$ is also a Cohen map of Itô type. We have to prove that for all $(y_0, v_0) \in TM \times TN$, $\tau_3\,T\phi(y_0, v_0, y_0)$ is semi-affine, i.e., if $t \mapsto y(t)$ is a $(\nabla^M)'$-geodesic in TM with $y(0) = y_0$, then the $(\nabla^N)'$-geodesic $t \mapsto v(t)$ in TN with $\dot{v}(0) = T_3T\phi(y_0, v_0, y_0)\,\dot{y}(0)$ satisfies $\ddot{v}(0) = \tau_3\,T\phi(y_0, v_0, y_0)\,\ddot{y}(0)$.

Let $(t,a) \mapsto x(t,a) \in M$ satisfy $\partial_a|_{a=0}\,x(t,a) = y(t)$ and such that $t \mapsto x(t,a)$ is a ∇^M-geodesic for all a. Note that this is possible because y is a Jacobi field. Let $(t,a) \mapsto z(t,a) \in N$ be such that for all a, $t \mapsto z(t,a)$ is a ∇^N-geodesic with

$$\dot{z}(0,a) = T_3\phi\big(x(0,a), z(0,a), x(0,a)\big)\,\dot{x}(0,a)$$

and $\partial_a|_{a=0}\,z(0,a) = v(0)$. Since $t \mapsto x(t,a)$ and $t \mapsto z(t,a)$ are geodesics and $T_3\phi(x,z,x)$ is semi-affine, we deduce that

$$\ddot{z}(0,a) = \tau_3\,\phi\big(x(0,a), z(0,a), x(0,a)\big)\,\ddot{x}(0,a).$$

Now we can apply Lemma 3.7 to obtain

$$(\partial_a z)^{\cdot\cdot}(0,a) = \tau_3\,T\phi(\partial_a x(0,a), \partial_a z(0,a), \partial_a x(0,a))\,(\partial_a x)^{\cdot\cdot}(0,a). \tag{3.9}$$

It remains to prove that $(\partial_a z)^{\cdot\cdot}(0,0) = \ddot{v}(0)$ and $(\partial_a x)^{\cdot\cdot}(0,0) = \ddot{y}(0)$. But $t \mapsto \partial_a z(t,a)$ and $t \mapsto \partial_a x(t,a)$ are geodesics for $(\nabla^N)'$ and $(\nabla^M)'$, hence it is sufficient to know that $(\partial_a z)^{\cdot}(0,0) = \dot{v}(0)$ and $(\partial_a x)^{\cdot}(0,0) = \dot{y}(0)$ (the last equality is already known). For this, we want to calculate $\langle dh, (\partial_a z)^{\cdot}(0,a)\rangle$ for $h = dg$ and $h = g \circ \pi$ with $g\colon N \to \mathbb{R}$ smooth (we will do the verification only for $h = dg$). Let $h = dg$, then

$$\begin{aligned}\langle dh, (\partial_a z)^{\cdot}(0,a)\rangle &= \partial_t|_{t=0}\langle dg, \partial_a z(t,a)\rangle \\ &= \partial_t|_{t=0}\partial_a(g\circ z)(t,a) = \partial_a\partial_t|_{t=0}(g\circ z)(t,a) \\ &= \partial_a\partial_t|_{t=0}(g\circ\phi)(x(0,a), z(0,a), x(t,a)) \\ &= \partial_t|_{t=0}\partial_a(g\circ\phi)(x(0,a), z(0,a), x(t,a)) \\ &= \partial_t|_{t=0}\big\langle dg, T\phi(\partial_a x(0,a), \partial_a z(0,a), \partial_a x(t,a))\big\rangle \\ &= \big\langle dh, T_3T\phi(\partial_a x(0,a), \partial_a z(0,a), \partial_a x(0,a))(\partial_a x)^{\cdot}(0,a)\big\rangle.\end{aligned}$$

In particular, for $a = 0$, this gives

$$\big\langle dh, (\partial_a z)^{\cdot}(0,0)\big\rangle = \big\langle dh, T_3T\phi\,(y_0, v_0, y_0)\,\dot{y}(0)\big\rangle.$$

Since $\dot{v}(0) = T_3T\phi(y_0, v_0, y_0)\,\dot{y}(0)$ we obtain $\dot{v}(0) = (\partial_a z)^{\cdot}(0,0)$ which finally gives with (3.9)

$$\ddot{v}(0) = (\partial_a z)^{\cdot\cdot}(0,0) = \tau_3\,T\phi(y_0, v_0, y_0)\,\ddot{y}(0).$$

This proves that $T\phi$ is a Cohen map of Itô type. □

Rewritten with Cohen maps of Itô type, we get the following result as a consequence of [E3 Theorem 12].

PROPOSITION 3.13. — *Let $k \geq 0$ and e be a C^k_{Lip} section of $T^*M \times TN$ over $M \times N$. Let ϕ be a $C^{k,\infty}_{\text{Lip}}$ Cohen map satisfying $e(x,z) = T_3\phi(x,z,x)$ for all $(x,z) \in M \times N$. Then the equations $d^{\nabla^N}Z = T_3\phi(X,Z,X)\,d^{\nabla^M}X$ and $\mathcal{D}Z = \tau_3\,\phi(X,Z,X)\,\mathcal{D}X$ are equivalent if and only if ϕ is a Cohen map of Itô type.*

The main motivation in our study of Cohen maps of Stratonovich and Itô type is the following result.

COROLLARY 3.14. — *1) Let $k \geq 0$ and e be a C^{k+1}_{Lip} section of the vector bundle $T^*M \times TN$ over $M \times N$. Assume that $a \mapsto X(a)$ is C^k in $\mathscr{S}(M)$, and $a \mapsto Z(a)$ is the maximal solution of*

$$\delta Z(a) = e\big(X(a), Z(a)\big)\,\delta X(a) \tag{3.10}$$

where $a \mapsto Z_0(a)$ is C^k in probability. Then $a \mapsto (X(a), Z(a))$ is C^k in $\hat{\mathscr{S}}(M \times N)$, and if $k \geq 1$, the derivative $\partial_a Z(a)$ is the maximal solution of

$$\delta\partial_a Z(a) = e'\big(\partial_a X(a), \partial_a Z(a)\big)\,\delta\partial_a X(a) \tag{3.11}$$

with initial condition $\partial_a Z_0(0)$ where e' is the C^k_{Lip} section of $T^*TM \times TTN$ over $TM \times TN$ defined as follows: if $e(x,z) = T_3\phi(x,z,x)$ with a $C^{k+1,\infty}_{\text{Lip}}$ Cohen map ϕ then $e'(u,v) = T_3T\phi(u,v,u)$ for $(u,v) \in TM \times TN$. If moreover $a \mapsto \xi_{X(a)}$ is continuous in probability, then $a \mapsto Z(a)$ is C^k in $\hat{\mathscr{S}}(N)$.

2) *Let $k \geq 0$ and e be a C^k_{Lip} section of the vector bundle $T^*M \times TN$ over $M \times N$. Assume that M (resp. N) is endowed with a connection ∇^M (resp. ∇^N), and denote by $(\nabla^M)'$ (resp. $(\nabla^N)'$) the complete lift of ∇^M (resp. ∇^N) in TM (resp. TN). Assume that $a \mapsto X(a)$ is C^k in $\hat{\mathscr{S}}(M)$, $a \mapsto Z(a)$ is the maximal solution of*

$$d^{\nabla^N} Z(a) = e\big(X(a), Z(a)\big)\, d^{\nabla^M} X(a) \tag{3.12}$$

where $a \mapsto Z_0(a)$ is C^k in probability. Then $a \mapsto (X(a), Z(a))$ is C^k in $\hat{\mathscr{S}}(M \times N)$, and the derivative $\partial_a Z(a)$ is the maximal solution of

$$d^{(\nabla^N)'} \partial_a Z(a) = e'\big(\partial_a X(a), \partial_a Z(a)\big)\, d^{(\nabla^M)'} \partial_a X(a) \tag{3.13}$$

with initial condition $\partial_a Z_0(0)$ where e' is the C^{k-1}_{Lip} section of $T^*TM \times TTN$ over $TM \times TN$ defined in 1). If moreover $a \mapsto \xi_{X(a)}$ is continuous in probability, then $a \mapsto Z(a)$ is C^k in $\hat{\mathscr{S}}(N)$.

REMARK. — We like to stress the pleasant point that both Stratonovich and Itô equations differentiate like equations involving smooth paths.

Proof of Corollary 3.14. — 1) We only have to consider the case $k \geq 1$. Let ϕ be a $C^{k,\infty}_{\text{Lip}}$ Cohen map of Stratonovich type such that $T_3\phi(x, z, x) = e(x, z)$ for all $(x, z) \in M \times N$. By Proposition 3.10, equation (3.10) is equivalent to

$$\mathcal{D}Z(a) = \tau_3\, \phi\big(X(a), Z(a), X(a)\big)\, \mathcal{D}X(a).$$

Applying Theorem 3.5, we can differentiate with respect to a and we get

$$\mathcal{D}\partial_a Z(a) = \tau_3\, T\phi\big(\partial_a X(a), \partial_a Z(a), \partial_a X(a)\big)\, \mathcal{D}\partial_a X(a). \tag{3.14}$$

But by Proposition 3.9, $T\phi$ is a $C^{k-1,\infty}_{\text{Lip}}$ Cohen map of Stratonovich type, and again by Proposition 3.10, equation (3.14) is equivalent to

$$\delta\partial_a Z(a) = T_3 T\phi\big(\partial_a X(a), \partial_a Z(a), \partial_a X(a)\big)\, \delta\partial_a X(a)$$

which is precisely equation (3.11).

2) The proof of 1) carries over verbatim, replacing Stratonovich by Itô, Proposition 3.10 by Proposition 3.13, and Proposition 3.9 by Proposition 3.12. □

We want to rephrase equation (3.13) in terms of covariant derivatives. For this we need some definitions and lemmas. Let R^M denote the curvature tensor of the connection ∇^M on M, which is assumed here to be torsion-free. If J is a semimartingale with values in TM endowed with the horizontal lift $(\nabla^M)^h$ of ∇^M (see [Y-I] for a definition), let DJ denote its covariant derivative, i.e. the projection of the vertical part of $d^{(\nabla^M)^h} J$, thus $DJ = v_J^{-1}(d^{(\nabla^M)^h} J)^v$ with $v_j \colon T_xM \to T_jTM$ denoting the vertical lift for $j \in T_xM$. We observe that also $DJ = /\!/_{0,\cdot}\, d(/\!/_{0,\cdot}^{-1} J)$ where $/\!/_{0,t}$ means parallel translation along $\pi(J)$. Indeed, this equality is verified if J is a smooth curve, and since by [Y-I] (9.2) p. 114, J is a geodesic if and only if $(\pi(J), /\!/_{0,\cdot}^{-1} J)$ is a geodesic in $M \times T_{\pi(J_0)}M$ for the product connection, using [E3] corollary 16, it extends to semimartingales as an Itô equation.

LEMMA 3.15. — *Let J be a semimartingale with values in TM, and $X = \pi(J)$ its projection to M. Then*

$$d^{(\nabla^M)'}J = d^{(\nabla^M)^h}J + \frac{1}{2}v_J\left(R^M(J,dX)dX\right) \tag{3.15}$$

where $v_j(u)$ is the vertical lift above $j \in T_xM$ of a vector $u \in T_xM$.

Proof. — Following [E3], if $\widetilde{\nabla}$ is a connection on TM, the Itô differential $d^{\widetilde{\nabla}}J$ may be written as $F^{\widetilde{\nabla}}(\mathcal{D}J)$ where $F^{\widetilde{\nabla}}\colon \tau TM \to TTM$ is the projection defined as follows: if A and B are vector fields on TM, then $F(A) = A$ and $F(AB) = \frac{1}{2}\left(\widetilde{\nabla}_A B + \widetilde{\nabla}_B A + [A,B]\right)$. The result is a direct consequence of the following Lemma. □

For $\ell \in \tau M$, let $b(\ell) \in TM \odot TM$ denote its symmetric bilinear part, i.e., $\langle df \otimes dg, b(\ell)\rangle = \frac{1}{2}\left[\ell(fg) - f\,\ell(g) - g\,\ell(f)\right]$ for f, g smooth functions on M.

LEMMA 3.16. — *Let L be an element of $\tau_u TM$ with $u \in T_xM$. Then*

$$\left(F^{(\nabla^M)'} - F^{(\nabla^M)^h}\right)(L) = v_u\left(\left(R^M(u,\cdot)\cdot\right)b(\pi_* L)\right)$$

where $\pi_*\colon \tau TM \to \tau M$ is induced by $\pi\colon TM \to M$.

Proof. — It is sufficient to prove this for $L_u = (AB)_u$ with A and B horizontal or vertical vector fields. But since among these possibilities $(\nabla^M)'_A B$ and $(\nabla^M)^h_A B$ coincide except if both A and B are horizontal, we can restrict to this case. Let A (resp. B) be the horizontal lift of $\bar{A}$ (resp. $\bar{B}$). Then by [Y-I],

$$(\nabla^M)'_{A_u} B - (\nabla^M)^h_{A_u} B = v_u\left(R^M(u,\bar{A}_x)\bar{B}_x\right)$$

where $x = \pi(u)$, and this gives the result, since $b(\pi_* L_u) = \frac{1}{2}\left(\bar{A}_x \otimes \bar{B}_x + \bar{B}_x \otimes \bar{A}_x\right)$. □

COROLLARY 3.17. — *Let $k \geq 0$ and e be a C^k_{Lip} section of the vector bundle $T^*M \times TN$ over $M \times N$. Assume that M (resp. N) is endowed with a torsion-free connection ∇^M (resp. ∇^N). Assume that $a \mapsto X(a)$ is C^k in $\hat{\mathscr{S}}(M)$, $a \mapsto Z(a)$ the maximal solution of*

$$d^{\nabla^N}Z(a) = e\big(X(a), Z(a)\big)\, d^{\nabla^M}X(a) \tag{3.16}$$

where $a \mapsto Z_0(a)$ is C^k in probability. Then $a \mapsto \big(X(a), Z(a)\big)$ is C^k in $\hat{\mathscr{S}}(M \times N)$, and the derivative $\partial_a Z(a)$ is the maximal solution of the covariant stochastic differential equation

$$\begin{aligned} D\partial_a Z &= e(X,Z)\, D\partial_a X + \nabla e(\partial_a X, \partial_a Z)\, d^{\nabla^M}X \\ &\quad + \frac{1}{2}\Big(e(X,Z)R^M(\partial_a X, dX)dX - R^N\big(\partial_a Z, e(X,Z)dX\big)e(X,Z)dX\Big) \end{aligned} \tag{3.17}$$

with initial condition $\partial_a Z_0(0)$. If moreover $a \mapsto \xi_{X(a)}$ is continuous in probability, then $a \mapsto Z(a)$ is C^k in $\hat{\mathscr{S}}(N)$.

REMARKS. — 1) If ∇^M and ∇^N are allowed to have torsion, one can write first a covariant equation of the form (3.17) with respect to the symmetrized connections $\bar\nabla^M$ and $\bar\nabla^N$. With the obvious notations, expressing $\bar D\partial_a X$, $\bar D\partial_a X$, $\bar R^M$ and $\bar R^N$ in terms of $D\partial_a X$, $D\partial_a X$, R^M, R^N and the torsion tensors, one obtains then a covariant equation with respect to ∇^M and ∇^N.

2) Starting with (3.11) in Corollary 3.14, one can also easily determine a Stratonovich covariant equation, identical to the equation for smooth processes.

Proof of Corollary 3.17. — Applying Lemma 3.15 to part 2) of Corollary 3.14 gives the following equation for $\partial_a Z$:

$$d^{(\nabla^N)^h}\partial_a Z = e'(\partial_a X, \partial_a Z)\left(d^{(\nabla^M)^h}\partial_a X + \frac{1}{2}v_{\partial_a X}\left(R^M(\partial_a X, dX)dX\right)\right) - \frac{1}{2}v_{\partial_a Z}\left(R^N(\partial_a Z, dZ)dZ\right).$$

But, if $u, w \in T_xM$, $z \in TN$, we have $e'(u,z)v_u(w) = v_z\left(e(\pi(u),\pi(z))w\right)$, and by definition, if $h_u^{\nabla^M}(w) \in T_uTM$ is the horizontal lift of w, then $v_z\left(\nabla e(u,z)w\right)$ is the vertical part of $e'(u,z)h_u^{\nabla^M}(w)$. These equalities applied to $u = \partial_a X$, $z = \partial_a Z$, and successively to $w = D\partial_a X$, $w = d^{\nabla^M}X$ and $w = \frac{1}{2}R^M(\partial_a X, dX)dX$, give the desired equation. □

As an application of Corollary 3.14, we get differentiability results for stochastic integrals, considered as particular instances of stochastic differential equations:

COROLLARY 3.18. — 1) *Let $k \geq 0$ and α be a C^{k+1}_{Lip} section of the vector bundle T^*M. Assume that $a \mapsto X(a)$ is C^k in $\hat{\mathscr{S}}(M)$.*

Then $a \mapsto \left(X(a), \int_0^\cdot \langle \alpha(X(a)), \delta X(a)\rangle\right)$ *is C^k in $\hat{\mathscr{S}}(M\times\mathbb{R})$.*

2) *Let $k \geq 0$ and α be a C^k_{Lip} section of the vector bundle $T^*M \times TN$ over $M\times N$. Assume that M (resp. N) is endowed with a connection ∇^M. Assume that $a \mapsto X(a)$ is C^k in $\hat{\mathscr{S}}(M)$.*

Then $a \mapsto \left(X(a), \int_0^\cdot \langle \alpha(X(a)), d^{\nabla^M}X(a)\rangle\right)$ *is C^k in $\hat{\mathscr{S}}(M\times\mathbb{R})$.*

4. Application to antidevelopment

If M is a manifold, we will denote by s: $TTM \to TTM$ the following canonical isomorphism: if $(t,a) \mapsto x(t,a)$ is a smooth M-valued map defined on some open subset of $\mathbb{R}^2$, then $\partial_t\partial_a x(t,a) = s\left(\partial_a\partial_t x(t,a)\right)$.

THEOREM 4.1. — *Let M be a manifold endowed with a connection ∇. Denote by ∇' the complete lift of ∇ on TM. Let $\mathcal{A}'$ denote the antidevelopment with respect to ∇'. Let $a \mapsto X(a) \in \hat{\mathscr{S}}(M)$ be a map of class C^1 defined on some interval I of $\mathbb{R}$. Then $a \mapsto (X(a), \mathcal{A}(X(a))) \in \hat{\mathscr{S}}(TM\times TM)$ is of class C^1 and*

$$s(\partial_a\mathcal{A}(X(a))) = \mathcal{A}'(\partial_a X(a)).$$

Moreover, if $a \mapsto \xi_{X(a)}$ is continuous in probability, then $a \mapsto \mathcal{A}(X(a))$ is C^1 in $\hat{\mathscr{S}}(TN)$.

Before proving this result we introduce some definitions and lemmas. Let M be a manifold of dimension m. If ∇ is a connection on M, we consider the complete lift ∇' of ∇ on TM, which is characterized by the relation $\nabla'_{X^c}Y^c = (\nabla_X Y)^c$ valid for all vector fields $X, Y \in \Gamma(TM)$. Here X^c denotes the complete lift of X, i.e. the vector field in $\Gamma(TTM)$ defined by $X^c_u = s(TX(u))$ (see [Y-I] for details). Recall that the geodesics for ∇' are the Jacobi fields for ∇.

Let $L(M)$ be the principal bundle of linear frames on TM: thus $L_x(M)$ is the set of linear isomorphisms $\mathbb{R}^m \to T_xM$ for each $x \in M$. There is a canonical embedding $\jmath\colon TL(M) \to L(TM)$ defined as follows: if $W \in TL(M)$ is equal to $(\partial_a U)(0)$ where $a \mapsto U(a)$ is a smooth path in $L(M)$, and if $v \in T\mathbb{R}^m = \mathbb{R}^{2m}$ is equal to $(\partial_a z)(0)$ where $a \mapsto z(a)$ is a smooth path in $\mathbb{R}^m$, then one has $\jmath(W)v = s\left((\partial_a(Uz))(0)\right)$. Let $(e_1, \dots, e_m, e_{\bar 1}, \dots, e_{\bar n})$ be the standard basis of $T\mathbb{R}^n$. Then $\jmath((\partial_a U)(0))e_\alpha = s\left((\partial_a(Ue_\alpha))(0)\right)$ and $\jmath((\partial_a U)(0))e_{\bar\alpha}$ is the vertical lift $(Ue_\alpha)^v(0)$ of $(Ue_\alpha)(0)$ above $\partial_a(\pi \circ U)(0)$ where $\pi\colon L(M) \to M$ is the canonical projection.

LEMMA 4.2. — *If $a \mapsto X(a) \in \hat{\mathscr{S}}(M)$ is C^1 and $U(a) \in \hat{\mathscr{S}}(L(M))$ is a horizontal lift of $X(a)$ such that $a \mapsto U_0(a)$ is C^1 in probability, then $a \mapsto U(a)$ is C^1 in $\hat{\mathscr{S}}(L(M))$ and $\jmath(\partial_a U(a))$ is a horizontal lift of $\partial_a X(a)$ with respect to the connection ∇'.*

Proof. — The fact that $a \mapsto U(a)$ is C^1 is a direct consequence of Corollary 3.14 and the fact that $\left\{\lim_{t\to\xi_{U(a)}} U_t(a) \text{ exists}\right\}$ is included in $\left\{\lim_{t\to\xi_{X(a)}} X_t(a) \text{ exists}\right\}$. Another consequence of Corollary 3.14 is that it suffices to prove the assertion with both $X_t(a) = x(t,a)$ and $a \mapsto U_0(a) = u(0,a)$ deterministic and smooth. Write $u(t,a) = U_t(a)$.

It is sufficient to verify that for all $i \in \{1, \dots, m\}$, $t \mapsto \jmath(\partial_a u(t,a))\, e_i$ and $t \mapsto \jmath(\partial_a u(t,a))\, e_{\bar\imath}$ are parallel transports. But by [Y-I chapt. I, prop. 6.3], we have

$$\nabla'_{(\partial_a x)^{\cdot}(t,a)} s\big(\partial_a(u(t,a)e_i)\big) = s\left(\partial_a\big(\nabla_{\dot x(t,a)} u(t,a)e_i\big)\right) = 0$$

and

$$\nabla'_{(\partial_a x)^{\cdot}(t,a)} \big(u(t,a)e_i\big)^v = \big(\nabla_{\dot x(t,a)} u(t,a)e_i\big)^v = 0.$$

This proves Lemma 4.2. □

Proof of Theorem 4.1. — The fact that $a \mapsto \big(X(a), \mathcal{A}(X(a))\big)$ is C^1 is a consequence of Corollary 3.14. We can calculate as if dealing with smooth deterministic paths. Let $a \mapsto U_0(a) \in L_{X_0(a)}(M)$ be C^1 in probability and denote by $U(a)$ the parallel transport of $U_0(a)$ along $X(a)$. Write $Z(a) = \mathcal{A}(X(a))$. Then we have the equation

$$U(a)U_0^{-1}(a)\, p(\delta Z(a)) = \delta X(a)$$

where if $z \in T_xM$ and $v \in T_zTM$ is a vertical vector, $p(v)$ denotes its canonical projection onto T_xM.

Denoting by m the dimension of M, we define a family of $\mathbb{R}^m$-valued processes $R(a)$ by $R(a) = U_0^{-1}(a)Z(a)$, hence

$$\delta R(a) = U_0^{-1}(a)\, p(\delta Z(a)). \tag{4.1}$$

We have $U(a)\delta R(a) = \delta X(a)$ and by differentiation with respect to a, using the definition of $\jmath$,

$$\jmath(\partial_a U(a))\, s(\delta\partial_a R(a)) = \delta\partial_a X(a). \tag{4.2}$$

On the other hand, differentiation of (4.1) gives

$$\jmath(\partial_a U_0(a))\, s(\delta\partial_a R(a)) = p'(\delta s(\partial_a Z(a))) \tag{4.3}$$

where p' on the vertical vectors to TTM is defined like p. Putting together (4.2) and (4.3) gives

$$\jmath(\partial_a U(a)) \left(\jmath(\partial_a U_0(a))\right)^{-1} p'\left(\delta s(\partial_a Z(a))\right) = \delta\partial_a X(a).$$

But $\jmath(\partial_a U(a))$ is the parallel transport of $\jmath(U_0(a))$ above $\partial_a X(a)$ by Lemma 4.2, hence $s(\partial_a Z(a)) = \mathcal{A}'(\partial_a X(a))$. □

COROLLARY 4.3. — *Let J be a TM-valued semimartingale with lifetime $\xi = \xi(0)$. There exists a C^1 family $(X(a))_{a\in\mathbb{R}}$ of elements in $\hat{\mathscr{S}}(M)$ such that the equality $J = \partial_a X(0)$ is satisfied. In particular, if $\xi(a)$ is the lifetime of $X(a)$, then $\xi(a) \wedge \xi(0)$ converges in probability to $\xi(0)$ as a tends to 0. The semimartingale J is a ∇'-martingale if and only if one can choose $(X(a))_{a\in\mathbb{R}}$ such that $X(a)$ is a ∇-martingale for each $a \in \mathbb{R}$.*

Proof. — With the notations of Theorem 4.1 define $V = \mathcal{A}'(J)$, and for $a \in \mathbb{R}$,

$$Z(a) = T\exp\left(s\left(as(V)\right)\right).$$

Note that the lifetime of $Z(a)$ can be 0 if $\exp aJ_0$ is not defined. A straightforward calculation shows that $s(\partial_a Z(0)) = V$. Define now $X(a)$ as the stochastic development of $Z(a)$. We have the relation $Z(a) = \mathcal{A}(X(a))$; by Corollary 3.14, the map $a \mapsto (Z(a), X(a))$ is C^∞ in $\hat{\mathscr{S}}(M)$ and in particular, $\xi(a) \wedge \xi(0)$ converges in probability to $\xi(0)$ as a tends to 0. By Theorem 4.1, the antidevelopment of the derivative of $a \mapsto X(a)$ at $a = 0$ is $s(\partial_a Z(0)) = V$. This implies that $\partial_a X(0) = J$.

If J is a martingale, then V is a local martingale (with possibly finite lifetime). It is easy to see that for each $a \in \mathbb{R}$, $Z(a)$ is also a local martingale, hence its development $X(a)$ is a martingale. □

REFERENCES

[C1] Cohen (S.) — *Géométrie différentielle stochastique avec sauts 1*, Stochastics and Stochastics Reports, t. **56**, 1996, p. 179–203.

[C2] Cohen (S.) — *Géométrie différentielle stochastique avec sauts 2: discrétisation et applications des EDS avec sauts*, Stochastics and Stochastics Reports, t. **56**, 1996, p. 205–225.

[E1] Emery (M.) — *Une topologie sur l'espace des semimartingales*, Séminaire de Probabilités XIII, Lecture Notes in Mathematics, Vol 721, Springer, 1979, p. 260–280.

[E2] Emery (M.) — *Equations différentielles stochastiques lipschitziennes: étude de la stabilité*, Séminaire de Probabilités XIII, Lecture Notes in Mathematics, Vol 721, Springer, 1979, p. 281–293.

[E3] Emery (M.) — *On two transfer principles in stochastic differential geometry*, Séminaire de Probabilités XXIV, Lecture Notes in Mathematics, Vol 1426, Springer, 1989, p. 407–441.

[E4] Emery (M.) — *Stochastic calculus in manifolds.* — Springer, 1989.

[K] Kendall (W.S.) — *Convex geometry and nonconfluent Γ-martingales II: Well-posedness and Γ-martingale convergence*, Stochastics, t. **38**, 1992, p. 135–147.

[Pi] Picard (J.) — *Convergence in probability for perturbed stochastic integral equations*, Probability Theory and Related Fields, t. **81**, 1989, p. 383–451.

[Pr] Protter (P.) — *Stochastic Integration and Differential Equations. A New Approach.* — Springer, 1990.

[Y-I] Yano (K.), Ishihara (S.) — *Tangent and Cotangent Bundles.* — Marcel Dekker, Inc., New York, 1973.

Propagation trajectorielle du chaos pour les lois de conservation scalaire

B.Jourdain [1]

Résumé

A l'aide d'un problème de martingales, nous donnons une interprétation probabiliste trajectorielle de la solution d'une équation cinétique associée aux lois de conservation scalaire. Puis nous montrons que l'unique solution de ce problème est la limite au sens de la propagation du chaos d'une suite de lois de systèmes de particules en interaction, étendant ainsi un résultat obtenu par Perthame et Pulvirenti [2] pour les marginales en temps.

L'équation cinétique non linéaire :

$$\left\{\begin{array}{l} \partial_t f(t,x,v) + a(v).\nabla_x f(t,x,v) + \lambda(f(t,x,v) - 1_{\{v \leq u(t,x)\}}) = 0 \\ \text{où } \ (t,x,v) \in \mathbb{R}_+ \times \mathbb{R}^d \times \mathbb{R}_+, \ \ u(t,x) = \int_{\mathbb{R}_+} f(t,x,v)dv \\ \text{et } \ \ f(0,x,v) = f_0(x,v) \end{array}\right. \tag{0.1}$$

$$\text{avec } \ \lambda > 0 \ \text{ et } \ a \ \text{ fonction continue de } \mathbb{R}_+ \text{ dans } \mathbb{R}^d$$

a été introduite par Perthame et Tadmor [3]. Ces auteurs ont montré que lorsque $w_0 \in L^1(\mathbb{R}^d)$ est positive et $f_0(x,v) = 1_{\{v \leq w_0(x)\}}$, dans le passage à la limite $\lambda \to +\infty$, u converge vers w, l'unique solution entropique de l'équation de conservation scalaire

$$\left\{\begin{array}{l} \partial_t w(t,x) + \nabla_x.A(w(t,x)) = 0, \ \ x \in \mathbb{R}^d, \ \ t \geq 0 \\ w(0,x) = w_0(x) \end{array}\right. \tag{0.2}$$

avec $A(v) = \int_0^v a(\tilde{v})d\tilde{v}$. Heuristiquement, on peut se convaincre de ce résultat de la façon suivante. Si on intègre (0.1) en v, on obtient $\partial_t u(t,x) + \nabla_x.\int_{\mathbb{R}_+} a(v)f(t,x,v)dv = 0$. Dans le passage à la limite, $f(t,x,v)$ et $1_{\{v\leq u(t,x)\}}$ deviennent très proches et remplaçant formellement $f(t,x,v)$ par $1_{\{v\leq u(t,x)\}}$ dans la dernière équation, on trouve que u vérifie l'équation de conservation scalaire.

Dans la première partie de ce travail, après avoir rappelé que pour $f_0 \in L^1(\mathbb{R}^d \times \mathbb{R}_+)$, l'équation (0.1) admet une unique solution faible dans $L^\infty([0,T], L^1(\mathbb{R}^d \times \mathbb{R}_+))$, nous cherchons à donner une interprétation probabiliste de cette solution lorsque f_0 est une densité de probabilité.
A cet effet, nous introduisons le problème de martingales non linéaire suivant : une

1. ENPC-CERMICS, 6-8 ave Blaise Pascal, Cité Descartes, Champs sur Marne, 77455 Marne la Vallée Cedex 2, FRANCE - e-mail : jourdain@cermics.enpc.fr

probabilité $P \in \mathcal{P}(D([0,T],\mathbb{R}^d \times \mathbb{R}_+))$ dont la marginale en t admet une densité $p(t,x,v)$ par rapport à la mesure de Lebesgue pour tout $t \in [0,T]$, est solution si $p(0,.,.) = f_0(.,.)$ et si, pour $u(t,x) = \int_{\mathbb{R}_+} p(t,x,v)dv$, $\forall \phi \in C_b^{1,0}(\mathbb{R}^d \times \mathbb{R}_+)$,

$$\begin{aligned}\phi(X_t,V_t) - \phi(X_0,V_0) - &\int_0^t a(V_s).\nabla_x\phi(X_s,V_s)ds \\ &- \lambda \int_0^t \frac{1_{\{u(s,X_s)>0\}}}{u(s,X_s)} \int_0^{u(s,X_s)} (\phi(X_s,v) - \phi(X_s,V_s))dv\, ds\end{aligned}$$

est une P-martingale locale (où (X,V) processus canonique sur $D([0,T],\mathbb{R}^d \times \mathbb{R}_+))$. Ce problème de martingales est associé à (0.1) au sens où si P est solution, alors $p(t,x,v)$ est solution faible de (0.1). Nous montrons qu'il admet une unique solution. La solution est construite de la façon suivante : la position X évolue suivant un mouvement de vitesse $a(V)$ tandis qu'aux instants de sauts (T_k) d'un processus de Poisson de paramètre λ, le paramètre de vitesse V se redistribue uniformément entre 0 et $u(T_k, X_{T_k})$ où u est la densité spatiale de la solution de (0.1).

Dans la seconde partie, nous généralisons un résultat de propagation du chaos dû à Perthame et Pulvirenti. Dans [2], ces auteurs restreignent la position x au tore $T^d = \mathbb{R}^d/\mathbb{Z}^d$, qu'ils subdivisent en cellules cubiques identiques disjointes de volume $|\Delta|$. Ils construisent un système à N particules à partir de N processus de Poisson indépendants de paramètre λ : l'évolution de la position de la ième particule est régie par son paramètre de vitesse et aux instants de sauts du ième processus de Poisson, ce paramètre de vitesse se redistribue uniformément entre 0 et la densité spatiale empirique dans la cellule où se trouve la particule (rapport entre le nombre de particules dans la cellule et $N|\Delta|$). Ils démontrent que pour $N \to +\infty$ avec $|\Delta| \to 0$ et $N|\Delta| \to +\infty$, les marginales en t des systèmes de particules sont $f(t,x,v)dxdv$-chaotiques où f est l'unique solution de (0.1). Le passage à la limite $|\Delta| \to 0, N|\Delta| \to +\infty, \lambda \to +\infty$ suffisamment doucement permet d'envisager la simulation de l'équation de conservation scalaire par une méthode de Monte-Carlo.
Nous étendons le résultat de Perthame et Pulvirenti en nous libérant de l'hypothèse technique de minoration de f_0 qui les contraignait à se limiter au tore et en travaillant sur $\mathcal{P}(D([0,T],\mathbb{R}^d \times \mathbb{R}_+))$ au lieu de considérer uniquement les marginales en temps. Nous introduisons un système de particules fortement inspiré du leur et nous montrons que pour f_0 satisfaisant une régularité en x précisée ultérieurement, il y a propagation du chaos en norme de variation sur $\mathcal{P}(D([0,T],\mathbb{R}^d \times \mathbb{R}_+))$ vers l'unique solution du problème de martingales partant de f_0.
Notre preuve reprend les deux étapes de la leur. Nous nous appuyons sur le couplage qu'ils utilisent mais avec un point de vue différent. Notre approche probabiliste nous permet de retrouver très facilement le résultat de leur première étape. Puis nous améliorons les majorations de la seconde étape en travaillant dans un cadre L^1 au lieu d'un cadre L^2.

Notations

Soit $D([0,T],\mathbb{R}^d \times \mathbb{R}_+)$ l'espace des fonctions càdlàg de $[0,T]$ dans $\mathbb{R}^d \times \mathbb{R}_+$. On note $\mathcal{P}(D([0,T],\mathbb{R}^d \times \mathbb{R}_+)$ l'espace des probabilités sur $D([0,T],\mathbb{R}^d \times \mathbb{R}_+)$ que l'on munit de la norme en variation

$$\|P-Q\| = \sup\{\int \phi\, dP - \int \phi\, dQ,\ \phi : D([0,T], \mathbb{R}^d \times \mathbb{R}_+) \to \mathbb{R} \ \text{ bornée par 1}\}$$

Si $P \in \mathcal{P}(D([0,T], \mathbb{R}^d \times \mathbb{R}_+)$, on note $(P_t)_{t\in[0,T]}$ les marginales en temps de P.
On désigne par $\tilde{\mathcal{P}}(D([0,T], \mathbb{R}^d \times \mathbb{R}_+)$ le sous ensemble de $\mathcal{P}(D([0,T], \mathbb{R}^d \times \mathbb{R}_+)$ constitué par les probabilités dont toutes les marginales en temps sont absolument continues par rapport à la mesure de Lebesgue sur $\mathbb{R}^d \times \mathbb{R}_+$. Si $P \in \tilde{\mathcal{P}}(D([0,T], \mathbb{R}^d \times \mathbb{R}_+)$, alors il existe une fonction mesurable $p(t,x,v)$ telle que pour tout $t \in [0,T]$, $p(t,.,.)$ est une densité de P_t par rapport à la mesure de Lebesgue (voir Meyer [1] p193-194). Une telle fonction est appelée version mesurable des densités pour P.
Le processus canonique sur $D([0,T], \mathbb{R}^d \times \mathbb{R}_+)$ est noté (X_s, V_s), $s \in [0,T]$ avec $X_s \in \mathbb{R}^d$ et $V_s \in \mathbb{R}_+$.

Remerciements: Je tiens à remercier Madame le Professeur Sylvie Méléard pour toute l'aide qu'elle m'a apportée au cours de ce travail.

1 Le problème de martingales non linéaire

1.1 L'équation cinétique (0.1)

Proposition 1.1 *Pour toute fonction $f_0 \in L^1(\mathbb{R}^d \times \mathbb{R}_+)$, l'équation cinétique (0.1) admet une unique solution faible f dans $L^\infty([0,T], L^1(\mathbb{R}^d \times \mathbb{R}_+))$. Si $f_0 \geq 0$, $f \geq 0$. En outre, si f' est la solution correspondant à la condition initiale f'_0,*

$$\|f - f'\|_\infty \leq \|f_0 - f'_0\|_{L^1} \tag{1.1}$$

où $\|\|_\infty$ désigne la norme de $L^\infty([0,T], L^1(\mathbb{R}^d \times \mathbb{R}_+))$.

Remarque 1.2 *Par invariance de l'équation cinétique par translation spatiale, si $f(t,x,v)$ est la solution associée à $f_0(x,v)$, $f(t,x+y,v)$ est la solution associée à $f_0(x+y,v)$. Avec la propriété de contraction (1.1), on en déduit que*

$$\left(\sup_{\substack{y\in\mathbb{R}^d\\ y\neq 0}} \frac{1}{|y|}\|f_0(.+y,.) - f_0(.,.)\|_{L^1} \leq K\right) \Rightarrow \left(\sup_{\substack{y\in\mathbb{R}^d\\ y\neq 0}} \frac{1}{|y|}\|f(.,.+y,.) - f(.,.,.)\|_\infty \leq K\right)$$

Preuve de la proposition 1.1 : Soit $f_0 \in L^1(\mathbb{R}^d \times \mathbb{R}_+)$, f une solution faible de (0.1) dans $L^\infty([0,T], L^1(\mathbb{R}^d \times \mathbb{R}_+))$ et $u(s,x) = \int_{\mathbb{R}_+} f(s,x,v)dv$. Ce qui suit est valable

pour t en dehors d'un borélien de $[0,T]$ de mesure de Lebesgue nulle.
Si $\psi \in C^\infty([0,T] \times \mathbb{R}^d \times \mathbb{R}_+)$ à support compact

$$\begin{aligned}\int_{\mathbb{R}^d\times\mathbb{R}_+} f(t,x,v)\psi(t,x,v)dxdv &= \int_{\mathbb{R}^d\times\mathbb{R}_+} f_0(x,v)\psi(0,x,v)dxdv \\ &+ \int_{(0,t]\times\mathbb{R}^d\times\mathbb{R}_+} f(s,x,v)\left(\partial_s\psi + a(v).\nabla_x\psi - \lambda\psi\right)(s,x,v)dsdxdv \\ &+ \lambda\int_{(0,t]\times\mathbb{R}^d\times\mathbb{R}_+} 1_{\{v\le u(s,x)\}}\psi(s,x,v)dsdxdv \qquad (1.2)\end{aligned}$$

Par densité, cette égalité reste vraie pour $\psi \in C^{1,1,0}([0,T] \times \mathbb{R}^d \times \mathbb{R}_+)$ à support compact.
Soit $\phi \in C^{1,0}(\mathbb{R}^d \times \mathbb{R}_+)$ à support compact et $\psi(s,x,v) = e^{\lambda(s-t)}\phi(x+(t-s)a(v),v)$.
La fonction ψ est $C^{1,1,0}$ à support compact dans $[0,T] \times \mathbb{R}^d \times \mathbb{R}_+$. En outre,

$$\forall(s,x,v) \in [0,T] \times \mathbb{R}^d \times \mathbb{R}_+,\ (\partial_s\psi + a(v).\nabla_x\psi - \lambda\psi)(s,x,v) = 0$$

En écrivant (1.2) pour ψ, on obtient alors par un simple changement de variables

$$\begin{aligned}\int_{\mathbb{R}^d\times\mathbb{R}_+} f(t,x,v)\phi(x,v)dxdv &= \int_{\mathbb{R}^d\times\mathbb{R}_+} e^{-\lambda t} f_0(x-ta(v),v)\phi(x,v)dxdv \\ &+ \lambda\int_0^t ds\int_{\mathbb{R}^d\times\mathbb{R}_+} 1_{\{v\le u(s,x-(t-s)a(v))\}} e^{\lambda(s-t)}\phi(x,v)dxdv\end{aligned}$$

Ainsi $t \in [0,T]$ p.p., (x,v)p.p.,

$$f(t,x,v) = e^{-\lambda t} f_0(x-ta(v),v) + \lambda\int_0^t e^{\lambda(s-t)} 1_{\{v\le u(s,x-(t-s)a(v))\}} ds \qquad (1.3)$$

Donc f est point fixe de l'application H qui à $g \in L^\infty([0,T], L^1(\mathbb{R}^d \times \mathbb{R}_+))$ associe,

$$H(g)(t,x,v) = e^{-\lambda t} f_0(x-ta(v),v) + \lambda\int_0^t e^{\lambda(s-t)} 1_{\{v\le u_g(s,x-(t-s)a(v))\}} ds$$

pour $u_g(s,x) = \int_{\mathbb{R}_+} g(s,x,v)dv$. On montre facilement que cette application est contractante de rapport $(1-e^{-\lambda T})$ dans $L^\infty([0,T], L^1(\mathbb{R}^d \times \mathbb{R}_+))$. Par le théorème du point fixe de Picard, elle admet un unique point fixe. On a donc obtenu l'unicité des solutions de (0.1) dans $L^\infty([0,T], L^1(\mathbb{R}^d \times \mathbb{R}_+))$.
Pour établir l'existence, nous allons montrer que le point fixe f est solution de (0.1). On pose $u(t,x) = \int_{\mathbb{R}_+} f(t,x,v)dv$. On se donne $t \in [0,T]$ et $\psi \in C^{1,1,0}([0,T]\times\mathbb{R}^d\times\mathbb{R}_+)$ à support compact. En utilisant notamment la formule d'intégration par parties pour les intégrales de Stieljes, on obtient

$$\begin{aligned}&\int_{(0,t]\times\mathbb{R}^d\times\mathbb{R}_+} \Big(\partial_s\psi + a(v).\nabla_x\psi - \lambda\psi\Big)(s,x,v)\int_0^s e^{\lambda(\tau-s)} 1_{\{v\le u(\tau,x-(s-\tau)a(v))\}} d\tau dsdxdv \\ &= \int_{\mathbb{R}^d\times\mathbb{R}_+}\int_0^t \partial_s\left(e^{-\lambda s}\psi(s,x+sa(v),v)\right)\int_0^s e^{\lambda\tau} 1_{\{v\le u(\tau,x+\tau a(v))\}} d\tau dsdxdv \\ &= \int_{\mathbb{R}^d\times\mathbb{R}_+} e^{-\lambda t}\psi(t,x+ta(v),v)\int_0^t e^{\lambda\tau} 1_{\{v\le u(\tau,x+\tau a(v))\}} d\tau dxdv \\ &- \int_{\mathbb{R}^d\times\mathbb{R}_+}\int_0^t e^{-\lambda s}\psi(s,x+sa(v),v)e^{\lambda s} 1_{\{v\le u(s,x+sa(v))\}} dsdxdv\end{aligned}$$

$$\int_{(0,t]\times\mathbb{R}^d\times\mathbb{R}_+}\Big(\partial_s\psi + a(v).\nabla_x\psi - \lambda\psi\Big)(s,x,v)\int_0^s e^{\lambda(\tau-s)}1_{\{v\leq u(\tau,x-(s-\tau)a(v))\}}d\tau dsdxdv$$

$$= \int_{\mathbb{R}^d\times\mathbb{R}_+}\psi(t,x,v)\int_0^t e^{\lambda(s-t)}1_{\{v\leq u(s,x-(t-s)a(v))\}}ds$$

$$- \int_{(0,t]\times\mathbb{R}^d\times\mathbb{R}_+}\psi(s,x,v)1_{\{v\leq u(s,x)\}}dsdxdv \qquad (1.4)$$

Un raisonnement analogue fournit

$$\int_{(0,t]\times\mathbb{R}^d\times\mathbb{R}_+}\Big(\partial_s\psi + a(v).\nabla_x\psi - \lambda\psi\Big)(s,x,v)e^{-\lambda s}f_0(x-sa(v),v)dsdxdv$$

$$= \int_{\mathbb{R}^d\times\mathbb{R}_+} e^{-\lambda t}f_0(x-ta(v),v)\psi(t,x,v)dxdv - \int_{\mathbb{R}^d\times\mathbb{R}_+} f_0(x,v)\psi(0,x,v)dxdv$$

En sommant cette équation et λ fois (1.4) puis en utilisant (1.3), on obtient que pour presque tout $t\in[0,T]$, (1.2) est vérifiée pour toute fonction $\psi\in C^{1,1,0}([0,T]\times\mathbb{R}^d\times\mathbb{R}_+)$ à support compact. Donc f est solution faible de l'équation cinétique (0.1).

Si $f_0\geq 0$, d'après (1.3), il est clair que $f\geq 0$. Si f' est la solution correspondant à la condition initiale f_0', toujours d'après (1.3),
$t\in[0,T]$ p.p., $\|f(t,.,.)-f'(t,.,.)\|_{L^1}\leq e^{-\lambda t}\|f_0-f_0'\|_{L^1}+(1-e^{-\lambda t})\|f-f'\|_\infty$.
On en déduit $\|f-f'\|_\infty\leq\|f_0-f_0'\|_{L^1}$. ■

1.2 Existence et unicité pour le problème de martingales

Définition 1.3 *Soit f_0 une densité de probabilité sur $\mathbb{R}^d\times\mathbb{R}_+$.*
On dit que $P\in\tilde{\mathcal{P}}(D([0,T],\mathbb{R}^d\times\mathbb{R}_+))$ est solution du problème de martingales (PM) partant de f_0 si P_0 admet f_0 comme densité et si pour $p(t,x,v)$ version mesurable des densités de P et $u(t,x)=\int_{\mathbb{R}_+}p(t,x,v)dv$,

$$\phi(X_t,V_t)-\phi(X_0,V_0)-\int_0^t a(V_s).\nabla_x\phi(X_s,V_s)ds$$
$$-\lambda\int_0^t\frac{1_{\{u(s,X_s)>0\}}}{u(s,X_s)}\int_0^{u(s,X_s)}(\phi(X_s,v)-\phi(X_s,V_s))dvds$$

est une P martingale locale pour toute fonction $\phi\in C_b^{1,0}(\mathbb{R}^d\times\mathbb{R}_+)$ (1.5)

Cette définition est indépendante du choix de la version mesurable des densités. En effet si p et p' sont deux telles versions et u et u' sont les densités spatiales associées,

$$P\text{p.s.},\ \forall t\in[0,T],\ \int_0^t\frac{1_{\{u(s,X_s)>0\}}}{u(s,X_s)}\int_0^{u(s,X_s)}(\phi(X_s,v)-\phi(X_s,V_s))dvds$$
$$=\int_0^t\frac{1_{\{u'(s,X_s)>0\}}}{u'(s,X_s)}\int_0^{u'(s,X_s)}(\phi(X_s,v)-\phi(X_s,V_s))dvds$$

Théorème 1.4 *Pour toute densité de probabilité f_0 sur $\mathbb{R}^d \times \mathbb{R}_+$, le problème de martingales (PM) partant de f_0 admet une unique solution P. De surcroît, toute version mesurable des densités pour P est solution de (0.1).*

Pour montrer le théorème, nous aurons besoin du lemme suivant :

Lemme 1.5 *Soit f_0 une densité de probabilité sur $\mathbb{R}^d \times \mathbb{R}_+$ et $h : [0,T] \times \mathbb{R}^d \to \mathbb{R}$ positive. On note (PM_h) le problème de martingales : $P \in \mathcal{P}(D([0,T], \mathbb{R}^d \times \mathbb{R}_+))$ est solution si P_0 admet une densité égale à f_0 et si (1.5) est vérifiée lorsque l'on y remplace u par h.*
Le problème (PM_h) admet une unique solution. En outre, la solution appartient à $\tilde{\mathcal{P}}(D([0,T], \mathbb{R}^d \times \mathbb{R}_+))$.

Preuve du lemme 1.5 : Soit P une solution de (PM_h). Pour montrer l'unicité pour ce problème, nous commençons par montrer que Pp.s., chaque trajectoire du processus X est une fonction déterministe de X_0 et de la trajectoire correspondante du processus V.
Soit $\phi \in C_b^1(\mathbb{R}^d)$. Le processus $M_t^\phi = \phi(X_t) - \phi(X_0) - \int_0^t a(V_s).\nabla_x\phi(X_s)ds$ est une martingale locale localement bornée et donc localement de carré intégrable. Par la formule d'intégration par parties, on obtient

$$\phi^2(X_t) = \phi^2(X_0) + 2\int_0^t \phi(X_{s-})dM_s^\phi + 2\int_0^t \phi(X_s)a(V_s).\nabla_x\phi(X_s)ds + [\phi(X)]_t$$

La fonction ϕ^2 étant dans $C_b^1(\mathbb{R}^d)$, $\phi^2(X_t) - \phi^2(X_0) - 2\int_0^t \phi(X_s)a(V_s).\nabla_x\phi(X_s)ds$ est une martingale locale. Donc $[\phi(X)]_t$ est une martingale locale. Puisque les sauts de M_t^ϕ et de $\phi(X_t)$ sont les mêmes, $[M^\phi]_t = [\phi(X)]_t$. Donc $[M^\phi]_t$ est une martingale locale et son compensateur $< M^\phi >_t$ est nul. Ainsi,

$$P\text{p.s.}, \ \forall t \in [0,T], \ \phi(X_t) = \phi(X_0) + \int_0^t a(V_s).\nabla_x\phi(X_s)ds$$

Avec des fonctions $\phi_{i,n} \in C_b^1(\mathbb{R}^d)$, $1 \le i \le d$, $n \in \mathbb{N}^*$ telles que pour x dans $[-n,n]^d$, $\phi_{i,n}(x) = x_i$, on en déduit

$$P\text{p.s.}, \ \forall t \in [0,T], \ X_t = X_0 + \int_0^t a(V_s)ds$$

On se ramène à un problème de martingales qui ne porte que sur X_0 et V. Pour se rapprocher des notations de l'appendice de Sznitman [4], on définit sur $[0,T] \times \mathbb{R}^d \times \mathbb{R}_+$ le noyau borélien à valeurs $\mathbb{R}$, $M(s,x,v,dv') = \lambda \frac{1_{\{h(s,x)>0\}}}{h(s,x)} 1_{\{-v \le v' \le -v+h(s,x)\}} dv'$.

$$\phi(V_t) - \phi(V_0) - \int_0^t \int_{\mathbb{R}} (\phi(V_s + v') - \phi(V_s)) M\left(s, X_0 + \int_0^s a(V_\tau)d\tau, V_s, dv'\right) ds$$

est une martingale locale pour toute fonction $\phi \in C_b(\mathbb{R}_+)$.
Par le théorème fonctionnel de classe monotone, cette propriété reste vraie pour ϕ borélienne et bornée. Par une adaptation immédiate de la peuve du Lemme 1 de l'appendice de Sznitman [4], on en déduit que Pp.s., V n'admet qu'un nombre fini de sauts sur chaque trajectoire et que $\forall t \in [0,T]$, $V_t = V_0 + \sum_{s\leq t} \Delta V_s$, la loi des instants et des valeurs des sauts sachant (X_0, V_0) étant déterminée de proche en proche (grâce à un choix judicieux de fonctions ϕ). D'où l'unicité pour (PM_h).

Pour l'existence, on se donne une variable aléatoire (χ_0, ν_0) distribuée suivant la loi de densité f_0, et, indépendamment, un processus de Poisson de paramètre λ de temps de sauts $(T_k)_{k\in\mathbb{N}^*}$ ainsi qu'une suite $(Z_k)_{k\in\mathbb{N}^*}$ de marques I.I.D. sur $[0,1]$. On construit (X, V) de la manière suivante :

- $(X_0, V_0) = (\chi_0, \nu_0)$
- sur $[0, T]$, entre les sauts du processus de Poisson, le paramètre de vitesse V est inchangé et la position X évolue suivant un mouvement libre de vitesse $a(V)$
- en T_k (si $T_k \leq T$), V prend la valeur $h(T_k, X_{T_k}) \times Z_k$ si $h(T_k, X_{T_k}) > 0$ et reste inchangé sinon. De cette façon, lorsque $h(T_k, X_{T_k}) > 0$, le paramètre de vitesse se redistribue uniformément sur $[0, h(T_k, X_{T_k})]$.

Nous allons vérifier que la loi P de ce processus est solution de (PM_h). A cet effet, pour $\phi \in C_b^{1,0}(\mathbb{R}^d \times \mathbb{R}_+)$, on pose $H(s,z) = 1_{\{s\leq T\}} 1_{\{h(s,X_s)>0\}}(\phi(X_s, h(s,X_s)z) - \phi(X_s, V_{s-}))$.
On définit $Q = \sum_k \delta_{\{T_k, Z_k\}}$ et on note $(\mathcal{G}_t)$ la filtration du processus $\sum_{k:T_k\leq t}(1 + Z_k)$.
Soit $\mathcal{F}_t = \mathcal{G}_t \vee \sigma(\chi_0, \nu_0)$ et $\mathcal{P}(\mathcal{F}_t)$ la tribu $\mathcal{F}_t$ prévisible. Pour $t \in [0,T]$,

$$\phi(X_t, V_t) = \phi(X_0, V_0) + \int_0^t a(V_s).\nabla_x \phi(X_s, V_s)ds + \int_{\mathbb{R}_+ \times [0,1]} H(s,z) 1_{\{s\leq t\}} Q(dsdz)$$

La mesure aléatoire Q est une $\mathcal{F}_t$-mesure de Poisson sur $\mathbb{R}_+ \times [0,1]$ de compensateur $\lambda dtdz$. Comme la fonction H est $\mathcal{P}(\mathcal{F}_t) \otimes \mathcal{B}([0,1])$ mesurable, on en déduit que

$$\phi(X_t, V_t) - \phi(X_0, V_0) - \int_0^t a(V_s).\nabla_x \phi(X_s, V_s)ds - \lambda \int_0^t \int_{[0,1]} H(s,z)dsdz$$

est une $\mathcal{F}_t$-martingale locale. Donc P est solution du problème de martingale (PM_h).

On utilise la construction précédente pour établir l'absolue continuité. On commence par supposer $\forall (t,x) \in [0,T] \times \mathbb{R}^d$, $h(t,x) > 0$. Soit $t \in (0,T]$ et $\phi : \mathbb{R}^d \times \mathbb{R}_+ \to \mathbb{R}$ mesurable bornée. Avec la convention $T_0 = 0$,

$$\text{si } T_k \leq t < T_{k+1},\ (X_t, V_t) = \left(\chi_0 + \sum_{j=1}^k (T_j - T_{j-1})a(V_{T_{j-1}}) + (t - T_k)a(V_{T_k}), V_{T_k}\right).$$

Ainsi $\mathbb{E}(\phi(X_t, V_t))$ est égal à

$$\sum_{k\in\mathbb{N}} e^{-\lambda t}\frac{(\lambda t)^k}{k!}\mathbb{E}\left(\phi\left(\chi_0+\sum_{j=1}^{k}(T_j-T_{j-1})a(V_{T_{j-1}})+(t-T_k)a(V_{T_k}), V_{T_k}\right)\Big| T_k\leq t< T_{k+1}\right)$$

et donc à

$$\sum_{k\in\mathbb{N}} e^{-\lambda t}\lambda^k \int_{0\leq t_1\leq t_2\leq\ldots\leq t_k\leq t} dt_1\ldots dt_k \int_{\mathbb{R}^d\times\mathbb{R}_+^{k+1}} \frac{1_{\{v_1\leq h(t_1,x_1)\}}}{h(t_1,x_1)}\cdots\frac{1_{\{v_k\leq h(t_k,x_k)\}}}{h(t_k,x_k)} f_0(x_0,v_0)\phi(x_t,v_k)dx_0dv_0dv_1\ldots dv_k \tag{1.6}$$

où $x_i = x_0+\sum_{j=1}^{i}(t_j-t_{j-1})a(v_{j-1})$ et $x_t = x_0+\sum_{j=1}^{k}(t_j-t_{j-1})a(v_{j-1})+(t-t_k)a(v_k)$ (par convention $t_0 = 0$).
En effectuant le changement de variable $[x_0 \to x_t$ et $\forall 0\leq i\leq k, v_i\to v_i]$ dans l'intégrale sur $\mathbb{R}^d\times\mathbb{R}_+^{k+1}$ et en utilisant le théorème de Fubini, on obtient $\mathbb{E}(\phi(X_t, V_t)) = \int_{\mathbb{R}^d\times\mathbb{R}_+}\phi(x,v)\times p(t,x,v)dxdv$ pour

$$p(t,x,v) = \sum_{k\in\mathbb{N}} e^{-\lambda t}\lambda^k \int_{0\leq t_1\leq t_2\leq\ldots\leq t_k\leq t} dt_1\ldots dt_k \int_{\mathbb{R}_+^k} \frac{1_{\{v_1\leq h(t_1,y_1)\}}}{h(t_1,y_1)}\cdots\frac{1_{\{v_{k-1}\leq h(t_{k-1},y_{k-1})\}}}{h(t_{k-1},y_{k-1})} \frac{1_{\{v\leq h(t_k,y_k)\}}}{h(t_k,y_k)} f_0(y_0,v_0)dv_0\ldots dv_{k-1}$$

avec $y_i = x-(t-t_k)a(v)-\sum_{j=i+1}^{k}(t_j-t_{j-1})a(v_{j-1})$.
Donc la loi de (X_t, V_t) est absolument continue.

Dans le cas où h peut s'annuler, au lieu de l'intégrale sur $\mathbb{R}^d\times\mathbb{R}_+^{k+1}$ de (1.6), on doit considérer 2^k intégrales suivant que $h(t_i,x_i)>0$ ou que $h(t_i,x_i)=0$ $(1\leq i\leq k)$.
On note I l'ensemble des i pour lesquels on prend $h(t_i,x_i)>0$ et $i_1,\ldots,i_n$ les éléments de I classés par ordre croissant. L'intégrale sur $\mathbb{R}^d\times\mathbb{R}_+^{k+1}$ est remplacée par la somme pour I variant dans l'ensemble des parties de $\{1,2,\ldots,k\}$ des intégrales :

$$\int_{\mathbb{R}^d\times\mathbb{R}_+^{n+1}}\prod_{i\notin I}1_{\{h(t_i,x_i)=0\}}\prod_{j=1}^{n}1_{\{h(t_{i_j},x_{i_j})>0\}}\frac{1_{\{v_j\leq h(t_{i_j},x_{i_j})\}}}{h(t_{i_j},x_{i_j})}f_0(x_0,v_0)\phi(x_t,v_n)dx_0dv_0dv_1\ldots dv_n$$

avec $t_{i_0}=0$, $x_i = x_0+\sum_{j=1}^{\max\{j:i_j\leq i\}}(t_{i_j}-t_{i_{j-1}})a(v_{j-1})+(t_i-t_{i_{\max\{j:i_j\leq i\}}})a(v_{\max\{j:i_j\leq i\}})$ et $x_t = x_0+\sum_{j=1}^{n}(t_{i_j}-t_{i_{j-1}})a(v_{j-1})+(t-t_{i_n})a(v_n)$. Et on conclut en effectuant un changement de variable dans chacun des 2^k termes. ∎

Preuve du théorème 1.4 : Soit f_0 une densité de probabilité sur $\mathbb{R}^d\times\mathbb{R}_+$, P une solution du problème de martingales (PM) partant de f_0, p une version mesurable des densités pour P et $u(t,x) = \int_{\mathbb{R}_+}p(t,x,v)dv$. On va montrer que p est solution faible de (0.1). La loi P est solution de (PM_u). En utilisant la construction de la solution de (PM_u) donnée dans la preuve du Lemme 1.5, on obtient que

$$\psi(t,X_t,V_t)-\psi(0,X_0,V_0)-\int_0^t(\partial_s\psi(s,X_s,V_s)+a(V_s).\nabla_x\psi(s,X_s,V_s))\,ds$$
$$-\lambda\int_0^t\frac{1_{\{u(s,X_s)>0\}}}{u(s,X_s)}\int_0^{u(s,X_s)}(\psi(s,X_s,v)-\psi(s,X_s,V_s))dvds \tag{1.7}$$

est une P-martingale pour $\psi\in C^{1,1,0}([0,T]\times\mathbb{R}^d\times\mathbb{R}_+)$ à support compact. On note A_t l'espérance du dernier terme. Par le théorème de Fubini,

$$\begin{aligned}
A_t &= \lambda\int_0^t\int_{\mathbb{R}^d\times\mathbb{R}_+}\frac{1_{\{u(s,x)>0\}}}{u(s,x)}\left(\int_0^{u(s,x)}\psi(s,x,v)dv\right)p(s,x,v')dxdv'ds\\
&\quad-\lambda\int_0^t\int_{\mathbb{R}^d\times\mathbb{R}_+}\frac{1_{\{u(s,x)>0\}}}{u(s,x)}\left(\int_0^{u(s,x)}\psi(s,x,v')dv\right)p(s,x,v')dxdv'ds\\
&=\lambda\int_0^t\int_{\mathbb{R}^d\times\mathbb{R}_+}\frac{1_{\{u(s,x)>0\}}}{u(s,x)}\psi(s,x,v)1_{\{v\le u(s,x)\}}\left(\int_{\mathbb{R}_+}p(s,x,v')dv'\right)dxdvds\\
&\quad-\lambda\int_0^t\int_{\mathbb{R}^d\times\mathbb{R}_+}1_{\{u(s,x)>0\}}\psi(s,x,v)p(s,x,v)dxdvds\\
&=\lambda\int_{(0,t]\times\mathbb{R}^d\times\mathbb{R}_+}\psi(s,x,v)\left(1_{\{v\le u(s,x)\}}-p(s,x,v)\right)dsdxdv
\end{aligned}$$

Avec la constance de l'espérance de la martingale (1.7), on en déduit facilement que p est solution faible de (0.1).

L'unicité pour (PM) est une conséquence de ce résultat et de la Proposition 1.1. En effet, si P et P' sont deux solutions de densités spatiales respectives $u(t,x)$ et $u'(t,x)$, le résultat d'unicité de la Proposition 1.1 implique (t,x)p.p., $u(t,x)=u'(t,x)$. On en déduit que P' est solution de (PM_u). Par l'unicité pour ce problème, $P'=P$.

Pour l'existence, on note f la solution de (0.1) et on pose $h(t,x)=\int_{\mathbb{R}_+}f(t,x,v)dv$. D'après la Proposition 1.1, $h\ge 0$. Soit P la solution de (PM_h), p une version mesurable des densités pour P et u la densité spatiale associée.

$$\psi(t,X_t,V_t)-\psi(0,X_0,V_0)-\int_0^t(\partial_s\psi(s,X_s,V_s)+a(V_s).\nabla_x\psi(s,X_s,V_s))\,ds$$
$$-\lambda\int_0^t\frac{1_{\{h(s,X_s)>0\}}}{h(s,X_s)}\int_0^{h(s,X_s)}(\psi(s,X_s,v)-\psi(s,X_s,V_s))dvds$$

est une P-martingale pour $\psi\in C^{1,1,0}([0,T]\times\mathbb{R}^d\times\mathbb{R}_+)$ à support compact. En prenant l'espérance, on obtient que p est solution faible de l'équation linéaire

$$\begin{aligned}
\int_{\mathbb{R}^d\times\mathbb{R}_+}\psi(t,x,v)p(t,x,v)dxdv &= \int_{\mathbb{R}^d\times\mathbb{R}_+}\psi(0,x,v)f_0(x,v)dxdv\\
&+\int_{(0,t]\times\mathbb{R}^d\times\mathbb{R}_+}(\partial_s\psi+a(v).\nabla_x\psi-\lambda\psi)(s,x,v)p(s,x,v)dsdxdv\\
&+\lambda\int_{(0,t]\times\mathbb{R}^d\times\mathbb{R}_+}\psi(s,x,v)\left(1_{\{h(s,x)>0\}}\frac{u(s,x)}{h(s,x)}1_{\{v\le h(s,x)\}}+1_{\{h(s,x)=0\}}p(s,x,v)\right)dsdxdv
\end{aligned}$$

Comme dans la preuve de la Proposition 1.1, on en déduit que $\forall t \in [0,T]$, (x,v)p.p.,

$$p(t,x,v) = e^{-\lambda t} f_0(x - ta(v), v)$$
$$+ \lambda \int_0^t e^{\lambda(s-t)} \left(1_{\{h(s,x_{t,s,v})>0\}} \frac{u(s,x_{t,s,v})}{h(s,x_{t,s,v})} 1_{\{v \le h(s,x_{t,s,v})\}} + 1_{\{h(s,x_{t,s,v})=0\}} p(s, x_{t,s,v}, v) \right) ds$$

avec $x_{t,s,v} = x - (t-s)a(v)$.
Donc p est point fixe de l'application H qui à $g \in L^\infty([0,T], L^1(\mathbb{R}^d \times \mathbb{R}_+))$ associe

$$H(g)(t,x,v) = e^{-\lambda t} f_0(x - ta(v), v)$$
$$+ \lambda \int_0^t e^{\lambda(s-t)} \left(1_{\{h(s,x_{t,s,v})>0\}} \frac{u_g(s,x_{t,s,v})}{h(s,x_{t,s,v})} 1_{\{v \le h(s,x_{t,s,v})\}} + 1_{\{h(s,x_{t,s,v})=0\}} g(s, x_{t,s,v}, v) \right) ds$$

où $u_g(t,x) = \int_{\mathbb{R}_+} g(t,x,v)dv$. On montre facilement que H est contractante de rapport $(1 - e^{-\lambda T})$ puis que f est l'unique point fixe de H. On en déduit que (t,x)p.p., $h(t,x) = u(t,x)$. Donc P est solution du problème de martingales (PM). ∎

2 Le résultat de propagation du chaos

2.1 Le système de particules en interaction

Soit $N \ge 2$. On fixe un hypercube D de $\mathbb{R}^d$ de mesure de Lebesgue $|D|$ centré en 0 que l'on subdivise en $\frac{|D|}{|\Delta|}$ cellules cubiques identiques de mesure de Lebesgue $|\Delta|$. Si $y \in D$, $\Delta(y)$ désigne la cellule dans laquelle se trouve y. Pour $y^N = (y^{N,1}, \ldots, y^{N,N}) \in \mathbb{R}^{Nd}$, on définit la densité empirique au point $y^{N,i}$ par

$$\rho_i(y^N) = 1_{\{y^{N,i} \in D\}} \frac{1}{(N-1)|\Delta|} \sum_{j \ne i} 1_{\Delta(y^{N,i})}(y^{N,j})$$

On note $(X_t^N, V_t^N) = (X_t^{N,1}, \ldots, X_t^{N,N}, V_t^{N,1}, \ldots, V_t^{N,N})$ le processus canonique sur $D([0,T], \mathbb{R}^{Nd} \times \mathbb{R}_+^N)$. On définit la loi du système de particules en interaction comme l'unique solution $P^{N,D,\Delta}$ du problème de martingales :
$P \in \mathcal{P}(D([0,T], \mathbb{R}^{Nd} \times \mathbb{R}_+^N))$ est solution si P_0 est égale à $(f_0 dx dv)^{\otimes N}$ et pour toute fonction $\phi \in C_b^{1,0}(\mathbb{R}^{Nd} \times \mathbb{R}_+^N)$,

$$\phi(X_t^N, V_t^N) - \phi(X_0^N, V_0^N) - \sum_{i=1}^N \int_0^t \left((a(V_s^{N,i}).\nabla_{x_i} \phi(X_s^N, V_s^N) ds + \lambda \frac{1_{\{\rho_i(X_s^N)>0\}}}{\rho_i(X_s^N)} \right.$$
$$\left. \int_0^{\rho_i(X_s^N)} (\phi(X_s^N, V_s^{N,1}, \ldots, V_s^{N,i-1}, v, V_s^{N,i+1}, \ldots, V_s^{N,N}) - \phi(X_s^N, V_s^N)) dv \right) ds$$

est une P-martingale locale. L'unicité pour ce problème de martingales s'obtient comme dans la preuve du lemme 1.5.

On se donne N processus de Poisson indépendants de paramètre λ (on note $(T_k^i)_{k\in\mathbb{N}^*}$ les instants de sauts du ième processus), N suites $(Z_k^i)_{k\in\mathbb{N}^*}$ de marques I.I.D. sur $[0,1]$ et N variables (χ_0^i, ν_0^i) I.I.D. suivant la loi de densité f_0. La probabilité $P^{N,D,\Delta}$ est la loi du processus (X^N, V^N) construit de la manière suivante :

- pour $1 \leq i \leq N$, $(X_0^{N,i}, V_0^{N,i}) = (\chi_0^i, \nu_0^i)$
- sur $[0,T]$, en dehors des sauts du ième processus de Poisson, $V^{N,i}$ le paramètre de vitesse de la ième particule est constant et sa position évolue suivant un mouvement libre de vitesse $a(V^{N,i})$
- à l'instant T_k^i (si $T_k^i \leq T$), $V^{N,i}$ prend la valeur $\rho_i(X^N_{T_k^i}) \times Z_k^i$ si $\rho_i(X^N_{T_k^i}) > 0$ et reste constant sinon.

Remarque 2.1 - *Comme on peut le voir sur la construction que l'on vient d'en donner, $P^{N,D,\Delta}$ est symétrique.*

- *Si on cherche seulement à approcher la solution de (0.1) dans un domaine contenu dans D, on peut arrêter de suivre toute particule qui sort de D. En effet, le paramètre de vitesse d'une telle particule ne peut plus changer. La particule part donc à l'infini et n'interagit plus avec les autres.*

2.2 Propagation du chaos

Théorème 2.2 *On se donne f_0 une densité de probabilité sur $\mathbb{R}^d \times \mathbb{R}_+$ qui vérifie*

$$\sup_{\substack{y\in\mathbb{R}^d \\ y\neq 0}} \frac{1}{|y|} \|f_0(.+y,.) - f_0(.,.)\|_{L^1} \leq K|y| \tag{2.1}$$

On note P la solution du problème de martingales (PM) partant de f_0 (théorème 1.4) et on pose $u(t,x) = \int_{\mathbb{R}_+} p(t,x,v)dv$ pour p version mesurable des densités de P. Pour $k \leq N$, on note $P_{(k)}^{N,D,\Delta}$ la loi des k premières particules sous $P^{N,D,\Delta}$. Alors en norme de variation sur $\mathcal{P}(D([0,T], \mathbb{R}^{kd} \times \mathbb{R}_+^k))$,

$$\left\| P_{(k)}^{N,D,\Delta} - P^{\otimes k} \right\| \leq 2k\lambda e^{3\lambda T} \left(\left(\sqrt{\frac{|D|}{(N-1)|\Delta|}} + 2^d \sqrt{d} K |\Delta|^{\frac{1}{d}} \right) T + \int_0^T \int_{D^c} u(s,z)dzds \right) \tag{2.2}$$

Ainsi, pour $N \to +\infty$ avec $|D| \to +\infty$, $|\Delta| \to 0$ et $\frac{N|\Delta|}{|D|} \to +\infty$, il y a propagation du chaos.

Preuve : La preuve est basée sur le couplage utilisé par Perthame et Pulvirenti dans [2]. On va construire un système à $2N$ particules tel que la loi du sous-système constitué par les N premières particules est $P^{N,D,\Delta}$ et que celle du sous-système constitué par les N dernières est $P^{\otimes N}$.

Le couplage
On note $(X^N, Y^N) \in \mathbb{R}^{2Nd}$ et $(V^N, W^N) \in \mathbb{R}_+^{2N}$ les positions et les paramètres de vitesse des $2N$ particules.
On se donne N variables (χ_0^i, ν_0^i) I.I.D. suivant la loi de densité $f_0(x,v)$ et N processus de Poisson de paramètre λ indépendants. On note $(T_k^i)_{k\in\mathbb{N}^*}$ les temps de sauts successifs du ième processus. On se donne également, indépendamment du reste, N suites indépendantes $(Z_k^{i,1}, Z_k^{i,2}, Z_k^{i,3})_{k\in\mathbb{N}^*}$ de marques I.I.D. suivant la loi uniforme sur $[0,1]^3$. On construit le couplage de la façon suivante :

- pour $1 \le i \le N$, $(X_0^{N,i}, V_0^{N,i}) = (Y_0^{N,i}, W_0^{N,i}) = (\chi_0^i, \nu_0^i)$
- sur $[0,T]$, en dehors des sauts du ième processus de Poisson, $V^{N,i}$ le paramètre de vitesse de la ième particule et $W^{N,i}$ celui de la N+ième restent constants et les positions de ces deux particules évoluent respectivement suivant des mouvements libres de vitesse $a(V^{N,i})$ et $a(W^{N,i})$
- à l'instant T_k^i (si $T_k^i \le T$), en notant $\rho_{i,k} = \rho_i(X_{T_k^i}^N)$ et $u_{i,k} = u(T_k^i, Y_{T_k^i}^{N,i})$ pour alléger les formules
  - si $\rho_{i,k} = u_{i,k} = 0$, alors $V^{N,i}$ et $W^{N,i}$ sont inchangés.
  - si $\rho_{i,k} > 0$ et $u_{i,k} = 0$ alors $W^{N,i}$ est inchangé et on pose $V_{T_k^i}^{N,i} = \rho_{i,k} \times Z_k^{2,i}$
  - si $\rho_{i,k} = 0$ et $u_{i,k} > 0$ alors $V^{N,i}$ est inchangé et on pose $W_{T_k^i}^{N,i} = u_{i,k} \times Z_k^{2,i}$
  - $\rho_{i,k} \ge u_{i,k} > 0$
    - si $Z_k^{1,i} \le \frac{u_{i,k}}{\rho_{i,k}}$, on pose $V_{T_k^i}^{N,i} = W_{T_k^i}^{N,i} = u_{i,k} \times Z_k^{2,i}$
    - sinon on pose $V_{T_k^i}^{N,i} = u_{i,k} + (\rho_{i,k} - u_{i,k}) \times Z_k^{2,i}$ et $W_{T_k^i}^{N,i} = u_{i,k} \times Z_k^{3,i}$
  - si $0 < \rho_{i,k} < u_{i,k}$
    - si $Z_k^{1,i} \le \frac{\rho_{i,k}}{u_{i,k}}$, on pose $V_{T_k^i}^{N,i} = W_{T_k^i}^{N,i} = \rho_{i,k} \times Z_k^{2,i}$
    - sinon on pose $W_{T_k^i}^{N,i} = \rho_{i,k} + (u_{i,k} - \rho_{i,k}) \times Z_k^{2,i}$ et $V_{T_k^i}^{N,i} = \rho_{i,k} \times Z_k^{3,i}$

Etape 1
Dans cette première étape qui utilise la compensation des mesures aléatoires de Poisson, nous majorons, pour $t \in [0,T]$, la probabilité pour que les trajectoires de la ième particule et de la (N+i)ème particule sur $[0,t]$ soient différentes. Par symétrie, cette probabilité est indépendante de i. On la note Q_t.

Soit $(\mathcal{F}_t^i)$ la filtration du processus $\sum_{k:T_k^i\le t}(1+Z_k^{1,i},Z_k^{2,i},Z_k^{3,i})$.
On pose aussi $\mathcal{F}_0=\sigma((\chi_0^i,\nu_0^i),\ 1\le i\le N)$ et $\mathcal{F}_t=\mathcal{F}_0\vee(\vee_{i=1}^N\mathcal{F}_t^i)$. La mesure aléatoire $M^i=\sum_k\delta_{\{T_k^i,Z_k^{1,i},Z_k^{2,i},Z_k^{3,i}\}}$ est une $\mathcal{F}_t^i-$ mesure de Poisson sur $\mathbb{R}_+\times[0,1]^3$ de compensateur $\lambda dtdz_1dz_2dz_3$. Par l'indépendance, M^i est encore une $\mathcal{F}_t$-mesure de Poisson de compensateur $\lambda dtdz_1dz_2dz_3$. On définit

$$G(s)=1_{\{\max(u(s,Y_s^{N,1}),\rho_1(X_s^N))>0\}}\frac{\min(u(s,Y_s^{N,1}),\rho_1(X_s^N))}{\max(u(s,Y_s^{N,1}),\rho_1(X_s^N))}+1_{\{\max(u(s,Y_s^{N,1}),\rho_1(X_s^N))=0\}}$$

$$H(s,z_1)=1_{\{s\in(0,t]\}}1_{\{z_1\ge G(s)\}}$$

Les trajectoires de la première particule et de la (N+1)ème particule sur $[0,t]$ sont différentes si et seulement si il y a avant t un saut du premier processus de Poisson tel que les paramètres de vitesse sont égaux avant ce saut et prennent des valeurs différentes à l'instant du saut. Pour que le kième saut vérifie cette propriété, il faut que $H(T_k^1,Z_k^1)=1$. Donc

$$Q_t\le\mathbb{E}\left(\int_{\mathbb{R}_+\times[0,1]^3}H(s,z_1)M^1(dsdz_1dz_2dz_3)\right)\tag{2.3}$$

On note $\mathcal{P}(\mathcal{F}_t)$ la tribu $\mathcal{F}_t$ prévisible. $H(s,z_1)$ est $\mathcal{P}(\mathcal{F}_t)\otimes\mathcal{B}([0,1]^3)$ mesurable. Donc

$$\begin{aligned}\mathbb{E}\left(\int_{\mathbb{R}_+\times[0,1]^3}H(s,z_1)M^1(dsdz_1dz_2dz_3)\right)&=\mathbb{E}\left(\int_{\mathbb{R}_+\times[0,1]^3}H(s,z_1)\lambda dsdz_1dz_2dz_3\right)\\&=\lambda\int_0^t\mathbb{E}\left(1_{\{\max(u(s,Y_s^{N,1}),\rho_1(X_s^N))>0\}}\frac{|u(s,Y_s^{N,1})-\rho_1(X_s^N)|}{\max(u(s,Y_s^{N,1}),\rho_1(X_s^N))}\right)ds\end{aligned}$$

Avec (2.3), on en déduit

$$Q_t\le\lambda\int_0^t\mathbb{E}\left(1_{\{\max(u(s,Y_s^{N,1}),\rho_1(X_s^N))>0\}}\frac{|u(s,Y_s^{N,1})-\rho_1(X_s^N)|}{\max(u(s,Y_s^{N,1}),\rho_1(X_s^N))}\right)ds$$

Or $\forall a,b,c\ge0,\ 1_{\{\max(a,b)>0\}}\dfrac{|a-b|}{\max(a,b)}\le1_{\{a=0\}}+1_{\{a>0\}}\dfrac{|a-c|}{a}+1_{\{\max(b,c)>0\}}\dfrac{|b-c|}{\max(b,c)}$

Donc

$$\begin{aligned}Q_t\le{}&\lambda\int_0^t\mathbb{E}\left(1_{\{u(s,Y_s^{N,1})=0\}}\right)ds+\lambda\int_0^t\mathbb{E}\left(1_{\{u(s,Y_s^{N,1})>0\}}\frac{|u(s,Y_s^{N,1})-\rho_1(Y_s^N)|}{u(s,Y_s^{N,1})}\right)ds\\&+\lambda\int_0^t\mathbb{E}\left(1_{\{\max(\rho_1(X_s^N),\rho_1(Y_s^N))>0\}}\frac{|\rho_1(X_s^N)-\rho_1(Y_s^N)|}{\max(\rho_1(X_s^N),\rho_1(Y_s^N))}\right)ds\end{aligned}\tag{2.4}$$

Comme $Y_s^{N,1}$ suit la loi de densité $u(s,y)$, le premier terme du second membre de (2.4) est nul.

Etape 2
Cette seconde étape est consacrée à la majoration du second et du troisième terme du second membre de (2.4).

Majoration de $A_s = \mathbb{E}\left(1_{\{u(s,Y_s^{N,1})>0\}}\frac{|u(s,Y_s^{N,1})-\rho_1(Y_s^N)|}{u(s,Y_s^{N,1})}\right)$
Nous travaillons directement dans le cadre L^1 au lieu de passer dans L^2 comme le font Perthame et Pulvirenti.

$$
\begin{aligned}
A_s &\leq \int_{\mathbb{R}^{Nd}} |u(s,y^1) - \rho_1(y^1,\ldots,y^N)|dy^1 u(s,y^2)dy^2 \ldots u(s,y^N)dy^N \\
&\leq \int_{D^c} u(s,y^1)dy^1 + \int_D \left| u(s,y^1) - \frac{1}{|\Delta|}\int_{\Delta(y^1)} u(s,z)dz \right| dy^1 \\
&+ \int_{D\times\mathbb{R}^{(N-1)d}} \left| \frac{1}{|\Delta|}\int_{\Delta(y^1)} u(s,z)dz - \rho_1(y^1,\ldots,y^N) \right| dy^1 u(s,y^2)dy^2 \ldots u(s,y^N)dy^N
\end{aligned}
\tag{2.5}
$$

On note A_s^2 et A_s^3 le second et le troisième terme du second membre. L'hypothèse (2.1) va nous permettre de contrôler A_s^2. Soit Γ l'hypercube de $\mathbb{R}^d$ de volume $2^d|\Delta|$ centré en 0.

$$
\begin{aligned}
A_s^2 &\leq \frac{1}{|\Delta|}\int_D \int_{\Delta(x)} |u(s,x) - u(s,z)|dzdx \leq \frac{1}{|\Delta|}\int_D \int_\Gamma |u(s,x) - u(s,x+y)|dydx \\
&\leq \frac{1}{|\Delta|}\int_\Gamma \|u(s,.) - u(s,.+y)\|_{L^1}dy
\end{aligned}
$$

En utilisant (2.1) et la Remarque 1.2, on en déduit

$$
A_s^2 \leq \frac{1}{|\Delta|}\int_\Gamma K|y|dy \leq \frac{1}{|\Delta|}\int_\Gamma K\sqrt{d}|\Delta|^{\frac{1}{d}}dy \leq 2^d\sqrt{d}K|\Delta|^{\frac{1}{d}}
\tag{2.6}
$$

La majoration de A_s^3 repose sur le fait que la variance d'une somme de variables indépendantes est égale à la somme des variances.

$$
\begin{aligned}
A_s^3 &= \sum_\Delta \mathbb{E}\left| \frac{1}{N-1}\sum_{k=2}^N \left(1_\Delta(Y_s^{N,k}) - \mathbb{E}\left(1_\Delta(Y_s^{N,k})\right)\right)\right| \\
&\leq \sum_\Delta \sqrt{\frac{1}{N-1}\mathbb{E}\left(1_\Delta(Y_s^{N,2}) - \mathbb{E}\left(1_\Delta(Y_s^{N,2})\right)\right)^2} \\
&\leq \frac{1}{\sqrt{N-1}}\sum_\Delta \sqrt{\int_\Delta u(s,z)dz\left(1 - \int_\Delta u(s,z)dz\right)} \\
&\leq \sqrt{\frac{|D|}{(N-1)|\Delta|}}\sqrt{\sum_\Delta \int_\Delta u(s,z)dz} \qquad \text{par l'inégalité de Schwarz} \\
&\leq \sqrt{\frac{|D|}{(N-1)|\Delta|}}
\end{aligned}
$$

Avec les inégalités (2.5) et (2.6), on obtient

$$
A_s \leq \sqrt{\frac{|D|}{(N-1)|\Delta|}} + 2^d\sqrt{d}K|\Delta|^{\frac{1}{d}} + \int_{D^c} u(s,z)dz
\tag{2.7}
$$

Majoration de $B_s = \mathbb{E}\left(1_{\{\max(\rho_1(X_s^N),\rho_1(Y_s^N))>0\}} \frac{|\rho_1(X_s^N)-\rho_1(Y_s^N)|}{\max(\rho_1(X_s^N),\rho_1(Y_s^N))}\right)$

Pour $y^N = (y^{N,1}, \ldots, y^{N,N}) \in \mathbb{R}^{Nd}$, on note $n_\Delta(y^N) = \sum_{i=1}^N 1_\Delta(y^{N,i})$.

$$\begin{aligned} B_s \leq\; & \mathbb{E}\left(1_{\{X_s^{N,1} \neq Y_s^{N,1}\}}\right) \\ & + \sum_\Delta \mathbb{E}\left(1_\Delta(X_s^{N,1}) 1_\Delta(Y_s^{N,1}) 1_{\{\max(n_\Delta(X_s^N),n_\Delta(Y_s^N))>1\}} \frac{|n_\Delta(X_s^N) - n_\Delta(Y_s^N)|}{\max(n_\Delta(X_s^N), n_\Delta(Y_s^N)) - 1}\right) \end{aligned} \tag{2.8}$$

Si $X_s^{N,1} \neq Y_s^{N,1}$, alors les trajectoires de la première particule et de la (N+1)ème particule sur $[0, s]$ sont différentes. Donc le premier terme du second membre est majoré par Q_s. On note B_s^2 le second terme. Sa majoration repose sur l'interchangeabilité des couples $(X_s^{N,i}, Y_s^{N,i})$, $1 \leq i \leq N$.

$$\begin{aligned} B_s^2 &= \frac{1}{N} \sum_\Delta \mathbb{E}\left(1_{\{\max(n_\Delta(X_s^N),n_\Delta(Y_s^N))>1\}} \frac{|n_\Delta(X_s^N) - n_\Delta(Y_s^N)| \sum_{i=1}^N 1_\Delta(X_s^{N,i}) 1_\Delta(Y_s^{N,i})}{\max(n_\Delta(X_s^N), n_\Delta(Y_s^N)) - 1}\right) \\ &\leq \frac{1}{N} \sum_\Delta \mathbb{E}\left(1_{\{\max(n_\Delta(X_s^N),n_\Delta(Y_s^N))>1\}} \frac{|n_\Delta(X_s^N) - n_\Delta(Y_s^N)| \min(n_\Delta(X_s^N), n_\Delta(Y_s^N))}{\max(n_\Delta(X_s^N), n_\Delta(Y_s^N)) - 1}\right) \\ &\leq \frac{1}{N} \sum_\Delta \mathbb{E}\left(|n_\Delta(X_s^N) - n_\Delta(Y_s^N)|\right) \\ &\leq \frac{1}{N} \sum_{i=1}^N \mathbb{E}\left(\sum_\Delta |1_\Delta(X_s^{N,i}) - 1_\Delta(Y_s^{N,i})|\right) \\ &\leq \frac{2}{N} \sum_{i=1}^N \mathbb{E}\left(1_{\{X_s^{N,i} \neq Y_s^{N,i}\}}\right) \leq 2Q_s \end{aligned}$$

L'inégalité (2.8) fournit alors

$$B_s \leq 3Q_s \tag{2.9}$$

<u>Conclusion</u>

En regroupant (2.9) et (2.7) dans (2.4), on obtient

$$Q_t \leq \lambda \left(\left(\sqrt{\frac{|D|}{(N-1)|\Delta|}} + 2^d \sqrt{d} K |\Delta|^{\frac{1}{d}} \right) t + \int_0^t \int_{D^c} u(s,z) dz ds \right) + 3\lambda \int_0^t Q_s ds$$

D'après le lemme de Gronwall,

$$Q_t \leq \lambda \left(\left(\sqrt{\frac{|D|}{(N-1)|\Delta|}} + 2^d \sqrt{d} K |\Delta|^{\frac{1}{d}} \right) t + \int_0^t \int_{D^c} u(s,z) dz ds \right) e^{3\lambda t} \tag{2.10}$$

Soit $k \leq N$. La probabilité pour que les processus $(X^{N,1}, V^{N,1}, \ldots, X^{N,k}, V^{N,k})$ et $(Y^{N,1}, W^{N,1}, \ldots, Y^{N,k}, W^{N,k})$ diffèrent sur $[0,T]$ est inférieure à kQ_T. Comme $P_k^{N,D,\Delta}$ et $P^{\otimes k}$ sont les lois de ces processus, la norme de variation de leur différence est inférieure au double de cette probabilité. Avec (2.10), on en déduit (2.2). ∎

Remarque 2.3 *En nous limitant au tore* $T^d \simeq [0,1]^d$ *pour les positions des particules, comme le font Perthame et Pulvirenti [2], et en prenant* $D = T^d$, *nous aurions obtenu à la place de (2.2) une majoration en* $2k\lambda\left(\sqrt{\frac{1}{(N-1)|\Delta|}} + 2^d\sqrt{d}K|\Delta|^{\frac{1}{d}}\right)Te^{3\lambda T}$ *que l'on peut comparer avec leur résultat en* $Ck\lambda e^{6\lambda T}\sqrt{|\Delta|^{\frac{1}{d}} + \frac{1}{N|\Delta|}}$.

Comme $\sqrt{\frac{1}{(N-1)|\Delta|}} + 2^d\sqrt{d}K|\Delta|^{\frac{1}{d}} \leq \sqrt{2\left(\frac{1}{(N-1)|\Delta|} + 4^d dK^2|\Delta|^{\frac{2}{d}}\right)}$, *notre majoration fournit une meilleure décroissance de la norme de variation avec* $|\Delta|$. *En outre, elle ne nécessite pas les hypothèses techniques de majoration et de minoraration de* f_0 *faites par Perthame et Pulvirenti.*

Références

[1] P.A. Meyer. *Probabilités et Potentiel.* Hermann, 1966.

[2] B. Perthame and M. Pulvirenti. On some large systems of random particles which approximate scalar conservation laws. *Asymt. Anal.*, 10(3):263–278, 1995.

[3] B. Perthame and E. Tadmor. A kinetic equation with kinetic entropy functions for scalar conservation laws. *Comm. Math. Phys.*, 136:501–517, 1991.

[4] A.S. Sznitman. Equations de type de Boltzmann spatialement homogènes. *Z. Warsch. Verw. Geb.*, 66:559–592, 1984.

SOME CALCULATIONS FOR PERTURBED BROWNIAN MOTION

R A DONEY

1. Introduction

If B is a standard Brownian motion starting from zero, and $\bar{B}_t = \sup_{0\le s\le t} B_s$, then the process X defined by

(1) $$X_t = B_t + \frac{\alpha}{1-\alpha}\bar{B}_t,$$

where $\alpha < 1$ is called an α-perturbed Brownian motion. It is immediate from (1) that if $\overline{X_t} = \sup_{0\le s\le t} X_s$ then

(2) $$\overline{X}_t = \frac{1}{1-\alpha}\bar{B}_t,$$

so that (1) shows that X is the unique pathwise solution of the functional equation

(3) $$X_t = B_t + \alpha\bar{X}_t.$$

This is a special case of the equation

$$X_t = B_t + \alpha\bar{X}_t + \beta\underline{X}_t\,,$$

where $\underline{X}_t = \inf_{0\le s\le t} X_s$,which has been studied by a number of authors; see ([3], [5], [4], and [8]). It should also be mentioned that, by the Lévy equivalence, (1) can be written as

$$-X_t = W_t - (1-\alpha)^{-1}L_t,$$

where W is a reflected Brownian motion whose local time at zero is L, so X is often referred to as "reflected Brownian motion perturbed by its local time". (See e.g.[11].)

From (1) it is clear that X is a non-Markov process which moves like Brownian motion except when it is at its maximum, and, moreover, X has the Brownian scaling property. Many other results known for Brownian motion have analogues for perturbed Brownian motion, including Lévy's Arc-sine law, the Ray-Knight theorems, and the solution to the two-sided exit problem. (See [7] , [2], [10], and [11].)

In this note we give an excursion theory approach, based on the excursions of X away from its maximum, which leads to simple proofs of some of these results, and to new ones. In particular, we give new proofs of the Ray-Knight theorems and extend the results known about the two-sided exit problem by computing the transition density of the bivariate Markov process $(X, \bar{X})$, killed when X exits the interval, at an exponential time. From these results we are able to deduce some information about "X conditioned to stay positive".

The basis for our calculations is the following observation; write $P^{(\alpha)}$ for the measure of X and $n^{(\alpha)}$ for the characteristic measure, under $P^{(\alpha)}$, of the excursions away from zero of $\bar{X} - X$. Note that $n = n^{(0)}$ coincides with the characteristic measure of excursions away from zero of reflected Brownian motion.

Proposition 1.1. *The measures $n^{(\alpha)}$ and n are related by*

(4) $$n^{(\alpha)} = (1-\alpha)n.$$

Proof. From (1)and(3) we have

$$\bar{X}_t - X_t = (1-\alpha)^{-1}\bar{B}_t - \{B_t + \alpha(1-\alpha)^{-1}\bar{B}_t\} = \bar{B}_t - B_t,$$

which tells us that $n^{(\alpha)}$ is a multiple of n. But (2) tells us that the local times at zero of $\bar{X} - X$ and $\bar{B} - B$ are related by $l^{(\bar{X}-X)} = (1-\alpha)^{-1} l^{(\bar{B}-B)}$, and this identifies the constant. ∎

2. Ray-Knight Theorems

Let L_t^x denote a jointly continuous version of the local time at level x and time t of X, and write Q_x^δ for the law of the square of a Bessel process of dimension δ starting from x.

Theorem 2.1. *For fixed $b > 0$ let $Z = \{Z(x), 0 \le x \le b\}$, where $Z_x = L_{T_b}^{b-x}$. Then the law of Z is the restriction to $[0, b]$ of $Q_0^{2\bar{\alpha}}$, where $\bar{\alpha} = 1 - \alpha$.*

Proof. Since the result is classical for $\alpha = 0$, it follows from the Lévy- Khintchine representation of Q_0^δ (see Theorem 3.2, p30 of [11]) that it suffices to show that for any Borel function $f \ge 0$

$$P^{(\alpha)}\{\exp - \int_0^b f(x)Z(x)dx\} = [P^{(0)}\{\exp - \int_0^b f(x)Z(x)dx\}]^{\bar{\alpha}}.$$

However, if we write $g(.) = f(b - .)$, the occupation density theorem gives

$$\int_0^b f(x)Z(x)dx = \int_0^{T_b} g(X_s)ds = \int_0^{\tau_b} g(l_s - Y_s)ds,$$

where $Y = \bar{X} - X$ and τ is the inverse of $l = l^{(Y)}$. Applying the master formula of excursion theory gives, with $\zeta = \zeta(\varepsilon)$ standing for the lifetime of a generic excursion ε,

$$\begin{aligned} P^{(\alpha)}\{\exp - \int_0^b f(x)Z(x)dx\} &= \exp -\{\int_0^b dt \int_\Omega n^{(\alpha)}(d\varepsilon)[1 - \exp - \int_0^\zeta g(t - \varepsilon(u))du]\} \\ &= [P^{(0)}\{\exp - \int_0^b f(x)Z(x)dx\}]^{\bar{\alpha}} \end{aligned}$$

by virtue of (4), and the result follows. ∎

Next, we deduce the second Ray-Knight theorem. We write σ for the inverse of L^0 and $\tilde{Q}_x^\delta$ for the measure of the square of a Bessel process of dimension δ, starting from x and killed on hitting zero.

Theorem 2.2. *For fixed $t > 0$ let $U^{(t)} = \{U_x^{(t)}, x \ge 0\}$, where $U_x^{(t)} = L^x(\sigma_t)$. Then under $P^{(\alpha)}$ the law of $U^{(t)}$ is $\tilde{Q}_t^{2\alpha}$.*

Proof. For $x_0 > 0$ it is clear that, given $L^{x_0}(\sigma_t) = t_0$, $\{L^{x_0+x}(\sigma_t), x \ge 0\}$ is independent of $\{L^y(\sigma_t), 0 \le y < x_0\}$, and is distributed as $U^{(t_0)}$. Thus $\{U_x^{(t)}, x \ge 0\}$ is Markov, and the result will follow if we can show that, for all Borel subsets A of $[0, \infty)$ and any $t > 0, x > 0$,

(5) $$P^{(\alpha)}\{U_x^{(t)} \in A\} = \tilde{Q}_t^{2\alpha}\{X_x \in A\}.$$

Now by Theorem 2,

$$P^{(\alpha)}\{U_x^{(t)}=0\}=P^{(\alpha)}\{L^0(T_x)>t\}=Q_0^{2\bar{\alpha}}\{X_x>t\},$$

whereas, writing $\lambda_t = \sup\{s : X_s = t\}$, it follows by time reversal(see e.g.Ex.1.23, p420 of [9]) that

$$\tilde{Q}_t^{2\alpha}\{X_x=0\}=Q_t^{2\alpha}\{T_0\le x\}=Q_0^{2+2\bar{\alpha}}\{\lambda_t\le x\}.$$

Finally, using the scaling property and the fact that the $Q_0^{2+2\bar{\alpha}}$ distribution of λ_1 coincides with the $Q_0^{2\bar{\alpha}}$ distribution of $\{X_1\}^{-2}$ (see Ex 1.18, p418 of [9]), we see that (5) holds for $A = \{0\}$. Next, on $\{U_x^{(t)} > 0\}$, we set $\tilde{T} = \inf\{s > T_x : X(s) = 0\}$,and write

$$U_x^{(t)}=L^x(\tilde{T})+\{L^x(\sigma_t)-L^x(\tilde{T})\}. \tag{6}$$

Since the excursions of X below x after time T_x have the same structure as the excursions below zero of a Brownian motion, it is clear that, given $L^0(T_x) = s$, the terms on the RHS of (6) are independent and, by the Ray-Knight theorems for Brownian motion, have the distribution of X_x under Q_0^2 and Q_{t-s}^0 respectively. Using the composition law for squares of Bessel processes (Theorem 1.2, p410 of [9]) and appealing again to Theorem 2 gives

$$\begin{aligned}P^{(\alpha)}\{U_x^{(t)}\in dy\} &= \int_0^t Q_0^{2\bar{\alpha}}\{X_x\in ds\}Q_{t-s}^2\{X_x\in dy\}\\ &= \frac{dy}{dt}\int_0^t Q_0^{2\bar{\alpha}}\{X_x\in ds\}Q_y\{X_x\in dt-s\}\\ &= \frac{dy}{dt}Q_y^{2+2\bar{\alpha}}\{X_x\in dt\}.\end{aligned}$$

Finally, time reversal gives

$$\frac{1}{dt}Q_y^{2+2\bar{\alpha}}\{X_x\in dt\}=\frac{1}{dy}Q_t^{2\alpha}\{X_x\in dy;T_0>x\}=\frac{1}{dy}\tilde{Q}_t^{2\alpha}\{X_x\in dy\},$$

which completes the proof of (5). ∎

3. The process killed on leaving $[-a, b]$.

We will write $S = S(a,b) = T_{-a}\wedge T_b$ for the first exit time of $[-a, b]$, and V_{θ^*} for an independent, exponentially distributed random variable with parameter $\theta^* = \theta^2/2$.

Theorem 3.1. *It holds that, for $a > 0, b > 0$,*

$$P^{(\alpha)}\{S\le V_{\theta^*};X(S)=b\}=\left(\frac{\sinh a\theta}{\sinh(a+b)\theta}\right)^{\bar{\alpha}}, \tag{7}$$

for $0 < y < b$,

$$P^{(\alpha)}\{S\le V_{\theta^*};X(S)=-a,\bar{X}(S)\in dy\}=\frac{\bar{\alpha}\theta(\sinh a\theta)^{\bar{\alpha}}}{\{\sinh(a+y)\theta\}^{\bar{\alpha}+1}}dy, \tag{8}$$

and for $-a < z < y, 0 < y < \infty$,

$$P^{(\alpha)}\{T_{-a}>V_{\theta^*};X(V_{\theta^*})\in dz,\bar{X}(V_{\theta^*})\in dy\}=\frac{\bar{\alpha}\theta^2(\sinh a\theta)^{\bar{\alpha}}\sinh(a+z)\theta}{\{\sinh(a+y)\theta\}^{\bar{\alpha}+1}}dydz. \tag{9}$$

Proof. Write $A(\theta^*, c)$ for $\{\varepsilon : \zeta(\varepsilon) > V_{\theta^*}\} \cup \{\varepsilon : \zeta(\varepsilon) \le V_{\theta^*}, \bar{\varepsilon}(\zeta) > c\}$, and recall that

$$n(A(\theta^*, c)) = \theta \coth c\theta. \tag{10}$$

Then $P^{(\alpha)}\{S \le V_{\theta^*}; X(S) = b\} = P^{(\alpha)}\{\phi > b\}$, where $\phi = \inf\{s : \varepsilon_s \in A(\theta^*, a+s)\}$. Thus

$$\begin{aligned} P^{(\alpha)}\{S &\le V_{\theta^*}; X(S) = b\} = \exp\{-\int_0^b n^{(\alpha)}(A(\theta^*, a+s))ds\} \\ &= \exp\{-\bar{\alpha} \int_0^b \theta \coth(a+s)\theta ds\}, \end{aligned}$$

from (4) and (10), and (7) follows. Also

$$\begin{aligned} P^{(\alpha)}\{S &\le V_{\theta^*}; X(S) = -a, \} = P^{(\alpha)}\{T_{-a} < T_b \wedge V_{\theta^*}\} \\ &= \int_0^b P^{(\alpha)}\{\phi > y\} n^{(\alpha)}\{\varepsilon : T_{a+y} < \zeta(\varepsilon) \wedge V_{\theta^*}\} dy \\ &= \int_0^b \left\{\frac{\sinh a\theta}{\sinh(a+y)\theta}\right\}^{\bar{\alpha}} \frac{\bar{\alpha}\theta}{\sinh(a+y)\theta} dy, \end{aligned}$$

where we have used another standard result for Brownian motion, and this is equivalent to (8). Similarly

$$\begin{aligned} P^{(\alpha)}\{S &> V_{\theta^*}; X(V_{\theta^*}) \in dz\} \\ &= \int_{z+}^b P^{(\alpha)}\{\phi > y\} n^{(\alpha)}\{\varepsilon : \zeta(\varepsilon) > V_{\theta^*}, \bar{\varepsilon}(V_{\theta^*}) \le a+y, \varepsilon(V_{\theta^*}) \in y - dz\} dy \end{aligned}$$

and since, for $0 < u < v$

$$\begin{aligned} n\{\varepsilon &: \zeta(\varepsilon) > V_{\theta^*}, \bar{\varepsilon}(V_{\theta^*}) \le v, \varepsilon(V_{\theta^*}) \in du\} \\ &= n\{\varepsilon : T_u < \zeta(\varepsilon) \wedge V_{\theta^*}\} P_u^{(0)}\{X(V_{\theta^*}) \in du, T_0 \wedge T_v > V_{\theta^*}\} \\ &= \left\{\frac{\theta}{\sinh u\theta}\right\} \cdot \left\{\frac{\theta \sinh u\theta \sinh(v-u)\theta du}{\sinh v\theta}\right\} = \frac{\theta^2 \sinh(v-u)\theta}{\sinh v\theta} du, \end{aligned}$$

(9) is also immediate. ∎

From this some known results in [2] and [8] follow immediately.

Corollary 3.2. *For α-perturbed Brownian motion we have*

$$P^{(\alpha)}\{X \text{ exits } [-a, b] \text{ at } b\} = \left(\frac{a}{a+b}\right)^{\bar{\alpha}}, \tag{11}$$

$$E^{(\alpha)}\{e^{-\theta^* T_b}\} = e^{-\bar{\alpha} b\theta},$$

and

$$E^{(\alpha)}\{e^{-\theta^* T_{-a}}\} = \int_0^\infty \frac{\bar{\alpha}\theta(\sinh a\theta)^{\bar{\alpha}}}{\{\sinh(a+y)\theta\}^{\bar{\alpha}+1}} dy.$$

We can also deduce some facts about X conditioned "to stay positive";

Corollary 3.3. *It holds that*

$$\lim_{a\downarrow 0}\lim_{k\uparrow\infty} E^{(\alpha)}\{e^{-\theta^* T_b} \mid X\,exits\,[-a,k]\;at\;k\} = \left(\frac{b\theta}{\sinh b\theta}\right)^{\bar{\alpha}}, \tag{12}$$

and

$$\begin{aligned}\lim_{a\downarrow 0}\lim_{k\uparrow\infty} P^{(\alpha)}\{X(V_{\theta^*}) &\in dz \mid X\,exits\,[-a,k]\;at\;k\} \\ &= \theta^{1+\bar{\alpha}} z\sinh z\theta . \int_z^\infty \frac{\bar{\alpha}\theta}{y^\alpha\{\sinh y\theta\}^{\bar{\alpha}+1}} dydz. \end{aligned}\tag{13}$$

Proof. Note that for $k > b$,

$$\begin{aligned} P^{(\alpha)}\{T_b &\le V_{\theta^*} \mid X\text{exits } [-a,k] \text{ at } k\} \\ &= \frac{P^{(\alpha)}\{S \le V_{\theta^*}, X(S) = b\}P^{(\alpha)}\{X \text{ exits } [-(a+b), k-b] \text{ at } k-b\}\}}{P^{(\alpha)}\{X \text{ exits } [-a,b] \text{ at } b\}} \\ &= \{\frac{\sinh a\theta}{\sinh(a+b)\theta}\}^{\bar{\alpha}}\{\frac{a+b}{a+k}\}^{\bar{\alpha}}\{\frac{a}{a+k}\}^{\bar{\alpha}}, \end{aligned}$$

which does not depend on k. So (12) follows by letting $a \downarrow 0$.

Similarly, we see that for $z < y < k$

$$\begin{aligned} &P^{(\alpha)}\{X(V_{\theta^*}) \in dz, \bar{X}(V_{\theta^*}) \in dy \mid \text{ exits } [-a,k] \text{ at } k\} \\ &= P^{(\alpha)}\{T_{-a} > V_{\theta^*}, X(V_{\theta^*}) \in dz, \bar{X}(V_{\theta^*}) \in dy\} P^{(0)}\{X \text{ exits } [-(a+z), y-z] \text{ at } y-z\} \\ &\times \left\{\frac{P^{(\alpha)}\{X \text{ exits } [-(a+y), k-y] \text{ at } k-y\}}{P^{(\alpha)}\{X \text{ exits } [-a,k] \text{ at } k\}}\right\} \\ &= \frac{\bar{\alpha}\theta^2(\sinh a\theta)^{\bar{\alpha}}\sinh(a+z)\theta}{\{\sinh(a+y)\theta\}^{\bar{\alpha}+1}} dydz . \frac{z+a}{y+a} . \left(\frac{a+y}{a=k}\right)^{\bar{\alpha}} . \left(\frac{a+k}{a}\right)^{\bar{\alpha}}, \end{aligned}$$

and this leads to (13). ∎

REMARK Using (13), it is not difficult to show that there is a probability measure $R^{(\alpha)}$ say, which is the weak limit of $P^{(\alpha)}(.\mid X \text{ exits } [-a,k])$ as $k \uparrow \infty$ and $a \downarrow 0$, and it would be interesting to describe X under $R^{(\alpha)}$. Of course $R^{(0)}$ corresponds to the BES(3) process, and one way to realize that is as $|B_t| + L_t$,where L is the local time at zero of $|B|$. This suggests the process $\Sigma^{(\delta)} = |B| + \frac{2}{\delta}L$, which has been studied in [11],chapter 4, as a candidate to have the $R^{(\alpha)}$ measure, for some suitable δ . Furthermore, when $\delta = 2(1-\alpha)$, one can check that, under $P^{(\alpha)}$, the time-reversed process $\{1 - X_{T_1 - t}, 0 \le t \le T_1\}$ has the same measure as $\{\Sigma_t^{(\delta)}, 0 \le t \le \lambda_1^{(\delta)}\}$, where $\lambda_1^{(\delta)} = \sup\{s : \Sigma_s^{(\delta)} = 1\}$. (I owe this observation, which extends a well-known connection between Brownian motion and BES(3), to Loic Chaumont.) However it follows from results in [1] that if $T^{(\delta)}$ is the hitting time process of Σ^δ, then

$$E\{e^{-\theta^* T_b^{(\delta)}}\} = \frac{\bar{\alpha}\theta}{(\sinh b\theta)^{\bar{\alpha}}}\int_0^b \frac{dy}{(\sinh y\theta)^\alpha}.$$

Since this disagrees with (12),we conclude that $\Sigma^{(\delta)}$ does not have $R^{(\alpha)}$ as its measure. This question is discussed further in [6].

References

[1] J.Azéma and M.Yor. Une solution simple au problème de Skorokhod. Sém. de Prob. XIII, Lecture notes in Mathematics, 721, 90-115, Springer, 1978.

[2] P.Carmona, F. Petit, and M.Yor. Some extensions of the arc-sine law as (partial) consequences of the scaling property of Brownian motion. Prob. Th. and Rel. Fields, 100, 1-29, 1994.

[3] P.Carmona, F.Petit, and M.Yor. Beta variables as the time spent in $[0, \infty)$ by certain perturbed Brownian motions. J.London Math. Soc.,(to appear,1997).

[4] L.Chaumont and R.A.Doney. Applications of a path decomposition for doubly perturbed Brownian motion. Preprint, 1997.

[5] B.Davis. Weak limits of perturbed random walks and the equation $Y_t = B_t + \alpha \sup_{s \leq t} Y_s + \beta \inf_{s \leq t} Y_s$. Ann. Prob. 24, 2007-2017, 1996.

[6] R.A.Doney, J.Warren, and M.Yor. Perturbed Bessel processes. This volume.

[7] F.Petit. Sur les temps passé par le mouvement brownien au dessus d'un multiple de son supremum, et quelques extensions de la loi de l'arcsinus. Thèse de doctorat de l'université Paris 7, 1992.

[8] M.Perman and W.Werner. Perturbed Brownian motions. Prob. Th. and Rel. Fields, 108, 357-383, 1997.

[9] D.Revuz and M. Yor. *Continuous Martingales and Brownian Motion.* .Springer-Verlag , Berlin, 1991.

[10] W.Werner. Some remarks on perturbed Brownian motion. Sém. de Prob., Lecture notes in Mathematics, 1613, 37-42, Springer, 1995.

[11] M.Yor.*Some aspects of Brownian motion, part I; some special functionals.Lectures in Mathematics, Birkhäuser*, ETH Zürich, 1992.

Mathematics Department, University of Manchester, Manchester M13 9PL, U.K.

Perturbed Bessel Processes

R.A.DONEY, J.WARREN, and M.YOR.

There has been some interest in the literature in Brownian motion perturbed at its maximum; that is a process $(X_t; t \geq 0)$ satisfying

$$X_t = B_t + \alpha M_t^X, \tag{0.1}$$

where $M_t^X = \sup_{0 \leq s \leq t} X_s$ and $(B_t; t \geq 0)$ is Brownian motion issuing from zero. The parameter α must satisfy $\alpha < 1$. For example arc-sine laws and Ray-Knight theorems have been obtained for this process; see Carmona, Petit and Yor [3], Werner [16], and Doney [7]. Our initial aim was to identify a process which could be considered as the process X conditioned to stay positive. This new process behaves like the Bessel process of dimension three except when at its maximum and we call it a perturbed three-dimensional Bessel process. We establish Ray-Knight theorems for the local times of this process, up to a first passage time and up to infinity (see Theorem 2.3), and observe that these descriptions coincide with those of the local times of two processes that have been considered in Yor [18]. We give an explanation for this coincidence by showing, in Theorem 2.2, that these processes are linked to the perturbed three dimensional Bessel process by space-time transformations and time-reversal.

A process which could be termed a perturbed one-dimensional Bessel process (or perturbed reflected Brownian motion) has already been studied, originally by Le Gall and Yor [11] in connection with windings of Brownian motion, and more recently by Chaumont and Doney [5] as a time change of the positive part of doubly perturbed Brownian motion. We are therefore motivated to introduce perturbed Bessel processes of dimension d, for any $d \geq 1$. Our fundamental result about these processes is Theorem 1.1, which shows how a perturbed Bessel process of dimension d is related to an ordinary Bessel process of dimension d via a space-time transformation. From this we deduce several extensions of results known for ordinary Bessel processes. Thus these processes have the Brownian scaling property, a power of a perturbed Bessel process is a time-change of another perturbed Bessel process (see Theorem 4.2), and there are descriptions of the local times of these processes which show that the Ciesielski-Taylor identity extends to this situation (see Theorem 5.2). On the other hand, some familiar properties of Bessel processes do not extend to perturbed Bessel processes. Thus they are not Markov processes, squares of perturbed Bessel processes do not have the additivity property, and the law of a perturbed 3-dimensional Bessel process up to a first hitting time is not invariant under time reversal (see Theorem 2.2).

We also show that some of these results extend to the case $0 < d < 1$ (see section 3) and to the case where the perturbation factor is replaced by a function of M_t^X (see section 6). Finally, in section 7 we discuss briefly a class of processes which can be thought of as Bessel processes of dimension $d > 2$ perturbed at their future minimum.

1. An h-transform of perturbed Brownian motion

We begin by observing that if X satisfies (0.1) then

$$M_t^X = \frac{1}{1-\alpha} M_t^B, \tag{1.1}$$

where $M_t^B = \sup_{0\le s\le t} B_s$, and consequently we can construct X from B thus,

$$X_t = B_t + \frac{\alpha}{1-\alpha} M_t^B. \tag{1.2}$$

From this we can see that the bivariate process $(X_t, M_t^X; t \ge 0)$ is Markov, and the classical theory of h-transforms of Markov processes tells us how to proceed in order to condition on X_t being positive for all time. We must look for a function h, strictly positive on $\{(x,m) : x > 0\}$ and zero on the set $\{(x,m) : x = 0\}$, such that $h(X_t, M_t^X)$ is a martingale for the bivariate process killed when X is first zero. Applying Itô's formula we find that h is given by

$$h(x,m) = cxm^{-\alpha}, \tag{1.3}$$

for some constant c. Consequently one introduces, for each $a > 0$,

$$\mathbb{P}_a^{3,\alpha}|_{\mathcal{F}_t} = \frac{1}{a^{1-\alpha}} \frac{X_{t\wedge T_0}}{(M_{t\wedge T_0}^X)^\alpha} \cdot \mathbb{P}_a^{(\alpha)}|_{\mathcal{F}_t}$$

where $T_0 = \inf\{u : X_u = 0\}$ and $\mathbb{P}_a^{(\alpha)}$ is the law of X started from a. We have

$$B_t = \tilde{B}_t + \int_0^t ds\, \frac{h'_x}{h}(X_s, M_s^X), \tag{1.4}$$

where $\tilde{B}$ is a $\mathbb{P}_a^{3,\alpha}$-Brownian motion, and so we find that under this latter law X has the following semimartingale decomposition,

$$X_t = a + \tilde{B}_t + \int_0^t \frac{ds}{X_s} + \alpha(M_t^X - a). \tag{1.5}$$

Of course, when $\alpha = 0$, this reduces to the equation which defines the ordinary Bessel process of dimension three, and it is well known that this has an extension to dimension $d \ge 1$. This motivates the following definition of the *perturbed Bessel processes of dimension* $d \ge 1$. We say that a continuous, $\mathbb{R}^+$-valued process $(R_{d,\alpha}(t); t \ge 0)$ is an α-perturbed Bessel process of dimension $d > 1$ starting from $a \ge 0$ if it satisfies

$$R_{d,\alpha}(t) = a + B_t + \frac{d-1}{2} \int_0^t \frac{ds}{R_{d,\alpha}(s)} + \alpha(M_t^R - a), \tag{1.6}$$

and an α-perturbed Bessel process of dimension 1 if it satisfies

$$R_{1,\alpha}(t) = a + B_t + \frac{1}{2} l^R(t) + \alpha(M_t^R - a), \tag{1.7}$$

where $M_t^R = \sup_{0\le s\le t} R_{d,\alpha}(s)$, B is a Brownian motion, and l_t^R is the semimartingale local time of $R_{1,\alpha}$ at zero, it being clear from (1.6) and (1.7) that $R_{d,\alpha}$ is a semimartingale for $d \ge 1$.

Theorem 1.1. *Let $d \ge 1$ and $\alpha < 1$. Suppose that $R_{d,\alpha}$ is defined from a given Bessel process $\tilde{R}$ of dimension d starting at $\tilde{a} \ge 0$ via the time-change*

$$\tilde{M}_u^{\alpha^*} \tilde{R}_u = R_{d,\alpha}\left(\int_0^u dv\, \tilde{M}_v^{2\alpha^*} \right), \tag{1.8}$$

where $\tilde{M}_t = \sup_{0\le s\le t} \tilde{R}_s$ and $\alpha^* = \alpha/1-\alpha$. Then $R_{d,\alpha}$ satisfies (1.6) when $d > 1$ and (1.7) when $d = 1$ with $a = R_{d,\alpha}(0) = \{\tilde{a}\}^{1/1-\alpha}$. Conversely, given a perturbed Bessel *process* $R_{d,\alpha}$ *starting from* a *the process* $\tilde{R}$ *defined via the time-change*

$$(1.9)\qquad \frac{R_{d,\alpha}(t)}{M_t^\alpha} = \tilde{R}\left(\int_0^t \frac{ds}{M_s^{2\alpha}}\right),$$

where $M_t = \sup_{0\le s\le t} R_{d,\alpha}(s)$, *is a Bessel process of dimension* d *starting from* $\tilde{a} = a^{1-\alpha}$.

Proof. Suppose $d > 1$ and (1.6) holds. Then, from an application of Itô's formula we see that

$$\frac{R_{d,\alpha}(t)}{M_t^\alpha} = a^{1-\alpha} + \int_0^t \frac{dB_s}{M_s^\alpha} + \frac{d-1}{2}\int_0^t \frac{ds}{M_s^{2\alpha}}\frac{M_s^\alpha}{R_{d,\alpha}(s)}.$$

Now replacing t by $a(t)$, the inverse of $A(t) = \int_0^t ds/M_s^{2\alpha}$, we see that $\tilde{R}$ is a Bessel process starting from $a^{1-\alpha}$. Moreover, we have

$$(1.10)\qquad \tilde{M}(A_t) = M_t^{1-\alpha},$$

so inverting the time-change we see that (1.9) and (1.8) are equivalent. If $d > 1$ and $\alpha > 0$ we start with the equation which $\tilde{R}$ satisfies (i.e. (1.6) with $\alpha = 0$), and the same argument shows that $R_{d,\alpha}$ defined by (1.8) satisfies (1.6). For $d = 1$ the argument is virtually the same. ∎

- It is a consequence of the representation (1.8) that equation (1.6) enjoys the uniqueness in law property.
- It is now known that in the case $d = 1$ equation (1.7) enjoys the pathwise uniqueness property: see [5]. However the corresponding question for (1.6) has not yet been resolved.
- Note also that if R is a Bessel process of dimension d starting from 0, and we use Theorem 1.1 with $\tilde{R}(.) = R_{T_a+.}$ to construct a family of $R_{d,\alpha}$ processes starting at $a \ge 0$ then these processes vary (in the uniform topology) continuously with a, and hence so do their laws. In particular, for $d = 3$ we see that
$$\mathbb{P}_a^{3,\alpha} \Rightarrow \mathbb{P}_0^{3,\alpha} \text{ as } a \downarrow 0,$$
so that one can also think of $\mathbb{P}_0^{3,\alpha}$ as the law of perturbed Brownian motion starting from zero conditioned to stay positive.
- An important deduction from (1.6) and (1.7) is that $R_{d,\alpha}$ *has the Brownian scaling property.*
- A further deduction is that, just as in the case $\alpha = 0$, the point 0 is instantaneously reflecting for $d < 2$ and polar for $d \ge 2$.
- We mention that although $R_{d,\alpha}$ does not have the Markov property when $\alpha \neq 0$, the pair $\{R_{d,\alpha}, M^{R_{d,\alpha}}\}$ is strong Markov.
- Henceforth we will write PBES(d,α) for a perturbed Bessel process of dimension d, and $\mathbb{P}_a^{d,\alpha}$ for its law if it starts from $a \ge 0$. For $\alpha = 0$ these will be abbreviated to BES(d) and $\mathbb{P}_a^d$.

2. Some Ray-Knight theorems on local time

We consider the perturbed Bessel processes of dimension $d = 3$, and write $\delta = 2(1-\alpha)$. Let $l_t^a(X)$ denote the (jointly continuous version of the) semimartingale local time attained before time t by a process X at the level a. As is now standard, $\mathbb{Q}_a^\delta$ denotes the law of the squared Bessel process of dimension δ starting from a, and

$\mathbb{Q}^{\delta}_{a\to b}$ the bridge of this process to a level b at time 1. Yor describes in [18], following Le Gall-Yor [10], the construction of two processes for which Ray-Knight theorems involving these squared Bessel processes are known. Specifically, given a Brownian motion B we define

$$\Sigma^{\delta}_t = |B_t| + \frac{2}{\delta} l^0_t(B), \tag{2.1}$$

and then define $(D^{\delta}_t; t < T_1)$ via the space-time change

$$\frac{\Sigma^{\delta}_t}{1+\Sigma^{\delta}_t} = D^{\delta}\left(\int_0^t \frac{ds}{(1+\Sigma^{\delta}_s)^4}\right). \tag{2.2}$$

Note that if (2.1) holds and $J^{\Sigma}_t = \inf_{s\geq t} \Sigma^{\delta}_s$, then $J^{\Sigma}_t = \frac{2}{\delta} l^0_t(B)$. It is then easy to see that (2.1) is equivalent to the existence of a Brownian motion $\hat{B}$ such that

$$\Sigma^{\delta}_t = \hat{B}_t + (1+\delta/2) J^{\Sigma}_t. \tag{2.3}$$

The result is

Theorem 2.1. *(Le Gall-Yor) The following descriptions of the local times of Σ^{δ} and D^{δ} hold.*

$$(l^a_{\infty}(\Sigma^{\delta}); a \geq 0) \textit{ has law } \mathbb{Q}^{\delta}_0,$$

and

$$(l^a_{T_1}(D^{\delta}); 0 \leq a \leq 1) \textit{ has law } \mathbb{Q}^{\delta}_{0\to 0}.$$

We are going to establish a similar result for the PBES$(3, \alpha)$ processes but first we need the following.

Theorem 2.2. *Suppose that $R_{3,\alpha}$ is a PBES$(3,\alpha)$ process starting from zero. Then the process Σ^{δ} defined by the space-time transform*

$$\frac{1}{R_{3,\alpha}(t)} = \Sigma^{\delta}\left(\int_t^{\infty} \frac{du}{(R_{3,\alpha}(u))^4}\right), \qquad \textit{for all } t > 0, \tag{2.4}$$

satisfies equation (2.1) with $\delta = 2(1-\alpha)$, and the local times of these processes are connected by;

$$l^a_{\infty}(R_{3,\alpha}) = a l^{1/a}_{\infty}(\Sigma^{\delta}) \textit{ for all } a \geq 0.$$

Moreover the process $X_{3,\alpha}$ defined from $R_{3,\alpha}$ via the space-time change

$$\frac{R_{3,\alpha}(t)}{1+R_{3,\alpha}(t)} = X_{3,\alpha}\left(\int_0^t \frac{ds}{(1+R_{3,\alpha}(s))^4}\right). \tag{2.5}$$

is a PBES(3,α) process, starting from zero, run until it first hits one, and it is related to the process D^{δ} defined from Σ^{δ} by equation (2.2) by time-reversal, i.e.

$$D^{\delta}_t = 1 - X_{3,\alpha}(T_1 - t) \qquad \textit{for all } 0 \leq t \leq T_1 = T_1^{X_{3,\alpha}}. \tag{2.6}$$

Finally it holds that

$$l^a_{T_1}(X_{3,\alpha}) = l^{1-a}_{T_1}(D^{\delta}) \qquad \textit{for all } 0 \leq a \leq 1.$$

Proof. Given a process $R = R_{3,\alpha}$ which is a solution of (1.6) with $d = 3$, we use (1.9) to define a BES(3) process $\tilde{R}$. Since the case $\alpha = 0$ of (2.4), when both $R_{3,\alpha}$ and Σ^δ become BES(3) processes, is a special case of representation results in [9] and [4], so also is the process $\hat{R}$ defined by

$$\frac{1}{\tilde{R}(t)} = \hat{R}\left(\int_t^\infty \frac{du}{(\tilde{R}(u))^4}\right).$$

From this it follows, using (1.9) again, that

$$\frac{\{M_t^R\}^\alpha}{R(t)} = \hat{R}\left(\int_t^\infty \frac{\{M_u^R\}^{2\alpha} du}{(R(u))^4}\right),$$

so that if Σ^δ is defined by (2.4) we have

$$\Sigma^\delta\left(\int_t^\infty \frac{du}{(R(u))^4}\right) \cdot \{M_t^R\}^\alpha = \hat{R}\left(\int_t^\infty \frac{\{M_u^R\}^{2\alpha} du}{(R(u))^4}\right).$$

Using the relation between M_t^R and $J_t^\Sigma = \inf\{\Sigma_u^\delta : u \geq t\}$ which follows from (2.4), we obtain the first in the following equivalent pair of representations

$$\text{(2.7)} \qquad \frac{\Sigma_t^\delta}{(J_t^\Sigma)^\alpha} = \hat{R}\left(\int_0^t \frac{ds}{(J_s^\Sigma)^{2\alpha}}\right), \qquad \hat{R}_t(\hat{J}_t)^{\alpha *} = \Sigma^\delta\left(\int_0^t \{\hat{J}_s\}^{2\alpha^*} ds\right),$$

where $\hat{J}_t = \inf\{\hat{R}_u : u \geq t\}$, and $\alpha^* = \alpha/(1-\alpha)$. The second follows by inverting the time change. Further, recalling that $2 - \alpha = 1 + \delta/2$, an application of Itô's formula shows that (2.7) is equivalent to the existence of a Brownian motion $\hat{B}$ such that (2.3) holds, and we have seen this is equivalent to (2.1). If we apply Itô's formula to $R_t/(1+R_t)$, and then make the time change, we see easily that the first assertion of the theorem about $X_{3,\alpha}$ is correct. To see that the relation (2.6) holds, we write $\hat{X}_t = 1 - X_{3,\alpha}(T_1 - t)$ for $t < T_1$ and note that $T_1 = \int_0^\infty \frac{du}{(1+R_u)^4}$, so that

$$\frac{1}{1+R_t} = \hat{X}\left\{\int_t^\infty \frac{du}{(1+R_u)^4}\right\}.$$

From (2.4) we see that

$$\int_t^\infty \frac{du}{(1+R_u)^4} = \int_0^{A_t} \frac{ds}{(1+\Sigma_s^\delta)^4},$$

where $A_t = \int_t^\infty \frac{du}{(R_u)^4}$. It follows that

$$\frac{\Sigma_t^\delta}{1+\Sigma_t^\delta} = \hat{X}\left\{\int_0^t \frac{ds}{(1+\Sigma_s^\delta)^4}\right\},$$

and comparing this to (2.2), we conclude that $X_{3,\alpha}$ and D^δ are related by time-reversal, as claimed. The results about the local times follow easily. ∎

Theorem 2.3. *The laws of the local times of $R_{3,\alpha}$ when it starts from zero, at times $T_1 = \inf\{t : R_{3,\alpha}(t) \geq 1\}$ and infinity, are respectively $\mathbb{Q}_{0\to 0}^\delta$ and $\mathbb{Q}_0^\delta$.*

Proof. These assertions follow from the statements about local times in Theorem 2.2, using the familiar properties of time inversion (for squared Bessel processes) and time reversal (for bridges of squared Bessel processes). ∎

We remark that in [18] there is presented the following additive decomposition,

$$\mathbb{Q}_0^\delta = \mathbb{Q}_{0\to 0}^\delta * R^\delta. \tag{2.8}$$

(We keep the notation R^δ from [18], hoping that it does not lead to any confusion with the various Bessel processes R_λ involved in our discussion.) The identification of the law of R^δ, except for the case $\delta = 2$, is not entirely satisfactory, involving a reweighting of the local times of the three dimensional Bessel process. We can now clarify this result by noting that if $R_{3,\alpha}$ starts from 1 and we define R^δ as the law of $(l_\infty^a(R_{3,\alpha}); a \geq 0)$, then (2.8) follows from the 'strong Markov' property of $R_{d,\alpha}$ at T_1.

Theorems 2.1 and 2.3 can be reformulated as statements about the *unperturbed* 3-dimensional Bessel process. These alternative presentations involve the local times of semimartingales whose martingale parts are not Brownian motions, and we stress that, if Y is such a semimartingale, then $l^a(Y)$ is an occupation density with respect to $d\langle Y\rangle_s$.

Theorem 2.4. *Let $\hat{R}$ be a BES(3) process starting from zero, put $\hat{M}_t = \sup_{s\leq t} \hat{R}_s$, $\hat{J}_t = \inf_{s\geq t} \hat{R}_s$, and define $Y_t^{(1)} = \{\hat{M}_t\}^{\alpha^*} \hat{R}_t$ and $Y_t^{(2)} = \{\hat{J}_t\}^{\alpha^*} \hat{R}_t$.*

Then for $i = 1, 2$

$$(l_\infty^a(Y^{(i)}); a \geq 0) \quad \textit{has law} \quad \mathbb{Q}_0^\delta, \tag{2.9}$$

and

$$(l_\infty^a((1+Y^{(i)})^{-1}); 0 \leq a \leq 1) \quad \textit{has law} \quad \mathbb{Q}_{0\to 0}^\delta \tag{2.10}$$

Furthermore, for $i = 1$, (2.10) is equivalent to

$$(l_{T_1}^a(Y^{(1)}); 0 \leq a \leq 1) \quad \textit{has law} \quad \mathbb{Q}_{0\to 0}^\delta. \tag{2.11}$$

Proof. From Theorem 1.1 we have the representation $Y_t^{(1)} = R(\Gamma_t)$, where R is a PBES$(3,\alpha)$ and $\Gamma_t = \int_0^t \hat{M}_s^{2\alpha^*} ds$. It follows that $l_t^a(Y^{(1)}) \equiv l_{\Gamma_t}^a(R)$, and since $T_1^R = \Gamma(T_1^{Y^{(1)}})$, statements (2.9) and (2.11) for $i = 1$ follow from Theorem 2.3. Also, by Theorem 2.2, we can write $1 - Y_t^{(1)} = R^*(\Theta(\Gamma_t))$, where R^* is a PBES$(3,\alpha)$ and $\Theta_t = \int_0^t \frac{ds}{(1+R_s)^4}$. It follows that $l_\infty^a(1 - Y^{(1)}) \equiv l_{\Gamma_t}^a(R^*)$, and (2.10) for $i = 1$ also follows from Theorem 2.3. For $i = 2$ we start with the representation $Y_t^{(2)} = \Sigma^\delta(\Phi_t)$ of (2.7), where $\Phi_t = \int_0^t \{\hat{J}_s\}^{2\alpha^*} ds$, and appeal to Theorem 2.1. But note that there is no analogue of (2.11), because $l_{T_1}^a(Y^{(2)}) \equiv l_{T_1}^a(\Sigma^\delta)$, and this is not $\mathbb{Q}_0^\delta$ distributed. ∎

3. Perturbed Bessel processes of dimension $0 < d < 1$

The problem of extending the definition of the perturbed Bessel processes to dimension $0 < d < 1$ can, as in the unperturbed case, be solved by defining first the *perturbed squared Bessel processes.* Note first that if $R_{d,\alpha}$ is a PBES(d, α) with $d \geq 1$, starting from $a \geq 0$, then one deduces easily from (1.6) and (1.7) that $Y = \{R_{d,\alpha}\}^2$ satisfies

$$Y_t = a^2 + 2\int_0^t \sqrt{Y_s} dB_s + (d.t) + \alpha(M_t^Y - a^2). \tag{3.1}$$

Of course, this equation makes sense for any $d > 0$, and in fact has a solution which is unique in law. This is a consequence of the following analogue of our basic representation result, Theorem 1.1.

Theorem 3.1. *Let $d > 0$ and $\alpha < 1$. Suppose that Y is defined from a given squared Bessel process $\tilde{Y}$ of dimension d starting at $\tilde{a}^2 \geq 0$ via the time-change*

$$\{\tilde{M}_u\}^{\alpha^*} \tilde{Y}_u = Y\left(\int_0^u dv\, \tilde{M}_v^{\alpha^*}\right), \tag{3.2}$$

where $\tilde{M}_u = \sup_{0\leq s\leq u} \tilde{R}_s$ and $\alpha^* = \alpha/1-\alpha$. Then Y satisfies (3.1) with $a = \{\tilde{a}\}^{1/1-\alpha}$. Conversely if Y solves (3.1) then the process $\tilde{Y}$ defined via the time-change

$$\frac{Y(t)}{M_t^\alpha} = \tilde{Y}\left(\int_0^t \frac{ds}{M_s^\alpha}\right), \tag{3.3}$$

where $M_t = \sup_{0\leq s\leq t} Y(s)$, is a squared Bessel process of dimension d starting from $\tilde{a}^2 = a^{2(1-\alpha)}$.

Proof. This follows the same lines as the proof of Theorem 1.1. Note that there is an analogue of (1.10), so that (3.2) and (3.3) are actually equivalent. ∎

We now define, for $0 < d < 1$ and $\alpha < 1$, the a-perturbed Bessel process of dimension d starting at $a \geq 0$ as the square root of an α-perturbed squared Bessel process of dimension d starting at a^2. Just as in the case $\alpha = 0$, for $0 < d < 1$ the perturbed Bessel process is not a semimartingale, although it is clear from (3.1) that its square is. However our next result shows that its expression as a Dirichlet process is exactly the α-perturbed version of the corresponding expression for BES(d), which is discussed in [1] and Chap. X of [19].

Theorem 3.2. *If $0 < d < 1$ and R is a PBES(d, α) process starting from $a \geq 0$, then it satisfies the equation*

$$R_t = a + B_t + (\frac{d-1}{2})K_t + \alpha(M_t^R - a), \qquad t \geq 0. \tag{3.4}$$

Here B is a Brownian motion and the drift term K is defined by

$$K_t = p.v. \int_0^t \frac{ds}{R_s} = \int_0^\infty a^{d-2}\{\lambda_t(a) - \lambda_t(0)\}da, \tag{3.5}$$

where the occupation density λ satisfies, for any Borel function $\phi \geq 0$,

$$\int_0^t \phi(R_s)ds = \int_0^\infty a^{d-1}\phi(a)\lambda_t(a)da. \tag{3.6}$$

Proof. Applying the Itô-Tanaka formula to $\phi_\varepsilon(Y)$, where $\varepsilon > 0$ and $\phi_\varepsilon(y) = \sqrt{y\wedge\varepsilon}$ and $Y = R^2$ solves (3.1), gives

(3.7)

$$\phi_\varepsilon(Y_t) = \sqrt{a\wedge\varepsilon} + \int_0^t 1_{\{Y_s\geq\varepsilon\}}dB_s + \frac{\alpha}{2}\int_0^t 1_{\{Y_s\geq\varepsilon\}}Y_s^{-\frac{1}{2}}dM_s^Y + \frac{d-1}{2}\int_0^t 1_{\{Y_s\geq\varepsilon\}}Y_s^{-\frac{1}{2}}ds + \frac{1}{4}\varepsilon^{-\frac{1}{2}}l_t^\varepsilon(Y).$$

From (3.6) we see that

$$l_t^a(Y) = 2a^{\frac{d}{2}}\lambda_t(\sqrt{a}), \tag{3.8}$$

and hence we have

$$\int_0^t 1_{\{Y_s\geq\varepsilon\}}Y_s^{-\frac{1}{2}}ds = \int_0^t 1_{\{Y_s\geq\varepsilon\}}\frac{d\langle Y\rangle_s}{4Y_s^{\frac{3}{2}}} = \int_\varepsilon^\infty \frac{l_t^x(Y)}{4x^{\frac{3}{2}}}dx = \frac{1}{2}\int_\varepsilon^\infty x^{\frac{d-3}{2}}\lambda_t(\sqrt{x})dx. \tag{3.9}$$

Hence

$$\int_\varepsilon^\infty x^{\frac{d-3}{2}}\lambda_t(\sqrt{x})dx+\frac{1}{2d-2}\varepsilon^{\frac{d-1}{2}}\lambda_t(\sqrt{\varepsilon}) = \frac{1}{2}\int_\varepsilon^\infty x^{\frac{d-3}{2}}\{\lambda_t(\sqrt{x})-\lambda_t(\sqrt{\varepsilon})dx$$

$$= \int_{\sqrt{\varepsilon}}^\infty a^{d-2}\{\lambda_t(a)-\lambda_t(\sqrt{\varepsilon})da. \tag{3.10}$$

Now it is not difficult to see, as in the proof of Theorem 4.2 below, that the process defined by $W_t=\{R_t\}^{2-d}=\{Y_t\}^{1-\frac{d}{2}}$ is a semimartingale whose local time satisfies

$$\int_0^\infty \phi(a)l_t^a(W)da=(2-d)^2\int_0^t R_s^{2-2d}\phi(R_s^{2-d})ds.$$

Comparing this to (3.6) yields the identity $\lambda_t(a)\equiv l_t^{a^{2-d}}(W)$. As a consequence of Kolmogorov's criterion we deduce that for any $\gamma\in(0,\frac{1}{2})$ we have $|\lambda_t(a)-\lambda_t(0)|\le ca^{\gamma(2-d)}$ for some positive constant c. This implies both that the final expression in (3.5) is finite, and that (3.4) results by letting $\varepsilon\downarrow 0$ in (3.7). ∎

Remark 3.1. *We can now see that Theorem 1.1, and its consequences, extends immediately to the case* $0<d<1$.

4. More time changes

In Theorem 1.1 we saw that for any fixed $d>0$ we can represent a perturbed Bessel process of dimension d with any perturbation parameter $\alpha<1$ in terms of another Bessel process of dimension d . It is therefore not surprising that there is a similar link between perturbed Bessel processes of dimension d with different perturbation parameters.

Theorem 4.1. *Suppose that R is an α-perturbed Bessel process of dimension $d>0$, $M_t=\sup_{s\le t}R_s$, and $\beta<1$. Then there exists a γ-perturbed Bessel process $\hat{R}$ of dimension d, with $\gamma=\frac{\alpha-\beta}{1-\beta}$, such that*

$$\frac{R(t)}{M_t^\beta}=\hat{R}\left(\int_0^t\frac{ds}{\{M_s\}^{2\beta}}\right).$$

Proof. Just use Theorem 1.1 twice, first to define a BES(3) process $\tilde{R}$ from R, and then to define a PBES(d,γ) process $\hat{R}$ from $\tilde{R}$. ∎

Perhaps more importantly, the fact that a power of a Bessel process is a time-change of a Bessel process of a different dimension (see e.g. Proposition 1.11, chap. XI of [15]) has an analogue for perturbed processes; note that the perturbation parameter is unchanged.

Theorem 4.2. *Suppose that R is a PBES(d,α) process with $d>0$, and β is such that $d_\beta:=2+\frac{d-2}{\beta}>0$. Then there exists a PBES(d_β,α) process $R^\#$ such that*

$$\{R(t)\}^\beta=R^\#\left(\beta^2\int_0^t ds\{R(s)\}^{2(\beta-1)}\right).$$

Proof. If $\beta = 1$ there is nothing to prove. If $\beta > 1$ it is straightforward to apply Itô's formula to Y^β, where $Y = R^2$, to deduce from (3.1) that $Y^\# = \{R^\#\}^2$ satisfies (3.1) with d replaced by d_β, and the result follows. If $\beta < 1$ we note that $d = 2 + \beta(d_\beta - 2)$, and repeat the previous argument with β replaced by $1/\beta$ and d and d_β interchanged. ∎

Remark 4.1. *The important cases of this result are when $d = 1$ and $d = 3$, since it allows us to express any PBES(d, α) process in terms of a PBES$(1, \alpha)$ process if $0 < d < 2$, and in terms of a PBES$(3, \alpha)$ process if $2 < d < \infty$.*

We also mention that just as a Bessel process of dimension d is given by Lamperti's representation as a time change of the exponential of Brownian motion with drift, so can a perturbed Bessel process be expressed in terms of a *perturbed Brownian motion with drift*.

Theorem 4.3. *Define the index $\nu = (d/2) - 1$, and let $\left(B_t^{(\nu,\alpha)}, t \geq 0\right)$ be equal in law to the solution of*

$$X_t = B_t + \nu t + \alpha M_t^X, t \geq 0,$$

where B is a Brownian motion. Then, for $d \geq 0$ and $\alpha < 1$ we have the representation

$$\exp\left(B_t^{(\nu,\alpha)}\right) = R_{d,\alpha}\left(\int_0^t ds \exp(2B_s^{(\nu,\alpha)})\right),$$

where $R_{d,\alpha}$ is a PBES(d, α) process.

Proof. Apply Itô's formula to $\exp(2B_t^{(\nu,\alpha)})$ to see that a time-change of this satisfies (3.1). ∎

Recalling that $\mathbb{P}_a^{d,\alpha}$ stands for the law of a PBES(d, α) starting from a, we now give an absolute continuity relationship between $\mathbb{P}_a^{d,\alpha}$ and $\mathbb{P}_a^{2,\alpha}$ which extends a result for BES(d) processes given as Exercise 1.22, Chap. XI of [15].

Theorem 4.4. *For $d \geq 2$ and $a > 0$ it holds that*

$$\mathbb{P}_a^{d,\alpha} \mid \mathcal{F}_t = \left(\frac{R_t}{a^{1-\alpha} M_t^\alpha}\right)^\nu \exp\left(-\frac{\nu^2}{2}\int_0^t \frac{ds}{R_s^2}\right) . \mathbb{P}_a^{2,\alpha} \mid \mathcal{F}_t.$$

Proof. This is a consequence of Girsanov's theorem, and the fact that, under $\mathbb{P}_a^{2,\alpha}$, $\log(R_t/M_t^\alpha)$ is a local martingale. Alternatively it could be deduced from the result for $\alpha = 0$, using the relationship (1.9). ∎

It is well-known that Spitzer's theorem on the asymptotics of planar Brownian windings (see, e.g., Thm. 4.1, Chap. X in [15]) may be deduced from some asymptotics for the BES(2) process. We now extend these results to PBES processes.

Theorem 4.5. *1. Assume $R_{2,\alpha}(0) > 0$. Then*

$$\frac{4}{(\log t)^2}\int_0^t \frac{ds}{R_{2,\alpha}^2(s)} \overset{(law)}{\to} \sigma_\alpha \equiv \inf\{t : B_t^{0,\alpha} = 1\} \text{ as } t \to \infty,$$

and, moreover, σ_α is equal in law to $\sigma = \inf\{t : B_t = 1\}$.
2. Assume $d > 2$ and $R_{d,\alpha}(0) > 0$. Then

$$\frac{1}{\log t}\int_0^t \frac{ds}{R_{d,\alpha}^2(s)} \overset{a.s.}{\to} E_0^{d,\alpha}\left(\frac{1}{R_1^2}\right) = \frac{1-\alpha}{d-2} \text{ as } t \to \infty.$$

Proof. The first result follows easily from the Lamperti relationship obtained in Theorem 4.3 above. (This deduction for the case $\alpha = 0$ is presented in Exercise 4.11, Chap. X in [15].)

The second result may be deduced from Birkhoff's theorem on path-space. (Again, for the case $\alpha = 0$, see Exercise 3.20, Chap. X in[15].) ∎

5. An extension of the Ciesielski-Taylor Theorem.

The classical version of the Ciesielski-Taylor theorem is the case $d = 3, \alpha = 0$, of the identity

$$\int_0^\infty ds 1\{R_{d,\alpha} \leq 1\} \stackrel{(\text{law})}{=} T_1\{R_{d-2,\alpha}\}, \tag{5.1}$$

which we will show below to be valid for any $d > 2, \alpha < 1$. For the case $\alpha = 0$ this has been established by several authors; see e.g. [2], [8] and [17]. The method used in [17] was to write both sides of (5.1) as integrals with respect to the local times $l^a_\infty\{R_{d,0}\}$ and $l^a_{T_1}\{R_{d-2,0}\}$ respectively, and then exploit the $\alpha = 0$ case of Theorem 4.2 to express these local times in terms of squares of Bessel processes. The same method can be applied to the case $\alpha \neq 0$, once we know that (5.1) holds with $d = 3$. This can be seen from the description of $l^a_\infty\{R_{3,\alpha}\}$ given in Theorem 2.3, together with the Ray-Knight theorems for perturbed Brownian motion (see [3], [16],or [7]), and the observation that a time change of the positive part of an α-perturbed Brownian motion is an α-perturbed Bessel process of dimension 1. (See [5]).

Alternatively we can express the integrals with respect to the local times in terms of integrals with respect to $\mathbf{n}^s_{d,\alpha} ds$, the intensity measure of the excursions away from 0 of the strong Markov process $M^{R_{d,\alpha}} - R_{d,\alpha}$. Then the validity of (5.1) for $\alpha \neq 0$ and $d > 2$ follows from its validity when $\alpha = 0$, the Lévy-Khintchine representations of $\mathbb{Q}^\delta_0$ and $\mathbb{Q}^\delta_{0\to 0}$ (see [18]), and the following result.

Lemma 5.1. *For any $\alpha < 1, d > 0$ it holds that*

$$\mathbf{n}^s_{d,\alpha} = (1-\alpha)\mathbf{n}^s_{d,0}.$$

Combining this with Theorem 4.1 of [18] gives the following result, in which $\{q^\delta(a), a \geq 0\}$ [respectively $\{\tilde{q}^\delta(a), 0 \leq a \leq 1\}$] denotes a process with the law $\mathbb{Q}^\delta_0$ [$\mathbb{Q}^\delta_{0\to 0}$], and again $\delta = 2(1-\alpha)$.

Theorem 5.2. *1. The Ciesielski-Taylor identity (5.1) is valid for any $\alpha < 1, d > 2$.*

2. For $d > 2$ we have

$$(l^a_\infty(R_{d,\alpha}), a \geq 0) \stackrel{(law)}{=} \left(\frac{a^{3-d}}{d-2} q^\delta(a^{d-2}), a \geq 0\right),$$

and

$$(l^a_{T_1}(R_{d,\alpha}), 0 \leq a \leq 1) \stackrel{(law)}{=} \left(\frac{a^{3-d}}{d-2} \tilde{q}^\delta(a^{d-2}), 0 \leq a \leq 1\right).$$

3. For $d = 2$ we have $(l^a_{T_1}(R_{2,\alpha}), 0 < a \leq 1) \stackrel{(law)}{=} (a\tilde{q}^\delta(\log 1/a), 0 < a \leq 1)$.

4. For $0 < d < 2$ we have

$$(l^a_{T_1}(R_{d,\alpha}), 0 < a \leq 1) \stackrel{(law)}{=} \left(\frac{1}{2-d} \tilde{q}^\delta(1 - a^{2-d}), 0 < a \leq 1\right).$$

6. Variable Perturbations

A process $(\Sigma_t^\Delta, t \geq 0)$ has been considered by Le-Gall and Yor, and others; see [19], section 18.3, and the references therein, and [6]. It is a simple generalization of the process Σ^δ defined via equation (2.1), where the constant δ has been replaced by a strictly increasing C^1 function $\Delta : \mathbb{R}^+ \to \mathbb{R}^+$, satisfying $\Delta(0) = 0$ and $\Delta(\infty) = \infty$. More precisely the process Σ^Δ satisfies

(6.1) $$\Sigma_t^\Delta = |B_t| + \Delta^{-1}(2l_t^0(B)),$$

for $t \geq 0$, where Δ^{-1} is the inverse of the function Δ. By taking $\Delta(t) = \delta t$ we recover Σ^δ.It was shown in [11] that the local time process $\left(l_\infty^a(\Sigma^\Delta); a \geq 0\right)$ has law, denoted by $\mathbb{Q}_0^\Delta$, which is that of a process $(Z_t; t \geq 0)$ satisfying

(6.2) $$Z_t = 2\int_0^t \sqrt{Z_s}d\beta_s + \Delta(t), \qquad t \geq 0,$$

for some Brownian motion β.

We now consider an analogous generalization of the perturbed Bessel process of dimension three. Suppose Δ is as above, and additionally Δ' is bounded away from zero and infinity (this condition could be weakened). The process $(R_{3,h}(t); t \geq 0)$ obtained from Σ^Δ by the space-time inversion

(6.3) $$\frac{1}{R_{3,h}(t)} = \Sigma^\Delta\left(\int_t^\infty \frac{du}{(R_{3,h}(u))^4}\right), \qquad \text{for all } t > 0,$$

satisfies

(6.4) $$R_{3,h}(t) = \beta_t + \int_0^t \frac{ds}{R_{3,h}(s)} + h(M_t),$$

where β is a Brownian motion and $M_t = \sup_{s \leq t} R_{3,h}(s)$. The C^1 function $h : \mathbb{R}^+ \to \mathbb{R}$ satisfies

(6.5) $$h(0) = 0, \quad 2(1 - h'(y)) = \Delta'(1/y), \qquad 0 < y < \infty,$$

which generalizes the relation between α and δ. The proof of this assertion follows a now familiar course. The process Σ^Δ can be obtained from a three-dimensional Bessel process R by a space-time change

(6.6) $$\frac{\Sigma_t^\Delta}{\sigma(J_t)} = R\left(\int_0^t \frac{ds}{(\sigma(J_s))^2}\right),$$

where $J_t = \inf_{u \geq t} \Sigma_u^\Delta$ and the function σ satisfies

(6.7) $$\frac{1}{2}\Delta'(y) = 1 - \frac{\sigma'(y)}{\sigma(y)}y, \qquad 0 < y < \infty.$$

Likewise the process $R_{3,h}$ satisfies (6.4) if and only if it is obtained from some three-dimensional Bessel process $\tilde{R}$ via

(6.8) $$\frac{R_{3,h}(t)}{k(M_t)} = \tilde{R}\left(\int_0^t \frac{ds}{\{k(M_s)\}^2}\right),$$

where the function k satisfies

(6.9) $$h'(x) = \frac{k'(x)}{k(x)}x, \qquad 0 < x < \infty.$$

Next, we observe that the space-time inversion (6.3) which connects $R_{3,h}$ and Σ^Δ corresponds exactly to

$$\frac{1}{R(t)} = \hat{R}\left(\int_t^\infty \frac{du}{(R(u))^4}\right). \tag{6.10}$$

Moreover the relation (6.5) is now a simple consequence of combining (6.7) and (6.9).

The local time process $(l_\infty^a(R_{3,h}); a \geq 0)$ is equal in law to $(\hat{Z}_t \equiv t^2 Z_{1/t}; t \geq 0)$ where Z satisfies (6.2). However it can be shown, although we do not pursue this here, that the law of $\hat{Z}$ is only of the form $\mathbb{Q}_0^{\hat{\Delta}}$ when $\Delta(y) = \hat{\Delta}(y) = \delta y$.

7. Perturbations at the future minimum.

Inspection of (2.3), rewritten as

$$\Sigma_t^\delta = B_t + (2-\alpha)J_t^\Sigma, \tag{7.1}$$

shows that a Σ^δ process can be thought of as a version of a BES(3) process, perturbed at its future minimum. In order to obtain analogous generalisations of BES(d), $d > 2$, we remark first that if R is a BES(d) process, starting from zero, and $d > 2$, then R satisfies

$$R_t = B_t + 2J_t^R + \frac{1}{2}(3-d)\int_0^t \frac{ds}{R_s}.$$

(See, e.g. [19], Chap. XII, p46, [13], [14], and [12].) This motivates the following definition. We say that Σ is a Bessel process of dimension $d > 2$, α-perturbed at its future minimum, and write Σ is a JBES(d,α), if it satisfies

$$\Sigma_t = B_t + (2-\alpha)J_t^\Sigma + \frac{1}{2}(3-d)\int_0^t \frac{ds}{\Sigma_s}. \tag{7.2}$$

It is easy to see that uniqueness of law of solutions of (7.2) holds, by establishing that the equivalent pair of relations given in (2.7) extend to the situation where $\hat{R}$ is a BES(d) process and Σ is a JBES(d,α) process. This is the "JBES version" of Theorem 1.1, and in fact many of our results have extensions to this situation. We conclude by giving some of these, without proofs.

First, the mapping given in (2.4) which maps a PBES($3,\alpha$) into a JBES($3,\alpha$) also maps a PBES(d,α) into a JBES(d,α) for any $d > 2$. Consequently, using Theorem 5.2, one can deduce the law of $\{l_\infty^a(\Sigma_{d,\alpha}), a \geq 0\}$. Moreover the mapping that extends (2.5) by mapping a PBES(d,α) R into a PBES(d,α) X killed on hitting 1 is given by

$$\frac{R_t}{\{1+R_t^{1/\beta}\}^\beta} = X\left(\int_0^t \frac{du}{\{1+R_u^{1/\beta}\}^{2+2\beta}}\right). \tag{7.3}$$

However it can not be true, for $d \neq 3$, that the process we get by applying this same transformation to Σ, the JBES(d,α) process which is the image under (2.4) of R, is related to X by time-reversal, as this is not true when $\alpha = 0$.

Finally the extension of Theorem 4.2 is the assertion that, if $d > 2$ and $\beta = 1/(d-2)$, then a JBES(3,α) process Σ and a JBES(d,α) process $\Sigma^\#$ are related by

$$\{\Sigma(t)\}^\beta = \Sigma^\#\left(\beta^2\int_0^t ds\{\Sigma(s)\}^{2(\beta-1)}\right). \tag{7.4}$$

References

[1] J.Bertoin. Excursions of a $BES_o(d)$ process ($0 < d < 1$). Prob. Th. and Rel. Fields, 84, 231-250, 1990.

[2] P.Biane.Comparaison entre temps d'atteinte et temps de séjour de certaines diffusions réelles. Sém. Prob.XIX, Lecture Notes in Mathematics, vol. 1123, Springer, Berlin Heidelberg New York, 291-296, 1985.

[3] P.Carmona, F. Petit, and M.Yor. Some extensions of the arc-sine law as (partial) consequences of the scaling property of Brownian motion. Prob. Th. and Rel. Fields, 100,1-29, 1994.

[4] J.Y.Calais and M.Génin. Sur les martingales locales continues indexées par $]0,\infty[$. Sém.Prob.XVII, LectureNotes in Mathematics, vol. 986 Springer. Berlin Heidelberg New York, 454-466, 1988.

[5] L.Chaumont and R.A.Doney. Pathwise uniqueness for pereturbed versions of Brownian motion and reflected Brownian motion. Preprint, 1997.

[6] C.Donati-Martin and M.Yor. Some Brownian functionals and their laws. Ann. Prob., 25, 1011-1058, 1997.

[7] R.A.Doney. Some calculations for perturbed Brownian motion. In this volume.

[8] R.K.Getoor and M.J.Sharpe. Excursions of Brownian motion and Bessel processes. Zeit. für Wahr. 47, 83-106, 1979.

[9] J.F.Le Gall. Sur la mesure de Haussdorff de la courbe Brownienne. Sém.Prob.XIX, Lecture Notes in Mathematics, vol. 1123, Springer, Berlin Heidelberg New York, 297-313, 1985.

[10] J.F.Le Gall and M.Yor. Excursions browniennes et carrés de processus de Bessel. Comptes Rendus Acad. Sci. I, 303, 73-76, 1986.

[11] J.F.Le Gall and M.Yor. Enlacements du mouvement brownien autour des courbes de l'espace.Trans. Amer. Math. Soc. 317, 687-722, 1990.

[12] B.Rauscher. Some remarks on Pitman's theorem. Sém.Prob.XXXI, Lecture Notes in Mathematics, vol. 1655, Springer, Berlin Heidelberg New York, 29, 1997.

[13] Y.Saisho and H.Tanemura. Pitman type theorems for one- dimensional diffusion. Tokyo J. Math., 2, 429-440, 1990.

[14] K.Takaoka. On the martingales obtained by an extension due to Saisho, Tanemura and Yor of Pitman's theorem. Sém.Prob.XXXI, Lecture Notes in Mathematics, vol. 1655, Springer, Berlin Heidelberg New York, 29, 1997.

[15] D.Revuz and M. Yor. *Continuous Martingales and Brownian Motion*. Springer-Verlag , Berlin, 1991.

[16] W.Werner. Some remarks on perturbed Brownian motion. Sém. Prob.XXIX, Lecture notes in Mathematics, vol. 1613, Springer, Berlin Heidelberg New York, 37-42, 1995.

[17] M.Yor. Une explication du théoreme de Ciesielski-Taylor. Ann.Inst.Henri Poincaré, Prob. et Stat.,27, 201-213,1991.

[18] M.Yor. *Some aspects of Brownian motion, part I; some special functionals.Lectures in Mathematics, Birkhaüser*, ETH Zürich, 1992.

[19] M.Yor. *Some aspects of Brownian motion, part II; some recent martingale problems. Lectures in Mathematics, Birkhaüser*, ETH Zürich, 1997.

R.A.DONEY, Mathematics Department, University of Manchester, Manchester M13 9PL, UK.

J.WARREN, Statistics Department, University of Warwick, Coventry, CV4 7AL, UK.

M.YOR, Laboratoire de Probabilités, Université Pierre et Marie Curie, tour 56, 4 place Jussieu, 75252 Paris cedex 05.

The Maximum Maximum of a Martingale

David G. Hobson

Department of Mathematical Sciences, University of Bath,
Claverton Down, Bath, BA2 7AY. UK.

Abstract

Let $(M_t)_{0\leq t\leq 1}$ be any martingale with initial law $M_0 \sim \mu_0$ and terminal law $M_1 \sim \mu_1$ and let $S \equiv \sup_{0\leq t\leq 1} M_t$. Then there is an upper bound, with respect to stochastic ordering of probability measures, on the law of S.

An explicit description of the upper bound is given, along with a martingale whose maximum attains the upper bound.

1 Introduction

Let μ_0 and μ_1 be probability measures on $\mathbb{R}$ with associated distribution functions $F_i[x] \equiv \mu_i((-\infty, x])$. Now let $\mathcal{M} \equiv \mathcal{M}(\mu_0, \mu_1)$ be the space of all martingales $(M_t)_{0\leq t\leq 1}$ with initial law μ_0 and terminal law μ_1. For such a martingale $M \in \mathcal{M}$ let $S \equiv \sup_{0\leq t\leq 1} M_t$ and denote the law of S by ν. In this short article we are interested in the set $\mathcal{P} \equiv \mathcal{P}(\mu_0, \mu_1) \equiv \{\nu; M \in \mathcal{M}\}$ of possible laws ν, and in particular we find a least upper bound for $\mathcal{P}$. The fact that M is a martingale imposes quite restrictive conditions on ν.

Clearly $\mathcal{M}$ is empty unless the random variables corresponding to the laws μ_i have the same finite mean, and henceforth we will assume without loss of generality that this mean is zero. Moreover a simple application of Jensen's inequality shows that a further necessary condition for the space to be non-empty is that

$$(1) \qquad \int_x^\infty (y-x)\mu_0(dy) \leq \int_x^\infty (y-x)\mu_1(dy) \qquad \forall x.$$

These conditions are also sufficient, see for example Strassen [16, Theorem 2] or Meyer [10, Chapter XI].

The question described in the opening paragraph is a special case of a problem first considered in Blackwell and Dubins [4] and Dubins and Gilat [7]. There the authors derive conditions on the possible laws $\tilde{\nu}$ of the supremum $\tilde{S}$ of a martingale $(\tilde{M}_t)_{0\leq t\leq 1}$ whose terminal distribution $\tilde{\mu}_1$ is given, but whose initial law $\tilde{\mu}_0$ is *not* specified. Let $\preceq$ denote stochastic ordering on probability measures, (so that $\rho \preceq \pi$ if and only if $F_\rho[x] \geq F_\pi[x]$ $\forall x$, with the obvious notational convention) and let ρ^* denote the Hardy transform of a probability measure ρ. Then it follows from [4] and [7] that

$$\tilde{\mu}_1 \preceq \tilde{\nu} \preceq \tilde{\mu}_1^*. \tag{2}$$

Indeed, Kertz and Rösler [9] have shown that the converse to (2) also holds: for any probability measure ρ satisfying $\tilde{\mu}_1 \preceq \rho \preceq \tilde{\mu}_1^*$, there is a martingale with terminal distribution $\tilde{\mu}_1$ whose maximum has law ρ. If, moreover, ρ is concentrated on $[0,\infty)$ then the martingale $\tilde{M}$ can be taken to have initial law consisting of the unit mass at 0. See also Rogers [14] for a proof of these results based on excursion theory, and Vallois [17] for a discussion of the case where $\tilde{M}$ is a continuous martingale. Thus if $\mu_0 \equiv \delta_0$ (the unit mass at 0) then our problem is solved and

$$\mathcal{P}(\delta_0, \mu_1) \equiv \{\nu : \delta_0 \vee \mu_1 \preceq \nu \preceq \mu_1^*\}.$$

Otherwise, as Kertz and Rösler [9, Remark 3.3] observe,

$$\mathcal{P}(\mu_0, \mu_1) \subseteq \{\nu : \mu_0 \vee \mu_1 \preceq \nu \preceq \mu_1^*\}. \tag{3}$$

In a sense Kertz and Rösler [9, Theorem 3.4] answer our question of interest also. They describe necessary and sufficient conditions for a candidate probability measure ν to be a member of $\mathcal{P}(\mu_0, \mu_1)$. These conditions involve displaying a pair of bivariate densities with marginals (μ_0, μ_1) and (μ_0, ν) and may be thought of as a restatement of the problem. In contrast the solution presented here is both explicit and constructive.

The main results of this article are that the set $\mathcal{P}(\mu_0, \mu_1)$ is bounded above by a probability measure $\mu_{0,1}^*$ (in the sense that if $\nu \in \mathcal{P}$ then $\nu \preceq \mu_{0,1}^*$), and that this upper bound is attained. Moreover we provide an explicit construction of this upper bound: we do so now for the nice special case where μ_1 has a continuous distribution. For $i = 0, 1$ define $\eta_i = \int_x^\infty (y - x)\mu_i(dy)$ and let $a(z)$ be the solution with $a(z) \geq z$ to the equation

$$\eta_0(a(z)) = \eta_1(z) + (a(z) - z)\eta_1'(z).$$

Pictorially $a(z)$ is the x-co-ordinate of the point where the tangent to η_1 at z intersects the graph of η_0. See Figure 1. Then $\mu_{0,1}^*$ is defined by $\mu_{0,1}^*((-\infty, x]) = \mu_1((-\infty, a^{-1}(x)])$. In Section 2 we prove this result and extend to arbitrary measures μ_1.

Clearly one non-constructive definition of $\mu_{0,1}^*$ is via its distribution function $F_{0,1}^*$:

$$F_{0,1}^*[x] \equiv \inf_{\nu \in \mathcal{P}} F_\nu[x]. \tag{4}$$

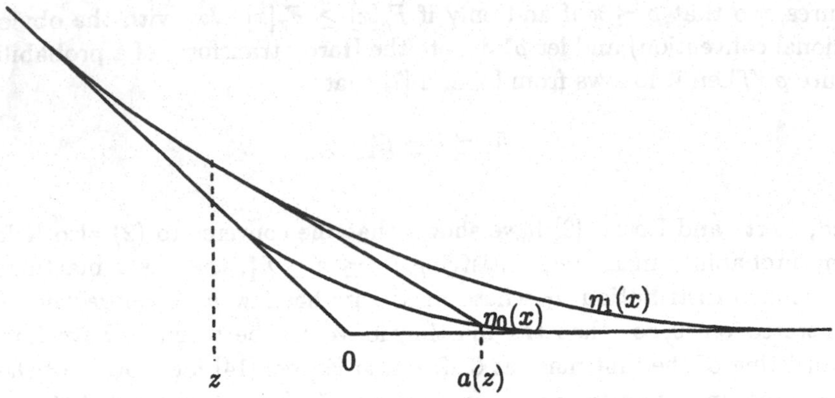

Figure 1: The functions $\eta_i(x)$ and $a(z)$.

There seems no reason *a priori* why the measure corresponding to (4) should be an element of $\mathcal{P}(\mu_0, \mu_1)$. Indeed if the greatest lower bound $\mu_*^{0,1}$ is defined via its distribution function $F_*^{0,1}$:

$$F_*^{0,1}[x] \equiv \sup_{\nu \in \mathcal{P}} F_\nu[x]; \tag{5}$$

then $\mu_*^{0,1}$ need not be an element of $\mathcal{P}$; in particular it is not in general true that $\mu_0 \vee \mu_1 \in \mathcal{P}$. See Section 3.4 for a simple example showing non-attainment of the lower bound.

The Skorokhod embedding theorem concerns the embedding of a given law in Brownian motion by construction of a suitably minimal stopping time. (Skorokhod embeddings for Brownian motion and other processes remains an active area of research; see the recent paper by Bertoin and Le Jan [3] for a new class of suitable stopping rules.) We show that one martingale whose maximum attains the upper bound is a (time-change of) Brownian motion, and this explains why the prescient reader will recognise in the arguments expounded below elements of the Chacon and Walsh [5] and Azéma and Yor [2] proofs of the Skorokhod theorem (and also the Rogers [12] excursion theoretic version of the Azéma-Yor argument). An incidental remark in Section 3.3 indicates how these alternative derivations of the Skorokhod theorem are in fact closely related.

Finally some brief words on a motivation for studying this problem. Let M_t be the price process of a financial asset, and suppose that interest rates are zero. Then standard arguments from the theory of complete markets show that when pricing contingent claims or derivative securities it is natural to treat M as if it were a martingale. The simplest and most liquidly traded contingent claims are European call options which, at maturity T, have payoff $(M_T - k)^+$.

Suppose now that instead of attempting to model M_t and thence to predict the prices of call options, we assume that the prices of calls are fairly determined by the market. From knowledge of call prices for all strikes k it is possible to infer the law (at least under the measure used for pricing derivatives) of M_T. Bounds on the prices of 'exotic' derivatives can be obtained by maximising the expected payoff of the exotic option over the space of martingales with the given (or rather the inferred) terminal distribution. These bounds depend on the market prices of call options, but they do not rely on any modelling assumptions which attempt to describe the underlying price process.

As an example, the lookback option is a security which at maturity T has value S_T, the maximum price attained by the asset over the interval $[0,T]$. Given the set of prices of call options with maturity T we can deduce the (implied) law of M_T under the pricing measure. Since M_0 is fixed and M_t is a martingale under the pricing measure, the problem of characterising the possible prices of a lookback security is solved once the possible laws of the maximum S_T have been determined. This is the problem under consideration in Blackwell and Dubins [4], and more generally here. See Hobson [8] for a more detailed analysis of the lookback option, the derivation of non-parametric bounds on the lookback price and the description of an associated hedging strategy.

Acknowledgement It is a pleasure to thank David Marles, Uwe Rösler and an anonymous referee for helpful and insightful comments on a previous version of this paper.

2 Main results

In this section we derive conditions relating the distribution of the maximum ν to the initial and terminal distributions μ_0, μ_1. First we recall some simple bounds which do not depend on the initial law μ_0.

Clearly $\mathbb{P}[M_1 > x] \leq \mathbb{P}[S > x]$ so that $\mu_1 \preceq \nu$. Define the non-decreasing barycentre function b_1 by

$$b_1(x) = \frac{\mathbb{E}[M_1; M_1 \geq x]}{\mathbb{P}[M_1 \geq x]}$$

for all x such that $\mathbb{P}[M_1 \geq x] > 0$, and $b_1(x) = x$ otherwise. By Doob's submartingale inequality

$$c\mathbb{P}[S \geq c] \leq \mathbb{E}[M_1; S \geq c].$$

Fixing c, then at least in the case where μ_1 has no atoms, there is some d with $\mathbb{P}[S \geq c] = \mathbb{P}[M_1 \geq d]$. Moreover it is trivial that $\mathbb{E}[M_1 - d; A] \leq \mathbb{E}[M_1 - d; M_1 \geq d]$ for all sets A, so that $\mathbb{E}[M_1; S \geq c] \leq \mathbb{E}[M_1; M_1 \geq d] = b_1(d)\mathbb{P}[M_1 \geq d]$. Thus $c \leq b_1(d)$ and it follows that $\nu \preceq \mu_1^*$, where μ_1^*, the Hardy transform of μ_1, has associated distribution function F_1^* given by $F_1^*[x] = F_1[b_1^{-1}(x)]$. The proof in the case where μ_1 contains atoms requires only minor modifications.

The above paragraph, which follows Blackwell and Dubins [4] closely, contains a proof of (2), and provides many of the essential arguments we will use in Proposition 2.1 to find an upper bound for $\mathcal{P}$. Our purpose is to consider the effect of fixing the initial law of the martingale.

For $i = 0, 1$ define the functions $\eta_i(x)$ by

$$\eta_i(x) = \int_x^\infty (1 - F_i[y])dy = \int_x^\infty (y - x)\mu_i(dy) = \mathbb{E}[(M_i - x)^+].$$

The functions η_i are positive, decreasing and convex with $\eta_i(x) \geq -x$. Recall from the introduction that a necessary and sufficient condition for $\mathcal{M}(\mu_0, \mu_1)$ to be non-empty is that $\eta_0(x) \leq \eta_1(x)$ for all x. Henceforth we assume that this condition is satisfied. Furthermore, if $\eta_1(x) = \eta_0(x)$ then

$$\begin{aligned}
\mathbb{E}((M_1 - x)^+; M_0 \leq x) &= \mathbb{E}((M_1 - x)^+) - \mathbb{E}((M_1 - x)^+; M_0 > x) \\
&\leq \eta_1(x) - \mathbb{E}((M_0 - x)^+; M_0 > x) \\
&= 0,
\end{aligned}$$

so that $\mathbb{P}(M_1 > x, M_0 \leq x) = 0$. Similarly $\mathbb{P}(M_1 < x, M_0 \geq x) = 0$ so that if μ_0 associates any mass with a point $x \in \{z : \eta_1(z) = \eta_0(z)\}$ then we must have that the martingale M is constant on the set $M_0 = x$, and μ_1 must include a corresponding atom. By considering such atoms separately, and by dividing the set $I = \{x : \eta_1(x) > \eta_0(x)\}$ into its constituent intervals we can reduce to the case where I takes the form of a single interval $I \equiv (i^-, i^+) \subseteq (-\infty, \infty)$. and assume further that μ_0 places no mass at the endpoints of I.

We now construct functions a, α and β which will play a crucial role in subsequent analysis. As motivation, suppose temporarily that μ_1 has no atoms. Note that the derivative of η_1 is given by $\eta_1'(x) = -\mathbb{P}[M_1 > x] \equiv F_1[x] - 1$. For $z \leq i^-$ and $z \geq i^+$ let $a(z) = z$ and otherwise define $a(z)$ to be the unique solution with $a(z) > z$ to the equation

$$\eta_0(a(z)) = \eta_1(z) + (a(z) - z)\eta_1'(z). \tag{6}$$

$a(z)$ is the x-coordinate of the point where the tangent to η_1 at z, taken in the direction of increasing z, intersects with the function $\eta_0(z)$. Recall Figure 1. The function a is non-decreasing, and on I it satisfies

$$a(dz)(\eta_0'(a(z)) - \eta_1'(z)) = \mu_1(dz)(a(z) - z).$$

Define $F_{0,1}^*$ to be the distribution function given by

$$F_{0,1}^*[x] = F_1[a^{-1}(x)].$$

(The composition $F_1 \circ a^{-1}$ is well defined since μ_1 assigns no mass to intervals where a is constant.) Let $\mu^*_{0,1}$ be the associated probability measure; $\mu^*_{0,1}$ will be the upper bound on $\mathcal{P}(\mu_0, \mu_1)$.

For the general case where μ_1 has atoms we let the above argument guide our intuition. For $u \in (0,1)$ define $\beta(u) = \inf\{x : F_1[x] \geq u\}$. The parameter u will play the role of defining the slope of the relevant tangent. Define h_u via

$$h_u(z) = \eta_0(z) + z(1-u) - \eta_1(\beta(u)) - \beta(u)(1-u).$$

Then h_u is a convex function; if $h_u(x) < 0$ then h_u has a unique root in (x, ∞). On $\eta_1(\beta(u)) = \eta_0(\beta(u))$ set $\alpha(u) = \beta(u)$, and on $\eta_1(\beta(u)) > \eta_0(\beta(u))$ let $\alpha(u)$ be the root in $(\beta(u), \infty)$ of h_u; then the function α satisfies

$$\eta_0(\alpha(u)) = \eta_1(\beta(u)) - (\alpha(u) - \beta(u))(1-u). \tag{7}$$

Informally, $\alpha(u)$ is the x-coordinate of the point where the tangent to η_1 with gradient $-(1-u)$ intersects the graph of η_0. If F_1 has an atom of mass v at i^+ then for all $u > 1-v$ we have $\alpha(u) = \beta(u)$; meanwhile $\alpha(u) > \beta(u)$ on $(0, 1-v)$. The function α is continuous and has (left)-derivative

$$\frac{d\alpha}{du} = \frac{\alpha(u) - \beta(u)}{\eta_0'(\alpha(u)) + 1 - u}, \tag{8}$$

where η_0' is again a left-derivative. By the convexity of η_0, and the definition of α as the x-coordinate of the point where a line with slope $-(1-u)$ intersects η_0, it is clear that for $u \in (0,1)$ the denominator must be positive. Finally define the measure $\mu^*_{0,1}$ via its distribution function

$$F^*_{0,1}[x] \equiv \inf\{u : \alpha(u) > x\}. \tag{9}$$

Where defined we have that $\alpha(u) = a(\beta(u))$, and the two definitions of $\mu^*_{0,1}$ agree.

Example 2.1 Examples always help to make things clearer...
Let μ_0, μ_1 be the uniform measures on $\{-1, 1\}$, $\{-2, 0, 2\}$ respectively. In particular μ_1 is discrete so that a is not well defined, and this example illustrates the general method. Then $\eta_0(x) = \max\{-x, 0, (1-x)/2\}$ and $\eta_1(x) = \max\{-x, 0, 2(1-x)/3, (2-x)/3\}$. Further

$$\alpha(u) = 2I_{(u>2/3)} + 2/(3(1-u))I_{(1/3<u\leq 2/3)} + (4u-1)/(1-2u)I_{(u\leq 1/3)}.$$

See Figure 2 for a pictorial representation of the functions α, β and F_0^{-1}. Note that α is continuous and non-decreasing, and that for $u \in (0,1)$, $\alpha(u) \geq \beta(u+) \vee F_0^{-1}[u+]$.

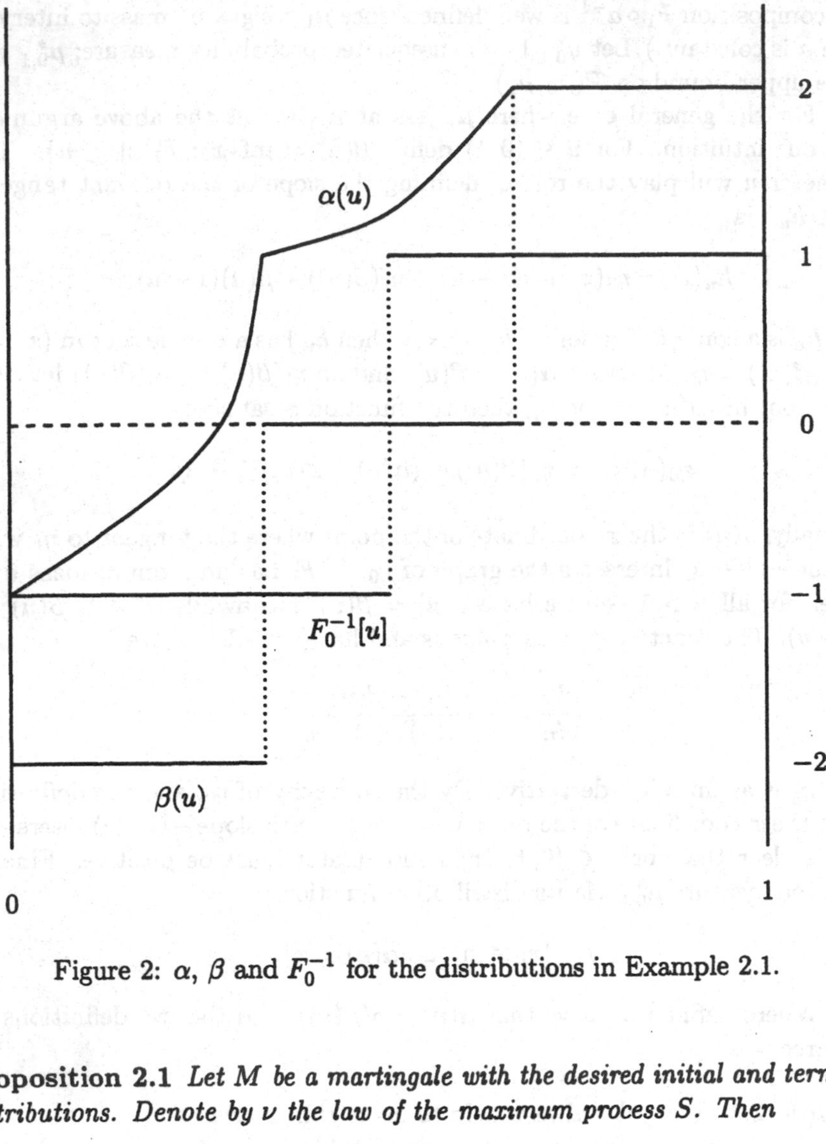

Figure 2: α, β and F_0^{-1} for the distributions in Example 2.1.

Proposition 2.1 *Let M be a martingale with the desired initial and terminal distributions. Denote by ν the law of the maximum process S. Then*

$$\nu \preceq \mu^*_{0,1} \qquad \forall \nu \in \mathcal{P}(\mu_0, \mu_1),$$

so that $\mu^*_{0,1}$ is an upper bound for $\mathcal{P}(\mu_0, \mu_1)$.

Proof
We prove that $F_\nu^{-1}[u] \leq \alpha(u)$ for all $u \in (0,1)$, where F_ν is the distribution function associated with the law ν. Since F_ν^{-1} and α are increasing functions it is sufficient to prove that if $\mathbb{P}(S > c) = 1 - F_\nu[c] = 1 - u$, then $c \leq \alpha(u)$.

Fix c, then by Doob's submartingale inequality

$$c\mathbb{P}[S > c, M_0 \leq c] \leq \mathbb{E}[M_1; S > c, M_0 \leq c].$$

Similarly, by the martingale property,

$$\mathbb{E}[M_0; M_0 > c] = \mathbb{E}[M_1; S > c, M_0 > c].$$

Adding these two expressions yields after some elementary manipulations

$$c\mathbb{P}[S > c] + \eta_0(c) \leq \mathbb{E}[M_1; S > c].$$

Now suppose $\mathbb{P}[S > c] = 1 - u$, then, since for any set A it is true that $\mathbb{E}[Y; A] \leq \mathbb{E}[Y; Y \geq 0]$,

$$\begin{aligned} \mathbb{E}[M_1; S > c] &= \mathbb{E}[M_1 - \beta(u); S > c] + (1-u)\beta(u) \\ &\leq \mathbb{E}[M_1 - \beta(u); M_1 \geq \beta(u)] + (1-u)\beta(u). \end{aligned}$$

Using (7) we can summarise these inequalities:

$$\begin{aligned} c(1-u) + \eta_0(c) &\leq \eta_1(\beta(u)) + (1-u)\beta(u) \\ &= \eta_0(\alpha(u)) + \alpha(u)(1-u). \end{aligned}$$

It is easy to see by the convexity of η_0 that $x(1-u) + \eta_0(x)$ is increasing on $(F_0^{-1}[u+], \infty)$. Since $\alpha(u) \geq F_0^{-1}[u+]$ it follows that $c \leq \alpha(u)$.

□

Inspection of the proof of Proposition 2.1 reveals that the upper bound is attained if both the martingale is continuous, (so that there is equality in Doob's inequality), and the sets $(S > c)$ can be identified with sets of the form $(M_1 > d)$. Guided by these observations our goal now is to construct a martingale M with initial law μ_0 and terminal law μ_1, whose maximum has law $\mu^*_{0,1}$. We ensure continuity of the martingale M by basing the construction on a stopping time for a Brownian motion. Moreover the stopping rule is a function of the current maximum of the Brownian motion and its current value.

Rösler [15] provides an alternative construction of a martingale whose maximum attains the upper bound. This construction is in the spirit of arguments given by Blackwell and Dubins [4] and Kertz and Rösler [9]. In some respects the Rösler construction is simpler than the methods presented below; the advantage of the methods we use is that they provide a solution of the Skorokhod problem.

Suppose that an initial point x_0 is chosen according to the distribution μ_0. For motivation consider first the case where μ_1 has a continuous distribution function and the function a and its inverse are well defined. Let B be a Brownian motion started at x_0 and define

$$S_t^B = \sup_{0\le u\le t} B_u$$
$$\tau = \inf\{u : B_u \le a^{-1}(S_u^B)\}$$

Note that τ is almost surely finite for if $y > x_0$ and H_y denotes the first hitting time of the Brownian motion B at level y then $\tau \le \inf\{t > H_y : B_t \le a^{-1}(y)\}$.

We show that, when averaged over the law of the starting point, B_τ has the law μ_1. Then also

$$\mathbb{P}[S_\tau^B > x] = \mathbb{P}[a(B_\tau) > x],$$

and the distribution function of S^B is given by

$$F_1[a^{-1}(x)] \equiv F_{0,1}^*[x]. \tag{10}$$

It will follow (in Corollary 2.1 below) that $\mu_{0,1}^*$ defined via (9) *is* an element of $\mathcal{P}(\mu_0, \mu_1)$. Our goal now is to prove the above claim that B_τ has law μ_1, in the setting of a general probability measure μ_1.

Proposition 2.2 *For a Brownian motion B with initial law μ_0 define $S_t^B = \sup_{0\le u\le t} B_u$ and $\tau = \inf\{u : F_1[B_u] \le F_{0,1}^*[S_u^B]\}$. Then $B_\tau \sim \mu_1$ and $S_\tau^B \sim \mu_{0,1}^*$.*

Proof

This follows directly from Chacon and Walsh [5] although here we provide a direct proof, similar in spirit to Azéma and Yor [2]. For a connection between these two approaches, see Section 3.3.

Suppose that μ_1 has an atom of size v at i^+, and thence that $\mu_{0,1}^*$ has an atom of at least this size there also.

With α, β and τ all as above define the random variable Z via $Z = F_{0,1}^*(S_\tau^B)$. Then $F_1(B_\tau) \ge Z \ge F_1(B_\tau -)$ so that $\beta(Z) \equiv B_\tau$ and, since α is continuous, $\alpha(Z) \equiv S_\tau^B$. We find the law of Z: it is sufficient to show that Z has the uniform distribution on $[0, 1]$, or more particularly, and to allow for atoms in $\mu_{0,1}^*$, it is sufficient to show that $\mathbb{P}(Z \le u) = u$, for $0 < u < 1 - v$. Then

$$\mathbb{P}(B_\tau \le x) = \mathbb{P}(\beta(Z) \le x) = \mathbb{P}(Z \le F_1[x]) = F_1[x],$$

and similarly $\mathbb{P}(S_\tau^B \le x) = F_{0,1}^*[x]$.

Return to the consideration of the law of Z: for a test function Φ (continuous and compactly supported) set $g = \Phi \circ \alpha^{-1}$ and define $G(x) = \int_0^x g(u)du$. Note that $G(\alpha(x)) = \int_0^x \Phi(u)d\alpha(u)$. Itô calculus shows that $N_t \equiv G(S_t^B) - (S_t^B - B_t)g(S_t^B)$ is a continuous local martingale, moreover the stopped martingale N^τ is bounded (see [2] for details). Therefore one has

$$\begin{aligned} 0 = \mathbb{E}[N_\tau - N_0] &= \mathbb{E}[G(S_\tau^B) - (S_\tau^B - B_\tau)g(S_\tau^B) - G(S_0^B)] \\ &= \mathbb{E}[G(\alpha(Z)) - G(B_0)] - \mathbb{E}[(\alpha(Z) - \beta(Z))g(\alpha(Z))]. \end{aligned} \tag{11}$$

Let π denote the law of Z, and let $\overline{F}_\pi[x] \equiv 1 - F_\pi[x] = \pi((x,\infty))$. Note that π has support contained in the interval $[0,1]$. Then

$$\mathbb{E}[G(\alpha(Z))] = \int_{\mathbb{R}} \pi(dz) \int_0^z \Phi(u)d\alpha(u) = \int_0^1 \overline{F}_\pi[u]\Phi(u)d\alpha(u).$$

For the second term a simple transformation of variable yields

$$\mathbb{E}[G(B_0)] = \int_0^1 \overline{F}_0[\alpha(u)]\Phi(u)d\alpha(u)$$

and (11) becomes

$$\int_0^1 (\overline{F}_\pi[u] - \overline{F}_0[\alpha(u)])\Phi(u)d\alpha(u) = \int_0^1 (\alpha(u) - \beta(u))\Phi(u)\pi(du).$$

Since Φ is arbitrary, π must satisfy the identity

$$(12) \qquad (\alpha(u) - \beta(u))\pi(du) = (\overline{F}_\pi[u] - \overline{F}_0[\alpha(u)])d\alpha(u).$$

Substituting from (8) gives that at least for $x < 1 - v$, (whence $\alpha(u) > \beta(u)$ for all $u \in (0,x)$)

$$F_\pi[x] - x = \int_0^x \left(\frac{F_0[\alpha(u)] - F_\pi[u]}{F_0[\alpha(u)] - u} - 1 \right) du = -\int_0^x \frac{F_\pi[u] - u}{F_0[\alpha(u)] - u} du.$$

and it follows that $F_\pi[x] = x$ for $x < 1 - v$. □

Corollary 2.1 *The measure $\mu_{0,1}^*$ is an element of $\mathcal{P}(\mu_0, \mu_1)$.*

Proof
It suffices to show that $M_t \equiv B(\tau \wedge (t/(1-t))$ is a true martingale and not just a local martingale, or equivalently that $(B_{t\wedge\tau})_{t\geq 0}$ is uniformly integrable. This follows by a straightforward extension to Lemma 2.3 in Rogers [14] and an appeal to Theorem 1 in Azéma, Gundy and Yor [1]. □

3 Remarks

3.1 The case $\mu_0 = \delta_0$ and reduction to previous results.

If $\mu_0 \equiv \delta_0$ then $\eta_0(x) = x^- \equiv (-x) \vee 0$. Suppose that μ_1 has a continuous distribution function then since $a(x) \geq 0$ the formula (6) becomes

$$a(x) = \frac{\eta_1(x) - x\eta_1'(x)}{|\eta_1'(x)|}.$$

The function $a(x)$ is easily shown to equal the barycentre function $b_1(x)$ and, for general μ_1 also, the results of the previous section become the twin statements that μ_1^* is both an element of, and an upper bound for, $\mathcal{P}(\delta_0, \mu_1)$.

3.2 Excursion-theoretic arguments

The defining equation for π given in (12) can be derived using excursion arguments. Readers who would like an introduction to excursion theory are referred to Rogers [13]. For simplicity consider the case where μ_1 has no atoms and as before let B be a Brownian motion with maximum process S^B such that the initial point B_0 is chosen according to the law μ_0. Define $\tau = \inf\{u : a(B_u) \le S_u^B\}$.

Consider splitting the Brownian path into excursions below its maximum. Imagine plotting the excursions from the maximum $S_t^B = s$, in xy space, in such a way that the x-component is always B_t, and the y-component is chosen to keep the two-dimensional process on the tangent to η_1 which joins $(s, \eta_0(s))$ with $(a^{-1}(s), \eta_1(a^{-1}(s)))$. As the maximum s increases, so the line along which excursions are plotted changes. See Figure 3. τ is then the first time that one of these excursions first meets the curve $(x, \eta_1(x))$.

By construction $\mathbb{P}[B_\tau \le y] = \mathbb{P}[B_0 \le a(y)] - \mathbb{P}[B_0 \le a(y), S_\tau^B \ge a(y)]$. Now for the event $(B_\tau \in dy)$ to occur it must be true that both $(B_0 \le a(y))$ and $(S_\tau^B \ge a(y))$, and then, before the maximum rises to $a(y+dy)$, there must be an excursion down from the maximum of relative depth at least $a(y) - y$. Using Lévy's identity in law of the pairs $(S^B, S^B - B)$ and $(L^B, |B|)$, and the fact that the local time rate of excursions of height in modulus at least x is x^{-1}, it follows that

$$\mathbb{P}[B_\tau \in dy] = \mathbb{P}[B_0 \le a(y), S_\tau^B \ge a(y)] \frac{a(dy)}{a(y) - y}.$$

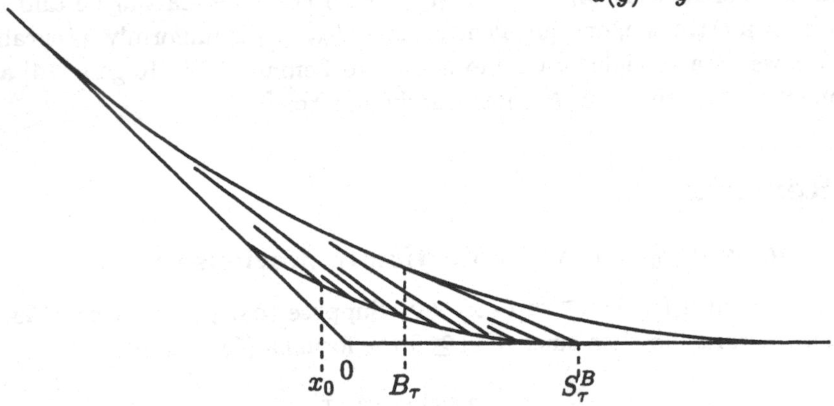

Figure 3: Some of the excursions below $S_t^B = s$ plotted along the tangents to η_1 at $a^{-1}(s)$

Define $\psi(y) = \mathbb{P}[B_\tau \le y] - F_1[y]$. We wish to show that $\psi(y) \equiv 0$. Now

$$\begin{aligned}\psi(y+dy) - \psi(y) &= \mathbb{P}[B_\tau \in dy] - \mu_1(dy) \\ &= (F_0[a(y)] - F_1[y] - \psi(y))\frac{a(dy)}{a(y)-y} - \mu_1(dy) \\ &= -\frac{\psi(y)}{F_0[a(y)] - F_1[y]}\mu_1(dy),\end{aligned}$$

where this last line follows from the identity

$$\frac{a(dy)}{a(y)-y} = \frac{\mu_1(dy)}{F_0[a(y)] - F_1[y]}.$$

It must follow that $\psi(y) \equiv 0$.

3.3 The Skorokhod Embedding Theorem

The proof of the Skorokhod embedding theorem given in Chacon and Walsh [5] can be expressed pictorially in a similar manner to Sections 2 and 3.2. One version of their algorithm, (see in particular Dubins [6]), involves a sequence of the following steps: firstly choose a value x and draw the tangent to η_1 at x; and secondly run a Brownian motion until it leaves some interval defined via this tangent (and the history of the construction to date). Our construction is a special case in which the values x chosen at each step are as small as possible. By the remarks in Section 3.1, when $\mu_0 \equiv \delta_0$ the function $a(x)$ is equivalent to the barycentre function $b_1(x)$ and the argument of Section 2 reduces to the Azéma-Yor proof. Thus the Azéma and Yor [2] and Rogers [14] proofs of the Skorokhod embedding theorem are seen to be special cases of the proof due to Chacon and Walsh [5].

3.4 The minimum maximum of a martingale

A *maximal* (respectively *minimal*) element of a set of measures $\mathcal{S}$ is a measure $\rho \in \overline{\mathcal{S}}$ for which there does *not* exist $\nu \in \mathcal{S}$ with $\nu \succ \rho$ (respectively $\nu \prec \rho$). The results of Section 2 show that $\mu^*_{0,1}$ is the *unique* maximal element of $\mathcal{P}(\mu_0, \mu_1)$. However it is not in general true that $\mathcal{P}$ has a unique minimal element.

It is clear that if M is any martingale with the desired initial and terminal laws, and if $\tilde{M}$ is the martingale consisting of a single jump such that $\tilde{M}_t \equiv M_0$ for $0 \le t < 1$ and $\tilde{M}_1 \equiv M_1$, then the law of the maximum of M stochastically dominates that of $\tilde{M}$. Thus the study of minimal elements reduces to a study of discrete parameter martingales at the timepoints 0 and 1.

Suppose the probability measures μ_0 and μ_1 are the Uniform measures on $\{-1, 1\}$ and $\{-2, 0, 2\}$ respectively. The joint laws of all (single jump) martingales with these initial and terminal distributions are parameterised by θ, $(0 < \theta < 1/12)$, in the following table:

		M_0^θ	
		-1	+1
M_1^θ	-2	$(1/4) + \theta$	$(1/12) - \theta$
	0	$(1/4) - 2\theta$	$(1/12) + 2\theta$
	2	θ	$(1/3) - \theta$

Table 1: The joint law of M^θ

If S^θ denotes the supremum of the martingale M^θ parameterised by θ then $\mathbb{P}[S^\theta > x] = \mathbb{P}[M_0^\theta \vee M_1^\theta > x]$. In particular $\mathbb{P}[S^\theta \geq 0] = (3/4) - \theta$ and $\mathbb{P}[S^\theta > 0] = (1/2) + \theta$. It is impossible to minimise both these expressions simultaneously. For this simple example there is a non-degenerate family of minimal elements of $\mathcal{P}(\mu_0, \mu_1)$ and the greatest lower bound of $\mathcal{P}(\mu_0, \mu_1)$ is not attained. Further $\mu_0 \vee \mu_1$ is not an element of $\mathcal{P}(\mu_0, \mu_1)$.

3.5 The minimum maximum of a continuous martingale

If the martingales M are further constrained to be continuous, then as Perkins [11] has shown, if the initial law is trivial, then there is a unique minimal element to the set of possible laws of the maximum.

Specifically, let μ_0 have zero mean and let $\mathcal{M}_C \equiv \mathcal{M}_C(\delta_0, \mu_1)$ be the space of all continuous martingales $(M_t)_{0 \leq t \leq 1}$ which are null at 0, and have terminal law μ_1. Let $\mathcal{P}_C \equiv \mathcal{P}_C(\delta_0, \mu_1)$ be the set of laws of the associated maxima. Then $\mathcal{P}_C$ has a unique minimal element, which can be represented as a Skorokhod embedding of a Brownian motion. See [11] for further details of this construction.

References

[1] Azéma, J., Gundy, R.F. and Yor, M; Sur l'integrabilité uniforme des martingales continues, *Séminaire de Probabilités*, **XIV**, 53–61, 1980.

[2] Azéma, J and Yor, M; Une solution simple au problème de Skorokhod, *Séminaire de Probabilités*, **XIII**, 90–115, 1979.

[3] Bertoin, J. and Le Jan, Y.; Representation of measures by balayage from a regular recurrent point, *Annals of Probability*, **20**, 538–548, 1992.

[4] BLACKWELL, D. AND DUBINS, L.D.; A Converse to the Dominated Convergence Theorem, *Illinois Journal of Mathematics*, **7**, 508–514, 1963.

[5] CHACON, R. AND WALSH, J.B.; One-dimensional potential embedding, *Séminaire de Probabilités*, **X**, 19–23, 1976.

[6] DUBINS, L.D.; On a theorem of Skorokhod, *Annals of Mathematical Statistics*, **39**, 2094–2097, 1968.

[7] DUBINS, L.D. AND GILAT, D.; On the distribution of maxima of martingales, *Proceedings of the American Mathematical Society*, **68**, 337–338, 1978.

[8] HOBSON, D.G.; Robust hedging of the lookback option, To appear in *Finance and Stochastics*, 1998.

[9] KERTZ, R.P. AND RÖSLER, U.; Martingales with given maxima and terminal distributions, *Israel J. Math.*, **69**, 173–192, 1990.

[10] MEYER, P.A.; *Probability and Potentials*, Blaidsell, Waltham, Mass., 1966.

[11] PERKINS, E.; The Cereteli-Davis solution to the H^1-embedding problem and an optimal embedding in Brownian motion, *Seminar on Stochastic Processes, 1985*, Birkhauser, Boston, 173–223, 1986.

[12] ROGERS, L.C.G.; Williams' characterisation of the Brownian excursion law: proof and applications, *Séminaire de Probabilités*, **XV**, 227–250, 1981.

[13] ROGERS, L.C.G.; A guided tour through excursions, *Bull. London Math. Soc.*, **21**, 305–341, 1989.

[14] ROGERS, L.C.G.; The joint law of the maximum and the terminal value of a martingale, *Prob. Th. Rel. Fields*, **95**, 451–466, 1993.

[15] RÖSLER, U.; Personal communication.

[16] STRASSEN, V.; The existence of probability measures with given marginals, *Ann. Math. Statist.*, **36**, 423–439, 1965.

[17] VALLOIS, P.; Sur la loi du maximum et du temps local d'une martingale continue uniformément intégrable, *Proc. London Math. Soc.*, **3**, 399–427, 1994.

AUTOUR D'UN THÉORÈME DE TSIRELSON SUR DES FILTRATIONS BROWNIENNES ET NON BROWNIENNES

par M.T. Barlow, M. Émery, F.B. Knight, S. Song et M. Yor

En 1979, l'un de nous a posé dans [23] le problème suivant : Si un mouvement brownien linéaire a la propriété de représentation prévisible dans une certaine filtration (qui peut être plus grosse que sa filtration naturelle mais pour laquelle ce mouvement brownien est une martingale), cette filtration est-elle nécessairement engendrée par un (autre) mouvement brownien? Dix ans plus tard, ce problème a été l'une des motivations de l'étude [4] sur les « mouvements browniens de Walsh »; si, comme on peut s'y attendre, la filtration naturelle de ces processus n'est pas brownienne, elle fournit un contre-exemple au problème initial. Mais le problème général est resté ouvert jusqu'en 1994, quand L. Dubins, J. Feldman, M. Smorodinsky et B. Tsirelson y ont apporté dans [8] une réponse négative au moyen d'un contre-exemple effroyablement compliqué (récemment simplifié par W. Schachermayer [19]). La question plus particulière de savoir si les mouvements browniens de Walsh, pourtant bien plus simples, ont aussi une filtration non-brownienne, restait pendante. Elle a été résolue en 1996 par B. Tsirelson, qui a donné une magnifique démonstration du caractère non brownien des filtrations walshiennes. Plus précisément, il montre qu'il n'existe aucun processus de Walsh dans la filtration naturelle d'un mouvement brownien, même si ce mouvement brownien est de dimension plus grande que 1, voire infinie!

L'outil fondamental introduit par Tsirelson est une propriété de rigidité des mouvements browniens : Si B et B' sont deux mouvements browniens tels que $\langle B, B' \rangle_t = (1-\varepsilon)\, t$, où ε est petit, leurs filtrations naturelles sont à la fois proches (au sens où tout objet dans l'une est approché par un objet analogue dans l'autre) et néanmoins bien distinctes (deux mouvements browniens de la forme $\int H\, dB$ et $\int H'\, dB'$ ont très peu de zéros communs; deux v. a. $U = f(B)$ et $U' = g(B')$ de lois diffuses vérifient nécessairement $\mathbf{P}[U = U'] = 0$). Tsirelson montre que les filtrations des processus de Walsh ne peuvent pas satisfaire simultanément ces deux contraintes, et en déduit qu'une filtration brownienne ne peut pas contenir de processus de Walsh.

Notre but initial était de traduire, pour les lecteurs du Séminaire, le travail de Tsirelson dans le langage de la théorie générale des processus. Cette traduction s'est peu à peu enrichie de remarques sur les idées de Tsirelson et de corollaires de sa méthode; elle est devenue un article autonome. De son côté, Tsirelson a continué à approfondir ces questions, et en comparant nos manuscrits en juin 1997, nous les avons trouvés très semblables sur le fond, bien que de style et de vocabulaire assez différents. Cette similitude n'est pas étonnante, les idées maîtresses, qui charpentent tout l'édifice, étant celles de Tsirelson. L'article [20] de Tsirelson paraîtra dans *Geometric and Functional Analysis.*

Notre première partie, assez fastidieuse, est consacrée à des généralités et à du vocabulaire. Elle est là surtout par souci de complétude et peut être omise sans inconvénient, à l'exception du joli théorème 1, puissant, simple et important pour la suite.

La seconde partie introduit une classe de processus, les martingales-araignées, qui nous paraissent être au cœur de la question (ce sont les martingales continues à $n-1$ dimensions qui prennent leurs valeurs dans n demi-droites issues de l'origine et non contenues dans un hyperplan, par exemple les martingales complexes Y telles que $\arg Y = 0 \bmod 2\pi/3$). Le résultat fondamental est le théorème 2 : les filtrations browniennes ne contiennent pas de martingales-araignées non triviales. C'est certainement le résultat le plus important de cette étude; la méthode est celle inventée par Tsirelson pour montrer que les filtrations browniennes ne contiennent pas de processus de Walsh non triviaux.

La troisième partie rappelle la définition des temps honnêtes (les instants de début des excursions d'une martingale-araignée sont des exemples de temps honnêtes) et déduit du théorème 2 la réponse (positive) à une question posée dans [4] : Si L est un temps honnête dans la filtration naturelle $\mathcal{F}$ d'un mouvement brownien, on a $\mathcal{F}_{L+} = \mathcal{F}_L \vee \sigma(A)$, où A est un événement; le supplément d'information qui apparaît à l'instant $L+$ est donc au maximum de un bit. Cette propriété est préservée par les changements (raisonnables!) de temps ou de probabilité.

La dernière partie s'intéresse aux processus de Walsh. En traduisant pour ces processus les résultats précédents, on voit que ces processus ne peuvent vivre dans une filtration brownienne; et en localisant la méthode de Tsirelson à des intervalles prévisibles, on obtient la diversité des filtrations walshiennes : deux processus de Walsh qui engendrent la même filtration, ou des filtrations isomorphes, ont la même loi.

Notre exposé ne prétend bien sûr pas se substituer à la lecture du travail original [20] de Tsirelson, dans lequel se trouvent entre autres les deux beaux théorèmes que voici et dont nous ne traitons pas :

a) *Si, dans une filtration brownienne,* Y *est une semimartingale continue à valeurs complexes telle que* $\arg Y = 0 \bmod 2\pi/3$, *et si la partie à variation finie de* Y *est portée par le fermé* $\{Y = 0\}$, *alors les parties à variation finie* $dA^{(j)}$ *des trois « composantes »* $Y^{(j)} = |Y|\,\mathbb{1}_{\{\arg Y = 2j\pi/3\}}$ *vérifient*

$$dA^{(1)} \wedge dA^{(2)} \wedge dA^{(3)} = 0\,.$$

b) *Soient* $U^{(1)}$, $U^{(2)}$, $U^{(3)}$ *trois ouverts connexes de* $\mathbf{R}^d$, *bornés et tels que* $U^{(j)} \cap U^{(k)} = \varnothing$ *pour* $j \neq k$ *; soient* $x_j \in U^{(j)}$ *et* $\mu^{(j)}$ *la mesure harmonique pour le couple* $(x_j, U^{(j)})$ (c'est-à-dire la loi du point d'atteinte de $\partial U^{(j)}$ par le brownien issu de x_j; la classe d'équivalence de $\mu^{(j)}$ ne dépend pas du choix de x_j dans $U^{(j)}$). *Tsirelson démontre que*

$$\mu^{(1)} \wedge \mu^{(2)} \wedge \mu^{(3)} = 0\,.$$

Notre principale source d'inspiration a été le travail de Tsirelson, que nous remercions pour nous avoir communiqué dès l'été 1996 une version préliminaire de [20], et pour sa sympathique correspondance. Merci aussi à Jacques Azéma pour ses remarques.

Les passages en petits caractères sont des digressions qui peuvent être omises sans nuire à la compréhension des résultats principaux (théorèmes 1 à 4).

1. Notations et préliminaires

Nous suivrons les conventions habituelles : les espaces probabilisés seront toujours complets, les sous-tribus considérées contiendront tous les négligeables et les filtrations seront continues à droite. Par exemple, la tribu dont est muni un produit $(\Omega, \mathcal{A}, \mathbf{P}) \times (\Omega', \mathcal{A}', \mathbf{P}')$ n'est pas la tribu produit, mais sa complétée pour la loi produit. Si $(\Omega, \mathcal{A}, \mathbf{P})$ est un espace probabilisé, $\mathrm{L}^0(\Omega, \mathcal{A}, \mathbf{P})$ désignera l'ensemble de toutes les classes d'équivalence de variables aléatoires presque sûrement finies, et sera muni de la topologie de la convergence en probabilité. Rappelons que $(\Omega, \mathcal{A}, \mathbf{P})$ est dit *essentiellement séparable* si l'espace vectoriel topologique $\mathrm{L}^0(\Omega, \mathcal{A}, \mathbf{P})$ est séparable; il existe alors dans $\mathcal{A}$ une algèbre de Boole $\mathcal{B}$ dénombrable et dense pour la distance non séparante $\mathbf{P}[A \triangle B]$; $\mathcal{A}$ est engendrée par $\mathcal{B}$ et par les événements négligeables. (Nous aurons besoin de cette hypothèse de séparabilité essentielle pour certains des espaces d'états dans lesquels vivront les processus, mais nous ne la ferons jamais pour les espaces d'épreuves Ω.)

Toutes les intégrales stochastiques et tous les crochets de semimartingales seront, par convention, nuls à l'origine. Le processus intégrale stochastique de X par rapport à Y, qui vaut $\int_0^t X_s\, dY_s$ à l'instant t, sera noté $\int X\, dY$; si Y est le processus croissant identité ($Y_t = t$ pour tout t), nous écrirons par abus de notation $\int X\, dt$.

La filtration naturelle (implicitement complétée et continue à droite) d'un processus X sera notée $\operatorname{Nat} X$.

Nous utiliserons la théorie des temps locaux des semimartingales continues, telle qu'elle est exposée, par exemple, dans le chapitre VI de [18]. Nous adopterons en particulier les conventions familières aux habitués du Séminaire : les temps locaux $(L_t^a)_{a\in\mathbf{R}, t\geqslant 0}$ sont continus à droite en la variable d'espace a; du coup, dans la formule du changement de variable pour les fonctions convexes, c'est une dérivée à gauche qui intervient, et $\operatorname{sgn} 0 = -1$.

Nous commençons par quelques définitions qui précisent la notion d'isomorphisme de filtrations. Il s'agit de choses que les probabilistes connaissent depuis fort longtemps et utilisent depuis bien plus longtemps encore; le lecteur est invité à les sauter pour passer directement au théorème 1, et à n'y revenir ensuite qu'en cas de nécessité, s'il a un doute sur la rigueur des opérations que nous ferons subir aux espaces probabilisés.

DÉFINITION. — *Si* $(\Omega, \mathcal{A}, \mathbf{P})$ *et* $(\Omega', \mathcal{A}', \mathbf{P}')$ *sont deux espaces probabilisés, on appellera* morphisme presque sûr de $(\Omega, \mathcal{A}, \mathbf{P})$ vers $(\Omega', \mathcal{A}', \mathbf{P}')$ *toute application* Ψ *de* $\mathrm{L}^0(\Omega, \mathcal{A}, \mathbf{P})$ *dans* $\mathrm{L}^0(\Omega', \mathcal{A}', \mathbf{P}')$ *telle que, pour tout* n, *toute fonction* f *borélienne sur* $\mathbf{R}^n$, *et tous* $U_1, \dots, U_n$ *dans* $\mathrm{L}^0(\Omega, \mathcal{A}, \mathbf{P})$, *on ait*

$$\Psi\big(f\circ(U_1, \dots, U_n)\big) = f\circ\big(\Psi(U_1), \dots, \Psi(U_n)\big) .$$

Parmi les exemples triviaux mais fort importants de morphismes presque sûrs figurent les changements absolument continus de probabilité (l'application « identique » de $\mathrm{L}^0(\Omega, \mathcal{A}, \mathbf{P})$ dans $\mathrm{L}^0(\Omega, \mathcal{A}, \mathbf{Q})$, où $\mathbf{Q}$ est une probabilité absolument continue par rapport à $\mathbf{P}$) et les grossissements de tribus (l'injection canonique de $\mathrm{L}^0(\Omega, \mathcal{A}, \mathbf{P})$ dans $\mathrm{L}^0(\Omega, \mathcal{A}', \mathbf{P})$, où $\mathcal{A}$ est une sous-tribu de $\mathcal{A}'$).

Un autre exemple intéressant est obtenu quand $\mathcal{A}$ est engendrée par une variable aléatoire X sur Ω et que l'on a sur Ω' une variable aléatoire X' de loi absolument continue par rapport à celle de X ; on peut alors définir Ψ par $\Psi(g \circ X) = g \circ X'$ pour toute g borélienne. En effet, Ψ est bien défini car si $g \circ X = h \circ X$ p. s., le borélien $\{g \neq h\}$ est négligeable pour la loi de X, donc pour celle de X', d'où $g \circ X' = h \circ X'$ p. s. ; et la propriété de morphisme presque sûr est vérifiée identiquement. (Les variables X et X' ne sont pas nécessairement réelles; nous aurons besoin du cas où elles sont à valeurs dans un espace métrique séparable.)

Comme leur nom l'indique, les morphismes presque sûrs ne correspondent pas nécessairement à une opération sur les espaces Ω et Ω', et ne peuvent en général pas être définis ω par ω. (Penser au cas où les deux tribus $\mathcal{A}$ et $\mathcal{A}'$ sont dégénérées.)

PROPOSITION 1. — *Les morphismes presque sûrs sont linéaires, préservent les constantes, sont croissants, et continus pour la convergence en probabilité.*

Si X est une variable aléatoire (ou un vecteur aléatoire) et Ψ un morphisme presque sûr, la loi de $\Psi(X)$ est absolument continue par rapport à celle de X ; en particulier, si X est l'indicatrice d'un événement, $\Psi(X)$ aussi.

Le composé de deux morphismes presque sûrs en est aussi un.

DÉMONSTRATION. — La linéarité et la préservation des constantes s'obtiennent en prenant f affine dans la définition. Si R est une relation binaire borélienne sur $\mathbf{R}$, $R(X,Y) \Rightarrow \mathbb{1}_R(X,Y) = 1 \Rightarrow \mathbb{1}_R(\Psi(X),\Psi(Y)) = 1 \Rightarrow R(\Psi(X),\Psi(Y))$ pour tout morphisme presque sûr Ψ; en particulier, Ψ est croissant.

Pour établir la continuité, il suffit de montrer que si une suite X_n tend vers 0 en probabilité, $\Psi(X_n)$ aussi. Pour le prouver, nous avons le droit de ne le vérifier que pour une sous-suite. Nous savons que $\mathbf{E}[|X_n| \wedge 1] \to 0$; en extrayant une sous-suite, on se ramène au cas où $\sum_n \mathbf{E}[|X_n| \wedge 1] < \infty$. On a alors $\sum_n (|X_n| \wedge 1) < \infty$ p. s., d'où $S = \sum_n |X_n| < \infty$ p. s. et, pour tout k, $\sum_{n \leqslant k} |X_n| \leqslant S$. La propriété de morphisme donne $\sum_{n \leqslant k} |\Psi(X_n)| \leqslant \Psi(S)$; il en résulte que $\sum_n |\Psi(X_n)| < \infty$ p. s. et $\Psi(X_n) \to 0$ p. s., donc en probabilité.

Enfin, si B est un borélien tel que $\mathbf{P}[X \in B] = 0$, on a $\mathbb{1}_B(X) = 0$ p. s., d'où $\mathbb{1}_B(\Psi(X)) = 0$, et $\mathbf{P}[\Psi(X) \in B] = 0$; la loi de $\Psi(X)$ est donc absolument continue par rapport à celle de X. ∎

DÉFINITION. — *On appellera* plongement de $(\Omega, \mathcal{A}, \mathbf{P})$ dans $(\Omega', \mathcal{A}', \mathbf{P}')$ *tout morphisme presque sûr Ψ tel que, pour tout $X \in \mathrm{L}^0(\Omega, \mathcal{A}, \mathbf{P})$, les variables aléatoires X et $\Psi(X)$ aient même loi.*

Les plongements transportent toute la structure probabiliste, sauf peut-être les épreuves ω. Il s'agit donc d'une notion pré-kolmogorovienne! Leur nom est justifié par l'injectivité que nous verrons ci-dessous.

Si $\mathcal{A}$ est engendrée par une variable aléatoire X et si X' est une variable aléatoire de même loi que X, la formule $\Psi(g \circ X) = g \circ X'$ définit un plongement. Un autre exemple, très utile, est l'apport d'information indépendante : $(\Omega_1, \mathcal{A}_1, \mathbf{P}_1)$ et $(\Omega_2, \mathcal{A}_2, \mathbf{P}_2)$ se plongent canoniquement dans $(\Omega_1, \mathcal{A}_1, \mathbf{P}_1) \times (\Omega_2, \mathcal{A}_2, \mathbf{P}_2)$.

PROPOSITION 2. — a) *Soit Ψ un plongement de $(\Omega, \mathcal{A}, \mathbf{P})$ dans $(\Omega', \mathcal{A}', \mathbf{P}')$. Il est linéaire, injectif et continu pour la convergence en probabilité; sa restriction à tout espace L^p est une isométrie. Si $\mathcal{B}$ est une sous-tribu de $\mathcal{A}$, il existe une unique sous-tribu $\Psi(\mathcal{B})$ de $\mathcal{A}'$ telle que Ψ soit une bijection entre $\mathrm{L}^0(\Omega, \mathcal{B}, \mathbf{P})$ et $\mathrm{L}^0(\Omega', \Psi(\mathcal{B}), \mathbf{P}')$, et l'on a $Y = \mathbf{E}[Z|\mathcal{B}]$ si et seulement si $\Psi(Y) = \mathbf{E}'[\Psi(Z)|\Psi(\mathcal{B})]$.*

Un processus M est une martingale (respectivement une semimartingale) pour une filtration $\mathcal{F}$ sur $(\Omega, \mathcal{A}, \mathbf{P})$ si et seulement si, avec des notations évidentes, le processus $\Psi(M)$ est une martingale (respectivement une semimartingale) pour la filtration $\Psi(\mathcal{F})$. La prévisibilité, les crochets et l'intégration stochastique se transfèrent de même.

Le composé de deux plongements est un plongement.

b) *Soit Ψ un morphisme presque sûr de $(\Omega, \mathcal{A}, \mathbf{P})$ dans $(\Omega', \mathcal{A}', \mathbf{P}')$. Il existe une probabilité $\mathbf{Q}$ absolument continue par rapport à $\mathbf{P}$ telle que Ψ soit un plongement de $(\Omega, \mathcal{A}, \mathbf{Q})$ dans $(\Omega', \mathcal{A}', \mathbf{P}')$. Deux variables X et Y de $\mathrm{L}^0(\Omega, \mathcal{A}, \mathbf{P})$ vérifient $\Psi(X) = \Psi(Y)$ si et seulement si $X = Y$ p. s. pour $\mathbf{Q}$.*

DÉMONSTRATION. — L'injectivité vient de ce que si $\Psi(X) = 0$, X et 0 ont même loi, donc $X = 0$. Pour l'unicité de $\Psi(\mathcal{B})$, rappelons que nous ne considérons que des sous-tribus contenant tous les événements négligeables. Le reste du a) résulte immédiatement des définitions des objets transportés. Pour le b), il suffit de définir $\mathbf{Q}$ par $\mathbf{Q}(X) = \mathbf{E}'[\Psi(X)]$ pour tout élément positif X de $\mathrm{L}^0(\Omega, \mathcal{A}, \mathbf{P})$. ∎

Il y a beaucoup de variables aléatoires, beaucoup de fonctions boréliennes, et donc beaucoup de conditions à vérifier pour s'assurer qu'une application de $\mathrm{L}^0(\mathcal{A})$ dans $\mathrm{L}^0(\mathcal{A}')$ est un plongement. Lorsque la tribu $\mathcal{A}$ est essentiellement séparable, le lemme ci-dessous permet de ne vérifier qu'une famille dénombrable de conditions. Rappelons que les opérations booléennes à n arguments sont les applications de $\{0,1\}^n$ dans $\{0,1\}$; par un petit abus de langage, nous les ferons opérer sur les événements.

LEMME 1. — *Soient $(\Omega, \mathcal{A}, \mathbf{P})$ et $(\Omega', \mathcal{A}', \mathbf{P}')$ deux espaces probabilisés, $\mathcal{B}^\circ$ une algèbre de Boole engendrant $\mathcal{A}$, et Ψ° une application de $\mathcal{B}^\circ$ dans $\mathcal{A}'$ qui commute avec les opérations booléennes et qui préserve les probabilités :*

(i) *pour tout n, tous $B_1, ..., B_n$ dans $\mathcal{B}^\circ$ et toute opération booléenne R à n arguments,*

$$\Psi^\circ\big[R(B_1, ..., B_n)\big] = R\big(\Psi^\circ(B_1), ..., \Psi^\circ(B_n)\big) \qquad p.\ s.\,;$$

(ii) *pour tout $B \in \mathcal{B}^\circ$, $\mathbf{P}'\big(\Psi^\circ(B)\big) = \mathbf{P}(B)$.*

Il existe un unique plongement Ψ de $(\Omega, \mathcal{A}, \mathbf{P})$ dans $(\Omega', \mathcal{A}', \mathbf{P}')$ tel que l'on ait $\Psi(\mathbb{1}_B) = \mathbb{1}_{\Psi^\circ(B)}$ pour tout $B \in \mathcal{B}^\circ$.

DÉMONSTRATION. — Soient $B_1, ..., B_n \in \mathcal{B}^\circ$ et $\alpha_1, ..., \alpha_n \in \mathbf{R}$ tels que $\sum \alpha_i \, \mathbb{1}_{B_i} = 0$ presque sûrement. Pour toute partie J de $\{1, ..., n\}$ telle que $\sum\limits_{j \in J} \alpha_j \neq 0$, l'événement

$$\Big(\bigcap_{j\in J} B_j\Big) \cap \Big(\bigcap_{k \in J^c} B_k^c\Big)\,,$$

sur lequel $\sum\limits_{i=1}^{n} \alpha_i \mathbb{1}_{B_i} = \sum\limits_{j\in J} \alpha_j$, est négligeable; son image par Ψ°,

$$\Big(\bigcap_{j\in J} \Psi^\circ(B_j)\Big) \cap \Big(\bigcap_{k \in J^c} \Psi^\circ(B_k^c)\Big)\,,$$

l'est donc aussi (préservation des probabilités), d'où $\sum \alpha_i \, \mathbb{1}_{\Psi^\circ(B_i)} = 0$ p. s. Il est donc possible de définir une application Φ sur l'ensemble $\mathrm{L}^\circ(\mathcal{B}^\circ)$ de toutes ces combinaisons par $\Phi\big[\sum \alpha_i \, \mathbb{1}_{B_i}\big] = \sum \alpha_i \, \mathbb{1}_{\Psi^\circ(B_i)}$; on a ainsi une application de $\mathrm{L}^\circ(\mathcal{B}^\circ)$

dans $L^0(\Omega', \mathcal{A}', \mathbf{P}')$ qui préserve les lois des v. a. et vérifie, pour tout n et toute fonction de $\mathbf{R}^n$ dans $\mathbf{R}$, l'identité

$$\Phi\big(f\circ(X_1,\dots,X_n)\big) = f\circ\big(\Phi(X_1),\dots,\Phi(X_n)\big)\,.$$

La convergence en probabilité sur Ω et Ω' peut être définie par les distances $d(X,Y) = \mathbf{E}\big[|X-Y| \wedge 1\big]$ et $d'(X',Y') = \mathbf{E}'\big[|X'-Y'| \wedge 1\big]$; pour ces distances, Φ est une injection isométrique de $L^\circ(\mathcal{B}^\circ)$ dans $L^0(\mathcal{A}')$, car il préserve les lois. Puisque $L^\circ(\mathcal{B}^\circ)$ est un sous-ensemble dense de $L^0(\mathcal{A})$ et que $L^0(\mathcal{A})$ et $L^0(\mathcal{A}')$ sont complets, Φ se prolonge de façon unique en une injection isométrique Ψ de $L^0(\mathcal{A})$ dans $L^0(\mathcal{A}')$; en particulier, Ψ est continue.

Si f est une fonction continue sur $\mathbf{R}^n$, l'identité

$$\Psi\big(f\circ(X_1,\dots,X_n)\big) = f\circ\big(\Psi(X_1),\dots,\Psi(X_n)\big)\,,$$

vraie sur $L^\circ(\mathcal{B}^\circ)$, s'étend par continuité à tous les $X_1,\dots,X_n \in L^0(\mathcal{A})$. Pour $X_1,\dots,X_n$ fixés dans $L^0(\mathcal{A})$, la classe des fonctions boréliennes f pour lesquelles cette identité est vraie est stable par les limites simples de suites; comme elle contient les fonctions continues, elle contient toutes les fonctions boréliennes. Ainsi, Ψ est un morphisme presque sûr.

Enfin l'unicité résulte de ce que, quand B décrit $\mathcal{B}^\circ$, $\mathbb{1}_B$ parcourt un ensemble total dans $L^0(\Omega, \mathcal{A}, \mathbf{P})$. ∎

DÉFINITIONS. — *Un* isomorphisme d'espaces probabilisés *est un plongement* Ψ *de* $(\Omega, \mathcal{A}, \mathbf{P})$ *dans* $(\Omega', \mathcal{A}', \mathbf{P}')$ *tel que l'application* $\Psi : L^0(\Omega, \mathcal{A}, \mathbf{P}) \to L^0(\Omega', \mathcal{A}', \mathbf{P}')$ *soit surjective.*

Un espace filtré $(\Omega, \mathcal{A}, \mathbf{P}, \mathcal{F})$ *est un espace probabilisé* $(\Omega, \mathcal{A}, \mathbf{P})$ *pourvu d'une filtration* $\mathcal{F}$ (qui vérifie les conditions habituelles).

Un isomorphisme d'espaces filtrés *entre* $(\Omega, \mathcal{A}, \mathbf{P}, \mathcal{F})$ *et* $(\Omega', \mathcal{A}', \mathbf{P}', \mathcal{F}')$ *est un isomorphisme* Ψ *de* $(\Omega, \mathcal{A}, \mathbf{P})$ *dans* $(\Omega', \mathcal{A}', \mathbf{P}')$ *tel que* $\Psi(\mathcal{F}) = \mathcal{F}'$.

Par exemple, si X et X' ont même loi, le plongement de $\big(\Omega, \sigma(X), \mathbf{P}\big)$ dans $\big(\Omega', \sigma(X'), \mathbf{P}'\big)$ qui transforme X en X' est un isomorphisme.

L'inverse d'un isomorphisme en est aussi un. Tout plongement Ψ de $(\Omega, \mathcal{A}, \mathbf{P})$ dans $(\Omega', \mathcal{A}', \mathbf{P}')$ est aussi un isomorphisme entre $(\Omega, \mathcal{A}, \mathbf{P})$ et $(\Omega', \Psi(\mathcal{A}), \mathbf{P}')$. Si, pour $i \in \{1,2\}$, $(\Omega_i, \mathcal{A}_i, \mathbf{P}_i)$ est isomorphe à $(\Omega'_i, \mathcal{A}'_i, \mathbf{P}'_i)$, alors $(\Omega_1, \mathcal{A}_1, \mathbf{P}_1)\times(\Omega_2, \mathcal{A}_2, \mathbf{P}_2)$ est isomorphe à $(\Omega'_1, \mathcal{A}'_1, \mathbf{P}'_1)\times(\Omega'_2, \mathcal{A}'_2, \mathbf{P}'_2)$.

La structure des espaces probabilisés essentiellement séparables, rappelée par la proposition 3 ci-dessous, est extrêmement simple : leur partie diffuse est isomorphe à un intervalle réel, muni de sa mesure de Lebesgue. Cette proposition peut être déduite des théorèmes de structure des algèbres de mesures homogènes; voir par exemple Maharan [13].

PROPOSITION 3. — a) *À isomorphisme près, un espace probabilisé essentiellement séparable est caractérisé par la liste des probabilités de ses atomes, rangées par ordre décroissant (avec des répétitions pour les atomes de même masse; la liste peut être vide, finie ou infinie). En particulier, tous les espaces probabilisés essentiellement séparables et sans partie atomique sont isomorphes.*

b) *Deux espaces probabilisés essentiellement séparables et plongeables chacun dans l'autre sont isomorphes.*

DÉMONSTRATION. — a) Si $(\Omega, \mathcal{A}, \mathbf{P})$ est un espace probabilisé essentiellement séparable, $\mathcal{A}$ est engendrée par une variable aléatoire réelle X ; soit F la partie continue de la fonction de répartition de X, de sorte que $F(-\infty) = 0$ et $F(+\infty)$ est la masse totale de la partie non atomique. Soit $(A_1, A_2, \ldots)$ une liste des atomes de $\mathcal{A}$ par ordre de masses décroissantes. La v. a. Y, égale à n sur l'atome A_n et à $F \circ X$ hors des atomes, engendre aussi $\mathcal{A}$, et sa loi ne dépend que de la liste $\big(\mathbf{P}(A_1), \mathbf{P}(A_2), \ldots\big)$. Si $(\Omega', \mathcal{A}', \mathbf{P}')$ et $(\Omega'', \mathcal{A}'', \mathbf{P}'')$ ont la même liste de masses atomiques, les v. a. Y' et Y'' ainsi obtenues ont même loi, et il existe donc un isomorphisme entre ces espaces, qui transforme Y' en Y''.

b) Nous allons utiliser le a) en caractérisant à isomorphisme près un espace probabilisé essentiellement séparable par la fonction $M : \,]0,1] \longrightarrow [0,1]$ ainsi définie : pour $0 < x \leqslant 1$, $M(x)$ est la somme des probabilités de tous les atomes A qui vérifient $\mathbf{P}(A) > x$. C'est une fonction décroissante et continue à droite; pour tout $x > 0$, le nombre d'atomes de masse x est égal à $(1/x)\big(M(x-) - M(x)\big)$; en particulier, les masses des atomes sont les abscisses des sauts de M.

Pour établir le b), il suffit de montrer que *s'il existe un plongement de* $(\Omega, \mathcal{A}, \mathbf{P})$ *dans* $(\Omega', \mathcal{A}', \mathbf{P}')$, *les fonctions* M *et* M' *associées à ces espaces vérifient* $M' \leqslant M$. L'existence d'un plongement dans l'autre sens montrera l'inégalité inverse, d'où $M = M'$, et le a) donnera l'isomorphisme.

Soit donc Ψ un plongement de $(\Omega, \mathcal{A}, \mathbf{P})$ dans $(\Omega', \mathcal{A}', \mathbf{P}')$. Appelons $\mathcal{H}$ (respectivement $\mathcal{H}'$) l'ensemble des atomes de $\mathcal{A}$ (respectivement $\mathcal{A}'$).

Puisque $\Psi(\mathcal{A})$ est une sous-tribu de $\mathcal{A}'$, tout atome de $\mathcal{A}'$ est inclus dans un atome de $\Psi(\mathcal{A})$; ceci définit une application F de $\mathcal{H}'$ dans $\mathcal{H}$ telle que l'on ait $\Psi\big(F(A')\big) \supset A'$ pour tout $A' \in \mathcal{H}'$. Cette F augmente les probabilités : $\mathbf{P}\big(F(A')\big) = \mathbf{P}'\big[\Psi\big(F(A')\big)\big] \geqslant \mathbf{P}'(A')$; plus généralement, si $\mathcal{K}'$ est une partie quelconque de $\mathcal{H}'$, en notant $F(\mathcal{K}') = \big\{F(A'),\ A' \in \mathcal{K}'\big\}$ et en remarquant que

$$\Psi\Big[\bigcup_{A \in F(\mathcal{K}')} A\Big] = \Psi\Big[\bigcup_{A' \in \mathcal{K}'} F(A')\Big] = \bigcup_{A' \in \mathcal{K}'} \Psi\big(F(A')\big) \supset \bigcup_{A' \in \mathcal{K}'} A' ,$$

on peut écrire

$$\mathbf{P}\Big[\bigcup_{A \in F(\mathcal{K}')} A\Big] \geqslant \mathbf{P}'\Big[\bigcup_{A' \in \mathcal{K}'} A'\Big] .$$

Pour $x > 0$, posons $\mathcal{K}_x = \{A \in \mathcal{H} : \mathbf{P}(A) > x\}$ et $\mathcal{K}'_x = \{A' \in \mathcal{H}' : \mathbf{P}'(A') > x\}$. Comme F augmente les probabilités, on a $\mathbf{P}\big(F(A')\big) > x$ pour tout A' tel que $\mathbf{P}'(A') > x$, d'où $F(\mathcal{K}'_x) \subset \mathcal{K}_x$, et l'inégalité ci-dessus entraîne

$$M(x) = \mathbf{P}\Big[\bigcup_{A \in \mathcal{K}_x} A\Big] \geqslant \mathbf{P}\Big[\bigcup_{A \in F(\mathcal{K}'_x)} A\Big] \geqslant \mathbf{P}'\Big[\bigcup_{A' \in \mathcal{K}'_x} A'\Big] = M'(x) . \qquad \blacksquare$$

PROPOSITION 4. — *Dans* $(\Omega, \mathcal{A}, \mathbf{P})$, *soient* $\mathcal{B}$ *et* $\mathcal{C}$ *deux sous-tribus indépendantes. Il existe un unique morphisme presque sûr* Ψ *de* $(\Omega, \mathcal{B} \vee \mathcal{C}, \mathbf{P})$ *dans* $(\Omega, \mathcal{B}, \mathbf{P}) \times (\Omega, \mathcal{C}, \mathbf{P})$ *tel que, si* B *(respectivement* C*) est une variable aléatoire mesurable pour* $\mathcal{B}$ *(respectivement* $\mathcal{C}$*), on ait* $\Psi(BC)(\omega_1, \omega_2) = B(\omega_1)\, C(\omega_2)$. *C'est un isomorphisme.*

DÉMONSTRATION. — Si X est une v. a. sur Ω, définissons des v. a. X' et X'' sur $\Omega \times \Omega$ par $X'(\omega_1, \omega_2) = X(\omega_1)$ et $X''(\omega_1, \omega_2) = X(\omega_2)$.

Toute v. a. mesurable pour $\mathcal{B} \vee \mathcal{C}$ est de la forme $f(B, C)$ où B et C sont respectivement mesurables pour $\mathcal{B}$ et $\mathcal{C}$, et où f est borélienne. Si Ψ existe, on doit avoir $\Psi(B) = B'$ et $\Psi(C) = C''$, d'où $\Psi\big(f(B,C)\big) = f(B',C'')$; ceci établit l'unicité.

Pour l'existence, remarquons que si B_1 et B_2 (respectivement C_1 et C_2) sont des v. a. mesurables pour $\mathcal{B}$ (respectivement $\mathcal{C}$), les vecteurs $\vec{B} = (B_1, B_2)$ et $\vec{B}' = (B_1', B_2')$ ont même loi, et de même $\vec{C} = (C_1, C_2)$ et $\vec{C}'' = (C_1'', C_2'')$. Comme de plus $\vec{B}$ et $\vec{C}$ sont indépendants, ainsi que $\vec{B}'$ et $\vec{C}''$, $(\vec{B}, \vec{C})$ a même loi que $(\vec{B}', \vec{C}'')$. Donc, si f et g sont deux fonctions boréliennes, $f(B_1, C_1) - g(B_2, C_2)$ a même loi que $f(B_1', C_1'') - g(B_2', C_2'')$. Et si la première différence est nulle, la seconde aussi; il est donc possible de définir Ψ par $\Psi\big(f(B,C)\big) = f(B',C'')$ sans créer d'ambiguïté; ceci définit un morphisme presque sûr. En prenant $g = 0$, on voit que Ψ préserve les lois, c'est donc un plongement. Enfin, la propriété d'isomorphisme vient de ce que toute v. a. mesurable pour $\mathcal{B}\otimes\mathcal{C}$ est de la forme $f(B', C'')$. ∎

COROLLAIRE 1. — *Sur $(\Omega, \mathcal{A}, \mathbf{P})$, soient X et Y deux variables aléatoires de même loi et $\mathcal{B}$ une sous-tribu indépendante de X et indépendante de Y. Il existe un unique isomorphisme entre $(\Omega, \mathcal{B}\vee\sigma(X), \mathbf{P})$ et $(\Omega, \mathcal{B}\vee\sigma(Y), \mathbf{P})$ qui soit l'identité sur $(\Omega, \mathcal{B}, \mathbf{P})$ et qui change X en Y.*

DÉMONSTRATION. — Par la proposition précédente, on se ramène à l'isomorphisme

$$(\Omega, \mathcal{B}, \mathbf{P}) \times (\Omega, \sigma(X), \mathbf{P}) \quad \simeq \quad (\Omega, \mathcal{B}, \mathbf{P}) \times (\Omega, \sigma(Y), \mathbf{P}) \,.$$ ∎

LEMME 2. — *Étant donné $(\Omega, \mathcal{A}, \mathbf{P})$, soient $\mathcal{B}$, $\mathcal{C}$ et $\mathcal{D}$ trois sous-tribus de $\mathcal{A}$. On suppose $\mathcal{D}$ essentiellement séparable et incluse dans $\mathcal{B}\vee\mathcal{C}$; on suppose aussi $\mathcal{B}$ indépendante de $\mathcal{C}$ et de $\mathcal{D}$. Il existe alors un plongement de $(\Omega, \mathcal{D}, \mathbf{P})$ dans $(\Omega, \mathcal{C}, \mathbf{P})$.*

REMARQUE. — Si l'on suppose en outre $\mathcal{B}$ indépendante de $\mathcal{C}\vee\mathcal{D}$, on a, sans utiliser la séparabilité, un résultat plus fort : $\mathcal{D}$ est incluse dans $\mathcal{C}$.

DÉMONSTRATION. — On ne restreint pas la généralité en supposant que $\mathcal{A} = \mathcal{B}\vee\mathcal{C}$. La proposition 4 permet alors, via un isomorphisme Φ, de travailler sur le produit $(\Omega', \mathcal{A}', \mathbf{P}') = (\Omega_1, \mathcal{B}, \mathbf{P}_1) \times (\Omega_2, \mathcal{C}, \mathbf{P}_2)$, où $\Omega_1 = \Omega_2 = \Omega$ et où $\mathbf{P}_1$ et $\mathbf{P}_2$ sont les restrictions de $\mathbf{P}$ à $\mathcal{B}$ et $\mathcal{C}$. Appelons $\mathcal{B}'$ la tribu $\Phi(\mathcal{B})$ sur Ω', c'est-à-dire la tribu produit $\mathcal{B}\otimes\{\varnothing, \Omega_2\}$ augmentée de tous les négligeables. Avec ces notations, la tribu $\mathcal{D}' = \Phi(\mathcal{D})$ est une tribu sur Ω', indépendante de $\mathcal{B}'$. Toute v. a. X sur l'espace produit admet une version mesurable pour la tribu produit $\mathcal{B}\otimes\mathcal{C}$ non complétée. Pour une telle version et pour tout $B \in \mathcal{B}$, si X est bornée le théorème de Fubini appliqué à $\mathbb{1}_B(\omega_1)\, X(\omega_1, \omega_2)$ dit que $\mathbb{1}_B(\omega_1)\, \mathbf{E}_2[X(\omega_1, .)]$ dépend mesurablement de ω_1 et admet $\mathbf{E}[\mathbb{1}_B X]$ pour intégrale en ω_1. Il en résulte que

$$\int X(\omega_1, \hat{\omega}_2)\, \mathbf{P}_2(d\hat{\omega}_2) = \mathbf{E}'[X|\mathcal{B}'](\omega_1, \omega_2) \quad \text{pour presque tout } (\omega_1, \omega_2).$$

Soit $\mathcal{D}^\circ$ une algèbre dénombrable dense dans $\mathcal{D}$. Pour chaque $D \in \mathcal{D}^\circ$, notons $D' = \Phi(D)$ et choisissons un représentant de D' mesurable pour la tribu produit $\mathcal{B}\otimes\mathcal{C}$ non complétée; ce représentant restera fixé pour toute la suite. Pour presque tout (ω_1, ω_2), on a $\mathbf{P}_2\{\hat{\omega}_2 : (\omega_1, \hat{\omega}_2) \in D'\} = \mathbf{P}'[D'|\mathcal{B}'](\omega_1, \omega_2) = \mathbf{P}'[D'] = \mathbf{P}[D]$, où la première égalité presque sûre s'obtient en prenant $X = \mathbb{1}_{D'}$ dans la formule vue ci-dessus, et où la deuxième égalité presque sûre vient de l'indépendance entre D' et $\mathcal{B}'$. Ainsi, pour presque tout ω_1, $\mathbf{P}_2\{\hat{\omega}_2 : (\omega_1, \hat{\omega}_2) \in D'\} = \mathbf{P}[D]$. Puisque $\mathcal{D}^\circ$ est dénombrable, il existe un événement négligeable $N \in \mathcal{B}$ tel que, pour tout $\omega_1 \in N^c$, les deux propriétés suivantes soient vérifiées :

(i) pour tout $D \in \mathcal{D}^\circ$, on a $\mathsf{P}_2\{\omega_2 : (\omega_1,\omega_2) \in D'\} = \mathsf{P}[D]$;
(ii) pour tout n, tous D_1, ..., D_n dans $\mathcal{D}^\circ$ et toute relation booléenne R à n arguments telle que $R(\mathbb{1}_{D_1},...,\mathbb{1}_{D_n})$ p. s., on a $R(\mathbb{1}_{D'_1}(\omega_1,\omega_2),...,\mathbb{1}_{D'_n}(\omega_1,\omega_2))$ pour presque tout ω_2.

Fixons un $\omega_1 \in N^c$ et posons $\Psi^\circ(D) = \{\omega_2 : (\omega_1,\omega_2) \in D'\}$ pour $D \in \mathcal{D}^\circ$. Ceci définit une application de $\mathcal{D}^\circ$ dans $\mathcal{C}$, qui préserve la loi grâce à la propriété (i). Pour des D_i dans $\mathcal{D}^\circ$ et si R est une opération booléenne, en posant $\Delta = R(D_1,...,D_n)$, la propriété (ii) appliquée à la relation à $n{+}1$ arguments $\delta = R(d_1,...,d_n)$ donne l'égalité presque sûre $\Psi^\circ(\Delta) = R(\Psi^\circ(D_1),...,\Psi^\circ(D_n))$. [Cette égalité presque sûre n'est en général pas une identité parce que nous n'avons pris aucune précaution de compatibilité en choisissant les représentants des éléments de $\mathcal{D}^\circ$.] Les hypothèses du lemme 1 étant vérifiées, Ψ° peut être prolongée en un plongement Ψ de $(\Omega,\mathcal{D},\mathsf{P})$ dans $(\Omega_2,\mathcal{C},\mathsf{P}_2) = (\Omega,\mathcal{C},\mathsf{P})$. ∎

La méthode de Tsirelson utilise, de façon tout à fait essentielle, le joli théorème de convergence ci-dessous; il semble classique chez les probabilistes russes sous le nom de « lemme de Slutsky », mais nous n'avons pas réussi à le trouver dans la littérature. Même dans le cas où $S = S' = \mathbf{R}$, il est nouveau pour nous; les énoncés voisins que nous connaissions, tels le lemme (5.7) du chapitre 0 de [18], donnent seulement des convergences en loi. Pour une généralisation de ce théorème, voir la note de F. Delbaen [5] dans ce volume.

THÉORÈME 1. — *Soient (S',δ') et (S,δ) deux espaces métriques séparables et h une application borélienne de S' dans S. Si $(X^k)_{k\in\mathbf{N}}$ est une suite de variables aléatoires à valeurs dans (S',δ') qui converge en probabilité vers une limite X, et si les X^k ont toutes même loi, alors $h\circ X^k$ tend en probabilité vers $h\circ X$.*

En d'autres termes, les plongements Ψ^k de $(\Omega,\sigma(X),\mathsf{P})$ dans $(\Omega,\sigma(X^k),\mathsf{P})$ tels que $\Psi^k(X) = X^k$ convergent vers l'identité.

REMARQUE. — L'hypothèse de séparabilité de S assure que la tribu borélienne sur $S{\times}S$ est le produit par elle-même de la tribu borélienne sur S; et de même sur S'. Lorsque cette hypothèse n'est pas réalisée, la fonction δ sur $S{\times}S$ peut ne pas être mesurable pour la tribu produit et $\delta(X,Y)$ n'est donc pas toujours une variable aléatoire. On pourrait aussi remplacer cete hypothèse par une définition, et exiger que toute variable aléatoire dans un espace métrique prenne ses valeurs dans un sous-espace séparable.

DÉMONSTRATION DU THÉORÈME 1. — Supposons d'abord h continue. De toute sous-suite de la suite $(h\circ X^k)_{k\in\mathbf{N}}$ on peut extraire une sous-sous-suite $(h\circ X^{k_j})_{j\in\mathbf{N}}$ telle que $X^{k_j} \to X$ p. s., donc aussi telle que $h\circ X^{k_j} \to h\circ X$ p. s. et a fortiori telle que $h\circ X^{k_j} \to h\circ X$ en probabilité; le résultat en découle (sans utiliser l'hypothèse que les X^k ont même loi).

Pour étendre ceci à toutes les fonctions boréliennes, il suffit de vérifier que l'ensemble des $h : S' \to S$ telles que $h\circ X^k$ converge en probabilité vers $h\circ X$ est stable par limites simples. Soient donc $(h_\ell)_{\ell\in\mathbf{N}}$ et h telles que, pour tout $x \in S'$, $h_\ell(x)$ tend vers $h(x)$ quand ℓ tend vers l'infini et que, pour chaque ℓ, $h_\ell\circ X^k$ tend vers $h_\ell\circ X$ en probabilité quand n tend vers l'infini. Dans la majoration

$$\begin{aligned} \mathsf{P}[\delta(h\circ X^k, h\circ X) > 3\varepsilon] \quad &\leqslant \quad \mathsf{P}[\delta(h\circ X^k, h_\ell\circ X^k) > \varepsilon] \\ &+ \mathsf{P}[\delta(h_\ell\circ X^k, h_\ell\circ X) > \varepsilon] \\ &+ \mathsf{P}[\delta(h_\ell\circ X, h\circ X) > \varepsilon]\,, \end{aligned}$$

le troisième terme est plus petit que ε pour un ℓ convenable (fixé dans la suite) car $h_\ell \circ X$ tend vers $h \circ X$ sûrement, donc en probabilité; le premier terme est égal au troisième car X^k a même loi que X; pour tout k assez grand, le deuxième terme est majoré par ε grâce à l'hypothèse sur h_ℓ, et on a donc $\mathbf{P}\big[\delta\big(h \circ X^k, h \circ X\big) > 3\varepsilon\big] \leqslant 3\varepsilon$. Ainsi, $h \circ X^k$ tend en probabilité vers $h \circ X$ et le théorème est établi. ∎

COROLLAIRE 2. — *Soit $(X^k)_{k\in\mathbf{N}}$ une suite de variables aléatoires, à valeurs dans un espace métrique séparable (S, δ), toutes de même loi, et convergeant en probabilité vers une limite X. Soit $\mathcal{B}$ une sous-tribu indépendante de chaque X^k et de X. Soit Ψ^k l'isomorphisme entre $(\Omega, \mathcal{B}\vee\sigma(X), \mathbf{P})$ et $(\Omega, \mathcal{B}\vee\sigma(X^k), \mathbf{P})$ qui préserve $\mathcal{B}$ et qui transforme X en X^k* (corollaire 1). *Si U est une variable aléatoire mesurable pour $\mathcal{B}\vee\sigma(X)$, les variables aléatoires $\Psi^k(U)$ convergent en probabilité vers U.*

DÉMONSTRATION. — La v. a. U, mesurable pour $\mathcal{B}\vee\sigma(X)$, est de la forme $f(B, X)$ pour une v. a. $B \in \mathrm{L}^0(\mathcal{B})$ et une f borélienne sur $\mathbf{R}\times S$; on a donc $\Psi^k(U) = f(B, X^k)$. Pour tout x, appelons $h(x)$ la v. a. $f(B, x) \in \mathrm{L}^0(\mathcal{B})$; h est une application borélienne de S dans l'espace métrique séparable $\mathrm{L}^0\big(\Omega, \sigma(B), \mathbf{P}\big)$, et le corollaire s'obtient en appliquant le théorème 1 à h et aux X^k. ∎

2. Martingales-araignées

DÉFINITIONS. — *Soient $n \geqslant 2$ un entier, $\mathbf{E}$ un espace vectoriel de dimension $n-1$ et $\mathbf{U}$ un ensemble de n vecteurs de $\mathbf{E}$, de somme nulle et engendrant $\mathbf{E}$. La* toile *de l'araignée sera la réunion $\mathbf{T} = \{\lambda u\,,\ \lambda \geqslant 0\,,\ u \in \mathbf{U}\}$ des demi-droites issues de l'origine dans les directions de $\mathbf{U}$.*

Nous appellerons martingale-araignée *(sur la toile $\mathbf{T}$) toute martingale locale continue à valeurs dans $\mathbf{T}$.*

L'entier n est appelé la multiplicité *de la toile et de la martingale-araignée; une martingale-araignée est dite* multiple *si $n \geqslant 3$.*

Une définition un peu plus contraignante, qui exige que les martingales-araignées soient de vraies martingales et pas seulement des martingales locales, figure dans le chapitre 17 de [26]; voir aussi le chapitre 5 de [21].

Lorsque $n = 2$, $\mathbf{T}$ est un espace vectoriel unidimensionnel; les martingales-araignées non multiples s'identifient dans ce cas aux martingales locales habituelles, à une dimension.

Lorsque $\mathbf{E}$ est le plan complexe $\mathbf{C}$ et $\mathbf{U}$ l'ensemble des racines cubiques de l'unité, $\mathbf{T}$ est l'ensemble des complexes y tels que $\arg y = 0 \bmod 2\pi/3$, et les martingales-araignées sont les martingales locales continues complexes Y telles que $Y^3 \in \mathbf{R}_+$.

Deux toiles $(\mathbf{E}', \mathbf{U}', \mathbf{T}')$ et $(\mathbf{E}'', \mathbf{U}'', \mathbf{T}'')$ ayant même multiplicité n se correspondent toujours (de $n!$ manières) par un isomorphisme entre les espaces vectoriels $\mathbf{E}'$ et $\mathbf{E}''$; le seul paramètre significatif dans la définition ci-dessus est la multiplicité n.

L'exemple complexe ci-dessus (où $n = 3$) s'étend à n quelconque en prenant $\mathbf{E}$ euclidien à $n-1$ dimensions et en prenant pour éléments de $\mathbf{U}$ les n sommets d'un simplexe régulier inscrit dans la sphère unité; ils vérifient $\|u\| = 1$ et $\langle u,v\rangle = -1/(n-1)$ pour $u \neq v$. Si $\mathbf{E}$ et $\mathbf{U}$ sont donnés, il est toujours possible de se ramener à ce cas par un choix convenable (unique) de la structure euclidienne.

Chaque point y de $\mathbf{T}$ peut être représenté par la famille de n nombres positifs $(y^u)_{u\in \mathbf{U}}$ dont un au plus est non nul et qui vérifient $y = \sum_{u\in\mathbf{U}} y^u u$; ces nombres y^u sont les *composantes* de y.

LEMME 3. — *Si M est la partie martingale d'une semimartingale continue Y,*

$$\int \mathbb{1}_{\{Y=0\}}\, dM = 0\,.$$

DÉMONSTRATION. — Cette intégrale est la partie martingale de $\int \mathbb{1}_{\{Y=0\}}\, dY$, qui est un processus à variation finie (voir Meyer [14], formule (12.4) page 365). ∎

PROPOSITION 5 (c'est la proposition 17.5 de [26]). — *Soit Y un processus continu, adapté, à valeurs dans $\mathbf{T}$. C'est une martingale-araignée si et seulement si les différences $Y^u - Y^v$ entre ses composantes sont des martingales locales.*

Chaque composante Y^u est alors une sous-martingale locale, de décomposition canonique $Y^u = Y^u_0 + M^u + L$, où M^u est une martingale locale vérifiant

$$M^u = \int \mathbb{1}_{\{Y^u>0\}}\, dM^u = \int \mathbb{1}_{\{Y\neq\vec{0}\}}\, dY^u\,,$$

et L un processus croissant ne dépendant pas de u et tel que

$$L = \int \mathbb{1}_{\{Y=\vec{0}\}}\, dL = \int \mathbb{1}_{\{Y^u=0\}}\, dY^u\,.$$

DÉMONSTRATION. — Soit $v \in \mathbf{U}$. L'ensemble $\mathbf{U}' = \mathbf{U} - \{v\}$ est une base de l'espace. La base duale est $\{f^v_u, u \in \mathbf{U}'\}$, où les formes f^v_u, définies par $f^v_u(u) = 1$ et $f^v_u(w) = 0$ pour $w \in \mathbf{U}' - \{u\}$, vérifient $f^v_u(v) = f^v_u(-\sum_{w\in\mathbf{U}'} w) = -\sum_{w\in\mathbf{U}'} f^v_u(w) = -1$ et donc $f^v_u(y) = y^u - y^v$ pour tout y de $\mathbf{T}$.

Le processus Y est une martingale-araignée si et seulement si chacune des formes f^v_u le transforme en une martingale locale réelle (ces formes engendrent en effet le dual), c'est-à-dire si et seulement si les $Y^u - Y^v$ sont des martingales locales.

Soit Y une martingale-araignée. Sa composante Y^u, qui vaut aussi $(Y^u - Y^v)^+$, est une sous-martingale locale; appelons $Y^u_0 + M^u + L^u$ sa décomposition canonique. Le processus $L^u - L^v$, partie à variation finie de la martingale locale $(Y^u - Y^v)$, est nul et L^u ne dépend donc pas de u; appelons L ce processus croissant.

Puisque M^u est la partie martingale de Y^u, on a (lemme 3) $\mathbb{1}_{\{Y^u=0\}}\, dM^u = 0$, d'où $M^u = \int \mathbb{1}_{\{Y^u>0\}}\, dM^u$. Par ailleurs, sur l'ouvert prévisible $\{Y^u > 0\}$, Y^u est égal à une martingale locale, par exemple $Y^u - Y^v$, d'où $\mathbb{1}_{\{Y^u>0\}}\, dL^u = 0$; en ajoutant cette égalité à la précédente, on a $dM^u = \mathbb{1}_{\{Y^u>0\}}\, dY^u$. En retranchant à dY^u chacun des deux membres de cette égalité, on obtient $L = \int \mathbb{1}_{\{Y^u=0\}}\, dY^u$. Pour $v \neq u$, on a $Y^u = 0$ sur l'ouvert prévisible $Y^v > 0$, donc $\mathbb{1}_{\{Y^v>0\}}\, dY^u = 0$. En sommant sur tous les v (y compris u) il vient $M^u = \int \mathbb{1}_{\{Y\neq\vec{0}\}}\, dY^u$. En prenant la partie à variation finie de $\mathbb{1}_{\{Y^v>0\}}\, dY^u = 0$, on obtient $\mathbb{1}_{\{Y^v>0\}}\, dL = 0$. Sommant en v, on voit que $\mathbb{1}_{\{Y\neq\vec{0}\}}\, dL = 0$ et on en tire $L = \int \mathbb{1}_{\{Y=\vec{0}\}}\, dL$. Les quatre égalités annoncées sont établies (elles contiennent les quatre signes $\int$ de ce paragraphe). ∎

REMARQUE. — Si Y est une martingale-araignée issue de l'origine, les n martingales locales $-M^u = L - Y^u$ sont orthogonales et ont le même supremum $L_t = \sup_{s\leqslant t}(-M^u_s)$. Cette propriété caractérise les martingales-araignées; en effet, *si $(N^i)_{1\leqslant i\leqslant n}$ est une famille de n martingales locales issues de 0, deux-à-deux orthogonales et ayant même processus de supremum S, alors $Y^i = S - M^i$ sont les n composantes d'une martingale-araignée.*

Pour le vérifier, il suffit d'établir que $Y^i = Y^j$ si $i \neq j$. Or le processus croissant S est porté par l'ensemble $\{N^i = S\} = \{Y^i = 0\}$, d'où $Y^i\,dS = 0$ et la formule d'intégration par parties donne $d(Y^iY^j) = Y^i\,d(S-N^j) + Y^j\,d(S-N^i) = -(Y^i dN^j + Y^j dN^i)$. Ainsi, Y^iY^j est une martingale locale positive et issue de 0, donc nulle.

On définit une distance sur la toile $\mathbf{T}$ par $\delta(y', y'') = \sum_{u\in \mathbf{U}} |y'^u - y''^u|$. Si y' et y'' sont sur le même fil u, ou si l'un des deux est à l'origine, leur distance est $|y'^u - y''^u|$; s'ils sont sur des fils distincts, on a $\delta(y', y'') = \delta(y', \vec{0}) + \delta(\vec{0}, y'')$. Bien entendu, la topologie définie par cette distance est la restriction à $\mathbf{T}$ de la topologie usuelle dans l'espace ambiant $\mathbf{E}$, et les martingales-araignées sont continues pour cette distance. Si $\mathbf{E}$ est euclidien et si tous les vecteurs de $\mathbf{U}$ sont unitaires, $\delta(y', y'')$ est aussi la longueur du chemin le plus court qui joint y' à y'' en restant dans $\mathbf{T}$ (très jacobine, notre araignée s'interdit de lancer un fil qui ne passerait pas par l'origine).

Si Y est une martingale-araignée dans $\mathbf{T}$, sa distance à l'origine $\delta(\vec{0}, Y)$ est simplement la somme de ses composantes $\Sigma = \sum_{u\in \mathbf{U}} Y^u = \Sigma_0 + \sum_u M^u + nL$; c'est une sous-martingale locale. La proposition suivante (qui ne sera pas utilisée ensuite) dit qu'il en va de même de sa distance à tout autre point.

PROPOSITION 6. — *Soient $v \in \mathbf{U}$ et $r > 0$; $y = rv$ est un point non nul de la toile. Soit Y une martingale-araignée, de composantes $Y^u = Y_0^u + M^u + L$; appelons Σ la somme des composantes de Y et Λ^r le temps local de Σ au point r. Le processus $\delta(y, Y)$ est une sous-martingale locale, de décomposition*

$$\delta(y, Y) = \delta(y, Y_0) + \sum_{u\neq v} M^u + \int \operatorname{sgn}(\Sigma - r)\, dM^v + (n-2)\, L + \int \mathbb{1}_{\{Y=y\}}\, d\Lambda^r .$$

DÉMONSTRATION. — Posons $V = Y^v - r$ et écrivons

$$\begin{aligned}\delta(y, Y) &= \sum_{u\in \mathbf{U}} \left|Y^u - y^u\right| = |Y^v - y^v| + \sum_{u\neq v} Y^u \\ &= \delta(y, Y_0) + |V| - |V_0| + \sum_{u\neq v} M^u + (n-1)\, L\,;\end{aligned}$$

il reste à calculer $|V| - |V_0|$. En appelant $\mathcal{L}$ le temps local de V à l'origine, on peut écrire $|V| = |V_0| + \int \operatorname{sgn} V\, dV + \mathcal{L}$. En y remplaçant dV par $dY^v = dM^v + dL$, on obtient $d|V| = \operatorname{sgn} V\, dM^v + \operatorname{sgn} V\, dL + d\mathcal{L}$; nous allons évaluer chacun de ces trois termes.

En utilisant la proposition 5, le premier vaut $\operatorname{sgn} V\, dM^v = \mathbb{1}_{\{Y^v>0\}} \operatorname{sgn} V\, dM^v$; mais, sur $\{Y^v > 0\}$, on a $V = \Sigma - r$ et $\operatorname{sgn} V = \operatorname{sgn}(\Sigma - r)$. Il reste $\operatorname{sgn} V\, dM^v = \operatorname{sgn}(\Sigma-r)\, dM^v$.

Pour le deuxième terme, on remarque que sur le support de L on a $Y = \vec{0}$, d'où $Y^v = 0$, $V = -r$ et $\operatorname{sgn} V = -1$. Ce terme vaut donc $-L$.

Pour le troisième terme, on observe que, sur l'ouvert $\{Y^v > 0\}$, qui contient l'ensemble $\{V = 0\} = \{Y^v = r\} = \{Y = y\}$ qui lui-même porte $\mathcal{L}$, on a $V = Y^v - r = \Sigma - r$; ceci permet d'écrire $d\mathcal{L} = \mathbb{1}_{\{Y^v>0\}}\, d\mathcal{L} = \mathbb{1}_{\{Y^v>0\}}\, d\Lambda^r = \mathbb{1}_{\{Y=y\}}\, d\Lambda^r$.

Il ne reste qu'à additionner tous ces morceaux pour parvenir à la décomposition annoncée; la propriété de sous-martingale locale en découle. ∎

REMARQUE. — Le terme de temps local à l'origine, $\int \mathbb{1}_{\{Y=\vec{0}\}}\, d\big(\delta(y, Y)\big)$, est égal à nL si $y = \vec{0}$ mais seulement à $(n-2)L$ si $y \neq \vec{0}$. Ceci est dû au fait que tous les débuts d'excursions contribuent à augmenter la distance à l'origine, alors que la distance à un $y \neq \vec{0}$ augmente lors des débuts des excursions dans $n-1$ directions mais diminue lors des débuts des excursions vers y, d'où le facteur $(n-1) - 1 = n - 2$.

Pour $y \in \mathbf{T} \setminus \{\vec{0}\}$, nous noterons $F(y)$ l'unique vecteur $u \in \mathbf{U}$ tel que $y^u > 0$ (c'est le fil sur lequel se trouve y); nous poserons aussi, par convention, $F(\vec{0}) = \vec{0}$, de sorte que F est une application de $\mathbf{T}$ dans $\mathbf{T}$.

PROPOSITION 7. — *Soient (dans une même filtration) Y et Y' deux martingales-araignées sur la même toile* $\mathbf{T}$. *Posons* $\Sigma = \delta(\vec{0}, Y) = \sum_u Y^u$ *et appelons M la partie martingale* $\sum_{u\in\mathbf{U}} M^u$ *de Σ (sa partie croissante est nL); définissons de même Σ' et M'.*

Le processus $\delta(Y, Y')$ est une sous-martingale locale; plus précisément, si l'on appelle Λ son temps local en 0, *on a*

$$\begin{aligned}\delta(Y,Y') = \delta(Y_0,Y_0') &+ \int \mathbb{1}_{\{F(Y)\neq F(Y')\}}\,(dM + dM')\\ &+ \int \mathbb{1}_{\{F(Y)=F(Y')\}}\,\mathrm{sgn}\,(\Sigma-\Sigma')\,(dM - dM')\\ &+ \tfrac{1}{2}\Lambda + (n-2)\left[\int \mathbb{1}_{\{Y'\neq\vec{0}\}}\,dL + \int \mathbb{1}_{\{Y\neq\vec{0}\}}\,dL'\right].\end{aligned}$$

DÉMONSTRATION. — Posons $Q = \delta(Y, Y')$. On a aussi $Q = \sum_{u\in\mathbf{U}} |Y^u - Y'^u|$, ce qui montre que Q est une semimartingale.

Tout d'abord, en appelant $(\ell_t^a)_{t\geqslant 0, a\in\mathbb{R}}$ le temps local de Q et ℓ^{a-} sa limite à gauche en a, le théorème (2 iv b) de [24] (voir aussi le théorème VI (1.7) de [18]) donne la formule $\int \mathbb{1}_{\{Q=a\}}\,dQ = \frac{1}{2}(\ell^a - \ell^{a-})$. Puisque Q est positive, $\ell^a = 0$ pour tout $a < 0$, d'où à la limite $\ell^{0-} = 0$, et $\int \mathbb{1}_{\{Y=Y'\}}\,dQ = \frac{1}{2}\ell^0 = \frac{1}{2}\Lambda$. Il reste à calculer le processus $\int \mathbb{1}_{\{Y\neq Y'\}}\,dQ$. On peut le décomposer en somme de quatre termes :

$$Q^1 = \sum_{\substack{u,v\in\mathbf{U}\\ u\neq v}} \int \mathbb{1}_{\{Y^u>0 \text{ et } Y'^v>0\}}\,dQ\,, \qquad Q^3 = \sum_{u\in\mathbf{U}} \int \mathbb{1}_{\{Y=\vec{0} \text{ et } Y'^u>0\}}\,dQ\,,$$

$$Q^2 = \sum_{u\in\mathbf{U}} \int \mathbb{1}_{\{Y^u>0\,,\,Y'^u>0 \text{ et } Y^u\neq Y'^u\}}\,dQ\,, \quad Q^4 = \sum_{u\in\mathbf{U}} \int \mathbb{1}_{\{Y'=\vec{0} \text{ et } Y^u>0\}}\,dQ\,.$$

Pour $u \neq v$, on a $Q = \Sigma + \Sigma'$ sur l'ouvert prévisible $G = \{Y^u > 0 \text{ et } Y'^v > 0\}$; d'où $\mathbb{1}_G\,dQ = \mathbb{1}_G\,d(\Sigma+\Sigma') = \mathbb{1}_G\,(dM+ndL+dM'+ndL')$. Mais G ne rencontre pas l'ensemble $\{Y = \vec{0} \text{ ou } Y' = \vec{0}\}$, qui porte L et L'; l'égalité précédente se réduit à $\mathbb{1}_G\,dQ = \mathbb{1}_G\,(dM+dM')$, et, en sommant sur tous les couples $u \neq v$,

$$Q^1 = \int \mathbb{1}_{\{Y\neq\vec{0}\,,\,Y'\neq\vec{0} \text{ et } F(Y)\neq F(Y')\}}\,(dM + dM')\,.$$

Pour calculer Q^2, fixons $u \in \mathbf{U}$ et posons $V = Y^u - Y'^u$. L'ouvert prévisible $G = \{Y^u > 0\,,\ Y'^u > 0 \text{ et } Y \neq Y'\}$ ne rencontre pas l'ensemble $\{Y = \vec{0} \text{ ou } Y' = \vec{0}\}$, qui porte L et L'. Sur G, on a $V = \Sigma - \Sigma'$, $dL = dL' = 0$ et $Q = |V| \neq 0$, d'où $\mathbb{1}_G\,dQ = \mathbb{1}_G\,\mathrm{sgn}\,V\,dV = \mathbb{1}_G\,\mathrm{sgn}(\Sigma-\Sigma')\,(dM-dM')$. Sommant en u, on obtient $dQ^2 = \mathbb{1}_{\{F(Y)=F(Y') \text{ et } Y\neq Y'\}}\,\mathrm{sgn}(\Sigma-\Sigma')\,(dM-dM')$. Et comme $M-M'$, partie martingale de $\Sigma-\Sigma'$, a selon le lemme 3 une intégrale nulle sur $\{\Sigma = \Sigma'\}$ et a fortiori sur l'ensemble plus petit $\{Y = Y'\}$,

$$Q^2 = \int \mathbb{1}_{\{F(Y)=F(Y')\}}\,\mathrm{sgn}(\Sigma-\Sigma')\,(dM - dM')\,.$$

Passons à Q^3. Pour u fixé, l'ouvert prévisible $G = \{\Sigma < Y'^u\}$ contient l'ensemble $\{Y = \vec{0} \text{ et } Y'^u > 0\}$. Sur G, on a $Y'^u = \Sigma'$ et $Q = \Sigma' + \Sigma - 2Y^u$. On en tire $\mathbb{1}_G\,dQ = \mathbb{1}_G\,[dM'+ndL'+dM+ndL-2dM^u-2dL]$, et, puisque $Y' > 0$ sur G, $\mathbb{1}_G\,dQ = \mathbb{1}_G\,[dM + dM' - 2dM^u + (n-2)dL]$. Intégrons $\mathbb{1}_{\{Y=\vec{0}\}}$ des deux côtés. Le terme dM^u disparaît grâce à la proposition 5, et il nous reste seulement

$\mathbb{1}_{\{Y=\vec{0} \text{ et } Y'^u>0\}}\, dQ = \mathbb{1}_{\{Y=\vec{0} \text{ et } Y'^u>0\}}\, [dM + dM' + (n-2)dL]$. Sommant en u, on trouve

$$Q^3 = \int \mathbb{1}_{\{Y=\vec{0} \text{ et } Y'\neq\vec{0}\}}\, [dM + dM' + (n-2)dL] \;;$$

et échangeant les rôles de Y et Y', on obtient

$$Q^4 = \int \mathbb{1}_{\{Y\neq\vec{0} \text{ et } Y'=\vec{0}\}}\, [dM + dM' + (n-2)dL'] \,.$$

La formule annoncée résulte de l'addition de Q_0, Q^1, Q^2, Q^3, Q^4 et $\frac{1}{2}\Lambda$. ∎

PROPOSITION 8. — *Les notations sont les mêmes que dans la proposition 7. Posons* $g_t = \sup\{s \leqslant t : Y_s = \vec{0}\}$ *et de même* $g'_t = \sup\{s \leqslant t : Y'_s = \vec{0}\}$, *avec la convention* $\sup \varnothing = 0$.

Le processus $n\,\delta(Y,Y') - (n-2)\left[\mathbb{1}_{\{Y'\circ g\neq\vec{0}\}}\Sigma + \mathbb{1}_{\{Y\circ g'\neq\vec{0}\}}\Sigma'\right]$ *est une sous-martingale locale.*

DÉMONSTRATION. — Posons $S = \big((n-2)/n\big)\left[\mathbb{1}_{\{Y'\circ g\neq\vec{0}\}}\Sigma + \mathbb{1}_{\{Y\circ g'\neq\vec{0}\}}\Sigma'\right]$. Puisque $\Sigma = 0$ sur l'ensemble $\{Y = \vec{0}\}$, la formule du balayage (voir par exemple [25]) dit que, si H est un processus prévisible borné, $H_g\Sigma = H_0\Sigma_0 + \int H_g\, d\Sigma$ (et c'est en particulier une semimartingale). On a donc $d(H_g\Sigma) = H_g\, d\Sigma = H_g\, dM + nH_g\, dL$; mais, sur le support de dL, on a $Y_t = 0$ d'où $g_t = t$ et $H_g = H$; finalement, $d(H_g\Sigma) = H_g\, dM + nH\, dL$. En prenant $H = \big((n-2)/n\big)\,\mathbb{1}_{\{Y'\neq\vec{0}\}}$, puis en échangeant les rôles de Y et Y', on voit que S est une sous-martingale locale, de partie à variation finie $(n-2)\left[\mathbb{1}_{\{Y'\neq\vec{0}\}}\, dL + \mathbb{1}_{\{Y\neq\vec{0}\}}\, dL'\right]$. D'après la proposition 7, ceci est un morceau de la partie croissante de la décomposition de la sous-martingale locale $\delta(Y,Y')$; le processus $\delta(Y,Y') - S$ est donc une sous-martingale locale. ∎

DÉFINITIONS. — *Soit I un ensemble dénombrable non vide. Un* mouvement brownien à valeurs dans $\mathbf{R}^I$ *est une famille indépendante* $X = (X_i)_{i\in I}$ *de mouvements browniens réels issus de l'origine.*

Une filtration $\mathcal{F}$ sur $(\Omega, \mathcal{A}, \mathbf{P})$ est dite brownienne *si elle est de la forme* $\mathcal{F} = \mathcal{F}_0 \vee \operatorname{Nat} X$, *où X est un mouvement brownien à valeurs dans $\mathbf{R}^I$, indépendant de $\mathcal{F}_0$.*

THÉORÈME 2. — *Dans une filtration brownienne, toute martingale-araignée multiple issue de l'origine est identiquement nulle.*

Nous rencontrerons plus bas des énoncés plus généraux (corollaires 3 et 4) : la probabilité peut être remplacée par une autre (absolument continue), et le mouvement brownien par une martingale pure. On peut aussi localiser le théorème 2 à une partie aléatoire de $\mathbf{R}_+$; nous le ferons dans la quatrième partie (proposition 24).

Il est essentiel de supposer que la martingale-araignée est multiple (ceci sera utilisé dans l'avant-dernière ligne de la démonstration); les martingales-araignées à deux composantes s'identifient aux martingales locales réelles et ne peuvent évidemment satisfaire un tel énoncé.

DÉMONSTRATION. — Par arrêt, il suffit de démontrer le théorème pour les martingales-araignées multiples bornées.

Soit Y une martingale-araignée bornée, issue de l'origine, à n composantes ($n \geqslant 3$), dans une filtration $\mathcal{F}$ vérifiant $\mathcal{F} = \mathcal{F}_0 \vee \operatorname{Nat} X$, où X est un mouvement brownien à valeurs dans $\mathbf{R}^I$, indépendant de $\mathcal{F}_0$.

Un grossissement indépendant (plus précisément, le plongement de Ω dans le produit de Ω par l'espace de Wiener) permet de supposer qu'il existe sur Ω un mouvement brownien $\tilde{X}$ à valeurs dans $\mathbf{R}^I$ et indépendant de $\mathcal{F}_\infty$. La filtration $\mathcal{G} = \mathcal{F}_0 \vee \operatorname{Nat}(X, \tilde{X})$ contient $\mathcal{F}$; $X = (X_i)_{i \in I}$ et $\tilde{X} = (\tilde{X}_i)_{i \in I}$ sont deux mouvements browniens indépendants sur $(\Omega, \mathcal{A}, \mathbf{P}, \mathcal{G})$.

Pour $-1 \leqslant r \leqslant 1$, posons $X^r = r\,X + \sqrt{1-r^2}\,\tilde{X}$. Les martingales réelles $X_i = X_i^1$ et X_i^r vérifient

$$d\langle X_i^r, X_j^r\rangle_t = r^2\,\delta_{ij}\,dt + (1-r^2)\,\delta_{ij}\,dt = \delta_{ij}\,dt\,,$$
$$d\langle X_i, X_j^r\rangle_t = r\,\delta_{ij}\,dt\,;$$

il en résulte en particulier que X^r est pour $\mathcal{G}$ un mouvement brownien à valeurs dans $\mathbf{R}^I$. Le corollaire 1 donne un isomorphisme Ψ^r entre $(\Omega, \mathcal{F}_\infty, \mathbf{P})$ et $(\Omega, \mathcal{F}_0 \vee \sigma(X^r), \mathbf{P})$, qui respecte $\mathcal{F}_0$ et qui transforme X en X^r.

Les tribus $\mathcal{F}_\infty = \mathcal{F}_0 \vee \sigma(X)$ et $\Psi^r(\mathcal{F}_\infty) = \mathcal{F}_0 \vee \sigma(X^r)$ sont liées par l'inégalité d'hypercontractivité, qui s'énonce ainsi : *si U (respectivement U') est une variable aléatoire positive et mesurable pour $\mathcal{F}_\infty$ (respectivement $\Psi^r(\mathcal{F}_\infty)$), et si $p, q \in [1, \infty]$ sont tels que $(p-1)(q-1) \geqslant r^2$, alors*

$$\mathbf{E}[UU'|\mathcal{F}_0] \leqslant \mathbf{E}[U^{\,p}|\mathcal{F}_0]^{\frac{1}{p}}\,\mathbf{E}[U'^{q}|\mathcal{F}_0]^{\frac{1}{q}}\,.$$

La démonstration de cette inégalité se copie *mutatis mutandis* sur celle donnée par Neveu [16] dans le cas où I n'a qu'un élément et où $\mathcal{F}_0$ est dégénérée; nous nous contenterons d'en rappeler le principe. Le résultat est en fait un peu plus général : *Soient, dans une même filtration, $X = (X_i)_{i \in I}$ et $X' = (X'_j)_{j \in J}$ des mouvements browniens; la formule $d\langle X_i, X'_j\rangle = r_{ij}\,dt$ définit (presque partout pour $dt \times d\mathbf{P}$) une matrice $R = (r_{ij})_{i \in I, j \in J}$ de processus prévisibles. On suppose que ${}^tRR$ existe (séries absolument convergentes) et vérifie $(p-1)(q-1)\mathrm{Id} - {}^tRR \geqslant 0$ au sens de la positivité des matrices symétriques. Si deux v. a. positives V et V' sont respectivement des intégrales stochastiques par rapport à X et X'* (ceci est plus général que des fonctionnelles de X et X'), *alors pour tout temps d'arrêt T,*

$$\mathbf{E}\big[V^{\frac{1}{p}}V'^{\frac{1}{q}}\,\big|\mathcal{F}_T\big] \leqslant \mathbf{E}[V|\mathcal{F}_T]^{\frac{1}{p}}\,\mathbf{E}[V'|\mathcal{F}_T]^{\frac{1}{q}}\,.$$

En effet, lorsque V et V' sont bornés, le membre de gauche (respectivement de droite) est la valeur à l'instant T d'une martingale (respectivement surmartingale; c'est ici qu'est utilisée l'hypothèse sur p, q et R) de valeur finale $V^{\frac{1}{p}}V'^{\frac{1}{q}}$.

Ici, $I = J$ et $r_{ij} = r\,\delta_{ij}$; ce cas est aussi traité dans Dellacherie-Maisonneuve-Meyer [7]. Par rapport à [16] ou à [7], les conditionnements par $\mathcal{F}_0$ ne sont pas une petite coquetterie supplémentaire, mais une contrainte : en prenant U et U' mesurables pour $\mathcal{F}_0$, on voit qu'il est en général impossible de s'en dispenser. Pour $|r| < 1$, l'inégalité habituelle (inconditionnelle) traduit en effet une sorte d'indépendance partielle entre les deux tribus; ici, les deux tribus ont une sous-tribu commune, $\mathcal{F}_0$, et l'inégalité conditionnelle ci-dessus exprime que cette indépendance partielle disparaît pour les variables aléatoires de $\mathcal{F}_0$.

L'hypercontractivité sera utilisée via le lemme suivant.

LEMME 4. — *Fixons* $r \in]-1,1[$. *Soient* G *et* G' *des variables aléatoires définies sur* $(\Omega,\mathcal{F}_\infty)$ *et* $(\Omega,\Psi^r(\mathcal{F}_\infty))$ *respectivement, à valeurs dans* $\mathbf{R}\cup\{+\infty\}$, *telles que, pour toute v. a. réelle* T *finie et mesurable pour* $\mathcal{F}_0$, *on ait* $\mathbf{P}[G=T]=\mathbf{P}[G'=T]=0$. *On a alors* $\mathbf{P}[G=G'<\infty]=0$.

Lorsque $\mathcal{F}_0$ est dégénérée, l'hypothèse se réduit à dire que les lois de G et G' sont diffuses, sauf peut-être des masses au point $+\infty$.

DÉMONSTRATION DU LEMME 4. — Quitte à remplacer G et G' par $\exp G$ et $\exp G'$, on peut les supposer positives. L'hypothèse dit que G évite les temps d'arrêt de la filtration constante égale à $\mathcal{F}_0$; cela entraîne que le processus croissant $A_t=\mathbf{P}[G<t|\mathcal{F}_0]$ a une version continue; il est nul en 0 et borné par 1. De même, $A'_t=\mathbf{P}[G'<t|\mathcal{F}_0]$ est continu, nul en 0 et borné par 1.

Pour m entier, posons $\tau_j=\inf\{t: A_t+A'_t\geqslant j/m\}$, de sorte que $\tau_{2m+1}=\infty$, et considérons les événements $E_j=\{\tau_j\leqslant G<\tau_{j+1}\}$ et $E'_j=\{\tau_j\leqslant G'<\tau_{j+1}\}$. On a

$$\mathbf{P}[E_j|\mathcal{F}_0]=\mathbf{P}[\tau_j\leqslant G<\tau_{j+1}|\mathcal{F}_0]=A_{\tau_{j+1}}-A_{\tau_j}\leqslant\frac{1}{m}$$

et de même $\mathbf{P}[E'_j|\mathcal{F}_0]=A'_{\tau_{j+1}}-A'_{\tau_j}\leqslant\dfrac{1}{m}$.

Écrivons l'inégalité hypercontractive avec $p=q=1+|r|<2$:

$$\mathbf{P}[E_j\cap E'_j|\mathcal{F}_0]\leqslant\mathbf{P}[E_j|\mathcal{F}_0]^{\frac1p}\mathbf{P}[E'_j|\mathcal{F}_0]^{\frac1p}\leqslant\Big(\frac1m\Big)^{\frac2p}.$$

On en tire $\mathbf{P}[E_j\cap E'_j]\leqslant m^{-\frac2p}$ et, en remarquant que $\{G=G'<\infty\}\subset\bigcup_{j=0}^{2m}E_j\cap E'_j$,

$$\mathbf{P}\big[G=G'<\infty\big]\leqslant\frac{2m+1}{m^{\frac2p}}.$$

Pour conclure, il ne reste qu'à faire tendre m vers l'infini : puisque $p<2$, le majorant tend alors vers zéro. ∎

REMARQUES. — Ce lemme reste vrai si l'on remplace l'hypothèse $\mathbf{P}[G=T]=\mathbf{P}[G'=T]=0$ par la condition plus faible $\mathbf{P}[G=G'=T]=0$. (Dans le cas où $\mathcal{F}_0$ est dégénérée, cela revient à remplacer « les lois de G et G' n'ont pas d'atomes sauf peut-être $\{+\infty\}$ » par « les lois de G et G' n'ont pas d'atomes communs sauf peut-être $\{+\infty\}$ ».) Cette extension, que nous n'utiliserons pas, est laissée au lecteur.

Ce lemme déduit une propriété qualitative d'une propriété quantitative qui semble bien plus forte (l'hypercontractivité) ; on pourrait espérer en donner une démonstration plus élémentaire. Voici un premier pas dans cette direction : La même démonstration marche encore si l'on remplace l'hypercontractivité par une inégalité affaiblie, en autorisant une constante au second membre. Or, à la fin de [16], Neveu observe que, contrairement à l'hypercontractivité proprement dite, cette hypercontractivité affaiblie peut être obtenue directement, à l'aide de majorations de convexité.

FIN DE LA DÉMONSTRATION DU THÉORÈME 2. — Toute martingale de la filtration $\mathcal{F}$ (respectivement $\Psi^r(\mathcal{F})$) est une intégrale par rapport à X (respectivement X^r); c'est donc aussi une martingale dans la filtration plus grosse $\mathcal{G}$. Notre martingale-araignée Y, dans $\mathcal{F}$, est transformée par Ψ^r en une martingale-araignée Y^r dans $\Psi^r(\mathcal{F})$; Y et Y^r sont toutes deux des martingales-araignées pour $\mathcal{G}$, issues de $\vec{0}$ et bornées (donc convergentes).

La proposition 8, appliquée à Y et Y^r dans $\mathcal{G}$, fournit la sous-martingale locale $S=n\,\delta(Y,Y^r)-(n-2)\big[\mathbb{1}_{\{Y^r\circ g\neq\vec{0}\}}\Sigma+\mathbb{1}_{\{Y\circ g^r\neq\vec{0}\}}\Sigma^r\big]$. Comme Y et Y^r et donc aussi

Σ et Σ^r sont bornés et issus de l'origine, S est une sous-martingale bornée, nulle en 0; on a donc la minoration (en posant $G = g_\infty$ et $G^r = g^r_\infty$ pour simplifier la typographie)

$$n\,\mathsf{E}[\delta(Y_\infty, Y^r_\infty)] \geqslant (n-2)\,\mathsf{E}\big[\mathbb{1}_{\{Y^r_G \neq \vec{0}\}}\Sigma_\infty + \mathbb{1}_{\{Y_{G^r}\neq\vec{0}\}}\Sigma^r_\infty\big]\,.$$

Sur l'événement $\{G^r < G\}$ on a $Y^r_G \neq \vec{0}$, et de même $Y_{G^r} \neq \vec{0}$ sur $\{G < G^r\}$; d'où

$$\mathbb{1}_{\{Y^r_G\neq\vec{0}\}}\Sigma_\infty + \mathbb{1}_{\{Y_{G^r}\neq\vec{0}\}}\Sigma^r_\infty \geqslant \mathbb{1}_{\{G^r<G\}}\Sigma_\infty + \mathbb{1}_{\{G<G^r\}}\Sigma^r_\infty \geqslant \mathbb{1}_{\{G\neq G^r\}}(\Sigma_\infty \wedge \Sigma^r_\infty)\,.$$

Mais G et G^r, derniers zéros des martingales Y et Y^r, évitent les temps d'arrêt et a fortiori les v. a. mesurables pour $\mathcal{F}_0$. On peut donc leur appliquer le lemme 4, et, sur l'événement $\{G = G^r\}$, on a $G = \infty$, donc $\Sigma_\infty = 0$. Ceci entraîne $\mathbb{1}_{\{G=G^r\}}(\Sigma_\infty \wedge \Sigma^r_\infty) = 0$, l'inégalité ci-dessus se simplifie en

$$\mathbb{1}_{\{Y^r_G\neq\vec{0}\}}\Sigma_\infty + \mathbb{1}_{\{Y_{G^r}\neq\vec{0}\}}\Sigma^r_\infty \geqslant \Sigma_\infty \wedge \Sigma^r_\infty\,,$$

et la minoration de $\delta(Y_\infty, Y^r_\infty)$ devient

$$n\,\mathsf{E}[\delta(Y_\infty, Y^r_\infty)] \geqslant (n-2)\,\mathsf{E}[\Sigma_\infty \wedge \Sigma^r_\infty]\,.$$

Prenons maintenant $r = r_k = 1 - \frac{1}{k}$ et faisons tendre k vers l'infini. Pour presque tout ω, $X^{r_k}(\omega)$ tend vers $X(\omega)$ dans l'espace $S = \mathrm{C}(\mathbf{R}_+, \mathbf{R}^I)$ muni de la convergence uniforme sur les compacts, coordonnée par coordonnée; le corollaire 2 permet d'affirmer que chaque v. a. $U \in \mathrm{L}^0(\mathcal{F}_\infty)$ est la limite en probabilité des $\Psi^{r_k}(U)$; en particulier, $Y^{r_k}_\infty$ tend vers Y_∞ et $\Sigma^{r_k}_\infty$ vers Σ_∞ ; $\Sigma_\infty \wedge \Sigma^{r_k}_\infty$ tend donc vers Σ_∞. Comme tout le monde est borné, les convergences ont lieu dans L^1, et l'inégalité ci-dessus, écrite pour chaque r_k, devient à la limite

$$0 \geqslant (n-2)\,\mathsf{E}[\Sigma_\infty]\,.$$

Puisque Y est une martingale-araignée multiple, $n-2$ est strictement positif. On a donc $\mathsf{E}[\Sigma_\infty] \leqslant 0$, d'où $\Sigma_\infty = 0$, c'est-à-dire $Y_\infty = \vec{0}$, et Y est identiquement nulle. ∎

3. Application aux temps honnêtes

La première application que nous allons donner du théorème 2 est la réponse à une question soulevée dans [4] : si L est la fin d'un ensemble optionnel dans une filtration brownienne, la tribu $\mathcal{F}_{L+}$ ne peut pas être beaucoup plus grosse que $\mathcal{F}_L$. Les définitions qui suivent nous permettront d'énoncer ceci de façon précise. Si $(\Omega, \mathcal{A}, \mathbf{P})$ est un espace probabilisé, la dimension de l'espace vectoriel $\mathrm{L}^0(\Omega, \mathcal{A}, \mathbf{P})$ est aussi le nombre maximal de valeurs différentes que peut prendre une v. a., ou encore le nombre maximal d'événements non négligeables et deux-à-deux disjoints. Si maintenant $\mathcal{B}$ est une sous-tribu de $\mathcal{A}$, ces équivalences subsistent conditionnellement à $\mathcal{B}$, c'est-à-dire en considérant comme constantes les v. a. mesurables pour $\mathcal{B}$; la dimension devient alors elle-même une v. a. mesurable pour $\mathcal{B}$; elle est introduite ci-dessous sous le nom de multiplicité conditionnelle de $\mathcal{A}$ par rapport à $\mathcal{B}$.

DÉFINITIONS ET NOTATIONS. — *Soit* $(\Omega, \mathcal{A}, \mathbf{P})$ *un espace probabilisé. Pour* $A \in \mathcal{A}$, *nous noterons* $\mathcal{A}_{|A}$ *la tribu trace* $\{B \in \mathcal{A} : B \subset A\}$ *de* $\mathcal{A}$ *sur* A.

Nous appellerons $\mathcal{Q}$ *l'ensemble de toutes les partitions finies et mesurables de* $(\Omega, \mathcal{A}, \mathbf{P})$. *Pour* $Q \in \mathcal{Q}$, $|Q|$ *est le nombre d'éléments de* Q *; si* $\mathcal{B}$ *est une sous-tribu de* $\mathcal{A}$, *l'événement* $S_{\mathcal{B}}(Q) = \{\forall A \in Q \;\; \mathbf{P}[A|\mathcal{B}] > 0\}$ *sera appelé le* support de Q relatif à $\mathcal{B}$ *; cet événement est dans* $\mathcal{B}$, *il est défini à un négligeable près.*

La multiplicité conditionnelle de $\mathcal{A}$ par rapport à $\mathcal{B}$ *est la variable aléatoire définie presque partout, mesurable pour* $\mathcal{B}$ *et à valeurs dans* $\mathbb{N}^* \cup \{\infty\}$

$$\mathrm{Mult}[\mathcal{A}|\mathcal{B}] = \operatorname*{ess\,sup}_{Q\in\mathcal{Q}} |Q|\, \mathbb{1}_{S_{\mathcal{B}}(Q)} .$$

La quantité déterministe $\|\mathrm{Mult}[\mathcal{A}|\mathcal{B}]\|_{L^\infty} = \inf\{n : \mathrm{Mult}[\mathcal{A}|\mathcal{B}] \leqslant n \text{ p. s.}\} \leqslant \infty$ est introduite dans [4] et désignée par *mult* $[\mathcal{A}|\mathcal{B}]$; nous ne l'utiliserons pas.

PROPOSITION 9. — *Soient* $(\Omega, \mathcal{A}, \mathbb{P})$ *un espace probabilisé et* $\mathcal{B}$ *une sous-tribu de* $\mathcal{A}$.

a) *On a* $\mathrm{Mult}[\mathcal{A}|\mathcal{B}] \geqslant 1$ *; l'égalité* $\mathrm{Mult}[\mathcal{A}|\mathcal{B}] = 1$ *a lieu si et seulement si* $\mathcal{A} = \mathcal{B}$.

b) *Soit* n *un entier non nul. Pour que* $\mathrm{Mult}[\mathcal{A}|\mathcal{B}] \leqslant n$ *p. s., il faut et il suffit que, pour tous événements* $A_1, ..., A_{n+1}$ *deux-à-deux disjoints, on ait* $\mathbb{P}[A_1|\mathcal{B}] \dots \mathbb{P}[A_{n+1}|\mathcal{B}] = 0$ *p. s.*

c) *Si l'on remplace* $\mathbb{P}$ *par une probabilité équivalente,* $\mathrm{Mult}[\mathcal{A}|\mathcal{B}]$ *ne change pas. Soient* $\mathbb{P}'$ *une probabilité absolument continue par rapport à* $\mathbb{P}$, *et* $\mathcal{A}'$ *et* $\mathcal{B}'$ *les complétées pour* $\mathbb{P}'$ *des tribus* $\mathcal{A}$ *et* $\mathcal{B}$. *Si, à l'aide de* $\mathbb{P}'$, *on définit* $\mathrm{Mult}'[\mathcal{A}'|\mathcal{B}']$ *presque partout pour* $\mathbb{P}'$, *alors* $\mathrm{Mult}'[\mathcal{A}'|\mathcal{B}'] \leqslant \mathrm{Mult}[\mathcal{A}|\mathcal{B}]$ *p. s. pour* $\mathbb{P}'$.

REMARQUE. — L'inégalité dans le c) peut être stricte. C'est par exemple le cas si $\Omega = \{1, 2\}$, $\mathcal{A} = \mathcal{P}(\Omega)$, $\mathcal{B} = \{\varnothing, \Omega\}$, $\mathbb{P} = \frac{1}{2}(\varepsilon_1 + \varepsilon_2)$ et $\mathbb{P}' = \varepsilon_1$. On a alors $\mathrm{Mult}[\mathcal{A}|\mathcal{B}] = 2$ p.s. pour $\mathbb{P}$, mais $\mathrm{Mult}'[\mathcal{A}'|\mathcal{B}'] = 1$ p.s. pour $\mathbb{P}'$.

DÉMONSTRATION DE LA PROPOSITION 9. — b) Par définition de $\mathrm{Mult}[\mathcal{A}|\mathcal{B}]$, on a $\mathrm{Mult}[\mathcal{A}|\mathcal{B}] \leqslant n$ p. s. si et seulement si, pour toute partition mesurable Q,

$$|Q| > n \quad \Longrightarrow \quad \prod_{A\in Q} \mathbb{P}[A|\mathcal{B}] = 0 \text{ p. s.}$$

Mais une partition plus fine que Q a un support plus petit, donc ceci équivaut à

$$(*) \qquad |Q| = n+1 \quad \Longrightarrow \quad \prod_{A\in Q} \mathbb{P}[A|\mathcal{B}] = 0 \text{ p. s.}$$

Si $A_1, ..., A_n, A_{n+1}$ sont des événements deux-à-deux disjoints, ou bien l'un d'eux est vide, ou bien $(A_1, ..., A_n, A')$ est une partition, où A' désigne $(A_1 \cup ... \cup A_n)^{\mathrm{c}}$. La condition ci-dessus peut donc se réécrire

$$\left.\begin{array}{l} A_1, ..., A_{n+1} \in \mathcal{A} \\ A_i \cap A_j = \varnothing \text{ pour } i \neq j \end{array}\right\} \quad \Longrightarrow \quad \prod_{i=1}^{n+1} \mathbb{P}[A_i|\mathcal{B}] = 0 \text{ p. s.}$$

et le b) est démontré.

a) Prenant $Q = \{\Omega\}$ dans la définition, on obtient $\mathrm{Mult}[\mathcal{A}|\mathcal{B}] \geqslant 1$ p. s. On a donc $\mathrm{Mult}[\mathcal{A}|\mathcal{B}] = 1$ si et seulement si $\mathrm{Mult}[\mathcal{A}|\mathcal{B}] \leqslant 1$, et la condition $(*)$ ci-dessus donne pour $n = 2$

$$\mathrm{Mult}[\mathcal{A}|\mathcal{B}] = 1 \text{ p. s.} \quad \Longleftrightarrow \quad \forall A \in \mathcal{A} \quad \mathbb{P}[A|\mathcal{B}]\mathbb{P}[A^{\mathrm{c}}|\mathcal{B}] = 0 \text{ p. s.}$$

Mais si $\mathbb{P}[A|\mathcal{B}]\mathbb{P}[A^{\mathrm{c}}|\mathcal{B}] = 0$, on a $A \subset \{\mathbb{P}[A|\mathcal{B}] > 0\} \subset \{\mathbb{P}[A^{\mathrm{c}}|\mathcal{B}] = 0\} \subset A$ p. s., d'où $A \in \mathcal{B}$. Réciproquement, si $A \in \mathcal{B}$, alors $\mathbb{P}[A|\mathcal{B}]\mathbb{P}[A^{\mathrm{c}}|\mathcal{B}] = 0$; le a) est établi.

c) Pour tout $A \in \mathcal{A}$, $\{\mathbb{P}[A|\mathcal{B}] > 0\}$ est le plus petit événement (modulo les négligeables) qui soit dans $\mathcal{B}$ et qui contienne A; il reste donc invariant lorsque l'on remplace $\mathbb{P}$ par une probabilité équivalente. Il en va de même de $S_{\mathcal{B}}(Q)$, puis de $\mathrm{Mult}[\mathcal{A}|\mathcal{B}]$.

Si maintenant $\mathbf{P}'$ est une probabilité absolument continue par rapport à $\mathbf{P}$, pour montrer que $\mathrm{Mult}'[\mathcal{A}'|\mathcal{B}'] \leqslant \mathrm{Mult}[\mathcal{A}|\mathcal{B}]$ p.s. pour $\mathbf{P}'$, on peut, quitte à remplacer $\mathbf{P}'$ par une probabilité équivalente, supposer que $\mathbf{P}'$ est la probabilité conditionnelle $\mathbf{P}[\ |C]$ pour un événement $C \in \mathcal{A}$ non négligeable. On a alors

$$\mathcal{A}' = \{A' \subset \Omega : A' \cap C \in \mathcal{A}\} \quad \text{et} \quad \mathcal{B}' = \{B' \subset \Omega : \exists B \in \mathcal{B}\ B' \cap C = B \cap C\}.$$

Nous devons établir que, si $Q' = (A'_1, ..., A'_k)$ est une partition de Ω mesurable pour $\mathcal{A}'$, alors $|Q'| \mathbb{1}_{S'_{\mathcal{B}'}(Q')} \leqslant \mathrm{Mult}[\mathcal{A}|\mathcal{B}]$ p. s. sur C, ou encore

$$|Q'| \mathbb{1}_{S'_{\mathcal{B}'}(Q') \cap C} \leqslant \mathrm{Mult}[\mathcal{A}|\mathcal{B}] \quad \text{p. s.}$$

Si $\mathbf{P}[A'_i \cap C] = 0$ pour un i, alors $\mathbf{P}'[A'_i] = 0$, $S'_{\mathcal{B}'}(Q')$ est négligeable et le premier membre est nul. Nous pouvons donc supposer $\mathbf{P}[A'_i \cap C] > 0$ pour tout i. Dans ce cas, $Q = (A'_1 \cap C, ..., A'_{k-1} \cap C, A'_k \cup C^c)$ est une partition mesurable pour $\mathcal{A}$, et il nous suffit de montrer que $S'_{\mathcal{B}'}(Q') \cap C \subset S_{\mathcal{B}}(Q)$ p. s., ou encore que, pour tout $A \in \mathcal{A}$,

$$\{\mathbf{P}'[A|\mathcal{B}'] > 0\} \cap C \subset \{\mathbf{P}[A|\mathcal{B}] > 0\} \quad \text{p. s.}$$

Or sur C (qui est dans $\mathcal{B}'$), on a $\mathbf{P}'[A|\mathcal{B}'] = \dfrac{\mathbf{P}[A \cap C|\mathcal{B}']}{\mathbf{P}[C|\mathcal{B}']} = \mathbf{P}[A|\mathcal{B}']$ p. s., et il suffit de vérifier que $\{\mathbf{P}[A|\mathcal{B}'] > 0\} \subset \{\mathbf{P}[A|\mathcal{B}] > 0\}$ p. s. Mais $\{\mathbf{P}[A|\mathcal{B}] > 0\}$ est un événement de $\mathcal{B}'$ (car $\mathcal{B}' \supset \mathcal{B}$) qui contient A; il contient donc aussi $\{\mathbf{P}[A|\mathcal{B}'] > 0\}$. ∎

La définition de la multiplicité conditionnelle présente une grande part d'arbitraire : aux nombreuses caractérisations de la dimension d'un espace vectoriel correspondent toute une famille de définitions équivalentes des multiplicités conditionnelles. Ceci fait l'objet des propositions 9, 10 et 11 ci-dessous, qui devraient être évidentes. Lorsque Ω est suffisamment bon, il admet une désintégration régulière mesurable pour $\mathcal{B}$ (voir par exemple Halmos [9]), et $\mathrm{Mult}[\mathcal{A}|\mathcal{B}](\omega)$ est simplement le nombre d'éléments de l'espace probabilisé qui constitue la fibre associée à ω; ceci raccourcit considérablement les démonstrations. Dans le cas général (où nous nous plaçons), il n'y a pas de désintégrations régulières, et l'on doit se contenter des espérances conditionnelles; les démonstrations sont alors élémentaires, mais longues et ennuyeuses.

LEMME 5. — *Soit* $n \geqslant 1$. *Si* $\mathbf{P}\big[\mathrm{Mult}[\mathcal{A}|\mathcal{B}] \geqslant n\big] > 0$, *il existe* $Q \in \mathcal{Q}$ *tel que* $|Q| = n$ *et que* $S_{\mathcal{B}}(Q) = \{\mathrm{Mult}[\mathcal{A}|\mathcal{B}] \geqslant n\}$ *(modulo les négligeables).*

DÉMONSTRATION. — Appelons B_n l'événement $\{\mathrm{Mult}[\mathcal{A}|\mathcal{B}] \geqslant n\}$. Il existe une partie dénombrable $\mathcal{R}$ de $\mathcal{Q}$ telle que $\bigcup_{R \in \mathcal{R}} S_{\mathcal{B}}(R) \supset B_n$ p. s. et $|R| \geqslant n$ pour tout $R \in \mathcal{R}$; on a donc $\bigcup_{R \in \mathcal{R}} S_{\mathcal{B}}(R) = B_n$ p. s. Il existe des événements $E(R) \in \mathcal{B}$ deux-à-deux disjoints tels que $E(R) \subset S_{\mathcal{B}}(R)$ p. s. et $\bigcup_{R \in \mathcal{R}} E(R) = B_n$ p. s. Pour chaque $R \in \mathcal{R}$, soit $\{A_1^R, ..., A_n^R\}$ une partition à n éléments, moins fine que R, et donc de support plus gros : $\mathbf{P}[A_i^R|\mathcal{B}] > 0$ p. s. sur $E(R)$. Lorsque i décrit $\{1, ..., n\}$, les événements $C_i^R = A_i^R \cap E(R)$ forment une partition de $E(R)$, et les événements $D_i = \bigcup_{R \in \mathcal{R}} C_i^R$ forment une partition de B_n. Puisque l'on a $D_i \cap E(R) = C_i^R = A_i^R \cap E(R)$, sur $E(R)$ on peut écrire $\mathbf{P}[D_i|\mathcal{B}] = \mathbf{P}[A_i^R|\mathcal{B}] > 0$, d'où $\mathbf{P}[D_i|\mathcal{B}] > 0$ sur B_n. Les événements $D_1 \cup B_n^c$, D_2, ... , D_n forment une partition $Q \in \mathcal{Q}$ telle que $|Q| = n$ et $S_{\mathcal{B}}(Q) \supset B_n$ p. s., donc $S_{\mathcal{B}}(Q) = B_n$ p. s. ∎

PROPOSITION 10. — *Soient* $\mathcal{B}$ *une sous-tribu sur* $(\Omega, \mathcal{A}, \mathbf{P})$, B *un événement de* $\mathcal{B}$ *et* n *un entier non nul. Les assertions* (i) *à* (iii) *ci-dessous sont équivalentes :*

(i) $\mathrm{Mult}[\mathcal{A}|\mathcal{B}] = n$ *sur* B *;*

(ii) *il existe des événements* $A_1, ..., A_n$ *deux-à-deux disjoints et de réunion* B, *tels que* $\mathbf{P}[A_i|\mathcal{B}] > 0$ *sur* B *et que* $\mathcal{A}_{|B} = \mathcal{B}_{|B} \vee \sigma(A_1, ..., A_n)$ *;*

(iii) *il existe des variables aléatoires* $Y_1, ..., Y_n$ *telles que, pour toute variable aléatoire* X, *il existe un unique vecteur aléatoire* $(v_1, ..., v_n)$ *mesurable pour* $\mathcal{B}$ *et vérifiant* $v_i = 0$ *sur* B^c *et* $X = \sum v_i Y_i$ *sur* B.

PROPOSITION 11. — *Soient $\mathcal{B}$ une sous-tribu sur $(\Omega, \mathcal{A}, \mathbb{P})$, B un événement de $\mathcal{B}$ et n un entier non nul. Les assertions* (iv) *à* (viii) *ci-dessous sont équivalentes :*

(iv) $\mathrm{Mult}[\mathcal{A}|\mathcal{B}] \geqslant n$ *sur* B ;

(v) *il existe des événements* $A_1, ..., A_n$ *deux-à-deux disjoints, de réunion* B *et tels que* $\mathbb{P}[A_i|\mathcal{B}] > 0$ *sur* B ;

(v') *il existe des événements* $A_1, ..., A_n$ *deux-à-deux disjoints et tels que* $\mathbb{P}[A_i|\mathcal{B}] > 0$ *sur* B ;

(vi) *il existe une variable aléatoire* X *et* n *variables aléatoires* $u_1, ..., u_n$ *mesurables pour* $\mathcal{B}$, *vérifiant* $u_i \neq u_j$ *p. s. sur* B *pour* $i \neq j$ *et* $\mathbb{P}[X = u_i|\mathcal{B}] > 0$ *sur* B ;

(vii) *il existe des variables aléatoires* $Y_1, ..., Y_n$ *telles que pour toutes variables aléatoires* $v_1, ..., v_n$ *mesurables pour* $\mathcal{B}$ *et vérifiant* $\sum_i v_i Y_i = 0$ *p. s. sur* B, *on ait* $v_1 = ... = v_n = 0$ *p. s. sur* B ;

(viii) *à toutes variables aléatoires* $Z_1, ..., Z_{n-1}$ *on peut associer une variable aléatoire* X *telle que, pour toutes variables aléatoires* $w_1, ..., w_{n-1}$ *mesurables pour* $\mathcal{B}$, *on ait* $\mathbb{P}[X \neq \sum_i w_i Z_i|\mathcal{B}] > 0$ *sur* B.

PROPOSITION 12. — *Soient $\mathcal{B}$ une sous-tribu sur $(\Omega, \mathcal{A}, \mathbb{P})$, B un événement de $\mathcal{B}$ et n un entier non nul. Les assertions* (ix) *à* (xiv) *ci-dessous sont équivalentes :*

(ix) $\mathrm{Mult}[\mathcal{A}|\mathcal{B}] \leqslant n$ *sur* B ;

(x) *il existe des événements* $A_1, ..., A_n$ *deux-à-deux disjoints, de réunion* B *et tels que* $\mathcal{A}_{|B} = \mathcal{B}_{|B} \vee \sigma(A_1, ..., A_n)$;

(xi) *pour tous événements* $A_1, ..., A_{n+1}$ *inclus dans* B *et deux-à-deux disjoints, on a* $\mathbb{P}[A_1|\mathcal{B}] ... \mathbb{P}[A_{n+1}|\mathcal{B}] = 0$;

(xii) *pour toute variable aléatoire* X *il existe des variables aléatoires* $u_1, ..., u_n$ *mesurables pour* $\mathcal{B}$ *et telles que* $(X-u_1)...(X-u_n) = 0$ *p. s. sur* B ;

(xiii) *il existe des variables aléatoires* $Y_1, ..., Y_n$ *telles que, pour toute variable aléatoire* X, *on ait* $X = \sum_i v_i Y_i$ *p. s. sur* B *pour des variables aléatoires* $v_1, ..., v_n$ *mesurables pour* $\mathcal{B}$;

(xiv) *pour toutes variables aléatoires* $Z_1, ..., Z_{n+1}$, *il y a un vecteur aléatoire* $(w_1, ..., w_{n+1})$ *mesurable pour* $\mathcal{B}$ *tel que, p. s. sur* B, *on ait* $\sum_i w_i Z_i = 0$ *et* $(w_1, ..., w_{n+1}) \neq (0, ..., 0)$.

REMARQUE. — Dans les assertions (ii) et (v), la condition $\mathbb{P}[A_i|\mathcal{B}] > 0$ sur B assure en général que les A_i ne sont pas vides; ils forment donc une partition. Mais il y a une exception : quand B est négligeable, les A_i ont le droit d'être vides; c'est d'ailleurs toujours le cas quand B est lui-même vide, et les énoncés deviendraient faux si l'on y exigeait des partitions (à moins de supposer partout $\mathbb{P}[B] > 0$). C'est pour la même raison que l'hypothèse $\mathbb{P}\big[\mathrm{Mult}[\mathcal{A}|\mathcal{B}]\leqslant n\big] > 0$ est introduite dans le lemme 5; quand elle n'est pas satisfaite, le lemme reste vrai à condition d'autoriser les partitions à avoir des éléments vides — ce qui n'affecte d'ailleurs aucunement la définition de $\mathrm{Mult}[\mathcal{A}|\mathcal{B}]$, puisque le support relatif à $\mathcal{B}$ d'une telle « partition » est négligeable.

DÉMONSTRATION DES TROIS PROPOSITIONS 10, 11 ET 12. — La structure logique de la démonstration est simple : nous allons montrer

(i) ⇒ (ii) ⇒ (iii) ⇒ (i);
(iv) ⇔ (v) ⇔ (v') ⇒ (vi) ⇒ (vii) ⇒ (viii) ⇒ (iv);
(x) ⇒ (xii) ⇒ (xiv) ⇒ (xi) ⇒ (ix) ⇒ (xiii) ⇒ (ix) ⇒ (x).

Malheureusement, la structure chronologique sera moins simple, car certaines de ces implications font appel à d'autres, que nous devrons donc avoir démontrées antérieurement, sous peine de nous égarer complètement.

(iv) ⇔ (v) ⇔ (v'). Si (iv) a lieu, $\mathrm{Mult}[\mathcal{A}|\mathcal{B}] \geqslant n$ sur B, et le lemme 5 donne une partition $Q = \{C_1, ..., C_n\} \in \mathcal{Q}$ telle que $S_{\mathcal{B}}(Q) \supset B$. Posons $A_i = C_i \cap B$. Sur B, on a $\mathbb{P}[A_i|\mathcal{B}] = \mathbb{P}[C_i|\mathcal{B}] > 0$, et (v) est établie. L'implication (v) ⇒ (v') est triviale. Supposons maintenant (v') vérifiée. Les événements $C_1 = (A_2 \cup ... \cup A_n)^c$, $C_2 = A_2$, ..., $C_n = A_n$ forment une partition Q telle que $|Q| = n$ et $S_{\mathcal{B}}(Q) \supset B$, donc $\mathrm{Mult}[\mathcal{A}|\mathcal{B}] \geqslant n\mathbb{1}_B$, d'où (iv).

(i) $\Rightarrow$ (ii). Supposons $\mathrm{Mult}[\mathcal{A}|\mathcal{B}] = n$ sur B. Nous venons voir que (iv) entraîne (v); il existe donc des événements $A_1, ..., A_n$ disjoints et de réunion B tels que $\mathbb{P}[A_i|\mathcal{B}] > 0$ sur B. Il reste à montrer que $\mathcal{A}_{|B} = \mathcal{B}_{|B} \vee \sigma(A_1, ..., A_n)$. Soit D un événement inclus dans A_i pour un certain i; on a $0 \leqslant \mathbb{P}[D|\mathcal{B}] \leqslant \mathbb{P}[A_i|\mathcal{B}]$. Sur $B' = \{0 < \mathbb{P}[D|\mathcal{B}] < \mathbb{P}[A_i|\mathcal{B}]\}$, on a à la fois $\mathbb{P}[D|\mathcal{B}] > 0$, $\mathbb{P}[A_i - D|\mathcal{B}] > 0$ et $\mathbb{P}[A_j|\mathcal{B}] > 0$ pour tout $j \neq i$. L'implication (v') $\Rightarrow$ (iv) vue plus haut, appliquée à B' et à $n+1$, fournit $\mathrm{Mult}[\mathcal{A}|\mathcal{B}] > n$ sur B'; comme $B' \subset \{\mathbb{P}[D|\mathcal{B}] > 0\} \subset B$, l'hypothèse (i) donne $\mathbb{P}[B'] = 0$. On a donc $\{\mathbb{P}[D|\mathcal{B}] > 0\} = \{\mathbb{P}[D|\mathcal{B}] = \mathbb{P}[A_i|\mathcal{B}]\}$; soit B'' cet événement, qui est dans $\mathcal{B}$. En prenant la trace sur A_i de l'inclusion

$$D \subset \{\mathbb{P}[D|\mathcal{B}] > 0\} = B'' = \{\mathbb{P}[A_i - D|\mathcal{B}] = 0\} = \{\mathbb{P}[D \cup A_i^c|\mathcal{B}] = 1\} \subset D \cup A_i^c ,$$

on obtient $D = B'' \cap A_i$, ce qui montre que D est dans la tribu $\mathcal{B}_{|B} \vee \sigma(A_i)$, et a fortiori dans $\mathcal{B}_{|B} \vee \sigma(A_1, ..., A_n)$. Si maintenant C est un événement quelconque inclus dans B, c'est une union finie d'événements chacun inclus dans un A_i, il est donc aussi dans $\mathcal{B}_{|B} \vee \sigma(A_1, ..., A_n)$; ceci établit (ii).

(x) $\Rightarrow$ (xii). Soient $A_1, ..., A_n$ des événements deux-à-deux disjoints, de réunion B et tels que $\mathcal{A}_{|B} = \mathcal{B}_{|B} \vee \sigma(A_1, ..., A_n)$. L'ensemble de toutes les v. a. X vérifiant $X\,\mathbb{1}_B = \sum u_i \mathbb{1}_{A_i}$, où $u_1, ..., u_n$ sont mesurables pour $\mathcal{B}$, est un espace vectoriel réticulé et stable par limites presque sûres de suites (si X^k converge p. s. et $X^k\,\mathbb{1}_B = \sum u_i^k \mathbb{1}_{A_i}$, poser $u_i = \mathbb{1}_B \limsup_k u_i^k$); il est donc de la forme $\mathrm{L}^0(\Omega, \mathcal{C}, \mathbb{P})$ pour une certaine une sous-tribu $\mathcal{C}$ de $\mathcal{A}$. Cette sous-tribu contient les indicatrices des A_i et toutes les v. a. X telles que $X\,\mathbb{1}_B$ soit mesurable pour $\mathcal{B}$; elle contient donc $\mathcal{A}_{|B^c}$ et $\mathcal{B}_{|B} \vee \sigma(A_1, ..., A_n)$. L'hypothèse (x) donne $\mathcal{C} = \mathcal{A}$, toute v. a. X vérifie $X\,\mathbb{1}_B = \sum u_i \mathbb{1}_{A_i}$, et (xii) en découle.

(ii) $\Rightarrow$ (iii). L'hypothèse (ii) nous fournit des événements $A_1, ..., A_n$ deux-à-deux disjoints, de réunion B, vérifiant $\mathbb{P}[A_i|\mathcal{B}] > 0$ sur B et $\mathcal{A}_{|B} = \mathcal{B}_{|B} \vee \sigma(A_1, ..., A_n)$. Nous allons montrer que les v. a. $Y_i = \mathbb{1}_{A_i}$ vérifient la condition (iii). La démonstration de (x) $\Rightarrow$ (xii) vue ci-dessus permet d'associer à toute v. a. X des $u_1, ..., u_n$ mesurables pour $\mathcal{B}$ et tels que $X\,\mathbb{1}_B = \sum u_i \mathbb{1}_{A_i}$; on peut évidemment imposer en plus aux u_i d'être nuls hors de B. Il ne nous reste qu'à montrer l'unicité des u_i. Si des v_i vérifient les mêmes conditions, on a $\sum (u_i - v_i)\,\mathbb{1}_{A_i} = 0$; comme les A_i sont disjoints, cela entraîne $A_i \subset \{u_i = v_i\}$, puis $\mathbb{P}[A_i|\mathcal{B}] \leqslant \mathbb{P}[u_i = v_i|\mathcal{B}] = \mathbb{1}_{\{u_i = v_i\}}$. Sur B, on a $\mathbb{P}[A_i|\mathcal{B}] > 0$, d'où $u_i = v_i$.

À ce stade de la démonstration, nous savons déjà que (i) $\Rightarrow$ (ii) $\Rightarrow$ (iii). L'implication (i) $\Rightarrow$ (iii), qui, sous une hypothèse de multiplicité constante, fournit une « base conditionnelle » du module $\mathrm{L}^0(\mathcal{A})$ sur l'anneau $\mathrm{L}^0(\mathcal{B})$, nous sera précieuse pour la suite. Nous allons maintenant établir la proposition 11, par la voie directe : (v) $\Rightarrow$ (vi) $\Rightarrow$ (vii) $\Rightarrow$ (viii) $\Rightarrow$ (iv).

(v) $\Rightarrow$ (vi). Prendre $X = \mathbb{1}_{A_1} + 2\mathbb{1}_{A_2} + ... + n\mathbb{1}_{A_n}$ et $u_i = i$.

(vi) $\Rightarrow$ (vii). Prendre $Y_i = \mathbb{1}_{\{X = u_i\}}$. Si $\sum v_i Y_i = 0$ p. s. sur B, les v. a. suivantes sont p. s. nulles sur B : $v_i \mathbb{1}_{\{X = u_i\}}$, puis $\mathbb{1}_{\{v_i \neq 0\}} \mathbb{1}_{\{X = u_i\}}$, $\mathbb{1}_{\{v_i \neq 0\}} \mathbb{P}[X = u_i|\mathcal{B}]$, $\mathbb{1}_{\{v_i \neq 0\}}$, et finalement v_i.

(vii) $\Rightarrow$ (viii). Fixons un vecteur aléatoire $Z = (Z_1, ..., Z_{n-1})$. Pour toute v. a. X, appelons $R(X)$ la réunion essentielle

$$R(X) = \bigcup_u \mathrm{ess} \left\{ \mathbb{P}[X = u{\cdot}Z\,|\mathcal{B}] = 1 \right\} ,$$

où u décrit tous les vecteurs aléatoires $(u_1, ..., u_{n-1})$ mesurables pour $\mathcal{B}$. On peut écrire $R(X) = \bigcup_{k \in \mathbb{N}} R^k$ où les R^k sont dans $\mathcal{B}$, deux-à-deux disjoints, et tels qu'il existe des vecteurs aléatoires u^k vérifiant $\mathbb{P}[X = u^k{\cdot}Z\,|\mathcal{B}] = 1$ sur R^k. Le vecteur aléatoire $u = \sum \mathbb{1}_{R^k} u^k$ vérifie $\mathbb{P}[X = u{\cdot}Z\,|\mathcal{B}] = \mathbb{P}[X = u^k{\cdot}Z\,|\mathcal{B}] = 1$ sur R^k, donc $\mathbb{P}[X = u{\cdot}Z\,|\mathcal{B}] = 1$ sur $R(X)$, et l'union essentielle ci-dessus est atteinte pour un u.

En appliquant ceci aux v. a. Y_i données par l'hypothèse (vii), nous obtenons des vecteurs $u_i \in \mathbb{R}^{n-1}$ tels que, pour chaque i, on ait $R(Y_i) = \left\{ \mathbb{P}[Y_i = u_i{\cdot}Z\,|\mathcal{B}] = 1 \right\}$. On a a fortiori $Y_i = u_i{\cdot}Z$ p. s.sur $R(Y_i)$, et, sur l'intersection $\underline{R} = \bigcap_i R(Y_i)$, on peut écrire $Y_i = \sum u_i^k Z_k$ pour tout $i \in \{1, ..., n\}$, où les u_i^k forment les coefficients d'une matrice aléatoire U, mesurable pour $\mathcal{B}$, à n lignes et $n-1$ colonnes.

Soit a une application borélienne qui à toute matrice m de dimensions $n\times(n-1)$ associe un vecteur ligne non nul v à n éléments tel que $vm = 0$.

Posons $V = a(U)$ et appelons $v_1, ..., v_n$ les éléments de V ; ce sont des v. a. mesurables pour $\mathcal{B}$. Sur $\underline{R}$, on a $\sum v_i Y_i = 0$; l'hypothèse (vii) donne $v_i \mathbb{1}_{\underline{R}} = 0$ p. s. sur B. Comme $V \neq 0$ par définition de a, il en résulte que $\mathbb{P}(B \cap \underline{R}) = 0$. Ceci signifie qu'il existe une v. a. I mesurable pour $\mathcal{B}$, à valeurs dans $\{1, ..., n\}$ et telle que $B \cap \{I=i\} \subset R_i^c$. La propriété (viii) a lieu avec $X = Y_I$. En effet, soient $w_1, ..., w_{n-1}$ mesurables pour $\mathcal{B}$. Par définition de R_i, on a $\mathbb{P}[Y_i = w\cdot Z\,|\mathcal{B}] < 1$ sur R_i^c, donc aussi $\mathbb{P}[Y_i \neq w\cdot Z\,|\mathcal{B}] > 0$ sur R_i^c. Sur $B \cap \{I = i\}$, qui est inclus dans R_i^c, on peut donc écrire $\mathbb{P}[Y_I \neq w\cdot Z\,|\mathcal{B}] = \mathbb{P}[Y_i \neq w\cdot Z\,|\mathcal{B}] > 0$; on en tire $\mathbb{P}[Y_I \neq w\cdot Z\,|\mathcal{B}] > 0$ sur B.

(viii) $\Rightarrow$ (iv). Nous supposons donc (viii). Pour tout $k < n$, soit B_k l'événement $\{\text{Mult}[\mathcal{A}|\mathcal{B}] = k\}$. L'implication (i) $\Rightarrow$ (iii) appliquée avec k et B_k donne une « base » $\{Y_1, ..., Y_k\}$. Appliquant l'hypothèse (viii) aux $n-1$ v. a. $Y_1, ..., Y_k, 0, ..., 0$; on obtient un X tel que $\mathbb{P}[X \neq w_1 Y_1 + ... + w_k Y_k\,|\mathcal{B}] > 0$ sur B pour tous $w_1, ..., w_k$ mesurables pour $\mathcal{B}$. La propriété (iii) vérifiée par les Y permet de construire $v_1, ..., v_k$ mesurables pour $\mathcal{B}$ tels que le X obtenu ci-dessus vérifie $X = \sum v_i Y_i$ sur B_k. En prenant alors $w_i = v_i$ dans la propriété satisfaite par X, on obtient $\mathbb{P}[X \neq \sum v_i Y_i|\mathcal{B}] > 0$ sur B. La comparaison de ces deux relations donne $\mathbb{P}[B \cap B_k] = 0$, et (iv) est établie.

Ceci achève la démonstration de la proposition 11 ; nous passons maintenant à la proposition 12. L'implication (x) $\Rightarrow$ (xii) a été démontrée plus haut; nous allons montrer successivement (xii) $\Rightarrow$ (xiv) $\Rightarrow$ (xi) $\Rightarrow$ (ix) $\Rightarrow$ (xiii) $\Rightarrow$ (ix) $\Rightarrow$ (x).

(xii) $\Rightarrow$ (xiv). Posons $E = \mathbb{R}^{n+1}$; soient α une bijection bimesurable de E dans $\mathbb{R}$ et a une application borélienne de E^n dans E telle que, pour tous $e_1, ..., e_n$ dans E, $a(e_1, ..., e_n)$ soit non nul et orthogonal à chacun des e_i. Si Z est un vecteur aléatoire à valeurs dans E, l'hypothèse (xii) appliquée à la v. a. $X = \alpha(Z)$ donne $u_1, ..., u_n$ mesurables pour $\mathcal{B}$ et telles que l'on ait $Z \in \{\alpha^{-1}u_1, ..., \alpha^{-1}u_n\}$ p. s. sur B. Le vecteur $(w_1, ..., w_{n+1}) = a(\alpha^{-1}u_1, ..., \alpha^{-1}u_n)$ est mesurable pour $\mathcal{B}$, p. s. non nul, et orthogonal à Z p. s. sur B.

(xiv) $\Rightarrow$ (xi). Soient $A_1, ..., A_{n+1}$ des événements inclus dans B et deux-à-deux disjoints. L'hypothèse (xiv) appliquée aux indicatrices des A_i donne $w_1, ..., w_{n+1}$ mesurables pour $\mathcal{B}$, telles que $B \subset \bigcup_i \{w_i \neq 0\}$ et vérifiant $\sum_i w_i \mathbb{1}_{A_i} = 0$ p. s. Sur $\{w_i \neq 0\}$, on a $\mathbb{1}_{A_i} = 0$ d'où $\mathbb{P}[A_i|\mathcal{B}] = 0$; on a donc $\prod_i \mathbb{P}[A_i|\mathcal{B}] = 0$ sur B.

(xi) $\Rightarrow$ (ix). Soit $Q = \{A_1, ..., A_k\}$ une partition mesurable de Ω telle que $|Q| = k > n$. Les événements $A_i' = A_i \cap B$ sont inclus dans B et deux-à-deux disjoints; en leur appliquant l'hypothèse (xi), on obtient $\mathbb{P}[A_1'|\mathcal{B}] ... \mathbb{P}[A_{n+1}'|\mathcal{B}] = 0$, donc $\mathbb{P}[A_1|\mathcal{B}] ... \mathbb{P}[A_{n+1}|\mathcal{B}] = 0$ sur B; en conséquence, $\mathbb{P}[S_{\mathcal{B}}(Q) \cap B] = 0$. Ainsi, seules les partitions vérifiant $|Q| \leqslant n$ peuvent avoir un support rencontrant B, et $\text{Mult}[\mathcal{A}|\mathcal{B}] \leqslant n$ sur B.

(ix) $\Rightarrow$ (xiii). Nous supposons $\text{Mult}[\mathcal{A}|\mathcal{B}] \leqslant n$ sur B, donc B est la réunion des événements $B_k = B \cap \{\text{Mult}[\mathcal{A}|\mathcal{B}] = k\}$ où k décrit $\{1, ..., n\}$. Pour chaque k, appliquons (i) $\Rightarrow$ (iii) à l'entier k et à l'événement B_k ; on obtient des v. a. $Y_1^k, ..., Y_k^k$ telles que toute v. a. X vérifie $X = \sum_{i\leqslant k} v_i^k Y_i^k$ sur B_k. Pour $1 \leqslant i \leqslant n$, posons $Y_i = \sum_{k\geqslant i} Y_i^k \mathbb{1}_{B_k}$ et $v_i = \sum_{k\geqslant i} v_i^k \mathbb{1}_{B_k}$. L'égalité $X = \sum v_i Y_i$ est en vigueur sur chaque B_k, donc sur B, et (xiii) a lieu.

(xiii) $\Rightarrow$ (ix). Soient $Y_1, ..., Y_n$ donnés par l'hypothèse (xiii); nous allons montrer que l'événement $B' = \{\text{Mult}[\mathcal{A}|\mathcal{B}] > n\}$ ne rencontre pas B. Par (iv) $\Rightarrow$ (viii) appliqué à B', à $n+1$ et aux Y_i, il existe X tel que, pour tous $w_1, ..., w_n$ mesurables pour $\mathcal{B}$, on ait $\mathbb{P}[X \neq \sum w_i Y_i|\mathcal{B}] > 0$ sur B'. Mais l'hypothèse (xiii) donne $v_1, ..., v_n$ tels que $X = \sum v_i Y_i$ p. s. sur B; comme on doit aussi avoir $\mathbb{P}[X \neq \sum v_i Y_i|\mathcal{B}] > 0$ sur B', la comparaison ce ces deux relations impose $\mathbb{P}[B \cap B'] = 0$.

(ix) $\Rightarrow$ (x). Nous supposons $\text{Mult}[\mathcal{A}|\mathcal{B}] \leqslant n$ sur B; B est donc la réunion, lorsque k décrit $\{1, ..., n\}$, des événements $B_k = B \cap \{\text{Mult}[\mathcal{A}|\mathcal{B}] = k\}$. En appliquant à k et B_k l'implication (i) $\Rightarrow$ (ii), on obtient pour chaque k des événements $A_1^k, ..., A_k^k$ deux-à-deux disjoints, de réunion B_k et tels que $\mathcal{A}_{|B_k} = \mathcal{B}_{|B_k} \vee \sigma(A_i^k, 1 \leqslant i \leqslant k)$. Les événements $A_k = A_k^k \cup ... \cup A_k^n$, sont deux-à-deux disjoints et ont pour union B. La tribu $\mathcal{B}_{|B} \vee \sigma(A_1, ..., A_n)$ sur B contient tous les $A_i \cap B_k = A_i^k$, donc tous les événements de $\mathcal{A}_{|B_k}$, et donc aussi toute la tribu $\mathcal{A}_{|B}$.

Ceci achève d'établir la proposition 12; il ne manque plus que la proposition 10, et plus précisément (iii) $\Rightarrow$ (i), la seule implication non encore prouvée. Mais (iii) est la conjonction des deux assertions (vii) et (xiii), cette dernière traduisant l'unicité des coefficients v_i. Supposant (iii), on a donc (vii) et (xiii), qui, nous l'avons vu, entraînent respectivement (iv) et (ix), d'où $\mathrm{Mult}[\mathcal{A}|\mathcal{B}] = n$ sur B. ∎

DÉFINITION. — *Si $\mathcal{F}$ est une filtration et τ une variable aléatoire à valeurs dans $[0,\infty]$, les tribus $\mathcal{F}_\tau$ et $\mathcal{F}_{\tau+}$ sont ainsi caractérisées : une variable aléatoire U est mesurable pour $\mathcal{F}_\tau$ (respectivement $\mathcal{F}_{\tau+}$) si et seulement s'il existe un processus optionnel (respectivement progressif) V, défini sur $[\![0,\infty]\!]$, tel que $U = V_\tau$.*

Il est bien connu que la continuité à droite des filtrations s'étend aux temps d'arrêt : on a $\mathcal{F}_{T+} = \mathcal{F}_T$ pour tout temps d'arrêt T. Nous allons nous intéresser à une classe de temps aléatoires ne partageant pas cette propriété, les temps honnêtes, introduits par P.A. Meyer, R.T. Smythe et J.B. Walsh dans [15].

DÉFINITION. — *Une variable aléatoire L à valeurs dans $[0,\infty]$ est un* temps honnête pour une filtration $\mathcal{F}$ *s'il existe pour chaque $t \geqslant 0$ une variable aléatoire ℓ_t mesurable pour $\mathcal{F}_t$ et telle que $L = \ell_t$ sur $\{L \leqslant t\}$.*

Les temps honnêtes jouent un rôle important dans la théorie du grossissement progressif d'une filtration; nous n'aborderons pas cet aspect.

Leurs principales propriétés sont rappelées, pour référence ultérieure, dans la proposition ci-dessous; nous n'en démontrerons que les parties triviales, le reste sera admis. Le a) caractérise les temps honnêtes comme étant les fins des ensembles optionnels (ou progressifs, c'est équivalent puisque l'adhérence d'un ensemble progressif est optionnelle).

PROPOSITION 13. — a) *Pour qu'une variable aléatoire L à valeurs dans $[0,\infty]$ soit un temps honnête, il faut et il suffit qu'il existe un ensemble optionnel $\Gamma \subset \mathbf{R}_+ \times \Omega$ tel que l'on ait $L(\omega) = \sup\{t : (t,\omega) \in \Gamma\}$ pour tout ω (avec la convention $\sup \varnothing = 0$).*

b) *Soit L un temps honnête pour une filtration $\mathcal{F}$.*

Si τ est une variable aléatoire telle que $L \leqslant \tau \leqslant \infty$, on a $\mathcal{F}_L \subset \mathcal{F}_\tau$.

La suite des tribus $\mathcal{F}_{L+1/n}$ est décroissante et sa limite est $\mathcal{F}_{L+}$.

Si L et L' sont deux temps honnêtes tels que $L \leqslant L'$, on a $\mathcal{F}_{L+} \subset \mathcal{F}_{L'+}$.

c) *On suppose que toutes les martingales de $\mathcal{F}$ sont continues* (ceci revient à dire que tous les optionnels sont prévisibles). *Dans $\mathcal{F}$, soient L un temps honnête et M une martingale uniformément intégrable. Pour que $\mathbb{E}[M_\infty|\mathcal{F}_L] = 0$ p. s., il faut et il suffit que $M_L = 0$ p. s.*

REMARQUE. — Le b) dit que pour L honnête, $\mathcal{F}_{L+}$ est la limite des $\mathcal{F}_{L+1/n}$. Ceci peut être précisé par les *théorèmes-quotient,* qui permettent de calculer explicitement, à l'aide de projections optionnelles, les espérances conditionnelles par rapport à $\mathcal{F}_{L+}$; voir [2] et [3].

DÉMONSTRATION. — a) C'est le théorème XX-13 de Dellacherie-Maisonneuve-Meyer [7].

b) Soient L un temps honnête et τ une variable aléatoire vérifiant $L \leqslant \tau \leqslant \infty$. Pour $U \in \mathrm{L}^\infty(\mathcal{F}_L)$, il existe un processus optionnel borné O, défini sur $[\![0,\infty]\!]$ et tel que $O_L = U$. Mais L est la fin d'un ensemble optionnel Γ, que l'on peut supposer fermé dans $[0,\infty]$ (voir le théorème IV-89 de Dellacherie-Meyer [6]). Le processus

O' défini par $O'_t = O_{\sup(\Gamma\cap[0,t])}$ est optionnel (si O est adapté et continu à droite, O' aussi ; le cas général s'en déduit par classes monotones). Comme $O' = U$ sur $[\![L,\infty]\!]$, on a $U = O_\tau \in \mathcal{F}_\tau$ et ceci montre que $\mathcal{F}_L \subset \mathcal{F}_\tau$.

Chaque $L+1/n$ est un temps honnête, car, sur $\{L+1/n \leqslant t\}$, $L+1/n$ est égal à une v. a. mesurable pour $\mathcal{F}_{t-1/n}$. La croissance, vue plus haut, de $L \mapsto \mathcal{F}_L$ entraîne que les tribus $\mathcal{F}_{L+1/n}$ décroissent ; il reste à démontrer que la limite est $\mathcal{F}_{L+}$. L'inclusion $\mathcal{F}_{L+} \subset \mathcal{F}_{L+1/n}$ résulte immédiatement de ce que, *si X_t est un processus progressif, $X_{(t-1/n)^+}$ est optionnel* pour la filtration constante par intervalles $\mathcal{F}'_t = \mathcal{F}_{[nt]/n}$ et a fortiori pour $\mathcal{F}$, qui est plus grosse. Enfin, pour avoir l'inclusion dans l'autre sens, $\bigcap_n \mathcal{F}_{L+1/n} \subset \mathcal{F}_{L+}$, il suffit d'établir que *si (X^n) est une suite uniformément bornée de processus progressifs, $\limsup_n X^n_{t+1/n}$ est progressif.* Or il est clair que cette limite supérieure est progressive pour la filtration plus grosse $(\mathcal{F}_{t+1/k})$; et l'intersection des tribus progressives des filtrations $(\mathcal{F}_{t+1/k})$ est la tribu progressive de $\mathcal{F}$ par le lemme de Lindvall-Rogers (voir l'appendice de [4]).

Enfin, l'inclusion $\mathcal{F}_{L+} \subset \mathcal{F}_{L'+}$ résulte facilement de ce qui précède, en passant à la limite dans $\mathcal{F}_{L+\frac{1}{n}} \subset \mathcal{F}_{L'+\frac{1}{n}}$.

c) Cela résulte aussitôt du théorème XX-35 de Dellacherie-Maisonneuve-Meyer [7] ; voir aussi [1]. ∎

LEMME 6. — *Soit L un temps honnête pour une filtration $\mathcal{F}$; on suppose que toutes les martingales de $\mathcal{F}$ sont continues. Appelons J la projection optionnelle de $\mathbb{1}_{[\![L,\infty[\![}$ (elle est càdlàg), J_- le processus des limites à gauche de J, K l'ensemble aléatoire $\{J_- = 0\}$ et G_t la v. a.* $\sup(K\cap[0,t])$.

(i) *Si U est une variable aléatoire mesurable pour $\mathcal{F}_{L+}$ et V un processus progressif (défini sur $[\![0,\infty]\!]$) tel que $V_L = U$, le processus $H_t = \mathbb{1}_{\{t\notin K\}} V_{G_t}$ est l'unique processus optionnel nul sur K et égal à U sur $]\!]L,\infty[\![$. L'application $\mathcal{O} : U \mapsto H$ ainsi définie est linéaire, positive et préserve les produits.*

(ii) *Si U est dans $\mathrm{L}^1(\mathcal{F}_{L+})$ et vérifie $\mathsf{E}[U|\mathcal{F}_L] = 0$, la martingale $\mathsf{E}[U|\mathcal{F}_t]$ est égale au processus $\mathcal{O}(U)_t J_{t-}$; elle est nulle à l'origine.*

(iii) *Étant donné $n \geqslant 2$, soient $A_1, \ldots, A_n$ des événements de $\mathcal{F}_{L+}$ deux-à-deux disjoints ; soit B l'événement $\{\forall i \in \{1,\ldots,n\},\ \mathsf{P}[A_i|\mathcal{F}_L] > 0\}$, qui est dans $\mathcal{F}_L$. Pour $1 \leqslant i \leqslant n$, posons*

$$U^i = \frac{\mathbb{1}_{A_i\cap B}}{\mathsf{P}[A_i|\mathcal{F}_L]} \qquad (\text{avec } \frac{0}{0} = 0)\,.$$

On a $\mathsf{E}[U^i|\mathcal{F}_L] = \mathbb{1}_B$. Les n processus $\mathcal{O}(U^i)J_-$ sont les composantes d'une martingale-araignée uniformément intégrable, issue de l'origine et de multiplicité n.

DÉMONSTRATION. — La projection optionnelle J de $\mathbb{1}_{[\![L,\infty[\![}$ est le processus adapté et càdlàg vérifiant $J_T \mathbb{1}_{\{T<\infty\}} = \mathsf{P}[L \leqslant T < \infty|\mathcal{F}_T]$ pour tout temps d'arrêt T ; avec la convention $J_{0-} = 0$, on a aussi $J_{T-}\mathbb{1}_{\{T<\infty\}} = \mathsf{P}[L < T < \infty|\mathcal{F}_T]$ (car $\mathcal{F}_T = \mathcal{F}_{T-}$ et tous les optionnels sont prévisibles), et $\Delta J_T \mathbb{1}_{\{T<\infty\}} = \mathsf{P}[L = T < \infty|\mathcal{F}_T]$. En particulier, les sauts de J sont positifs. Le processus J est une sous-martingale bornée, et sa limite à l'infini est $J_\infty = \mathbb{1}_{\{L<\infty\}}$.

Positif et à sauts positifs, J ne peut s'annuler lors d'un saut, et $\{J = 0\}$ est inclus dans $\{J_- = 0\}$; l'ensemble $K = \{J_- = 0\}$ est donc un fermé optionnel. Pour tout temps d'arrêt T, on peut écrire

$$[\![T]\!]\subset K \quad\Leftrightarrow\quad J_{T-}\mathbb{1}_{\{T<\infty\}}=0 \quad\Leftrightarrow\quad \mathsf{P}[L<T<\infty|\mathcal{F}_T]=0$$
$$\Leftrightarrow\quad \mathsf{P}[L<T<\infty]=0 \quad\Leftrightarrow\quad [\![T]\!]\subset[\![0,L]\!]\ ;$$

en conséquence, par le théorème de section, tout ensemble optionnel inclus dans $[\![0,L]\!]$ est inclus dans K.

En outre, K est lui-même inclus dans $[\![0,L]\!]$ (c'est ici qu'intervient l'honnêteté de L; c'est d'ailleurs comme cela que l'on construit Γ dans la proposition 13 a; voir [7], XX-13); K est donc le plus grand optionnel inclus dans $[\![0,L]\!]$, et L est aussi la fin de K. La formule $G_t=\sup\big(K\cap[0,t]\big)$ définit un processus croissant, continu à droite; on a $L=G_\infty$ et, pour tout temps d'arrêt T, on a aussi $L=G_T$ sur $\{L\leqslant T\}$.

(i) Pour $\varepsilon>0$, le processus $\mathbb{1}_{\{t\geqslant G_t+\varepsilon\}}V_{G_t}$ est adapté et continu à droite, donc optionnel; à la limite, $\mathbb{1}_{\{t>G_t\}}V_{G_t}$ est optionnel. Enfin, $\mathbb{1}_{\{t\notin K\}}V_{G_t}$, qui est égal à $\mathbb{1}_{\{t\notin K\}}\mathbb{1}_{\{t>G_t\}}V_{G_t}$, est lui aussi optionnel.

Pour $t>L$, on a $G_t=L$ et $t\notin K$, donc $H_t=U$. L'unicité de H vient de ce que deux processus optionnels égaux sur $]\!]L,\infty[\![$ diffèrent sur un optionnel inclus dans $[\![0,L]\!]$, donc dans K. Elle entraîne que $\mathcal{O}$ préserve les relations boréliennes à un nombre quelconque d'arguments, en particulier la linéarité, la positivité et les produits.

(ii) Supposons $U\in\mathrm{L}^1(\mathcal{F}_{L+})$ et $\mathsf{E}[U|\mathcal{F}_L]=0$. Soit T un temps d'arrêt fini. Si z est dans $\mathrm{L}^\infty(\mathcal{F}_T)$, le processus $Z=z\mathbb{1}_{[\![T,\infty[\![}$ est optionnel, la v. a. $z\mathbb{1}_{\{L\geqslant T\}}=Z_L$ est dans $\mathrm{L}^\infty(\mathcal{F}_L)$ et, puisque $\mathsf{E}[U|\mathcal{F}_L]=0$, $\mathsf{E}[Uz\mathbb{1}_{\{L\geqslant T\}}]=0$. On en déduit $\mathsf{E}[U\mathbb{1}_{\{L\geqslant T\}}|\mathcal{F}_T]=0$, ce qui permet d'écrire, avec la même notation V qu'en (i),

$$\mathsf{E}[U|\mathcal{F}_T]=\mathsf{E}[U\mathbb{1}_{\{L<T\}}|\mathcal{F}_T]=\mathsf{E}[V_L\mathbb{1}_{\{L<T\}}|\mathcal{F}_T]=\mathsf{E}[V_{G_T}\mathbb{1}_{\{L<T\}}|\mathcal{F}_T]$$
$$=V_{G_T}\,\mathsf{P}[L<T|\mathcal{F}_T]=V_{G_T}J_{T-}=\mathcal{O}(U)_TJ_{T-}\,.$$

Ceci établit que le processus $\mathcal{O}(U)_tJ_{t-}$ est la martingale $\mathsf{E}[U|\mathcal{F}_t]$; sa valeur initiale est $\mathsf{E}[U|\mathcal{F}_0]=\mathsf{E}[\mathsf{E}[U|\mathcal{F}_L]|\mathcal{F}_0]=0$.

(iii) Les U^i vérifient

$$\mathsf{E}[U^i|\mathcal{F}_L]=\mathbb{1}_{\{\mathsf{P}[A_i|\mathcal{F}_L]>0\}}\,\frac{1}{\mathsf{P}[A_i|\mathcal{F}_L]}\,\mathsf{P}[A_i\cap B|\mathcal{F}_L]=\mathbb{1}_{\{\mathsf{P}[A_i|\mathcal{F}_L]>0\}}\,\mathbb{1}_B=\mathbb{1}_B\ ;$$

en particulier, $U^i\in\mathrm{L}^1(\mathcal{F}_{L+})$. Posons $H^i=\mathcal{O}(U^i)$.

Soit $\mathbf{U}=\{\vec{u}_1,\ldots,\vec{u}_n\}$ un système de n vecteurs de somme nulle engendrant un espace de dimension $n-1$. Posons $\vec{U}=\sum U^i\vec{u}_i$ et $\vec{H}=\sum H^i\vec{u}_i$. Le vecteur aléatoire $\vec{U}$ vérifie $\mathsf{E}[\vec{U}|\mathcal{F}_L]=\sum\mathbb{1}_B\vec{u}_i=\vec{0}$. Le (ii) permet d'affirmer que $Y=\vec{H}J_-$ est la martingale issue de l'origine $\mathsf{E}[\vec{U}|\mathcal{F}_t]$. De plus, pour $i\neq j$, on a $A_i\cap A_j=\varnothing$ par hypothèse, d'où $U^iU^j=0$; cette propriété et la positivité des U^i se transfèrent par $\mathcal{O}$, donnant $H^i\geqslant0$ et $H^iH^j=0$, et Y est ainsi une martingale-araignée de multiplicité n, à valeurs dans la toile $\{\lambda\vec{u}_i,\ \lambda\geqslant0,\ 1\leqslant i\leqslant n\}$. ∎

DÉFINITION (empruntée à [4]). — *La* multiplicité de scindage *d'une filtration $\mathcal{F}$ est le plus petit entier n tel que* $\mathrm{Mult}[\mathcal{F}_{L+}|\mathcal{F}_L]\leqslant n$ *pour tout temps L honnête pour $\mathcal{F}$; elle est infinie s'il n'existe aucun tel entier.* On la note $\mathrm{Sp\,Mult}\,(\mathcal{F})$.

On a donc $1\leqslant\mathrm{Sp\,Mult}\,(\mathcal{F})\leqslant\infty$, et $\mathrm{Sp\,Mult}\,(\mathcal{F})\leqslant n$ si et seulement si $\mathrm{Mult}[\mathcal{F}_{L+}|\mathcal{F}_L]\leqslant n$ pour tout temps honnête L.

PROPOSITION 14. — *On suppose que, sur l'espace probabilisé filtré* $(\Omega, \mathcal{A}, \mathbf{P}, \mathcal{F})$, *toutes les martingales sont continues. Soit* $n \geqslant 1$ *un entier. Il y a équivalence entre les quatre conditions suivantes :*

(i) *toute martingale-araignée de multiplicité* $n+1$*, bornée et issue de l'origine est nulle ;*

(i') *toute martingale-araignée de multiplicité* $n+1$ *est nulle à partir de son premier zéro ;*

(ii) *tout temps honnête* L *vérifie* $\mathrm{Mult}[\mathcal{F}_{L+}|\mathcal{F}_L] \leqslant n$ *;*

(ii') $\mathrm{Sp\,Mult}\,(\mathcal{F}) \leqslant n$.

Pour $n = 1$, la condition (i) dit que toute martingale (usuelle) bornée issue de l'origine est nulle. Ceci se produit si et seulement si $\mathcal{F}_0 = \mathcal{F}_\infty$, c'est-à-dire si et seulement si la filtration est constante. Les filtrations constantes sont donc caractérisées par la relation $\mathrm{Sp\,Mult}\,(\mathcal{F}) = 1$.

DÉMONSTRATION DE LA PROPOSITION 14. — L'équivalence entre (ii) et (ii') ne fait que rappeler la définition de $\mathrm{Sp\,Mult}\,(\mathcal{F})$.

(i) $\Rightarrow$ (i'). Supposant (i), on en déduit par arrêt que toute martingale araignée de multiplicité $n+1$ et issue de l'origine est nulle; puis la condition (i') s'obtient en remarquant que, si Y est une martingale-araignée de premier zéro T, alors $Y\,\mathbb{1}_{[\![T,\infty[\![} = \int \mathbb{1}_{]\!]T,\infty[\![}\,dY$ est, sur la même toile et dans la même filtration, une martingale-araignée, nulle à l'origine et égale à Y sur $[\![T,\infty[\![$.

(i') $\Rightarrow$ (ii). L'hypothèse entraîne que toute martingale-araignée de multiplicité $n+1$ et issue de l'origine est identiquement nulle. Considérons un temps honnête L. Soient $A_1, ..., A_{n+1}$ des événements dans $\mathcal{F}_{L+}$ et deux-à-deux disjoints; appelons B l'événement $\{\forall i \in \{1,...,n+1\},\ \mathbf{P}[A_i|\mathcal{F}_L] > 0\}$. Le (iii) du lemme 6 fournit une martingale-araignée Y de multiplicité $n+1$, issue de l'origine et dont les composantes Y^i vérifient $\mathbf{E}[Y^i_\infty|\mathcal{F}_L] = \mathbb{1}_B$. L'hypothèse (i') donne $Y = 0$. Il en résulte $\mathbf{E}[Y^i_\infty] = 0$, c'est-à-dire $\mathbf{P}[B] = 0$, et $\mathrm{Mult}[\mathcal{F}_{L+}|\mathcal{F}_L] \leqslant n$ par la proposition 9 b). Comme L est arbitraire, (ii) est établie.

(ii) $\Rightarrow$ (i). Supposons (ii) vérifiée, et soit Y une martingale-araignée de multiplicité $n+1$, issue de l'origine et bornée (donc convergente). Pour montrer que $Y = 0$, considérons le temps honnête $L = \sup\{Y = \vec{0}\}$. En notant Y^u les composantes de Y, les $n+1$ événements $A_u = \{Y^u_\infty > 0\}$ sont deux-à-deux disjoints et dans $\mathcal{F}_{L+}$. En utilisant l'hypothèse $\mathrm{Mult}[\mathcal{F}_{L+}|\mathcal{F}_L] \leqslant n$, la proposition 9 b) fournit $\prod_u \mathbf{P}[A_u|\mathcal{F}_L] = 0$, et la réunion des $n+1$ événements $B_u = \{\mathbf{P}[A_u|\mathcal{F}_L] = 0\}$ a probabilité 1. Mais, par ailleurs, la martingale $Y^u - Y^v$ est nulle à l'instant L, donc $\mathbf{E}[Y^u_\infty - Y^v_\infty|\mathcal{F}_L] = 0$ par la proposition 13 c); et sur B_u, on a $\mathbf{E}[Y^u_\infty|\mathcal{F}_L] = 0$, d'où $\mathbf{E}[Y^v_\infty|\mathcal{F}_L] = 0$ et $\mathbf{P}[A_v|\mathcal{F}_L] = 0$; ceci montre que $B_u \subset B_v$. Tous les B_u sont donc égaux, et chacun d'eux a probabilité 1. En conséquence, chaque A_u est négligeable, Y_∞ est p. s. nulle, et $Y = \vec{0}$. Ainsi, (ii) $\Rightarrow$ (i). ∎

THÉORÈME 3. — *Toute filtration brownienne* $\mathcal{F}$ *vérifie* $\mathrm{Sp\,Mult}\,(\mathcal{F}) = 2$.

DÉMONSTRATION. — Le théorème 2 dit que toute martingale-araignée dans $\mathcal{F}$ de multiplicité 3 et issue de l'origine est nulle. La proposition 14 en déduit que $\mathrm{Sp\,Mult}\,(\mathcal{F}) \leqslant 2$. Puisque $\mathcal{F}$ n'est pas constante, on a $\mathrm{Sp\,Mult}\,(\mathcal{F}) > 1$. En fin de compte, $\mathrm{Sp\,Mult}\,(\mathcal{F}) = 2$. ∎

Si L est un temps honnête dans une filtration brownienne $\mathcal{F}$, on a donc $\mathrm{Mult}[\mathcal{F}_{L+}|\mathcal{F}_L] \leqslant 2$. Il serait intéressant de savoir reconnaître les L tels que $\mathcal{F}_{L+} = \mathcal{F}_L$ (c'est-à-dire tels que $\mathrm{Mult}[\mathcal{F}_{L+}|\mathcal{F}_L] = 1$). C'est toujours le cas si L est un temps d'arrêt, mais cela peut aussi se produire pour bien d'autres L : par exemple, si $\mathcal{F}$ est la filtration naturelle d'un mouvement brownien unidimensionnel X, l'instant $L \in [0,1]$ tel que $X_L = \sup_{t \leqslant 1} X_t$ vérifie $\mathcal{F}_{L+} = \mathcal{F}_L$. En effet, L est alors le dernier zéro avant 1 du mouvement brownien réfléchi $R_t = \sup_{s \leqslant t} X_s - X_t$, $\mathcal{F}$ est la filtration naturelle de R, et l'on est ainsi ramené à la situation étudiée dans l'appendice de [4].

Plus généralement, pour un temps honnête L, comment peut-on construire l'événement $\{\mathrm{Mult}[\mathcal{F}_{L+}|\mathcal{F}_L] = 1\}$, sur lequel les deux tribus coïncident ? Comment passe-t-on de $\mathcal{F}_L$ à $\mathcal{F}_{L+}$? Dans le cas brownien où $\mathrm{Mult}[\mathcal{F}_{L+}|\mathcal{F}_L] \leqslant 2$, les deux petites propositions qui suivent constituent un embryon d'attaque de ces questions.

PROPOSITION 15. — *Étant donné* $(\Omega, \mathcal{A}, \mathbf{P})$, *soit* $\mathcal{B}$ *une sous-tribu de* $\mathcal{A}$. *Pour que* $\mathrm{Mult}[\mathcal{A}|\mathcal{B}] \leqslant 2$ *p. s., il faut et il suffit qu'il existe un événement* $A \in \mathcal{A}$ *tel que* $\mathcal{A} = \mathcal{B} \vee \sigma(A)$. *Lorsque c'est le cas, les événements* A' *tels que l'on ait* $\mathcal{A} = \mathcal{B} \vee \sigma(A')$ *sont exactement les événements de la forme* $A' = A \triangle B$, *où* B *décrit* $\mathcal{B}$.

DÉMONSTRATION. — L'équivalence entre (ix) et (x) dans la proposition 12, avec $n = 2$ et $B = \Omega$, dit que $\mathrm{Mult}[\mathcal{A}|\mathcal{B}] \leqslant 2$ si et seulement s'il existe $A \in \mathcal{A}$ tel que $\mathcal{A} = \mathcal{B} \vee \sigma(A, A^c)$, c'est-à-dire $\mathcal{A} = \mathcal{B} \vee \sigma(A)$.

Pour $A \in \mathcal{A}$ et $B \in \mathcal{B}$, on a $A \triangle B \in \mathcal{B} \vee \sigma(A)$ et $A = (A \triangle B) \triangle B \in \mathcal{B} \vee \sigma(A \triangle B)$; d'où $\mathcal{B} \vee \sigma(A) = \mathcal{B} \vee \sigma(A \triangle B)$. Pour établir la proposition, il reste seulement à vérifier que, si $\mathcal{A} = \mathcal{B} \vee \sigma(A_1) = \mathcal{B} \vee \sigma(A_2)$, alors il existe $B \in \mathcal{B}$ tel que $A_2 = A_1 \triangle B$.

Supposons donc $\mathcal{A} = \mathcal{B} \vee \sigma(A_1) = \mathcal{B} \vee \sigma(A_2)$. L'événement $A_1 \backslash A_2$ est inclus dans A_1 et mesurable pour $\mathcal{B} \vee \sigma(A_1)$, il est donc de la forme $B_1 \cap A_1$ pour un B_1 dans $\mathcal{B}$; inclus dans A_2^c et mesurable pour $\mathcal{B} \vee \sigma(A_2)$, il est aussi de la forme $B_2 \cap A_2^c$ où $B_2 \in \mathcal{B}$. De $A_1 \cap A_2^c = B_1 \cap A_1$, on déduit $B_1 \subset A_1^c \cup A_2^c$; de $A_1 \cap A_2^c = B_2 \cap A_2^c$, on tire $B_2 \subset A_1 \cup A_2$; on peut donc écrire

$$A_1 \backslash A_2 \subset B_1 \cap B_2 \subset (A_1^c \cup A_2^c) \cap (A_1 \cup A_2) = A_1 \triangle A_2 .$$

En posant $B = B_1 \cap B_2$, ceci met en évidence un événement B dans $\mathcal{B}$ tel que

$$A_1 \backslash A_2 \subset B \subset A_1 \triangle A_2 ;$$

par symétrie, il existe de même B' dans $\mathcal{B}$ tel que $A_2 \backslash A_1 \subset B' \subset A_1 \triangle A_2$, et l'on a

$$A_1 \triangle A_2 = (A_1 \backslash A_2) \cup (A_2 \backslash A_1) \subset B \cup B' \subset A_1 \triangle A_2 .$$

Ceci entraîne $A_1 \triangle A_2 = B \cup B'$, puis $A_2 = A_1 \triangle (B \cup B')$. ∎

Même lorsque L est un temps honnête dans la filtration naturelle $\mathcal{F}$ d'un brownien réel X, nous ne savons pas s'il est possible d'écrire explicitement l'un des événements A qui figurent dans la proposition 15 en termes du comportement germe des trajectoires de X à l'instant $L+$. Une formule telle que

$$A = \limsup_{n \to \infty} \{X_{L+1/n} - X_L > 0\}$$

semble plausible, mais, pour qu'elle soit vraie en toute généralité, il faudrait déjà, d'après la proposition précédente, qu'y remplacer X par un autre brownien engendrant la même filtration revienne à y remplacer A par $A \triangle B$ pour un événement $B \in \mathcal{F}_L$; or nous ignorons si cette propriété de stabilité a lieu...

PROPOSITION 16. — *Dans une filtration* $\mathcal{F}$ *dont toutes les martingales sont continues, soit* L *un temps honnête tel que* $\mathrm{Mult}[\mathcal{F}_{L+}|\mathcal{F}_L] \leqslant 2$. *Il existe une martingale uniformément intégrable* M *nulle en* 0, *vérifiant les trois propriétés suivantes :*

(i) $\mathcal{F}_{L+} = \mathcal{F}_L \vee \sigma(\operatorname{sgn} M_\infty)$;

(ii) $\{M_\infty = 0\} = \{\mathrm{Mult}[\mathcal{F}_{L+}|\mathcal{F}_L] = 1\}$;

(iii) *sur l'événement* $\{M_\infty \neq 0\}$, L *est le dernier zéro de* M.

D'après (ii), l'événement $\{M_\infty = 0\}$ est dans $\mathcal{F}_L$, et l'on peut indifféremment convenir dans le (i) que $\operatorname{sgn} 0 = 0$, ou 1, ou -1 sans que cela change la tribu $\mathcal{F}_L \vee \sigma(\operatorname{sgn} M_\infty)$ qui y figure.

DÉMONSTRATION. — Par la proposition 15, il existe un événement $A \in \mathcal{F}_{L+}$ tel que $\mathcal{F}_{L+} = \mathcal{F}_L \vee \sigma(A)$. Soit

$$B = \{\mathbb{P}[A|\mathcal{F}_L] > 0 \text{ et } \mathbb{P}[A^c|\mathcal{F}_L] > 0\} = \{0 < \mathbb{P}[A|\mathcal{F}_L] < 1\} \in \mathcal{F}_L ,$$

de sorte que $B^c = \{\mathbb{P}[A|\mathcal{F}_L] = \mathbb{1}_A\}$. Sur B^c, la v. a. $\mathbb{1}_A$ est égale à $\mathbb{P}[A|\mathcal{F}_L]$, qui est mesurable pour $\mathcal{F}_L$, donc $(\mathcal{F}_{L+})_{|B^c} = (\mathcal{F}_L)_{|B^c}$ et $\mathrm{Mult}[\mathcal{F}_{L+}|\mathcal{F}_L] = 1$ sur B^c par le (ii) de la proposition 10. Sur B, on a $\mathrm{Mult}[\mathcal{F}_{L+}|\mathcal{F}_L] \geqslant 2$ par le point (v) de la proposition 11, donc, sur B, $\mathrm{Mult}[\mathcal{F}_{L+}|\mathcal{F}_L] = 2$ en utilisant l'hypothèse. Posons

$$M_\infty = \frac{\mathbb{1}_{A\cap B}}{\mathbb{P}[A|\mathcal{F}_L]} - \frac{\mathbb{1}_{A^c\cap B}}{\mathbb{P}[A^c|\mathcal{F}_L]} \quad \text{où } \frac{0}{0} = 0 .$$

Le lemme 6 (iii), appliqué avec $n = 2$, dit que M_∞ est la valeur finale de la martingale uniformément intégrable et nulle en zéro $M = \mathcal{O}(M_\infty)J_-$. On a

$$A\cap B = \{M_\infty > 0\} , \quad A^c\cap B = \{M_\infty < 0\} , \quad B^c = \{M_\infty = 0\} .$$

La propriété (i) résulte de la proposition 15 et de

$$\{M_\infty > 0\} = A \cap B = A \,\triangle\, (A\backslash B) = A \,\triangle\, \{\mathbb{P}[A|\mathcal{F}_L] = 1\} .$$

La propriété (ii) vient de $\{M_\infty = 0\} = B^c$; pour vérifier (iii), il suffit de remarquer que sur $]\!]L, \infty[\![$, on a $J_- > 0$ et $\mathcal{O}(M_\infty) = M_\infty$, donc M a sur cet intervalle le même signe (-1, 0 ou 1) que M_∞. ∎

Les corollaires 3 et 4 vont étendre le théorème 3 à des filtrations un peu plus générales que les filtrations browniennes, les filtrations browniennes changées de probabilité (elles ne sont pas nécessairement browniennes : voir [8]) et les filtrations browniennes changées de temps.

PROPOSITION 17. — *Soient $(\Omega, \mathcal{A}, \mathbb{P}, \mathcal{F})$ un espace filtré et $\mathbb{Q}$ une probabilité telle que $\mathbb{Q} \ll \mathbb{P}$. Appelons $\mathcal{A}'$ et $\mathcal{F}'$ les tribu et filtration complétées par tous les événements négligeables pour $\mathbb{Q}$. On a* $\mathrm{Sp\,Mult}(\Omega, \mathcal{A}', \mathbb{Q}, \mathcal{F}') \leqslant \mathrm{Sp\,Mult}(\Omega, \mathcal{A}, \mathbb{P}, \mathcal{F})$. *Si $\mathbb{Q}$ est équivalente à $\mathbb{P}$, on a* $\mathrm{Sp\,Mult}\,(\Omega, \mathcal{A}, \mathbb{Q}, \mathcal{F}) = \mathrm{Sp\,Mult}\,(\Omega, \mathcal{A}, \mathbb{P}, \mathcal{F})$.

DÉMONSTRATION. — Tout ensemble optionnel pour $(\Omega, \mathcal{A}', \mathbb{Q}, \mathcal{F}')$ est indistinguable pour $\mathbb{Q}$ d'un ensemble optionnel pour $(\Omega, \mathcal{A}, \mathbb{P}, \mathcal{F})$; tout temps honnête L pour $(\Omega, \mathcal{A}', \mathbb{Q}, \mathcal{F}')$ est donc égal presque sûrement pour $\mathbb{Q}$ à un temps honnête ℓ pour $(\Omega, \mathcal{A}, \mathbb{P}, \mathcal{F})$. La tribu $\mathcal{F}'_L$ est engendrée par $\mathcal{F}_\ell$ et par les événements négligeables pour $\mathbb{Q}$; de même pour $\mathcal{F}'_{L+}$ et $\mathcal{F}_{\ell+}$. Si $\mathrm{Sp\,Mult}\,(\Omega, \mathcal{A}, \mathbb{P}, \mathcal{F}) = n < \infty$, on a $\mathrm{Mult}[\mathcal{F}_{\ell+}|\mathcal{F}_\ell] \leqslant n$, la proposition 9 c) permet d'en déduire $\mathrm{Mult}[\mathcal{F}'_{L+}|\mathcal{F}'_L] \leqslant n$ (avec égalité si $\mathbb{Q}$ est équivalente à $\mathbb{P}$), et l'on a donc $\mathrm{Sp\,Mult}\,(\Omega, \mathcal{A}', \mathbb{Q}, \mathcal{F}') \leqslant n$, (égalité si $\mathbb{Q} \approx \mathbb{P}$). ∎

COROLLAIRE 3. — *Soient $\mathcal{F}$ une filtration brownienne sur un espace probabilisé $(\Omega, \mathcal{A}, \mathbb{P})$ et $\mathbb{Q}$ une probabilité absolument continue par rapport à $\mathbb{P}$. Appelons $\mathcal{A}'$ et $\mathcal{F}'$ les tribu et filtration complétées par tous les événements négligeables pour $\mathbb{Q}$.*

On a $\mathrm{Sp\,Mult}\,(\Omega, \mathcal{A}', \mathbb{Q}, \mathcal{F}') = 2$.

DÉMONSTRATION. — La proposition 17 et le théorème 3 entraînent immédiatement $\mathrm{Sp\,Mult}\,(\Omega, \mathcal{A}', \mathbb{Q}, \mathcal{F}') \leqslant 2$. Il existe un mouvement brownien X sur $(\Omega, \mathcal{A}, \mathbb{P}, \mathcal{F})$; son transformé de Girsanov est un mouvement brownien sur $(\Omega, \mathcal{A}', \mathbb{Q}, \mathcal{F}')$, et la filtration $\mathcal{F}'$ n'est donc pas constante. D'où $\mathrm{Sp\,Mult}(\Omega, \mathcal{A}', \mathbb{Q}, \mathcal{F}') \neq 1$, et finalement $\mathrm{Sp\,Mult}\,(\Omega, \mathcal{A}', \mathbb{Q}, \mathcal{F}') = 2$. ∎

DÉFINITION. — *Soit I un ensemble dénombrable non vide. Un processus X défini sur un espace filtré $(\Omega, \mathcal{A}, \mathbf{P}, \mathcal{F})$ et à valeurs dans $\mathbf{R}^I$ est une* martingale pure *si*

(i) *chaque coordonnée X^i de X est une martingale locale continue issue de l'origine;*

(ii) *le crochet $C = \langle X^i, X^i \rangle$ ne dépend pas de la coordonnée i (on notera τ l'inverse continu à droite de C ; τ est infini sur l'intervalle aléatoire $[\![C_\infty, \infty[\![$);*

(iii) *il existe un plongement Ψ de $(\Omega, \mathcal{A}, \mathbf{P})$ dans un espace $(\Omega', \mathcal{A}', \mathbf{P}')$ et un mouvement brownien ξ sur $(\Omega', \mathcal{A}', \mathbf{P}')$, indépendant de $\Psi(\mathcal{F}_0)$, tels que $\Psi(C_t)$ soit pour chaque $t \in [0, \infty]$ un temps d'arrêt de la filtration $\Psi(\mathcal{F}_0) \vee \mathrm{Nat}\, \xi$, et que ξ soit égal à $\Psi(X \circ \tau)$ sur l'intervalle stochastique $[\![0, \Psi(C_\infty)[\![$.*

REMARQUES. — Par rapport à la définition usuelle des martingales pures (voir par exemple [18], chapitre V, § 4), il y a ici deux changements : premièrement, X peut être multidimensionnelle, et le changement de temps qui la rend brownienne est alors le même pour tous les indices i; deuxièmement, ce changement de temps peut dépendre de $\mathcal{F}_0$, et pas seulement de ξ lui-même. La pureté ainsi définie ne dépend donc pas uniquement de la loi de X. Par exemple, avec notre définition, si T est une v. a. uniforme sur l'intervalle $[1, 2]$ et si B est un mouvement brownien pour $\mathcal{F}$, indépendant de $\sigma(T) \vee \mathcal{F}_0$, le brownien arrêté B^T est une martingale pure si et seulement si T est mesurable pour $\mathcal{F}_0$, et n'est donc pas pure si T est dans $\mathcal{F}_1$ mais pas dans $\mathcal{F}_0$, ou si T est totalement inaccessible. La raison du rôle un peu bizarre joué par $\mathcal{F}_0$ est purement technique; elle apparaîtra plus bas, lorsque nous parlerons de pureté dans un intervalle $[\![S, T]\!]$. Il sera alors naturel d'autoriser les changements de temps à dépendre de ce qui s'est passé avant S, et, travaillant dans la filtration décalée $\mathcal{G}_t = \mathcal{F}_{S+t}$, nous aurons besoin pour $\mathcal{G}$ d'une notion de pureté qui tienne compte de l'information initiale.

Toute martingale pure X a une limite X_∞ sur l'événement $\{C_\infty < \infty\}$; le processus $X \circ \tau$ est un mouvement brownien arrêté au temps C_∞. Ce processus est défini sur l'espace Ω, mais, lorsque $\mathbf{P}[C_\infty < \infty] > 0$, il se peut que Ω ne soit pas assez riche pour contenir tout un brownien coïncidant avec $X \circ \tau$ sur $[\![0, C_\infty]\!]$. Le rôle de Ψ est de remplacer Ω par un espace Ω' suffisamment gros pour que $X \circ \tau$ ait la place d'y être prolongé en un brownien ξ défini sur $[\![0, \infty[\![$. Lorsque $C_\infty = \infty$ ce plongement n'est pas nécessaire, et l'on a dans ce cas une définition équivalente en prenant dans (iii) $\Omega' = \Omega$ et $\Psi = \mathrm{Id}$.

Remarquer l'orthogonalité des coordonnées d'une martingale pure : $\langle X^i, X^j \rangle = 0$ pour $i \neq j$.

Une situation plus générale est décrite et étudiée dans [12] : On n'y suppose plus (ii) mais on demande que les coordonnées X^i soient orthogonales et que les C_t soient des temps d'arrêt vectoriels de $\mathcal{F}_0 \vee \mathrm{Nat}\,(X \circ \tau)$.

DÉFINITION. — *Une filtration $\mathcal{F}$ sur un espace probabilisé filtré $(\Omega, \mathcal{A}, \mathbf{P})$ est* pure *s'il existe dans $\mathcal{F}$ une martingale pure X telle que l'on ait $\mathcal{F} = \mathcal{F}_0 \vee \mathrm{Nat}\, X$.*

Le petit lemme suivant est bien connu; il intervient dans toutes les utilisations de la pureté. La présence du plongement Ψ dans l'énoncé et dans la démonstration rappelle que nous avons dû grossir l'espace pour que les browniens arrêtés, obtenus par changement de temps des martingales pures, soient de vrais browniens, définis sur $[0, \infty[$ avant d'être arrêtés; le lecteur gêné par ces Ψ peut se munir de blanc et les effacer!

LEMME 7. — *Si $\mathcal{F}$ est pure, en conservant les notations des deux définitions précédentes et en appelant $\hat{\mathcal{F}}$ la filtration* $\Psi(\mathcal{F}_0)\vee\mathrm{Nat}\,\xi$, *on a* $\Psi(\mathcal{F}_t)=\hat{\mathcal{F}}_{\Psi(C_t)}$. *De plus, si M est dans $\mathcal{F}$ une martingale uniformément intégrable, la martingale* $\mathsf{E}'[\Psi(M_\infty)|\hat{\mathcal{F}}_t]$ *de $\hat{\mathcal{F}}$ est égale à* $\Psi(M\circ\tau)$.

DÉMONSTRATION. — Notant d'abord que $\Psi(X_t)=\xi_{\Psi(C_t)}$ est mesurable pour $\hat{\mathcal{F}}_{\Psi(C_t)}$, on en déduit $\Psi(\mathcal{F}_t)\subset\hat{\mathcal{F}}_{\Psi(C_t)}$. Pour établir l'inclusion inverse, observons que $\xi_{s\wedge\Psi(C_t)}=\Psi(X_{\tau_s\wedge t})$ est mesurable pour $\Psi(\mathcal{F}_t)$, donc

$$\hat{\mathcal{F}}_{\Psi(C_t)}=\Psi(\mathcal{F}_0)\vee\sigma(\xi^{\Psi(C_t)})\subset\Psi(\mathcal{F}_t)\,.$$

Si maintenant M est une martingale uniformément intégrable pour la filtration $\mathcal{F}$, la martingale u. i. $\hat{M}_t=\mathsf{E}'[\Psi(M_\infty)|\hat{\mathcal{F}}_t]$ vérifie

$$\begin{aligned}\hat{M}_{\Psi(C_t)}&=\mathsf{E}'[\Psi(M_\infty)|\hat{\mathcal{F}}_{\Psi(C_t)}]=\mathsf{E}'[\Psi(M_\infty)|\Psi(\mathcal{F}_t)]\\&=\Psi\big(\mathsf{E}[M_\infty|\mathcal{F}_t]\big)=\Psi(M_t)\,.\end{aligned}$$

Il en résulte $\hat{M}_{\Psi(C_{\tau_t})}=\Psi(M_{\tau_t})$; en remarquant que $C_{\tau_t}=t\wedge C_\infty$, on en déduit que $\Psi(M_{\tau_t})=\hat{M}_{t\wedge\Psi(C_\infty)}$ et $\Psi(M\circ\tau)$ est une martingale u. i. pour $\hat{\mathcal{F}}$. Sa valeur à l'infini est $\Psi(M_{\tau_\infty})=\Psi(M_\infty)$. ∎

COROLLAIRE 4. — *Toute filtration pure $\mathcal{F}$ vérifie* $\mathrm{Sp\,Mult}\,(\mathcal{F})\leqslant 2$.

DÉMONSTRATION. — Dans $\mathcal{F}$, soit Y une martingale-araignée multiple bornée. Avec les notations des deux définitions qui précèdent, le lemme 7 dit que $\Psi(Y\circ\tau)$ est, dans la filtration $\mathcal{G}=\Psi(\mathcal{F}_0)\vee\mathrm{Nat}\,\xi$, la martingale de valeur finale $\Psi(Y_\infty)$; c'est en particulier une martingale-araignée multiple. Comme $\mathcal{G}$ est brownienne, le théorème 2 donne $\Psi(Y_\infty)=0$. On a donc $Y_\infty=0$, et $Y=0$; ainsi, la condition (i) de la proposition 14 est satisfaite pour $n=2$, d'où $\mathrm{Sp\,Mult}\,(\mathcal{F})\leqslant 2$. ∎

4. Application aux processus de Walsh

Il s'agit de processus introduits par J. Walsh dans l'épilogue de [22]. De nombreuses définitions de ces processus sont possibles; le lecteur trouvera l'une d'entre elles, ainsi que beaucoup de références, dans [4]; il pourra aussi consulter C. Rainer [17] ou E. Vernocchi [21]. La définition proposée ci-dessous est bien sûr équivalente à toutes les autres.

Puisque nous nous intéressons tout spécialement aux filtrations, nous allons définir un processus de Walsh, non pas seulement dans sa filtration naturelle, mais dans une filtration donnée a priori (comme on parle d'un mouvement brownien dans une filtration donnée); le point de vue le plus commode pour cela est la définition par un problème de martingales, en demandant que certaines fonctionnelles du processus soient des martingales dans la filtration considérée. Une fois la définition posée, nous en déduirons une description du processus (propositions 18 et 19); mais nous ne tenterons pas d'en établir l'existence, renvoyant à [4], et aux références qui s'y trouvent, le lecteur intéressé par une construction (ou seulement sceptique quant à l'existence). Bien entendu, l'existence n'est possible que si la filtration s'y prête, de même qu'il n'existe pas toujours de mouvement brownien dans une filtration donnée. Une conséquence directe du théorème 2 est précisément qu'il n'existe pas de processus de Walsh dans une filtration brownienne (corollaire 5).

Soit $(\mathbf{V}, \mathcal{V}, \pi)$ un espace probabilisé (appelé ensemble des rayons) ; pour des raisons de commodité, on supposera que le réel 0 n'est pas élément de $\mathbf{V}$. L'espace dans lequel vivront les processus de Walsh est $\mathbf{S} = \{c\} \cup \big(]0,\infty[\times \mathbf{V}\big)$, où la réunion est disjointe; le point c est appelé centre. L'espace $\mathbf{S}$ est muni de la tribu engendrée par $\{c\}$ et par la tribu produit sur $]0,\infty[\times \mathbf{V}$. On désignera par ρ l'application de $\mathbf{S}$ dans $\mathbb{R}_+$ définie par $\rho(c) = 0$ et $\rho\big((r,v)\big) = r$ pour $r > 0$ et $v \in \mathbf{V}$; et par θ l'application de $\mathbf{S}$ dans $\mathbf{V} \cup \{0\}$ donnée par $\theta(c) = 0$ et $\theta\big((r,v)\big) = v$ pour $r > 0$ et $v \in \mathbf{V}$. Il faut imaginer $\mathbf{S}$ comme une réunion de demi-droites toutes issues du même point c; chacune d'entre elles correspond à un rayon v et est paramétrée par $r \in]0,\infty[$. Les fonctions ρ et θ jouent le rôle de coordonnées polaires sur $\mathbf{S}$.

DÉFINITION. — *Un processus Z, défini sur un espace filtré $(\Omega, \mathcal{A}, \mathbb{P}, \mathcal{F})$ et à valeurs dans* $\mathbf{S}$, *est un processus de Walsh (plus précisément : un processus de Walsh associé à $(\mathbf{V}, \pi)$) si, en posant $R = \rho \circ Z$ et $\Theta = \theta \circ Z$,*

(i) *le processus R est, pour la filtration $\mathcal{F}$, un mouvement brownien réfléchi issu de l'origine ;*

(ii) *pour toute fonction f sur* $\mathbf{V}$, *mesurable, bornée et d'intégrale nulle, le processus $f(\Theta)\,R$ est une martingale de $\mathcal{F}$, continue à droite et pourvue de limites à gauche.*

Le choix de $R_0 = 0$ (et donc $Z_0 = c$) fait en (i) n'est pas indispensable, mais sera commode et nous évitera dans la suite quelques conditionnements par $\mathcal{F}_0$. On pourrait s'en dispenser, par exemple pour s'intéresser aux aspects markoviens de Z; il faudrait alors remplacer « martingale » par « martingale locale » dans le (ii), pour tenir compte du cas où R_0 ne serait pas intégrable.

Les propositions 18 et 19 ci-dessous donneront quelques propriétés des processus de Walsh; voici leur interprétation intuitive. Un processus de Walsh est markovien; sa distance au centre c est le brownien réfléchi R. Conditionnellement à R, la partie angulaire Θ est constante dans chaque excursion de R, a pour loi π, et ses valeurs dans des excursions différentes sont indépendantes. Au début de chaque excursion de R, le processus Z se trouve au centre et choisit au hasard suivant π, sans tenir compte du passé, un rayon qui sera conservé tout au long de l'excursion; et à la fin de l'excursion, Z revient au centre et le rayon qui avait été choisi n'a plus aucune influence sur le comportement ultérieur.

On prend souvent pour $\mathbf{V}$ le cercle unité du plan euclidien; ceci permet d'identifier $\mathbf{S}$ avec le plan au moyen des coordonnées polaires (ρ, θ) et le processus se déplace alors de façon brownienne sur des demi-droites issues de l'origine, en ne passant de l'une à l'autre que via l'origine.

D'importants exemples de processus de Walsh sont les mouvements browniens gauches (skew Brownian motions), introduits par K. Itô et H.P. McKean dans [11] ; ils correspondent au cas où $\mathbf{V} = \{-1, 1\}$ et $\mathbf{S} = \mathbb{R}$. Ils ont même valeur absolue qu'un brownien usuel, et les signes des excursions sont choisis dans $\{-1,1\}$ indépendamment et selon la loi π. Lorsque π est la loi uniforme, on retrouve bien sûr un mouvement brownien ordinaire mais si $\pi(\{1\}) < \frac{1}{2}$ (respectivement $\pi(\{1\}) > \frac{1}{2}$), on a seulement une surmartingale (respectivement une sous-martingale). J.M. Harrison et L.A. Shepp ont montré dans [10], à l'aide du théorème d'unicité trajectorielle de Nakao, que la filtration d'un mouvement brownien gauche est toujours celle d'un brownien linéaire usuel.

Rappelons qu'un brownien réfléchi R admet pour décomposition canonique $R = \beta + L$, où la partie martingale β est un mouvement brownien et où la partie à variation finie L, qui vérifie $L_t = -\inf_{s\leqslant t} \beta_s$, est aussi le demi-temps local en zéro de R, c'est-à-dire son temps local symétrique en zéro, ou encore le temps local en zéro de tout brownien ayant R pour valeur absolue. Réciproquement, tout brownien β est la partie martingale d'un unique brownien réfléchi R, donné par $R_t = \beta_t - \inf_{s\leqslant t} \beta_s$.

PROPOSITION 18. — *Soit Z un processus de Walsh, de coordonnées polaires R et Θ, défini sur $(\Omega, \mathcal{A}, \mathsf{P}, \mathcal{F})$. Appelons L le demi-temps local en 0 de R, c'est-à-dire le processus croissant nul à l'origine tel que $\beta = R - L$ soit un mouvement brownien linéaire.*

(i) *Pour toute fonction f mesurable sur* $\mathbf{V}$, *le processus $f{\circ}\Theta$ reste p. s. constant lors de chaque excursion de R; il est prévisible.*

(ii) *Le processus Z est aussi un processus de Walsh pour la filtration* Nat Z, *et, plus généralement, pour toute filtration intermédiaire entre $\mathcal{F}$ et* Nat Z.

(iii) *Si f est une fonction mesurable bornée sur* $\mathbf{V}$, *la martingale*

$$M^f = \int \mathbb{1}_{\{Z\neq c\}}\, f(\Theta)\, d\beta$$

vaut $f(\Theta)R - \pi(f)L$. En particulier, si $\pi(f) = 0$, la martingale $f(\Theta)R$, qui figure dans la définition des processus de Walsh, est continue.

REMARQUE. — Si l'on suppose en outre que $\mathbf{V}$ est essentiellement séparable et que deux points de $\mathbf{V}$ sont toujours séparés par deux éléments non négligeables de $\mathcal{V}$, le (i) dit que Θ lui-même est constant lors des excursions de R et prévisible. On peut toujours se ramener à ce cas en remplaçant $\mathbf{V}$ par un quotient convenable; nous préférons ne pas le faire ici pour éviter d'alourdir la proposition 20, plus bas.

DÉMONSTRATION. — (i) Soit $s > 0$; posons $D_s = \inf\{t \geqslant s : R_t = 0\}$. Pour $A \in \mathcal{V}$, la fonction $a = \mathbb{1}_A - \pi(A)$ est de moyenne nulle et $M = a(\Theta)\,R$ est donc une martingale. En raison de l'expression de a, les sauts de M ne peuvent prendre que les valeurs $\pm R$. Soient S le temps d'arrêt $\inf\{t > s : \Delta M_t \neq 0\}$ et $T = S\wedge D_s\wedge n$, où n est un entier plus grand que s. Sur l'intervalle $[\![s, T[\![$, M est continue, R est continu et strictement positif, $a{\circ}\Theta = M/R$ est continue, donc constante en raison de l'expression de a, et $M_t = a(\Theta_s)\,R_t$. Ceci permet d'écrire

$$\begin{aligned} M_T - M_s &= a(\Theta_s)\,(R_T - R_s) + \Delta M_T \\ &= a(\Theta_s)\,(R_T - R_s) + \mathbb{1}_{\{\Delta M_T\neq 0\}}(-1)^{\mathbb{1}_A\circ\Theta_s} R_T\,. \end{aligned}$$

Puisque T est majoré par D_s et borné, $\mathsf{E}[M_T - M_s|\mathcal{F}_s]$ et $\mathsf{E}[R_T - R_s|\mathcal{F}_s]$ sont nuls; la formule ci-dessus entraîne $\mathsf{E}[\mathbb{1}_{\{\Delta M_T\neq 0\}} R_T|\mathcal{F}_s] = 0$, donc $\mathbb{1}_{\{\Delta M_T\neq 0\}} R_T = 0$ p. s. et $T = D_s$ sur $\{\Delta M_T \neq 0\}$. Mais $\Delta M_{D_s} = \pm R_{D_s} = 0$; sur $\{\Delta M_T \neq 0\}$ on devrait avoir $\Delta M_T = \Delta M_{D_s} = 0$; ainsi, $\Delta M_T = 0$ p. s., d'où $S \geqslant D_s\wedge n$, $T \geqslant D_s\wedge n$, et $a{\circ}\Theta$ est constant sur $[\![s, D_s\wedge n[\![$. En faisant varier s parmi les rationnels et n parmi les entiers, on obtient que $\mathbb{1}_A{\circ}\Theta$ est constant sur chaque excursion de R; l'extension des ensembles aux fonctions est immédiate.

La prévisibilité de $f{\circ}\Theta$ s'obtient en écrivant $f{\circ}\Theta = \lim_{n\to\infty} f{\circ}\Theta\,\mathbb{1}_{\{R\geqslant 1/n\}}$.

(ii) Si l'on remplace $\mathcal{F}$ par une filtration plus petite, mais à laquelle Z est encore adapté, les processus β et $f{\circ}\Theta\, R$ restent adaptés, donc restent des martingales (pour $\pi(f) = 0$), et Z reste un processus de Walsh.

(iii) Lorsque f est constante, la formule $\int f\circ\Theta\, d\beta = f\circ\Theta\, R - \pi(f)\,L$ découle aussitôt de $R = \beta + L$; il suffit donc de la vérifier lorsque $\pi(f) = 0$. Dans ce cas, le processus $M = f\circ\Theta\, R$ est une martingale, et même une martingale continue : elle est continue dans les excursions de R car $f\circ\Theta$ y est constant, et elle est continue en tout zéro de R car $|M| \leqslant (\sup|f|)\,|R|$. On a donc, par le lemme 3, $M = \int \mathbb{1}_{\{M\neq 0\}} dM$. Mais l'ouvert prévisible $\{M \neq 0\}$ est inclus dans la réunion des ouverts $]\!]t, D_t[\![$, où t décrit les rationnels et où $D_t = \inf\{u \geqslant t : R_u = 0\}$. Sur un tel ouvert, on peut écrire $dM = f(\Theta_t)\, dR = f(\Theta_t)\, d\beta = f\circ\Theta\, d\beta$. Ainsi, $\mathbb{1}_{\{M\neq 0\}} dM = \mathbb{1}_{\{M\neq 0\}} f\circ\Theta\, d\beta = f\circ\Theta\, \mathbb{1}_{\{R\neq 0\}}\, d\beta$ et $M = \int f\circ\Theta\, d\beta$. ∎

Les quelques propriétés des processus de Walsh décrites dans la proposition 18 sont celles dont nous aurons besoin dans la suite, mais elles ne permettent guère de se représenter ces processus. La proposition 19 ci-dessous déduit la loi d'un processus de Walsh de la définition donnée plus haut, et établit donc l'équivalence entre cette définition et celles données par d'autres auteurs. Compte tenu du sujet qui nous occupe, il nous semble intéressant de décrire le comportement de la filtration aux instants ou le processus de Walsh quitte le centre c; ceci nécessitera le lemme 8 ci-dessous. Ce lemme propose des variations sur un thème bien connu (voir par exemple l'exercice V-(4.16) de [18]). Si nous le rappelons ici, c'est parce que le b) et sa conséquence (c.iii), bien que très faciles, ne sont pas classiques, et surtout parce que le point (c.iv) nous servira dans la proposition 19 pour construire le semi-groupe de transition d'un processus de Walsh (sans recourir à la théorie des excursions).

La condition (c.i) du lemme est l'*hypothèse* H de la littérature; il est bien connu, et élémentaire, que dans (c.i) on peut remplacer « martingale bornée » par « martingale » ou par « martingale locale ».

LEMME 8. — *L'espace probabilisé* $(\Omega, \mathcal{A}, \mathbf{P})$ *est fixé. Nous appelons* $\mathbf{E}^{\mathcal{B}}$ *l'opérateur d'espérance conditionnelle par rapport à une sous-tribu* $\mathcal{B}$ *et* $\mathbb{O}^{\mathcal{F}}$ *l'opérateur de projection optionnelle associé à une filtration* $\mathcal{F}$; $\mathrm{Opt}_{\mathrm{b}}(\mathcal{F})$ *désignera l'espace de tous les processus bornés et optionnels pour* $\mathcal{F}$.

a) *Soient* $\mathcal{B}$, $\mathcal{C}$ *et* $\mathcal{D}$ *trois sous-tribus de* $\mathcal{A}$ *telles que* $\mathcal{B} \subset \mathcal{C} \cap \mathcal{D}$. *Les trois conditions suivantes sont équivalentes (et traduisent toutes l'indépendance conditionnelle de* $\mathcal{C}$ *et* $\mathcal{D}$ *par rapport à* $\mathcal{B}$) :

(a.i) $$\mathbf{E}^{\mathcal{C}}\mathbf{E}^{\mathcal{D}} = \mathbf{E}^{\mathcal{B}}\ ;$$

(a.ii) $$\mathbf{E}^{\mathcal{D}}\mathbf{E}^{\mathcal{C}} = \mathbf{E}^{\mathcal{B}}\ ;$$

(a.iii) $$\forall C \in \mathrm{L}^\infty(\mathcal{C}) \quad \forall D \in \mathrm{L}^\infty(\mathcal{D}) \quad \mathbf{E}^{\mathcal{B}}(CD) = \mathbf{E}^{\mathcal{B}}(C)\,\mathbf{E}^{\mathcal{B}}(D)\,.$$

Quand ces conditions sont satisfaites, $\mathcal{B} = \mathcal{C} \cap \mathcal{D}$.

b) *Soient* $\mathcal{F}$, $\mathcal{G}$ *et* $\mathcal{H}$ *trois filtrations sur* $(\Omega, \mathcal{A}, \mathbf{P})$ *telles que* $\mathcal{F} \subset \mathcal{G} \cap \mathcal{H}$. *Les trois conditions suivantes sont équivalentes :*

(b.i) $$\mathbb{O}^{\mathcal{G}}\mathbb{O}^{\mathcal{H}} = \mathbb{O}^{\mathcal{F}}\ ;$$

(b.ii) $$\mathbb{O}^{\mathcal{H}}\mathbb{O}^{\mathcal{G}} = \mathbb{O}^{\mathcal{F}}\ ;$$

(b.iii) $$\forall G \in \mathrm{Opt}_{\mathrm{b}}(\mathcal{G}) \quad \forall H \in \mathrm{Opt}_{\mathrm{b}}(\mathcal{H}) \quad \mathbb{O}^{\mathcal{F}}(GH) = \mathbb{O}^{\mathcal{F}}(G)\,\mathbb{O}^{\mathcal{F}}(H)\,.$$

Quand ces conditions sont satisfaites, $\mathcal{F} = \mathcal{G} \cap \mathcal{H}$.

c) *Soient* $\mathcal{F}$ *et* $\mathcal{G}$ *deux filtrations telles que* $\mathcal{F} \subset \mathcal{G}$. *Les cinq conditions suivantes sont équivalentes :*

(c.i) *toute martingale bornée de* $\mathcal{F}$ *est aussi une martingale bornée de* $\mathcal{G}$;

(c.ii) *pour tout* $t \geqslant 0$ *les tribus* $\mathcal{G}_t$ *et* $\mathcal{F}_\infty$ *sont conditionnellement indépendantes par rapport à* $\mathcal{F}_t$;

(c.iii) *pour toute variable aléatoire* τ *mesurable pour* $\mathcal{F}_\infty$ *et à valeurs dans* $[0,\infty]$, *les tribus* $\mathcal{G}_\tau$ *et* $\mathcal{F}_\infty$ *sont conditionnellement indépendantes par rapport à* $\mathcal{F}_\tau$;

(c.iv) *pour tout temps* L *honnête pour* $\mathcal{F}$ *(donc aussi pour* $\mathcal{G}$*), les tribus* $\mathcal{G}_{L+}$ *et* $\mathcal{F}_\infty$ *sont conditionnellement indépendantes par rapport à* $\mathcal{F}_{L+}$;

(c.v) *en appelant* $\mathcal{H}$ *la filtration constante telle que* $\mathcal{H}_t = \mathcal{F}_\infty$ *pour tout* t, *les conditions* (b.i), (b.ii) *et* (b.iii) *sont satisfaites.*

DÉMONSTRATION DU LEMME 8. — Exercice tout à fait classique, le a) rappelle trois définitions équivalentes de l'indépendance conditionnelle; il serait d'autant plus superflu de le redémontrer ici qu'il se déduit facilement du b).

b) Nous commençons par (b.i) $\Rightarrow$ (b.iii). Soient $G \in \mathrm{Opt_b}(\mathcal{G})$ et $H \in \mathrm{Opt_b}(\mathcal{H})$; on a $\mathbf{O}^{\mathcal{G}}(H) = \mathbf{O}^{\mathcal{G}}\mathbf{O}^{\mathcal{H}}(H) = \mathbf{O}^{\mathcal{F}}(H)$. Si F est un processus croissant, borné et optionnel pour $\mathcal{F}$, le processus $\int G\,dF$ est optionnel pour $\mathcal{G}$ et de variation totale finie, et l'on a donc l'égalité $\mathbb{E}\big[\int_0^\infty HG\,dF\big] = \mathbb{E}\big[\int_0^\infty \mathbf{O}^{\mathcal{G}}(H)\,G\,dF\big] = \mathbb{E}\big[\int_0^\infty \mathbf{O}^{\mathcal{F}}(H)\,G\,dF\big]$. Ceci ayant lieu pour tout F, il en résulte $\mathbf{O}^{\mathcal{F}}(GH) = \mathbf{O}^{\mathcal{F}}\big(G\mathbf{O}^{\mathcal{F}}(H)\big) = \mathbf{O}^{\mathcal{F}}(G)\,\mathbf{O}^{\mathcal{F}}(H)$, c'est-à-dire (b.iii).

(b.iii) $\Rightarrow$ (b.ii). Soient T un temps d'arrêt pour $\mathcal{H}$ et G un processus borné, adapté à $\mathcal{G}$ et continu à droite. Le processus $H = \mathbb{1}_{\{T>s\}}\mathbb{1}_{[\![T,\infty[\![}$ est optionnel pour $\mathcal{H}$ et l'hypothèse (b.iii) donne, pour $s < t$,

$$\begin{aligned}\mathbb{E}[G_t\mathbb{1}_{\{s<T\leqslant t\}}] &= \mathbb{E}[G_tH_t] = \mathbb{E}\big[\mathbf{O}^{\mathcal{F}}(GH)_t\big] = \mathbb{E}\big[\mathbf{O}^{\mathcal{F}}(G)_t\,\mathbf{O}^{\mathcal{F}}(H)_t\big]\\ &= \mathbb{E}\Big[\mathbf{O}^{\mathcal{F}}\big(\mathbf{O}^{\mathcal{F}}(G)H\big)_t\Big] = \mathbb{E}\big[\mathbf{O}^{\mathcal{F}}(G)_tH_t\big] = \mathbb{E}\big[\mathbf{O}^{\mathcal{F}}(G)_t\mathbb{1}_{\{s<T\leqslant t\}}\big]\,.\end{aligned}$$

En écrivant ceci pour les instants $s = (k-1)2^{-n}$ et $t = k2^{-n}$, en sommant sur $k \geqslant 0$ et en appelant T_n le plus petit dyadique d'ordre n supérieur ou égal à T, on a $\mathbb{E}[G_{T_n}\mathbb{1}_{\{T<\infty\}}] = \mathbb{E}\big[\mathbf{O}^{\mathcal{F}}(G)_{T_n}\mathbb{1}_{\{T<\infty\}}\big]$. Puisque G est continu à droite, $\mathbf{O}^{\mathcal{F}}(G)$ l'est également (voir le théorème VI-47 de Dellacherie-Meyer [6]), et l'on a à la limite $\mathbb{E}[G_T\mathbb{1}_{\{T<\infty\}}] = \mathbb{E}\big[\mathbf{O}^{\mathcal{F}}(G)_T\mathbb{1}_{\{T<\infty\}}\big]$. Comme T est un temps d'arrêt arbitraire de $\mathcal{H}$, on en déduit $\mathbf{O}^{\mathcal{F}}(G) = \mathbf{O}^{\mathcal{H}}(G)$. Par classes monotones, ceci s'étend à tout $G \in \mathrm{Opt_b}(\mathcal{G})$, d'où $\mathbf{O}^{\mathcal{F}} = \mathbf{O}^{\mathcal{H}}\mathbf{O}^{\mathcal{G}}$.

Nous venons de voir (b.i) $\Rightarrow$ (b.iii) $\Rightarrow$ (b.ii); on en tire (b.i) $\Rightarrow$ (b.ii), et (b.ii) $\Rightarrow$ (b.i) s'en déduit en échangeant les rôles de $\mathcal{G}$ et $\mathcal{H}$.

Supposons satisfaites les trois conditions (b). Pour $A \in \mathcal{G}_t \cap \mathcal{H}_t$, la v. a. $T = t + \mathbb{1}_A$ est un temps d'arrêt pour $\mathcal{G}$ et pour $\mathcal{H}$, le processus $\mathbb{1}_{[\![T,\infty[\![}$ est optionnel pour ces deux filtrations, donc aussi, par (b.i), pour $\mathcal{F}$, et T est un temps d'arrêt pour $\mathcal{F}$. En conséquence, $A \in \mathcal{F}_t$; et l'on a $\mathcal{F}_t = \mathcal{G}_t \cap \mathcal{H}_t$.

c) Tout d'abord (c.iv) $\Rightarrow$ (c.ii) est trivial, car tout temps fixe t est honnête, et vérifie $\mathcal{F}_{t+} = \mathcal{F}_t$.

(c.ii) $\Rightarrow$ (c.i) est classique et facile : Si M est une martingale bornée pour $\mathcal{F}$, il existe $U \in \mathrm{L}^\infty(\mathcal{F}_\infty)$ tel que $M_t = \mathbb{E}[U|\mathcal{F}_t]$; en utilisant l'indépendance conditionnelle sous la forme (a.i), on a $M_t = \mathbb{E}^{\mathcal{F}_t}(U) = \mathbb{E}^{\mathcal{G}_t}\mathbb{E}^{\mathcal{F}_\infty}(U) = \mathbb{E}^{\mathcal{G}_t}(U)$ et M est aussi une martingale pour $\mathcal{G}$.

(c.i) $\Rightarrow$ (c.v). Supposons que toute martingale bornée pour $\mathcal{F}$ est une martingale pour $\mathcal{G}$; appelons $\mathcal{H}$ la filtration constante telle que $\mathcal{H}_t = \mathcal{F}_\infty$. Soit $U \in \mathrm{L}^\infty(\mathcal{F}_\infty)$; le processus $M_t = \mathbb{E}[U|\mathcal{F}_t]$ est une martingale continue à droite pour $\mathcal{F}$ et donc aussi pour $\mathcal{G}$, d'où $M_t = \mathbb{E}[U|\mathcal{G}_t]$; si f est une fonction borélienne bornée, le processus $X_t(\omega) = f(t)U(\omega)$ vérifie $\mathbf{O}^{\mathcal{F}}(X) = fM = \mathbf{O}^{\mathcal{G}}(X)$. Par classes monotones, on en tire $\mathbf{O}^{\mathcal{F}}(X) = \mathbf{O}^{\mathcal{G}}(X)$ pour tout processus X borné et mesurable pour la tribu-produit $\mathrm{Borel}(\mathbb{R}_+)\otimes\mathcal{F}_\infty$. La condition (b.i) est ainsi satisfaite, (b.ii) et (b.iii) en découlent.

(c.v) $\Rightarrow$ (c.iii). Si τ est une v. a. mesurable pour $\mathcal{F}_\infty$ et à valeurs dans $[0,\infty]$, c'est un temps d'arrêt de $\mathcal{H}$ et l'on a bien sûr $\mathcal{H}_\tau = \mathcal{F}_\infty$. Tout $U \in \mathrm{L}^\infty(\mathcal{G}_\tau)$ est de la forme $G_\tau\mathbb{1}_{\{\tau<\infty\}} + V\mathbb{1}_{\{\tau=\infty\}}$ où $G \in \mathrm{Opt_b}(\mathcal{G})$ et $V \in \mathrm{L}^\infty(\mathcal{G}_\infty)$. On peut donc écrire, en appliquant (b.ii) au processus G,

$$\begin{aligned}\mathbb{E}^{\mathcal{F}_\infty}(U) &= \mathbb{E}^{\mathcal{F}_\infty}[G_\tau\mathbb{1}_{\{\tau<\infty\}}] + \mathbb{E}^{\mathcal{F}_\infty}[V\,\mathbb{1}_{\tau=\infty}] = \mathbb{E}^{\mathcal{H}_\tau}[G_\tau\mathbb{1}_{\{\tau<\infty\}}] + \mathbb{E}^{\mathcal{F}_\infty}[V\,\mathbb{1}_{\tau=\infty}]\\ &= \big(\mathbf{O}^{\mathcal{H}}(G)\big)_\tau\,\mathbb{1}_{\{\tau<\infty\}} + \mathbb{E}^{\mathcal{F}_\infty}(V)\,\mathbb{1}_{\tau=\infty} = \big(\mathbf{O}^{\mathcal{F}}(G)\big)_\tau\,\mathbb{1}_{\{\tau<\infty\}} + \mathbb{E}^{\mathcal{F}_\infty}(V)\,\mathbb{1}_{\tau=\infty}\,.\end{aligned}$$

Ceci montre que $\mathbb{E}^{\mathcal{F}_\infty}(U)$ est mesurable pour $\mathcal{F}_\tau$, d'où $\mathbb{E}^{\mathcal{F}_\infty}(U) = \mathbb{E}^{\mathcal{F}_\tau}(U)$, c'est-à-dire $\mathbb{E}^{\mathcal{F}_\infty}\mathbb{E}^{\mathcal{G}_\tau} = \mathbb{E}^{\mathcal{F}_\tau}\mathbb{E}^{\mathcal{G}_\tau} = \mathbb{E}^{\mathcal{F}_\tau}$. Ainsi, la condition (a.ii) est satisfaite par les trois tribus $\mathcal{F}_\tau$, $\mathcal{G}_\tau$ et $\mathcal{F}_\infty$.

(c.iii) $\Rightarrow$ (c.iv). Ceci résulte aussitôt de la proposition 13 b), en prenant $\tau = L + \frac{1}{n}$ dans la relation $\mathbb{E}^{\mathcal{F}_\tau} = \mathbb{E}^{\mathcal{G}_\tau}\mathbb{E}^{\mathcal{F}_\infty}$ et en faisant tendre n vers l'infini. ∎

PROPOSITION 19. — *Soit Z un processus de Walsh défini sur $(\Omega, \mathcal{A}, \mathbb{P}, \mathcal{F})$, associé à $(\mathbf{V}, \mathcal{V}, \pi)$. Appelons R et Θ ses coordonnées polaires, et β la partie martingale de R.*

(i) *Le processus Z est markovien par rapport à $\mathcal{F}$. Pour $0 \leqslant s < t$, si f et g sont des fonctions mesurables bornées (ou toutes deux positives) sur $\mathbf{V} \cup \{0\}$ et $\mathbb{R}_+$ respectivement, en posant $D_s = \inf\{t \geqslant s : R_t = 0\}$ et*[1]

$$\gamma_t(x) = \frac{1}{\sqrt{2\pi t}}\, e^{-\frac{x^2}{2t}} ,$$

$$a(x,g) = \int_0^\infty g(x)\left[\gamma_{t-s}(x-R_s) - \gamma_{t-s}(x+R_s)\right] dx ,$$

$$b(x,g) = \int_0^\infty g(x)\, 2\,\gamma_{t-s}(x+R_s)\, dx ,$$

on a

$$\begin{aligned}\mathbb{E}[f(\Theta_t)g(R_t)|\mathcal{F}_s] &= f(\Theta_s)\,\mathbb{E}[g(R_t)\mathbb{1}_{\{D_s \geqslant t\}}|\mathcal{F}_s] + \pi(f)\,\mathbb{E}[g(R_t)\mathbb{1}_{\{D_s < t\}}|\mathcal{F}_s] \\ &= f(\Theta_s)\, a(R_s, g) + \pi(f)\, b(R_s, g) .\end{aligned}$$

(ii) *Le mouvement brownien β possède la propriété de représentation prévisible dans la filtration* Nat Z. *En particulier, toutes les martingales de* Nat Z *sont continues.*

(iii) *Pour $t > 0$, posons $G_t = \sup\{s \leqslant t : Z_s = c\} = \sup\{s \leqslant t : R_s = 0\}$. La variable aléatoire Θ_T est indépendante de $\mathcal{F}_{G_t}$ et a pour loi π.*

DÉMONSTRATION. — (i) Comme R est un mouvement brownien réfléchi pour $\mathcal{F}$, Les formules $\mathbb{E}[g(R_t)\mathbb{1}_{\{D_s \geqslant t\}}|\mathcal{F}_s] = a(R_s, g)$ et $\mathbb{E}[g(R_t)\mathbb{1}_{\{D_s < t\}}|\mathcal{F}_s] = b(R_s, g)$ résultent du principe de réflexion. Prenons f et g bornées. Pour établir la formule de semi-groupe

$$(*) \qquad \mathbb{E}[f(\Theta_t)g(R_t)|\mathcal{F}_s] = f(\Theta_s)\,\mathbb{E}[g(R_t)\mathbb{1}_{\{D_s \geqslant t\}}|\mathcal{F}_s] + \pi(f)\,\mathbb{E}[g(R_t)\mathbb{1}_{\{D_s < t\}}|\mathcal{F}_s] ,$$

nous allons procéder en plusieurs étapes. Posons $G = \sup\big([0,t] \cap \{R=0\}\big)$ et appelons $\mathcal{R}$ la filtration naturelle de R; c'est aussi la filtration naturelle de β.

1) En décomposant R en un pont sur $[0, G]$ et un méandre sur $[G, t]$, et en utilisant le lemme de Lindvall-Rogers, on vérifie que $\mathcal{R}_{G+} = \mathcal{R}_G$ (voir les détails dans l'appendice de [4]).

2) Toute martingale de $\mathcal{R}$ est une intégrale stochastique par rapport à β, lui-même mouvement brownien de $\mathcal{F}$; les martingales de $\mathcal{R}$ sont donc des martingales de $\mathcal{F}$, et, par le (c.iv) du lemme 8, les tribus $\mathcal{F}_{G+}$ et $\mathcal{R}_\infty$ sont conditionnellement indépendantes par rapport à $\mathcal{R}_{G+}$, c'est-à-dire par rapport à $\mathcal{R}_G$ (étape précédente). Ainsi, pour $V \in \mathrm{L}^1(\mathcal{R}_\infty)$, on a $\mathbb{E}[V|\mathcal{F}_{G+}] = \mathbb{E}[V|\mathcal{R}_G]$; puisque $\mathcal{R}_G \subset \mathcal{F}_G \subset \mathcal{F}_{G+}$, cela entraîne $\mathbb{E}[V|\mathcal{F}_{G+}] = \mathbb{E}[V|\mathcal{F}_G]$. Pour $U \in \mathrm{L}^\infty(\mathcal{F}_{G+})$ et $V \in \mathrm{L}^1(\mathcal{R}_\infty)$, on a donc

$$\mathbb{E}[UV|\mathcal{F}_G] = \mathbb{E}\big[U\mathbb{E}[V|\mathcal{F}_{G+}]\,\big|\,\mathcal{F}_G\big] = \mathbb{E}\big[U\mathbb{E}[V|\mathcal{F}_G]\,\big|\,\mathcal{F}_G\big] = \mathbb{E}[U|\mathcal{F}_G]\,\mathbb{E}[V|\mathcal{F}_G] .$$

3) Si maintenant f est une fonction mesurable et bornée sur $\mathbf{V}$, de moyenne $\pi(f)$ nulle, le processus $M = f\circ\Theta\, R$, égal à $\int f\circ\Theta\, d\beta$ par la proposition 18, est une martingale continue. Cette martingale vérifie $M_G = 0$, donc aussi $\mathbb{E}[M_t|\mathcal{F}_G] = 0$ (voir [2], [3] ou [25]). Par l'étape précédente, ceci s'écrit $\mathbb{E}[f(\Theta_t)|\mathcal{F}_G]\mathbb{E}[R_t|\mathcal{F}_G] = 0$. Mais $\mathbb{E}[R_t|\mathcal{F}_G] = c\sqrt{t-G_t} > 0$ p. s.; il reste $\mathbb{E}[f(\Theta_t)|\mathcal{F}_G] = 0$. En utilisant à nouveau l'étape précédente, on obtient $\mathbb{E}[f(\Theta_t)\, V|\mathcal{F}_G] = 0$ pour tout $V \in \mathrm{L}^1(\mathcal{R}_\infty)$.

4) Toujours pour $\pi(f) = 0$, si g est bornée on a

$$\begin{aligned}\mathbb{E}[f(\Theta_t)\, g(R_t)\mathbb{1}_{\{G>s\}}|\mathcal{F}_s] &= \mathbb{E}\big[\mathbb{1}_{\{G>s\}}\mathbb{E}[f(\Theta_t)\, g(R_t)|\mathcal{F}_{G\vee s}]\,\big|\,\mathcal{F}_s\big] \\ &= \mathbb{E}\big[\mathbb{1}_{\{G>s\}}\mathbb{E}[f(\Theta_t)\, g(R_t)|\mathcal{F}_G]\,\big|\,\mathcal{F}_s\big] ;\end{aligned}$$

ceci est nul par l'étape précédente, et

$$\begin{aligned}\mathbb{E}[f(\Theta_t)\, g(R_t)|\mathcal{F}_s] &= \mathbb{E}[f(\Theta_t)\, g(R_t)\mathbb{1}_{\{G\leqslant s\}}|\mathcal{F}_s] \\ &= \mathbb{E}[f(\Theta_s)\, g(R_t)\mathbb{1}_{\{G\leqslant s\}}|\mathcal{F}_s] = f(\Theta_s)\,\mathbb{E}[\, g(R_t)\mathbb{1}_{\{G\leqslant s\}}|\mathcal{F}_s] .\end{aligned}$$

1. Dans la formule donnant γ_t, le symbole π ne désigne pas la probabilité sur $\mathbf{V}$!

Ceci établit (*) lorsque $\pi(f) = 0$; le cas où f est constante est trivial, le cas général s'en déduit par addition.

Le caractère markovien de Z dans la filtration $\mathcal{F}$ (et a fortiori dans sa filtration naturelle) se déduit immédiatement de cette formule de semi-groupe par un argument de classes monotones.

(ii) Pour établir la propriété de représentation prévisible, appelons $\mathcal{Z}$ la filtration Nat Z et désignons par $\mathcal{C}(\Omega, \mathcal{Z}_\infty, \mathbb{P}, \mathcal{Z}, \beta)$ l'ensemble de toutes les probabilités $\mathbb{Q}$ sur $(\Omega, \mathcal{Z}_\infty)$ telles que $\mathbb{Q} \ll \mathbb{P}$ et que β est une martingale pour $\mathbb{Q}$. Pour une telle $\mathbb{Q}$, le processus $R_t = \beta_t - \inf_{s\leqslant t} \beta_s$ est un brownien réfléchi sur $(\Omega, \mathcal{Z}_\infty, \mathbb{Q}, \mathcal{Z})$; pour f telle que $\pi(f) = 0$, le processus $f\circ\Theta\, R$, qui est aussi égal à $\int f\circ\Theta\, d\beta$, est une martingale pour $(\Omega, \mathcal{Z}_\infty, \mathbb{Q}, \mathcal{Z})$. Sur cet espace filtré, Z est donc un processus de Walsh et sa loi est donnée par la formule de semi-groupe ci-dessus. Ceci entraîne que $\mathbb{Q}$ est la restriction de $\mathbb{P}$ à $\mathcal{Z}_\infty$. Ainsi, $\mathcal{C}(\Omega, \mathcal{Z}_\infty, \mathbb{P}, \mathcal{Z}, \beta)$ contient un seul élément, $\mathbb{P}_{|\mathcal{Z}_\infty}$, qui est évidemment un point extrémal de $\mathcal{C}(\Omega, \mathcal{Z}_\infty, \mathbb{P}, \mathcal{Z}, \beta)$; ceci montre que β possède la propriété de représentation prévisible dans $\mathcal{Z}$.

(iii) Fixons $t > 0$. Soit f bornée telle que $\pi(f) = 0$. En prenant $g = 1$ dans le (i), on voit que la martingale continue et bornée $M_s = \mathbb{E}[f(\Theta_t)|\mathcal{F}_s]$ est nulle à l'instant G_t. La proposition 13 c) donne donc $\mathbb{E}[f(\Theta_t)|\mathcal{F}_{G_t}] = \mathbb{E}[M_t|\mathcal{F}_{G_t}] = 0$, d'où le résultat. ∎

PROPOSITION 20. — *Soient $(\mathbf{V}, \mathcal{V})$ et $(\mathbf{V}', \mathcal{V}')$ deux espaces mesurables et ϕ une application mesurable de $\mathbf{V}$ vers $\mathbf{V}'$. Définissons une application mesurable $\tilde{\phi}$ de $\mathbf{S}$ vers $\mathbf{S}'$ par $\tilde{\phi}(c) = c'$ et $\tilde{\phi}\big((r,v)\big) = \big(r, \phi(v)\big)$, et appelons π' la mesure image $\pi\circ\phi^{-1}$ de π par ϕ. Si Z est, dans une filtration, un processus de Walsh associé à $(\mathbf{V}, \pi)$, $\tilde{\phi}\circ Z$ est, dans cette filtration, un processus de Walsh associé à $(\mathbf{V}', \pi')$.*

DÉMONSTRATION. — Cela résulte facilement de la définition des processus de Walsh : si f' est une fonction bornée sur $\mathbf{V}'$ telle que $\pi'(f') = 0$, la fonction $f = f'\circ\phi$ vérifie $\pi(f) = 0$ par définition de π', et $f'(\Theta')R' = f'\big(\phi\circ\Theta\big)R = f(\Theta)R$ est donc une martingale càdlàg. ∎

MOUVEMENTS WALSHIENS. — Il existe une variante des processus de Walsh, légèrement différente; pour la distinguer de la précédente, on pourrait l'appeler *mouvement walshien.* On prend pour $\mathbf{V}$ un espace vectoriel, que nous supposerons de dimension finie pour éviter des détails techniques sans intérêt ici. Si $Z = (R, \Theta)$ est un processus de Walsh associé à $(\mathbf{V}, \pi)$ au sens précédent, le mouvement walshien associé à $(\mathbf{V}, \pi)$ est le processus à valeurs dans $\mathbf{V}$ lui-même (et non pas dans $\mathbf{S}$), égal à $R\Theta$ quand $R \neq 0$ et au vecteur nul $\vec{0} \in \mathbf{V}$ quand $R = 0$ $(= \Theta)$. Attention! Le processus Θ, à valeurs dans $\mathbf{V}$, est bien un vecteur, et non pas une variable angulaire. Les mouvements walshiens sont des processus continus dans $\mathbf{V}$.

Si X est un mouvement brownien gauche mais pas un mouvement brownien, $|X|$ est un exemple de mouvement walshien qui n'est pas un processus de Walsh

Quand π est portée par une sphère de $\mathbf{V}$ (c'est le cas considéré par J. Walsh, ainsi que dans [4]), les deux notions de mouvement walshien et de processus de Walsh sont équivalentes. Mais quand $\pi\times\pi$ charge l'ensemble des couples $(x, y) \in \mathbf{V}\times\mathbf{V}$ tels que les vecteurs x et y sont colinéaires et de même sens, comme dans le cas de $|X|$ ci-dessus, cela fait une grosse différence : les processus walshiens ne sont en général pas markoviens car la seule position du processus détermine la demi-droite sur laquelle il se trouve mais non le coefficient de diffusion (variation quadratique) valable pour toute l'excursion en cours. Par contre, pour les questions de filtration qui nous intéresseront plus bas, la distinction ne s'impose plus puisque les filtrations sont les mêmes (du moins si $\pi(\{\vec{0}\}) = 0$); en effet, la position et la variation quadratique instantanées d'un mouvement walshien $R\Theta$ déterminent alors sans équivoque le processus de Walsh (R, Θ) dont il est issu.

PROPOSITION 21. — *Dans un espace vectoriel $\mathbf{V}$ de dimension finie, soit W un mouvement walshien associé à une probabilité π.*

Pour que W soit une semimartingale, il faut et il suffit que les moments d'ordre 1 de π existent. En ce cas, l'intégrale stochastique $\int \Theta\, d\beta$ existe et est une martingale, et W a pour décomposition canonique $\int \Theta\, d\beta + b(\pi)L$, où $b(\pi) = \int_{\mathbf{V}} x\, \pi(dx)$ est le barycentre de π.

En particulier, W est une martingale locale si et seulement si π est centrée; et W est alors une martingale.

DÉMONSTRATION. — Supposant que π a des moments d'ordre 1, nous allons d'abord montrer que $W - b(\pi)L$ est une martingale.

Pour $0 \leqslant s \leqslant t$ et pour ϕ mesurable bornée sur $\mathbf{V}$, puisque $\phi(\Theta)R - \pi(\phi)L$ est une martingale, on peut écrire $\mathbb{E}[\phi(\Theta_t)R_t|\mathcal{F}_s] = \phi(\Theta_s)R_s + \pi(\phi)\mathbb{E}[L_t - L_s|\mathcal{F}_s]$; cette formule s'étend par limites croissantes à toute fonction ϕ mesurable positive. Prenant d'abord pour ϕ une norme sur $\mathbf{V}$, on en déduit $\mathbb{E}[\|W_t\|] = \int \|x\|\,\pi(dx)\,\mathbb{E}[L_t] < \infty$ et W_t est donc intégrable. Étendant ensuite la formule aux fonctions ϕ mesurables bornées à valeurs dans $\mathbf{V}$, et prenant $\phi_n(x) = \mathbb{1}_{\{\|x\|\leqslant n\}}\, x$, on écrit $\mathbb{E}[\mathbb{1}_{\{\|\Theta_t\|\leqslant n\}} W_t|\mathcal{F}_s] = \mathbb{1}_{\{\|\Theta_s\|\leqslant n\}} W_s + \mathbb{E}[L_t - L_s|\mathcal{F}_s]\pi(\phi_n)$; lorsque n tend vers $+\infty$, $\pi(\phi_n)$ tend vers $b(\pi)$ et, W_t étant intégrable, on obtient $\mathbb{E}[W_t|\mathcal{F}_s] = W_s + \mathbb{E}[L_t - L_s|\mathcal{F}_s]\, b(\pi)$; ainsi, $W - b(\pi)L$ est une martingale.

Il reste à établir deux points : que cette martingale s'écrit $\int \Theta\, d\beta$, et que réciproquement, si W est une semimartingale, la probabilité π a un barycentre.

Supposons donc que W est une semimartingale; notons $W = M + A$ sa décomposition canonique. Appelons $[\![S_k, T_k]\!]$ les intervalles de montées de 0 à ε du processus R, de sorte que $S_0 = 0$, $R_{S_0} = 0$, $R_{T_0} = \varepsilon$, $R_{S_1} = 0$, etc. Sur $[\![T_k, S_{k+1}[\![$ on a identiquement $\Theta = \Theta_{T_k}$ et $W - W_{T_k} = \Theta(\beta - \beta_{T_k})$; donc, sur cet intervalle, $dM = \Theta\, d\beta$ et $dA = 0$. Ces deux égalités sont vraies sur l'ensemble $\{R > \varepsilon\}$, qui est inclus dans la réunion des $[\![T_k, S_{k+1}[\![$, et, à la limite en ε, on a $dM = \Theta\, d\beta$ et $dA = 0$ sur $\{R > 0\}$. Comme $\{R = 0\}$ est négligeable pour $d\beta$, donc aussi (représentation prévisible) pour dM, on en déduit que Θ est intégrable par rapport à β (au sens où le processus croissant $\int \|\Theta\|^2 dt$ est fini) et que $M = \int \Theta\, d\beta$. Le premier point est donc établi.

On en déduit aussi $A = \int \mathbb{1}_{\{R=0\}}\, dW$. Puisque l'ensemble prévisible $\bigcup_k [\![S_k, T_k]\!]$ tend vers $\{R = 0\}$ quand ε tend vers 0, A_t est la limite en probabilité

$$A_t = \lim_{\varepsilon\to 0} \sum_{k\geqslant 0} (W_{T_k\wedge t} - W_{S_k\wedge t}) = \lim_{\varepsilon\to 0} \sum_{k\geqslant 0} W_{T_k\wedge t}\, \mathbb{1}_{\{S_k\leqslant t\}}\,.$$

Pour terminer la démonstration, c'est-à-dire établir l'existence de $b(\pi)$, il suffit de montrer que $\int \|x\|\,\pi(dx)$ est finie, autrement dit que la mesure image de π par l'application norme a elle-même un barycentre. En travaillant avec $\|W\|$, qui est aussi un mouvement walshien (proposition 20) et une semimartingale, *on est ainsi ramené au cas où* $\mathbf{V} = \mathbb{R}$ *et* $\Theta \geqslant 0$. Pour $a > 0$, la probabilité π^a image de π par $x \mapsto x\mathbb{1}_{[0,a]}(x)$ est portée par $[0, a]$, donc, d'après la première partie de la démonstration, le mouvement walshien $W^a = W\mathbb{1}_{\{\Theta\leqslant a\}}$ est une semimartingale de partie à variation finie $A^a = b(\pi^a)L$. Mais on a aussi, par la formule ci-dessus,

$$A^a_t = \lim_{\varepsilon\to 0} \sum_{k\geqslant 0} W^a_{T_k\wedge t}\, \mathbb{1}_{\{S_k\leqslant t\}}\,.$$

Puisque $W \geqslant W^a$, on en tire $A \geqslant A^a = b(\pi^a)L$, d'où la majoration presque sûre $b(\pi^a) \leqslant A_\tau$, où τ est l'instant tel que $L_\tau = 1$. En faisant tendre a vers l'infini, on obtient $b(\pi) \leqslant A_\tau$, donc $b(\pi) < \infty$. ∎

Lorsque $\mathbf{V}$ est un ensemble fini et π la loi uniforme, les processus de Walsh asssociés à $(\mathbf{V}, \pi)$ s'identifient à des martingales-araignées. Ceci se généralise à un espace $(\mathbf{V}, \pi)$ quelconque, à condition de regrouper des rayons pour n'en garder qu'un nombre fini, et de compenser la non-uniformité de la loi par des homothéties sur chaque rayon :

PROPOSITION 22. — *Soient $n \geqslant 2$ un entier et Q une partition finie de $\mathbf{V}$ en n parties mesurables, non négligeables. Si Z est un processus de Walsh associé à $(\mathbf{V}, \pi)$ dans une filtration $\mathcal{F}$, les n processus positifs définis pour $q \in Q$ par*

$$Y^q = \frac{1}{\pi(q)}\, \mathbb{1}_{\{\Theta\in q\}}\, R$$

sont les composantes d'une martingale-araignée pour $\mathcal{F}$.

DÉMONSTRATION. — Dans un espace de dimension $n-1$, soit $\mathbf{T}$ une toile à n fils. Les Y^q sont les n composantes d'un processus à valeurs dans $\mathbf{T}$ car pour tout (t,ω), un au plus des nombres $Y_t^q(\omega)$ n'est pas nul. La fonction f définie sur $\mathbf{V}$ par $f(v) = \mathbb{1}_{\{v\in q_1\}}/\pi(q_1) - \mathbb{1}_{\{v\in q_2\}}/\pi(q_2)$ vérifie $\pi(f) = 0$; le processus $Y^{q_1} - Y^{q_2} = f\circ\Theta\, R$ est une martingale de la filtration $\mathcal{F}$, et, par la proposition 5, Y est une martingale-araignée. ∎

DÉFINITION. — *Un processus de Walsh associé à un espace probabilisé* $(\mathbf{V},\mathcal{V},\pi)$ *sera dit* multiple *s'il existe une partition mesurable de* $\mathbf{V}$ *en trois parties non négligeables.*

Les processus de Walsh qui ne sont pas multiples sont les mouvements browniens réfléchis, obtenus quand $\mathbf{V}$ est grossier, et les mouvements browniens gauches (skew Brownian motions) qui correspondent au cas où $\mathbf{V}$ est formé de deux atomes, et qui contiennent comme cas particulier les mouvements browniens linéaires. La filtration naturelle d'un processus de Walsh non multiple est toujours brownienne.

COROLLAIRE 5. — *Dans une filtration brownienne, il n'existe aucun processus de Walsh multiple.*

DÉMONSTRATION. — S'il existait un processus de Walsh multiple Z dans une filtration brownienne, la martingale-araignée multiple Y issue de $\vec{0}$ que la proposition 22 associe à Z serait identiquement nulle par le théorème 2 ; en ce cas le brownien réfléchi $R = \sum_{q\in Q} \pi(q)\, Y^q$ devrait aussi être nul, ce qui est absurde. ∎

Une variation de cette méthode va permettre l'étude des filtrations naturelles des processus de Walsh. Il faut pour cela localiser la notion de filtration brownienne à un ensemble aléatoire ; ceci nécessite de travailler dans des filtrations browniennes arrêtées. Il n'est techniquement pas plus difficile d'utiliser des changements de temps plus généraux que le simple arrêt, ce qui amène à se placer dans des filtrations engendrées par des martingales pures ; mais nous devrons généraliser la notion de filtration pure, introduite plus haut, avant le corollaire 4.

NOTATIONS. Si T est un temps d'arrêt d'une filtration $\mathcal{F}$, nous noterons $\mathcal{F}^T$ la filtration arrêtée, définie par $\mathcal{F}_t^T = \mathcal{F}_{T\wedge t}$; nous noterons $\vartheta_T\mathcal{F}$ la filtration décalée, définie par $(\vartheta_T\mathcal{F})_t = \mathcal{F}_{T+t}$.

DÉFINITIONS. — *Étant donné un espace probabilisé filtré* $(\Omega,\mathcal{A},\mathsf{P},\mathcal{F})$, *soient* S *et* T *deux temps d'arrêt tels que* $S \leqslant T$, *et* $A \subset \mathbf{R}_+\times\Omega$ *un ensemble aléatoire.*

a) *On dira que la filtration* $\mathcal{F}$ *est* pure entre S et T *s'il existe dans la filtration décalée* $\vartheta_S\mathcal{F}$ *une martingale pure* X *telle que, en appelant* $\mathcal{G}$ *la sous-filtration de* $\mathcal{F}$ *définie par*

$$\mathcal{G}^S = \mathcal{F}^S \qquad \text{et} \qquad \vartheta_S\mathcal{G} = \mathcal{F}_S \vee \operatorname{Nat} X\,,$$

T soit un temps d'arrêt de $\mathcal{G}$ et vérifie $\mathcal{F}^T = \mathcal{G}^T$.

b) *On dira que* $\mathcal{F}$ *est* pure dans A *s'il existe deux suites* $(S_k)_{k\in\mathbf{N}}$ *et* $(T_k)_{k\in\mathbf{N}}$ *de temps d'arrêt telles que* $S_k \leqslant T_k$, *que les intervalles* $[\![S_k,T_k]\!]$ *recouvrent* A *et que, pour chaque* k, $\mathcal{F}$ *soit pure entre* S_k *et* T_k.

Dans l'utilisation que nous ferons de cette notion, l'ensemble A sera un ouvert prévisible; il n'y aurait donc aucun inconvénient à remplacer les intervalles fermés $[\![S_k, T_k]\!]$ par des intervalles ouverts $]\!]S_k, T_k[\![$. Nous n'avons pas cherché à rendre la définition raisonnable pour des ensembles A plus généraux que les ouverts prévisibles (par exemple, avec cette définition, toute filtration est pure dans n'importe quel ensemble optionnel à coupes dénombrables, ce qui est certainement déraisonnable!).

PROPOSITION 23. — *La filtration naturelle d'un processus de Walsh Z est pure dans l'ouvert aléatoire $\{Z \neq c\}$.*

DÉMONSTRATION. — Appelons $\mathcal{F}$ la filtration, R le brownien réfléchi $\rho \circ Z$, et β la partie martingale de R. Pour $t > 0$, soit D_t le début de $\{R = 0\} \cap [\![t, \infty[\![$. Lorsque t décrit les rationnels, les intervalles $[\![t, D_t]\!]$ recouvrent l'ouvert $\{Z \neq c\} = \{R > 0\}$; il suffit donc de vérifier que $\mathcal{F}$ est pure entre t et D_t. Si l'on prend pour X le mouvement brownien réel $X_s = \beta_{t+s} - \beta_t$, la filtration $\mathcal{G}$ figurant dans la définition a) ci-dessus n'est autre que la filtration naturelle du couple (Z^t, β), où Z^t désigne l'arrêté de Z à t. Elle contient le brownien réfléchi R (déjà adapté à $\operatorname{Nat}\beta$) et donc le temps d'arrêt D_t, premier zéro de R après t.

Il reste à voir que $\mathcal{F}^{D_t} \subset \mathcal{G}^{D_t}$. Comme toutes les martingales de $\mathcal{F}$ sont continues, D_t est annoncé par des temps d'arrêt antérieurs, donc $\mathcal{F}^{D_t}$ est engendrée par Z^{D_t}. Or ce dernier vérifie $\rho \circ Z^{D_t} = R^{D_t}$ et $\theta \circ Z^{D_t} = \theta \circ Z^t$; il est donc adapté à $\mathcal{G}^{D_t}$. ∎

PROPOSITION 24. — *Sur l'espace filtré $(\Omega, \mathcal{A}, \mathbf{P}, \mathcal{F})$, soient A un ensemble aléatoire tel que la filtration $\mathcal{F}$ soit pure dans A, et Y une martingale-araignée multiple.*

L'ensemble $\mathbf{G} = \{t \geqslant 0 : Y_t = \vec{0}$ *et* $\exists \varepsilon > 0 : Y \neq \vec{0}$ *sur* $]t, t+\varepsilon[\}$ *des débuts des excursions de Y vérifie* $\mathbf{G} \cap A = \varnothing$ *p. s.*

DÉMONSTRATION. — Par définition de la pureté de $\mathcal{F}$ dans A, il suffit de démontrer que si S et T sont deux temps d'arrêt vérifiant $S \leqslant T$ et si $\mathcal{F}$ est pure entre S et T, alors $\mathbf{G} \cap [\![S, T]\!]$ est évanescent.

Si $S = \infty$ p. s., le résultat est trivial. Sinon, le processus $\vartheta_S Y$ défini par $(\vartheta_S Y)_t = Y_{S+t}$ est une martingale-araignée dans la filtration décalée $\vartheta_S \mathcal{F}$ et pour la probabilité conditionnée $\mathbf{P}[\,\cdot\,|S < \infty]$. Ce décalage de S, qui respecte la pureté entre deux instants, permet de se ramener au cas où $S = 0$, ce que nous supposons dorénavant.

Puisque $\mathcal{F}$ est pure entre 0 et T, il existe une martingale pure X telle que la filtration $\mathcal{G} = \mathcal{F}_0 \vee \operatorname{Nat} X$ coïncide avec $\mathcal{F}$ sur $[\![0, T]\!]$ et que T soit un temps d'arrêt de $\mathcal{G}$. Nous devons montrer que $\mathbf{G}$ ne rencontre pas $[\![0, T]\!]$; nous savons qu'il ne rencontre pas $[\![T]\!]$ (car les débuts des excursions des martingales évitent les temps d'arrêt) et il reste à établir qu'il ne rencontre pas $[\![0, T[\![$. Nous pouvons donc remplacer Y par l'arrêtée Y^T, qui est une martingale-araignée pour la filtration $\mathcal{F}^T = \mathcal{G}^T$, et a fortiori pour $\mathcal{G}$.

Le point (iii) de la définition des martingales pures permet alors, par plongement et changement de temps, de se ramener au cas où X est un mouvement brownien dans $\mathbf{R}^I$, $\mathcal{F}$ est la filtration $\mathcal{F}_0 \vee \operatorname{Nat} X$, T est un temps d'arrêt de $\mathcal{F}$ et Y est une martingale-araignée vérifiant $Y = Y^T$. Ce cas brownien est traité par le théorème 2, qui montre que Y, et donc aussi Y^T, est absorbée par l'origine. ∎

COROLLAIRE 6. — *Soit Y une martingale-araignée multiple dans la filtration naturelle d'un processus de Walsh Z. L'ensemble des débuts des excursions de Y est inclus dans l'ensemble $\{Z = c\}$.*

C'est une conséquence immédiate des propositions 23 et 24.

PROPOSITION 25. — *Soit Z un processus de Walsh, de filtration naturelle $\mathcal{F}$, associé à un espace probabilisé $(\mathbf{V}, \mathcal{V}, \pi)$.*

a) *Si $\mathcal{V}'$ est une sous-tribu de $\mathcal{V}$, il existe dans la filtration $\mathcal{F}$ un processus de Walsh associé à $(\mathbf{V}, \mathcal{V}', \pi)$.*

b) *Réciproquement, soit Z' un processus de Walsh multiple, dans la filtration $\mathcal{F}$, associé à un espace probabilisé essentiellement séparable $(\mathbf{V}', \mathcal{V}', \pi')$. Les parties radiales $R = \rho \circ Z$ et $R' = \rho' \circ Z'$ de Z et Z' sont égales (et les processus Z et Z' ont donc les mêmes zéros : $\{Z = c\} = \{Z' = c'\}$); en outre, il existe un plongement de $(\mathbf{V}', \mathcal{V}', \pi')$ dans $(\mathbf{V}, \mathcal{V}, \pi)$. En particulier, Z est lui aussi un processus de Walsh multiple.*

REMARQUE. — Le cas b) recouvre des situations beaucoup plus générales que celles du type $\Theta' = f \circ \Theta$ données par la proposition 20 : on peut par exemple avoir $\Theta'_t(\omega) = f(g_t(\omega), \omega, \Theta_t(\omega))$, où f est une fonction qui dépend prévisiblement des deux premiers arguments et g_t le dernier zéro de R avant t.

DÉMONSTRATION. — Le a) est trivial : c'est simplement la proposition 20 dans le cas de l'application identique de $(\mathbf{V}, \mathcal{V})$ dans $(\mathbf{V}, \mathcal{V}')$, qui est mesurable puisque $\mathcal{V}'$ est une sous-tribu de $\mathcal{V}$.

b) Soient Q' une partition mesurable de $\mathbf{V}'$ en trois parties non négligeables, et Y' la martingale-araignée à trois composantes associée par la proposition 22 à Z' et Q'. Le corollaire 6 dit que l'ensemble $\mathbf{G}'$ des débuts des excursions de Y' est inclus dans le fermé $\{Z = c\}$. Puisque Q' est une partition, les débuts d'excursions de Y' et de Z' sont les mêmes (ce sont ceux du brownien réfléchi R'); puisque $\{Z' = c'\}$ est d'intérieur vide, $\mathbf{G}'$ y est dense. Son adhérence est aussi incluse dans $\{Z = c\}$, et tout zéro de Z' est donc un zéro de Z. Pour tout t, le début D_t de $\{Z = c\} \cap [\![t, \infty[\![$ et le début D'_t de $\{Z' = c'\} \cap [\![t, \infty[\![$ vérifient donc $D_t \leqslant D'_t$; mais D_t et D'_t ont même loi (car R et R' sont tous deux des browniens réfléchis issus de l'origine); ceci implique $D_t = D'_t$ p. s. Comme R et R' sont des browniens réfléchis dans la filtration $\mathcal{F}$, on a p. s. pour t fixé

$$\mathsf{P}[D_t < t+1 | \mathcal{F}_t] = 2\,\mu\big([R_t, \infty[\big) \quad \text{et} \quad \mathsf{P}[D'_t < t+1 | \mathcal{F}_t] = 2\,\mu\big([R'_t, \infty[\big)\,,$$

où μ est la loi normale centrée et réduite; on en tire $\mu\big([R_t, \infty[\big) = \mu\big([R'_t, \infty[\big)$, puis $R_t = R'_t$ p. s. Ainsi, $R = R'$; la première assertion est établie.

Soient $T = \inf\{t : R_t = 1\}$ et G le dernier zéro de R avant T. Si f et f' sont des fonctions mesurables sur $\mathbf{V}$ et $\mathbf{V}'$, bornées et telles que $\pi(f) = \pi'(f') = 0$, les processus $M^f = f(\Theta)\,R$ et $N^{f'} = f'(\Theta')\,R$ sont des martingales de $\mathcal{F}$ par définition des processus de Walsh; elles sont bornées sur $[\![0, T]\!]$. De $M^f_G = N^{f'}_G = 0$, la proposition 13 c) permet de déduire que $\mathsf{E}[M^f_T | \mathcal{F}_G] = \mathsf{E}[N^{f'}_T | \mathcal{F}_G] = 0$, c'est-à-dire

$$\mathsf{E}[f(\Theta_T) | \mathcal{F}_G] = \mathsf{E}[f'(\Theta'_T) | \mathcal{F}_G] = 0\,.$$

Si l'on ne suppose plus que f et f' sont d'intégrale nulle, ces égalités deviennent $\mathsf{E}[f(\Theta_T) | \mathcal{F}_G] = \pi(f)$ et $\mathsf{E}[f'(\Theta'_T) | \mathcal{F}_G] = \pi'(f')$; ceci établit que Θ_T et Θ'_T sont indépendants de $\mathcal{F}_G$ et de lois respectives π et π'.

Ils sont en outre mesurables pour $\mathcal{F}_{G+}$ car $\Theta_T = \Theta_{(G+\varepsilon)\wedge T}$ (et de même pour Θ'_T) pour tout $\varepsilon > 0$. L'appendice de [4] montre que $\mathcal{F}_{G+} = \mathcal{F}_G \vee \sigma(\Theta_T)$. En prenant $\mathcal{B} = \mathcal{F}_G$, $\mathcal{C} = \sigma(\Theta_T)$ et $\mathcal{D} = \sigma(\Theta'_T)$ dans le lemme 2, on obtient l'existence d'un plongement de $(\Omega, \sigma(\Theta'_T), \mathbf{P})$ dans $(\Omega, \sigma(\Theta_T), \mathbf{P})$. Ceci termine la démonstration, puisque, les lois de Θ_T et Θ'_T étant π et π', $(\Omega, \sigma(\Theta_T), \mathbf{P})$ est isomorphe à $(\mathbf{V}, \mathcal{V}, \pi)$ et $(\Omega, \sigma(\Theta'_T), \mathbf{P})$ à $(\mathbf{V}', \mathcal{V}', \pi')$. ∎

THÉORÈME 4. — *Soient Z' et Z'' deux processus de Walsh multiples, définis sur des espaces probabilisés $(\Omega', \mathcal{A}', \mathbf{P}')$ et $(\Omega'', \mathcal{A}'', \mathbf{P}'')$, et associés à des espaces probabilisés essentiellement séparables $(\mathbf{V}', \mathcal{V}', \pi')$ et $(\mathbf{V}'', \mathcal{V}'', \pi'')$.*

Les deux espaces filtrés $(\Omega', \sigma(Z'), \mathbf{P}', \mathrm{Nat}\, Z')$ et $(\Omega'', \sigma(Z''), \mathbf{P}'', \mathrm{Nat}\, Z'')$ sont isomorphes si et seulement si les espaces probabilisés $(\mathbf{V}', \mathcal{V}', \pi')$ et $(\mathbf{V}'', \mathcal{V}'', \pi'')$ le sont.

DÉMONSTRATION. — S'il existe un isomorphisme $\Phi : (\mathbf{V}', \mathcal{V}', \pi') \longrightarrow (\mathbf{V}'', \mathcal{V}'', \pi'')$, on peut définir $\Psi : \mathrm{L}^0(\Omega', \sigma(Z'), \mathbf{P}') \longrightarrow \mathrm{L}^0(\Omega'', \sigma(Z''), \mathbf{P}'')$ à l'aide du lemme 1, comme étant l'unique isomorphisme qui transforme tout événement de la forme $\{R' \in C, \Theta'_{t_1} \in D_1, \ldots, \Theta'_{t_n} \in D_n\}$ en $\{R'' \in C, \Theta''_{t_1} \in \Phi(D_1), \ldots, \Theta''_{t_n} \in \Phi(D_n)\}$. Ces deux événements ont en effet même probabilité, car les vecteurs aléatoires $(\mathbb{1}_{D_1}, \ldots, \mathbb{1}_{D_n})$ et $\big(\Phi(\mathbb{1}_{D_1}), \ldots, \Phi(\mathbb{1}_{D_n})\big)$ ont même loi; et la condition de compatibilité est satisfaite, car Φ est un isomorphisme. On a $\Psi(Z') = Z''$, d'où $\Psi(\mathrm{Nat}\, Z') = \mathrm{Nat}\, Z''$, et les deux espaces filtrés $(\Omega', \sigma(Z'), \mathbf{P}', \mathrm{Nat}\, Z')$ et $(\Omega'', \sigma(Z''), \mathbf{P}'', \mathrm{Nat}\, Z'')$ sont isomorphes.

Réciproquement, s'il existe un isomorphisme Ψ de $(\Omega', \sigma(Z'), \mathbf{P}', \mathrm{Nat}\, Z')$ vers $(\Omega'', \sigma(Z''), \mathbf{P}'', \mathrm{Nat}\, Z'')$, la filtration naturelle de $\Psi(Z')$ est $\Psi(\mathrm{Nat}\, Z') = \mathrm{Nat}\, Z''$; chacun des deux processus de Walsh multiples $\Psi(Z')$ et Z'' vit dans la filtration naturelle de l'autre, et, par la proposition 25, chacun des deux espaces $(\mathbf{V}', \mathcal{V}', \pi')$ et $(\mathbf{V}'', \mathcal{V}'', \pi'')$ peut être plongé dans l'autre. Ils sont donc isomorphes d'après la proposition 3. ∎

RÉFÉRENCES

[1] J. Azéma, T. Jeulin, F. Knight, G. Mokobodzki & M. Yor. Sur les processus croissants de type injectif. *Séminaire de Probabilités XXX*, Lecture Notes in Mathematics 1626, Springer 1996.

[2] J. Azéma, P. A. Meyer & M. Yor. Martingales relatives. *Séminaire de Probabilités XXVI*, Lecture Notes in Mathematics 1526, Springer 1992.

[3] J. Azéma & M. Yor. Sur les zéros des martingales continues. *Séminaire de Probabilités XXVI*, Lecture Notes in Mathematics 1526, Springer 1992.

[4] M.T. Barlow, J.W. Pitman & M. Yor. On Walsh's Brownian motions. *Séminaire de Probabilités XXIII*, Lecture Notes in Mathematics 1372, Springer 1989.

[5] F. Delbaen. A remark on Slutsky's theorem. Dans ce volume.

[6] C. Dellacherie & P. A. Meyer. Probabilités et Potentiel. Chapitres V à VIII. Hermann, 1980.

[7] C. Dellacherie, B. Maisonneuve & P. A. Meyer. Probabilités et Potentiel. Chapitres XVII à XXIV : Processus de Markov (fin), Compléments de calcul stochastique. Hermann, 1992.

[8] L.E. Dubins, J. Feldman, M. Smorodinsky & B. Tsirelson. Decreasing sequences of σ-fields and a measure change for Brownian motion. *Ann. Prob. 24*, 882–904, 1996.

[9] P.R. Halmos. The decomposition of measures. *Duke Math. J. 9*, 386–392, 1941.

[10] J.M. Harrison & L.A. Shepp. On skew Brownian motion. *Ann. Prob. 9*, 309–313, 1981.

[11] K. Itô & H.P. McKean, Jr. Diffusion Processes and their Sample Paths. Springer, 1965.

[12] F.B. Knight. Poisson representation of strict regular step filtrations. *Séminaire de Probabilités XX*, Lecture Notes in Mathematics 1204, Springer 1986.

[13] D. Maharan. On homogeneous measure algebras. *Proc. Nat. Acad. Sc. 28*, 108–111, 1942.

[14] P.A. Meyer. Un cours sur les intégrales stochastiques. *Séminaire de Probabilités X*, Lecture Notes in Mathematics 511, Springer 1976.

[15] P.A. Meyer, R.T. Smythe & J.B. Walsh. Birth and death of Markov processes. *Proc. 6th Berkeley Sympos. Math. Statist. Probab. 3*, 295–305, 1972.

[16] J. Neveu. Sur l'espérance conditionnelle par rapport à un mouvement brownien. *Ann. Inst. Henri Poincaré, Section B, 12*, 105–109, 1976.

[17] C. Rainer. Fermés marqués, filtrations lentes, équations de structure. Thèse de Doctorat de l'Université de Paris VI, 1994.

[18] D. Revuz & M. Yor. Continuous Martingales and Brownian Motion. Springer, 1991.

[19] W. Schachermayer. On certain probabilities equivalent to Wiener measure, d'après Dubins, Feldman, Smorodinsky and Tsirelson. À paraître au *Séminaire de Probabilités XXXIII.*

[20] B. Tsirelson. Triple points: From non-Brownian filtrations to harmonic measures. *GAFA, Geom. funct. anal. 7*, 1096–1142, 1997.

[21] E. Vernocchi. Considerazioni sul moto browniano di Walsh. Tesi di Laurea, Università degli Studi di Milano, 1993.

[22] J.B. Walsh. A diffusion with a discontinuous local time. *Temps Locaux, Astérisque 52-53*, 37–45, 1978.

[23] M. Yor. Sur l'étude des martingales continues extrémales. *Stochastics 2*, 191–196, 1979.

[24] M. Yor. Sur la continuité des temps locaux associés à certaines semimartingales. *Temps Locaux, Astérisque 52-53*, 23–35, 1978.

[25] M. Yor. Sur le balayage des semimartingales continues. *Séminaire de Probabilités XIII*, Lecture Notes in Mathematics 721, Springer 1979.

[26] M. Yor. Some Aspects of Brownian Motion. Part II: Some Recent Martingale Problems. *Lectures in Mathematics*, ETH Zürich. Birkhäuser, 1997.

Martin T. Barlow
Department of Mathematics
University of British Columbia
Vancouver, B. C. V6T 1Y4
Canada
`barlow@math.ubc.ca`

Michel Émery
Université Louis Pasteur et C.N.R.S.
I.R.M.A.
7 rue René Descartes
67 084 Strasbourg Cedex
`emery@math.u-strasbg.fr`

Frank B. Knight
Department of Mathematics
University of Illinois
1409 West Green Street
Urbana, Illinois 61801
U.S.A.

Shiqi Song
Département de Mathématiques
Université d'Évry
Boulevard des Coquibus
91025 Évry Cedex
`Shiqi.Song@lami.univ-evry.fr`

Marc Yor
Université Pierre et Marie Curie
Laboratoire de Probabilités
4 place Jussieu
75 252 Paris Cedex 05

SUR UN THÉORÈME DE TSIRELSON RELATIF À DES MOUVEMENTS BROWNIENS CORRÉLÉS ET À LA NULLITÉ DE CERTAINS TEMPS LOCAUX

par M. Émery et M. Yor

B. Tsirelson a démontré dans [9] que la filtration naturelle d'un mouvement brownien (à une ou plusieurs dimensions) ne contient pas de processus de Walsh. Des variations sur la méthode de Tsirelson et sur ce résultat font l'objet de l'exposé précédent [1], auquel nous renvoyons le lecteur pour un bref résumé de l'histoire de cette question ainsi que pour des références. La superbe démonstration de Tsirelson repose de façon essentielle sur l'argument suivant (parmi d'autres) : Si deux mouvements browniens X et Y vérifient $\frac{d}{dt}\langle X,Y\rangle_t \leqslant 1-\varepsilon < 1$, alors les processus croissants $M_t = \sup_{s\leqslant t} X_s$ et $N_t = \sup_{s\leqslant t} Y_s$ ont peu de points de croissance communs; plus précisément, les mesures dM et dN sont presque sûrement étrangères (ce qui ne signifie pas que leurs supports soient presque sûrement disjoints). Tsirelson le démontre en exhibant dans le quart de plan une fonction vérifiant une inéquation aux dérivées partielles et certaines conditions au bord; nous en proposons ici aux habitués des semimartingales une autre démonstration, dans le cadre du calcul stochastique.

Ceci fait, nous nous pencherons sur les méthodes utilisées pour les affiner un peu et terminer par des estimations plus précises que celles qui suffisent à établir le théorème.

THÉORÈME (Tsirelson). — *Soient, dans une même filtration, X et Y deux mouvements browniens réels, issus de l'origine et vérifiant $d\langle X,Y\rangle_t = \Gamma_t\,dt$, où Γ est un processus prévisible tel que $\Gamma \leqslant r < 1$. En posant $M_t = \sup_{s\leqslant t} X_s$ et $N_t = \sup_{s\leqslant t} Y_s$, on a*

$$\int \mathbb{1}_{\{Y=N\}}\,dM = 0\,.$$

Si l'on remplace l'hypothèse $\Gamma \leqslant r$ par $\Gamma = r$, le processus $(M-X, N-Y)$ devient une diffusion dans le premier quadrant, réfléchie normalement aux bords; une transformation affine en fait un mouvement brownien dans un angle de mesure $\alpha = \text{Arc}\cos(-r)$, avec, sur chaque côté de l'angle, une réflexion oblique parallèle à l'autre côté. Dans ce cas, l'ensemble $\{X=M \text{ et } Y=N\}$, intersection des supports de dM et dN, est $\{0\}$ si $r\in[-1,0]$, et est l'adhérence de l'image d'un subordinateur d'indice $1-\frac{\pi}{2\alpha}$ si $r\in\,]0,1[$ (sur tout ceci, voir le chapitre IV de Le Gall [5]); on a ainsi un résultat quantitatif bien plus fin que la simple propriété qualitative de nullité de la mesure dM sur le support de dN. Toujours dans le cas où $\Gamma = r$, si l'on ne s'intéresse qu'à l'existence d'instants où $X=M$ et $Y=N$, le critère de S.R.S. Varadhan et R. Williams ([10]) montre déjà qu'il existe de tels instants (non nuls) si et seulement si $r>0$.

COROLLAIRE 1. — *Avec les notations du théorème, on appelle L le temps local de X en zéro et on suppose $|\Gamma| \leqslant r < 1$. On a alors*

$$\int \mathbb{1}_{\{Y=N\}}\, dL = \int \mathbb{1}_{\{Y=0\}}\, dM = \int \mathbb{1}_{\{Y=0\}}\, dL = 0\,.$$

DÉMONSTRATION. — Le mouvement brownien $\hat{X} = -\int \operatorname{sgn} X\, dX = L - |X|$ vérifie $\sup_{s\leqslant t} \hat{X}_s = L_t$ et donc aussi $\{\hat{X}_t = \sup_{s\leqslant t} \hat{X}_s\} = \{X_t=0\}$; de même, le brownien $\hat{Y} = -\int \operatorname{sgn} Y\, dY$ vérifie $\{\hat{Y}_t = \sup_{s\leqslant t} \hat{Y}_s\} = \{Y_t=0\}$. Les crochets de X, $\hat{X}$, Y et $\hat{Y}$ valent $d\langle \hat{X}, Y\rangle = -\operatorname{sgn} X\, \Gamma\, dt$, $d\langle X, \hat{Y}\rangle = -\operatorname{sgn} Y\, \Gamma\, dt$ et $d\langle \hat{X}, \hat{Y}\rangle = \operatorname{sgn}(XY)\, \Gamma\, dt$. La nullité des trois intégrales du corollaire peut s'obtenir en appliquant le théorème à chacun des trois couples $(\hat{X}, Y)$, $(X, \hat{Y})$ et $(\hat{X}, \hat{Y})$. ∎

Nous commençons par la démonstration du théorème. Nous suivons la convention de Meyer sur les temps locaux : si l'on note $(L^a_t)_{t\geqslant 0,\, a\in\mathbb{R}}$ le processus des temps locaux d'une semimartingale continue U, et si $\phi : \mathbb{R} \to \mathbb{R}$ est une différence de fonctions convexes, c'est la dérivée à gauche ϕ'_{g} qui intervient dans la formule du changement de variable

$$\phi \circ U_t = \phi(U_0) + \int_0^t \phi'_{\mathrm{g}}(U_s)\, dU_s + \tfrac{1}{2} \int_{a\in\mathbb{R}} L^a_t\, \phi''(da)\,,$$

où ϕ'' est la mesure dérivée seconde de ϕ (voir par exemple [8], théorème (1.5) du chapitre VI). Du coup, le temps local est continu à droite en la variable a avec des limites à gauche L^{a-}_t, et l'on a la formule

$$\int \mathbb{1}_{\{U=a\}}\, dU = \tfrac{1}{2}(L^a - L^{a-})$$

(*ibid.*, théorème (1.7)) ; cette formule va être généralisée par le lemme 2, plus bas.

LEMME 1. — *Soient U une semimartingale continue positive et H un ensemble prévisible. Si $U = 0$ sur H, le processus $\int \mathbb{1}_H\, dU$ est croissant.*

DÉMONSTRATION DU LEMME. — Puisque U est positive, $L^a = 0$ pour tout $a < 0$, d'où $L^{0-} = 0$ et la formule ci-dessus devient $\int \mathbb{1}_{\{U=0\}}\, dU = \frac{1}{2} L^0$. On en déduit $\int \mathbb{1}_H\, dU = \int \mathbb{1}_H \mathbb{1}_{\{U=0\}}\, dU = \frac{1}{2} \int \mathbb{1}_H\, dL^0$. ∎

COROLLAIRE 2. — *Soient f et g deux fonctions différences de convexes sur $\mathbb{R}^d$, vérifiant $f \leqslant g$ et $f(0) = g(0)$. Si Z est une semimartingale continue dans $\mathbb{R}^d$, $\int \mathbb{1}_{\{Z=0\}}\, d(f \circ Z) \leqslant \int \mathbb{1}_{\{Z=0\}}\, d(g \circ Z)$.*

C'est immédiat en prenant $H = \{Z = 0\}$ et $U = (g-f)\circ Z$ dans le lemme 1. ∎

LEMME 2. — *Soient $\phi : \mathbb{R} \to \mathbb{R}$ une différence de deux fonctions convexes, ϕ'_{g} et ϕ'_{d} ses dérivées à gauche et à droite, et a un réel. Si U est une semimartingale continue et $(L^b_t)_{t\geqslant 0,\, b\in\mathbb{R}}$ le processus de ses temps locaux, on a*

$$\int \mathbb{1}_{\{U=a\}}\, d(\phi \circ U) = \tfrac{1}{2} \left(\phi'_{\mathrm{d}}(a) L^a - \phi'_{\mathrm{g}}(a) L^{a-} \right).$$

DÉMONSTRATION DU LEMME. — Puisque le temps local L^b, qui figure dans la formule du changement de variable

$$\phi \circ U_t = \phi(U_0) + \int_0^t \phi'_{\mathrm{g}}(U_s)\, dU_s + \tfrac{1}{2} \int_{\mathbb{R}} L^b_t\, \phi''(db)\,,$$

est un processus croissant porté par l'ensemble $\{U = b\}$, on peut écrire

$$\begin{aligned}\int_0^t \mathbb{1}_{\{U_s=a\}}\, d(\phi\circ U)_s &= \int_0^t \mathbb{1}_{\{U_s=a\}}\, \phi'_{\mathrm{g}}(U_s)\, dU_s + \tfrac12 \int_{\mathbb{R}} \phi''(db) \int_0^t \mathbb{1}_{\{U_s=a\}} dL_s^b \\ &= \phi'_{\mathrm{g}}(a) \int_0^t \mathbb{1}_{\{U_s=a\}}\, dU_s + \tfrac12\, \phi''(\{a\}) \int_0^t \mathbb{1}_{\{U_s=a\}} dL_s^a \\ &= \tfrac12\, \phi'_{\mathrm{g}}(a)\, (L_t^a - L_t^{a-}) + \tfrac12\, (\phi'_{\mathrm{d}}(a) - \phi'_{\mathrm{g}}(a))\, L_t^a \\ &= \tfrac12\, (\phi'_{\mathrm{d}}(a) L_t^a - \phi'_{\mathrm{g}}(a) L_t^{a-})\,. \qquad \blacksquare\end{aligned}$$

COROLLAIRE 3. — *Soit V un processus continu, adapté et positif. Si, pour un exposant $p \in\,]0,1[$, le processus V^p est une semimartingale, V est aussi une semimartingale et $\int \mathbb{1}_{\{V=0\}}\, dV = 0$.*

DÉMONSTRATION DU COROLLAIRE. — En posant $U = V^p$ et $\phi(x) = |x|^{1/p}$, on a $V = \phi\circ U$. Puisque ϕ est convexe, V est une semimartingale; et puisque $\phi'_{\mathrm{g}}(0) = \phi'_{\mathrm{d}}(0) = 0$, le lemme 2 donne $\int \mathbb{1}_{\{V=0\}}\, dV = \int \mathbb{1}_{\{U=0\}}\, d(\phi\circ U) = 0$. ■

DÉMONSTRATION DU THÉORÈME. — Sans perdre de généralité, nous supposerons $r > 0$. Les processus $\xi = M - X$ et $\eta = N - Y$ sont des sous-martingales.

Pour $\varepsilon > 0$, la semimartingale $U^\varepsilon = (\varepsilon + \xi^2 + \eta^2)^{\frac{r}{(1+r)}}$ peut être calculée par la formule du changement de variable; on trouve

$$dU^\varepsilon = \frac{2r}{1+r}\, (\varepsilon+\xi^2+\eta^2)^{-\frac{1}{1+r}} \Big[\xi\, d\xi + \eta\, d\eta + \Big(1 - \frac{1}{1+r}\, \frac{\xi^2+\eta^2+2\xi\eta\Gamma}{\varepsilon+\xi^2+\eta^2}\Big)\, dt\Big]\,.$$

Le coefficient de dt est positif car $0 \leqslant \dfrac{2\xi\eta}{\varepsilon+\xi^2+\eta^2} \leqslant \dfrac{\xi^2+\eta^2}{\varepsilon+\xi^2+\eta^2} \leqslant 1$ et $\Gamma \leqslant r$; les termes $\xi\, d\xi$ et $\eta\, d\eta$ donnent eux aussi des sous-martingales locales; ainsi, U^ε est une sous-martingale locale. Comme $\sup_{s\leqslant t} (U^\varepsilon)_s$ est dans tous les L^p, la sous-martingale locale U^ε est une vraie sous-martingale; en faisant tendre ε vers zéro, on voit que *le processus $U = (\xi^2 + \eta^2)^{\frac{r}{1+r}}$ est une sous-martingale* (et donc une semimartingale).

Posons $\rho = \sqrt{\xi^2 + \eta^2}$ et $p = 2r/(1+r) < 1$. Comme $\rho^p = U$, le corollaire 3 dit que la semimartingale ρ vérifie $\int \mathbb{1}_{\{\rho=0\}}\, d\rho = 0$. Et en appliquant le corollaire 2 à la semimartingale bidimensionnelle $Z = (\xi, \eta)$ et aux trois fonctions convexes nulles à l'origine $f(x,y) = 0$, $g(x,y) = |x|$ et $h(x,y) = \sqrt{x^2 + y^2}$, qui vérifient $f \leqslant g \leqslant h$, on obtient $0 \leqslant \int \mathbb{1}_{\{\rho=0\}}\, d\xi \leqslant \int \mathbb{1}_{\{\rho=0\}}\, d\rho$. Il s'ensuit que $\int \mathbb{1}_{\{\rho=0\}}\, d\xi = 0$, et, en prenant la partie à variation finie, $\int \mathbb{1}_{\{\rho=0\}}\, dM = 0$. En remplaçant dans cette égalité $\mathbb{1}_{\{\rho=0\}}$ par $\mathbb{1}_{\{Y=N\}}\mathbb{1}_{\{X=M\}}$ puis $\mathbb{1}_{\{X=M\}}\, dM$ par dM, il reste $\int \mathbb{1}_{\{Y=N\}}\, dM = 0$. Le théorème est établi. ■

REMARQUES. — En remplaçant ξ par X et η par Y, le même calcul permet d'établir, si $|\Gamma| \leqslant r < 1$, que $(X^2 + Y^2)^{r/(1+r)}$ est une sous-martingale. En outre, l'exposant $r/(1+r)$ est le plus petit possible; en effet, le coefficient de dt dans le calcul de $d(X^2+Y^2)^q$ vaut, au facteur positif $2q(X^2+Y^2)^{q-1}$ près,

$$1 - (1-q)\Big(1 + \frac{2XY}{X^2+Y^2}\,\Gamma\Big)\,,$$

donc, lorsque $\Gamma \equiv r$ et $q < r/(1+r)$, ce coefficient est négatif dans un angle contenant la diagonale $\{x = y\}$.

Dans le cas $r = 0$ où (X, Y) est un mouvement brownien plan, ce qui précède montre que $(X^2 + Y^2)^q$ est une sous-martingale pour tout $q > 0$. Ce résultat est dû à Getoor et Sharpe, qui l'ont établi dans [4] pour toutes les martingales conformes. On pourrait de même étendre l'argument ci-dessus au cas d'une martingale (X, Y) « r-conforme », c'est-à-dire vérifiant une inégalité de Kunita-Watanabe renforcée du facteur $r < 1$.

Le cas $r = 1$ est en particulier celui du mouvement brownien unidimensionnel ($X = Y$), pour lequel $|X|^q$ n'est une semimartingale pour aucun exposant $q < 1$.

Le cas général interpole de façon continue entre ces deux cas extrêmes. Inspirée de Neveu [6], sa démonstration rappelle celle de l'inégalité de sous-harmonicité de Janson (voir par exemple le paragraphe 6.7 de [3]).

Des thèmes proches de celui-ci sont aussi abordés dans [2] et [7]; cette dernière référence contient tout un formulaire bien intéressant sur les temps locaux.

Voici pour finir deux extensions de ce qui précède. La première généralise le corollaire 3, et donne un critère un peu plus opératoire de nullité de $\int \mathbb{1}_{\{V=0\}}\, dV$.

LEMME 3. — *Soit V une semimartingale continue. Pour que $\int \mathbb{1}_{\{V=0\}}\, dV = 0$, il suffit qu'il existe une fonction impaire $h : \mathbf{R}-\{0\} \to \mathbf{R}$, strictement positive sur $]0, \infty[$, telle que*

$$\lim_{\varepsilon \to 0+} \int_0^t \frac{d\langle V, V\rangle_s}{h(V_s)}\, \mathbb{1}_{\{|V_s|>\varepsilon\}} \quad \textit{existe,}$$

mais

$$\int_0 \frac{da}{h(a)} = \infty\,.$$

DÉMONSTRATION. — Il suffit d'écrire la formule de densité d'occupation :

$$\int_0^t \frac{d\langle V, V\rangle_s}{h(V_s)}\, \mathbb{1}_{\{|V_s|>\varepsilon\}} = \int_\varepsilon^\infty \frac{da}{h(a)}\,(L_t^a - L_t^{-a})\,.$$

Lorsque ε tend vers $0+$, le premier membre converge; lorsque a tend vers $0+$, $L_t^a - L_t^{-a}$ tend vers $\int_0^t \mathbb{1}_{\{V=0\}}\, dV$. Ceci n'est possible que si $\int_0^t \mathbb{1}_{\{V=0\}}\, dV = 0$. ∎

PROPOSITION 1. — *Soit $\phi : \mathbf{R}_+ \to \mathbf{R}_+$ une différence de deux fonctions convexes telle que ${\phi'}^2/\phi$ soit intégrable sur tout intervalle $[0, a]$ (où $a > 0$). Si U est une semimartingale positive, la semimartingale $V = \phi o U$ vérifie $\int \mathbb{1}_{\{V=0\}}\, dV = 0$.*

Le corollaire 3 est un cas particulier de cette proposition, puisque la fonction $\phi(x) = x^{\frac{1}{p}}$ vérifie pour $0 < p < 1$ l'hypothèse de la proposition.

DÉMONSTRATION. — En prenant $h(x) = x$ dans le lemme 3, il suffit de vérifier que l'intégrale $\int_0^t d\langle V, V\rangle_s / V_s$ est finie. En appelant L le temps local de U, on écrit

$$\int_0^t \frac{d\langle V, V\rangle_s}{V_s} = \int_0^t \frac{(\phi' o U_s)^2\, d\langle U, U\rangle_s}{\phi o U_s} = \int_0^\infty \frac{da\, \phi'(a)^2}{\phi(a)}\, L_t^a$$

et il ne reste qu'à remarquer que les trajectoires de $a \mapsto L_t^a$ sont continues sur $[0, \infty[$ et à support compact. ∎

Nous terminons par une extension du théorème, ou plutôt de la dernière égalité du corollaire 1.

PROPOSITION 2. — *Soient r et p tels que $\frac{1}{3} < r < 1$ et $p > \dfrac{r+1}{2r}$. Soient X et Y deux semimartingales continues, de la forme*

$$X_t = B_t + \int_0^t H_s\,ds \qquad \text{et} \qquad Y_t = C_t + \int_0^t K_s\,ds\,,$$

où B et C sont des mouvements browniens réels et H et K des processus prévisibles tels que

$$\int_0^t |H_s|^p\,ds < \infty \qquad \text{et} \qquad \int_0^t |K_s|^p\,ds < \infty\,.$$

Appelons L le temps local de X à l'origine. Si l'on a $d\langle X,Y\rangle_t = \Gamma_t\,dt$ où Γ est un processus prévisible vérifiant $|\Gamma| \leqslant r$, alors

$$\int \mathbb{1}_{Y=0}\,dL = 0\,.$$

REMARQUE. — Pour $r \leqslant \frac{1}{3}$, le résultat subsiste, mais n'est pas intéressant car l'exposant p est alors supérieur à 2; on peut dans ce cas obtenir mieux (le même résultat pour $p = 2$) par un simple changement de probabilité, en se ramenant au corollaire 1 par le théorème de Girsanov. En revanche, pour $r > \frac{1}{3}$, l'exposant p peut être choisi plus petit que 2; il peut d'ailleurs être pris proche de 1 si r lui-même est proche de 1.

DÉMONSTRATION. — Il suffit d'établir que le processus $U = (X^2+Y^2)^{\frac{r}{1+r}}$ est une semimartingale; un argument calqué sur la fin de la démonstration du théorème permettra alors de conclure.

Comme plus haut, nous allons approcher U par $U^\varepsilon = (\varepsilon + X^2 + Y^2)^{\frac{r}{(1+r)}}$. Posons $\rho = \sqrt{X^2+Y^2}$. Le même calcul que pour le théorème donne $U^\varepsilon = \frac{2r}{1+r}(I^\varepsilon + A^\varepsilon)$, où l'on a posé

$$I^\varepsilon = \int \frac{X\,dX + Y\,dY}{(\varepsilon+\rho^2)^{\frac{1}{1+r}}} \quad \text{et} \quad A^\varepsilon = \int \frac{1}{(\varepsilon+\rho^2)^{\frac{1}{1+r}}}\Big(1 - \frac{1}{1+r}\,\frac{X^2+Y^2+2\Gamma XY}{\varepsilon+X^2+Y^2}\Big)\,dt\,.$$

L'hypothèse $|\Gamma| \leqslant r$ entraîne que A^ε est un processus croissant. Pour achever la démonstration, il suffit de montrer que I^ε converge, lorsque $\varepsilon \to 0$, vers l'intégrale stochastique

$$I = \int \frac{X\,dX + Y\,dY}{\rho^{\frac{2}{1+r}}}\,;$$

comme U^ε tend vers U, les A^ε devront alors par différence converger vers un processus A, qui sera croissant comme limite de processus croissants, et l'on aura ainsi établi que U est la semimartingale $\frac{2r}{1+r}\,(I + A)$.

Pour montrer que l'intégrale stochastique I a bien un sens et est la limite des I^ε, il suffit de vérifier que le processus prévisible $|X|\,\rho^{-\frac{2}{1+r}}$ (qui est positif et peut prendre des valeurs infinies) est intégrable par rapport à la semimartingale X. En échangeant X et Y, on en déduira à la fois que I existe et, par le théorème de convergence dominée pour les semimartingales, que les I^ε convergent vers I.

Prenant en compte la décomposition de X (le cas de Y est analogue), nous sommes ramenés à vérifier la convergence des intégrales

$$\int_0^t \frac{X_s}{\rho_s^{\frac{2}{1+r}}}\,dB_s \quad \text{et} \quad \int_0^t \frac{|X_s|}{\rho_s^{\frac{2}{1+r}}}\,|H_s|\,ds\,;$$

en minorant ρ par $|X|$, il suffit de montrer que

$$\int_0^t \frac{1}{|X_s|^{2\frac{1-r}{1+r}}}\,ds < \infty \quad \text{et} \quad \int_0^t \frac{1}{|X_s|^{\frac{1-r}{1+r}}}\,|H_s|\,ds < \infty\,.$$

Puisque $r > \frac{1}{3}$, l'exposant $\alpha = 2\frac{1-r}{1+r}$ vérifie $\alpha < 1$, la fonction $|x|^{2-\alpha}$ est convexe, et la première intégrale apparaît dans la formule

$$|X_t|^{2-\alpha} = (2-\alpha)\int_0^t |X_s|^{1-\alpha}\,\mathrm{sgn}\,X_s\,dX_s + \frac{(2-\alpha)(1-\alpha)}{2}\int_0^t \frac{ds}{|X_s|^{\alpha}}\,,$$

ce qui montre qu'elle converge. Pour l'autre intégrale, introduisons l'exposant q conjugué de p et posons $\beta = q\frac{1-r}{1+r}$. On peut écrire

$$\int_0^t \frac{1}{|X_s|^{\frac{1-r}{1+r}}}\,|H_s|\,ds \leqslant \left[\int_0^t \left(\frac{1}{|X_s|^{\frac{1-r}{1+r}}}\right)^q ds\right]^{\frac{1}{q}} \left[\int_0^t |H_s|^p\,ds\right]^{\frac{1}{p}}$$

$$= \left[\int_0^t \frac{1}{|X_s|^{\beta}}\,ds\right]^{\frac{1}{q}} \left[\int_0^t |H_s|^p\,ds\right]^{\frac{1}{p}}\,.$$

L'hypothèse $p > \frac{r+1}{2r}$ donne $q < \frac{1+r}{1-r}$ d'où $\beta < 1$, et le premier facteur est fini par le même argument que pour α ci-dessus; l'autre facteur est fini par hypothèse. La proposition est ainsi démontrée. ∎

Références

[1] M. Barlow, M. Émery, F. Knight, S. Song & M. Yor. Autour d'un théorème de Tsirelson sur des filtrations browniennes et non browniennes. Dans ce volume.

[2] J.-M. Bismut. On the set of zeros of certain semimartingales. *Proc. London Math. Soc. 49*, 73–86, 1984.

[3] R. Durrett. Brownian Motion and Martingales in Analysis. Wadsworth Mathematics Series, Wadsworth, 1984.

[4] R.K. Getoor & M.J. Sharpe. Conformal martingales. *Invent. Math. 16*, 271–308, 1972.

[5] J.-F. Le Gall. Some properties of planar Brownian motion. École d'Été de Probabilités de Saint-Flour XX, Lecture Notes in Mathematics 1527, Springer 1992.

[6] J. Neveu. Sur l'espérance conditionnelle par rapport à un mouvement brownien. *Ann. Inst. Henri Poincaré, Section B, 12*, 105–109, 1976.

[7] Y. Ouknine & M. Rutkovski. Local times of functions of continuous semimartingales. *Stochastic Analysis and Applications 13*, 211–232, 1995.

[8] D. Revuz & M. Yor. Continuous Martingales and Brownian Motion. Springer, 1991.

[9] B. Tsirelson. Triple points: From non-Brownian filtrations to harmonic measures. *GAFA, Geom. funct. anal. 7*, 1096–1142, 1997.

[10] S.R.S. Varadhan & R.J. Williams. Brownian motion in a wedge with oblique reflection. *Comm. Pure Appl. Math. 38*, 405–443, 1984.

[11] M. Yor. Sur la continuité des temps locaux associés à certaines semimartingales. *Temps Locaux, Astérisque 52-53*, 23–35, 1978.

Michel Émery
Université Louis Pasteur et C.N.R.S.
I.R.M.A.
7 rue René Descartes
67 084 STRASBOURG Cedex
emery@math.u-strasbg.fr

Marc Yor
Université Pierre et Marie Curie
Laboratoire de Probabilités
4 place Jussieu
75 252 PARIS Cedex 05

A REMARK ON SLUTSKY'S THEOREM

FREDDY DELBAEN

Departement für Mathematik, ETH Zürich

1. Introduction and Notation.

In Theorem 1 of the paper by [BEKSY] a generalisation of a theorem of Slutsky is used. In this note I will present a necessary and sufficient condition that assures that whenever X_n is a sequence of random variables that converges in probability to some random variable X, then for each Borel function f we also have that $f(X_n)$ tends to $f(X)$ in probability. The abstract way of formulating the result has the advantage that it shows how to decompose the problem. The key result is the Dunford-Pettis characterisation of relatively weakly compact subsets of the space L^1. Because of this immediate relationship I believe that the result is known. However I could not find a reference.
In the sequel $(\Omega, \mathcal{A}, \mathbb{P})$ is a fixed probability space and $(E, \mathcal{E})$ is a measurable space. The sequence $(X_n)_{n\geq 1}$ denotes a sequence of measurable functions of Ω into E. Also X denotes a measurable function of Ω into E. The distributions (image measures) of X_n, resp. X are denoted by μ_n, resp. μ.
The symbol $\mathcal{H}$, a subset of the space of measurable functions from E into $\mathbb{R}$, denotes the set which consists of those functions g such that $g(X_n)$ tends to $g(X)$ in probability. It is clear that $\mathcal{H}$ satifies some stability properties. First of all it is clear that $\mathcal{H}$ is a vector space stable for multiplication, i.e. an algebra. Also if $\phi\colon \mathbb{R} \to \mathbb{R}$ is continuous and $f \in \mathcal{H}$, then $\phi(f) \in \mathcal{H}$. It follows that for each $m \geq 0$ and $f \in \mathcal{H}$, the truncation f^m of f is also in $\mathcal{H}$, f^m is defined as $f^m(x) = f(x)$ if $|f(x)| \leq m$, $f^m(x) = m$ if $f(x) > m$ and $f^m(x) = -m$ if $f(x) < -m$. Conversely if all the truncations f^m are in $\mathcal{H}$, then also $f \in \mathcal{H}$. It is also obvious that $\mathcal{H}$ is closed for uniform convergence.
Let $M(E, \mathcal{E})$, M for short, be the space of all signed measures defined on the space $(E, \mathcal{E})$. A subset K of $M(E, \mathcal{E})$ is said to be relatively weakly compact if it is relatively weakly compact for the weak topology (i.e. $\sigma(M, M^*)$) on M. The Dunford-Pettis theorem states that K is relatively weakly compact if and only if there is a probability measure $\lambda \in M$ such that every element $\nu \in K$ is absolutely continuous with respect to λ and such that the set $\left\{\frac{d\nu}{d\lambda} \mid \nu \in K\right\}$ of Radon-Nikodym derivatives, is uniformly integrable in $L^1(\lambda)$. For information on weak compactness and related topics I refer to [G], last chapter.

1991 *Mathematics Subject Classification.* 28A20, 28A33 60B10, 60B12.
Key words and phrases. Slutsky's Theorem, weak compactness.

Theorem 1. *Let us assume that the set $(\mu_n)_{n\geq 1}$ of distributions of X_n is relatively weakly compact. If $(f_k)_{k\geq 1}$ is a sequence of functions in $\mathcal{H}$ that converges pointwise to a function f then also $f \in \mathcal{H}$, i.e. $\mathcal{H}$ is stable for taking pointwise convergent limits.*

Proof. Let $K = \{\mu_n \mid n \geq 1\} \cup \{\mu\}$. Clearly K is relatively weakly compact. Because of the stability properties of $\mathcal{H}$, we may and do assume that the sequence f_k is uniformly bounded, e.g. for each k, we have $|f_k| \leq 1$. Since the measures in K have uniformly integrable RN derivatives, we immediately obtain that $\sup_{\nu\in K} \int_E |f - f_k|\, d\nu$ tends to zero. For given $\epsilon > 0$ we now take k_0 big enough to assure that $\sup_{\nu\in K} \int_E |f - f_{k_0}|\, d\nu < \epsilon$. Now we take n_0 so that for $n \geq n_0$, $\int_\Omega |f_{k_0}(X_n) - f_{k_0}(X)|\, d\mathbb{P} < \epsilon$. For $n \geq n_0$ we then have $\int_\Omega |f(X_n) - f(X)|\, d\mathbb{P} < 3\epsilon$. This reasoning shows that $f(X_n)$ tends to $f(X)$ in $L^1(\mathbb{P})$ and hence in probability. □

By a standard argument on monotone classes we can now deduce the next theorem, which I give without proof.

Theorem 2. *If the set $(\mu_n)_{n\geq 1}$ of distributions of X_n is relatively weakly compact and if $\mathcal{H} \supset \mathcal{G}$, then $\mathcal{H}$ contains all measurable functions with respect to the sigma algebra $\mathcal{B}$, generated by $\mathcal{G}$.*

In the paper by [BEKSY], the functions X_n take values in a separable metric space S and X_n tend to X in probability. Since in their case, all the X_n have the same distribution, it immediately follows that for every Borel function h on S, we have that $h \circ X_n$ tend to $h \circ X$ in probability. More precisely we have the following.

Theorem 3. *Let S be a metric space and suppose that the sequence of S-valued random variables X_n converges to X in probability. In order that for each Borel measurable function f, the sequence $f(X_n)$ converges to $f(X)$ in probability, it is necessary and sufficient that the sequence of distributions $(\mu_n)_{n\geq 1}$ is relatively weakly compact.*

Proof. The sufficency is dealt with in Theorem 1 and 2 above. The necessity of the weak compactness condition is rather trivial. Suppose that the sequence of distributions, $(\mu_n)_{n\geq 1}$, is not weakly compact. Then there is a bounded measurable function such that $\int g\, d\mu_n$ does not converge to $\int g\, d\mu$. It follows that $g(X_n)$ cannot converge to $g(X)$ in probability. □

If in the previous theorem we replace convergence in probability by convergence almost surely, then the statement is wrong. To see this we will give a counterexample. We start with the circle $\mathbb{T} = \mathbb{R}/\mathbb{Z}$ equipped with the usual normalised Lebesgue measure m. Let O be an open subset of $\mathbb{T}$ such that $m(O) < 1/2$ and such that O is dense in $\mathbb{T}$. I will construct a sequence X_n, defined on some probability space, such that X_n converges to a random variable X almost surely. All the variables will be distributed uniformly on $\mathbb{T}$, i.e. $\mu_n = m$ for all n. However it will turn out that the almost sure covergence of $\mathbf{1}_O(X_n)$ to $\mathbf{1}_O(X)$ is false. The construction goes as follows. For each $\delta > 0$ and $x \in \mathbb{T}$, we put $g(\delta, x) = \frac{1}{2\delta} m(O \cap I_\delta^x)$, where I_δ^x is the symmetric interval around x with length 2δ. Since

O is dense we obtain that $g(\delta, x) > 0$ for all $x \in \mathbb{T}$ and all $\delta > 0$. It is now easy to find integers $(k_l)_{l\geq 1}$ such that for almost every $x \in \mathbb{T}$ we have that

$$\sum_{l\geq 1} k_l\, g\left(x, \frac{1}{l+1}\right) = \infty.$$

To construct the variables X_n, we need a sequence of independent variables $(V_n)_{n\geq 1}$, uniformly distributed on $[-1,1]$. The variable X is taken to be independent of the sequence V_n and to have a distribution equal to m. Let us put $K_0 = 0$ and $K_{l+1} = K_l + k_{l+1}$. For each n, $K_l < n \leq K_{l+1}$, we define $X_n = X + \frac{1}{l+1} V_n$. The distributions of the X_n are easily seen to be equal to m. Since for almost every $x \in \mathbb{T}$ we have that

$$\sum_n \mathbb{P}[X_n \in O \mid X = x] = \sum_{l\geq 1} k_l\, g\left(x, \frac{1}{l+1}\right) = \infty,$$

it follows from independence and the Borel Cantelli lemma that for almost every $\omega \in X^{-1}(O^c)$, $X_n(\omega) \in O$ infinitely often. The construction of the counterexample is therefore complete.

Acknowledgement. The author thanks the participants in the Mathematical Finance workshop in Oberwolfach, September 97 for discussions on this topic.

References

[BEKSY] M.T. Barlow, M.Émery, F.B. Knight, S. Song, M. Yor, *Autour d'un théorème de Tsirelson sur des filtrations browniennes et non browniennes*, Séminaire de Probabilités XXXII (1998).

[G] A. Grothendieck, *EVT*, Publicação da Sociedade de Matemática e S. Paulo (1954).

Eidgenössische Technische Hochschule Zürich, CH-8092 Zürich, Switzerland
E-mail address: delbaen@math.ethz.ch

Quelques calculs de compensateurs impliquant l'injectivité de certains processus croissants.

J. Azéma[1] T. Jeulin[2] F. Knight[3] M. Yor[1]

[1] Laboratoire de Probabilités - Université Paris 6 et CNRS URA 224
4 place Jussieu - Tour 56 - 3ème étage - Couloir 56-66
75272 PARIS CEDEX 05.
[2] Université Paris 7 et CNRS URA 1321 - 2 place Jussieu
Tour 45 - 5ème étage - Couloir 45-55 - 75251 PARIS CEDEX 05.
[3] University of Illinois - Department of Mathematics - 273 Altgeld Hall
1409 West Green Street - URBANA, IL 61801 - U.S.A.

1 Introduction

Soit $(\Omega, \mathcal{A}, (\mathcal{G}_t)_{t\geq 0}, \mathbb{P})$ un espace de probabilité filtré satisfaisant aux conditions habituelles. Dans l'article [1], est introduite et étudiée la notion suivante de *processus injectif* :

un processus $\mathcal{G}$-prévisible, à variation bornée $(A_t)_{t\geq 0}$ est appelé processus injectif (on dit aussi qu'il possède la propriété d'injectivité) s'il satisfait la condition suivante :

si H est un processus prévisible tel que $\int_{]0,\infty[} |H_s|\,|dA_s| < \infty$ et $\int_{]0,\infty[} H_s dA_s = 0$,

alors $\int_{]0,t]} H_s\,dA_s = 0$, pour tout $t \geq 0$.

On montre, dans le même article, que si A^L est la projection duale prévisible du processus $1_{[L,\infty[}$, où L est une fin d'ensemble $\mathcal{G}$-prévisible, alors A^L est un processus injectif.

Toutefois, il existe d'autres processus injectifs que les processus A^L ci-dessus, et une caractérisation de la propriété d'injectivité de A est donnée dans [1], en terme du support gauche et des sauts de A.

Finalement, toujours dans le même article [1], nous donnons les exemples explicites suivants de processus injectifs :

Théorème 1 *Soit μ une mesure de Radon sur $\mathbb{R}$ et $(L_t^x)_{x\in\mathbb{R},\,t\geq 0}$ la famille bicontinue des temps locaux du mouvement brownien réel. Le processus*

$$V_t = \int_{\mathbb{R}} L_t^x \, d\mu(x)$$

a la propriété d'injectivité si, et seulement si, le support de μ est d'intérieur vide.

Dans dans le paragraphe *2* du présent article, nous calculons explicitement, les processus A^L lorsque

$$L = g_\alpha^F \equiv \sup\{t < \alpha \mid X_t \in F\}$$

pour $\alpha \in \mathbb{R}_+$, F fermé de $\mathbb{R}$ et X mouvement brownien réel, ce qui nous fournit de nombreux exemples de processus croissants injectifs. En particulier, pour tout ensemble fini $F_n = \{x_1, .., x_n\}$, le processus $\left(\sum_{1\leq j\leq n} L_{t\wedge\alpha}^{x_j}\right)_{t\geq 0}$ est injectif, car équivalent, en tant que mesure aléatoire sur $\mathbb{R}_+$ à A^L où $L = g_\alpha^{F_n}$ (cela résulte aussi du théorème *1*).

Pour calculer les projections duales prévisibles mentionnées ci-dessus, nous avons besoin d'une formule d'Itô-Tanaka qui ne semble pas se trouver dans la littérature ; nous l'établissons au paragraphe *3*.

2 Derniers temps de passage du brownien [linéaire]-espace temps et propriété d'injectivité

Proposition 2 *Soit $\alpha > 0$, F un sous-ensemble fermé de $\mathbb{R}$ et g_α^F le dernier temps de passage du mouvement brownien X dans F avant l'instant α :*

$$g_\alpha^F = \sup\{t < \alpha \mid X_t \in F\}.$$

1) Soit ρ une variable exponentielle, de paramètre p, indépendante de $\mathcal{G}_\infty$. On note $\mathcal{G}^\rho$ la plus petite filtration contenant $\mathcal{G}$, pour laquelle ρ est un temps d'arrêt. La projection duale $\mathcal{G}^\rho$-prévisible de $1_{\{g_\rho^F>0\}}1_{[g_\rho^F,\infty[}$ est

$$\frac{1}{2}\left(\sqrt{2p}\sum_{\substack{]h,k[\text{ composante}\\ \text{connexe de } F^c}} \tanh\left(\sqrt{2p}\frac{k-h}{2}\right)\left(L_{t\wedge\rho}^h + L_{t\wedge\rho}^k\right)\right) + p\int_0^{t\wedge\rho} 1_F(X_s)\, ds.$$

2) La projection duale prévisible de $1_{\{g_\alpha^F>0\}}1_{[g_\alpha^F,\infty[}$ est

$$A_t^{g_\alpha^F} = 1_{\{X_\alpha\in F\}}1_{\{\alpha\leq t\}} + \frac{1}{2}\sum_{\substack{]h,k[\text{ composante}\\ \text{connexe de } F^c}}\int_0^{t\wedge\alpha}\frac{1}{\sqrt{2\pi(\alpha-s)}}\vartheta\left(\frac{k-h}{\sqrt{\alpha-s}}\right)\left(dL_s^h + dL_s^k\right)$$

où

$$\vartheta(x) = \frac{\sqrt{2\pi}}{x}\sum_{k\in\mathbb{Z}}\exp\left(-\frac{(2k+1)^2\pi^2}{2x^2}\right) = \sum_{k\in\mathbb{Z}}(-1)^k\exp\left(-\frac{k^2x^2}{2}\right)$$

(on pose $\vartheta(0) = 0$ et $\vartheta(\infty) = 1$).

Démonstration. 1) Soit pour $-\infty < u < v < \infty$, $J(u, v, .)$ la solution de l'équation différentielle :

$$\frac{1}{2}\varphi'' - p\varphi = 0, \text{ avec condition frontière : } \varphi(u) = \varphi(v) = 1\ ;$$

une expression explicite de J est :

$$J(u, v, x) = \frac{\cosh\left(\sqrt{2p}\left(x - \frac{u+v}{2}\right)\right)}{\cosh\left(\sqrt{2p}\frac{u-v}{2}\right)}\ ;$$

on notera aussi

$$\begin{aligned} J(-\infty, v, x) &= \exp\left(\sqrt{2p}(x-v)\right) \qquad (v \in \mathbb{R}), \\ J(u, \infty, x) &= \exp\left(\sqrt{2p}(u-x)\right) \qquad (u \in \mathbb{R}). \end{aligned}$$

Soit enfin, pour $x \in \mathbb{R}$,

$$\delta_F(x) = \inf(y > x \mid y \in F), \ \gamma_F(x) = \sup(y < x \mid y \in F).$$

$$x \to j(x) = 1_F(x) + 1_{F^c}(x) J(\gamma_F(x), \delta_F(x), x)$$

est continue sur $\mathbb{R}$; $\{j = 1\} = F$. j vérifie sur F^c l'équation différentielle $\frac{1}{2}j'' - pj = 0$. Pour $]u, v[$ composante connexe de F^c, j a une dérivée à gauche en v [resp. à droite en u] donnée par

$$\begin{aligned} & j'(v-) = -j'(u+) = \sqrt{2p}\tanh\left(\sqrt{2p}\tfrac{u-v}{2}\right) && (-\infty < u < v < \infty) \\ & j'(-\infty) = 0, \ -j'(v-) = \sqrt{2p} && (-\infty = u < v < \infty) \\ & j'(v-) = j'(u+) = -\sqrt{2p}, \ j'(\infty-) = 0 && (-\infty < u < v = \infty) \end{aligned}$$

on en déduit que j est dérivable à droite et à gauche en tout point de $\mathbb{R}$ (en un point $c \in F$, non isolé à droite [resp. à gauche] dans F, $j'(c+) = 0$ (resp. $j'(c-) = 0$) et que pour $x < y$

$$j'(y+) - j'(x+) = 2p \int_x^y (j 1_{F^c})(s)\, ds - \sum_{\substack{]h,k[\text{ composante} \\ \text{connexe de } F^c}} \sqrt{2p}\tanh\left(\sqrt{2p}\tfrac{k-h}{2}\right)\left(1_{\{x<h\le y\}} + 1_{\{x<k\le y\}}\right) ;$$

ainsi,

$$j(x) = \frac{1}{\sqrt{2p}} \int_{\mathbb{R}} e^{-\sqrt{2p}|x-y|}\, d\mu_{p,F}(y)$$

$$\text{où} \quad \mu_{p,F} = \sqrt{2p} \sum_{\substack{]h,k[\text{ composante} \\ \text{connexe de } F^c}} \tanh\left(\sqrt{2p}\tfrac{k-h}{2}\right)\left(\tfrac{\delta_h + \delta_k}{2}\right) + p 1_F . \lambda$$

est la mesure d'équilibre du fermé F pour le mouvement brownien tué à temps exponentiel [indépendant] de paramètre p (λ est la mesure de Lebesgue sur $\mathbb{R}$; $\tanh(\infty) = 1$).

$\left(1_{\{\rho \le t\}} - p\inf(t, \rho)\right)$ est une $\mathcal{G}^\rho$-martingale et X est un $\mathcal{G}^\rho$-mouvement brownien. Par suite, pour U $\mathcal{G}^\rho$-temps d'arrêt,

$$\mathbb{P}\left[U < g_\rho^F\right] = \mathbb{P}\left[U < \rho, \ X_U \in F\right] + \mathbb{P}\left[U' < \rho, \ X_U \in F^c\right]$$

où $U' = \inf\{t > U \mid X_t \in F\}$;

$$j(X_{t \wedge U'})\, 1_{\{X_U \in F^c, \ U \le t \wedge U' < \rho\}}$$

étant une $\mathcal{G}^\rho$-martingale de variable terminale $1_{\{X_U \in F^c, \ U' < \rho\}}$,

$$\mathbb{P}\left[U' < \rho, \ X_U \in F^c\right] = \mathbb{E}\left[j(X_U) \ ; \ X_U \in F^c, \ U < \rho\right].$$

La projection $\mathcal{G}^\rho$-optionnelle de $1_{[0,g_\rho^F[}$ est donc

$$Z_t^{g_\rho^F} = 1_{\{t<\rho\}} j(X_t) = j(X_{t\wedge\rho}) - 1_{\{\rho\le t\}} j(X_\rho).$$

D'après la formule d'Itô-Tanaka, sa partie martingale est

$$-1_{\{\rho\le t\}} j(X_\rho) + p\int_0^{t\wedge\rho} j(X_s)\,ds + \int_0^{t\wedge\rho} j'(X_s)\,dX_s,$$

sa partie à variation finie $\mathcal{G}^\rho$-prévisible est

$$\frac{1}{2}\left(\sqrt{2p}\sum_{\substack{]h,k[\text{ composante}\\ \text{connexe de }F^c}} \tanh\left(\sqrt{2p}\frac{k-h}{2}\right)\left(L_{t\wedge\rho}^h + L_{t\wedge\rho}^k\right)\right) + p\int_0^{t\wedge\rho} 1_F(X_s)\,ds.$$

2) Il est bien connu que pour $z\in\mathbb{R}$

$$\tanh z = \sum\nolimits_{n\in\mathbb{Z}} \frac{z}{z^2+(2n+1)^2\frac{\pi^2}{4}},$$

ce qui permet d'écrire pour $\tau>0$,

$$\begin{aligned}\sqrt{2p}\tanh\left(\sqrt{2p}\tfrac{\tau}{2}\right) &= \tfrac{p}{\tau}\sum\nolimits_{n\in\mathbb{Z}} \tfrac{2}{p+(2n+1)^2\frac{\pi^2}{2\tau^2}}\\ &= \tfrac{2p}{\tau}\int_0^\infty e^{-p\alpha}\sum\nolimits_{n\in\mathbb{Z}}\exp\left(-\tfrac{(2n+1)^2\pi^2\alpha}{2\tau^2}\right)d\alpha\\ &= 2p\int_0^\infty e^{-p\alpha}\tfrac{1}{\sqrt{2\pi\alpha}}\vartheta\left(\tfrac{\tau}{\sqrt{\alpha}}\right)d\alpha\end{aligned}$$

(*nota bene* : cette formule est vraie aussi pour $\tau=\infty$) .
Pour K processus prévisible, positif, borné, on a d'une part

$$\int_0^\infty e^{-ps}\mathbb{E}[K_s\,;\,X_s\in F]\,ds = \mathbb{E}\left[\int_0^\rho K_s 1_F(X_s)\,ds\right]$$

et, d'autre part, pour tout $\tau\in\overline{\mathbb{R}}_+^*$ et $h\in\mathbb{R}$,

$$\begin{aligned}p&\int_0^\infty e^{-p\alpha}\mathbb{E}\left[\int_0^\alpha \tfrac{1}{\sqrt{2\pi(\alpha-s)}}\vartheta\left(\tfrac{\tau}{\sqrt{\alpha-s}}\right)K_s\,dL_s^h\right]d\alpha\\ &= \mathbb{E}\left[\int_0^\infty\left(p\int_s^\infty e^{-p\alpha}\tfrac{1}{\sqrt{2\pi(\alpha-s)}}\vartheta\left(\tfrac{\tau}{\sqrt{\alpha-s}}\right)d\alpha\right)K_s\,dL_s^h\right]\\ &= \mathbb{E}\left[\int_0^\infty e^{-ps}K_s\,dL_s^h\right]\left(p\int_0^\infty e^{-p\alpha}\tfrac{1}{\sqrt{2\pi\alpha}}\vartheta\left(\tfrac{\tau}{\sqrt{\alpha}}\right)d\alpha\right)\\ &= \tfrac{1}{2}\sqrt{2p}\tanh\left(\sqrt{2p}\tfrac{\tau}{2}\right)\mathbb{E}\left[\int_0^\rho K_s\,dL_s^h\right]\end{aligned}$$

Par sommation, on obtient :

$$\mathbb{E}\left[K_{g_\rho^F}\right] = p\int_0^\infty e^{-p\alpha}\mathbb{E}\left[K_{g_\alpha^F}\right]d\alpha = p\int_0^\infty e^{-p\alpha}\mathbb{E}\left[\int_{]0,\alpha]} K_s\,dA_s^{g_\alpha^F}\right]d\alpha$$

ce qui permet d'obtenir *2*) par inversion de transformation de Laplace.

3) Nous donnons également une démonstration directe de *2*), utilisant une formule d'Itô-Tanaka adéquate.

Notons, pour $t > 0$ et $-\infty < a \leq 0 < b \leq \infty$, $\Phi(a,b,t)$ la probabilité que X quitte l'intervalle $]a,b[$ avant l'instant t ; soit enfin, pour $(t,x) \in \mathbb{R}_+ \times \mathbb{R}$,

$$\Psi_F(x,t) = 1_{F^c}(x)\,\Phi(\gamma_F(x) - x, \delta_F(x) - x, t) + 1_F(x).$$

Soit U un temps d'arrêt ; vu la régularité des points pour le mouvement brownien réel,

$$\begin{aligned}\mathbb{P}\left[U < g_\alpha^F\right] &= \mathbb{P}[U < \alpha,\ X_U \in F] + \mathbb{P}\left[U < g_\alpha^F,\ X_U \in F^c\right] \\ &= \mathbb{P}[U < \alpha,\ X_U \in F] + \mathbb{P}[U' < \alpha,\ X_U \in F^c]\end{aligned}$$

où $U' = \inf\{t > U \mid X_t \in F\}$; par application de la propriété de Markov au temps U,

$$\mathbb{P}\left[U < g_\alpha^F\right] = \mathbb{E}[\Psi_F(X_U, \alpha - U)\ ;\ U < \alpha]\ ;$$

En conséquence, la projection optionnelle $Z^{g_\alpha^F}$ de $1_{[0,g_\alpha^F[}$ est

$$Z_t^{g_\alpha^F} = 1_{\{t<\alpha\}}\Psi_F(X_t, \alpha - t) = \Psi_F\left(X_{t\wedge\alpha}, (\alpha - t)_+\right) - 1_{\{X_\alpha \in F\}}1_{\{\alpha \leq t\}}.$$

Remarquons que :

$$\begin{gathered} Z_{t-}^{g_\alpha^F} = 1_{\{t\leq\alpha\}}\Psi_F(X_t, \alpha - t) \text{ et} \\ \left\{g_\alpha^F = \alpha\right\} = \{X_\alpha \in F\},\ \left\{Z_{g_\alpha^F}^{g_\alpha^F} = 1\right\} = \{X_\alpha \notin F\}. \end{gathered}$$

La partie purement discontinue de $A^{g_\alpha^F}$ est donc bien $1_{\{X_\alpha\in F\}}1_{[\alpha,\infty[}$.

4) Pour identifier la partie continue de $A^{g_\alpha^F}$, nous nous appuierons sur le

Lemme 3 *Pour* $-\infty \leq a < 0 < b \leq \infty$,

$$1 - \Phi(a,b,t) = \mathbb{P}\left[\inf_{s\leq t} X_s \geq a,\ \sup_{s\leq t} X_s \leq b\right].$$

Par suite, si $-\infty < a < 0 < b < \infty$,

$$\begin{aligned}\Phi(-\infty, b, t) &= b\int_0^t e^{-\frac{b^2}{2\rho}}\frac{1}{\sqrt{2\pi\rho^3}}\,d\rho \\ 1 - \Phi(a,b,t) &= \frac{1}{\sqrt{2\pi t}}\sum_{k\in\mathbb{Z}}\int_a^b \left(e^{-\frac{(\xi+2k(b-a))^2}{2t}} - e^{-\frac{(\xi-2b+2k(b-a))^2}{2t}}\right)d\xi.\end{aligned}$$

Ces formules se trouvent dans la plupart des traités classiques sur le mouvement brownien (voir, par exemple, [2]-*X.5* ou [3]) et s'obtiennent par intégration sur $]a,b[$ de l'égalité

$$\begin{aligned}&\mathbb{P}\left[B_t \in dx\ ;\ \inf_{s\leq t} B_s \geq a,\ \sup_{s\leq t} B_s \leq b\right] \\ &= \frac{1}{\sqrt{2\pi t}}\sum_{k\in\mathbb{Z}}\left(e^{-\frac{(x+2k(b-a))^2}{2t}} - e^{-\frac{(x-2b+2k(b-a))^2}{2t}}\right)dx \quad (a < x < b).\end{aligned}$$

$(x,t) \in F^c \times \mathbb{R}_+^* \to \Psi_F(x,t)$ est de classe C^2 ; si $]a,b[$ est une composante connexe de F^c, on a, pour $(x,t) \in\,]a,b[\, \times \mathbb{R}_+^*$:

$$\frac{\partial \Psi_F}{\partial x}(x,t) = \begin{cases} \sqrt{\frac{2}{\pi t}} e^{-\frac{(b-x)^2}{2t}} & \text{si } -\infty = a < b < \infty \\ \sqrt{\frac{2}{\pi t}} e^{-\frac{(x-a)^2}{2t}} & \text{si } -\infty < a < b = \infty \\ \frac{1}{\sqrt{2\pi t}} \sum_{k\in\mathbb{Z}} \left(\begin{array}{l} e^{-\frac{(a-x+2kb-a)^2}{2t}} - e^{-\frac{(b-x+2k(b-a))^2}{2t}} \\ \quad - e^{-\frac{(b+x-2a+2k(b-a))^2}{2t}} + e^{-\frac{(x-a+2k(b-a))^2}{2t}} \end{array} \right) \\ \qquad \text{si } -\infty < a < b < \infty \end{cases}$$

en outre,

$$\frac{1}{2}\frac{\partial^2 \Psi_F}{\partial x^2} - \frac{\partial \Psi_F}{\partial t} = 0 \text{ sur } F^c \times \mathbb{R}_+^*.$$

Si $a \in F$ est isolé à gauche dans F, $\frac{1}{2}\frac{\partial \Psi_F}{\partial x}(a+,t)$ existe et

$$\frac{1}{2}\frac{\partial \Psi_F}{\partial x}(a+,t) = \frac{1}{\sqrt{2\pi t}} \vartheta\left(\frac{b-a}{\sqrt{t}}\right);$$

de même, si $b \in F$ est isolé à droite dans F, $\frac{1}{2}\frac{\partial \Psi_F}{\partial x}(b-,t)$ existe et

$$\frac{1}{2}\frac{\partial \Psi_F}{\partial x}(b-,t) = \frac{-1}{\sqrt{2\pi t}} \vartheta\left(\frac{b-a}{\sqrt{t}}\right).$$

Comme $\vartheta(x) \underset{x\to 0}{\longrightarrow} 0$, si $c \in F$ n'est pas isolé à droite [resp. à gauche] dans F, $\frac{\partial \Psi_F}{\partial x}(c+,t)$ $\left[\text{resp. } \frac{\partial \Psi_F}{\partial x}(c-,t)\right]$ existe et est nulle.

La généralisation de la formule d'Itô-Tanaka donnée au paragraphe suivant permet d'écrire la décomposition canonique de la semimartingale

$$(\Psi_F(X_{t\wedge\alpha}, \alpha - t))_{0\le t\le\alpha},$$

sous la forme :

$$\begin{aligned} \Psi_F(X_{t\wedge\alpha}, \alpha - t) &= \Psi_F(0,\alpha) + \int_0^t \tfrac{\partial \Psi_F}{\partial x}(X_s, s)\, dX_s \\ &+ \sum_{\substack{]h,k[\text{ composante} \\ \text{connexe de } F^c}} \int_0^{t\wedge\alpha} \tfrac{1}{\sqrt{2\pi(\alpha-s)}} \vartheta\left(\tfrac{k-h}{\sqrt{\alpha-s}}\right) \left(dL_s^h + dL_s^k\right), \end{aligned}$$

ce qui donne *2)* □

3 Formule d'Itô-Tanaka pour le brownien [linéaire]-espace temps.

N'ayant pas trouvé dans la littérature de *formule d'Itô-Tanaka* à notre convenance pour le couple (X_t, t) où X est un mouvement brownien réel issu de 0, nous établissons

Proposition 4
Soit $H : \mathbb{R} \times \mathbb{R}_+ \to \mathbb{R}$, continue vérifiant les quatre propriétés suivantes :
i) pour presque tout $x \in \mathbb{R}$, $t \to H(x,t)$ est absolument continue, de dérivée [de Radon-Nikodym] $\frac{\partial H}{\partial t}(x,t)$ telle que

$$\forall A,\ t \in \mathbb{R}_+, \int_{[-A,A]\times[0,t]} \left|\frac{\partial H}{\partial t}(x,s)\right| dx \frac{ds}{\sqrt{s}} < \infty ;$$

ii) il existe une mesure de Radon ν sur $\mathbb{R}$ et une fonction h borélienne sur $\mathbb{R}\times\mathbb{R}_+$ avec

$$\forall A,\ t \in \mathbb{R}_+, \int_{[-A,A]\times[0,t]} |h(x,s)|\, d\nu(x) \frac{ds}{\sqrt{s}} < \infty$$

et telles que, pour presque tout $t \in \mathbb{R}_+$, $x \to H(x,t)$ a pour dérivée seconde (au sens des distributions) la mesure $h(x,t)\, d\nu(x)$
$\left(\text{en particulier, } x \to H(x,t) \text{ a une dérivée à droite } \frac{\partial H}{\partial x}(x+,t)\right)$.

$$iii) \qquad \forall A,\ t \in \mathbb{R}_+, \int_{[-A,A]\times[0,t]} \left(\frac{\partial H}{\partial x}(x+,s)\right)^2 dx \frac{ds}{\sqrt{s}} < \infty ;$$

iv) pour presque tout t, $x \to h(x,t)$ est ρ-presque sûrement continue, où ρ est la partie de ν étrangère à la mesure de Lebesgue.
On a alors :

$$\begin{aligned} H(X_t,t) &= H(X_0,0) + \int_0^t \tfrac{\partial H}{\partial x}(X_s,s)\, dX_s \\ &+ \int_0^t \tfrac{\partial H}{\partial t}(X_s,s)\, ds + \tfrac{1}{2}\int_{\mathbb{R}} \left(\int_0^t h(z,s)\, d_sL_s^z\right) d\nu(z). \end{aligned}$$

Démonstration. Quitte à remplacer H par HK_n où K_n est de classe C^∞ sur $\mathbb{R}\times\mathbb{R}_+$ avec $0 \le K_n \le 1$ et $K_n = 1$ sur $[-n,n]\times[0,n]$ (cela revient à localiser) on peut supposer :

$$\int_{\mathbb{R}\times\mathbb{R}_+} \left|\tfrac{\partial H}{\partial t}(x,s)\right| dx \tfrac{ds}{\sqrt{s}} + \int_{\mathbb{R}\times\mathbb{R}_+} |h(x,s)|\, d\nu(x) \tfrac{ds}{\sqrt{s}}$$
$$+ \int_{\mathbb{R}\times\mathbb{R}_+} \left(\frac{\partial H}{\partial x}(x+,t)\right)^2 dx \tfrac{ds}{\sqrt{s}} < \infty$$

On a alors :

$$\mathbb{E}\left[\int_0^t \left(\tfrac{\partial H}{\partial x}(X_s,s)\right)^2 ds\right] < \infty,$$
$$\mathbb{E}\left[\int_0^t \left|\tfrac{\partial H}{\partial t}(X_s,s)\right| ds + \tfrac{1}{2}\int_{\mathbb{R}} \left(\int_0^t |h(z,s)|\, d_sL_s^z\right) d\nu(z)\right] < \infty,$$

$\left(\int_0^t \frac{\partial H}{\partial x}(X_s,s)\,dX_s\right)_{t\geq 0}$ est une martingale de carré intégrable,

$\left(\int_0^t \frac{\partial H}{\partial t}(X_s,s)\,ds + \frac{1}{2}\int_{\mathbb{R}}\left(\int_0^t h(z,s)\,d_sL_s^z\right)d\nu(z)\right)_{t\geq 0}$ est un processus continu, à variation finie. Il suffit en outre de montrer, pour $0 < \varepsilon < t$, la formule

$$(\natural)\quad \left\{\begin{aligned} H(X_t,t) &= H(X_\varepsilon,\varepsilon) + \int_\varepsilon^t \frac{\partial H}{\partial x}(X_s,s)\,dX_s \\ &\quad + \int_\varepsilon^t \frac{\partial H}{\partial t}(X_s,s)\,ds + \frac{1}{2}\int_{\mathbb{R}}\left(\int_\varepsilon^t h(z,s)\,d_sL_s^z\right)d\nu(z). \end{aligned}\right.$$

Nous allons montrer ($\natural$) par régularisation de H (prolongée par 0 sur $\mathbb{R}\times\mathbb{R}_-^*$).

Soit $f,\ g:\mathbb{R}\to\mathbb{R}_+$ de classe C^∞, à support compact, avec

$$\int_{\mathbb{R}} f(x)\,dx = \int_{\mathbb{R}} g(x)\,dx = 1.$$

Pour $m,\ n\in\mathbb{N}^*$,

$$f_n(x) = nf(nx)\,,\ g_m(x) = mg(mx)$$

$$H_{n,m}(x,t) = \int_{\mathbb{R}^2} H(y,s)\,f_n(x-y)\,g_m(t-s)\,dyds.$$

$H_{n,m}$ étant de classe C^∞, on peut lui appliquer la formule d'Itô "ordinaire" :

$$\begin{aligned} H_{n,m}(X_t,t) &= H_{n,m}(X_\varepsilon,\varepsilon) + \int_\varepsilon^t \frac{\partial H_{n,m}}{\partial x}(X_s,s)\,dX_s \\ &\quad + \int_\varepsilon^t \left(\frac{1}{2}\frac{\partial^2 H_{n,m}}{\partial x^2} + \frac{\partial H_{n,m}}{\partial t}\right)(X_s,s)\,ds. \end{aligned}$$

▷ $H_{n,m}(X_\varepsilon,\varepsilon) \underset{m,n\to\infty}{\longrightarrow} H(X_\varepsilon,\varepsilon)$.

▷ En ce qui concerne l'intégrale stochastique par rapport à X, remarquons ("intégration par parties" ou Fubini) que, pour presque tout t,

$$\begin{aligned} \frac{\partial H_{n,m}}{\partial x}(x,t) &= \int_{\mathbb{R}^2} H(y,s)\,f_n'(x-y)\,g_m(t-s)\,dy\,ds \\ &= \int_{\mathbb{R}^2} \frac{\partial H}{\partial x}(y,s)\,f_n(x-y)\,g_m(t-s)\,dy\,ds \end{aligned}$$

et que (résultats classiques sur les approximations de l'unité) :

$$\frac{\partial H_{n,m}}{\partial x} \underset{m,n\to\infty}{\longrightarrow} \frac{\partial H}{\partial x} \text{ dans } L^2(\mathbb{R}^2).$$

Comme

$$\mathbb{E}\left[\sup_{\varepsilon\le r\le t}\left|\int_\varepsilon^r \frac{\partial H_{n,m}}{\partial x}(X_s,s)\,dX_s-\int_\varepsilon^r \frac{\partial H}{\partial x}(X_s,s)\,dX_s\right|^2\right]$$

$$\le 4\mathbb{E}\left[\left(\int_\varepsilon^r\left(\frac{\partial H_{n,m}}{\partial x}(X_s,s)-\frac{\partial H}{\partial x}(X_s,s)\right)^2 ds\right)\right]$$

$$=4\int_{\mathbb{R}^2} dy\left(\int_\varepsilon^r \frac{1}{\sqrt{2\pi s}}e^{-\frac{a^2}{2s}}\left(\frac{\partial H_{n,m}}{\partial x}(a,s)-\frac{\partial H}{\partial x}(a,s)\right)^2 ds\right)$$

$$\le \frac{4}{\sqrt{2\pi\varepsilon}}\left\|\frac{\partial H_{n,m}}{\partial x}-\frac{\partial H}{\partial x}\right\|^2_{L^2(\mathbb{R}^2)},$$

il y a convergence (en probabilité, uniformément sur les compacts de $\mathbb{R}_+$) de $\int_\varepsilon^t \frac{\partial H_{n,m}}{\partial x}(X_s,s)\,dX_s$ vers $\int_\varepsilon^t \frac{\partial H}{\partial x}(X_s,s)\,dX_s$.

▷ De même, pour presque tout x,

$$\frac{\partial H_{n,m}}{\partial t}(x,t)=\int_{\mathbb{R}^2}\frac{\partial H}{\partial t}(y,s)\,f_n(x-y)\,g_m(t-s)\,dy\,ds$$

et

$$\frac{\partial H_{n,m}}{\partial t}\underset{m,n\to\infty}{\longrightarrow}\frac{\partial H}{\partial t}\text{ dans }L^1(\mathbb{R}^2),$$

si bien que :

$$\mathbb{E}\left[\int_\varepsilon^t\left|\frac{\partial H_{n,m}}{\partial t}-\frac{\partial H}{\partial t}\right|(X_s,s)\,ds\right]\underset{m,n\to\infty}{\longrightarrow}0.$$

▷ Pour le dernier terme,

$$\begin{aligned}\frac{\partial^2 H_{n,m}}{\partial x^2}(x,t)&=\int_{\mathbb{R}^2}\frac{\partial H}{\partial x}(y,s)\,f_n'(x-y)\,g_m(t-s)\,dy\,ds\\&\underset{m\to\infty}{\longrightarrow}\int_{\mathbb{R}}\frac{\partial H}{\partial x}(y,t)\,f_n'(x-y)\,dy\text{ dans }L^1(\mathbb{R}^2).\end{aligned}$$

Il reste à s'occuper de la convergence, quand $n\to\infty$, de

$$\int_\varepsilon^t\left(\int_{\mathbb{R}}\frac{\partial H}{\partial x}(y,s)\,f_n'(X_s-y)\,dy\right)ds=\int_{\mathbb{R}}\left(\int_\varepsilon^t f_n(X_s-y)\,h(y,s)\,ds\right)d\nu(y).$$

Soit ν_a la partie absolument continue de ν, ϕ sa densité par rapport à la mesure de Lebesgue ;

$$\int_{\mathbb{R}}\left(\int_\varepsilon^t f_n(X_s-z)\,h(z,s)\,ds\right)d\nu_a(z)$$

$$=\int_\varepsilon^t\left(\int_{\mathbb{R}} f_n(X_s-z)\,h(z,s)\,\phi(z)\,dz\right)ds\,;$$

pour tout $y \in \mathbb{R}$, tout $s > 0$,

$$\int_{\mathbb{R}} f_n(y-z)\,h(z,s)\,\phi(z)\,dz \underset{n\to\infty}{\longrightarrow} h(y,s)\,\phi(y) \text{ dans } L^1(\mathbb{R}^2) \quad \text{et}$$

$$\mathbb{E}\left[\int_\varepsilon^t \left|\int_{\mathbb{R}} f_n(X_s-z)\,h(z,s)\,\phi(z)\,dz - h(X_s,s)\,\phi(X_s)\right| ds\right]$$
$$\leq \frac{1}{\sqrt{2\pi\varepsilon}} \int_{\mathbb{R}\times[\varepsilon,t]} \left|\int_{\mathbb{R}} f_n(y-z)\,h(z,s)\,\phi(z)\,dz - h(y,s)\,\phi(y)\right| dy\,ds \underset{n\to\infty}{\longrightarrow} 0.$$

A l'aide de la continuité de $(L_t^x)_{(x,t)\in\mathbb{R}\times\mathbb{R}_+}$, pour tout $z \in \mathbb{R}$, la suite des mesures

$$\left(A \in \mathcal{R}_+ \to \int_A f_n(X_s - z)\,ds\right)_{n\geq 1}$$

converge (presque sûrement) étroitement vers la mesure

$$A \in \mathcal{R}_+ \to \int_A d_s L_s^z.$$

Vu la condition *iv*), pour ρ-presque tout z,

$$\int_\varepsilon^t f_n(X_s-z)\,h(z,s)\,ds \underset{n\to\infty}{\longrightarrow} \int_\varepsilon^t h(z,s)\,dL_s^z \text{ et}$$

$$\int_{\mathbb{R}}\left(\int_\varepsilon^t f_n(X_s-z)\,h(z,s)\,ds\right) d\rho(z) \underset{n\to\infty}{\longrightarrow} \int_{\mathbb{R}}\left(\int_\varepsilon^t h(z,s)\,d_sL_s^z\right) d\rho(z).$$

On peut en effet se limiter au cas où h est positif, auquel cas (lemme de Scheffé) il suffit de vérifier :

$$\mathbb{E}\left[\int_{\mathbb{R}}\left(\int_\varepsilon^t f_n(X_s-z)\,h(z,s)\,ds\right) d\rho(z)\right]$$
$$\underset{n\to\infty}{\longrightarrow} \mathbb{E}\left[\int_{\mathbb{R}}\left(\int_\varepsilon^t h(z,s)\,d_sL_s^z\right) d\rho(z)\right].$$

Or

$$\mathbb{E}\left[\int_{\mathbb{R}}\left(\int_\varepsilon^t f_n(X_s-z)\,h(z,s)\,ds\right) d\rho(z)\right]$$
$$= \int_{\mathbb{R}^2\times[\varepsilon,t]} f_n(y-z)\,h(z,s)\,e^{-\frac{y^2}{2s}}\,dy\,d\rho(z)\,\frac{ds}{\sqrt{2\pi s}}$$
$$= \int_{\mathbb{R}\times[\varepsilon,t]} h(z,s)\left(\int_{\mathbb{R}} f(y)\,e^{-\frac{(z+\frac{y}{n})^2}{2s}}\,dy\right) d\rho(z)\,\frac{ds}{\sqrt{2\pi s}}$$
$$\underset{\substack{n\to\infty\\ \text{Lebesgue}}}{\longrightarrow} \int_{\mathbb{R}\times[\varepsilon,t]} h(z,s)\,e^{-\frac{z^2}{2s}}\,d\rho(z)\,\frac{ds}{\sqrt{2\pi s}}$$

tandis que

$$\mathbb{E}\left[L_t^z\right] = \int_0^t \frac{1}{\sqrt{2\pi s}} \exp\left(-\frac{z^2}{2s}\right) ds$$

est dérivable par rapport à t et $\sqrt{2\pi}\frac{\partial}{\partial t}\left(\mathbb{E}\left[L_t^z\right]\right) = \frac{1}{\sqrt{t}}e^{-\frac{z^2}{2t}}$.

Remarque 5

Soit F un sous-ensemble fermé de $\mathbb{R}$ et Ψ_F la fonction introduite dans la démonstration de la proposition *2* ;

$$(x,t) \to \Psi_F(x,\alpha-t) = 1_F(x) + 1_{F^c}(x)\,\Phi(\gamma_F(x)-x,\delta_F(x)-x,\alpha-t)$$

vérifie (sur $\mathbb{R}\times[0,\alpha[$) les hypothèses de la proposition *4*. En effet :

▷ si $]-\infty,b[$ $(b<\infty)$ est une composante connexe de F^c, on a, pour $x<b$,

$$\Psi_F(x,t) = (b-x)\int_0^t e^{-\frac{(b-x)^2}{2u}}\frac{du}{\sqrt{2\pi u^3}} = \sqrt{\tfrac{2}{\pi}}\int_{\frac{b-x}{\sqrt{t}}}^{\infty} e^{-\frac{v^2}{2}}\,dv\,;$$

$$\frac{\partial\Psi_F}{\partial x}(x,t) = \sqrt{\frac{2}{\pi t}}e^{-\frac{(b-x)^2}{2t}},\quad \left|\frac{\partial\Psi_F}{\partial x}(x,t)\right| \le \sqrt{\frac{2}{\pi t}}\,;$$

$$\frac{\partial^2\Psi_F}{\partial x^2}(x,t) = \sqrt{\frac{2}{\pi t^3}}(b-x)\,e^{-\frac{(b-x)^2}{2t}},\quad \left|\frac{\partial^2\Psi_F}{\partial x^2}(x,t)\right| \le \sqrt{\frac{2}{e\pi t}}\,;$$

▷ si $]a,b[$ $(-\infty<a<b<\infty)$ est une composante connexe de F^c, soit

$$\ell = b-a \text{ et } L = \bigcup_{k\in\mathbb{Z}}[2k\ell,(2k+1)\ell]\,;$$

on a, pour $x\in\,]a,b[$,

$$\begin{aligned}
\frac{\partial\Psi_F}{\partial x}(x,t) &= \frac{1}{\sqrt{2\pi t}}\left(\int_{x-a+L}\frac{u}{t}e^{-\frac{u^2}{2t}}\,du + \int_{-(x-a)+L}\frac{u}{t}e^{-\frac{u^2}{2t}}\,du\right),\\
\left|\frac{\partial\Psi_F}{\partial x}(x,t)\right| &\le \frac{1}{\sqrt{2\pi t}}\int_{\mathbb{R}}\left(1_{\{u-x+a\in L\}}+1_{\{u+x-a\in L\}}\right)\frac{|u|}{t}e^{-\frac{u^2}{2t}}\,du\\
&\le \sqrt{\frac{2}{\pi t}}\int_{\mathbb{R}}\frac{|u|}{t}e^{-\frac{u^2}{2t}}\,du = 2\sqrt{\frac{2}{\pi t}}.
\end{aligned}$$

On montre de même :

$$\begin{aligned}
\frac{\partial^2\Psi_F}{\partial x^2}(x,t) &= \frac{1}{\sqrt{2\pi t^3}}\int_{\mathbb{R}}\left(1_{\{u-x+a\in L\}}+1_{\{u+x-a\in L\}}\right)\left(1-\frac{u^2}{t}\right)e^{-\frac{u^2}{2t}}\,du\\
\left|\frac{\partial^2\Psi_F}{\partial x^2}(x,t)\right| &\le \sqrt{\frac{2}{\pi t^3}}\int_{\mathbb{R}}\left|1-\frac{u^2}{t}\right|e^{-\frac{u^2}{2t}}\,du \le \frac{4}{t}.
\end{aligned}$$

Par ailleurs,

$$\frac{\partial\Psi_F}{\partial t}(x,t) = 1_{F^c}(x)\,\frac{1}{2}\frac{\partial^2\Psi_F}{\partial x^2}(x,t)\,.$$

En outre, pour $t>0$, $x\to\Psi_F(x,t)$ a pour dérivée seconde, au sens des distributions,

$$\begin{aligned}
&1_{F^c}\frac{\partial^2\Psi_F}{\partial x^2}(.,t)\cdot\lambda + \sum_{\substack{]h,k[\text{ composante}\\ \text{connexe de } F^c}}\left(\frac{\partial\Psi_F}{\partial x}(h+,t)-\frac{\partial\Psi_F}{\partial x}(k-,t)\right)(\delta_h+\delta_k)\\
&= 1_{F^c}\tfrac{\partial^2\Psi_F}{\partial x^2}(.,t)\cdot\lambda + \tfrac{1}{\sqrt{2\pi t}}\sum_{\substack{]h,k[\text{ composante}\\ \text{connexe de } F^c}}\vartheta\left(\tfrac{k-h}{\sqrt{t}}\right)(\delta_h+\delta_k)\,;
\end{aligned}$$

enfin, ϑ est continue sur $\mathbb{R}_+$ et, si $M = \sup_{x>0} \frac{\vartheta(x)}{x}$, on a pour tout $A > 0$:

$$\sum_{\substack{]h,k[\text{ composante connexe de } F^c, \\ h \text{ ou } k \in]-A,A[}} \vartheta\left(\frac{k-h}{\sqrt{t}}\right) \leq 2 + \frac{M}{\sqrt{t}} \lambda\left[F^c \cap]-A, A[\right] \quad \square$$

Bibliographie.

[1] AZEMA J., JEULIN T., KNIGHT F., MOKOBODZKI G., YOR M. Sur les processus croissants de type injectif. Séminaire de Probabilités XXX, 312-343, Lect. Notes in Math 1626, Springer 1996.

[2] FELLER W. : An Introduction to Probability Theory and its applications. Tome 2. Wiley 1971.

[3] FREEDMAN D. : Brownian Motion and Diffusion. Holden Day, 1971.

The Brownian Burglar: conditioning Brownian motion by its local time process

J. WARREN[1] and M. YOR[2]

Imagine a Brownian crook who spent a month in a large metropolis. The number of nights he spent in hotels A,B,C...etc. is known; but not the order, nor his itinerary. So the only information the police has is total hotel bills.....

Let $(W_t; t \geq 0)$ be reflecting Brownian motion issuing from zero, and let $l(t,y)$, for $y \in \mathbb{R}^+$ and $t \geq 0$, denote the local time that W has accrued at level y by time t. Throughout this paper our normalisation of local time is such that it is an occupation density with respect to Lebesgue measure. Let T_1 be the first time t such that $W_t = 1$. The celebrated Ray-Knight theorem describes the law of $\left(l(T_1, 1-y); 0 \leq y \leq 1\right)$ as being that of a diffusion; specifically a squared Bessel process of dimension two, started from zero. The question now naturally arises of obtaining some description of W conditional on these local times.

To begin one may look for functionals of W for which we can describe the conditional law. Such a functional is the process $\left(l(T_a, y); 0 \leq y \leq a\right)$ for some $a \in (0,1)$. Specifically we find that

$$\frac{l(T_a, y)}{l(T_1, y)} = Y^{2,0}_{\int_y^a \frac{dz}{l(T_1,z)}}, \tag{0.1}$$

where $Y^{2,0}$ is a diffusion, independent of $\left(l(T_1, 1-y); 0 \leq y \leq 1\right)$, and with generator $2y(1-y)D^2 + 2(1-y)D$. It belongs to a class of diffusions on $[0,1]$ known as Jacobi diffusions. We were then motivated to try and obtain a process $\hat{W}$, which we call the burglar, whose local times would give rise to such diffusions. Define $\hat{W}$ via the space-time change

$$\theta(W_t) = \hat{W}_{A_t}, \tag{0.2}$$

for $0 \leq t < T_1$, where

$$A_t = \int_0^t \frac{ds}{\left(l(T_1, W_s)\right)^2} \qquad \text{and} \qquad \theta(y) = \int_0^y \frac{dz}{l(T_1, z)}.$$

With probability one, the function θ maps $[0,1)$ onto $\mathbb{R}^+$, and A maps $[0, T_1)$ onto $[0, \infty)$; both being continuous and strictly increasing. The main result of this paper is to show that the burglar, so defined, is independent of the local times accrued by W at time T_1.

Theorem 1. *The reflecting Brownian motion* $\left(W_t; 0 \leq t < T_1\right)$ *admits the representation* (0.2) *in terms of its local times* $\left(l(T_1, y); 0 \leq y \leq 1\right)$ *and an independent burglar* $\left(\hat{W}_u; 0 \leq u < \infty\right)$.

[1]University of Warwick, United Kingdom
[2]Université Pierre et Marie Curie, Paris, France

Although we leave to a future article a thorough study of the law of the burglar, including Markovian properties and martingale characterisations, already (0.1) may be interpreted as a Ray-Knight theorem for the local times of $\hat{W}$. In fact, we now present burglar variants of the two classical Ray-Knight theorems for Brownian local times. We first observe that the burglar $\hat{W}$ possesses a jointly continuous local time process $(\rho(t,y); t \geq 0, y \geq 0)$. Recall the definitions of θ and A.

Lemma 2. *Define, for $0 \leq t < T_1$, and $0 \leq y < 1$,*

$$\rho(A_t, \theta(y)) = \frac{l(t,y)}{l(T_1,y)},$$

then ρ is a jointly continuous local time process for the burglar, in that for any positive measurable function f on $\mathbb{R}^+$,

$$\int_0^t f(\hat{W}_s)\, ds = \int_0^\infty f(y)\rho(t,y)\, dy.$$

Theorem 3. *The local times $(\rho(t,y); y \in \mathbb{R}^+, t \in \mathbb{R}^+)$ of a Brownian burglar $\hat{W}$ admit the following descriptions.*

For $a > 0$ let $\hat{T}_a = \inf\{t : \hat{W}_t = a\}$. Then we have

$$(\rho(\hat{T}_a, a-y); 0 \leq y \leq a) \stackrel{law}{=} (Y_y^{2,0}; 0 \leq y \leq a),$$

where $Y_0^{2,0} = 0$.

For $0 \leq s \leq 1$ let $\hat{\tau}_s = \inf\{u : \rho(u,0) \geq s\}$. Then we have

$$(\rho(\hat{\tau}_s, y); y \geq 0) \stackrel{law}{=} (Y_y^{0,2}; y \geq 0),$$

where $Y_0^{0,2} = s$.

In the above $Y^{0,2}$ denotes a Jacobi diffusion with generator $2y(1-y)D^2 - 2yD$.

Our paper is organised as follows.

- Section 1 contains a discussion showing how a group action on a probability space can induce a factorisation. This is illustrated with reference to the standard Brownian bridge and the skew-product representation of planar Brownian motion.

- In Section 2 we prove the independence of the burglar and $(l(T_1,y); 0 \leq y \leq 1)$ as an application of the general method presented in the previous section.

- Section 3 contains a proof of the Ray-Knight theorems for the burglar (Theorem 3 above).

- In Section 4 we give an application of the burglar to the problem of describing W conditional on $(l(\tau_1,y); y \in \mathbb{R}^+)$, where $\tau_1 = \inf\{t : l(t,0) = 1\}$. In order to do this we must decompose the path of W at its maximum. The result of this section can be seen as describing a contour process for the Fleming-Viot process, and should be compared with the skew-decomposition of super-Brownian motion in terms of its total mass process and an independent Fleming-Viot process achieved by March and Etheridge [7]; see also Dawson [5].

1 Group actions and factorisations

Suppose that $(\Omega, \mathcal{F}, \mu)$ is a probability space and that $\mathcal{G}$ a sub-σ-algebra of $\mathcal{F}$. It is well known (and the cause of much grief!) that, in general, there are many different independent complements to $\mathcal{G}$. That is sub-σ-algebras $\mathcal{H}$ such that $\mathcal{F} = \mathcal{G} \vee \mathcal{H}$ and such that $\mathcal{G}$ and $\mathcal{H}$ are independent. However it is usual to have some additional structure on Ω which allows one to single out some distinguished complement in a natural manner. Here we will be concerned with cases in which this additional structure arises from the action of a group G on Ω.

We suppose that we have a second probability space $(E, \mathcal{E}, \nu)$ on which there is a G-action also defined; and a measurable map $\phi : \Omega \mapsto E$ so that ν is the image of μ under ϕ, and so that ϕ is a homomorphism of G-spaces, that is,

$$\phi(g\omega) = g\phi(\omega), \tag{1.1}$$

for all $g \in G$ and $\omega \in \Omega$. We assume that the measures μ and ν are quasi-invariant under the action of G on Ω and E, and denote their images under the transformation associated with an element $g \in G$ by μ^g and ν^g respectively. In the dynamical systems literature the space E is known as a *factor* of Ω, see, for example, Cornfeld, Fomin and Sinai [4].

We are interested in conditions under which the space $(\Omega, \mathcal{F}, \mu)$ can be identified with the product of $(E, \mathcal{E}, \nu)$ and some complementary space. To develop this fully would involve us in much measure theoretic detail, as described by Rohlin [16]. This would be, for our purpose, unnecessary, and we can be satisfied with the following elementary lemma. Recall that the action of G on E is said to be ergodic if every $\mathcal{E}$-measurable, G-invariant function on E is constant modulo a set of measure zero.

Lemma 4. *Consider a probability space $(\Omega, \mathcal{F}, \mu)$ upon which a group G acts. Let ϕ be a homomorphism between this space and a factor $(E, \mathcal{E}, \nu)$. Let $\mathcal{G} = \sigma(\phi)$ and $\mathcal{H}$ be the σ-algebra of G-invariant subsets of Ω, thus*

$$\mathcal{H} = \{H \in \mathcal{F} : \mu(H \Delta gH) = 0 \text{ for all } g \in G\}.$$

Then if the Radon-Nikodým densities $d\mu^g/d\mu$ are $\mathcal{G}$-measurable for all $g \in G$, and if the action of G on E is ergodic, the σ-algebras $\mathcal{G}$ and $\mathcal{H}$ are independent.

Proof. Let $H \in \mathcal{H}$ and consider the conditional expectation $\mu[\mathcal{I}_H|\mathcal{G}]$. For any $g \in G$, we find that for μ-almost all $\omega \in \Omega$,

$$\mu[\mathcal{I}_H|\mathcal{G}](\omega) = \mu^g[\mathcal{I}_{gH}|g\mathcal{G}](g\omega) = \mu^g[\mathcal{I}_H|\mathcal{G}](g\omega).$$

But, since $d\mu^g/d\mu$ is $\mathcal{G}$-measurable, we have

$$\mu^g[\mathcal{I}_H|\mathcal{G}] = \mu[\mathcal{I}_H|\mathcal{G}].$$

Thus $\mu[\mathcal{I}_H|\mathcal{G}](\omega) = \mu[\mathcal{I}_H|\mathcal{G}](g\omega)$ for μ-almost all ω. Now the ergodicity of the G-action implies that $\mu[\mathcal{I}_H|\mathcal{G}]$ is equal to some constant μ-almost always, and thus the σ-algebras $\mathcal{G}$ and $\mathcal{H}$ are independent. □

This lemma says nothing to guarantee that $\mathcal{F} = \mathcal{G} \vee \mathcal{H}$; indeed in general this is manifestly false. However in the examples we consider it will be evident that $\mathcal{G}$ and $\mathcal{H}$ do generate everything.

We now illustrate the above discussion with some concrete examples. It is important to stress that there are alternative treatments of these than the one we will describe. However we hope that our rather unusual approach will pay dividends in demonstrating that the Burglar is constructed in a very natural way.

Our first example is that of the Brownian bridge. Let $(X_t; 0 \leq t \leq 1)$ be a Brownian motion issuing from zero. We wish to find an independent complement to X_1. We take $(\Omega, \mathcal{F}, \mu)$ to be Wiener space and consider $(X_t; 0 \leq t \leq 1)$ to be the co-ordinate projection maps on Ω. Introduce the action of the group $G \equiv (\mathbb{R}, +)$ on Ω by defining, for any $a \in G$ and $\omega \in \Omega$,

$$X_t(a\omega) = X_t(\omega) + at, \tag{1.2}$$

for $0 \leq t \leq 1$. Of course the measure μ^a is the law of Brownian motion with drift a, and the absolute continuity relation,

$$\frac{d\mu^a}{d\mu} = \exp\{aX_1 - \tfrac{1}{2}a^2\}, \tag{1.3}$$

holds. For the factor we take the map $\phi \equiv X_1$, and the probability space $(E, \mathcal{E}, \nu)$ is just the real line equipped with its Borel σ-algebra and standard Gaussian measure, with G acting by translation. Now consider the bridge $(\hat{X}_t; 0 \leq t \leq 1)$, defined by

$$\hat{X}_t(\omega) = X_t(\omega) - X_1(\omega)t, \tag{1.4}$$

for $0 \leq t \leq 1$. Note $\hat{X}$ is invariant under the action of G:

$$\hat{X}_t(a\omega) = \hat{X}_t(\omega), \tag{1.5}$$

for all $\omega \in \Omega$ and $a \in G$. Thus, since G acts transitively on E, and the Radon-Nikodým derivatives (1.3) are $\sigma(\phi)$-measurable, Lemma 4 is applicable and the bridge $\hat{X}$ is independent of X_1. Bridges of the gamma process, see Vershik and Yor [17], may be treated in exactly the same manner.

Our second example, slightly more involved, is that of the decomposition of planar Brownian motion into its radial and angular parts. However before we present this, let us introduce a group and a probability space that will play a central role both in this example and in the construction of the burglar. Let G be the group of increasing C^2-diffeomorphisms of $[0,1]$. Let $(E, \mathcal{E}, \nu)$ be the space of continuous paths indexed by $[0,1]$, together with the usual Borel σ-algebra, and ν the law of the squared Bessel process of dimension two starting from zero. Let $(Z_t; 0 \leq t \leq 1)$ be the co-ordinate projection maps on E. We can define an action of G on E as follows. For any $g \in G$ define for each $\omega \in E$ its image $g\omega$ under the action of g via,

$$Z_{g(t)}(g\omega) = g'(t) Z_t(\omega), \tag{1.6}$$

for all $0 \leq t \leq 1$. Under ν^g, $(Z_t; 0 \leq t \leq 1)$ is a time-inhomogeneous diffusion with generator

$$2z\frac{d^2}{dz^2} + (2F_g(t)z + 2)\frac{d}{dz}, \tag{1.7}$$

where $F_g(t) = -h''(t)/2h'(t)$, if $h = g^{-1}$. Such diffusions are considered by Pitman and Yor [13].

Lemma 5. *The action of G on $(E, \mathcal{E}, \nu)$ is ergodic.*

Proof. The law ν^g is absolutely continuous with respect to ν with Radon-Nikodým derivative

$$\frac{d\nu^g}{d\nu} = \exp\left\{\int_0^1 F_g(t)dM_t - \tfrac{1}{2}\int_0^1 F_g^2(t)Z_t dt\right\},$$

where $2M_t = Z_t - 2t$. Let Ξ be the collection of random variables which are proportional to $d\nu^g/d\nu$ for some $g \in G$, and lie in $\mathcal{L}^2(\nu)$. We will show that Ξ is total in $\mathcal{L}^2(\nu)$. The ergodicity of the G-action follows easily from this, for if Φ is a bounded random variable that is invariant under the action of G, then $\Phi - \int_E \Phi d\nu$ is orthogonal to each member of Ξ, and thus almost surely zero.

Suppose that F_g is continuously differentiable, then

$$F_g(1)Z_1 = \int_0^1 F_g(t)dZ_t + \int_0^1 Z_t dF_g(t),$$

and we may write

$$\frac{d\nu^g}{d\nu} = \exp\left\{\tfrac{1}{2}F_g(1)Z_1 - \int_0^1 F_g(t)dt - \tfrac{1}{2}\int_0^1 Z_t\left[dF_g(t) + F_g^2(t)dt\right]\right\}.$$

So the $\mathcal{L}^2$-closure of Ξ contains every random variable of the form

$$\exp\left\{-\tfrac{1}{2}\int_0^1 Z_t\alpha(dt)\right\},$$

where α is a positive measure on $[0,1]$; and the collection of all such random variables is closed under multiplication and generates $\mathcal{E}$. Thus if Φ is orthogonal to Ξ, on application of the monotone class lemma, we may deduce that Φ is orthogonal to all of $\mathcal{L}^2(\nu)$ and must be zero. □

Returning to the planar Brownian motion let $(X_t; t \in [0,1])$ be the co-ordinate projection maps on Ω, the space of all continuous paths in the plane $\mathbb{R}^2$, issuing from the origin, indexed by time in $[0,1]$. We may take μ to be the law of planar Brownian motion issuing from zero. We define the action of G on Ω as follows. For $g \in G$ the image of $\omega \in \Omega$ under the action of g satisfies

$$X_{g(t)}(g\omega) = \sqrt{g'(t)}X_t(\omega), \tag{1.8}$$

for all $t \in [0,1]$. Under μ^g the process X has a radial drift. In fact it satisfies

$$X_t = \beta_t + \int_0^t F_g(s)X_s\, ds, \tag{1.9}$$

where β, under μ^g, is a planar Brownian motion. The measure μ^g is absolutely continuous with respect to μ with the Radon-Nikodým derivative being

$$\frac{d\mu^g}{d\mu} = \exp\left\{\int_0^1 F_g(t)dM_t - \tfrac{1}{2}\int_0^1 F_g^2(t)|X_t|^2 dt\right\}, \tag{1.10}$$

where $2M_t = |X_t|^2 - 2t$.

The factor space is to be the radial part of X_t. Specifically define the operator $\phi : \Omega \mapsto E$ by,

$$Z_t(\phi\omega) = |X_t(\omega)|^2, \tag{1.11}$$

for all $t \in [0,1]$. It is well known that the measure ν, the law of the squared Bessel process of dimension two starting from zero, is the image of μ under ϕ. Observe that the Radon-Nikodým derivative (1.10) is $\sigma(\phi)$-measurable. Now consider the angular part of X appropriately time-changed:

$$\hat{X}_{\int_t^1 \frac{dv}{|X_v|^2}} = \frac{X_t}{|X_t|}. \tag{1.12}$$

With probability one, this defines a process $(\hat{X}_u; 0 \le u < \infty)$, which is, in fact, a Brownian motion on the unit circle. It is easy to check that $\hat{X}$ is invariant under the action of G on Ω, and consequently we deduce, with the aid of Lemma 4, that it is independent of the radial process.

Define $E_0 \subset E$ by,

$$E_0 = \left\{ \eta \in E : Z_t(\eta) > 0 \text{ for } 0 < t \le 1 \quad \text{and} \quad \int_{0+}^1 \frac{dt}{Z_t(\eta)} = \infty \right\}. \tag{1.13}$$

Observe that $\nu(E_0) = 1$. For $\eta \in E_0$ define a process $(\hat{X}_t^\eta; 0 \le t \le 1)$ on the probability space $(\Omega, \mathcal{F}, \mu)$, via,

$$\hat{X}_t^\eta(\omega) = Z_t(\eta) \hat{X}_{\int_t^1 \frac{dv}{Z_v(\eta)}}(\omega). \tag{1.14}$$

$\hat{X}^\eta$ is a process with radial part $Z_t(\eta)$ and (time-changed) angular part $\hat{X}_u$. Its law μ_η is supported on the fibre $\phi^{-1}(\eta) \subset \Omega$. The family of laws $(\mu_\eta; \eta \in E_0)$ form a regular probability distribution for μ given ϕ, see Parthasarathy [12].

2 The Burglar

We shall now apply the technique we have demonstrated in the previous section to our original problem of conditioning with respect to local times. For presentational reasons it is convenient to reverse the rôles that zero and one take in the introduction. Thus our Brownian motion is reflected down from level one and stopped on first reaching level zero.

We take Ω to be the space of continuous paths taking values in the interval $[0,1]$, starting from 1, and stopped at $T_0(\omega)$ which is the first time the path reaches 0. The σ-algebra $\mathcal{F}$ is the Borel σ-algebra generated by the uniform topology, and the measure μ will be the law of Brownian motion on $[0,1]$ with reflection at the boundaries, stopped on hitting level 0. Denote the co-ordinate projections by $(X_t; 0 \le t \le T_0)$. A path admits a bicontinuous local time $(l(t,y); 0 \le t \le T_0, 0 \le y \le 1)$ with probability one, and we extend l to the whole of Ω, defining it to be identically zero otherwise. We will be concerned with exactly the same group G as that featured in the second example of the preceding section, but this time the group action is defined very differently. For $g \in G$ define the image of $\omega \in \Omega$ under the action of g via,

$$g(X_t(\omega)) = X_{H_t}(g\omega), \tag{2.1}$$

where

$$H_t = \int_0^t g'(X_s(\omega))^2 ds. \tag{2.2}$$

It is simple to check that this does indeed define a G-action. Under μ^g the co-ordinate process satisfies

$$X_t = \beta_t - \tfrac{1}{2}l(t,1) + \int_0^t F_g(X_s)ds, \tag{2.3}$$

where β is a μ^g-Brownian motion. Thus μ^g is absolutely continuous with respect to μ and the Radon-Nikodým derivative is given by,

$$\frac{d\mu^g}{d\mu} = \exp\left\{\int_0^{T_0} F_g(X_t)dB_t - \tfrac{1}{2}\int_0^{T_0} F_g^2(X_t)dt\right\}, \tag{2.4}$$

where $B_t = X_t + \frac{1}{2}l(t,1)$.

We retain the space $(E, \mathcal{E}, \nu)$ as before; but $\phi : \Omega \mapsto E$ now takes the form:

$$Z_y(\phi\omega) = l(T_0, y)(\omega), \tag{2.5}$$

for $y \in [0,1]$. The Ray-Knight theorem states that ν is indeed the image of μ under ϕ. That ϕ is a homomorphism of G-spaces follows immediately from the following lemma.

Lemma 6. *If the path ω admits a bicontinuous local time $\left(l(t,y); 0 \le t \le T_0, 0 \le y \le 1\right)$, then the path $g\omega$ defined by equation (2.1) does also, and denoting this by $\left(l^g(t,y); 0 \le t \le T_0^g, 0 \le y \le 1\right)$, the two are related by,*

$$l^g(H_t, g(y)) = g'(y)l(t,y),$$

for all $y \in [0,1]$ and $0 \le t \le T_0$.

Proof. For any bounded, measurable test function f we have, for $0 \le t \le T_0$,

$$\int_0^{H_t} f(X_v(g\omega))dv = \int_0^t f(X_{H_s}(g\omega))dH_s = \int_0^t f(g(X_s(\omega)))g'(X_s(\omega))^2 ds$$
$$= \int_0^1 f(g(y))g'(y)^2 l(t,y)dy.$$

Thus, defining $l^g(H_t, g(y)) = g'(y)l(t,y)$, we find, substituting $u = H_t$ and $z = g(y)$,

$$\int_0^u f(X_v(g\omega))dv = \int_0^1 f(z)l^g(u,z)dz,$$

and so l^g forms a bicontinuous local time for $g\omega$. □

The arguments of Bouleau and Yor, [3], show that,

$$-\int_0^1 F_g(y)dy = \int_0^{T_0} F_g(X_t)dB_t - \tfrac{1}{2}\int_0^1 F_g(y)d_y l(T_0, y). \tag{2.6}$$

Using this, the Radon-Nikodým derivative (2.4) may be re-written in the form:

$$\frac{d\mu^g}{d\mu} = \exp\left\{\int_0^1 F_g(y)dM_y - \tfrac{1}{2}\int_0^1 F_g^2(y)l(T_0, y)dy\right\}, \tag{2.7}$$

where $2M_y = l(T_0, y) - 2y$, and is now evidently $\sigma(\phi)$-measurable. Define $\hat{X}$ via the space-time change

$$\theta(X_t) = \hat{X}_{A_t}, \tag{2.8}$$

for $0 \leq t < T_0$, where

$$A_t = \int_0^t \frac{ds}{\left(l(T_0, X_s)\right)^2} \qquad \text{and} \qquad \theta(y) = \int_y^1 \frac{dz}{l(T_0, z)}. \tag{2.9}$$

With probability one, $(\hat{X}_u; 0 \leq u < \infty)$ is well-defined, and satisfies $\lim_{u\to\infty} \hat{X}_u = \infty$. The process $(\hat{X}_u; u \geq 0)$, or more generally any process with the same law, will be called a Brownian burglar. The following invariance property is the key to proving the independence claimed in Theorem 1.

Lemma 7. *The burglar $\hat{X}$ defined by equations* (2.8) *and* (2.9) *is invariant under the action of G on Ω,*

Proof. Fix $g \in G$. Write A^g and θ^g for the functions A and θ translated by the action of g; that is

$$A_t^g(\omega) = A_t(g\omega) \qquad \text{and} \qquad \theta^g(y)(\omega) = \theta(y)(g\omega).$$

Now we make repeated use of the previous lemma. For any $y \in (0, 1]$,

$$\theta^g(g(y)) = \int_{g(y)}^1 \frac{dz}{l^g(T_0^g, z)} = \int_y^1 \frac{g'(x)dx}{l^g(T_0^g, g(x))} = \int_y^1 \frac{dx}{l(T_0, x)} = \theta(y).$$

Similarly we find,

$$A_{H_t}^g = \int_0^{H_t} \frac{ds}{l^g(T_0^g, X_s(g\omega))^2} = \int_0^t \frac{dH_u}{l^g(T_0^g, X_{H_u}(g\omega))^2}$$

$$= \int_0^t \frac{g'(X_u(\omega))^2 du}{l^g(T_0^g, g(X_u(\omega)))^2} = \int_0^t \frac{du}{l(T_0, X_u(\omega))^2} = A_t.$$

Now consider the definition of the burglar,

$$\theta(X_t(\omega)) = \hat{X}_{A_t}(\omega),$$

and the same relationship at $g\omega$,

$$\theta^g(X_t(g\omega)) = \hat{X}_{A_t^g}(g\omega).$$

Replacing t by H_t, this latter equation becomes,

$$\theta^g(X_{H_t}(g\omega)) = \hat{X}_{A^g_{H_t}}(g\omega),$$

and now, using equation (2.1) and the relationships that we have just derived, we obtain,

$$\theta(X_t(\omega)) = \hat{X}_{A_t}(g\omega).$$

But comparing this with the original equation defining the burglar, we deduce,

$$\hat{X}_u(g\omega) = \hat{X}_u(\omega),$$

for all u, and the invariance s proven. □

In order to prove Theore n 1 we apply Lemma 4, having now confirmed that its premises hold.

Recall the definition of E , made in the previous section. For $\eta \in E_0$, let $k : \mathbb{R}^+ \mapsto (0,1]$ be the function

$$k(\ \iota) = \sup\left\{y : \int_y^1 \frac{dz}{\eta(z)} \geq a\right\}, \tag{2.10}$$

and then define a process $\hat{X}^\eta$ on $(\Omega, \mathcal{F}, \mu)$ via the space-time change,

$$k(\hat{X}_t) = \hat{X}^\eta_{K_t}, \tag{2.11}$$

where

$$K_t = \int_0^t k'(\hat{X}_s)^2 ds.$$

The process $(\hat{X}^\eta_t; 0 \leq t < T_0)$ can be thought of as X conditioned on $l(T_0, \cdot) = \eta(\cdot)$. In fact, if we denote the law of $\hat{X}^\eta$, which is supported on $\phi^{-1}(\eta) \subset \Omega$, by μ_η, then the family $(\mu_\eta : \eta \in E_0)$ form a regular probability distribution for μ given ϕ.

3 Some Ray-Knight Theorems

We denote by $Y^{d,d'}$ a diffusion on $[0,1]$ with infinitesimal generator

$$2y(1-y)D^2 + (d - (d+d')y)D. \tag{3.1}$$

These diffusions, called Jacobi diffusions with dimensions d and d', have been well studied, particularly in relations to models in genetics, see for example Ethier-Kurtz [8], Karlin-Taylor [9] and Kimura [10], or more recently in financial models, see Delbaen-Shirakawa [6]. For other studies and motivations, including hypercontractivity, see Bakry [1] and Mazet [11]. Some further results are given by the authors of this paper in [18], where the Jacobi diffusions are introduced via the following proposition.

Proposition 8. *Let $(Z_t; t \geq 0)$ and $(Z'_t; t \geq 0)$ be two independent squared Bessel processes of dimensions d and d′ starting from z and z′ respectively, with $z + z' > 0$, and let $T = \inf\{u : Z_u + Z'_u = 0\}$. Then there exists a Markov process $(Y_u^{d,d'} : u \geq 0)$, a diffusion on $[0,1]$ with infinitesimal generator given by (3.1) such that*

$$\frac{Z_t}{Z_t + Z'_t} = Y^{d,d'}_{\left(\int_0^t \frac{ds}{Z_s + Z'_s}\right)}, \qquad \text{for } 0 \leq t < T,$$

with $Y^{d,d'}$ being independent of $(Z_t + Z'_t; t \geq 0)$.

The above skew-product decomposition also holds when Z and Z′ are replaced by processes $\tilde{Z}$ and $\tilde{Z}'$ obtained from Z and Z′ via,

$$\tilde{Z}_t = \frac{1}{u'(t)} Z_{u(t)} \qquad \text{and} \qquad \tilde{Z}'_t = \frac{1}{u'(t)} Z'_{u(t)},$$

with $u : [0, \infty) \mapsto [0, \infty)$ a strictly increasing, C^1-function, and $u(0) = 0$.

Proof. We give a proof based on the stochastic calculus. The squared Bessel processes Z and Z' satisfy

$$Z_t = z + \int_0^t 2\sqrt{Z_s}\, d\beta_s + dt,$$
$$Z'_t = z' + \int_0^t 2\sqrt{Z'_s}\, d\beta'_s + d't,$$

where β and β' are independent Brownian motions. Now we sum these two expressions and use 'Pythagoras':

$$Z_t + Z'_t = z + z' + \int_0^t 2\sqrt{Z_s + Z'_s}\, d\gamma_s + (d + d')t,$$

where γ is the Brownian motion:

$$\gamma_t = \int_0^t \frac{\sqrt{Z_s}\, d\beta_s + \sqrt{Z'_s}\, d\beta'_s}{\sqrt{Z_s + Z'_s}},$$

defined up to the time T. Now one introduces,

$$\xi_t = \frac{Z_t}{Z_t + Z'_t},$$

for $0 \leq t < T$. We deduce with the aid of Itô's formula that

$$\xi_t = \xi_0 + 2\int_0^t \sqrt{\xi_s(1 - \xi_s)} \frac{\sqrt{1 - \xi_s}\, d\beta_s - \sqrt{\xi_s}\, d\beta'_s}{\sqrt{Z_s + Z'_s}} + \int_0^t (d - (d + d')\xi_s) \frac{ds}{Z_s + Z'_s}.$$

The process $Y^{d,d'}$ is obtained as a time-change of ξ, thus $Y_u^{d,d'} = \xi_{\alpha_u}$ where $\alpha_u = \inf\{t : \int_0^t ds/(Z_s + Z'_s) \geq u\}$. Applying this time-change to the above equation we obtain

$$Y_u^{d,d'} = Y_0^{d,d'} + 2\int_0^u \sqrt{Y_v^{d,d'}(1 - Y_v^{d,d'})}\, d\hat{\beta}_v + \int_0^u (d - (d + d')Y_v^{d,d'})\, dv,$$

where $\hat{\beta}$ is the time-change of

$$\int_0^t \frac{\sqrt{1-\xi_s}d\beta_s - \sqrt{\xi_s}d\beta'_s}{\sqrt{Z_s + Z'_s}}.$$

Observe that this martingale is orthogonal to γ, and thus we deduce using Knight's Theorem on continuous orthogonal martingales that the Brownian motions $\hat{\beta}$ and γ are independent. Now $Y^{d,d'}$ is adapted to the filtration of $\hat{\beta}$, and consequently independent of $(Z_t + Z'_t; t \geq 0)$.

The extension to $\tilde{Z}$ and $\tilde{Z}'$ follows immediately on making the deterministic time change $t \mapsto u(t)$. □

Before proceeding to the proofs of the Ray-Knight theorems for the burglar, we must prove Lemma 2 which confirms that the burglar possesses local times.

Proof of Lemma 2. This is really the same argument as for Lemma 6, but it bears repeating. For any bounded and compactly supported, measurable test function f on $[0,1)$, we have,

$$\int_0^{A_t} f(\hat{W}_v)dv = \int_0^t f(\hat{W}_{A_s})dA_s = \int_0^t f(\theta(W_s))\frac{ds}{l(T_1,W_s)^2}$$
$$= \int_0^1 f(\theta(y))\frac{l(t,y)}{l(T_1,y)^2}dy.$$

Consequently, if ρ is defined by,

$$\rho(A_t, \theta(y)) = \frac{l(t,y)}{l(T_1,y)},$$

for $y \in [0,1)$ and $0 \leq t < T_1$, then, substituting $z = \theta(y)$ and $u = A_t$, we obtain,

$$\int_0^u f(\hat{W}_v)dv = \int_0^\infty f(z)\rho(u,z)dz.$$

□

Proof of first part of Theorem 3. Fix $x \in (0,1)$. Consider the local times of the reflecting Brownian motion $(W_t; 0 \leq t \leq T_1)$, stopped when it first hits level 1, split into a contribution from before time $T_x = \inf\{u : W_u = x\}$ and a contribution from between times T_x and T_1. It follows from the Ray-Knight theorems for Brownian motion, and arguments familiar in excursion theory, that

$$\left(l(T_x,y), l(T_1,y) - l(T_x,y); 0 \leq y \leq x\right) \stackrel{law}{=} \left(Z^2(x-y), Z^0(x-y); 0 \leq y \leq x\right),$$

where Z^2 is a squared Bessel process of dimension two starting from zero and Z^0 is an independent squared Bessel process of dimension zero starting from $l(T_1,x)$. By Lemma 2, proved above,

$$\rho(\hat{T}_{\theta(x)}, \theta(y)) = \frac{l(T_x,y)}{l(T_1,y)},$$

and so we deduce from Proposition 8 that

$$(\rho(\hat{T}_{\theta(x)}, \theta(x) - y); 0 \le y \le \theta(x)) \stackrel{law}{=} (Y_y^{2,0}; 0 \le y \le \theta(x)),$$

where $Y_0^{2,0} = 0$. Now ρ is independent of $\theta(x)$, since the burglar is independent of $(l(T_1, y); y \in \mathbb{R}^+)$, whence for each fixed a we must have,

$$(\rho(\hat{T}_a, a - y); 0 \le y \le a) \stackrel{law}{=} (Y_y^{2,0}; 0 \le y \le a),$$

as desired. □

Proof of second part of Theorem 3. Fix $s \in [0,1]$. Let $\sigma = l(T_1, 0)$, and then $\tau_{s\sigma} = \inf\{t : l(t,0) \ge sl(T_1, 0)\}$. This time we split the local times that W has attained by time T_1 into a contribution from before time $\tau_{s\sigma}$, and a contribution from between times $\tau_{s\sigma}$ and T_1. We find that,

$$(l(\tau_{s\sigma}, y), l(T_1, y) - l(\tau_{s\sigma}, y); 0 \le y \le 1) \stackrel{law}{=} (Z^0_{s\sigma \to 0}(y), Z^2_{(1-s)\sigma \to 0}(y); 0 \le y \le 1),$$

where $Z^d_{x \to 0}$ denotes the bridge of the squared Bessel process of dimension d, from x to 0. The two bridges appearing in the above equation are taken to be independent. By virtue of Lemma 2 we have, since $A_{\tau_{s\sigma}} = \hat{\tau}_s$,

$$\rho(\hat{\tau}_s, \theta(x)) = \frac{l(\tau_{s\sigma}, x)}{l(T_1, x)},$$

and since, see Revuz and Yor, [15],

$$(Z^d_{x \to 0}(t); 0 \le t \le 1) \stackrel{law}{=} ((1-t)^2 Z^d_{t/(1-t)}; 0 \le t \le 1)$$

we may apply Proposition 8 to obtain the result. □

4 Stopping at τ_1

In this section we describe $(W_t; 0 \le t \le \tau_1)$ conditional on $(l(\tau_1, y); y \in \mathbb{R}^+)$ where, as usual, $\tau_1 = \inf\{u : l(u, 0) \ge 1\}$.

Theorem 9. *We consider a reflecting Brownian motion,* $(W_t; 0 \le t \le \tau_1)$, *with its maximum* $M = \sup_{0 \le t \le \tau_1} W_t$ *attained at time* T_M. *Then, there exists a Jacobi diffusion,* $Y^{2,2}$, *independent of* $(l(\tau_1, y); y \in \mathbb{R}^+)$, *such that, for* $0 \le y < M$,

$$\frac{l(T_M, y)}{l(\tau_1, y)} = Y^{2,2}_{\left(\int_0^y \frac{dz}{l(\tau_1, z)}\right)},$$

with $Y_0^{2,2} = l(T_M, 0)/l(\tau_1, 0)$ *having uniform distribution.*

We define a process $(\hat{W}_u^{(1)}; u \ge 0)$ *by*

$$\int_0^{W_t} \frac{dy}{l(T_M, y)} = \hat{W}^{(1)}_{\left(\int_0^t \frac{ds}{(l(T_M, W_s))^2}\right)},$$

for $0 \le t < T_M$, *and a process* $(\hat{W}^{(2)}_u; u \ge 0)$ *by*

$$\int_0^{W_t} \frac{dy}{l(\tau_1, y) - l(T_M, y)} = \hat{W}^{(2)}_{\left(\int_{\tau_1 - t}^{\tau_1} \frac{ds}{(l(\tau_1, W_s) - l(T_M, W_s))^2}\right)},$$

for $T_M < t \le \tau_1$. *Then* $\hat{W}^{(1)}$ *and* $\hat{W}^{(2)}$ *both have the law of the Brownian burglar. The four processes* $\hat{W}^{(1)}$, $\hat{W}^{(2)}$, $Y^{2,2}$ *and* $l(\tau_1, \cdot)$ *are independent, and from them we can reconstruct* $(W_t; 0 \le t \le \tau_1)$.

We use the following lemma which is a combination of the agreement formula of Pitman-Yor [14] and the relationship between the bridge and the pseudo-bridge given by Biane-Le Gall-Yor [2].

Lemma 10. *Let* $R^{(1)}$ *and* $R^{(2)}$ *be two independent* $BES(1)$ *processes starting from* 0, *and let* $T^{(1)}$ *and* $T^{(2)}$ *be their respective hitting times of level* 1. *Define* $R^{(+)}$ *by connecting the paths of* $R^{(1)}$ *and* $R^{(2)}$ *back to back:*

$$R^{(+)}_t = \begin{cases} R^{(1)}_t & \text{if } t \le T^{(1)} \\ R^{(2)}_{T^{(1)}+T^{(2)}-t} & \text{if } T^{(1)} \le t \le T^{(1)} + T^{(2)}. \end{cases}$$

Now finally let R *be obtained by scaling* $R^{(+)}$ *so as to normalise its local time:*

$$R_t = \frac{1}{l^{(1)} + l^{(2)}} R^{(+)}\big((l^{(1)} + l^{(2)})^2 t\big), \qquad \text{for } t < \frac{T^{(1)} + T^{(2)}}{\big(l^{(1)} + l^{(2)}\big)^2},$$

where $l^{(1)}$ *is the local time at level* 0 *that* $R^{(1)}$ *has accrued by time* $T^{(1)}$ *and* $l^{(2)}$ *is similarly defined. Then the law of* R *is equivalent to the law of the reflecting Brownian motion* W *run until its local time at level* 0 *first reaches* 1, *and for any suitable path-functional* F

$$\mathbf{E}[F(R)] = \mathbf{E}\left[\frac{1}{2M} F(W)\right],$$

where $M = \sup_{0 \le t \le \tau_1} W_t$.

We will be satisfied with sketching the proof of this lemma. The above mentioned references give some more detail. Let L^W be the local time at zero that W has accrued when it attains its maximum M^W. Begin by observing that,

$$\mathbb{P}\big(M^W \in dm, L^W \in dl\big) = \frac{e^{-1/2m}}{2m^2}\, dm dl. \tag{4.1}$$

Using the law of $l^{(1)}$ and $l^{(2)}$, a simple calculation confirms that,

$$\mathbb{P}\big(M^R \in dm, L^R \in dl\big) = \frac{e^{-1/2m}}{4m^3}\, dm dl, \tag{4.2}$$

where L^R and M^R have the obvious meaning. Thus the conclusion of the lemma holds for F depending only on the maximum level attained and the local time at zero when this occurs. In order to lift the result to an equality of laws on path space, we condition on these two quantities. We can then easily confirm, using Brownian scaling and Williams' description of the Itô excursion measure, that the excursions from zero have identical conditional law under the two regimes.

Proof of Theorem 9. Consider the construction of the preceding lemma. Define burglars $\hat{R}^{(1)}$ and $\hat{R}^{(2)}$ from the processes $R^{(1)}$ and $R^{(2)}$ in the usual manner. Define $\tilde{Y}^{2,2}$ via,

$$\frac{l^{(1)}(y)}{l^{(1)}(y)+l^{(2)}(y)} = Y^{2,2}_{\int_0^y \frac{dz}{l^{(1)}(z)+l^{(2)}(z)}}, \tag{4.3}$$

where $l^{(1)}(y)$ is the local time at level y accrued by $R^{(1)}$ before $T^{(1)}$ and $l^{(2)}(y)$ is similarly defined. As is used in the proof of Theorem 3 we have

$$(l^{(1)}(y); 0 \leq y \leq 1) \overset{law}{=} (Z^2(1-y); 0 \leq y \leq 1) \overset{law}{=} (Z^2_{l(1)\to 0}(y); 0 \leq y \leq 1),$$

and $Y^{2,2}$ is a Jacobi diffusion by virtue of Proposition 8. The four processes $\hat{R}^{(1)}$, $\hat{R}^{(2)}$, $Y^{2,2}$ and $l(\tau_1, \cdot)$ must be independent as a consequence of the independence of the two $BES(1)$ processes, and the results of Theorem 1 and Proposition 8.

Now let the reflecting Brownian motion W be obtained by completing the construction of R, and then making the appropriate change of measure. It is simple to check that $\hat{R}^{(1)} = \hat{W}^{(1)}$ and $\hat{R}^{(2)} = \hat{W}^{(2)}$. The process $Y^{2,2}$ just defined by (4.3) is also identical to that defined in the statement of the theorem. Since the change of measure we have made affects only the marginal law of $l(\tau_1, \cdot)$ the theorem is proved. □

Acknowledgements. The first author was supported by an EPSRC research grant.

References

[1] D. Bakry. Remarques sur les semi-groupes de Jacobi. In *Hommage à P.A. Meyer et J. Neveu. Asterisque 236*, pages 23–40. Soc. Math. France, 1996.

[2] Ph. Biane, J.F. Le Gall, and M. Yor. Un processus qui ressemble au pont brownien. In *Séminaire de Probabilités XXI, Lecture notes in Mathematics 1247*, pages 270–275. Springer, 1987.

[3] N. Bouleau and M. Yor. Sur la variation quadratique des temps locaux de certaines semimartingales. *Comptes Rendus Acad. Sci., Paris*, 292:491–494, 1981.

[4] I.P. Cornfeld, S.V. Fomin, and Ya.G. Sinai. *Ergodic theory.* Springer-Verlag, Berlin, 1982.

[5] D.A. Dawson. Measure-valued Markov processes. In *Ecole d'Eté de Probabilitiés de Saint-Flour, 1991, Lecture Notes in Mathematics 1541.* Springer, Berlin, 1993.

[6] F. Delbaen and H. Shirakawa. Interest rate model with upper and lower bounds. Preprint, 1996.

[7] A. Etheridge and P. March. A note on superprocesses. *Probability Theory and Related Fields*, 89:141–147, 1991.

[8] S.N. Ethier and T.G. Kurtz. *Markov processes: characterization and convergence.* Wiley, New York, 1986.

[9] S. Karlin and H.M. Taylor. *A second course in stochastic processes.* Academic Press, New York, 1981.

[10] M. Kimura. Diffusion models in population genetics. *Journal of Applied Probability*, 1:177–232, 1964.

[11] O. Mazet. Classification des semigroupes de diffusion sur $\mathbb{R}$ associés à une famille de polynomes orthogonaux. In *Séminaire de Probabilités XXXI, Lecture notes in Mathematics 1655*, pages 40–53. Springer, Berlin, 1997.

[12] K.R. Parthasarathy. *Introduction to probability and measure.* Macmillan, London, 1977.

[13] J. Pitman and M. Yor. Sur une décomposition des ponts de Bessel. In *Functional analysis in Markov processes, Lecture notes in Mathematics 923*, pages 276–285. Springer, Berlin, 1982.

[14] J. Pitman and M. Yor. Decomposition at the maximum for excursions and bridges of one-dimensional diffusions. In M. Fukushima, N. Ikeda, H. Kunita, and S. Watanabe, editors, *Itô's stochastic calculus and probability theory*, pages 293–310. Springer, 1996.

[15] D. Revuz and M. Yor. *Continuous martingales and Brownian motion.* Springer, Berlin, 1991.

[16] V.A. Rohlin. Selected topics in the metric theory of dynamical systems. *American Mathematical Society Translations Series 2*, 49:171–240, 1966.

[17] A. Vershik and M. Yor. Multiplicativité du processus gamma et étude asymptotique des lois stables d'indice α, lorsque α tend vers 0. Technical Report 289, Laboratoire de Probabilités, Université Pierre et Marie Curie, Paris, 1995.

[18] J. Warren and M. Yor. Skew-products involving Bessel and Jacobi processes. Technical report, Statistics group, University of Bath, 1997.

ON THE UPCROSSING CHAINS OF STOPPED BROWNIAN MOTION

FRANK B. KNIGHT

INTRODUCTION

We follow the notations of [6], although that paper is not required for the present work. $B'(t)$ denotes $B(t \wedge T(-1))$ where $B(t)$ is a Brownian motion on R, $B(0) = 0$, and $T(-1) = \inf\{t : B(t) = -1\}$ (we assume the paths of B are unbounded above and below, so that $T(-1) < \infty$).

We set $\alpha_n = 2^{-n}$, and define a random walk R_n by $R_n(0) = 0$, $R_n(k\alpha_n^2) = B'(T_k)$ where $T_0 = 0$ and inductively $T_{k+1} = \inf\{t > T_k : B'(t) - B'(T_k) = \pm\alpha_n\}$; $\inf \emptyset = \infty$, $B'(\infty) = -1$.

Our main objects of concern are the upcrossing chains $N_n(k) = \#\{j < M_n : R_n(j\alpha_n^2) = k\alpha_n,\ R_n((j+1)\alpha_n^2) = (k+1)\alpha_n\}, -2^n \le k$, where $M_n = \inf\{k : R_n(k\alpha_n^2) = -1\}$. Clearly, $N_n(k) = 0$ for $k \le -2^n$, or for $k \ge K(n) := \inf\{j \ge 0 : N_n(j) = 0\}$, and it is known [4] that, for each $n \ge 0$, $N_n(k)$ is a Markov chain in k with negative binomial one-step transition function

$$p(i,j) = \begin{cases} \binom{i+j-1}{j}\alpha_{i+j};\ k \ge 0;\ i,j \ge 0 \\ \binom{i+j}{j}\alpha_{i+j+1};\ -2^n \le k < 0;\ i,j \ge 0. \end{cases}$$

Thus, in the parameter range $k \ge 0$ we have a Galton-Watson branching chain with geometric offspring ($p = \frac{1}{2}$) while for $-2^n < k < 0$ there is a superimposed geometric immigration ($p = \frac{1}{2}$). Only the parameter range depends on n (since our definition of N_n does not include the scaling used for R_n). These Markov chains are both elementary and much-studied, and they are not the subject here. What is not as well understood, and will be our principle concern, is the dependence of $N_n(\cdot)$ on $n (\ge 0)$. It turns out that N_n is also a Markov chain with parameter n. Its "upward" ($n \downarrow$) and "downward" ($n \uparrow$) one-step transition functions will be investigated (Section 2), and it turns out that they are "almost" homogeneous in n.

The original motivation for this work was a question of J. Pitman and M. Yor (unpublished) which we understand as follows. Let $L(x)$, $-1 \le x$, denote the (continuous) local time of B' : $L(x) := \frac{d}{dx}\int_0^{T(-1)} I_{(-\infty,x)}(B'(s))ds$. The problem is to construct the law of $B'(\cdot)$ conditional on $\sigma(L(\cdot))$. Now our approach is to introduce N_n into the problem, so that it has two stages: first, one constructs the law of $\{N_n(\cdot), 0 \leq n\}$ given $L(\cdot)$; second, one constructs the law of $B'(\cdot)$ given $\{N_n(\cdot), 0 \leq n\}$. Since we have $L(x) = \lim_{n\to\infty} 2\alpha_n N_n[2^n x]$ uniformly x, P-a.s. (see [1], [4]), it is not necessary to include $L(\cdot)$ explicitly in the given data for the second stage. This convergence was recently studied in [6], where other references are given. There, we obtained the law of $L(\cdot)$ given $N_n(\cdot)$ for fixed n. The principle obstacle in stage one is to reverse this to find the law of N_n given $L(\cdot)$. We emphasize that a simple Bayes rule application does not succeed in the function space setting. Nor does it seem possible to find the higher order transition functions of $N_n(\cdot)$ and pass to a limit as $n \to \infty$. A very plausible conjecture, for example, is that $\sigma(L(\cdot)) = \lim_{N\to\infty} \sigma(N_n, n \geq N)$ up to P-nullsets, but we could not prove it.

Nevertheless, we are able by means of a comparison theorem (Theorem 2.5 below) to get solid information about the law of $N_n(\cdot)$ given $L(\cdot)$. This has emboldened us to write down, in the following Sections 2 and 3, what we have found along these lines, in the hope that it may be useful for some more skillful subsequent treatment elsewhere. Indeed, the subject seems to us attractive, both because of its relevance to Brownian motion and because of its combinatorial overtones.

Before entering into the dependence of N_n on n, however, we give in Section 1 a construction of the law of $B'(\cdot)$ given N_n for a single n. This is easy and no doubt known, but it gives the first step in the solution of stage 2 of the Pitman-Yor problem, and we imagine it may take the place of Section 4 for all but the more diligent readers. Thus the outline of the paper is as follows. In Section 1 we construct the law of B' given N_n, n fixed. In Section 2 we discuss the dependence of N_n on n, and study the explicit transition functions. In Section 3 we obtain estimates for the law of N_n given L, and carry out stage 1 of the Pitman-Yor problem as far as we are able. Then, in Section 4, we carry out stage 2 of the Pitman-Yor problem, which does not depend on stage 1.

After completion of this paper, we received a first draft of a paper [9] by J. Warren and M. Yor which gives a "solution" to the problem when B' is replaced by a reflected Brownian motion. This paper has virtually nothing in common with ours, which we regard as a paper on the random walk approximation as much as on the Pitman-Yor problem per se. Nevertheless, the solution of [9] is remarkable, both as to completeness and conciseness. From the standpoint of the present paper, its main implication is that it suffices only to treat the case $L(\cdot) \equiv 1$ — the general case follows from this by changes of scale and time. It remains to be seen whether the conditional law of $N_n(k)$ given $L \equiv 1$ can be given explicitly.

Section 1. Construction of the law of B' given N_n.

For the rest of this section, $n(\geq 0)$ is fixed. For $i > -2^n$, a "random walk path" starting at i is a sequence $(i_0, i_1, \ldots, i_{M_n})$ where $0 < M_n < \infty$, $i_0 = i$, $i_{M_n} = -2^n < i_k$ and $i_{k+1} = i_k \pm 1$ for $0 \leq k < M_n$. We define $N_n^i(k)$ as the number of upcrossings $k \uparrow (k+1)$, as a function of paths starting at i, $-2^n \leq k$, where $N_n^i(-2^n) = 0$ (thus $N_n^0 \equiv N_n$ in an obvious sense). For $-2^n \leq k$, let

$X^i_k(j)$, $0 \le j \le N^i_n(k)$, denote the number of upcrossings $(k+1) \uparrow (k+2)$ occurring after the jth upcrossing $k \uparrow k+1$ but before the $(j+1)$-st, where if $j = 0$ the first requirement is absent, and if $j = N^i_n(k)$ the second requirement is absent (for example, if $j = 0 = N^i_n(k)$ then $X^i_k(0) = 0$ unless $k \le i-1$, but if $k \le i-1$ then $X^i_k(0)$ is the total number of upcrossings $k \uparrow k+1$).

Lemma 1.1. *The function $X^i_k(j)$ determines the path starting at i uniquely. Indeed, there is an algorithm for its determination.*

Remark We assume that at least one path for $X^i_k(j)$ exists.

Proof. The algorithm is as follows. If $X^i_{i-1}(0) > 0$, the first step is to $i+1$, and the X-function for the subsequent path (starting at $i+1$) is obtained from $X^i_k(j)$ as follows:

(a) $X^{i+1}_{i-1}(0) = X^i_{i-1}(0) - 1$
(b) $X^{i+1}_i(j) = X^i_i(j+1)$ for $0 \le j \le N^{i+1}_n(i) = N^i_n(i) - 1$
(c) $X^{i+1}_k(j) = X^i_k(j)$ for all other (k, j).

On the other hand, if $X^i_{i-1}(0) = 0$, the first step is to $i-1$, and the X-function of the subsequent path is obtained simply by changing superscript i to $i-1$ (if $i = -2^n + 1$, we leave X^{i-1} undefined).

It is an elementary task to see that the first step is the only possibility consistent with the given $X^i_k(j)$, and that the modified X-function is actually the X-function uniquely determined by the subsequent path. This being true, our proof of uniqueness is by induction on the total number of steps M_n. If $M_n = 2^n + i$, then there is only one possible path, namely all steps are down, and $X^i_k(0) = 0$, $k < i$. Clearly this is the path determined by repetition of the algorithm. So we make the induction assumption that, for a certain $k \ge 1$ and all $i > -2^n$, the uniqueness has been proved for $M_n < 2^n + i + k$. Then if $M_n = 2^n + i + k$ for a certain path, either $X^i_{i-1}(0) > 0$, the first step is up and M_n is reduced by 1, so that the induction assumption applies to the subsequent path (which is thus uniquely determined) or $X^i_{i-1}(0) = 0$, the first step is down and M_n is reduced by 1, so that the equality is maintained for the subsequent path starting at $i-1$. Since this step is uniquely determined, we can repeat the procedure on the subsequent path, leading eventually to a path for which the first step is up (since there were more than $2^n + i$ steps). At that point the induction assumption applies to the subsequent path, and the whole path is thus uniquely determined.

We now determine the law of R_n given N_n. In view of Lemma 1.1, it is enough to determine the joint law of the random vectors $(X_k(j),\ 0 \le j \le N^0_n(k))$, $-2^n \le k$; where we omit the superscript 0 in $X^0_k(j)$. We observe that $\Sigma_j X_k(j) = N_n(k+1)$. It turns out that this is the only restriction imposed on an X-function $X_k(j)$ when N_n is given. Following the terminology of W. Feller [2] we introduce

Definition 1.2. Non-negative, integer-valued random variables $X_1, \ldots, X_n$ are said to be determined by Bose-Einstein sampling of size (n, N), $N \ge 0$, if all distinct $(x_1, \ldots, x_n)$ with $\Sigma_1^n x_j = N$ are equally likely. Then $P\{(X_1, \ldots, X_n) = (x_1, \ldots, x_n)\} = \binom{N+n-1}{N}^{-1}$.

Now we have

Lemma 1.3. *Given N_n, the $(X_k(j),\ 0 \le j \le N_n(k))$ are independent over $k \ge -2^n$. For $-2^n \le k < 0$ they have the law of Bose-Einstein sampling of size $(N_n(k)+1,\ N_n(k+1))$, whereas, for $0 \le k$, $X_k(0) = 0$ and $(X_k(j),\ 1 \le j \le N_n(k))$ is either vacuous (if $N_n(k) = 0$) or Bose-Einstein of size $(N_n(k),\ N_n(k+1))$ (if $N_n(k) > 0$).*

Proof. For fixed $k < 0$, it follows from the transition function of N_n in case $i = 0$, namely $p(0,j) = \alpha_{j+1}$, that $X_k(0)$ is geometric with $p = \frac{1}{2}$ (apply the Markov property of R_n at its passage time to $(k+1)\alpha_n$). Similarly, by the Markov property of R_n at its subsequent returns to $(k+1)\alpha_n$ from $k\alpha_n$, given that they occur, each $X_k(j)$, $j \le N_n(k)$, is geometric ($p = \frac{1}{2}$) and they are independent conditionally on $N_n(k)$. Thus, given $N_n(k)$, $P\{\bigcap_{0 \le j \le N_n(k)} X_k(j) = x_j\} = \alpha_{N_n(k)+1+\Sigma x_j}$, and when $N_n(k+1) = \Sigma_j x_j$ is also given, the $X_k(j)$, $0 \le j \le N_n(k)$, are Bose-Einstein of size $(N_n(k)+1,\ N_n(k+1))$, as asserted. For fixed $k \ge 0$ an analogous reasoning applies to $(X_k(j),\ 1 \le j \le N_n(k))$ based on the passage times of R_n to $(k+1)\alpha_n$ from $k\alpha_n$.

It remains to see that the vectors $(X_k(j), 0 \le j \le N_n(k))$ are mutually independent given $N_n(\cdot)$, and that each is conditionally dependent only on $(N_n(k),\ N_n(k+1))$. This follows by Proposition 1.1, p. 92, of J. B. Walsh [8]. In brief, for each k we introduce the upcrossing field $\mathcal{U}_{k,k+1}$ generated during the successive upcrossings $k\alpha_n \uparrow (k+1)\alpha_n$. We also introduce downcrossing field $\mathcal{V}_{k+1,k}$ generated during the downcrossings of $(k+1)\alpha_n \downarrow k\alpha_n$, say $Z_1(t), \ldots, Z_N(t)$, where $N = N_n(k)+1$ for $k < 0$, $N = N_n(k)$ for $k \ge 0$. Then $\sigma\{N_n(j),\ j \le k\} \subset \mathcal{U}_{k,k+1}$, $\sigma\{N_n(j),\ j > k\} \subset \mathcal{V}_{k+1,k}$, and $\mathcal{U}_{k,k+1}$ is conditionally independent of $\mathcal{V}_{k+1,k}$ given $N_n(k)$. Hence $N_n(k+1)$ is independent of $\sigma\{N_n(j),\ j \le k\}$ given $N_n(k)$, and we have therefore obtained, as above, the conditional law of $(X_k(j),\ 0 \le j \le N_n(k))$ given $\sigma\{N_n(j),\ j \le k+1\}$ (note that for every $i \ge 0$, $(\{N_n(k) = i\} \bigcap \sigma\{X_k(j),\ 0 \le j \le i\}) \subset (\{N_n(k) = i\} \bigcap \mathcal{V}_{k+1,k})$, so that when $N_n(k)$ is given, $\sigma(X_k(j),\ 0 \le j \le N_n(k))$ is independent of $\sigma\{N_n(j),\ j \le k\}$). On the other hand, $\sigma\{N_n(j),\ j > k+1\} \subset \mathcal{V}_{k+2,k+1}$, and given $N_n(k+1)$, $\mathcal{V}_{k+2,k+1}$ is independent of $\mathcal{U}_{k+1,k+2}$, which contains not only $\sigma\{N_n(j),\ j \le k+1\}$ but also $\sigma(X_k(j),\ 0 \le j \le N_n(k))$ (as a picture will show). Thus the law of $(X_k(j),\ 0 \le j \le N_n(k))$ given $\sigma\{N_n(k),\ N_{n+1}(k)\}$ is not only the same as given $\sigma\{N_n(j),\ j \le k+1\}$ but also the same as given $\sigma\{N_n(j),\ -2^n < j\}$, as asserted.

Corollary 1.4. *Given N_n, every family $x_k(j)$ with $0 \le x_k(j)$, $x_k(0) = 0$ for $k \ge 0$, and $\sum_{j \le N_n(k)} x_k(j) = N_n(k+1)$, $-2^n \le k$, corresponds to a unique random walk path starting at 0, and all such families are equally likely.*

Proof. Obvious.

Remark. It is also easy to see that any sequence $0 \leq N_n(k)$, $-2^n < k$, with $N_n(k) = 0$ for all $k \geq K(n) := \inf\{j \geq 0 : N_n(j) = 0\} < \infty$, has positive probability for N_n. Indeed, the probability follows by iteration of the transition function $p(i,j)$. The total probability of all such sequences is one. The number of such paths is included below in Theorem 4.5.

To complete the construction of B' conditional on N_n, it now remains only to fill in B' given R_n (the conditional law of R_n given N_n being identified by Lemmas 1.1 and 1.3, where $\sigma(N_n) \subset \sigma(R_n) \subset \sigma(B')$). For this we need one more lemma, which appears as Lemma 1.1 of [6] and will be used repeatedly in the sequel. For the reader's convenience we repeat it here with a different (and simpler) proof.

Lemma 1.5. *For $0 \leq k < M_n$, set*

$$Y_k(t) = sgn_k(B((T_k + t) \wedge T_{k+1}) - B(T_k)); \; 0 \leq t,$$

where sgn_k is the choice of sign in the definition of T_{k+1}. Then conditional on $\sigma(R_n)$ (which includes $\sigma(M_n)$), $Y_0, \ldots, Y_{M_n-1}$ are independent and identically distributed with the law of a $BES^3_{\alpha_n}(t \wedge T(2\alpha_n)) - \alpha_n$.

Terminology. We call $Y_k(t)$ an "n-insert", $k < M_n$, and the result "(conditional) independence of n- inserts".

Proof. Instead of stopping at M_n, we continue the sequence T_k by using $B(t)$ instead of $B'(t)$, and in this way define $R_n(k\alpha_n^2)$ for all k so that it becomes an unstopped symmetric random walk. Then we obtain a sequence of processes Y_k of which the first M_n are those of the lemma. The strong Markov property of B at T_k, together with the symmetry $B \longleftrightarrow -B$, shows that for each k, given $\{R_n(j\alpha_n^2),\ j \leq k\}$, Y_k is conditionally independent of $Y_0, \ldots, Y_{k-1}$ and has the same law as Y_0. It is plain that the law of Y_0 is that of B starting at 0, stopped at α_n, and conditioned to reach α_n before $-\alpha_n$. Then it is a familiar fact that this is the asserted law of a BES^3 starting at α_n and translated by $-\alpha_n$ (indeed, this is the law of the h-path transform of B killed at $\pm\alpha_n$ with $h(x) = x + \alpha_n$, which has the BES^3 generator by a simple calculation). Now the same law holds if $R_n((k+1)\alpha_n^2)$ is also given (considering the two possibilities separately), and then the strong Markov property at T_{k+1} shows that we may as well be given $R_n(j\alpha_n^2)$ for all j. Finally, since $T(-1)(= T_{M_n})$ is a stopping time, the same reasoning shows that $\{R_n((j + M_n)\alpha_n^2),\ 1 \leq j\}$ is independent of $\sigma(B'(t),\ 0 \leq t)$, so it suffices to be given only $\{R_n((k \wedge M_n)\alpha_n^2),\ 1 \leq k\}$, proving the Lemma.

We summarize these findings as

Theorem 1.6. *To construct the law of B' given N_n, since $M_n = (2^n + 2\Sigma_k N_n(k)) \in \sigma(N_n)$, we may begin with a random sequence $Y_0, \ldots, Y_{M_n-1}$ of independent n-inserts, with absorption times $\zeta_0, \ldots, \zeta_{M_N-1}$, respectively. Then we determine R_n from N_n by Bose-Einstein sampling (Lemma 1.3), and, setting $T_k = \zeta_0 + \cdots + \zeta_{k-1} (0 < k)$, we define $B'((T_k + t) \wedge T_{k+1}) = (sgn_k) Y_k(t) + R_n(k\alpha_n^2)$,*

$0 \le t$, $0 \le k$, *where* $sgn_k = sgn(R_n((k+1)\alpha_n^2) - R_n(k\alpha_n^2))$. *This defines* $B'(t)$, $T_k \le t \le T_{k+1}$, *for all* k, *as required.*

Section 2. The N_n-chain, with n as parameter.

The random walks R_n are nested, in such a way that $\sigma(R_n) \subset \sigma(R_{n+1})$, but that is not true for $N_n : \sigma(N_n)$ is not comparable to $\sigma(N_{n+1})$ although, of course, $\sigma(N_n) \subset \sigma(R_n)$. We do have a Markov property, as consequence of

Theorem 2.1. *Given* N_n, *for* n *fixed*, $R_n(\cdot)$ *is conditionally independent of* $\{N_{n+k}, 1 \le k\}$. *In particular, since* $\sigma\{N_0, \ldots, N_n\} \subset \sigma(R_n)$, N_n *is a Markov chain in* n.

Proof. Intuitively, the assertion is reasonably obvious from independence of inserts (Lemma 1.5). Indeed, to go from N_n to R_n involves only an ordering of the n-inserts at each level, while to go to N_{n+k} involves only interpolation of $(n+k)$-inserts into the n-inserts without changing level and counting those at each sub-level, a result which is not affected by the reordering. As to a proof, it is enough to show conditional independence, for all $n \ge 0$, of R_n and N_{n+1} given N_n, because then, for every k, $\sigma\{R_n, N_n, \ldots, N_{n+k}\}$ is conditionally independent of $\sigma(N_{n+k+1})$ given $\sigma(N_{n+k})$. Then if the law of $N_{n+1}, \ldots, N_{n+k}$ given $\sigma(R_n)$ depends only on $\sigma(N_n)$ (by induction assumption), the same is true for $N_{n+1}, \ldots, N_{n+k+1}$. To write this out, let f_j, $1 \le j \le k+1$, be bounded Borel functions. Then we have

$$
\begin{aligned}
&E\left(\prod_1^{k+1} f_j(N_{n+j}) \middle| R_n\right) \\
&= E\left[E(f_{k+1}(N_{n+k+1}) | R_n, N_{n+1}, \ldots, N_{n+k}) \prod_1^k f_j(N_{n+j}) \middle| R_n\right] \\
&= E\left[E(f_{k+1}(N_{n+k+1}) \middle| N_{n+k}) \prod_1^k f_j(N_{n+j}) \middle| R_n\right] \\
&= E\left[E(f_{k+1}(N_{n+k+1}) \middle| N_{n+k}) \prod_1^k f_j(N_{n+j}) \middle| N_n\right] \\
&\in \sigma(N_n).
\end{aligned}
$$

The argument is completed by appeal to the monotone class theorem applied to the linear algebra generated by such products, and then by letting $k \to \infty$.

Now for $n \ge 0$, we observe that if N_n is given, so are the numbers of downcrossings at each level $(k+1) \downarrow k$ (namely, $N_n(k)$ for $k \ge 0$, and $N_n(k) + 1$ for $-2^n \le k < 0$). The law of N_{n+1} given $\sigma(R_n)$ may be constructed by independently interpolating R_{n+1} into each of the independent n-inserts, and then adding the upcrossings over those n-inserts which can contribute to a given level for N_{n+1} (for example, the level $2k$ is only contributed to from upcrossings $k \uparrow k+1$ and downcrossings $k \downarrow k-1$ of R_n). All that matters is the number of summands of each type, which in turn depends only on $\sigma(N_n)$ when R_n is given. This suffices for the proof (the enumeration procedure will be explained in full detail below when we obtain the transition mechanism).

We take the point of view that a "step" $N_n \to N_{n+1}$ is "down" (toward $L(\cdot)$), and discuss first the downward transition mechanism. There are two approaches, leading to different (but of course equivalent) descriptions, and (since we cannot iterate either one to obtain explicitly the $N_n \to N_{n+k}$ transition function) we shall present them both quite briefly.

The first approach is based on symmetry. Let $Y(t)$ be an n-insert (Lemma 1.5), and let us interpolate a random walk of step size α_{n+1} into Y, just as we did R_{n+1} into B'. Clearly the total number of steps $0 \to \pm\alpha_{n+1}$ has the same law as if we were interpolating into $B'(t \wedge T(1))$ (since the condition $R_n(\alpha_n^2) = \alpha_n$ is analogous to $R_n(\alpha_n^2) = -\alpha_n$). Thus the law is that of $1 + \text{geo}(\frac{1}{2})$, where for brevity we shall write geo(p) for (a) random variable X with $P\{X = k\} = p(1-p)^k$, $0 \le k$ with an analogous interpretation for bin(n,p) and neg. bin.(n,p). Moreover, given this variable geo$(\frac{1}{2})$, the number of passages $0 \to \alpha_{n+1}$ has the law of $1 + \text{bin}(\text{geo}(\frac{1}{2}), \frac{1}{2})$, namely 1 plus a binomial variable with $p = \frac{1}{2}$, which determines the number of passages $0 \to -\alpha_{n+1}$ as $\text{geo}(\frac{1}{2}) - \text{bin}(\text{geo}(\frac{1}{2}), \frac{1}{2})$. Again, this is obvious enough when we interpolate into $B'(t \wedge T(1))$ instead of into Y, and condition on $B'(T(1)) = \alpha_n$. In more detail, we can use the strong Markov property of $B'(t \wedge T(1))$ to write for $k > 0$

$$\begin{aligned}
&P^0\{B' \text{ reaches } -\alpha_{n+1} \text{ before } \alpha_{n+1} | k \text{ returns to } 0, \text{ then to } \alpha_n\} \\
&= 2^{+(k+1)} P^0\{-\alpha_{n+1} \text{ before } \alpha_{n+1} \text{ then } k \text{ returns to } 0, \text{ then to } \alpha_n\} \\
&= 2^{+(k+1)} 2^{-2} P^0\{(k-1) \text{ returns to } 0, \text{ then to } \alpha_n\} \\
&= \frac{1}{2}.
\end{aligned}$$

Thus, conditional on $k > 0$ returns to 0, the first step of R_{n+1} interpolated into the n-insert Y goes to $\pm\alpha_{n+1}$ each with probability $p = \frac{1}{2}$. Then, given this return to 0, by the strong Markov property given $k - 1 > 0$, the second exit is to α_{n+1} with $p = \frac{1}{2}$, independently of the first, and so forth to the kth exit from 0, giving the law of bin$(k, \frac{1}{2})$.

We observe that the same law applies to an interpolation into $-Y$ since $(\text{geo}\,\frac{1}{2}) - \text{bin}(\text{geo}\,\frac{1}{2}, \frac{1}{2}) \stackrel{d}{=} \text{bin}(\text{geo}\,\frac{1}{2}, \frac{1}{2})$, except that the extra 1 adds to $\#(0 \to -\alpha_{n+1})$. Secondly, the set of all n-inserts which are in a position to contribute interpolated steps from $j\alpha_n \to j\alpha_n \pm \alpha_{n+1}$ for given j are precisely those with $R_n(T_k) = j\alpha_n$. By independence of inserts it follows that we have

Theorem 2.2. *For $j > 0$ (resp. $-2^n < j \le 0$) the law of $N_{n+1}(2j-1) + N_{n+1}(2j)$ conditional on N_n is that of $(N_n(j-1) + N_n(j) + \text{neg. bin.}(N_n(j-1) + N_n(j), \frac{1}{2}))$ (resp. replace $N_n(j-1)$ by $N_n(j-1)+1$), and conditional on this* neg. bin.*$(= k$, say)*

(2.1) $$(N_{n+1}(2j-1), N_{n+1}(2j)) \stackrel{d}{=}$$

$$= (N_n(j-1) + \text{bin}(k, \frac{1}{2}), N_n(j) + k - \text{bin}(k, \frac{1}{2}))$$

(where the two variables $\text{bin}(k, \frac{1}{2})$ *are identical). Given* N_n*, these conditional pairs of random variables are mutually independent in* j*, and thus determine the conditional law of* N_{n+1} *given* N_n*.*

Proof. For $j > 0$, we have only to sum the interpolated steps over all contributing inserts and appeal to their mutual independence. The sum of independent $\text{geo}(\frac{1}{2})$ terms gives the neg. bin.$(N_n(j-1) + N_n(j), \frac{1}{2})$ terms, $j > 0$, and the sum over the conditionally independent $\text{bin}(\text{geo}(\frac{1}{2}), \frac{1}{2})$ terms then gives the $\text{bin}(k, \frac{1}{2})$. For $j \leq 0$ there are $1 + N_n(j-1)$ steps $j\alpha_n \to (j-1)\alpha_n$, and the result is the joint law of $(1 + N_{n+1}(2j-1),\ N_{n+1}(2j))$. But, for given $k(=$ neg. bin.$(1 + N_n(j-1) + N_n(j), \frac{1}{2}))$ the two ones drop out, and no change in (2.1) is needed.

Remark. The transition mechanism described by Theorem 2.2 does not depend on n *except* for the fact that the state of N_n is a sequence $(N_n(k); -2^n \leq k)$. This dependence cannot be avoided by defining $N_n(k) = 0$ for $k < -2^n$ since then 0 is preserved only for $k < -2^{n+1}$, which depends on n.

Another way to present the result of Theorem 2.2 is to exhibit $N_{n+1}(k)$ given N_n as a (conditional) Markov chain in k. Clearly there is only a 2-step dependence on N_n, and it is equivalent to use either increasing or decreasing k. In terms of increasing k, the result is

Theorem 2.3. *For* $k > 0$*, given* N_n*, we have* $N_{n+1}(2k-1) \stackrel{d}{=} N_n(k-1) +$ neg. bin.$(N_n(k-1) + N_n(k), \frac{2}{3})$*, and given both* N_n *and* $N_{n+1}(2k-1)$*, we have* $N_{n+1}(2k) \stackrel{d}{=} N_n(k) +$ neg. bin.$(N_{n+1}(2k-1) + N_n(k), \frac{3}{4})$*. For* $k \leq 0$*, analogous facts hold after replacing* $N_n(k-1)$ *by* $1 + N_n(k-1)$ *and* $N_{n+1}(2k-1)$ *by* $1 + N_n(2k-1)$*.*

Proof. It is possible to prove this by deriving the corresponding result for an n-insert, and then adding over contributing inserts as we did for Theorem 2.2, but it is somewhat long. For brevity, we derive the result directly from Theorem 2.2. For $j > 0$, we had $N_{n+1}(2j-1) \stackrel{d}{=} N_n(j-1) + \text{bin}(k, \frac{1}{2})$ where $k \stackrel{d}{=}$ neg. bin.$(N_n(j-1)+N_n(j), \frac{1}{2})$, while for $j \leq 0$ this last becomes $k \stackrel{d}{=}$ neg. bin.$(1+ N_n(j-1) + N_n(j), \frac{1}{2})$. Writing "bin" for $\text{bin}(k, \frac{1}{2})$ with the random k, we have for $j > 0$ $P\{\text{bin} = i\} =$

$$\sum_{k=i}^{\infty} \binom{k}{i} \binom{N_n(j-1) + N_n(j) + k - 1}{k} 2^{-(2k+N_n(j-1)+N_n(j))}$$

$$= \frac{2^{-(2i)}}{i!(N_n(j-1) + N_n(j) - 1)!} \Bigg[\sum_{\ell=0}^{\infty} \binom{N_n(j-1) + N_n(j) + i + \ell - 1}{\ell}$$

$$4^{-\ell} (\frac{3}{4})^{N_n(j-1)+N_n(j)+i} \Bigg] \cdot (\frac{2}{3})^{(N_n(j-1)+N_n(j))} (\frac{4}{3})^i (etc.)$$

$$= P\{\text{neg. bin.}(N_n(j-1) + N_n(j); \frac{2}{3}) = i\},$$

as asserted for the marginal distribution, where for $j \le 0$ we replace $N_n(j-1)$ by $(1+N_n(j-1))$ throughout. [Of course, the marginal law of $N_{n+1}(2j)$ follows by replacing $N_n(j-1)$ by $N_n(j)$, using (2.1)].

Clearly $N_{n+1}(j)$ is a conditional Markov chain given N_n. It remains to derive the asserted transition function for $N_{n+1}(2k)$ given $N_{n+1}(2k-1)$ and N_n. But with i and k as before, for $j > 0$ we have

$$P(N_{n+1}(2j) = N_n(j) + k - i \big| N_{n+1}(2j-1))$$

$$= P\left(\text{neg. bin.}\,(N_n(j-1) + N_n(j))\,, \frac{1}{2}\right) = k \big| \text{bin}\,(\text{neg. bin.}(etc.) = i)$$

$$= P(\text{bin} = i \big| \text{neg. bin.} = k) \cdot P(\text{neg. bin.} = k) P^{-1}(\text{bin} = i)$$

$$= \binom{k}{i} 2^{-k} \frac{\binom{N_n(j-1)+N_n(j)+k-1}{k} 2^{-(N_n(j-1)+N_n(j)+k)}}{\binom{N_n(j-1)+N_n(j)+i-1}{i} (\frac{2}{3})^{N_n(j-1)+N_n(j)} (\frac{1}{3})^i}$$

$$= (\frac{3}{4})^{N_n(j-1)+N_n(j)+i} (\frac{1}{4})^{k-i} \frac{(N_n(j-1) + N_n(j) + k - 1)!}{(N_n(j-1) + N_n(j) + i - 1)!(k-i)!}$$

$$= (\frac{3}{4})^{N_n(2j-1)+N_n(j)} (\frac{1}{4})^{k-i} \binom{N_n(j-1)+N_n(j)+k-1}{k-i}$$

$$= P\{\text{neg. bin.} \left(N_n(2j-1) + N_n(j), \frac{3}{4}\right) = k - i\},$$

as required for the second assertion. For $k \le 0$, as before, we need to add 1 to $N_n(k-1)$ and $N_n(2k-1)$ in the argument of the neg. bin.

In view of Theorems 2.2 and 2.3, let us note that while the downward mechanism is perhaps more tractable than we had a right to expect, it does not seem tractable enough to iterate explicitly to 2 or more steps. As indicated in the Introduction, what we really need is information about iteration to k steps as k becomes large, and this seems to be out of reach for the downward transitions. It turns out, thanks to some fortuitous comparisons, that it is not entirely out of reach for the "upward" transitions $N_n \to N_{n-1}$. Accordingly, we turn our attention now to these.

Since $\sigma(N_0, \ldots, N_n) \subset \sigma(R_n)$, the conditional law of N_{n-1} given N_n is implicit in the construction of R_n from N_n in Section 1. Indeed, referring to Lemma 1.1, given N_n we see that $N_{n-1}(k)$ equals the number of $X_{2k}(j)$, $1 \le j \le N_n(2k)$, which are non-zero. We note that, even for $k < 0$, we do not count $X_{2k}(0)$ since it only gives the steps of N_n from $(2k+1)$ to $2(k+1)$ before reaching $2k$, which do not yield steps of N_{n-1} from k to $k+1$. On the other hand, for $1 \le j$, even if $X_{2k}(j) > 1$ there is at most one step of N_{n-1} from k to $k+1$ starting with the jth of N_{n+1} from $2k$ to $2k+1$ and before the $(j+1)^{st}$ (void if $j = N_n(2k)$). Now according to Lemma 1.3, for $k \ge 0$ when N_n is given, $\{X_{2k}(j),\ 1 \le j \le N_n(2k)\}$ are determined by Bose-Einstein statistics of size $(N_n(2k),\ N_n(2k+1))$, and for $k < 0$, $\{X_{2k}(j),\ 0 \le j \le N_n(2k)\}$ are determined by Bose-Einstein statistics of size $(1 + N_n(2k),\ N_n(2k+1))$. Thus, for $k \geq 0$, the number of $X_{2k}(j) > 0$ is equal to the *number of non-empty boxes* in Bose-Einstein statistics with $N_n(2k)$ boxes and $N_n(2k+1)$ balls. This is a familiar combinatorial problem, and the answer is easily derivable from Exercise 17 of W. Feller [2, II.11] (see below). On the other hand, for $k < 0$ we need the law of the number of non-empty boxes *excluding the first box*, in Bose-Einstein statistics with $1 + N_n(2k)$ boxes and $N_n(2k+1)$ balls. This is a mixture of the former. The probability of i balls in box 1 is

$$\frac{\binom{N_n(2k)+(N_n(2k+1)-i)-1}{N_n(2k+1)-i}}{\binom{N_n(2k)+N_n(2k+1)}{N_n(2k+1)}}, 0 \leq i \leq N_n(2k+1),$$

and given i the problem reduces to the former with $N_n(2k+1)$ replaced by $N_n(2k+1) - i$. Thus we can prove

Theorem 2.4. *For $n > 0$, given $\sigma(N_n)$ the variables $N_{n-1}(k), -2^{n-1} < k$, are conditionally independent. Moreover*

(a) *For $k \geq 0$, $P\left(N_{n-1}(k) = j | N_n\right) = \frac{\binom{N_n(2k)}{j}\binom{N_n(2k+1)-1}{j-1}}{\binom{N_n(2k)+N_n(2k+1)-1}{N_n(2k+1)}}$; $1 \leq j \leq N_n(2k) \wedge N_n(2k+1)$, $(= 1$ if $j = N_n(2k) \wedge N_n(2k+1) = 0$ or $1)$.*

(b) *For $-2^{n-1} < k < 0$, $P\left(N_{n-1}(k) = j | N_n\right) = \frac{\binom{N_n(2k)}{j}\binom{N_n(2k+1)}{j}}{\binom{N_n(2k)+N_n(2k+1)}{N_n(2k+1)}}$; $0 \leq j \leq N_n(2k) \wedge N_n(2k+1)$ $(= 1$ if $j = N_n(2k) \wedge N_n(2k+1) = 0)$.*

Proof. We first show (a) $\Rightarrow$ (b). Indeed, by the remarks before the theorem (b) is a mixture of (a) in which $\binom{N_n(2k)+N_n(2k+1)-i-1}{N_n(2k+1)-i}$ cancels out leaving (for $k < 0$)

$$P\left(N_{n-1}(k) = j | N_n\right) =$$
$$= \binom{N_n(2k)}{j}\left(\sum_{i=0}^{N_n(2k+1)-j}\binom{N_n(2k+1)-i-1}{j-1}\right)$$
$$\cdot \binom{N_n(2k)+N_n(2k+1)}{N_n(2k+1)}^{-1}$$

and the sum reduces to $\binom{N_n(2k+1)}{j}$ by identity (12.8a) of [2, II. 12, p. 64]. Now to prove (a), we write for Bose-Einstein statistics with $N_n(2k)$ boxes and $N_n(2k+1)$ balls

$$\begin{aligned}
&P\,\{\text{exactly } j \text{ boxes are not empty}\}\\
&= P\,\{\text{exactly } N_n(2k) - j \text{ boxes remain empty}\}\\
&= \binom{N_n(2k)}{j} P\{N_n(2k) - j \text{ given boxes are empty, the other } j \text{ nonempty}\}\\
&= \binom{N_n(2k)}{j}\binom{N_n(2k+1)-1}{j-1}\binom{N_n(2k)+N_n(2k+1)-1}{N_n(2k+1)}^{-1},
\end{aligned}$$

as required, where we used [2, II, (5.2), p. 38] at the last step.

Remark. As with the downward transitions, the upward transition mechanism is practically free of n, but this time there is no exception. We need only extend it to $-\infty < k < \infty$ by the convention $P\left(N_{n-1}(k) = 0 \middle| N_n\right) = 1$ if $N_n(2k) \wedge N_n(2k+1) = 0$, and also define $P\{N_n(k) = 0,\ k \leq -2^n\} = 1$. This preserves the necessary zeros of $N_{n-1}(k)$ for
$k \leq -2^{n-1}$ and for $k > -2^{n-1}$ we note that neither (a) nor (b) depends on n explicitly.

In view of Theorem 2.4 (a), we are led to study the (hypergeometric) transition kernels

$$P\{Y = k | X_1 = j_1,\ X_2 = j_2\} = \binom{j_1}{k}\binom{j_2 - 1}{k - 1} \Big/ \binom{j_1 + j_2 - 1}{j_2}, \tag{2.2}$$
$$1 \leq k \leq j_1 \wedge j_2,$$

for $1 \leq j_1 \wedge j_2$, while $P\{Y = 0 | X_1 = j_1,\ X_2 = j_2\} = 1$ if $0 = j_1 \wedge j_2$. It turns out that when L is given the joint law of X_1 and X_2, when $X_1 = N_n(2k)$ and $X_2 = N_n(2k+1)$, is that of (conditionally) independent random variables. Consequently, we only examine the case when X_1 and X_2 are independent. In fact, we shall mainly be interested in the additional assumption that X_1 and X_2 also are identically distributed. Then we can prescribe a distribution $F = F_{X_1} = F_{X_2}$ for X_1 and X_2, and study the iteration of the transition mechanism with $F_Y = F_{Y_1} = F_{Y_2}$ in place of F (Y_1 and Y_2 being taken independent) and so on to the higher iterates (this applies also for $k < 0$, where there is an analogous iteration which one derives from Theorem 2.4 (b)). To be sure, $N_n(2k)$ and $N_n(2k+1)$ are not identically distributed given L except in special cases, but we aim for a comparison theorem with the identically distributed case. Anyway, this extra assumption is not needed for the basic comparison (Theorem 2.5). We introduce the familiar ordering of distribution functions:

Notation 2.5. For integer-valued random variables $X_1 \geq 0$ and $X_2 \geq 0$, we write $X_1 << X_2$ if, for all $k \geq 0$, $F_{X_1}(k) \geq F_{X_2}(k)$.

Now we will derive

(Comparison) Theorem 2.5. If F_{Y_1} is determined from (independent) $X_{1,1}$ and $X_{1,2}$ by (2.2), and F_{Y_2} similarly from (independent) $X_{2,1}$ and $X_{2,2}$, then if $X_{1,1} << X_{2,1}$ and $X_{1,2} << X_{2,2}$, it follows that $Y_1 << Y_2$.

Note. The probability space and joint distribution of (Y_1, Y_2) is irrelevant and unspecified.

Proof. The proof is rather long, but simple in outline. Viewing (2.2) as a transformation of pairs of point-probability distributions $(\delta_{j_1}, \delta_{j_2}) \to F_Y$, we first observe that it suffices that this be monotone increasing in both variables (j_1, j_2) in the sense of the order of Notation 2.5. Indeed, the relation $X_1 << X_2$ means that F_{X_1} may be obtained from F_{X_2} by transfer of probability mass from larger to smaller values. In brief, we can first obtain $F_{X_1}(0)$ by successive transfer from $\{k > 0 : F_{X_2}(k+1) - F_{X_2}(k) > F_{X_1}(k+1) - F_{X_1}(k)\}$ to 0, adding the surplus mass to $F_{X_2}(0)$ starting with the smallest possible k. Then with $F_{X_1}(0) = F_{X_2}(0)$ having been obtained for the resultant distribution F_{X_2}, we proceed analogously to obtain $F_{X_1}(1) - F_{X_1}(0)$ by transfers from $\{k > 1\}$; and so forth to $F_{X_1}(k+1) - F_{X_1}(k)$ for all $k \geq 0$. Obviously such transfers of mass reduce the distribution F_{X_2} in the sense of Notation 2.5 without destroying the relation $X_1 << X_2$ for the new distributions F_{X_2}. To see that such an operation also reduces F_Y (whether performed for $F_{X_{1,1}} << F_{X_{1,2}}$ or for $F_{X_{2,1}} << F_{X_{2,2}}$) it obviously suffices to show that (2.2) is monotone increasing in each variable when $Y = k$ is replaced by $Y \geq k$. Evidently (see Theorem 2.4 (a)) there is no difficulty if $j_1 \wedge j_2 = 0$ or if $j_1 \wedge j_2 = 1$. To examine the other cases, it is useful intuitively to regard (2.2) as a special case of the hypergeometric distribution, in which Y is the number of objects of "type one" chosen at random from j_1+j_2-1 objects in j_2 choices when there are initially j_1 of type 1 and $j_2 - 1$ of type 2 (to see this, note that $\binom{j_2-1}{k-1} = \binom{j_2-1}{j_2-k}$; this interpretation also makes clear why $Y \equiv 1$ if $j_1 \wedge j_2 = 1$).

We now examine the dependence on the variable j_2 (with j_1 fixed) by looking at the difference of (2.2) at j_2 and at $j_2 + 1$, for $1 \leq k \leq j_1 \wedge j_2$. We obtain by routine cancellations

$$\binom{j_1}{k}\binom{j_2-1}{k-1}\Big/\binom{j_1+j_2-1}{j_2} - \binom{j_1}{k}\binom{j_2}{k-1}\Big/\binom{j_1+j_2}{j_2+1}$$
$$:= D_1(j_1, j_2, k) = C_1(j_1, j_2, k)(j_1 j_2 - (j_1+j_2)k + j_1),$$

where $C_1(j_1, j_2, k)$ is a *non-negative* common factor whose complicated exact expression need not concern us further. The lesson derived from this is that it suffices, in order to prove monotonicity in j_2, to observe that $\sum_{k=1}^{j_1 \wedge j_2} D_1(j_1, j_2, k) \geq 0$. Indeed, since the sums $\sum_{k=1}^{j} D_1(j_1, j_2, k)$ are manifestly unimodal in j (i.e., increasing to a positive maximum and decreasing thereafter), we need only show that the last is non-negative, which follows since for $j = j_1 \wedge j_2$ the sum of the first terms is 1 and that of the second is ≤ 1.

A similar argument applies to the dependence on j_1. We obtain, for $1 \leq k \leq j_1 \wedge j_2$,

$$\binom{j_1}{k}\binom{j_2-1}{k-1}\Big/\binom{j_1+j_2-1}{j_2} - \binom{j_1+1}{k}\binom{j_2-1}{k-1}\Big/\binom{j_1+j_2}{j_2}$$
$$:= D_2(j_1, j_2, k) = C_2(j_1, j_2, k)(j_1 j_2 - (j_1+j_2)k)$$

where C_2 is non-negative. Again the partial sums in k are unimodal, and to show that they are all non-negative it suffices to observe that the sum from 1 to $j_1 \wedge j_2$ is obviously non-negative.

Corollary 2.5. *The comparison of Theorem 2.5 remains valid if, in place of (2.2), we replace $N_n(2k)$ by X_1 and $N_n(2k+1)$ by X_2 in the conditional distribution of Theorem 2.4 (b).*

Proof. As observed before Theorem 2.4, part (b) is a mixture, over i, of part (a) with $N_n(2k+1)$ replaced by $N_n(2k+1) - i$, where i is the number of balls in box 1 for Bose-Einstein statistics of size $(1 + N_n(2k),\ N_n(2k+1))$. We need only realize that if the distribution of $N_n(2k)$ is increased in the sense of Notation 2.5, that of i is decreased, hence that of $N_n(2k+1)-i$ is again increased. Obviously it also increases along with the distribution of $N_n(2k+1)$, so our assertion follows from that of Theorem 2.5.

Theorem 2.5 of course carries over to iterations of the transition mechanism (2.2): in particular if $X_{1,1}$ and $X_{1,2}$ (resp. $X_{2,1}$ and $X_{2,2}$) are i.i.d. with $X_{1,1} << X_{2,1}$, then $Y_1 << Y_2$ generate i.i.d. pairs to which the same operation applies, preserving the order.

A key to applying this is to identify distributions for which the operation and its iterates can be more or less explicitly calculated to serve as a basis for comparisons. If we take the obvious $F_X = \delta_k$ it turns out that F_Y is too complicated to iterate again (even once?). However, as sometimes happens in such situations, the Poisson distributions provide a better candidate for iteration.

Lemma 2.6. *If X_1 and X_2 in (2.2) are i.i.d. with the Poisson distribution, parameter $\lambda > 0$, then*

$$P\{Y=k\} = \begin{cases} e^{-\lambda}(2-e^{-\lambda}); & k=0 \\ 2e^{-\lambda}\lambda^{2k}/(2k)!; & k \geq 1. \end{cases}$$

In particular, if X_1 and X_2 are i.i.d. with law

$$P\{X_1=k\} = \begin{cases} 0; k=0 \\ \frac{e^{-\lambda}}{1-e^{-\lambda}}\lambda^k/k!; & k \geq 1, \text{ then} \end{cases}$$

$$P\{Y=k\} = \begin{cases} 0; & k=0 \\ (\cosh\lambda - 1)^{-1}\lambda^{2k}/(2k)!; & 1 \leq k. \end{cases}$$

In other words, given $X_1 > 0$ and $X_2 > 0$, $2Y$ has a $\cosh(\lambda)$-distribution conditioned to be non-zero.

Proof. The second assertion follows from the first by observing that the condition $X_1 > 0$ and $X_2 > 0$ is equivalent to $Y > 0$. As to the first, we need only calculate for $k > 0$

$$\begin{aligned} P\{Y=k\} &= e^{-2\lambda}\sum\sum_{i,j\geq k}\frac{\lambda^{i+j}}{i!j!}\binom{i}{k}\binom{j-1}{k-1}\Big/\binom{i+j-1}{j} \\ &= e^{-2\lambda}\sum_{n=2k}^{\infty}\frac{\lambda^n}{(n-1)!}\sum_{i=k}^{n-k}\binom{i-1}{k-1}\binom{n-i-1}{k-1} \\ &= e^{-2\lambda}k^{-1}\lambda^{2k}\sum_{n=2k}^{\infty}\frac{\lambda^{n-2k}}{(n-1)!}\binom{2k+(n-2k)-1}{n-2k} \\ &= e^{-2\lambda}k^{-1}\frac{\lambda^{2k}}{(2k-1)!}e^{\lambda} = 2e^{-\lambda}\lambda^{2k}/(2k)!. \end{aligned}$$

We now turn to a comparison of the law of $2Y$ from Lemma 2.6 with a Poisson law. Let us set

Notation 2.7. For any integer-valued random variable $X \geq 0$, let $\mathcal{O}X$ denote (any) random variable with the law derived from (2.2) when X_1 and X_2 are independent with the law of X. Further, let $\mathcal{P}_\lambda$ denote any random variable with the Poisson law, parameter $\lambda > 0$.

Noting that from Lemma 2.6 we have

$$P\{2\mathcal{O}\mathcal{P}_\lambda = k\} = \begin{cases} 0; \ 1 \leq k \text{ odd} \\ e^{-\lambda}(2-e^{-\lambda}); \ k=0\,, \\ 2e^{-\lambda}\frac{\lambda^k}{k!}; \ 0 < k \text{ even} \end{cases} \quad \text{we prove}$$

Theorem 2.8. *For $\lambda > 6$, set $c = (\lambda-1)^{-1}$. Then for $n \geq 1$, $(2\mathcal{O})^n\mathcal{P}_\lambda << \mathcal{P}_{\mu_n}$, where*
$\mu_n = \lambda\prod_{k=o}^{n-1}(1 + c(1+c)^k\ell n\lambda)$.

Remark. This result can probably be improved, however it represents a compromise. Due to the nonlinearity of (2.2) it does not seem possible to extract the factors 2 from Theorem 2.8 to get an estimate of $\mathcal{O}^n\mathcal{P}_\lambda$. Consequently, Theorem 2.8 is not used in the sequel, and an uninterested reader can skip to Theorem 2.9.

Proof. We first note that $2\mathcal{O}\mathcal{P}_\lambda << X$, where $P\{X=0\} = e^{-\lambda} - e^{-2\lambda}$, $P\{X = 1\} = e^{-\lambda}$, and for $1 \leq k$, $P\{X = 2k\} = P\{X = 2k+1\} = e^{-\lambda}\frac{\lambda^{2k}}{(2k)!}$. We wish to minimize μ such that $X << \mathcal{P}_\mu$. We set $T_k = P\{X = k\} - e^{-\mu}\frac{\mu^k}{k!}$, so we want $\sum_{j=0}^k T_j \geq 0$ for $0 \leq k$. $T_0 \geq 0$ means $e^{-\lambda} - e^{-2\lambda} - e^{-\mu} \geq 0$, and $T_1 \geq 0$ means $e^{-\lambda} - \mu e^{-\mu} \geq 0$. Now if $T_1 \geq 0$ holds, then $T_0 \geq \mu e^{-\mu} - \mu^2 e^{-2\mu} - e^{-\mu}$, so it suffices that $\mu - \mu^2 e^{-\mu} \geq 1$. Here the left side is increasing for $\mu > 1$, and exceeds 1 for $\mu = \lambda(1 + \frac{\ell n\lambda}{\lambda-1})$. Indeed, logarithmic differentiation shows this last is increasing for $\lambda > 1$ with limit 2 at $\lambda = 1$, and indeed $2 - 4e^{-2} \geq 1$. Hence to ensure that $T_0 \geq 0$ it suffices to require $T_1 \geq 0$. We need to show that this will also imply that $\sum_{j=o}^n T_j \geq 0$ for every n. Treating the even and odd terms separately, we first show that both $\sum_{j=1}^n T_{2j}$ and $\sum_{j=0}^n T_{2j+1}$ are unimodal, in the sense that the signs $sgnT_{2j}$ are decreasing in j (once -1, then remaining -1), and likewise $sgnT_{2j+1}$. Indeed, for $T_{2k} > 0$ it is equivalent that $\ell n\left(e^{-\lambda}\frac{\lambda^{2k}}{(2k)!}\right) > \ell n\left(e^{-\mu}\frac{\mu^{2k}}{(2k)!}\right)$, or $-\lambda + (2k)\ell n\lambda > -\mu + 2k\ell n\mu$, and for $1 < \lambda < \mu$ this holds if and only if $2k < (\mu - \lambda)(\ell n\frac{\mu}{\lambda})^{-1}$, $1 \leq k$. Similarly, for $0 \leq k$, $T_{2k+1} > 0$ is equivalent to $\ell n\left(e^{-\lambda}\frac{\lambda^{2k}}{(2k)!}\right) > \ell n\left(e^{-\mu}\frac{\mu^{2k+1}}{(2k+1)!}\right)$, which reduces to $-\lambda + (2k)\ell n\lambda > -\mu + (2k+1)\ell n\mu - \ell n(2k+1)$. Replacing $2k+1$ by x, we find

$$\frac{d}{dx}[(\mu - \lambda) - x(\ell n\mu - \ell n\lambda) - \ell n\lambda + \ell nx] = -\ell n(\mu/\lambda) + \frac{1}{x},$$

decreasing in x, so the inequality holds only in an initial interval. Setting $0 = (\mu - \lambda) - x(\ell n\mu - \ell n\lambda) - \ell n\lambda + \ell nx$, we get a unique root of $x = ((\mu - \lambda) +$

$\ell n\frac{x}{\lambda})(\ell n\frac{\mu}{\lambda})^{-1}$. Now if $\mu-\lambda=\frac{\lambda}{\lambda-1}\ell n\lambda$ from the Theorem at $n=1$, then at $x=\lambda$ the right side becomes $\frac{\lambda}{\lambda-1}\ell nx/\ell n(1+\frac{1}{\lambda-1}\ell n\lambda)$ which exceeds λ. Thus the root x exceeds λ, and since $\ell n\frac{x}{\lambda}>0$ we see that it exceeds $(\mu-\lambda)(\ell n\frac{\mu}{\lambda})^{-1}$ as well. On the other hand, at $x=\mu=\lambda+\frac{\lambda}{\lambda-1}\ell n\lambda$ the right side is $(\mu-\lambda)(\ell n\frac{\mu}{x})^{-1}+1$, which equals $(\frac{\lambda}{\lambda-1}\ell n\lambda)/\ell n(1+\frac{\ell n\lambda}{\lambda-1})+1$. Setting $\frac{\ell n\lambda}{\lambda-1}=\epsilon$, this is $\lambda(\frac{\epsilon}{\ell n(1+\epsilon)})+1<$ $(\frac{\lambda}{1-\frac{\epsilon}{2}})+1=\lambda\left(1+\frac{\epsilon/2}{1-\epsilon/2}\right)+1$. Now for $\epsilon\le\frac{1}{2}$ this is at most $\lambda(1+\frac{\epsilon}{3})+1$, and if $\frac{\lambda}{6}\ge 1$ this is less than μ. A check shows that for $\lambda>6$ the condition on ϵ is met, so that the root x is less than μ. Since the range of the right side in $\lambda\le x\le\mu$ is 1, we see that the root exceeds $(\mu-\lambda)(\ell n\frac{\mu}{\lambda})^{-1}$ by less than 1.

Now we have shown that the first $2k>0$ for which $T_{2k}<0$ is the first even integer exceeding $(\mu-\lambda)(\ell n\frac{\mu}{\lambda})^{-1}$, while the first $2k+1$ for which $T_{2k+1}<0$ exceeds this by less than 1. Thus, whether the first integer exceeding $(\mu-\lambda)(\ell n\frac{\mu}{\lambda})^{-1}$ is even or odd, if $2k_0$ is the first even integer exceeding it, then $T_{2k_0+1}<0$. In all cases, the index of the first even negative term differs from the index of the first odd negative term by 1. This implies, since $T_0\ge 0$ and $T_1\ge 0$, and $\sum_{j=0}^{\infty}T_j=0$, that $\sum_{j=0}^{n}T_j\ge 0$ for all n, which will complete the proof for $n=1$.

It remains to examine the requirement $T_1\ge 0$, i.e. $e^{-\lambda}-\mu e^{-\mu}\ge 0$, and to iterate it over n. With λ fixed we estimate the root of $-\lambda=\ell n\mu-\mu$ by using Newton's method starting with $\mu_0=\lambda>1$. Then $\mu_1=\lambda+(\frac{\lambda}{1-\lambda})\ell n\lambda$ over-estimates the root, since $\frac{d}{d\mu}(\ell n\mu-\mu)=\frac{1-\mu}{\mu}$ is decreasing, so Theorem 2.8 is proved for $n=1$. For $n=2$ we replace λ by μ_1 and repeat the procedure to get, setting $c=(\lambda-1)^{-1}$,

$$\begin{aligned}\mu_2&=\lambda(1+c\,\ell n\lambda)(1+c\ell n[\lambda(1+c\ell n\lambda])\\&<\lambda(1+c\ell n\lambda)(1+c(\ell n\lambda+c\ell n\lambda))=\lambda(1+c\ell n\lambda)(1+c+c^2)\ell n\lambda,\end{aligned}$$

using the estimate $\ell n(1+x)<x$ for $x>0$. In the same way, we get for induction step

$$\begin{aligned}&\mu_n(1+c\,\ell n\mu_n)=\\&=\lambda\prod_0^{n-1}(1+c(1+c)^k\ell n\lambda)(1+c(\ell n\lambda+\sum_0^{n-1}\ell n(1+c(1+c)^k\ell n\lambda))\\&<\lambda\prod_0^{n-1}(1+c(1+c)^k\ell n\lambda)(1+c\,\ell n\lambda+c^2\sum_0^{n-1}(1+c)^k\ell n\lambda)\\&=\lambda\prod_0^{n}(1+c(1+c)^k\ell n\lambda),\quad\text{as required.}\end{aligned}$$

While we do not have a reasonable bound on the law of $\mathcal{O}^nX$, even if $X=\mathcal{P}_\lambda$, we do obtain good upper bounds on the first two moments $E\mathcal{O}^nX$ and $E(\mathcal{O}^nX)^2$ for general X. These will lead in Section 3 to bounds on the two moments of the law of $(N_n\mid L)$. With more work, the following for $n>1$ may also be sharpened to some extent, but it seems quite satisfactory as stated.

Theorem 2.9. *For any random variable $X \geq 0$ (integer-valued), we have*
(a) $E(\mathcal{O}X) \leq \frac{1}{2}(EX+1)$; $E(\mathcal{O}^n X) \leq 2^{-n}EX + 1$ *for* $1 \leq n$.
(b) $0 \leq 2E(\mathcal{O}X) - E(\mathcal{O}(2X)) \leq \frac{2}{3}$, *and*
(c) $E(\mathcal{O}X)^2 \leq \frac{1}{4}EX^2 + \frac{5}{8}(EX+1)$; $E(\mathcal{O}^n X)^2 \leq 4^{-n}EX^2 + 5\cdot 2^{-1(n+2)}EX + \frac{5}{6}$; $1 \leq n$.

Proof.

(a) As noted in the proof of Theorem 2.5, given $X_1 = i \geq 1$ and $X_2 = j \geq 1$, $\mathcal{O}X$ has a hypergeometric distribution with mean $j(\frac{i}{i+j-1})$. Thus if X has distribution $p_i = P(X=i)$, $0 \leq i$, we have

$$
\begin{aligned}
E\mathcal{O}X &= \frac{1}{2}\sum_{i,j\geq 1}\sum p_i p_j 2ij/i+j-1 \\
&\leq \frac{1}{4}\sum_{i,j\geq 1}\sum p_i p_j \left(\frac{(i+j)^2}{i+j}\right)\left(\frac{i+j}{i+j-1}\right) \\
&= \frac{1}{4}\sum\sum p_i p_j (i+j)\left(1+\frac{1}{i+j-1}\right) \\
&= \frac{1}{4}\sum_i p_i(i+EX) + \frac{1}{4}\sum\sum p_i p_j \left(1+\frac{1}{i+j-1}\right) \\
&\leq \frac{1}{2}EX + \frac{1}{4} + \frac{1}{4}\sum_{i,j\geq 1}\sum p_i p_j / i+j-1 \\
&\leq \frac{1}{2}(EX+1),
\end{aligned}
$$

proving the first assertion. Now by iteration, setting $e_0 = EX$, $e_{n+1} = \frac{1}{2}(e_n+1)$, we obtain easily $e_n = 2^{-n}e_0 + (\frac{1}{2}+\frac{1}{4}+\cdots+2^{-n}) \leq 2^{-n}e_0 + 1$, as required.

(b) Evidently we can assume again, without loss of generality, that $1 \leq X$, so that, given (i,j), $\mathcal{O}X$ has a hypergeometric distribution with mean $\frac{ij}{i+j-1}$. Then we need only note for $1 \leq i \wedge j$

$$
\begin{aligned}
0 &\leq 2(ij/i+j-1) - \frac{(2i)2j)}{2(i+j)-1} \\
&= 2ij\left(\frac{1}{i+j-1} - \frac{1}{i+j-\frac{1}{2}}\right) \\
&= ij\frac{1}{(i+j-1)(i+j-\frac{1}{2})} \\
&= \frac{ij}{(i+j)^2 - \frac{3}{2}(i+j) + \frac{1}{2}} \\
&< \frac{(i+j)^2}{4((i+j)^2 - \frac{3}{2}(i+j)) + \frac{1}{2})} \\
&< \frac{2}{3}
\end{aligned}
$$

(the last expression is monotone decreasing in $i+j>1$ by routine differentiation, so obviously stronger inequalities hold for $i+j>K$, depending on K, but $\frac{1}{4}$ is a lower bound obtained by letting $i+j\to\infty$).

(c) As to the square, it is well-known that the variance of a hypergeometric distribution is less than that of the corresponding binomial $b(n,p)$ where, in our case (given $X_1=i$, $X_2=j$), we have $n=j$, $p=\frac{i}{i+j-1}$. Thus (for $1\leq i,j$) $(VarOX\mid i,j)\leq j\frac{i}{i+j-1}\left(1-\frac{i}{i+j-1}\right)$. Thus $E((OX)^2\mid i,j)\leq(E(OX\mid i,j))^2+E(OX\mid i,j)-\frac{ji^2}{(i+j-1)^2}$, and taking expectations gives

$$\begin{aligned}E(OX)^2&\leq E(OX)+\sum\sum p_ip_j\left(\frac{ij}{i+j-1}\right)\frac{(i-1)j}{(i-1)+j}\\&\leq E(OX)+\sum\sum p_ip_j\frac{(ij)^2}{(i+j-1)(i+j)},\end{aligned}$$

where we used the observation that $\frac{d}{dx}\left(\frac{xy}{x+y}\right)=\frac{y^2}{(x+y)^2}>0$ for $0<x,y$. Now

$$\begin{aligned}\sum\sum p_ip_j\frac{(ij)^2}{(i+j-1)(i+j)}&=\frac{1}{4}\sum\sum p_ip_j\frac{(2ij)^2}{(i+j-1)(i+j)}\\&\leq\frac{1}{16}\sum\sum p_ip_j(i+j)^2\left(\frac{i+j}{i+j-1}\right)\\&\leq\frac{1}{4}EX^2+\frac{1}{16}\sum\sum p_ip_j\frac{(i+j)^2}{(i+j-1}\\&\leq\frac{1}{4}EX^2+\frac{1}{8}(EX+1),\end{aligned}$$

where we used the inequality from the proof of (a) at the last step. Combining with EOX now gives the first assertion of (c).

As to the second, we will prove by induction that $E(O^nX)^2\leq 4^{-n}EX^2+5(\sum_{k=n+2}^{2n+3}2^{-k})EX+\frac{5}{2}\sum_{k=1}^{n}4^{-k}$. For $n=1$ it follows by the first assertion, so we assume it for n. Then

$$\begin{aligned}E(O^{n+1}X)^2&\leq\frac{1}{4}E(O^nX)^2+\frac{5}{8}E(O^nX+1)\\&\leq 4^{-(n+1)}EX^2+\left[\frac{5}{4}\left(\sum_{k=n+2}^{2n+3}2^{-k}\right)+\frac{5}{8}2^{-n}\right]EX\\&\quad+\frac{5}{8}\sum_{k=1}^{n}4^{-k}+\frac{5}{8}\\&=4^{-(n+1)}EX^2+\frac{5}{4}\left(\sum_{k=n+1}^{2n+3}2^{-k}\right)EX+\frac{5}{2}\left(\sum_{k=1}^{n+1}4^{-k}\right),\end{aligned}$$

and the proof is completed by summing the series.

Remark. For brevity, and because it is not needed for Theorem 3.6 below, we do not attempt any analog of Theorem 2.9 using the operation of Theorem 2.4 (b) in place of (2.2).

Section 3. The law of $(N_n \mid L), n$ fixed.

We start by admitting that we do not have any closed expression for the above law. One approach would be to examine $\lim_{m\to\infty} (N_n \mid N_m)$ in law; however, as stated in the Introduction, we could not prove that $\sigma(L) \equiv \lim_{m\to\infty} \sigma(N_k, k \geq m)$, and besides, an explicit expression for the $(m-n)^{th}$ iterate of the transition mechanism is lacking. The approach via Bayes Formula also seems doomed to failure, although the ingredients are known from [6]. A more rewarding method is to find, first of all, the law of N_n given $\{L(k\alpha_n), -2^n < k\}$. This is done explicitly below. Then we can use the fact that $(N_n \mid L) = \lim_{m\to\infty} (N_n \mid L(k\alpha_m), -2^m < k)$ (in law) together with the results of Section 2, to obtain information about $(N_n \mid L)$.

Theorem 3.1. *Choose $0 < x_k;\ -2^n < k \leq \kappa$, and set $0 = x_k$ for $k > \kappa$, for some $\kappa \geq 0$. Then for $-2^n < j < \kappa$*

$$P(N_n(j) = n_j \mid L(k\alpha_n) = x_k,\ -2^n < k) = \tag{3.1}$$

$$\begin{cases} \left(\frac{x_j x_{j+1}}{4\alpha_n^2}\right)^{n_j} (n_j!)^{-2} I_0^{-1}\left(\sqrt{\frac{x_j x_{j+1}}{\alpha_n^2}}\right); j < 0,\ 0 \leq n_j \\ \left(\frac{x_j x_{j+1}}{4\alpha_n^2}\right)^{n_j - \frac{1}{2}} (n_j!(n_j-1)!)^{-1} I_1^{-1}\left(\sqrt{\frac{x_j x_{j+1}}{\alpha_n^2}}\right); j \geq 0,\ 0 < n_j \end{cases}$$

where I_0 and I_1 are the modified Bessel functions (so that either sum over n_j equals 1).

Remark 1. For $j \geq \kappa$ the conditional probability is 1 for $n_j = 0$. The exact meaning of the conditioning is that, for $\kappa := \inf\{k \geq 0 : L((k+1)\alpha_n) = 0\}$ (note that $\{N_n(k) = 0\} \equiv \{N_n(j) = 0 \text{ for } j \geq k\} \equiv \{L((k+1)\alpha_n) = 0\}$) the expression gives a regular conditional distribution of $N_n(j)$ given $\{L(k\alpha_n),\ -2^n < k\}$. Moreover, as seen easily from the proof below, the variables $N_n(j)$ are conditionally independent given $\{L(k\alpha_n),\ -2^n < k\}$, hence by multiplication this also gives the regular conditional joint distributions.

Remark 2. Following J. Pitman and M. Yor [7] we call the first case the (discrete) Bessel I_0-distribution with parameter $z = \sqrt{\frac{x_j x_{j+1}}{\alpha_n^2}}$ and the second case the I_1-distribution with the same z. This appearance of these Bessel distributions has a long history, the case of I_1 going back at least to F. Knight [5, p. 180] and the case of I_0 to Pitman and Yor [loc sit., p. 449] where the other Bessel distributions are also discussed. But the present derivation, which includes also the joint distributions, seems to be new.

Proof. By Lemma 1.5, if $N_n(\cdot)$ is given, $L(k\alpha_n)$ equals a sum over inserts which "start" at $k\alpha_n$ (i.e. start at some T_j with $B(T_j) = k\alpha_n$), and these are all independent. On the other hand, a well-known fact about the local time of B at 0 and time $T(\pm\alpha_n)$ is that it has the exponential law with $\lambda = \alpha_n^{-1}$ (it suffices to check the case $n = 0$ and apply Brownian scaling). This is clearly the same for the local time at 0 of an n-interest. Thus we see that, given $N_n(\cdot)$, the $L(k\alpha_n)$ are mutually independent with marginal law having the gamma densities $\Gamma(N_n(k-1) + N_n(k);\ \alpha_n^{-1})$ for $k > 0$ and $\Gamma(N_n(k-1) + N_n(k) + 1;\ \alpha_n^{-1})$ for $-2^n < k \le 0$ (note that for $k \le 0$ there is a first trip from $k\alpha_n$ to $(k-1)\alpha_n$ not counted in $N_n(k-1)$).

We now work out the joint distribution

$$P(N_n(k) = n_k,\ 1 \le k < \kappa \mid N_n(0) = n_0,\ L(k\alpha_n) = x_k,\ 1 \le k \tag{3.2}$$

for $n_k > 0$, $0 \le k < \kappa$ fixed. Applying Bayes' rule, this is proportional (for n_0 fixed, $n_\kappa = 0$) to $P(L(k\alpha_n) = x_k,\ 1 \le k \le \kappa \mid N_n(k) = n_k,\ 0 \le k \le \kappa) \cdot P(N_n(k) = n_k,\ 1 \le k < \kappa \mid N_n(0) = n_0,\ \kappa)$ with a factor of proportionality such that the sum over $n_k > 0$, $1 \le k < \kappa$, equals 1, and by the Markov property of $N_n(k)$ this would not be changed if $\{N_n(k),\ k \le 0\}$ were given instead of only $N_n(0)$. Referring to the transition function of $N_n(k)$ from the Introduction, and setting $n_\kappa = 0$, the last product equals

$$\begin{aligned}
&\prod_{k=1}^{\kappa} \frac{x_k^{n_{k-1}+n_k-1} e^{-\alpha_n^{-1}x_k}}{\alpha_n^{n_{k-1}+n_k-1}\Gamma(n_{k-1}+n_k)} \prod_{k=1}^{\kappa} \binom{n_{k-1}+n_k-1}{n_k} \alpha_{n_{k-1}+n_k} \\
&= \prod_{k=1}^{\kappa} \left(\frac{x_k}{2\alpha_n}\right)^{n_{k-1}+n_k-1} ((n_{k-1}-1)!n_k!)^{-1} e^{-\alpha_n^{-1}x_k} \\
&= \frac{\left(\frac{x_1}{2\alpha_n}\right)^{n_0-\frac{1}{2}}}{(n_0-1)!} \left(\prod_{k=1}^{\kappa-1} \sqrt{\frac{x_k x_{k+1}}{4\alpha_n^2}}^{\,2n_k-1} ((n_k-1)!n_k!)^{-1} e^{-\alpha_n^{-1}x_k}\right).
\end{aligned} \tag{3.3}$$

Thus, in so far as the dependence on $(n_1, \ldots, n_{\kappa-1})$ is concerned, we recognize for each k the $(n_k-1)^{th}$ term in the series expansion of $I_1\left(\sqrt{\frac{x_k x_{k+1}}{\alpha_n^2}}\right)$. Hence by normalization the whole expression (3.2) must reduce to

$$\frac{\prod_{k=1}^{\kappa-1} \sqrt{\frac{x_k x_{k+1}}{4\alpha_n^2}}^{\,2n_k-1} ((n_k-1)!n_k!)^{-1}}{\prod_{k=1}^{\kappa-1} I_1\left(\sqrt{\frac{x_k x_{k+1}}{\alpha_n^2}}\right)}.$$

We note the curious fact that there is no dependence on the given n_0. Moreover, since $N_n(k) = n_k > 0$ implies $k < \kappa$, we can fix $K \ge 1$ and sum over all κ and $n_k : K < k < \kappa$ to obtain, for $1 \le n_k$, $1 \le k \le K$,

$$\begin{aligned}
&P(N_n(k) = n_k,\ 1 \le k \le K \mid N_n(0),\ L(k\alpha_n) = x_k,\ 1 \le k \le K+1) \\
&= \prod_{k=1}^{K} \sqrt{\frac{x_k x_{k+1}}{4\alpha_n^2}}^{\,2n_k-1} \left[(n_k-1)!n_k! I_1\left(\sqrt{\frac{x_k x_{k+1}}{\alpha_n^2}}\right)\right]^{-1},
\end{aligned} \tag{3.4}$$

where $\{\kappa > K\}$ is also given since $x_{K+1} > 0$.

With this as model, let us now work out

$$P(N_n(k) = n_k,\ -2^n < k < \kappa \mid L(k\alpha_n) = x_k,\ -2^n < k \le \kappa) \tag{3.5}$$

where $n_k \ge 0$ for $k \le 0$; $n_k > 0$ for $0 < k < \kappa$, and $x_k > 0$ for $2^n < k \le \kappa$ (and we set $x_{-2^n} = 0 = n_{-2^n}$, $x_{\kappa+1} = 0 = n_\kappa$). Applying Bayes' Rule we get a result proportional to (as the n_k vary with κ fixed) the product of (3.4) with

$$P(L(k\alpha_n) = x_k,\ -2^n < k \le 0 \mid N_n(\cdot))P(N_n(k) = n_k,\ -2^n < k \le 0)$$

$$= \prod_{-2^n+1}^{0} \frac{x_k^{n_{k-1}+n_k-1} e^{-\alpha_n^{-1}x_k}}{\alpha_n^{n_{k-1}+n_k}\Gamma(n_{k-1}+n_k)} \prod_{-2^n+1}^{0} \binom{n_{k-1}+n_k}{n_k} \alpha_{n_{k-1}+n_k-1}$$

$$= \prod_{-2^n+1}^{0} \left(\frac{x_k}{2\alpha_n}\right)^{n_{k-1}+n_k} (n_{k-1}!n_k!)^{-1}\, e^{-\alpha_n^{-1}x}$$

$$= \frac{\left(\frac{x_0^{n_0}}{2\alpha_n}\right)}{n_0!} \prod_{-2^n+1}^{-1} \sqrt{\frac{x_k x_{k+1}}{4\alpha_n^2}}^{\,2n_k} (n_k!)^{-2}\,.$$

Combining the first factor on the right with the factor $\frac{\left(\frac{x_1}{2\alpha_n}\right)^{n_0-\frac{1}{2}}}{(n_0-1)!}$ from the right side of (3.3) we recognize the general term in the expansion of $I_1\left(\sqrt{\frac{x_0x_1}{\alpha_n^2}}\right)$, for $n_0 > 0$, while the terms of the subsequent product are those of $I_0\left(\sqrt{\frac{x_kx_{k+1}}{\alpha_n^2}}\right)$, where we permit $n_k = 0$. On the other hand, if $n_0 = 0$ then $\kappa = 0$ and (3.3) is vacuous. In that case we have only the terms from $I_0\left(\sqrt{\frac{x_kx_{k+1}}{\alpha_n^2}}\right)$; $-2^n < k < 0$. This completes the proof of Theorem 3.1. Moreover, since our conditional probability given $\{L(k\alpha_n) = x_k,\ -2^n < k\}$ factors into a product of terms in n_j, we see that the events $\{N_n(j) = n_j\}$ are all conditionally independent given $\{L(k\alpha_n);\ -2^n < k\}$. Indeed, since the conditional law of $N_n(k)$ depends only on $L(k\alpha_n)$ and $L((k+1)\alpha_n)$, we can state

Corollary 3.1. *For $-2^n < k_1 < \cdots < k_m$, $\{N_n(k_j);\ 1 \le j \le m\}$ are conditionally independent given $\{L(k_j\alpha_n),\ L(k_j+1)\alpha_n;\ 1 \le j \le m\}$ and have the marginal distributions of Theorem 3.1.*

Remark 1. The distribution of κ is not contained in Theorem 3.1. However, it is easy to find:

$$P\{\kappa \le K\} = P\{\max B'(t) < (K+1)\alpha_n\}$$
$$= (K+1)\alpha_n\,(1+(K+1)\alpha_n)^{-1}\,,\ 0 \le K.$$

Remark 2. The independence assertion of Corollary 3.1 can be seen as a consequence of the conditional independence, given $\{L(k_j\alpha_n),\ L((k_j+1)\alpha_n)\}$, of the processes in $[k_j\alpha_n, (k_j+1)\alpha_n]$ obtained by excising the excursions of B' outside the interval (not proved here).

For application of Theorem 2.9 to these distributions, we need to work out the first two moments of the second marginal in (3.1) (namely, of the I_1-distribution).

Rappelons qu'un brownien réfléchi R admet pour décomposition canonique $R = \beta + L$, où la partie martingale β est un mouvement brownien et où la partie à variation finie L, qui vérifie $L_t = -\inf_{s\leqslant t} \beta_s$, est aussi le demi-temps local en zéro de R, c'est-à-dire son temps local symétrique en zéro, ou encore le temps local en zéro de tout brownien ayant R pour valeur absolue. Réciproquement, tout brownien β est la partie martingale d'un unique brownien réfléchi R, donné par $R_t = \beta_t - \inf_{s\leqslant t} \beta_s$.

PROPOSITION 18. — *Soit Z un processus de Walsh, de coordonnées polaires R et Θ, défini sur $(\Omega, \mathcal{A}, \mathbf{P}, \mathcal{F})$. Appelons L le demi-temps local en 0 de R, c'est-à-dire le processus croissant nul à l'origine tel que $\beta = R - L$ soit un mouvement brownien linéaire.*

(i) *Pour toute fonction f mesurable sur $\mathbf{V}$, le processus $f\circ\Theta$ reste p. s. constant lors de chaque excursion de R; il est prévisible.*

(ii) *Le processus Z est aussi un processus de Walsh pour la filtration* Nat Z, *et, plus généralement, pour toute filtration intermédiaire entre $\mathcal{F}$ et* Nat Z.

(iii) *Si f est une fonction mesurable bornée sur $\mathbf{V}$, la martingale*

$$M^f = \int \mathbb{1}_{\{Z\neq c\}} f(\Theta)\, d\beta$$

vaut $f(\Theta)R - \pi(f)L$. En particulier, si $\pi(f) = 0$, la martingale $f(\Theta)R$, qui figure dans la définition des processus de Walsh, est continue.

REMARQUE. — Si l'on suppose en outre que $\mathbf{V}$ est essentiellement séparable et que deux points de $\mathbf{V}$ sont toujours séparés par deux éléments non négligeables de $\mathcal{V}$, le (i) dit que Θ lui-même est constant lors des excursions de R et prévisible. On peut toujours se ramener à ce cas en remplaçant $\mathbf{V}$ par un quotient convenable; nous préférons ne pas le faire ici pour éviter d'alourdir la proposition 20, plus bas.

DÉMONSTRATION. — (i) Soit $s > 0$; posons $D_s = \inf\{t \geqslant s : R_t = 0\}$. Pour $A \in \mathcal{V}$, la fonction $a = \mathbb{1}_A - \pi(A)$ est de moyenne nulle et $M = a(\Theta)\, R$ est donc une martingale. En raison de l'expression de a, les sauts de M ne peuvent prendre que les valeurs $\pm R$. Soient S le temps d'arrêt $\inf\{t > s : \Delta M_t \neq 0\}$ et $T = S\wedge D_s\wedge n$, où n est un entier plus grand que s. Sur l'intervalle $[\![s, T[\![$, M est continue, R est continu et strictement positif, $a\circ\Theta = M/R$ est continue, donc constante en raison de l'expression de a, et $M_t = a(\Theta_s)\, R_t$. Ceci permet d'écrire

$$\begin{aligned} M_T - M_s &= a(\Theta_s)\,(R_T - R_s) + \Delta M_T \\ &= a(\Theta_s)\,(R_T - R_s) + \mathbb{1}_{\{\Delta M_T \neq 0\}}(-1)^{\mathbb{1}_A\circ\Theta_s} R_T\,. \end{aligned}$$

Puisque T est majoré par D_s et borné, $\mathbf{E}[M_T - M_s | \mathcal{F}_s]$ et $\mathbf{E}[R_T - R_s | \mathcal{F}_s]$ sont nuls; la formule ci-dessus entraîne $\mathbf{E}[\mathbb{1}_{\{\Delta M_T\neq 0\}} R_T | \mathcal{F}_s] = 0$, donc $\mathbb{1}_{\{\Delta M_T\neq 0\}} R_T = 0$ p. s. et $T = D_s$ sur $\{\Delta M_T \neq 0\}$. Mais $\Delta M_{D_s} = \pm R_{D_s} = 0$; sur $\{\Delta M_T \neq 0\}$ on devrait avoir $\Delta M_T = \Delta M_{D_s} = 0$; ainsi, $\Delta M_T = 0$ p. s., d'où $S \geqslant D_s\wedge n$, $T \geqslant D_s\wedge n$, et $a\circ\Theta$ est constant sur $[\![s, D_s\wedge n[\![$. En faisant varier s parmi les rationnels et n parmi les entiers, on obtient que $\mathbb{1}_A\circ\Theta$ est constant sur chaque excursion de R; l'extension des ensembles aux fonctions est immédiate.

La prévisibilité de $f\circ\Theta$ s'obtient en écrivant $f\circ\Theta = \lim_{n\to\infty} f\circ\Theta\mathbb{1}_{\{R\geqslant 1/n\}}$.

(ii) Si l'on remplace $\mathcal{F}$ par une filtration plus petite, mais à laquelle Z est encore adapté, les processus β et $f\circ\Theta\, R$ restent adaptés, donc restent des martingales (pour $\pi(f) = 0$), et Z reste un processus de Walsh.

(b) *(Local Independence). Apart from a P-null set, for every* $-1 < x_1 < x_2$, *and* $\forall(k,n)$ *with* $(k\alpha_n, (k+1)\alpha_n) \subset (x_1, x_2)$, $N_n(k)$ *under* $\mathcal{L}$ *is jointly independent of all* $N_m(j)$ *with* $(j\alpha_m, (j+1)\alpha_m)$ *outside* $[x_1, x_2]$, *and its conditional law depends only on* $\{L(x);\ x_1 \le x \le x_2\}$. *In words: the* $N_n(k)$ *which count upcrossings of disjoint intervals are mutually independent given L, and their laws depends only on L inside the intervals counted.*

(c) *(Local monotonicity). Apart from* $w_i, i = 1$ *or 2, in a P-null set, for every* $-1 < x_1 < x_2$, $L(x, w_1) < L(x, w_2)$ *for* $x_1 \le x \le x_2$ *implies that* $(N_n(k) \mid L(\cdot, w_1)) << (N_n(k) \mid L(\cdot, w_2))$ *for* $\forall(k,n)$ *with* $x_1 \le k\alpha_n$ *and* $(k+1)\alpha_n \le x_2$, *where* $(N_n(k) \mid L(\cdot, w))$ *is any choice of* $N_n(\cdot)$ *having the conditional law* $\mathcal{L}$ *at* w.

Proof. Since $L(x)$ is a known diffusion process (we may assume continuous paths and generated filtrations continuous in x, so that "fringe effects" do not play a role: $\sigma(L(y),\ y \le x) \equiv \bigcap_{\epsilon>0} \sigma(L(y),\ y < x + \epsilon))$ we have $\sigma(L) = \bigvee_n \sigma(L(k\alpha_n),\ -2^n < k)$, the right term being monotone in n. For each n, let $\mathcal{L}_n(w)$ be a regular conditional probability for $(N_n,\ N_{n-1}, \ldots, N_0)$ given $(L(k\alpha_n),\ -2^n < k)$. Since $\sigma(L) \subset \sigma(N_{n+m},\ 0 \le m)$, it follows easily by Theorem 2.1 that, with probability one, $\mathcal{L}_n(w)$ makes $(N_n, N_{n-1}, \ldots, N_0)$ a Markov chain with the same transition mechanism as (N_n, P) (Theorem 2.4) and initial distribution of N_n that of Corollary 3.1. Indeed, we may and do take this as *definition* of $\mathcal{L}_n$.

Now we define the limit in distribution

$$\mathcal{L}(N_n(\cdot),\ 0 \le n \mid L(w)) = \lim_{N\to\infty} \mathcal{L}_N(w) \tag{3.6}$$

if, for every finite subset $(N_{n_1}(k_1), \ldots, N_{n_j}(k_j))$, the joint law for $\mathcal{L}_N$ converges (hence the limit is uniquely extendible to a probability on a product of discrete spaces by Kolmogorov's Extention Theorem), and $L(N_n(\cdot),\ 0 \le n \mid L(w)) = \delta_{N_n(\cdot, w')}$ elsewhere, where $w' \in \Omega$ is fixed but arbitrary. By a simple consideration of martingale convergence the first case has probability 1 in $\sigma(L(x), -1 < x)$, and defines a regular conditional probability of $\sigma(N_n, 0 \le n)$ given $\sigma(L(x), -1 < x)$. To see that (a) is true, we need only observe that the transition mechanism is obviously continuous under convergence in law of the marginal distributions. As $N \to \infty$ with n fixed, the Markov chain law $\mathcal{L}_N$ applied to $(N_n, N_{n-1}, \ldots, N_0)$ yields a Markov chain limit law when we condition on $\sigma(N_k;\ n \le k \le N)$ (which converges to $\sigma(N_k; n \le k)$). Since n is arbitrary, (a) follows.

Turning to (b), we have first to observe that it suffices to prove that, for every n, $\{N_n(k),\ -2^n < k\}$ are mutually independent for $\mathcal{L}$ (with probability 1) and that their conditional laws (as w varies) depend only on L inside the intervals covered (we refer to $(k\alpha_n, (k+1)\alpha_n)$ as the interval "covered by" $N_n(k)$). Indeed, any finite collections $\{N_{n_j,i}(k_j^i),\ 1 \le j \le \ell_i\}$, $1 \le i \le I$, such that the covered intervals are disjoint over i has a law built up by applying the transition mechanism to disjoint subsets of $\{N_{\hat{n}}(k),\ -2^{\hat{n}} < k\}$ where $\hat{n} = \max_{i,j} n_{j,i}$. Given that these disjoint subsets are mutually independent for $\mathcal{L}$, it follows that so are the $\{N_{n_j}(k_j)\}$ as i varies. This will prove the independence assertion, and the dependence assertion also follows (indeed, it is clear by independence that, if the law of each $N_{\hat{n}}(k)$ depends only on L in the covered interval, then any finite subset covering a subset of $[x_1, x_2]$ has joint law depending only on L in $[x_1, x_2]$).

Now to prove the conditional independence of the $N_n(k)$ for fixed n we start with $\mathcal{L}_n$, for which the independence is part of Corollary 3.1. More generally, for $M \geq n$, since $\mathcal{L}_M$ makes $\{N_M(k),\ -2^{+M} < k\}$ independent, and given these the $\mathcal{L}_M$-law of $\{N_n(k),\ -2^n < k\}$ is built up by applying the transition mechanism $M - N$ times to the disjoint subsets of $\{N_M(k)\}$ which cover subintervals of the intervals covered by $\{N_n(k)\}$, we see that $\mathcal{L}_M$ makes these last independent. Similarly, since the $\mathcal{L}_M$-law of $N_M(k)$ depends only on L at the endpoints of the covered interval (by Corollary 3.1 again), the $\mathcal{L}_M$-law of $N_n(k)$ depends only on L at the $N_M(j)$-endpoints contained in the interval covered by $N_n(k)$, hence on L in the covered interval. Then as $M \to \infty$, the $\mathcal{L}_M$-law of the $\{N_n(k),\ -2^n < k\}$ converges to the $\mathcal{L}$-law (with probability 1) preserving both the independence and the individual limits of dependence on L to the covered intervals. This finishes (b).

As to (c), it follows by Lemma 3.3 that if $L(x, w_1) < L(x, w_2), x_1 \leq x \leq x_2$, and if $N_n(k)$ covers a subinterval of (x_1, x_2), then the $\mathcal{L}_n$-laws of $N_n(k)$ (given by Corollary 3.1) are ordered in the same direction (in the sense of $<<$). Therefore since by Theorem 2.5 and Corollary 2.5 the transition mechanism preserves this ordering, and the $\mathcal{L}_M$-law of $N_n(k)$ is developed from that of $\{N_M(j); -2^M < j\}$ by applying the transition mechanism to those $N_M(j)$ covering subintervals of (x_1, x_2), the ordering is also preserved by $\mathcal{L}_M$, $M \geq n$. Letting $M \to \infty$, and noting again that the order $<<$ is conserved under convergence in law, we see that (apart from w_1 or w_2 in the set where (3.6) fails) the ordering is also preserved by $\mathcal{L}$. This finishes (c).

We come now to our upper bounds for the first two moments of $N_n(k)$ under $\mathcal{L}$ from Theorem 3.5.

Theorem 3.6. *Apart from the P-null set where (3.6) of Theorem 3.5 (Proof) fails, for all $n \geq 0$ and $-2^n < k$, if $L^* := \max_{k\alpha_n \leq x \leq (k+1)\alpha_n} L(x)$, then under $\mathcal{L}$ we have*

(a) $E(N_n(k) \mid L) \leq 2^{n-1}L^* + 1$, *and*
(b) $E(N_n^2(k) \mid L) \leq 4^{n-1}(L^*)^2 + 5 \cdot 2^{n-3}L^* + \frac{5}{6}$.

Proof. We may and do define both conditional expectations using $\mathcal{L}$. Now for $M \geq n$, since $L(j\alpha_M) < L^*$ for $k\alpha_n \leq j\alpha_M \leq (k+1)\alpha_m$, by Lemma 3.3 the two conditional expectations computed under $\mathcal{L}_M$ instead of $\mathcal{L}$ (see proof of Theorem 3.5) will not be decreased if we replace L by L^* in the interval. Suppose, first, that $0 \leq k$. Then the $L(j\alpha_M)$ under $\mathcal{L}_M$ have the I_1-distributions with parameter $\beta = \frac{L(j\alpha_M)L((j+1)\alpha_m)}{4\alpha_M^2}$. If we replace L by L^*, β becomes $\frac{(L^*)^2}{4\alpha_M^2}$ and the distribution is increased in the sense of the order <<. By Lemma 3.2, the corresponding moments become respectively $\frac{L^*}{2\alpha_M} I_0(\frac{L^*}{\alpha_M})/I_1(\frac{L^*}{\alpha_M})$, and this plus $\frac{(L^*)^2}{4\alpha_M^2}$. Now applying (a) and (c) of Theorem 2.9, respectively, to the $(M-m)^{th}$ iterate of the transition mechanism $\mathcal{O}$ (which preserves <<) it follows that under $\mathcal{L}_M$,

$$E(N_n(k) \mid L(j\alpha_M), \forall_j) \leq 2^{+n-1} L^* I_0\left(\frac{L^*}{\alpha_M}\right) / I_1\left(\frac{L^*}{\alpha_M}\right) + 1 \quad \text{and}$$

$$E(N_n^2(k) \mid L(j\alpha_M), \forall_j) \leq 4^{+n-1}(L^*)^2 + 4^{-(M-n)} \frac{L^*}{2\alpha_M} \frac{I_0}{I_1}\left(\frac{L^*}{\alpha_M}\right)$$

$$+ 5 \cdot 2^{-(M-n)-2} \frac{L^*}{2\alpha_M} \frac{I_0}{I_1}\left(\frac{L^*}{\alpha_M}\right) + \frac{5}{6}$$

$$= 4^{+n-1}(L^*)^2 + [2^{-M-1}(4^n) + 5 \cdot 2^{n-3}] L^* \frac{I_0}{I_1}\left(\frac{L^*}{\alpha_M}\right) + \frac{5}{6}$$

Then as $M \to \infty$, except on the P-null set we have $\mathcal{L}_M \to \mathcal{L}$, and convergence in law implies that the moments for $\mathcal{L}$ are bounded by the liminf of those for $\mathcal{L}_M$ (approximate x or x^2 from below by bounded continuous functions). Meanwhile, we have $\lim_{x\to\infty} I_0(x)/I_1(x) = 1$ [10, p. 343]. Combining these observations gives the asserted bounds when $0 \leq k$.

It remains to discuss the case $-2^n < k < 0$. This reduces to the former case. We have only to decompose $L(x) = L_1(x) + L_2(x)$, $k\alpha_n \leq x$, where $L_1(x) := L(x, T(k\alpha_n))$ is the local time before reaching $k\alpha_n$, and $L_2(x)$ is the local time in $T(k\alpha_n) \leq t \leq T(-1)$. By the strong Markov property, L_1 and L_2 are independent processes (we need not use their characterization as diffusions) and $L(k\alpha_n) = L_2(k\alpha_n)$. We claim that the case $k < 0$ reduces to $k \geq 0$ with L_2 in place of L. Indeed, just as we can treat $k \geq 0$ in terms of the process after reaching $k\alpha_n$ (which does not change L in $(k\alpha_n, (k+1)\alpha_n)$), we can do the same for $k < 0$, but then we have to reduce L by L_1. The salient facts here are

(a) the process before reaching $k\alpha_n$ has no effect on $N_n(k)$ (apart from the case $T(k\alpha_n) = \infty$ when $0 < k$) and
(b) $B'(T(k\alpha_n)+t) - B'(T(k\alpha_n))$ looks like B' with $T(-1)$ replaced by $T(-1+k\alpha_n)$, and this reduces to B' by Brownian scaling. Using these facts, for $k < 0$ the inequalities of Theorem 3.6 are seen to hold with L_2 in place of L. But then they also hold given $\sigma(L_1, L_2)$, and with L^* in place of $\max_{k\alpha_n \leq x \leq (k+1)\alpha_n} L_2(x)$ (since L_1 is independent of $N_n(k)$, and L^* is larger than $\max L_2$). Since L^* depends only on L, we may as well only assume L, and the case $k < 0$ follows as stated.

Remark 1. In order to accurately evaluate the sharpness of these inequalities, one could develop reverse inequalities in Theorem 2.9 (a), (c), but it looks complicated. Meanwhile, at least for n reasonably large (where the proof of these inequalities is tightest) there is reason to believe that they are quite sharp. Indeed, as n becomes large, $2\alpha_n N_n(k) \to L$ and $L^* \to L$, in such a way that after multiplying (a) by $2\alpha_n$ and replacing $E(N_n(k) \mid L)$ by $N_n(k)$, it converges to the tautology $L \leq L$ and similarly for (b), using $4\alpha_n^2 N_n^2(k) \to L^2$. Moreover, without multiplying by $2\alpha_n$, if we replace L^* by L on the right and take expectations, using the identities $EN_n(k) = 2^n$, $EN_n^2(k) = 2^{2n+1} + k2^{n+1} + 2^n$, and $EL(x) = 2$, $EL^2(x) = 8(x+1)$ for $0 \leq x = k\alpha_n$ (by standard calculations), then for $0 \leq x = k\alpha_n$ (a) becomes $2^n \leq 2^n + 1$, while (b) becomes $2^{2n+1} + k2^{n+1} + 2^n \leq 2^{2n+1} + k2^{n+1} + 5 \cdot 2^{n-2} + \frac{5}{6}$. This verifies that (a) and (b) hold "on the average", and at the same time shows that they are quite sharp on the average, particularly if n and k are large with $x = k\alpha_n$ fixed.

Remark 2. It may be of interest to calculate the law of $\mathcal{O}X$ explicitly when X has the I_1-distribution (for $\mathcal{O}X$ see Notation 2.8 — when $X = \mathcal{P}_\lambda$ this was done in Lemma 2.6). The result, after simplification, is

$$P\{\mathcal{O}X = k\} = \left(\frac{\beta}{2}\right)^{k-\frac{1}{2}} (k!(k-1)!)^{-1} I_{2k-1}(2\sqrt{2\beta})/I_1^2(2\sqrt{\beta}).$$

Thus modified Bessel functions of all odd orders appear, and it looks to us hopeless to simplify the law of $\mathcal{O}^2 X$, much less that of $\mathcal{O}^n X$ for general n.

Remark 3. In terms of $\mathcal{L}(N_n \mid L)$, the unverified conjecture of the Introduction that $\sigma(L(\cdot)) \equiv \lim_{N\to\infty} \sigma(N_n,\ n \geq N)$ is equivalent to asserting that, for every m, $\lim_{n\to\infty} \mathcal{L}(N_m \mid N_n) = \mathcal{L}(N_m \mid L)$, P-a.s., where the convergence is to be in law. It seems that a proof would require additional results along the general lines of Theorem 3.6.

4. Construction of the law of B' given $(N_n;\ 0 \leq n)$.

Here we assume that $(N_n(\cdot),\ 0 \leq n)$ has one of the entrance laws $\mathcal{L}(N \mid L)$ with properties (a)—(c) of Theorem 3.5 and also (a)—(b) of Theorem 3.6 (i.e. we exclude the P-nullset where (3.6) fails), and we seek to define the law of B' consistent with this $\mathcal{L}(N \mid L)$, to obtain $\mathcal{L}(B' \mid L)$. Let us then also impose one further condition, namely that $\lim_{n\to\infty} 2\alpha_n N_n[2^n x] = L(x)$ uniformly in x, $P^{\mathcal{L}(N|L)}$-a.s. Since this holds P-a.s., it also holds $P^{\mathcal{L}(N|L)}$-a.s. for P-almost all $w \in \Omega$, so the condition holds except on a P-nullset. Then if we define the law $\mathcal{L}(B' \mid N_.)$ for P-almost all laws $\mathcal{L}(N \mid L)$, we can write

$$\mathcal{L}(B' \mid L) = E[\mathcal{L}(B' \mid \sigma(L, N_.)) | \sigma(L)] = E(\mathcal{L}(B' \mid N) | \sigma(L)), \tag{4.1}$$

where E is to be calculated using $\mathcal{L}(N \mid L)$. This gives the solution to our problem.

In fact, we define $\mathcal{L}(B' \mid N_.)$ for all paths of $N_.$ having the *consistency properties* that for all $0 \leq n$,

(a) $N_n(k) > 0 \quad \text{for} \quad 0 \leq k < K(n) := \inf\{j \geq 0 : N_n(j) = 0\}$,
(b) $N_{n+1}(2k) \wedge N_{n+1}(2k+1) \geq N_n(k) \quad \text{for} \quad -2^n < k$.

This conditional law is entirely free of the given L. We do this by defining $\mathcal{L}(R_n, 0 \leq n \mid N_.)$, which solves our problem since $P\{\lim_{n\to\infty} R_n(t) = B'(t)$ uniformly in $0 \leq t \leq T(-1) \mid N(\cdot)\} = 1$, P-a.s. The first step is to define $\mathcal{L}(R_0 \mid N_.)$ which, in accordance with Theorem 2.1 at $n = 0$ is simply $\mathcal{L}(R_0 \mid N_0)$. But this has already been done in Section 1 (taking again $n = 0$). For induction, suppose we have defined $\mathcal{L}(R_n \mid N_.)$ (since we have $\sigma(R_0, \ldots, R_n) = \sigma(R_n)$, this gives $\mathcal{L}(R_k,\ 0 \leq k \leq n \mid N_.)$, and by Theorem 2.1 it depends only on $(N_0, \ldots, N_n)$). The induction step reduces to defining the law $\mathcal{L}(R_{n+1} \mid R_n, N_{n+1})$ (since $\sigma(N_0, \ldots, N_n) \subset \sigma(R_n)$). Once this conditional law is defined for general (and hence by induction for every) n our problem is reduced to a last appeal to the Kohmogorov-Bochner extension theorem for the consistent families of laws $\mathcal{L}(R_n \mid N_.)$. Here the canonical space of each R_n is countable, and there is no serious problem in topologizing the projective limit space so as to apply the extension theorem. This was done in [3] for the unconditional case, and deserves no further mention here. So it remains only to define $\mathcal{L}(R_{n+1} \mid R_n, N_{n+1})$, and here we may take $n = 1$ for convenience.

The picture to keep in mind is the reduction of B' to its 1-inserts (Lemma 1.5), which are independent when R_1 is given. The effect of being given also N_2 is to enumerate how many 2-inserts there are with given starting values $k\alpha_2$ and given sign ± 1 (i.e. up or down). The problem is simply to interpolate into the 1-inserts (whose starting and ending values are dictated by R_1) the internal R_2-steps, subject to the total multiplicities N_2.

To derive the law of this interpolation (which in non-explicit terms is just the law that all possible R_2-paths are equally likely) we need to examine the law of the $(n+1)$-upcrossings embedded into an n-insert. This is the same for every n, so taking $n = 1$, we introduce

Notation 4.1. Let V denote the number of upcrossings of $(-\alpha_2, 0)$ of a 1-insert, and let $U + 1$ be the number of upcrossings of $(0, \alpha_2)$ for the same 1-insert.

Lemma 4.2. *The pair (V, U) has a symmetric joint distribution. In terms of the notations* $\text{geo}(p)$ *and* $\text{neg. bin.}(n, p)$ *of Section 2, the marginal law of V and U is* $\text{geo}(\frac{2}{3})$ *and given $\{V = k\}$, the conditional law of U is* $\text{neg. bin.}(k+1, \frac{3}{4})$, *as is that of V given $\{U = k\}$.*

Proof. This is mostly a consequence of Theorem 2.3 when $k = 0$, $n = 1$, and $N_1(0) = 0$, $N_1(-1) = 0$. This gives $N_2(-1) \stackrel{d}{=} \text{neg. bin.}(1, \frac{2}{3}) = \text{geo}(\frac{2}{3})$, and, given $N_2(-1)$, $N_2(0) \stackrel{d}{=} \text{neg. bin.}(1 + N_2(-1), \frac{3}{4})$. Since in this case $N_2(-1)$ and $N_2(0)$ are entirely contained in a single insert starting at 0 and ending at -1, we have the identification $(V, U) \stackrel{d}{=} (N_2(0), N_2(-1))$, where we noted that since this insert has sign -1 the roles of V and U are interchanged. It remains only to note that the sign of the first 1-insert described by B' is determined at the last exit of B' from 0 before reaching $\{\pm 1\}$, independently of the crossings from 0 to $\{\pm\frac{1}{2}\}$ before that time. If we discount the single crossing to $\{\pm\frac{1}{2}\}$ after this last exit (as we did in defining U), the remaining crossings are determined by symmetric Beromoulli trials, hence the law of (V, U) is also symmetric (from an analytic viewpoint, this was already treated before Theorem 2.2).

Remark. It follows easily, although not needed below, that $P\{V = i,\ U = j\} = \frac{1}{2}\binom{i+j}{j}\frac{1}{4}^{i+j};\ 0 \le i, j$.

We return to our problem of the interpolation of the 2- inserts into the 1-inserts when (R_1, N_2) is given. Actually we need only determine the joint law of the total numbers of 2-inserts of sign $+1$ in each 1-insert. Indeed, for a 1-insert of sign $+1$ starting at $k/2$ there are no 2-inserts $\frac{(k+1)}{2} \downarrow \frac{2k+1}{4}$; the number of 2-inserts $\frac{2k+1}{4} \downarrow \frac{k}{2}$ is 1 less than the number $\frac{k}{2} \uparrow \frac{2k+1}{4}$, the number $\frac{k}{2} \downarrow \frac{2k-1}{4}$ equals the number $\frac{2k-1}{4} \uparrow \frac{k}{2}$, and there are none $\frac{2k-1}{4} \downarrow \frac{k-1}{2}$. Similar reasoning applies to a 1-insert of sign -1. Now if R_1 is given but not N_2, the numbers of 2-inserts of sign $+1$ at each level in each 1-insert are determined independently (by independence of 1-inserts; Lemma 1.5) with the law of (V, U) from Lemma 4.2 (if the 1-insert has sign $+1$, we must add 1 to U). When N_2 is also given, for each k we are given the total number of 2-inserts from $\frac{k}{2} - \frac{1}{4} \uparrow \frac{k}{2}$, and the total number from $\frac{k}{2} \uparrow \frac{k}{2} + \frac{1}{4}$. Now 2-inserts $\frac{k}{2} - \frac{1}{4} \uparrow \frac{k}{2}$ can only occur during 1-inserts $\frac{k}{2} \uparrow \frac{k+1}{2}$, or during 1-inserts $\frac{k}{2} \downarrow \frac{k-1}{2}$, *except that* exactly one occurs during each 1-insert $\frac{k-1}{2} \uparrow \frac{k}{2}$, and the same holds for 2-inserts $\frac{k}{2} \uparrow \frac{k}{2} + \frac{1}{4}$ without exception. Conversely, these two types of 1-insert have all their embedded 2-inserts of sign 1 partitioned among these two types, *except for* one extra 2-insert from $\frac{k}{2} + \frac{1}{4} \uparrow \frac{k+1}{2}$ in each 1-insert $\frac{k}{2} \uparrow \frac{k+1}{2}$. It follows that the conditional law of the numbers of embedded 2-inserts of sign 1 in the 1-inserts, given (R_1, N_2), is determined independently for each k, by the conditional law of the embedded 2-inserts of sign 1 in the one inserts starting at $\frac{k}{2}$. This is slightly different for $k > 0$ and for $-2 < k \le 0$. We introduce a notation for the random frequencies whose conditional law is to be determined. Let $1+U_i^+(k)$, $1 \le i \le N_1(k)$, denote the numbers of upcrossings $\frac{k}{2} \uparrow \frac{k}{2} + \frac{1}{4}$ in the successive 1-inserts of sign 1 starting at $\frac{k}{2}$ (the order being that determined by R_1), and let $V_i^+(k)$, $1 \le i \le N_1(k)$, denote the analogous numbers of upcrossings $\frac{k}{2} - \frac{1}{4} \uparrow \frac{k}{2}$. Similarly, for $k > 0$, let $V_i^-(k)$, $1 \le i \le N_n(k-1)$, denote the numbers of upcrossings of $\frac{k}{2} \uparrow \frac{k}{2} + \frac{1}{4}$ in the successive 1-inserts of sign -1 starting at $\frac{k}{2}$ (and for $-2 < k \le 0$, the same notation with $1 \le i \le N_n(k-1) + 1$), and similarly let $U_i^-(k)$ denote the numbers of upcrossings $\frac{k}{2} - \frac{1}{4} \uparrow \frac{k}{2}$ (since the 1-inserts have sign -1, U^- counts those farthest in the direction of advance). Then for $0 < k \le K(1)$, we have easily

$$(4.2) \qquad \sum_{i=1}^{N_1(k)} U_i^+(k) + \sum_{i=1}^{N_1(k-1)} V_i^-(k) = N_2(2k) - N_1(k);$$

$$(4.3) \qquad \sum_{i=1}^{N_1(k)} V_i^+(k) + \sum_{i=1}^{N_1(k-1)} U_i^-(k) = N_2(2k-1) - N_1(k-1),$$

where we subtracted $N_1(k)$ on both sides of the first equation for reasons of symmetry. For $-2 < k \le 0$, the analogous equations hold except that $N_1(k-1)$ in the two upper limits is replaced by $N_1(k-1) + 1$.

We observe that if R_1 is given but N_2 is not given, then by independence of inserts and Lemma 4.2, the left sides of (4.2) and (4.3) have the same law, namely a neg. bin.$(N_1(k)+N_1(k-1),\frac{2}{3})$ if $k>0$, or a neg. bin.$(N_1(k)+N_1(k-1)+1,\frac{2}{3})$ if $-2<k\le 0$. But suppose the collection $S:=\{U_i^+(k),\ 1\le i\le N_1(k);\ V_i^-(k),\ 1\le i\le N_1(k-1)\}$ for $k>0$ is also given (replace $N_1(k-1)$ by $N_1(k-1)+1$ if $k\le 0$). Then by Lemma 4.2 the left side of (4.3) has the law neg. bin.$(N_2(2k)-N_1(k)+N_1(k)+N_1(k-1),\frac{3}{4})=$ neg. bin.$(N_2(2k)+N_1(k-1),\frac{3}{4})$, depending only on the single parameter $N_2(2k)$ from N_2. Therefore, given $N_2(2k)$ (as well as R_1), the collection S is independent of $N_2(2k-1)$. So if (R_1,N_2) is given, the law of S is the same as if only $(R_1,N_2(2k))$ were given, i.e. the sum $N_2(2k)-N_1(k)$ is the only restraint on the otherwise independent terms (4.2). It follows that the joint law of S is determined by Bose-Einstein sampling of size $(N_1(k)+N_1(k-1),\ N_2(2k)-N_1(k))$ if $0<k\le K(1)$, or of size $(N_1(k)+N_1(k-1)+1,\ N_2(2k)-N_1(k))$ if $-2<k\le 0$ (see Definition 1.2 for the probabilities). Of course, by symmetry an analogous fact holds for the collection on the left of (4.3). It remains only to discuss the joint law of the two collections. Suppose, to this effect, that the collection S is given. Since the pairs $(V_i^+(k),U_i^+(k))$ and $(V_j^-(k),U_j^-(k))$ are jointly independent over i and j when S is not given, it follows from Lemma 4.2 that when S is given (but not $N_2(2k-1)$) the variables $V_i^+(k)$ and $U_j^-(k)$ are mutually independent over i and j (they all pertain to different 1-inserts), and the conditional law of $V_i^+(k)$ is neg. bin.$(U_i^+(k)+1,\frac{3}{4})$ while that of $U_j^-(k)$ is neg. bin.$(V_j^-(k)+1,\frac{3}{4})$. Let us introduce a corresponding

Definition 4.3. Integer-valued random variables $X_i\ge 0$, $1\le i\le n$, have the law of a *Bose-Einstein partition* of size $(x_1,\dots,x_n;N)$ if the joint law is obtained by defining $(Y_1,\dots,Y_K)$ by Bose-Einstein sampling of size (K,N), where $K=\sum_{i=1}^n x_i$, and then combining to get $X_1=\sum_{i=1}^{x_1}Y_i$; $X_2=\sum_{i=x_1+1}^{x_1+x_2}Y_i,\dots,X_n=\sum_{i=x_1+\dots+x_{n-1}+1}^{x_1+\dots+x_n}Y_i$.

Lemma 4.4. *In a Bose-Einstein partition of size $(x_1,\dots,x_n;N)$ we have (setting again $K=\sum_{i=1}^n x_i$)*

$$P\{X_1=k_1,\dots,X_n=k_n\}=\binom{K+N-1}{N}^{-1}\prod_{j=1}^n\binom{x_j+k_j-1}{k_j};$$

$$\sum_1^n k_j=N.$$

Proof. The first factor on the right is the probability of a sample point in Bose-Einstein sampling of size (K,N). For the same reason, the factors in the product are the numbers of sample points in Bose-Einstein sampling of size (x_j,k_j), which is the same as the number of occupancy x_j-tuples of x_j cells by k_j balls. The event $\bigcap_1^n\{X_j=k_j\}$ occurs if and only if each consecutive set of x_j cells receives k_j balls. Clearly the total number of such occupancies is just the product, as asserted.

Now the gist of the discussion preceding Definition 4.3 for $k > 0$ is

(a) the conditional joint law of S given (R_1, N_2) is Bose-Einstein of size $(N_1(k) + N_1(k-1),\ N_2(2k) - N_1(k))$, and

(b) given S as well as (R_1, N_2), the summands on the left of (4.3) have law determined as that of independent, neg. bin.$(U_i^+(k)+1, \frac{3}{4})$ and neg. bin.$(V_j^-(k)+1, \frac{3}{4})$ random variables whose total sum is given by $N_2(2k-1) - N_1(k-1)$.

It is easy to see by writing neg. bin.(n,p) as a sum of n independent geo(p) terms in each of these negative binomials that the effect is to determine the conditional law of $\{V_i^+(k), U_j^-(k)\}$ by Bose-Einstein partitioning of size $(U_i^+(k) + 1\ (1 \le i \le N_1(k)),\ V_j^-(k) + 1\ (1 \le j \le N_1(k-1));\ N = N_2(2k-1) - N_1(k-1)$ if $0 < k \le K(1)$ and for $-2 < k \le 0$ one need only replace $N_1(k-1)$ by $N_1(k-1)+1$ in the range of j. Of course, a parallel argument applies for every $n \ge 0$ to the conditional law of the numbers of $(n+1)$-inserts interpolated into the n-inserts when (R_n, N_{n+1}) is given. To determine, now, the conditional law of R_{n+1}, it remains only to condition also on these numbers of $(n+1)$-inserts (as we noted before, the numbers of sign -1 are determined by those of sign 1). Then by the independence of n-inserts given R_n, the *ordering* of the two types of random $(n+1)$-inserts (namely, those from $k\alpha_n$ to $k\alpha_n + \alpha_{n+1}$, and those from $k\alpha_n - \alpha_{n+1}$ to $k\alpha_n$) is made independently in each n-insert, given the totals of each type, and moreover this ordering determines the R_{n+1}-path uniquely. Indeed for an n-insert of sign 1, each step $k\alpha_n$ to $k\alpha_n + \alpha_{n+1}$ except the last (hence, a total of $U_i^+(k)$ if $n = 1$) is followed by a step from $k\alpha_n + \alpha_{n+1}$ back to $k\alpha_n$, while each from $k\alpha_n - \alpha_{n-1}$ to $k\alpha_n$ is preceded by one from $k\alpha_n$ to $k\alpha_n - \alpha_{n+1}$ (a total of $V_i^+(k)$ if $n = 1$), and an analogous argument applies to n-inserts of sign -1. This obviously determines the R_{n+1}-path uniquely. Finally, it is clear from the symmetry of R_{n+1} that these orderings are all equally likely, given the totals (for example, given $(U_i^+(k),\ V_i^-(k))$ if $n = 1$ and the 1-insert in question is the ith of sign $+1$ starting from $\frac{k}{2}$). For given totals (U, V), the number of such orderings is $\binom{U+V}{V}$.

To write the conditional probability of a choice of $\{U_i^+(k),\ V_i^-(k)\}$ and $\{V_i^+(k),\ U_j^-(k)\}$, $1 \le i \le N_1(k)$ and $1 \le j \le N_1(k-1)$ for $0 < k \le K(1)$ (resp.$1 \le j \le N_1(k-1)+1$ if $-2 < k \le 0$), given (R_1, N_2), we need only multiply the Bose-Einstein probabilities of the former by the Bose-Einstein partition probabilities of the later, and then divide by the corresponding product of the order counts $\binom{U+V}{V}$. When we carry this out, it is seen that the order counts precisely cancel the products in the numerator of the Bose-Einstein partition probabilities (i.e. the $\prod_{j=1}^n \binom{x_j+k_j-1}{k_j}$ in Lemma 4.4) leaving only, for $0 < k \le K(1)$, the expression

$$(4.4)\qquad \left[\binom{N_2(2k) + N_1(k-1) - 1}{N_2(2k) - N_1(k)}\binom{N_2(2k) + N_2(2k-1) - 1}{N_2(2k-1) - N_1(k-1)}\right]^{-1}$$

$$= \left[\binom{N_2(2k) + N_1(k-1) - 1}{N_2(2k) - N_1(k)}\binom{N_2(2k) + N_2(2k-1) - 1}{N_2(2k) + N_1(k-1) - 1}\right]^{-1}$$

$$= [(N_2(2k) + N_2(2k-1) - 1)!]^{-1} \cdot$$
$$(N_1(k) + N_1(k-1) - 1)!(N_2(2k) - N_1(k))!(N_2(2k-1) - N_1(k-1))! \,.$$

An analogous calculation $-2 < k \leq 0$ gives the result

$$[(N_2(2k) + N_2(2k-1))!]^{-1} \cdot \tag{4.5}$$
$$(N_1(k) + N_1(k-1))!(N_2(2k) - N_1(k))!(N_2(2k-1) - N_1(k-1))! \,.$$

Thus, the probability of a sample point of R_2 given (R_1, N_2) is the product of (4.4) over $k > 0$ times (4.5) for $-2 < k \leq 0$. To be sure, precisely the same result holds for any R_{n+1} given (R_n, N_{n+1}), replacing the subscripts and letting the product for $k \leq 0$ range from $-2^n < k \leq 0$. This may not appear especially simple, until one takes into account the amount of cancellation which has already occurred. Nevertheless, we shall state

Theorem 4.5. *For every $n \geq 0$, the conditional probability of a sample path of R_{n+1} given (R_n, N_{n+1}) is (either 0, if the point is inconsistent with (R_n, N_{n+1}), or) the product of (4.4) (with subscripts $(1,2)$ replaced by $(n, n+1)$) over $0 < k \leq K(n)$, and of (4.5) (with the same replacement) over $-2^n < k \leq 0$.*

Remark. One fact which does not seem quite obvious, but follows from this expression, is that for given (R_n, N_{n+1}), all possible paths of R_{n+1} are equally likely, their total number being the reciprocal of this product. Of course, if we change the given R_n, we get an entirely disjoint but equinumerous set of possible R_{n+1}-paths. Indeed, since the conditional result depends only on (N_n, N_{n+1}), when we change R_n keeping N_n fixed the cardinality of the set of possible R_{n+1}-paths does not change.

We want to consider, finally, the law of R_{n+1} given $(N_0, \ldots, N_{n+1})$. We obtain these probabilities by multiplying the conditional laws, viz $\mathcal{L}(R_0 \mid N_0) \cdot \mathcal{L}(R_1 \mid R_0, N_1) \cdot \mathcal{L}(R_2 \mid R_1, N_2) \ldots \mathcal{L}(R_{n+1} \mid R_n, N_{n+1})$.

Now by Lemma 1.3 $\mathcal{L}(R_0 \mid N_0)$ is determined by equally likely cases with the probability of a case being

$$p_0 = \left[\prod_{k=0}^{K(0)-1} \binom{N_0(k) + N_0(k+1) - 1}{N_0(k+1)} \right]^{-1} . \tag{4.6}$$

Next, by Theorem 4.5 $\mathcal{L}(R_1 \mid R_0, N_1)$ has equally likely cases with probability

$$\begin{aligned} &[(N_1(0) + N_1(-1)!]^{-1} N_0(0)!(N_1(0) - N_0(0))!(N_1(-1) - N_0(-1))! \\ &\cdot \prod_{k=1}^{K(0)} [(N_1(2k) + N_1(2k-1) - 1)!]^{-1} (N_0(k) + N_0(k-1) - 1)!(N_1(2k) \\ &- N_0(k))!(N_1(2k-1) - N_0(k-1))! \end{aligned} \tag{4.7}$$

The product of (4.6) and (4.7) gives the probability of a case for $\mathcal{L}(R_1 \mid N_0, N_1)$ as

$$p_1 = [(N_1(0) + N_1(-1))!]^{-1} N_0(0)! \Bigg[\prod_{k=0}^{K(0)} (N_1(2k) - N_0(k))!(N_1(2k-1) - N_0(k-1))!(N_0(k)-1)!N_0(k+1)! \Bigg] \cdot \prod_{k=1}^{K(0)} [(N_1(2k) + N_1(2k-1) - 1)!]^{-1} \quad \text{(where } (-1)! := 1)$$

(4.8)
$$= [(N_1(0) + N_1(-1))!]^{-1} \left(\prod_{k=0}^{K(0)} (N_0(k)-1)!N_0(k)! \right) \left(\prod_{k=0}^{K(0)} (N_1(2k) - N_0(k))!(N_1(2k-1) - N_0(k-1))! \right) \cdot \left[\prod_{k=1}^{K(0)} (N_1(2k) + N_1(2k-1) - 1)! \right]^{-1}$$

Now to write $\mathcal{L}(R_2 \mid N_0, N_1, N_2)$ we have to multiply (4.8) by the product of (4.4) over $0 < k \leq K(1)$ and of (4.5) over $-2 < k \leq 0$. We get for the probabilities

(4.9)
$$p_2 = \left(\prod_{k=0}^{K(0)} (N_0(k)-1)!N_0(k)! \right) \left[N_1(-1) \prod_{k=1}^{K(0)} (N_1(2k-1) + N_1(2k-1) - 1)! \right] \cdot \left[\prod_{k=0}^{K(0)} (N_1(2k) - N_0(k))!(N_1(2k-1) - N_0(k-1))! \prod_{k=-1}^{K(0)} (N_2(2k) - N_1(k))!(N_2(2k-1) - N_1(k-1))! \right] \cdot \left[\prod_{k=-1}^{0} (N_2(2k) + N_2(2k-1))! \prod_{k=1}^{K(1)} (N_2(2k) + N_2(2k-1) - 1)! \right]^{-1} .$$

To detect the general rule, it is necessary only to write the next case $\mathcal{L}(R_3 \mid N_0, N_1, N_2, N_3)$ by the same method. Omitting the details, we obtain

Theorem 4.6. *For every $n \geq 0$, $\mathcal{L}(R_n \mid N_{(\cdot)}, L) = \mathcal{L}(R_n \mid N_0, \ldots, N_n)$ is determined by equally likely cases over the set of consistent R_n-paths. The probabilities are given by (4.6) for $n = 0$, by (4.8) for $n = 1$, by (4.9) for $n = 2$, and*

for $n \geq 2$ they have the form

$$p_n = T_0 \left(\prod_{j=1}^{n-1} T_1(j) \right) \left(\prod_{j=0}^{n-1} T_2(j) \right) T_n^{-1}, \quad \textit{where}$$

$$T_0 := \prod_{k=1}^{K(0)} (N_0(k)-1)!(N_0(k))!; T_1(j) := \prod_{-2^{j-1}}^{-1} (N_j(2k+1)+N_j(2k))!.$$

$$\prod_{k=1}^{K(j-1)} (N_j(2k-1)+N_j(2(k-1))-1)!;$$

$$T_2(j) := \prod_{-2^j+1}^{K(j)} (N_{j+1}(2k)-N_j(k))!(N_{j+1}(2k-1)-N_j(k-1))!, T_n :=$$

$$\prod_{-2^{n-1}+1}^{0} (N_n(2k)+N_n(2k-1))! \prod_{k=1}^{K(n-1)} (N_n(2k)+N_n(2k-1)-1)!\ .$$

We observe here that, except for $T_2(j)$, each factor depends only on N_j for a simple j, while $T_2(j)$ depends only on (N_j, N_{j+1}). The terms $T_1(j)$ and T_n are quite similar, but $T_1(j)$ combines pairs $(2k, 2k+1)$ while T_n combines pairs $(2k-1, 2k)$. The terms $T_2(j)$ represent the product of the extra $(j+1)$-upcrossings beyond those necessitated by the j-upcrossings of the α_j-intervals containing them. Note, finally, that it is the reciprocal of T_n that figures in p_n.

Final Remark. We have considered B stopped at $T(-1)$, but of course B stopped at any $c \neq 0$ can be covered by scale changes if we replace α_n by $|c|\alpha_n$. It takes only a little more thought to see that B stopped at $\inf\{t : L(t,0) \geq c\}$, $c > 0$, is also covered. Actually, for this we need only adapt the arguments for $k \geq 0$ to construct B' on $[0,\infty)$ reflected at 0 (the excursions into $(-\infty, 0)$ are spliced out). To see this note that, by excursion theory from 0, B' on $[0,\infty)$ may be constructed by replacing $T(-1)$ by $\inf\{t : L(t,0) = e\}$ where e is exponential, independent of B (since $L(T(-1),0) \overset{(d)}{=} e$). Now given $e = c$, the law of $\{L(x), x \geq 0\}$ is the same as its law for c in the other case, and the same is true of the reflected B'.

Furthermore, the dependence of N_n on $L(x)$ is local, in such a way that $\mathcal{L}(N_n(k), 0 \leq k \mid L\}$ depends only on $\{L(x), 0 \leq x\}$. Thus stage one of our construction carries over. Stage 2, to construct $\mathcal{L}(R_n, 0 \leq n \mid N)$, also carries over if R_n is replaced by its analog on $[0,\infty)$ (splicing out the negative excursions). For $n = 0$, we need only determine the variables $X_k(j)$ for $0 \leq k$ and $1 \leq j \leq N_0(k)$. The same argument gives a Bose-Einstein law when N_0 is given, and these determine R_0 in each positive excursion by the (classical) branching process argument. The induction step, to define the law $\mathcal{L}(R_{n+1} \mid R_n, N_{n+1})$, relied on the independence of inserts, which is equally valid in the reflected case. The only change needed beyond ignoring $k < 0$ is that, for $k = 0$, we again have a simple Bose-Einstein distribution of the $N_{n+1}(0)$ upcrossings among the $N_n(0)$

n-inserts, since only the crossings $0 \uparrow \alpha_n$ can contribute. For $k > 0$, however, the argument of Section 4 remains applicable. Thus we again have equally likely cases, and the precise expressions are analogous to those of Theorem 4.6 but somewhat simpler.

References

1. Chacon, R., LeJan, Y., Perkins, E. and Taylor, S. J., *Generalized arc length for Brownian motion and Lévy processes*, Z. Wahrsch. Verw. Gebiete **57** (1981), 197–211.
2. Feller, W., *An Introduction to Probability Theory and its Applications*, Vol. I., 3rd Edition. John Wiley and Sons, New York (1968).
3. Knight, F. B., *On the random walk and Brownian motion*, TAMS **103** (1962), 218–228.
4. ———, *Random walks and a sojourn density process of Brownian motion*, TAMS **109** (1963), 56–86.
5. ———, *Brownian local times and taboo processes*, TAMS **143** (1969), 173–185.
6. ———, *Approximation of stopped Brownian local time by dyadic crossing chains*, Stochastic Processes and their Applications Vol. 66, No 2 (1997), 253–271.
7. Pitman, J., and Yor, M., *A decomposition of Bessel bridges*, Z. Wahrsch. Verw. Gebiete **59** (1982), 425–457.
8. Walsh, J. B., *Downcrossings and the Markov property of local time*, In Temps Locaux, Asterisque **52-53** (1978), 89–116.
9. Warren, J., and Yor, M., *The Brownian burglar: conditioning Brownian motion by its local time process*, Séminaire de Prob. XXXII (?).
10. Whittaker, E. T., and Watson, G. N., *A Course of Modern Analysis*, 4th Edition Cambridge (1965).

Department of Mathematics
University of Illinois
1409 West Green Street
Urbana, Illinois 61801
U.S.A.

LE THÉOREME DE RAY-KNIGHT À TEMPS FIXE

Christophe Leuridan
Institut Fourier, UMR 5582 CNRS-Université de Grenoble I
BP 74, F-38402 St Martin d'Hères Cedex

Introduction

L'objet de cet article est de décrire la loi du processus des temps locaux browniens à instant fixe (théorème 1) et à un instant "inverse" du temps de séjour dans $\mathbf{R}_+$ (théorème 2). Une version moins précise du théorème 1 a déjà été obtenue indépendamment par R. van der Hofstad, F. den Hollander, W. König : voir [3], lemme 1. Ce lemme 1, que les auteurs justifient de façon proche mais incomplète, ne constitue pas le but de l'article [3]. Il est semblable à notre théorème 1, hormis le fait que certaines lois ne sont pas complètement explicitées. Le lecteur trouvera au paragraphe IV.3 plus de précisions sur les liens et les différences entre les arguments et les résultats de [3] et ceux du présent article.

L'un des avantages de notre présentation est que la méthode permet théoriquement d'obtenir des théorèmes de Ray-Knight pour tous les instants inverses d'une fonctionnelle additive croissante du mouvement brownien. Mais en fait, les résultats ne s'explicitent bien que dans quelques cas, comme celui du théorème 2 (où la fonctionnelle additive est le temps de séjour dans $\mathbf{R}_+$).

On considère donc un mouvement brownien $B = (B_t)_{t\in\mathbf{R}_+}$, dans $\mathbf{R}$, issu de 0, et une version continue de ses temps locaux $(L_t^x)_{t\in\mathbf{R}_+}^{x\in\mathbf{R}}$. Lorsque l'on se donne un temps aléatoire τ, les temps locaux L_τ^x définissent un processus indexé par la variable d'espace x. En 1963, D. B. Ray [8] et F. B. Knight [5] ont obtenu la loi du processus $(L_\tau^x)^{x\in\mathbf{R}}$ pour certains instants τ comme :

- le temps d'atteinte d'un point a par le mouvement brownien B :

$$\sigma_a = \inf\left\{t \in \mathbf{R}_+ \mid B_t = a\right\};$$

- le temps d'atteinte d'une valeur r par le temps local en 0 :

$$\tau_r^0 = \inf\left\{t \in \mathbf{R}_+ \mid L_t^0 = r\right\};$$

– un temps exponentiel indépendant de B.

Depuis, la loi du processus $(L_\tau^x)^{x\in\mathbb{R}}$ a été décrite pour d'autres temps τ, comme :

$$\alpha_r = \inf\left\{t \in \mathbb{R}_+ \mid \sup_{x\in\mathbb{R}_+} L_t^x \geq r\right\},$$

par N. Eisenbaum [2]. On trouvera dans [10] quelques extensions, ainsi qu'un résumé des principaux résultats obtenus à ce sujet.

Dans les deux premiers exemples ci-dessus ($\tau = \sigma_a$ ou $\tau = \tau_r^0$), le processus $(L_\tau^x)^{x\in\mathbb{R}}$ est markovien et la position B_τ est constante. Dans les deux exemples suivants (τ est un temps exponentiel indépendant de B ou $\tau = \alpha_r$), le processus $(L_\tau^x)^{x\in\mathbb{R}}$ est markovien conditionnellement à B_τ.

La situation est plus complexe lorsque τ est un temps fixe t, car le processus $(L_t^x)^{x\in\mathbb{R}}$ n'est pas markovien, même après conditionnement par B_t. La principale obstruction empêchant le processus $(L_t^x)^{x\in\mathbb{R}}$ d'être markovien est l'égalité :

$$\int_{\mathbb{R}} L_t^x\, dx = t.$$

En effet, cette égalité entraîne par exemple :

$$E\left[\int_x^{+\infty} L_t^y\, dy \mid \sigma\left(B_t, (L_t^y)^{y\leq x}\right)\right] = t - \int_{-\infty}^x L_t^y\, dy,$$

ce qui montre que les processus $(L_t^y)^{y\leq x}$ et $(L_t^y)^{y\geq x}$ ne sont pas indépendants conditionnellement à (B_t, L_t^x).

En 1981, E. Perkins [7] a démontré, au prix de longs calculs, que le processus $(L_1^x)^{x\in\mathbb{R}}$ est une semi-martingale. Il a même donné sa décomposition comme somme d'une martingale (relativement à la filtration $(\mathcal{E}_x)_{x\in\mathbb{R}}$ des excursions au dessous d'un niveau donné) et d'un processus à variation finie. Il conjecture enfin que le processus $\left(L_1^x, \int_x^{+\infty} L_1^y\, dy, \mathbb{1}_{\{x<B_1\}}\right)_{x\in\mathbb{R}}$ est markovien (inhomogène) relativement à la filtration $(\mathcal{E}_x)_{x\in\mathbb{R}}$.

En 1985, T. Jeulin donne dans [4] la décomposition de la semi-martingale $(L_t^x)^{x\in\mathbb{R}}$ dans la filtration $(\widetilde{\mathcal{E}}_x)_{x\in\mathbb{R}}$, où $\widetilde{\mathcal{E}}_x = \sigma\left(\mathcal{E}_x, B_t, \inf_{s\leq t} B_s\right)$, en utilisant la théorie du grossissement de filtrations. La méthode qu'il a employée consiste à conditionner par $\tau = t$ la loi de $(L_\tau^x)^{x\in\mathbb{R}}$, où τ est un temps exponentiel indépendant de B.

Il signale (page 260) que le calcul stochastique permet de montrer que le processus $\left(L_t^x, \int_{-\infty}^x L_t^y\, dy\right)_{x\in\mathbb{R}}$ est markovien conditionnellement à $\left(B_t, \inf_{s\leq t} B_s\right)$ et d'expliciter son générateur ; mais –écrit-il– "le résultat obtenu est assez compliqué et plutôt long à écrire".

L'objet de ce travail est de donner une description plus simple de la loi du processus $(L_t^x)_{x\in\mathbb{R}}$, en conditionnant par rapport au 5-uplet $\left(B_t, L_t^{B_t}, L_t^0, V_t, V_t'\right)$, où V_t

et V'_t désignent le temps passé à l'instant t par le mouvement brownien B hors du segment $[0, B_t]$, du côté de B_t et du côté de 0 :

$$V_t = \int_0^t \mathbb{1}_{\{B_s/B_t>1\}}\, ds = \int_{\mathbf{R}} \mathbb{1}_{\{y/B_t>1\}} L_t^y\, dy$$

$$V'_t = \int_0^t \mathbb{1}_{\{B_s/B_t<0\}}\, ds = \int_{\mathbf{R}} \mathbb{1}_{\{y/B_t<0\}} L_t^y\, dy$$

La différence $U_t = t - V_t - V'_t$ représente alors le temps passé à l'instant t par le mouvement brownien B dans le segment $[0, B_t]$.

La description que nous donnons a l'intérêt d'être symétrique, c'est-à-dire de faire apparaître l'identité en loi :

$$\left(B_t, \left(L_t^{B_t-x}\right)^{x\in\mathbf{R}}\right) \stackrel{\mathcal{L}}{=} \left(B_t, \left(L_t^x\right)^{x\in\mathbf{R}}\right) ,$$

qui provient de l'identité en loi :

$$(B_t - B_{t-s})_{0\le s\le t} \stackrel{\mathcal{L}}{=} (B_s)_{0\le s\le t} .$$

La méthode que nous employons est un peu plus simple que celles de E. Perkins et T. Jeulin. Mais avant de la présenter, énonçons les principaux résultats.

Notations et énoncés des résultats.

Suivant l'usage, on note pour $\delta \in \mathbf{R}_+$ et $a, r, r' \in \mathbf{R}_+^*$:

Q_r^δ la loi d'un carré de Bessel de dimension δ issu de r

$Q_{r,r'}^{\delta,a}$ la loi d'un pont de carré de Bessel de dimension δ, de longueur a et de r à r'.

Les densités de certaines variables aléatoires vont jouer un grand rôle dans la suite. En appelant $(X_z)_{z\in\mathbf{R}_+}$ le processus canonique sur $\mathcal{C}_c(\mathbb{R}_+, \mathbb{R}_+)$, on note :

$f(r,\cdot)$ la densité de la variable aléatoire $\int_0^{+\infty} X_z\, dz$ sous Q_r^0 ,

$q_a(r,\cdot)$ la densité de la variable aléatoire X_a sous Q_r^2 ,

$g_a(r,\cdot)$ la densité de la variable aléatoire $\int_0^a X_z\, dz$ sous Q_r^2 ,

$g_a(r,r',\cdot)$ la densité de la variable aléatoire $\int_0^a X_z\, dz$ sous $Q_{r,r'}^{2,a}$.

Les deux premières densités ont des expressions relativement simples (voir [9] au chapitre XI)

$$f(r,v) = \frac{r}{\sqrt{8\pi}v^{3/2}} \exp\left(-\frac{r^2}{8v}\right) \quad \text{pour } v > 0$$

$$q_a(r,r') = \frac{1}{2a} \exp\left(-\frac{r+r'}{2a}\right) I_0\left(\frac{\sqrt{rr'}}{a}\right) \quad \text{pour } r' > 0 .$$

Les deux autres n'ont pas d'expression simple. Cependant, on connaît bien leurs transformées de Laplace. Pour tout $\theta \in \mathbf{R}_+$, on a en effet :

$$\int_0^{+\infty} \exp\left(-\frac{\theta^2}{2}u\right) g_a(r,u)\, du = Q_r^2\left[\exp\left(-\frac{\theta^2}{2}\int_0^a X_z\, dz\right)\right]$$
$$= \exp\left(-\frac{r\theta}{2}\operatorname{th}(a\theta)\right)\frac{1}{\operatorname{ch}(a\theta)}$$

$$\int_0^{+\infty} \exp\left(-\frac{\theta^2}{2}u\right) g_a(r,r',u)\, du = Q_{r,r'}^{2,a}\left[\exp\left(-\frac{\theta^2}{2}\int_0^a X_z\, dz\right)\right]$$
$$= \frac{\theta a}{\operatorname{sh}(\theta a)}\exp\left(\frac{r+r'}{2a}(1-\theta a\coth(\theta a))\right) I_0\left(\frac{\theta\sqrt{rr'}}{\operatorname{sh}(\theta a)}\right) \Big/ I_0\left(\frac{\sqrt{rr'}}{a}\right).$$

On obtient les transformées de Fourier en remplaçant $\frac{\theta^2}{2}$ par $i\lambda$ et θ par $(1 + i\operatorname{sgn}(\lambda))\sqrt{|\lambda|}$ dans les formules ci-dessus. La formule d'inversion de Fourier montre que l'on peut prendre les densités continues sur $\mathbf{R}_+^*$, et de classe C^1 vis à vis du paramètre r.

Nous pouvons maintenant énoncer :

Théorème 1. — *La loi conjointe de B_t et du processus $(L_t^z)^{z\in\mathbf{R}}$ est décrite par les propriétés suivantes :*

- *le 5-uplet $(B_t, L_t^{B_t}, L_t^0, V_t, V_t')$ admet comme densité sur $\mathbf{R}\times(\mathbf{R}_+^*)^4$:*
$$(a,r,r',v,v') \longmapsto q_{|a|}(r,r')f(r,v)f(r',v')g_{|a|}(r,r',t-v-v')\,;$$
- *pour $a>0$, conditionnellement à $(B_t, L_t^{B_t}, L_t^0, V_t, V_t') = (a,r,r',v,v')$:*
  - *les processus $(L_t^{a+z})_{z\geq 0}$, $(L_t^{-z})_{z\geq 0}$ et $(L_t^{a-z})_{0\leq z\leq a}$ sont indépendants ;*
  - *les processus $\left(L_t^{a+z}, \int_{a+z}^{+\infty} L_t^y\, dy\right)_{z\geq 0}$ et $\left(L_t^{-z}, \int_{-\infty}^{-z} L_t^y\, dy\right)_{z\geq 0}$ sont markoviens de générateur :*
  $$2x\frac{\partial^2}{\partial x^2} + \left(4 - \frac{x^2}{y}\right)\frac{\partial}{\partial x} - x\frac{\partial}{\partial y}\,;$$
  - *le processus $\left(L_t^{a-z}, \int_0^{a-z} L_t^y\, dy, z\right)_{0\leq z\leq a}$ est markovien de générateur :*
  $$2x\frac{\partial^2}{\partial x^2} + \left(2+4x\left(\frac{\partial_1 q_{a-z}}{q_{a-z}}(x,r') + \frac{\partial_1 g_{a-z}}{g_{a-z}}(x,r',y)\right)\right)\frac{\partial}{\partial x} - x\frac{\partial}{\partial y} + \frac{\partial}{\partial z}\,;$$
- *pour $a<0$, on a des formules semblables par symétrie.*

On dispose d'un résultat similaire pour l'instant τ_s^+ où le temps de séjour dans $\mathbf{R}_+$ du mouvement brownien B atteint la valeur s :

$$\tau_s^+ = \inf\left\{t\in\mathbf{R}_+ \mid \int_{\mathbf{R}_+} L_t^y\, dy \geq s\right\}.$$

THÉORÈME 2. — *La loi conjointe de $B_{\tau_s^+}$ et du processus $(L^z_{\tau_s^+})^{z\in\mathbf{R}_+}$ est donnée par les propriétés suivantes :*

- *le triplet* $\left(B_{\tau_s^+}, L_{\tau_s^+}^{B_{\tau_s^+}}, U_{\tau_s^+}\right)$ *admet comme densité sur* $(\mathbf{R}_+^*)^3$:

$$(a, r, u) \longmapsto g_a(r, u) f(r, s-u) ;$$

- *conditionnellement à* $\left(B_{\tau_s^+}, L_{\tau_s^+}^{B_{\tau_s^+}}, U_{\tau_s^+}\right) = (a, r, u)$:

  * *les processus* $(L^{a+z}_{\tau_s^+})_{z\geq 0}$ et $(L^{a-z}_{\tau_s^+})_{z\geq 0}$ *sont indépendants ;*

  * *le processus* $\left(L^{a+z}_{\tau_s^+}, \int_{a+z}^{+\infty} L^y_{\tau_s^+}\, dy\right)_{z\geq 0}$ *est markovien de générateur :*

$$2x\frac{\partial^2}{\partial x^2} + \left(4 - \frac{x^2}{y}\right)\frac{\partial}{\partial x} - x\frac{\partial}{\partial y} ;$$

  * *le processus* $\left(L^{a-z}_{\tau_s^+}, \int_0^{a-z} L^y_t\, dy\right)_{z\geq 0}$ *est markovien inhomogène ;*

  * *le processus* $\left(L^{a-z}_{\tau_s^+}, \int_0^{a-z} L^y_t\, dy, z\right)_{0\leq z\leq a}$ *a pour générateur :*

$$2x\frac{\partial^2}{\partial x^2} + \left(2 + 4x\frac{\partial_1 g_{a-z}}{g_{a-z}}(x,y)\right)\frac{\partial}{\partial x} - x\frac{\partial}{\partial y} + \frac{\partial}{\partial z} ;$$

  * *le processus* $(L^{-z}_{\tau_s^+})_{z\geq 0}$ *est un carré de Bessel de dimension 0.*

Remarques.

• Dans l'énoncé du théorème 1, on peut remplacer le processus

$$\left(L_t^{a-z}, \int_0^{a-z} L_t^y\, dy, z\right)_{0\leq z\leq a} \quad \text{par} \quad \left(L_t^z, \int_z^a L_t^y\, dy, z\right)_{0\leq z\leq a}$$

à condition d'intervertir r et r' dans l'expression de son générateur.

• Dans l'énoncé du théorème 2, on peut conditionner par rapport à $L^0_{\tau_s^+}$, ce qui fournit une formulation proche de celle du théorème 1.

Des théorèmes 1 et 2, on déduit immédiatement :

COROLLAIRE 3. — *Les processus*

$$\left(L_t^{(B_t\vee 0)+z}, \int_{(B_t\vee 0)+z}^{+\infty} L_t^y\, dy\right)_{z\geq 0}, \qquad \left(L_t^{(B_t\wedge 0)-z}, \int_{-\infty}^{(B_t\wedge 0)-z} L_t^y\, dy\right)_{z\geq 0}$$

et

$$\left(L_{\tau_s^+}^{B_{\tau_s^+}+z}, \int_{B_{\tau_s^+}+z}^{+\infty} L_t^y\, dy\right)_{z\geq 0}$$

sont tous trois markoviens de générateur

$$2x\frac{\partial^2}{\partial x^2} + \left(4 - \frac{x^2}{y}\right)\frac{\partial}{\partial x} - x\frac{\partial}{\partial y}.$$

Les trois processus ci-dessus sont tous des couples formés par le temps local et sa primitive vis à vis de la variable d'espace. On peut se ramener à un processus markovien unidimensionnel en faisant un changement de paramètre lié à la primitive du temps local. En effet, on a le résultat suivant :

PROPOSITION 4. — *Soit $(X_z, Y_z)_{z\geq 0}$ une diffusion dans $\mathbf{R}_+^2$ issue de $(x_0, y_0) \in (\mathbf{R}_+^*)^2$ et de générateur*

$$2x\frac{\partial^2}{\partial x^2} + \left(4 - \frac{x^2}{y}\right)\frac{\partial}{\partial x} - x\frac{\partial}{\partial y}.$$

Pour $y \in]0, y_0[$, notons $\zeta(y)$ l'unique valeur de z telle que $Y_z = y$. Alors le processus ρ défini par :

$$\rho_t = (y_0 e^{-t})^{-1/2} X_{\zeta(y_0 e^{-t})} \quad pour\ t \geq 0$$

est markovien de générateur :

$$2\frac{\partial^2}{\partial r^2} + \left(\frac{4}{r} - \frac{r}{2}\right)\frac{\partial}{\partial r}.$$

Présentation de la méthode suivie.

Expliquons maintenant la méthode employée pour démontrer les théorèmes 1 et 2.

Étant donné un temps aléatoire τ, fini presque sûrement, on note P^τ la loi du mouvement brownien tué à l'instant τ, c'est-à-dire du processus $(B_u)_{0\leq u\leq\tau}$. Cette loi est une probabilité sur l'espace des trajectoires continues à durée de vie finie :

$$\mathcal{W} = \bigcup_{t\in\mathbf{R}_+} \mathcal{C}([0,t], \mathbf{R}).$$

À une trajectoire $w \in \mathcal{W}$, on associe sa durée de vie $\zeta(w)$ définie par $\zeta(w) = t$ si $w \in \mathcal{C}([0,t], \mathbf{R})$, et la famille de ses temps locaux à l'instant final $\zeta(w)$ définie P^τ-presque sûrement par :

$$L^z(w) = \lim_{\varepsilon\downarrow 0} \frac{1}{2\varepsilon}\int_0^{\zeta(w)} \mathbb{1}_{\{|w(u)-z|\leq\varepsilon\}}\, du.$$

On note $\mathcal{L}(P^\tau)$ la loi de $(B_\tau, (L_\tau^z)^{z\in\mathbf{R}})$, qui n'est autre que la mesure image de P^τ par la fonction $\mathcal{L} : w \longmapsto (w(\zeta(w)), L^\cdot(w))$. Cette loi définit une probabilité sur

$\mathbf{R} \times \mathcal{C}_c(\mathbf{R}, \mathbf{R}_+)$ où $\mathcal{C}_c(\mathbf{R}, \mathbf{R}_+)$ désigne l'ensemble des fonctions continues à support compact de $\mathbf{R}$ dans $\mathbf{R}_+$.

Notre but est de décrire les lois $\mathcal{L}(P^t)$ et $\mathcal{L}(P^{\tau_s^+})$ pour $t, s \in \mathbf{R}_+^*$. Les théorèmes de Ray-Knight permettent d'obtenir facilement les lois $\mathcal{L}(P^{\tau_r^a})$, où τ_r^a est le premier instant où le temps local en a atteint la valeur r :

$$\tau_r^a = \inf\{t \in \mathbf{R}_+ | L_t^a \geq r\}.$$

Nous passons des lois $\mathcal{L}(P^{\tau_r^a})$ aux lois $\mathcal{L}(P^t)$ et $\mathcal{L}(P^{\tau_s^+})$ en utilisant les identités suivantes :

$$(1) \qquad \int_0^{+\infty} P^t \, dt = \int_{\mathbf{R}} \left(\int_0^{+\infty} P^{\tau_r^a} \, dr \right) da$$

$$(2) \qquad \int_0^{+\infty} P^{\tau_s^+} \, ds = \int_0^{+\infty} \left(\int_0^{+\infty} P^{\tau_r^a} \, dr \right) da.$$

Les identités (1) et (2) entraînent les identités analogues pour les mesures images par $\mathcal{L}$:

$$(1') \qquad \int_0^{+\infty} \mathcal{L}(P^t) \, dt = \int_{\mathbf{R}} \left(\int_0^{+\infty} \mathcal{L}(P^{\tau_r^a}) \, dr \right) da$$

$$(2') \qquad \int_0^{+\infty} \mathcal{L}(P^{\tau_s^+}) \, ds = \int_0^{+\infty} \left(\int_0^{+\infty} \mathcal{L}(P^{\tau_r^a}) \, dr \right) da.$$

Chacune de ces identités représente deux désintégrations d'une même mesure sur $\mathbf{R} \times \mathcal{C}_c(\mathbf{R} \times \mathbf{R}_+)$:

— dans le membre de gauche, la mesure est désintégrée par rapport à la fonctionnelle $\Lambda_\mu : (a, \ell) \longmapsto \int_{\mathbf{R}} \ell(z)\mu(dz)$, où μ est la mesure de Lebesgue sur $\mathbf{R}$ ou sur $\mathbf{R}_+$;

— dans le membre de droite, la mesure est désintégrée par rapport à la fonctionnelle $(a, \ell) \longmapsto (a, \ell(a))$ de $\mathbf{R} \times \mathcal{C}_c(\mathbf{R}, \mathbf{R}_+)$ dans $\mathbf{R} \times \mathbf{R}_+$.

On passe donc des lois $\mathcal{L}(P^{\tau_r^a})$ aux lois $\mathcal{L}(P^t)$ et $\mathcal{L}(P^{\tau_s^+})$ en intégrant les lois $\mathcal{L}(P^{\tau_r^a})$ par rapport à a et r et en désintégrant par rapport à la fonctionnelle Λ_μ. Nous sommes ainsi conduits à conditionner le processus $(L^z_{\tau_r^a})^{z \in \mathbf{R}}$ par son intégrale sur $\mathbf{R}$ ou sur $\mathbf{R}_+$. Pour cela, il est commode de séparer les restrictions des temps locaux aux trois intervalles délimités par 0 et $B_{\tau_r^a} = a$ pour obtenir trois processus indépendants conditionnellement à $L^0_{\tau_r^a}$.

Pour expliciter les lois obtenues (carrés ou ponts de carrés de Bessel conditionnés par leur intégrale), nous utilisons la théorie des h-processus de Doob.

La suite de notre travail s'organise de la façon suivante :

— dans la première partie ("préliminaires") nous démontrons les identités 1 et 2, nous donnons une description de la loi $\mathcal{L}(P^{\tau_r^a})$ et nous démontrons la proposition 4 (qui se prouve indépendamment du reste) ;

— dans la deuxième partie, nous démontrons en détail le théorème 2 ;

— dans la troisième partie, nous donnons les grandes lignes de la démonstration du théorème 1, très voisine de celle du théorème 2, mais plus lourde à écrire ;

— enfin, dans la quatrième partie, nous terminons par quelques remarques.

I. Préliminaires

1. Démonstration des identités (1) et (2).

Dans l'article [6], j'ai démontré l'identité (1) pour retrouver de façon élémentaire une identité que P. Biane et M. Yor ont déduit de la théorie des excursions. Au cours d'un exposé que j'ai donné à ce sujet aux Journées de Probabilités de 1995, J. Azema m'a suggéré la généralisation suivante :

Soit μ une mesure positive localement finie sur $\mathbf{R}_+$. Considérons la fonctionnelle additive L^μ et son "inverse" τ^μ définis par :

$$L_t^\mu = \int_{\mathbf{R}} L_t^x \, \mu(dx) \text{ pour } t \in \mathbf{R}_+$$
$$\tau_s^\mu = \inf\{t \in \mathbf{R}_+ \mid L_t^\mu \geq s\} \text{ pour } s \in \mathbf{R}_+ .$$

On définit une mesure σ-finie M sur $\mathcal{W}$ en posant, pour toute fonctionnelle $\mathcal{F}$-mesurable positive sur $\mathcal{W}$:

$$M(F) = E\left[\int_0^{+\infty} F\left((B_u)_{0\leq u\leq t}\right) dL_t^\mu\right].$$

Cette mesure M ainsi construite vérifie :

$$M = \int_0^{+\infty} P^{\tau_s^\mu} \, ds = \int_{\mathbf{R}} \left(\int_0^{+\infty} P^{\tau_r^a} \, dr\right) \mu(da). \tag{3}$$

Les identités (1) et (2) sont les cas particuliers de l'identité (3) obtenus en prenant la mesure de Lebesgue sur R et la mesure de Lebesgue sur $\mathbf{R}_+$ comme mesure

μ. Il suffit donc de montrer l'identité (3), qui est une conséquence de l'observation suivante :

LEMME 5. — *Presque sûrement la mesure image de la mesure $\mu(da)\, dr$ par l'application $(a, r) \longmapsto \tau_r^a$ est égale à la mesure de Stieltjes dL_t^μ (qui est la mesure image de la mesure ds par l'application $s \mapsto \tau_s^\mu$).*

Une fois le lemme connu, il suffit en effet d'écrire que pour toute fonctionnelle F mesurable positive sur $\mathcal{W}$, on a presque sûrement :

$$\int_0^{+\infty} F((B_u)_{0\le u\le t})\, dL_t^\mu = \int_0^{+\infty} F((B_u)_{0\le u\le \tau_s^\mu})\, ds$$
$$= \int_{\mathbf{R}} \left(\int_0^{+\infty} F((B_u)_{0\le u\le \tau_r^a})\, dr \right) \mu(da).$$

En passant aux espérances, on démontre l'identité (3).

Démonstration du lemme 5. — Il suffit de constater que pour tout $t \in \mathbf{R}_+$

$$\int_{\mathbf{R}} \left(\int_0^{+\infty} \mathbb{1}_{\{\tau_r^a \le t\}}\, dr \right) \mu(da) = \int_{\mathbf{R}} \left(\int_0^{+\infty} \mathbb{1}_{\{r \le L_t^a\}}\, dr \right) \mu(da)$$
$$= \int_{\mathbf{R}} L_t^a\, \mu(da) = L_t^\mu = \int_0^t \mathbb{1}_{\{s \le L_t^\mu\}} ds$$
$$= \int_0^t \mathbb{1}_{\{\tau_s^\mu \le t\}} ds.$$

2. Description de la loi $\mathcal{L}(P_0^{\tau_r^a})$.

Par symétrie, on peut se contenter de décrire la loi $\mathcal{L}(P_0^{\tau_r^a})$ pour $a \in \mathbf{R}_+$ et $r > 0$. On a bien sûr $B_{\tau_r^a} = a$ presque sûrement. Il suffit donc de décrire la loi du processus $(L_{\tau_r^a}^z)^{z\in\mathbf{R}}$. En appliquant la propriété de Markov au premier instant d'atteinte de a par le mouvement brownien B et en utilisant l'additivité des carrés de Bessel, on déduit, des théorèmes de Ray et Knight, la description suivante :

COROLLAIRE (Ray-Knight). — *Soient $a \in \mathbf{R}_+$ et $r > 0$. Les processus $(L_{\tau_r^a}^{a+z})_{z\ge 0}$ et $(L_{\tau_r^a}^{a-z})_{z\ge 0}$ sont indépendants et markoviens, le premier étant homogène mais pas le second :*

- *le processus $(L_{\tau_r^a}^{a+z})_{z\ge 0}$ est un carré de Bessel de dimension 0 issu de r ;*
- *le processus $(L_{\tau_r^a}^{a-z})_{0\le z\le a}$ est un carré de Bessel de dimension 2 issu de r ;*
- *le processus $(L_{\tau_r^a}^{-z})_{z\ge 0}$ est un carré de Bessel de dimension 0.*

3. Démonstration de la proposition 4.

Soit $(X_z, Y_z)_{z\geq 0}$ une diffusion sur $(\mathbf{R}_+)^2$ de générateur $2x\frac{\partial^2}{\partial x^2} + \left(4 - \frac{x^2}{y}\right)\frac{\partial}{\partial x} - x\frac{\partial}{\partial y}$ et issue de $(x_0, y_0) \in (\mathbf{R}_+^*)^2$. On peut supposer que $(X_z, Y_z)_{z\geq 0}$ est la solution du système différentiel stochastique :

$$\begin{cases} dX_z = 2\sqrt{X_z}d\beta_z + (4 - \frac{X_z^2}{Y_z})\,dz \\ dY_z = -X_z\,dz \end{cases}$$

vérifiant $(X_0, Y_0) = (x_0, y_0)$, où β est un mouvement brownien.

Soit $\zeta_0 = \inf\{z \in \mathbf{R}_+ \mid Y_z = 0\}$. Pour $0 \leq z < \zeta_0$, on a :

$$d\left(\frac{X_z}{\sqrt{Y_z}}\right) = \left(\frac{dX_z}{\sqrt{Y_z}}\right) - \frac{1}{2}\left(\frac{X_z\,dY_z}{Y_z^{3/2}}\right) = 2\sqrt{\frac{X_z}{Y_z}}\,d\beta_z + \left(4 - \frac{X_z^2}{2Y_z}\right)\frac{dz}{\sqrt{Y_z}}.$$

Donc en notant $\zeta(y)$ la valeur de z telle que $Y_z = y$ pour $0 < y \leq y_0$ et en posant $\rho_t = (y_0e^{-t})^{-1/2}X_{\zeta(y_0e^{-t})}$ pour $t \geq 0$, on a :

$$\rho_t = \rho_0 + 2W_t + \int_0^{\zeta(y_0e^{-t})}\left(4 - \frac{X_z^2}{2Y_z}\right)\frac{dz}{\sqrt{Y_z}} \quad \text{où } W_t = \int_0^{\zeta(y_0e^{-t})}\sqrt{\frac{X_z}{Y_z}}\,d\beta_z.$$

Le processus W est un mouvement brownien puisque :

$$\langle W, W\rangle_t = \int_0^{\zeta(y_0e^{-t})}\frac{X_z}{Y_z}\,dz = [-\ln Y_z]_{z=0}^{z=\zeta(y_0e^{-t})} = t.$$

En effectuant les changements de variable $y = Y_z$ puis $y = y_0e^{-s}$, on obtient :

$$\begin{aligned}\rho_t - \rho_0 - 2W_t &= \int_{y_0}^{y_0e^{-t}}\left(4 - \frac{X_{\zeta(y)}^2}{2y}\right)\frac{dy}{-X_{\zeta(y)}\sqrt{y}} \\ &= \int_0^t\left(4 - \frac{\rho_s^2}{2}\right)\frac{y_0e^{-s}\,ds}{X_{\zeta(y_0e^{-s})}(y_0e^{-s})^{1/2}} \\ &= \int_0^t\left(\frac{4}{\rho_s} - \frac{\rho_s}{2}\right)ds,\end{aligned}$$

ce qui prouve la proposition.

II. Démonstration du théorème 2

Dans cette partie, on prend la mesure de Lebesgue sur $\mathbf{R}_+$ comme mesure μ. La mesure M correspondante est

$$M = \int_0^{+\infty} P^{\tau_s^+}\,ds = \int_0^{+\infty}\left(\int_0^{+\infty} P^{\tau_r^a}\,dr\right)da,$$

et la fonctionnelle Λ_μ est l'application $(a, \ell) \longmapsto \int_{\mathbf{R}_+} \ell(y)\, dy$ de $\mathbf{R} \times \mathcal{C}_c(\mathbf{R}, \mathbf{R}_+)$ dans $\mathbf{R}_+$. L'égalité

$$\mathcal{L}(M) = \int_0^{+\infty} \mathcal{L}(P_0^{\tau_s^+})\, ds$$

n'est autre que la désintégration de la mesure $\mathcal{L}(M)$ sur $\mathbf{R} \times \mathcal{C}_c(\mathbf{R}, \mathbf{R}_+)$ par rapport à la fonctionnelle Λ_μ. Nous allons obtenir une autre expression de cette désintégration en décomposant pour $a, r \in \mathbf{R}_+^*$ fixés la probabilité $\mathcal{L}(P^{\tau_r^a})$. Comme $B_{\tau_r^a} = a$ presque sûrement, cela revient à conditionner le processus $(L^z_{\tau_r^a})^{z \in \mathbf{R}}$ par rapport à son intégrale $\int_0^{+\infty} L^y_{\tau_r^a}\, dy$ sur $\mathbf{R}_+$. On commence par le conditionner par le couple $(U_{\tau_r^a}, V_{\tau_r^a})$, vu que $U_{\tau_r^a} = \int_0^a L^y_{\tau_r^a}\, dy$ et $V_{\tau_r^a} = \int_a^{+\infty} L^y_{\tau_r^a}\, dy$.

1. Conditionnement de $(L^z_{\tau_r^a})^{z \in \mathbf{R}}$ par rapport au couple $\left(U_{\tau_r^a}, V_{\tau_r^a}\right)$.

Par indépendance des processus $(L^{a-z}_{\tau_r^a})_{z \geq 0}$ et $(L^{a+z}_{\tau_r^a})_{z \geq 0}$, il suffit de conditionner séparément le premier par $U_{\tau_r^a}$ et le second par $V_{\tau_r^a}$. D'après les théorèmes de Ray-Knight, le processus $(L^{a-z}_{\tau_r^a})_{z \geq 0}$ est markovien inhomogène. Or la variable $U_{\tau_r^a}$ ne dépend que de la restriction de ce processus à l'intervalle $[0, a]$. Donc après conditionnement par rapport à $U_{\tau_r^a}$, le processus $(L^{-z}_{\tau_r^a})_{z \geq 0}$ reste un carré de Bessel de dimension 0 indépendant de $(L^{a-z}_{\tau_r^a})_{0 \leq z \leq a}$ conditionnellement à $L^0_{\tau_r^a}$.

La loi du processus $(L^z_{\tau_r^a})^{z \in \mathbf{R}}$ sachant que $(U_{\tau_r^a}, V_{\tau_r^a}) = (u, v)$ est donc la loi $Q_{a,r,u,v}$ définie comme suit. En notant $\ell = (\ell(z))_{z \in \mathbf{R}}$ le processus canonique sur $\mathcal{C}_c(\mathbf{R}, \mathbf{R}_+)$, sa loi sous $Q_{a,r,u,v}$ est donnée par les propriétés suivantes :

- les restrictions de ℓ aux intervalles $]-\infty, a]$ et $[a, +\infty[$ sont indépendantes ;
- les restrictions de ℓ aux intervalles $]-\infty, 0]$ et $[0, a]$ sont indépendantes conditionnellement à $\ell(0)$;
- le processus $(\ell(a+z))_{z \geq 0}$ est un carré de Bessel de dimension 0 issu de r conditionné à ce que son intégrale vaille v ;
- le processus $(\ell(a-z))_{0 \leq z \leq a}$ est un carré de Bessel de dimension 2 issu de r conditionné à ce que son intégrale vaille u ;
- le processus $(\ell(-z))_{z \geq 0}$ est un carré de Bessel de dimension 0.

Nous expliciterons les lois des carrés de Bessel conditionnés par leur intégrale aux paragraphes II.3 et II.4. Mais voyons d'abord en quoi les lois $Q_{a,r,u,v}$ sont utiles pour notre problème.

2. Application à la description des lois $\mathcal{L}(P^{\tau_s^+})$.

Comme les variables $U_{\tau_r^a}$ et $V_{\tau_r^a}$ sont indépendantes et de densité $g_a(r,\cdot)$ et $f(r,\cdot)$, le conditionnement ci-dessus fournit la désintégration suivante :

$$\mathcal{L}(P^{\tau_r^a}) = \int_0^{+\infty}\int_0^{+\infty} (\delta_a \otimes Q_{a,r,u,v}) g_a(r,u) f(r,v)\, du\, dv\,.$$

Intégrons cette formule par rapport à a et r. En admettant la mesurabilité de l'application

$$(a,r,u,v) \longmapsto (\delta_a \otimes Q_{a,r,u,v})[F] g_a(r,u) f(r,v),$$

pour toute fonctionnelle F mesurable positive sur $\mathbf{R} \times \mathcal{C}_c(\mathbf{R},\mathbf{R}_+)$ on peut intervertir les intégrales et écrire :

$$\begin{aligned}\mathcal{L}(M) &= \int_0^{+\infty}\int_0^{+\infty} \mathcal{L}(P^{\tau_r^a})\, da\, dr \\ &= \int_0^{+\infty}\int_0^{+\infty}\int_0^{+\infty}\int_0^{+\infty} (\delta_a \otimes Q_{a,r,u,v}) g_a(r,u) f(r,v)\, da\, dr\, du\, dv\,.\end{aligned}$$

Sous la loi $Q_{a,r,u,v}$, les paramètres u et v représentent les intégrales du processus canonique ℓ sur les intervalles $[0,a]$ et $[a,+\infty[$. La quantité $s = u + v$ représente donc l'intégrale sur $\mathbf{R}_+$. Écrivons :

$$\mathcal{L}(M) = \int_0^{+\infty}\left(\int_0^{+\infty}\int_0^{+\infty}\int_0^{+\infty} (\delta_a \otimes Q_{a,r,u,s-u}) g_a(r,u) f(r,s-u)\, da\, dr\, du\right) ds$$

(on peut intégrer par rapport à u sur l'intervalle $[0,+\infty[$ tout entier car $f(r,v) = 0$ pour $v \le 0$).

Cette égalité constitue une désintégration par rapport à la fonctionnelle Λ_μ de la mesure $\mathcal{L}(M)$, dont on connaissait la désintégration :

$$\mathcal{L}(M) = \int_0^{+\infty} \mathcal{L}(P^{\tau_s^+})\, ds\,.$$

Par unicité essentielle de la désintégration, on a donc pour presque tout $s \in \mathbf{R}_+^*$:

$$\mathcal{L}(P^{\tau_s^+}) = \int_0^{+\infty}\int_0^{+\infty}\int_0^{+\infty} (\delta_a \otimes Q_{a,r,u,s-u}) g_a(r,u) f(r,s-u)\, da\, dr\, du\,.$$

En fait, l'égalité a lieu pour tout $s \in \mathbf{R}_+^*$ par continuité, ce que nous justifierons au paragraphe II.5.

Remarque. — Les égalités

$$\begin{aligned}\int_0^{+\infty} \mathcal{L}(P^{\tau_s^+})\, ds &= \mathcal{L}(M) \\ &= \int_0^{+\infty}\left(\int_0^{+\infty}\int_0^{+\infty}\int_0^{+\infty} (\delta_a \otimes Q_{a,r,u,s-u}) g_a(r,u) f(r,s-u)\, da\, dr\, du\right) ds\end{aligned}$$

et le fait que $\Lambda_\mu = s$ presque sûrement sous les lois $\mathcal{L}(P^{T_s^+})$ et $\delta_a \otimes Q_{a,r,u,s-u}$ montrent en particulier que :

$$\int_0^{+\infty}\int_0^{+\infty}\int_0^{+\infty} g_a(r,u) f(r,s-u)\, da\, dr\, du = 1$$

pour (presque) tout $s \in \mathbf{R}_+^*$, c'est-à-dire que l'application $(a,r,u) \longmapsto g_a(r,u)f(r,s-u)$ est bien une densité de probabilité sur $(\mathbf{R}_+^*)^3$. Cette propriété n'est en fait qu'une conséquence de l'identité :

$$\int_0^{+\infty} P^{T_s^+}\, ds = \int_0^{+\infty}\left(\int_0^{+\infty} P^{T_r^a}\, dr\right) da\,.$$

Nous allons maintenant donner une version plus explicite des carrés de Bessel conditionnés par leur intégrale qui interviennent dans la loi $Q_{a,r,u,v}$.

On note (X_z) le processus canonique sur $\mathcal{C}([0,a],\mathbf{R}_+)$ ou sur $\mathcal{C}_c(\mathbf{R}_+,\mathbf{R}_+)$, $(\mathcal{F}_z)$ la filtration canonique, et $Y_\cdot = \int_0^\cdot X_w\, dw$ la primitive de $X_\cdot$ nulle en 0.

Nous allons exhiber des versions régulières (et mêmes continues) des lois conditionnelles $Q_r^2[\,\cdot \mid Y_a = u]$ et $Q_r^0[\,\cdot \mid Y_\infty = v]$.

3. Description de la loi $Q_r^2[\,\cdot \mid Y_a = u]$.

Soient $z \in [0,a[$, $A \in \mathcal{F}_z$ et B un borélien de $\mathbf{R}_+$. D'après la propriété de Markov, on a :

$$\begin{aligned} Q_r^2[A\,;Y_a \in B] &= Q_r^2\left[A\,;Y_z + \int_0^{a-z} X_{z+w}\,dw \in B\right] \\ &= Q_r^2\left[\mathbf{1}_A \cdot \int_{\mathbf{R}} \mathbf{1}_B(Y_z+y) g_{a-z}(X_z,y)\,dy\right] \\ &= \int_B Q_r^2\left[\mathbf{1}_A\, g_{a-z}(X_z, u - Y_z)\right] du \\ &= \int_B Q_r^2\left[\mathbf{1}_A\, h_u(X_z, Y_z, z)\right] g_a(r,u)\, du \end{aligned}$$

où $h_u(x,y,z) = \frac{g_{a-z}(x,u-y)}{g_a(r,u)}$ pour $x,y,z \in \mathbf{R}_+$.

On peut donc prendre comme version régulière des lois conditionnelles $Q_r^2[\,\cdot \mid Y_a = u]$ la famille de probabilités définies par :

$$Q_r^2[A \mid Y_a = u] = Q_r^2[\mathbf{1}_A\, h_u(X_z,Y_z,z)] \text{ pour } A \in \mathcal{F}_z\,.$$

Sous la loi Q_r^2, le processus $(X_z, Y_z, z)_{0 \le z \le a}$ est markovien de générateur :

$$L = 2x\frac{\partial^2}{\partial x^2} + 2\frac{\partial}{\partial x} + x\frac{\partial}{\partial y} + \frac{\partial}{\partial z}\,.$$

Donc, d'après la théorie des h-processus (voir proposition 3.9 de [9]), sous la loi $Q_r^2[\cdot \mid Y_a = u]$, le processus $(X_z, Y_z, z)_{0\leq z<a}$ est markovien de générateur :

$$\begin{aligned} h_u^{-1}L(h_u\cdot) &= L + h_u^{-1}4x\frac{\partial h_u}{\partial x}\frac{\partial}{\partial x} \\ &= 2x\frac{\partial^2}{\partial x^2} + \left(2 + 4x\frac{\partial_1 g_{a-z}}{g_{a-z}}(x, u-y)\right)\frac{\partial}{\partial x} + x\frac{\partial}{\partial y} + \frac{\partial}{\partial z}. \end{aligned}$$

Sous la loi $Q_r^2[\cdot \mid Y_a = u]$, le processus $(X_z, \int_z^a X_w\, dw, z)_{0\leq z<a}$ est donc markovien de générateur :

$$2x\frac{\partial^2}{\partial x^2} + \left(2 + 4x\frac{\partial_1 g_{a-z}}{g_{a-z}}(x, y)\right)\frac{\partial}{\partial x} - x\frac{\partial}{\partial y} + \frac{\partial}{\partial z}.$$

4. Description de la loi $Q_r^0[\cdot \mid Y_\infty = v]$.

Notons $\zeta_0 = \inf\{z \in \mathbf{R}_+ \mid X_z = 0\}$. D'après la propriété de Markov, on a pour tout $z > 0$, $A \in \mathcal{F}_z$ et tout borélien B de $\mathbf{R}_+^*$:

$$\begin{aligned} Q_r^0[A\,; z < \zeta_0\,; Y_\infty \in B] &= Q_r^0\left[A\,; z < \zeta_0\,; Y_z + \int_0^{+\infty} X_{z+w}dw \in B\right] \\ &= Q_r^0\left[\mathbf{1}_A\mathbf{1}_{\{z<\zeta_0\}}\int_{\mathbf{R}}\mathbf{1}_B(Y_z + y)f(X_z, y)\, dy\right] \\ &= \int_B Q_r^0\left[\mathbf{1}_A\mathbf{1}_{\{z<\zeta_0\}}f(X_z, v - Y_z)\right]\, dv \\ &= \int_B Q_r^0\left[\mathbf{1}_A\mathbf{1}_{\{z<\zeta_0\}}h_v(X_z, Y_z)\right] f(r, v)\, dv \end{aligned}$$

où $h_v(x, y) = \frac{f(x, v-y)}{f(r, v)}$ pour $x, y \in \mathbf{R}_+^* \times \mathbf{R}_+^*$.

En choisissant convenablement la version régulière des lois conditionnelles par rapport à Y_∞, on a donc pour tout $v \in \mathbf{R}_+^*$:

$$Q_r^0[A\,; z < \zeta_0 \mid Y_\infty = v] = Q_r^0[\mathbf{1}_A\mathbf{1}_{\{z<\zeta_0\}}h_v(X_z, Y_z)].$$

Or sous la loi Q_r^0, le processus $(X_z, Y_z)_{0\leq z<\zeta_0}$ est markovien de générateur $L = 2x\frac{\partial^2}{\partial x^2} + x\frac{\partial}{\partial y}$. Donc sous la loi $Q_r^0[\cdot \mid Y_\infty = v]$, le processus $(X_z, Y_z)_{0\leq z<\zeta_0}$ est markovien de générateur

$$\begin{aligned} h_v^{-1}L(h_v\cdot) &= L + h_v^{-1}4x\frac{\partial h_v}{\partial x}\frac{\partial}{\partial x} = L + 4x\frac{\partial_1 f}{f}(x, v-y)\frac{\partial}{\partial x} \\ &= 2x\frac{\partial^2}{\partial x^2} + \left(4 - \frac{x^2}{v-y}\right)\frac{\partial}{\partial x} + x\frac{\partial}{\partial y}. \end{aligned}$$

Donc sous la loi $Q_r^0[\cdot \mid Y_\infty = v]$, le processus $(X_z, \int_z^{+\infty} X_w\, dw)_{0\leq z<\zeta_0}$ est markovien de générateur $2x\frac{\partial^2}{\partial x^2} + \left(4 - \frac{x^2}{y}\right)\frac{\partial}{\partial x} - x\frac{\partial}{\partial y}$. Le processus $\left(X_z, \int_z^{+\infty} X_w\, dw\right)_{z\geq 0}$ est donc une diffusion issue de (r, v), de générateur $2x\frac{\partial^2}{\partial x^2} + \left(4 - \frac{x^2}{y}\right)\frac{\partial}{\partial x} - x\frac{\partial}{\partial y}$, arrêtée lorsqu'elle atteint $(0, 0)$.

5. Justification des points manquants du II.2.

Dans le paragraphe II.2, nous avons admis la mesurabilité de l'application :

$$(a, r, u, v) \longmapsto (\delta_a \otimes Q_{a,r,u,v})[F] g_a(r, u) f(r, v)$$

pour toute fonctionnelle mesurable positive F sur $\mathbf{R} \times \mathcal{C}_c(\mathbf{R}, \mathbf{R}_+)$, ainsi que la continuité de l'application :

$$s \longmapsto \int_0^{+\infty}\int_0^{+\infty}\int_0^{+\infty} (\delta_a \otimes Q_{a,r,u,s-u}) g_a(r, u) f(r, s-u)\, da\, dr\, du\,,$$

pour obtenir l'égalité pour tout $s > 0$:

$$P^{\mathrm{T}_s^+} = \int_0^{+\infty}\int_0^{+\infty}\int_0^{+\infty} (\delta_a \otimes Q_{a,r,u,s-u}) g_a(r, u) f(r, s-u)\, da\, dr\, du\,.$$

En fait, pour obtenir cette égalité, il suffit de montrer l'égalité des lois finidimensionnelles correspondantes. On fixe donc n points $z_1, \ldots, z_n$, une fonction continue bornée ψ_0 de $\mathbf{R}$ dans $\mathbf{R}_+$, et $n+1$ fonctions bornées $\psi_1, \ldots, \psi_n, \varphi$ de $\mathbf{R}_+$ dans $\mathbf{R}_+$. On note F et G les fonctionnelles positives sur $\mathbf{R} \times \mathcal{C}_c(\mathbf{R}, \mathbf{R}_+)$ définies par :

$$G(a, \ell) = \psi_0(a)\psi_1(\ell(z_1)) \cdots \psi_n(\ell(z_n))$$
$$F(a, \ell) = G(a, \ell)\varphi\left(\int_0^{+\infty} \ell(z)\, dz\right)\,.$$

On a alors :

$$\begin{aligned}(\delta_a \otimes Q_{a,r,u,v})[F] &= (\delta_a \otimes Q_{a,r,u,v})[G]\varphi(u+v)\\ &= \psi_0(a) Q_{a,r,u,v}\left[\psi_1(\ell(z_1)) \cdots \psi_n(\ell(z_n))\right] \varphi(u+v)\,,\end{aligned}$$

en notant $(\ell(z))_{z \in \mathbf{R}}$ le processus canonique sur $\mathcal{C}_c(\mathbf{R}, \mathbf{R}_+)$. On vérifie que la quantité $Q_{a,r,u,v}[\psi_1(\ell(z_1)) \cdots \psi_n(\ell(z_n))]$ dépend continûment de chacune des variables a, r, u, v, ainsi que la quantité $g_a(r, u) f(r, v)$.

L'application $(a, r, u, v) \longmapsto (\delta_a \otimes Q_{a,r,u,v})[F] g_a(r, u) f(r, v)$ est donc mesurable ce qui permet d'appliquer le théorème de Fubini-Tonnelli. On a ainsi :

$$\begin{aligned}&\int_0^{+\infty} \mathcal{L}(P^{\mathrm{T}_s^+})[G]\varphi(s)\, ds\\ &= \int_0^{+\infty} \mathcal{L}(P^{\mathrm{T}_s^+})[F]\, ds = \mathcal{L}(M)[F]\\ &= \int_0^{+\infty}\int_0^{+\infty} \mathcal{L}(P^{\mathrm{T}_r^a})[F]\, da\, dr\\ &= \int_0^{+\infty}\int_0^{+\infty}\int_0^{+\infty}\int_0^{+\infty} (\delta_a \otimes Q_{a,r,u,v})[F] g_a(r, u) f(r, v)\, da\, dr\, du\, dv\\ &= \int_0^{+\infty}\left(\int_0^{+\infty}\int_0^{+\infty}\int_0^{+\infty} (\delta_a \otimes Q_{a,r,u,s-u})[G] g_a(r, u) f(r, s-u)\, da\, dr\, du\right)\varphi(s)\, ds\,.\end{aligned}$$

Comme cette égalité est vraie pour toute fonction φ continue bornée de $\mathbf{R}_+$ dans $\mathbf{R}_+$, on a donc pour presque tout $s \in \mathbf{R}_+^*$:

$$\mathcal{L}(P^{\tau_s^+})[G] = \int_0^{+\infty}\int_0^{+\infty}\int_0^{+\infty} (\delta_a \otimes Q_{a,r,u,s-u})[G] g_a(r,u) f(r,s-u)\, da\, dr\, du\,.$$

Il reste à montrer la continuité de chaque membre vis à vis de s pour obtenir l'égalité pour tout $s \in \mathbf{R}_+^*$. Le membre de gauche est égal à :

$$E\left[\psi_0(B_{\tau_s^+})\psi_1(L^{z_1}_{\tau_s^+})\cdots\psi_n(L^{z_n}_{\tau_s^+})\right]\,.$$

Il suffit d'appliquer le théorème de convergence dominée, compte tenu de ce que $\tau_s^+ \to \tau_{s_0}^+$ presque sûrement quand $s \to s_0$.

Pour le membre de droite, on remarque que les fonctions

$$h_s : (a,r,u) \mapsto g_a(r,u) f(r,s-u)$$

convergent simplement vers la fonction h_{s_0} quand $s \to s_0$. Toutes ces fonctions étant positives et d'intégrale 1, il y a également convergence dans $L^1((\mathbf{R}_+^*)^3)$. En effet, comme la fonction $h_s - h_{s_0}$ est d'intégrale nulle, on a :

$$\|h_s - h_{s_0}\|_1 = 2\int_0^{+\infty}\int_0^{+\infty}\int_0^{+\infty} (h_{s_0} - h_s)_+(a,r,u)\, da\, dr\, du\,,$$

et on peut utiliser la domination $(h_{s_0} - h_s)_+ \le h_{s_0}$. On écrit alors :

$$\begin{aligned}
&\int_0^{+\infty}\int_0^{+\infty}\int_0^{+\infty} (\delta_a \otimes Q_{a,r,u,s-u})[G] h_s(a,r,u)\, da\, dr\, du \\
&\quad - \int_0^{+\infty}\int_0^{+\infty}\int_0^{+\infty} (\delta_a \otimes Q_{a,r,u,s_0-u})[F] h_{s_0}(a,r,u)\, da\, dr\, du \\
&= \int_0^{+\infty}\int_0^{+\infty}\int_0^{+\infty} (\delta_a \otimes Q_{a,r,u,s-u})[G](h_s(a,r,u) - h_{s_0}(a,r,u))\, da\, dr\, du \\
&\quad + \int_0^{+\infty}\int_0^{+\infty}\int_0^{+\infty} ((\delta_a \otimes Q_{a,r,u,s-u})[G] - (\delta_a \otimes Q_{a,r,u,s_0-u})[G])\, h_{s_0}(a,r,u)\, da\, dr\, du
\end{aligned}$$

et on termine en utilisant la continuité de l'application $s \mapsto Q_{a,r,u,s-u}[G]$ et le théorème de convergence dominée de Lebesgue.

III. Démonstration du théorème 1

Dans cette partie, on prend la mesure de Lebesgue sur $\mathbf{R}$ comme mesure μ. La mesure M correspondante M est :

$$M = \int_0^{+\infty} P^t\, dt = \int_{\mathbf{R}} \left(\int_0^{+\infty} P^{\tau_r^a}\, dr\right) da,$$

et la fonctionnelle Λ_μ est l'application $(a, \ell) \longmapsto \int_{\mathbf{R}} \ell(y)\, dy$ de $\mathbf{R} \times \mathcal{C}_c(\mathbf{R}, \mathbf{R}_+)$ dans $\mathbf{R}_+$. L'égalité :

$$\mathcal{L}(M) = \int_0^{+\infty} \mathcal{L}(P^t)\, dt$$

représente la désintégration de la mesure $\mathcal{L}(M)$ par rapport à la fonctionnelle Λ_μ. Nous allons obtenir une autre expression de cette désintégration en décrivant, pour tout $a \in \mathbf{R}^*$ et $r \in \mathbf{R}_+^*$, les lois conditionnelles du processus $(L^z_{\tau^a_r})^{z \in \mathbf{R}}$ par rapport à son intégrale $\int_{\mathbf{R}} L^y_{\tau^a_r}\, dy$. La première étape consiste à conditionner le processus $(L^z_{\tau^a_r})^{z \in \mathbf{R}}$ par rapport au quadruplet $(L^0_{\tau^a_r}, U_{\tau^a_r}, V_{\tau^a_r}, V'_{\tau^a_r})$

1. Conditionnement du processus $(L^z_{\tau^a_r})^{z \in \mathbf{R}}$ par rapport au quadruplet $(L^0_{\tau^a_r}, U_{\tau^a_r}, V_{\tau^a_r}, V'_{\tau^a_r})$.

Par symétrie, on peut se contenter de décrire les lois conditionnelles pour $a \in \mathbf{R}_+^*$. D'après les théorèmes de Ray-Knight, la variable $L^0_{\tau^a_r}$ admet comme densité $r' \mapsto q_a(r, r')$, et conditionnellement à $L_{\tau^a_r} = r'$:

- les processus $(L^{a+z}_{\tau^a_r})_{z \geq 0}$, $(L^{-z}_{\tau^a_r})_{z \geq 0}$ et $(L^{a-z}_{\tau^a_r})_{0 \leq z \leq a}$ sont indépendants ;
- les processus $(L^{a+z}_{\tau^a_r})_{z \geq 0}$ et $(L^{-z}_{\tau^a_r})_{z \geq 0}$ sont des carrés de Bessel de dimension 0 respectivement issus de r et r' ;
- le processus $(L^{a-z}_{\tau^a_r})_{0 \leq z \leq a}$ est un pont de carré de Bessel de dimension 2, de longueur a, et de r à r'.

La loi du processus $(L^z_{\tau^a_r})^{z \in \mathbf{R}}$ sachant que $(L^0_{\tau^a_r}, U_{\tau^a_r}, V_{\tau^a_r}, V'_{\tau^a_r}) = (r', u, v, v')$ est donc la loi $Q_{a,r,r',u,v,v'}$ sous laquelle le processus canonique $\ell = (\ell(z))_{z \in \mathbf{R}}$ sur $\mathcal{C}_c(\mathbf{R}, \mathbf{R}_+)$ possède les propriétés suivantes :

- les restrictions de ℓ aux intervalles $]-\infty, 0]$, $[0, a]$ et $[a, +\infty[$ sont indépendantes ;
- le processus $(\ell(a+z))_{z \geq 0}$ est un carré de Bessel de dimension 0 issu de r conditionné à ce que son intégrale vaille v ;
- le processus $(\ell(-z))_{z \geq 0}$ est un carré de Bessel de dimension 0 issu de r' conditionné à ce que son intégrale vaille v' ;
- le processus $(\ell(a-z))_{0 \leq z \leq a}$ est un pont de carré de Bessel de dimension 2, de longueur a, de r à r', conditionné à ce que son intégrale vaille u.

Les lois de carrés de Bessel de dimension 0 conditionnés par leur intégrale ont été explicitées au paragraphe II.4. Les lois des ponts de carrés de Bessel de dimension 2 conditionnés par leur intégrale s'explicitent par une méthode semblable à celle que nous avons utilisée au paragraphe II.3.

En notant X le processus canonique sur $\mathcal{C}([0,a],\mathbf{R})$ et $Y = \int_0^{\cdot} X_z\,dz$, on trouve que sous la loi $Q^{2,a}_{r,r'}$, le processus $(X_z, Y_z, z)_{0\leq z<a}$ est markovien de générateur :

$$2x\frac{\partial^2}{\partial x^2} + \left(2 + 4x\frac{\partial_1 q_{a-z}}{q_{a-z}}(x,r')\right)\frac{\partial}{\partial x} + x\frac{\partial}{\partial y} + \frac{\partial}{\partial z},$$

sous la loi conditionnelle $Q^{2,a}_{r,r'}[\,\cdot\mid Y_a = u]$, le processus $(X_z, Y_z, z)_{0\leq z<a}$ est markovien de générateur :

$$2x\frac{\partial^2}{\partial x^2} + \left(2 + 4x\left(\frac{\partial_1 q_{a-z}}{q_{a-z}}(x,r') + \frac{\partial_1 g_{a-z}}{g_{a-z}}(x,r',u-y)\right)\right)\frac{\partial}{\partial x} + x\frac{\partial}{\partial y} + \frac{\partial}{\partial z}.$$

2. Application à la description des lois $\mathcal{L}(P^t)$.

Conditionnellement à $L^0_{\tau^a_r} = r'$, les variables aléatoires $U_{\tau^a_r}$, $V_{\tau^a_r}$ et $V'_{\tau^a_r}$ sont indépendantes de densités $g_{|a|}(r,r',\cdot)$, $f(r,\cdot)$ et $f(r',\cdot)$. Le conditionnement du processus $(L^z_{\tau^a_r})^{z\in\mathbf{R}}$ par rapport au quadruplet $(L^0_{\tau^a_r}, U_{\tau^a_r}, V_{\tau^a_r}, V'_{\tau^a_r})$ fournit la désintégration suivante :

$$\mathcal{L}(P^{\tau^a_r}) = \int_{(\mathbf{R}_+)^4} (\delta_a \otimes Q_{a,r,r',u,v,v'})\,q_{|a|}(r,r')\,g_{|a|}(r,r',u)\,f(r,v)\,f(r',v')\,dr'\,du\,dv\,dv'.$$

En intégrant cette formule par rapport à $da\,dr$ sur $\mathbf{R}\times\mathbf{R}_+$, on obtient :

$$\mathcal{L}(M) = \int_{\mathbf{R}\times(\mathbf{R}_+)^5} (\delta_a \otimes Q_{a,r,r',u,v,v'})\,q_{|a|}(r,r')\,g_{|a|}(r,r',u)\,f(r,v)\,f(r',v')\,da\,dr\,dr'\,du\,dv\,dv'.$$

Sous la loi $Q_{a,r,r',u,v,v'}$, l'intégrale sur $\mathbf{R}$ du processus canonique ℓ vaut $u+v+v'$. Effectuons donc le changement de variables $t = u+v+v'$ dans l'intégrale ci-dessus. On obtient :

$$\mathcal{L}(M) = \int_{\mathbf{R}\times(\mathbf{R}_+)^5} (\delta_a \otimes Q_{a,r,r',t-v-v',v,v'})\,q_{|a|}(r,r')\,g_{|a|}(r,r',t-v-v')\,f(r,v)\,f(r',v')\,da\,dr\,dr'\,dt\,dv\,dv'.$$

On peut garder $\mathbf{R}\times(\mathbf{R}_+)^5$ comme domaine d'intégration compte-tenu de ce que $g_{|a|}(r,r',u) = 0$ pour $u<0$.

Par unicité essentielle de la désintégration de la mesure $\mathcal{L}(M)$ par rapport à la fonctionnelle Λ_μ, on a ainsi :

$$\mathcal{L}(P^t) = \int_{\mathbf{R}\times(\mathbf{R}_+)^4} (\delta_a \otimes Q_{a,r,r',t-v-v',v,v'})\,q_{|a|}(r,r')\,g_{|a|}(r,r',t-v-v')\,f(r,v)\,f(r',v')\,da\,dr\,dr'\,dv\,dv'$$

pour presque tout $t\in\mathbf{R}^*_+$. On montre comme au paragraphe II.5 que cette identité est en fait vraie pour tout $t\in\mathbf{R}^*_+$, par continuité.

IV. Remarques

1. Une généralisation possible de la méthode.

Dans les parties II et III, nous nous sommes servis des identités :

$$\int_0^{+\infty} P^t\,dt = \int_{\mathbf{R}} \left(\int_0^{+\infty} P^{\tau_r^a}\,dr \right) da \tag{1}$$

$$\int_0^{+\infty} P^{\tau_s^+}\,ds = \int_0^{+\infty} \left(\int_0^{+\infty} P^{\tau_r^a}\,dr \right) da \tag{2}$$

pour décrire les lois $\mathcal{L}(P^t)$ et $\mathcal{L}(P^{\tau_s^+})$. Il est tout à fait possible d'obtenir des résultats proches de ceux de T. Jeulin [4], en décrivant plus généralement les lois $\mathcal{L}(P^{\tau_s^\mu})$ pour une large classe de mesures μ. Il suffit pour cela d'appliquer la même méthode avec l'identité plus générale suggérée par J. Azema :

$$\int_0^{+\infty} P^{\tau_s^\mu}\,ds = \int_{\mathbf{R}} \left(\int_0^{+\infty} P^{\tau_r^a}\,dr \right) \mu(da) \tag{3}$$

Mais, dans cette généralité, les formules obtenues sont alors fort lourdes, peu explicites, et perdent une partie de leur intérêt.

On obtient des formules explicites dans le cas particulier où μ est une combinaison des mesures de Lebesgue sur $\mathbf{R}_+$ et sur $\mathbf{R}_-$. Les deux lois que nous avons étudiées en sont les exemples les plus simples et les plus intéressants.

2. Lien entre les théorèmes de Ray-Knight "à temps fixe" et à temps exponentiel indépendant.

Une idée pour décrire les lois $\mathcal{L}(P^t)$ consiste à partir de la loi $\mathcal{L}(P^T)$ où T est un temps exponentiel indépendant de B, donnée par le théorème de Ray, et à conditionner par $T = \int_{\mathbf{R}} L_T^x\,dx$. T. Jeulin a utilisé cette méthode ainsi que la théorie du grossissement de filtrations pour décomposer la semi-martingale $(L_t^x)^{x \in \mathbf{R}}$.

Il est possible de retrouver le théorème 1 en déduisant les lois $\mathcal{L}(P^t)$ de la loi $\mathcal{L}(P^T)$: on commence par conditionner le couple formé de la variable B_T et du processus $(L_T^z)^{z \in \mathbf{R}}$ par le 6-uplet $(B_T, L_T^{B_T}, L_T^0, U_T, V_T, V_T')$, puis on conditionne par $T = U_T + V_T + V_T'$ seulement.

Cette méthode a l'inconvénient de faire intervenir des lois moins "classiques" que celles des carrés de Bessel. Mais en fait, elle n'est pas vraiment différente de celle que nous avons menée au paragraphe III.2. En effet, la formulation simple du théorème de Ray, due à Biane et Yor [1] se démontre à partir des deux théorèmes de Ray et Knight les plus connus, et d'une identité en loi très voisine de l'identité (1) que nous utilisons (voir à ce sujet [6]).

3. Lien avec les travaux de R. van der Hofstad, F. den Hollander, W. König.

Sous une formulation différente, mais équivalente à celle de notre théorème 1, le lemme 1 de [3] indique :

- la densité du 5-uplet $(B_t, L_t^{B_t}, L_t^0, V_t, V_t')$;
- l'indépendance des processus $(L_t^{B_t+z})_{z\geq 0}, (L_t^{-z})_{z\geq 0}$ et $(L_t^{B_t-z})_{0\leq z\leq B_t}$ sachant que $(B_t, L_t^{B_t}, L_t^0, V_t, V_t') = (a, r, r', v, v')$ pour $(a, r, r', v, v') \in \mathbf{R}_+{}^5$.

En revanche, R. van der Hofstad, F. den Hollander et W. König se contentent de décrire la loi conditionnelle de ces processus comme celle de carrés, ou de ponts de carrés de Bessel conditionnés par leur intégrale. Ils n'ont pas explicité le générateur des couples formés des temps locaux et de leur primitive sur les trois intervalles délimités par 0 et B_t, mais ils n'en avaient pas besoin pour obtenir les théorèmes limites qui étaient le but de leur article.

Pour justifier le lemme 1, ils ont fait les observations suivantes :

- l'application $(a, r) \longmapsto \tau_r^a$ de $\mathbf{R} \times \mathbf{R}_+^*$ dans $\mathbf{R}_+^*$ est injective et admet comme inverse à gauche l'application $t \mapsto (B_t, L_t^{B_t})$. Elle est "presque" bijective au sens où pour tout $t > 0$:

$$\tau^{B_t}_{L_t^{B_t}} = t \text{ presque sûrement}.$$

- On a l'identité en loi suivante :

$$(*) \qquad P[\tau_r^a \in dt]\, da\, dr = P[B_t \in da\,;\, L_t^{B_t} \in dr]\, dt$$

Ils ont ensuite affirmé que la loi du processus $(L_t^x)^{x\in\mathbf{R}}$ sachant que $(B_t, L_t^{B_t}) = (a, r)$ est égale à la loi du processus $(L_{\tau_r^a}^x)^{x\in\mathbf{R}}$ sachant que $\tau_r^a = t$. Comme il n'y a pas de notion intrinsèque de conditionnement par des événements de mesure nulle, cette affirmation nécessite une démonstration, et les arguments précédents ne suffisent pas.

Néanmoins, l'identité (1) que nous avons utilisée, et qui se trouve déjà dans [6] permet de donner une justification rigoureuse, en exprimant de deux manières différentes la désintégration de la mesure sur $\mathbf{R} \times \mathcal{C}_c(\mathbf{R}, \mathbf{R}_+)$:

$$\int_0^{+\infty} \mathcal{L}(P^t)\, dt = \int_{\mathbf{R}} \left(\int_0^{+\infty} \mathcal{L}(P^{\tau_r^a})\, dr \right) da$$

par rapport à la fonctionnelle :

$$(a, \ell) \longmapsto \left(a, \ell(a), \int_{\mathbf{R}} \ell(z)\, dz \right).$$

Signalons enfin que l'identité en loi $(*)$ de R. van der Hofstad, F. den Hollander et W. König peut être vue comme un corollaire de l'identité trajectorielle :

$$\mathbb{1}_{\{\tau_r^a \in dt\}}\, da\, dr = \mathbb{1}_{\{B_t \in da\,;\, L_t^{B_t} \in dr\}}\, dt,$$

que l'on démontre comme le lemme 5.

Bibliographie

[1] BIANE P., YOR M. — *Sur la loi des temps locaux pris en un temps exponentiel*, Sém. Prob. XXII, LNM 1321, Springer (1988), 454–466.

[2] EISENBAUM N. — *Un théorème de Ray-Knight lié au supremum des temps locaux browniens*, PTRF 87 (1990), 79–95.

[3] VAN DER HOFSTAD R., DEN HOLLANDER F., KÖNIG W. — *Central limit theorem for the Edwards model*, Annals of Probability 25 (2) (1997), 573–597.

[4] JEULIN T., YOR M. — *Grossissement de filtrations: exemples et applications*, LNM 1118, Springer, 1985.

[5] KNIGHT F. B. — *Random walks and a sojourn density process of brownian motion*, Trans. Am. Math. Soc. 109 (1963), 56–86.

[6] LEURIDAN C. — *Une démonstration élémentaire d'une identité de Biane et Yor*, Sém. Prob. XXX, LNM 1626, Springer (1996), 255–260.

[7] PERKINS E. — *Local time is a semimartingale*, Z.W. 60 (1982), 79–117.

[8] RAY D. B. — *Sojourn times of a diffusion process*, Ill. J. Math. 7 (1963), 615–630.

[9] REVUZ D., YOR M. — *Continuous martingales and brownian motion*, Springer, 1991.

[10] VALLOIS P. — *Une extension des théorèmes de Ray et Knight sur les temps locaux browniens*, PTRF 88, vol 4 (1991), 445–482.

– ◇ –

Christophe LEURIDAN
Université de Grenoble I
Institut Fourier
UMR 5582
UFR de Mathématiques
B.P. 74
38402 ST MARTIN D'HÈRES Cedex (France)
e-mail: Christophe.Leuridan@ujf-grenoble.fr

Point le plus visité par un mouvement brownien avec dérive.

Jean Bertoin et Laurence Marsalle
Laboratoire de Probabilités, Université Pierre et Marie Curie,
4 Place Jussieu, F-75252 Paris Cedex 05, France

1 Introduction et principaux résultats

Soit $(B_t, t \geq 0)$ un mouvement brownien réel, et $(\ell_t^x, t \geq 0, x \in \mathbb{R})$ la famille bicontinue de ses temps locaux. Le processus $(\overline{V}_t, t \geq 0)$ défini pour tout $t \geq 0$ par

$$\overline{V}_t := \inf\{x \geq 0 : \ell_t^x \vee \ell_t^{-x} = \sup\{\ell_t^y : y \in \mathbb{R}\}\}$$

représente, en valeur absolue, le point le plus visité par B sur l'intervalle de temps $[0, t]$. Bass et Griffin [1] ont montré qu'il était transient, et ont également obtenu une loi du logarithme itéré

$$\limsup_{t \to +\infty} \frac{\overline{V}_t}{(2t \log\log t)^{1/2}} = 1 \ p.s.$$

Voir également Eisenbaum [3] et Leuridan [4] pour des travaux proches.

Nous nous intéressons ici au point le plus visité par un mouvement brownien réel avec coefficient de dérive $\mu > 0$, $(X_t = B_t + \mu t, t \geq 0)$. Considérons le processus continu des densités d'occupation $(\ell^x, x \in \mathbb{R})$, où

$$\ell^x := \lim_{\varepsilon \to 0+} \frac{1}{\varepsilon} \int_0^{+\infty} \mathbb{1}_{\{x \leq X_t \leq x+\varepsilon\}} dt,$$

puis définissons le point le plus visité par X sur l'intervalle d'espace $[0, y]$

$$V(y) := \inf\{x \geq 0 : \ell^x = m_y\}, \quad \text{avec } m_y = \max\{\ell^x : 0 \leq x \leq y\}.$$

Notons encore que, comme $(\ell^x, x \leq 0)$ est borné et $\sup\{\ell^x : x \geq 0\} = +\infty$, $V(y)$ est également le point de $(-\infty, y]$ le plus visité par X, pourvu que y soit assez grand. L'objet principal de ce travail est d'établir les deux résultats asymptotiques suivants.

Proposition 1.1
$V(y)/y$ converge en distribution vers la loi uniforme sur $[0, 1]$ quand y tend vers $+\infty$.

Théorème 1.2
Soit $f : (0, +\infty) \to (0, +\infty)$ *une fonction croissante. Alors*

$$\liminf_{y \to +\infty} \frac{V(y)f(y)}{y} = 0 \quad ou \quad +\infty \quad p.s.$$

selon que l'intégrale $\int_1^{+\infty} \frac{dt}{tf(t)}$ *diverge ou converge.*

Remarque : Il est facile de voir que $V(y) \leq y$ pour tous les $y \geq 0$, et que $V(y) = y$ pour certains y arbitrairement grands. En effet, si H_n désigne le temps d'atteinte par $(\ell^x, x \geq 0)$ du niveau n, alors pour tout $n \geq 1$, $V(H_n) = H_n$. C'est la raison pour laquelle on ne regarde que le comportement de la limite inférieure.

Ce papier est organisé de la façon suivante. La deuxième partie est consacrée à l'étude de $(\ell^x, x \geq 0)$, qui, d'après un théorème de Ray-Knight, est une diffusion récurrente positive. La théorie des excursions d'Itô ramène alors l'étude de $V(y)$ à celle des temps de record d'un processus de Poisson ponctuel. Dans la troisième partie, nous précisons le comportement asymptotique du point où un processus de Poisson ponctuel réalise son maximum, ce qui nous permet dans la quatrième partie de prouver les résultats énoncés ci-dessus. Enfin, la cinquième partie contient quelques résultats supplémentaires sur le comportement asymptotique du point le moins visité par X, ainsi que sur le point le plus visité par X sur l'intervalle de temps $[0, t]$, qui est défini de façon analogue à celle de Bass et Griffin pour le mouvement brownien.

2 Préliminaires

2.1 Processus des densités d'occupation

Nous commençons par rappeler que, d'après un théorème de type Ray-Knight, $(\ell^x, x \geq 0)$ a la loi du carré d'un processus d'Ornstein-Uhlenbeck plan de paramètre μ (voir par exemple Borodin et Salminen [2], page 78). Le processus $(\ell^x, x \geq 0)$ est en particulier une diffusion ergodique, le point 1 est régulier et récurrent. Le processus markovien $(\ell^x, x \geq 0)$ admet donc un temps local en 1.
Nous savons d'autre part que le carré d'un processus d'Ornstein-Uhlenbeck plan de paramètre μ admet pour probabilité invariante la loi exponentielle de paramètre μ, qu'on note m. Nous notons L le temps local en 1 normalisé par $\mathbb{E}_m[L_1] = 1$, où $\mathbb{E}_m$ désigne la loi du carré d'un processus d'Ornstein-Uhlenbeck plan sous la loi initiale m. Cette normalisation entraîne que la mesure associée à la fonctionnelle additive L est la masse de Dirac en 1 (cf. Revuz et Yor [7], chapitre X, section 2), et le théorème ergodique pour les fonctionnelles additives (cf. Revuz et Yor [7], chapitre X, section 3) donne $\lim_{t \to +\infty} L_t/t = 1$ *p.s.*

Au temps local L on associe son inverse continu à droite

$$\lambda_u := \inf\{t > 0 : L_t > u\}, \qquad u \geq 0.$$

On a immédiatement

$$\lim_{t\to+\infty} \lambda_t/t = 1 \ p.s. \tag{1}$$

Pour simplifier, nous travaillerons conditionnellement à $\ell^0 = 1$, et étudierons les excursions du processus des temps locaux en dehors de 1. Ceci se généralise *mutis mutandi* à $\ell^0 = y$, pour $y > 0$, en considérant les excursions en dehors de y. Rappelons d'abord la définition du processus des excursions de $(\ell^x, x \geq 0)$ en dehors de 1. On désigne par $\mathcal{U}$ l'ensemble des fonctions continues $u : [0, +\infty) \to \mathbb{R}$ telles que

$$0 < H(u) < +\infty \quad \text{et} \quad u(t) = 1 \quad \text{pour tout } t \geq H(u),$$

où $H(u) := \inf\{t > 0 : u(t) = 1\}$. Par abus, on note 1 la fonction identiquement égale à 1. Pour tout $t \geq 0$, on pose

$$e_t = \begin{cases} 1 & \text{si } \lambda_{t-} = \lambda_t \\ (\ell^{\lambda_{t-}+s}, 0 \leq s < \lambda_t - \lambda_{t-}) & \text{si } \lambda_t > \lambda_{t-}. \end{cases}$$

D'après la théorie des excursions d'Itô (voir par exemple Rogers et Williams [8], chapitre VI, section 8), nous savons que le processus $(e_t, t \geq 0)$ est un processus de Poisson ponctuel à valeurs dans $\mathcal{U} \cup \{1\}$. Notons n sa mesure caractéristique (mesure d'Itô).

Soit maintenant $x > 1$, nous considérons

$$\mathcal{U}^x := \{u \in \mathcal{U} : \sup_{t\geq 0} u(t) > x\},$$

l'ensemble des excursions de hauteur supérieure à x. On sait alors qu'il existe une constante $c > 0$ telle que

$$n(\mathcal{U}^x) = \frac{c}{s(x)},$$

où s désigne une fonction d'échelle du processus $(\ell^x, x \geq 0)$ satisfaisant $s(1) = 0$ (cf. Pitman et Yor [5], section 3). La fonction d'échelle étant continue, et comme $s(+\infty) = +\infty$, il existe une hauteur $h > 1$ telle que $n(\mathcal{U}^h) = 1$.

Pour tout $t \geq 0$, on définit alors γ_t de la façon suivante :

- Si $\lambda_{t-} = \lambda_t$, alors $\gamma_t = 0$.
- Si $\lambda_t > \lambda_{t-}$, alors

$$\gamma_t = \begin{cases} 0 & \text{si } e_t \notin \mathcal{U}^h \\ \sup\{e_t(s) : 0 \leq s < \lambda_t - \lambda_{t-}\} & \text{sinon.} \end{cases}$$

On déduit immédiatement le lemme suivant.

Lemme 2.1
Le processus $(\gamma_t, t \geq 0)$ *est un processus de Poisson ponctuel discret à valeurs dans* $[0, +\infty)$, *dont la mesure caractéristique est diffuse et a une masse totale égale à* 1.

2.2 Temps de record d'un processus de Poisson ponctuel

Soit $(Y, \mathbb{P})$ un processus de Poisson ponctuel discret à valeurs dans $[0, +\infty)$, de mesure caractéristique une loi de probabilité ν qu'on suppose diffuse. Le point 0 est pris comme point isolé. On note $(\mathcal{F}_t, t \geq 0)$ la filtration naturelle du processus Y et on définit l'opérateur de translation θ de la façon usuelle : pour tout $t \geq 0$, pour tout $s \geq 0$,

$$(\theta_t Y)_s := Y_{t+s}.$$

On introduit également la loi du processus issu de x, $\mathbb{P}^x$, définie comme la loi Y^x, où

$$Y^x := \begin{cases} x & \text{si } t = 0 \\ Y_t & \text{si } t > 0. \end{cases}$$

Ceci nous permet d'énoncer la propriété forte de Markov sous la forme suivante. Soit T un $(\mathcal{F}_t)$-temps d'arrêt, H une fonctionnelle du processus Y, alors

$$\mathbb{E}(H \circ \theta_T \mid \mathcal{F}_T) = \mathbb{E}^{Y_T}(H).$$

Notre but est d'obtenir des renseignements sur le comportement asymptotique de l'instant en lequel Y réalise son maximum sur l'intervalle $[0, t]$

$$G_t = \inf\{s \geq 0 : M_s = M_t\}, \qquad \text{avec } M_t = \max\{Y_s : 0 \leq s \leq t\}. \tag{2}$$

Afin d'obtenir des résultats trajectoriels, nous introduisons la notion de *temps de record* du processus Y. On dit que le processus Y présente un *record* à l'instant t si Y_t dépasse toutes les valeurs précédentes du processus, i.e. $Y_t > M_{t-}$. Nous notons $T = (T_n, n \geq 0)$ la suite des *temps de record* de Y, définie par $T_0 = 0$ et pour $n \geq 1$ par

$$T_n = \inf\{t > T_{n-1} : Y_t > M_{T_{n-1}}\}.$$

L'intérêt de cette notion pour l'étude de G_t repose sur le fait que $G_t = T_n$ pour tout $t \in [T_n, T_{n+1})$.

Il est bien connu que la loi de la suite des temps de record ne dépend pas de la loi caractéristique du processus Y. Plus précisément, si Y' est un second processus de Poisson ponctuel dont la mesure caractéristique est également une loi de probabilité diffuse sur $[0, +\infty)$, alors la suite des temps de record de Y' a même loi que celle de Y.

Pour étudier les temps de record du processus Y, nous pouvons donc nous ramener au cas où

$$\nu(dx) = \mathrm{e}^{-x} \mathbb{1}_{\{x \geq 0\}} dx.$$

Le choix de la loi exponentielle est motivée par la propriété d'absence de mémoire de cette loi, ce qui va nous permettre de préciser la structure des records (voir les propositions 4.1 et 4.8 dans Resnick [6] pour des résultats très proches). Plus précisément, introduisons

$$R_n := Y_{T_n}, \qquad n \geq 0$$

la suite des records de Y, ainsi que celle des inter-records

$$\Delta_n = T_{n+1} - T_n , \qquad n \geq 0.$$

Notons que $T_n = \Delta_0 + \cdots + \Delta_{n-1}$.

Proposition 2.2
(i) La suite $(R_n, n \geq 0)$ est une marche aléatoire sur $\mathbb{R}_+$, de pas la loi exponentielle de paramètre 1.
(ii) Conditionnellement à $R_0, R_1, \ldots, R_n$, les variables aléatoires $\Delta_0, \Delta_1, \ldots, \Delta_n$ sont indépendantes et suivent des lois exponentielles de paramètres respectifs $1, e^{-R_1}, \ldots, e^{-R_n}$.

Preuve : Elle repose essentiellement sur la propriété forte de Markov que nous allons appliquer au $(\mathcal{F}_t)$-temps d'arrêt T_n. Ecrivons d'abord

$$T_{n+1} = T_n + T_1 \circ \theta_{T_n},$$

où nous rappelons que $T_1 = \inf\{t > 0 : Y_t > Y_0\}$. Ceci entraîne les deux identités suivantes

$$R_{n+1} - R_n = (Y_{T_1} - Y_0) \circ \theta_{T_n} \quad , \quad \Delta_n = T_1 \circ \theta_{T_n}.$$

On déduit de la propriété de Markov des processus de Poisson ponctuels que, si on pose $Y_{T_n} = y$, alors

$$\mathbb{P}(R_{n+1} - R_n \in dx,\ \Delta_n \in dt \mid \mathcal{F}_{T_n}) = \mathbb{P}^y(R_1 - y \in dx,\ T_1 \in dt) = \mathbb{P}^y(R_1 - y \in dx)\mathbb{P}^y(T_1 \in dt)$$

(en utilisant pour la dernière égalité le fait que l'instant du premier saut d'un processus de Poisson ponctuel est indépendant de la valeur de ce saut).
Or nous savons (voir par exemple Revuz et Yor [7], chapitre XII, section 1) d'une part que

$$\mathbb{P}^y(R_1 - y \in dx) = e^{-x} \mathbb{1}_{\{x>0\}} dx,$$

et d'autre part que

$$\mathbb{P}^y(T_1 \in dt) = e^{-y} \exp\{-e^{-y} t\} \mathbb{1}_{\{t \geq 0\}} dt,$$

et on peut alors compléter la preuve par récurrence. □

Corollaire 2.3
La suite des inter-records $(\Delta_n, n \geq 0)$ admet la représentation

$$\Delta_n = e^{R_n} Z_n , \qquad n \geq 0,$$

où $(Z_n, n \geq 0)$ est une suite de variables i.i.d., de loi exponentielle de paramètre 1, et indépendante de la suite $(R_n, n \geq 0)$.

Preuve : Posons $Z_n = e^{-R_n}\Delta_n$ pour tout $n \geq 0$. D'après la proposition 2.2 (ii), nous savons que conditionnellement à $R_1, \ldots, R_n$, les variables $Z_1, \ldots, Z_n$ sont indépendantes, et de plus nous pouvons calculer leur loi conditionnelle : soit h borélienne, i compris entre 1 et n fixé,

$$\begin{aligned} \mathbb{E}[h(Z_i) \mid R_1, \ldots, R_n] &= \mathbb{E}[h(e^{-R_i}\Delta_i) \mid R_1, \ldots, R_n] \\ &= \int_0^{+\infty} h(e^{-R_i}x)e^{-R_i}\exp\{-xe^{-R_i}\}dx \\ &= \int_0^{+\infty} h(u)e^{-u}du. \end{aligned}$$

Le corollaire est établi. □

3 Comportement asymptotique du dernier temps de record

Rappelons que G_t, le dernier temps de record sur l'intervalle de temps $[0, t]$, a été défini par l'équation (2). On connaît explicitement la loi de G_t. Si $M_t = 0$ (évènement de probabilité e^{-t}), alors $G_t = 0$, et conditionnellement à $M_t > 0$, G_t suit une loi uniforme sur $[0, t]$. On a donc

$$\mathbb{P}(G_t/t \leq u) = (1 - e^{-t})u + e^{-t}, \qquad 0 \leq u \leq 1, \tag{3}$$

ce qui entraîne immédiatement le lemme suivant.

Lemme 3.1
G_t/t converge en distribution vers la loi uniforme sur $[0, 1]$ quand t tend vers $+\infty$.

Les résultats sur les temps de record nous permettent d'étudier le comportement trajectoriel de G_t; nous commençons par établir un lemme technique.

Lemme 3.2
Pour tout $\beta > 2$, on a $\mathbb{P}(T_{n+1} \leq \beta e^{R_n}) \geq (1 - 2/\beta)$.

Preuve : Elle repose sur l'inégalité de Markov. Nous avons, grâce à la proposition 2.2 et au corollaire 2.3,

$$\mathbb{E}[T_{n+1}e^{-R_n}] = \sum_0^n \mathbb{E}[\Delta_k e^{-R_n}] = \sum_0^n \mathbb{E}[e^{R_k - R_n} Z_k] = \sum_0^n \mathbb{E}[\exp(-R_1)]^{n-k} = \sum_0^n 2^{-k}.$$

Il suffit alors d'écrire que

$$\mathbb{P}(T_{n+1}e^{-R_n} \geq \beta) \leq (1/\beta)\mathbb{E}[T_{n+1}e^{-R_n}] \leq 2/\beta,$$

et le lemme est prouvé. □

Dans le but d'obtenir un test intégral pour G_t, nous allons considérer la suite des évènements

$$\Lambda_n^{(\beta)} = \{T_{n+1} \leq \beta e^{R_n}, \Delta_{n+1} \geq \beta f(e^{2n}) e^{R_n}\}, \qquad n \geq 0$$

où $f : (0, +\infty) \to (0, +\infty)$ est une fonction croissante et $\beta > 2$.

Lemme 3.3
Si l'intégrale $\int_1^{+\infty} \frac{dt}{tf(t)}$ *diverge, alors pour tout* $\beta > 2$,

$$\mathbb{P}(\limsup_{n\to+\infty} \Lambda_n^{(\beta)}) \geq (1 - 2/\beta)^2.$$

Preuve : Nous allons utiliser une version généralisée du lemme de Borel-Cantelli (cf. Spitzer [9] page 317); nous commençons donc par évaluer $\mathbb{P}(\Lambda_n^{(\beta)})$. Nous utilisons le corollaire 2.3 pour écrire

$$\begin{aligned}
\mathbb{P}(\Lambda_n^{(\beta)}) &= \mathbb{P}(T_{n+1} \leq \beta e^{R_n}, e^{R_{n+1}} Z_{n+1} \geq \beta f(e^{2n}) e^{R_n}) \\
&= \mathbb{P}(T_{n+1} \leq \beta e^{R_n}, e^{R_{n+1} - R_n} Z_{n+1} \geq \beta f(e^{2n})) \\
&= \mathbb{P}(T_{n+1} \leq \beta e^{R_n}) \mathbb{P}(e^{R_1} Z_1 \geq \beta f(e^{2n})).
\end{aligned}$$

Le lemme 3.1 nous permet de minorer le premier terme du produit, regardons le second. On a

$$\begin{aligned}
\mathbb{P}(e^{R_1} Z_1 \geq \beta f(e^{2n})) &= \mathbb{P}(R_1 \geq \log \beta + \log(f(e^{2n})) - \log(Z_1)) \\
&= \int_0^{+\infty} \exp\{\log x - \log \beta - \log(f(e^{2n}))\} e^{-x} dx \\
&= \frac{1}{\beta f(e^{2n})} \int_0^{+\infty} x e^{-x} dx \\
&= \frac{1}{\beta f(e^{2n})}.
\end{aligned}$$

On a donc obtenu la minoration

$$\mathbb{P}(\Lambda_n^{(\beta)}) \geq \frac{(1 - 2/\beta)}{\beta f(e^{2n})}. \tag{4}$$

On vérifie ensuite que la série de terme général $1/f(e^{2n})$ et l'intégrale $\int^{+\infty} dt/tf(t)$ sont de même nature. Ceci découle simplement du fait que la fonction f est croissante, donc

$$\frac{(1 - 1/e)}{f(e^{2(k+1)})} \leq \int_{e^k}^{e^{k+1}} \frac{dt}{tf(t^2)} \leq \frac{(e - 1)}{f(e^{2k})},$$

et du fait que les intégrales $\int^{+\infty} dt/tf(t)$ et $\int^{+\infty} dt/tf(t^2)$ sont de même nature. En conclusion, la série $\sum_0^{+\infty} \mathbb{P}(\Lambda_n^{(\beta)})$ diverge.

Pour pouvoir utiliser la version généralisée du lemme de Borel-Cantelli, nous allons montrer que pour $|n - m| \geq 2$,

$$\mathbb{P}(\Lambda_n^{(\beta)} \cap \Lambda_m^{(\beta)}) \leq (\beta/(\beta - 2))^2 \mathbb{P}(\Lambda_n^{(\beta)}) \mathbb{P}(\Lambda_m^{(\beta)}).$$

Pour cela, on écrit d'abord

$$\begin{aligned}
\mathbb{P}(\Lambda_n^{(\beta)} \cap \Lambda_m^{(\beta)}) &= \mathbb{P}\left(T_{n+1} \le \beta e^{R_n}, T_{m+1} \le \beta e^{R_m}, \Delta_{n+1} \ge \beta f(e^{2n}) e^{R_n}, \Delta_{m+1} \ge \beta f(e^{2m}) e^{R_m}\right) \\
&\le \mathbb{P}\left(\Delta_{n+1} \ge \beta f(e^{2n}) e^{R_n}, \Delta_{m+1} \ge \beta f(e^{2m}) e^{R_m}\right) \\
&\le \mathbb{P}\left(e^{(R_{n+1}-R_n)} Z_{n+1} \ge \beta f(e^{2n})\right) \mathbb{P}\left(e^{(R_{m+1}-R_m)} Z_{m+1} \ge \beta f(e^{2m})\right),
\end{aligned}$$

car la condition d'écartement sur n et m nous assure l'indépendance des deux évènements considérés. Leurs probabilités ont déjà été calculées ci-dessus, ce qui entraîne

$$\mathbb{P}(\Lambda_n^{(\beta)} \cap \Lambda_m^{(\beta)}) \le \frac{1}{\beta f(e^{2n})} \times \frac{1}{\beta f(e^{2m})},$$

et pour conclure on utilise le fait que, d'après (4), pour tout $n \ge 0$

$$\frac{1}{\beta f(e^{2n})} \le (\beta/(\beta-2)) \mathbb{P}(\Lambda_n^{(\beta)}),$$

ce qui donne la majoration voulue. Le lemme de Borel-Cantelli généralisé complète la preuve.

□

Nous sommes maintenant en mesure d'énoncer et de prouver un test intégral sur le comportement asymptotique de G_t.

Théorème 3.4
Soit $f : (0, +\infty) \to (0, +\infty)$ une fonction croissante. Alors

$$\liminf_{t \to +\infty} \frac{G_t f(t)}{t} = 0 \quad ou \quad +\infty \ p.s.$$

selon que l'intégrale $\displaystyle\int_1^{+\infty} \frac{dt}{t f(t)}$ *diverge ou converge.*

Preuve : Nous commençons par supposer que l'intégrale $\int_1^{+\infty} dt/tf(t)$ converge. Notre but est d'appliquer le lemme de Borel-Cantelli. Par commodité, nous introduisons $g : (0, +\infty) \to (0, +\infty)$, définie par $g(t) = tf(t)$. La fonction g étant strictement croissante, nous lui associons sa fonction réciproque, φ. Grâce à (3), nous avons

$$\begin{aligned}
\mathbb{P}\left(G_{2^n} f(G_{2^n}) \le 2^{n+1}\right) &= \mathbb{P}\left(G_{2^n} \le \varphi(2^{n+1})\right) \\
&\le 2^{-n}(1 - e^{-2^n}) \varphi(2^{n+1}) + e^{-2^n} \\
&\le 2^{-n} \varphi(2^{n+1}) + e^{-2^n}.
\end{aligned}$$

On note ensuite que

$$\sum_0^{+\infty} \frac{\varphi(2^{n+1})}{2^n} = 8 \sum_0^{+\infty} \frac{\varphi(2^{n+1})}{2^{n+3}} \le 8 \sum_0^{+\infty} \int_{2^{n+1}}^{2^{n+2}} \frac{\varphi(t)}{t^2} dt = 8 \int_2^{+\infty} \frac{\varphi(t)}{t^2} dt.$$

Etudions la convergence de l'intégrale de droite :

$$\int^{+\infty} \frac{\varphi(t)}{t^2} dt = \int^{+\infty} \frac{u}{g(u)^2} dg(u) = \int^{+\infty} \frac{du}{uf(u)} + \int^{+\infty} \frac{df(u)}{f(u)^2}.$$

Le premier terme de la somme est fini par hypothèse. Le second l'est également, car, f étant croissante et l'intégrale $\int_1^{+\infty} du/uf(u)$ convergente, on a nécessairement $\lim_{+\infty} f = +\infty$. On en déduit la convergence de la série $\sum_0^{+\infty}(2^{-n}\varphi(2^{n+1}) + e^{-2^n})$, le lemme de Borel-Cantelli entraîne qu'avec probabilité 1 il existe n_0 tel que $G_{2^n} f(G_{2^n}) > 2^{n+1}$ pour tout $n \geq n_0$.

Un argument standard de monotonie entraîne alors $\liminf_{t\to+\infty} G_t f(t)/t \geq 1$. Comme le résultat du test intégral est inchangé quand on remplace f par εf, où ε est un réel positif arbitraire, on conclut que

$$\liminf_{t\to+\infty} \frac{G_t f(t)}{t} = +\infty.$$

On suppose maintenant que l'intégrale $\int_1^{+\infty} dt/tf(t)$ diverge. On introduit $\tilde{f}(x) = f(x^2)$. La fonction $\tilde{f}$ est croissante et l'intégrale $\int_1^{+\infty} dt/t\tilde{f}(t)$ diverge. Posons

$$\Lambda_n^{(\beta)}(\tilde{f}) = \left\{T_{n+1} \leq \beta e^{R_n},\ \Delta_{n+1} \geq \beta \tilde{f}(e^{2n}) e^{R_n}\right\};$$

nous allons montrer que pour tout $\beta > 2$ fixé,

$$\limsup_{n\to+\infty} \Lambda_n^{(\beta)}(\tilde{f}) \subset \left\{\liminf_{t\to+\infty} \frac{G_t f(t)}{t} \leq 1\right\}. \tag{5}$$

Nous travaillons avec ω fixé appartenant à $\limsup_{+\infty} \Lambda_n^{(\beta)}(\tilde{f})$. Il existe donc une suite $(n_k, k \geq 0)$ d'entiers tendant vers $+\infty$ telle que pour tout $k \geq 0$:

$$T_{n_k+1} \leq \beta e^{R_{n_k}},\ \ \Delta_{n_k+1} \geq \beta \tilde{f}(e^{2n_k}) e^{R_{n_k}}.$$

D'après la loi forte des grands nombres et la proposition 2.2 (i), $R_n/n \to 1$ *p.s.* et nous pouvons supposer que $\beta e^{R_{n_k}} \leq e^{2n_k}$ pour tout $k \geq 0$.
Nous en déduisons la série d'inégalités suivantes

$$T_{n_k+2} > \Delta_{n_k+1} \geq \beta \tilde{f}(e^{2n_k}) e^{R_{n_k}} \geq \beta \tilde{f}(\beta e^{R_{n_k}}) e^{R_{n_k}} \geq \tilde{f}(T_{n_k+1}) T_{n_k+1}.$$

Ensuite, on remarque que pour tout $t \in [T_{n_k+1}, T_{n_k+2}[$, $G_t = T_{n_k+1}$, et *a fortiori* $G_t\tilde{f}(G_t) < t$, quitte à prendre t suffisamment proche de T_{n_k+2}. Plus précisément, pour tout $k \geq 0$, il existe un réel $t_k \in [T_{n_k+1}, T_{n_k+2}[$ tel que

$$G_{t_k} \tilde{f}(G_{t_k}) \leq t_k.$$

Mais d'après la première partie de la preuve, nous savons que $\lim_{t\to+\infty} G_t/\sqrt{t} = +\infty$ *p.s.* Il existe donc k_0 tel que pour tout $k \geq k_0$,

$$G_{t_k} \geq \sqrt{t_k} \quad \text{et en conséquence} \quad \tilde{f}(G_{t_k}) \geq \tilde{f}(\sqrt{t_k}) = f(t_k).$$

Ainsi $G_{t_k} f(t_k) \leq t_k$ pour tout $k \geq k_0$, ce qui établit (5).

En combinant (5), le lemme 3.3 et le fait que β peut être choisi arbitrairement grand, on obtient que $\liminf_{t\to+\infty} G_t f(t)/t \leq 1$ *p.s.* Le résultat du test intégral n'étant pas modifié lorsqu'on remplace la fonction f par f/ε, où $\varepsilon > 0$ est un réel arbitraire, on en déduit que $\liminf_{t\to+\infty} G_t f(t)/t = 0$ *p.s.* Le théorème est prouvé. □

4 Preuve des résultats

Nous allons appliquer les résultats obtenus dans la partie 3 au processus de Poisson ponctuel $Y = \gamma$ introduit dans la section 2.1. Nous rappelons que, pour simplifier, on travaille conditionnellement à $\ell^0 = 1$, et que λ désigne l'inverse de L, le temps local au niveau 1 de ℓ.

Lemme 4.1
Sur $\{G_y \neq 0\}$, *nous avons l'encadrement*

$$\lambda_{G_y -} \leq V(\lambda_y) \leq \lambda_{G_y}.$$

Preuve : Regardons l'information que donne G_y.
• Si $G_y = 0$, alors le processus $(\ell^x, x \geq 0)$ est resté sous le niveau h sur l'intervalle $[0, \lambda_y]$.
• Si $G_y = x > 0$, alors le supremum de $\{\ell^v : 0 \leq v \leq \lambda_y\}$ est réalisé sur l'intervalle de temps $[\lambda_{x-}, \lambda_x]$, ce qui revient exactement à dire que

$$\lambda_{G_y -} \leq V(\lambda_y) \leq \lambda_{G_y},$$

le lemme est donc prouvé. □

Par la suite, nous utiliserons implicitement le fait que $\lim_{y\to+\infty} G_y = +\infty$ *p.s.*, de sorte que l'hypothèse du lemme 4.1 est vérifiée dès que y est suffisamment grand. Nous allons maintenant nous servir du fait que $\lambda_t \sim t$ quand $t \to +\infty$, et de l'encadrement du lemme 4.1 pour montrer la proposition 1.1.

Preuve de la proposition 1.1 : Lorsque $t \to +\infty$, $G_t \to +\infty$ et $\lambda_t/t \to 1$ (d'après (1)), ce qui entraîne la convergence en probabilité de λ_{G_t}/G_t et $\lambda_{G_t -}/G_t$ vers 1. Pour tout $\varepsilon > 0$, il existe t_0 tel que pour tout $t \geq t_0$

$$\mathbb{P}(|\lambda_t/t - 1| > \varepsilon) \leq \varepsilon, \quad \mathbb{P}(|\lambda_{G_t}/G_t - 1| > \varepsilon) \leq \varepsilon \quad \text{et} \quad \mathbb{P}(|\lambda_{G_t -}/G_t - 1| > \varepsilon) \leq \varepsilon.$$

Par ailleurs, nous savons que $\mathbb{P}(G_t = 0) = e^{-t}$, nous pouvons donc supposer que pour tout $t \geq t_0$, $\mathbb{P}(G_t = 0) \leq \varepsilon$. Ceci nous permet alors d'écrire l'inégalité suivante : pour tout $\beta \in [0,1]$, pour tout $t \geq t_0$,

$$\mathbb{P}(V(t)/t \leq \beta) \leq \mathbb{P}(\{V(t)/t \leq \beta\} \cap \{|\lambda_t/t - 1| \leq \varepsilon\} \cap \{|\lambda_{G_t-}/G_t - 1| \leq \varepsilon\} \cap \{G_t \neq 0\}) + 3\varepsilon.$$

Fixons $\omega \in \{|\lambda_t/t - 1| \leq \varepsilon\} \cap \{|\lambda_{G_t-}/G_t - 1| \leq \varepsilon\} \cap \{G_t \neq 0\}$. Alors

$$V((1+\varepsilon)t) \geq V(\lambda_t) \geq \lambda_{G_t-} \geq (1-\varepsilon)G_t,$$

ce qui donne, si on ne garde plus que les deux extrémités des inégalités

$$\frac{V((1+\varepsilon)t)}{(1+\varepsilon)t} \geq \frac{(1-\varepsilon)}{(1+\varepsilon)} \frac{G_t}{t}.$$

On pose alors $u = (1+\varepsilon)t$, donc pour tout $u \geq (1+\varepsilon)t_0$

$$\frac{V(u)}{u} \geq (1-\varepsilon) \frac{G_{u/(1+\varepsilon)}}{u}.$$

Ainsi, pour tout $\beta \in [0,1]$

$$\mathbb{P}(V(u)/u \leq \beta) \leq \mathbb{P}\left(\frac{G_{u/(1+\varepsilon)}}{u/(1+\varepsilon)} \leq \beta \frac{(1+\varepsilon)}{(1-\varepsilon)} \right) + 3\varepsilon,$$

ce qui entraîne d'après le lemme 3.1 que pour tout $\beta \in [0,1]$, pour tout $\varepsilon > 0$,

$$\limsup_{u \to +\infty} \mathbb{P}(V(u)/u \leq \beta) \leq \beta \frac{(1+\varepsilon)}{(1-\varepsilon)} + 3\varepsilon,$$

et finalement $\limsup_{u \to +\infty} \mathbb{P}(V(u)/u \leq \beta) \leq \beta$.
Pour obtenir la minoration de la limite inférieure, il suffit de procéder de manière analogue en considérant cette fois-ci $\mathbb{P}(V(u)/u > \beta)$. Ceci termine la preuve de la proposition. □

Preuve du théorème 1.2 : Nous commençons par supposer que l'intégrale $\int_1^{+\infty} dt/tf(t)$ diverge, ce qui équivaut à la divergence de $\int_1^{+\infty} dt/tf(t/2)$. Le théorème 3.4 donne

$$\mathbb{P}\left(\liminf_{t \to +\infty} \frac{G_t f(t/2)}{t} = 0 \right) = 1.$$

On fixe alors ω dans $\{\lim_{t \to +\infty} \lambda_t/t = 1\} \cap \{\liminf_{t \to +\infty} G_t f(t/2)/t = 0\}$, évènement de probabilité 1. Pour tout $\eta > 0$, on peut choisir t arbitrairement grand tel que

$$\frac{G_t f(t/2)}{t} \leq \eta \quad , \quad |\frac{\lambda_t}{t} - 1| \leq 1/2 \quad \text{et} \quad |\frac{\lambda_{G_t}}{G_t} - 1| \leq 1.$$

Grâce au lemme 4.1, ceci entraîne

$$V(t/2) \leq V(\lambda_t) \leq \lambda_{G_t} \leq 2G_t,$$

et donc

$$\frac{V(t/2)f(t/2)}{t/2} \leq 4\frac{G_t f(t/2)}{t} \leq 4\eta.$$

Nous avons ainsi établi que $\liminf_{t\to+\infty} V(t)f(t)/t = 0$ avec probabilité 1.

On suppose ensuite que l'intégrale $\int_1^{+\infty} dt/tf(t)$ converge. La convergence de $\int_1^{+\infty} dt/tf(2t)$ en découle, et ainsi, d'après le théorème 3.4,

$$\mathbb{P}\left(\lim_{t\to+\infty} \frac{G_t f(2t)}{t} = +\infty\right) = 1.$$

Comme précédemment, on fixe $\omega \in \{\lim_{t\to+\infty} G_t f(2t)/t = +\infty\}$. Alors pour tout $A > 0$ et tout t suffisamment grand

$$\frac{G_t f(2t)}{t} \geq A, \ |\frac{\lambda_t}{t} - 1| \leq 1 \ \text{ et } \ |\frac{\lambda_{G_t -}}{G_{t_k}} - 1| \leq 1/2.$$

En conséquence $V(2t) \geq V(\lambda_t) \geq \lambda_{G_t-} \geq (1/2)G_t$, et donc $V(2t)f(2t)/t \geq A/2$; la preuve du théorème est complète. □

5 Quelques compléments

5.1 Point le moins visité

Si on définit le point le moins visité par X sur $[0, y]$,

$$v(y) := \inf\{x \geq 0 : \ell^x = \min_{0\leq u\leq y} \ell^u\},$$

nous pouvons adapter la méthode utilisée pour l'étude de $V(y)$ à celle de $v(y)$. On considère les excursions de $(\ell^x, x \geq 0)$ en-dessous de 1, il existe une hauteur $h' \in (0, 1)$ telle que

$$n(u \in \mathcal{U} : \inf_{t\geq 0} u(t) < h') = 1.$$

On définit un nouveau processus de Poisson ponctuel, γ', de façon analogue à γ. Nous pouvons alors appliquer les résultats de la partie 3 à γ', puis passer à $v(y)$.
D'autre part, les processus des excursions au-dessus et en-dessous de 1 étant indépendants, γ et γ' sont indépendants. Ceci nous permet de renforcer la proposition 1.1 et le théorème 1.2 de la façon suivante.

Proposition 5.1
Le couple $(V(y)/y, v(y)/y)$ converge en distribution vers le produit de deux lois uniformes sur $[0, 1]$ quand y tend vers $+\infty$.

Pour énoncer la seconde proposition, nous adoptons les notations suivantes :
Si $f:(0,+\infty)\to(0,+\infty)$ et $g:(0,+\infty)\to(0,+\infty)$ sont deux fonctions croissantes,

$$\mathbf{C}=\left\{\int_1^{+\infty} dt/tf(t) \text{ converge}\right\} \quad , \quad \mathbf{D}=\left\{\int_1^{+\infty} dt/tf(t) \text{ diverge}\right\},$$

$$\mathbf{c}=\left\{\int_1^{+\infty} dt/tg(t) \text{ converge}\right\} \quad , \quad \mathbf{d}=\left\{\int_1^{+\infty} dt/tg(t) \text{ diverge}\right\}.$$

Proposition 5.2
On a p.s.

$$\bullet\mathbf{Cc} \Longrightarrow \lim_{y\to+\infty}\frac{V(y)f(y)}{y}=+\infty \quad et \quad \lim_{y\to+\infty}\frac{v(y)g(y)}{y}=+\infty.$$

$$\bullet\mathbf{Cd} \Longrightarrow \lim_{y\to+\infty}\frac{V(y)f(y)}{y}=+\infty \quad et \quad \liminf_{y\to+\infty}\frac{v(y)g(y)}{y}=0.$$

$$\bullet\mathbf{Dc} \Longrightarrow \liminf_{y\to+\infty}\frac{V(y)f(y)}{y}=0 \quad et \quad \lim_{y\to+\infty}\frac{v(y)g(y)}{y}=+\infty.$$

$$\bullet\mathbf{Dd} \Longrightarrow \liminf_{y\to+\infty}\frac{V(y)f(y)}{y}=0 \quad et \quad \liminf_{y\to+\infty}\frac{v(y)g(y)}{y}=0.$$

5.2 Point le plus visité sur l'intervalle de temps $[0,t]$

Les résultats asymptotiques obtenus pour $V(y)$, point de l'intervalle d'espace $[0,y]$ le plus visité par X, permettent facilement de déduire le comportement asymptotique du point le plus visité par X pendant l'intervalle de temps $[0,t]$.
On considère $(\ell_t^x : x\in\mathbb{R},\, t\geq 0)$ la famille bicontinue des temps locaux de X, et on introduit

$$M_t=\max\{\ell_t^x : x\geq 0\} \quad \text{et} \quad \overline{V}_t=\inf\{x\geq 0 : \ell_t^x=M_t\}.$$

Corollaire 5.3
$\overline{V}_t/\mu t$ converge en distribution vers la loi uniforme sur $[0,1]$ quand t tend vers $+\infty$.

Corollaire 5.4
Soit $f:(0,+\infty)\to(0,+\infty)$ une fonction croissante. Alors

$$\liminf_{t\to+\infty}\frac{\overline{V}_t f(t)}{t}=0 \quad ou \quad +\infty \ p.s.$$

selon que l'intégrale $\displaystyle\int_1^{+\infty}\frac{dt}{tf(t)}$ *diverge ou converge.*

La démonstration des corollaires repose sur le fait que $X_t/t\to\mu$, ce qui permet de ramener l'étude de $\overline{V}_t$ à celle de $V(\mu t)$ au moyen du lemme suivant.

Lemme 5.5
Pour tout $\varepsilon \in (0,1)$,

$$\mathbb{P}\left(V(\mu s(1-\varepsilon)) \leq \overline{V}_s \leq \frac{1+\varepsilon}{1-\varepsilon} V(\mu s(1+\varepsilon)) \text{ pour tout } s \geq t\right)$$

tend vers 1 *lorsque* t *tend vers* $+\infty$.

Preuve : On introduit $S_t = \sup\{X_s : 0 \leq s \leq t\}$ et $J_t = \inf\{X_s : s \geq t\}$, respectivement le supremum passé et l'infimum futur de X. Fixons $\varepsilon > 0$ et $\eta > 0$. Comme $X_t/t \to \mu$, on peut choisir t suffisamment grand pour que la probabilité de l'évènement

$$\{\mu s(1-\varepsilon) \leq J_s \text{ et } S_s \leq \mu s(1+\varepsilon) \text{ pour tout } s \geq t\}$$

vaille au moins $1-\eta$. On travaille désormais sur cet évènement.
Fixons $s \geq t$. Le fait que $J_s \geq \mu s(1-\varepsilon)$ entraîne

$$\ell^x = \ell^x_s \quad \text{pour tout } x < \mu s(1-\varepsilon). \tag{6}$$

L'inégalité $V(\mu s(1-\varepsilon)) \leq \overline{V}_s$ en découle aisément.

Vérifions ensuite que $(1-\varepsilon)\overline{V}_s \leq (1+\varepsilon)V(\mu s(1+\varepsilon))$; pour cela, distinguons deux cas. Dans un premier temps, supposons que $\overline{V}_s < \mu s(1-\varepsilon)$. L'identité (6) entraîne alors que $\overline{V}_s = V(\mu s(1-\varepsilon))$, et *a fortiori* $(1-\varepsilon)\overline{V}_s \leq (1+\varepsilon)V(\mu s(1+\varepsilon))$.
Supposons ensuite que $\overline{V}_s \geq \mu s(1-\varepsilon)$. Pour tout $y \in [0, \mu s(1-\varepsilon))$, on a clairement

$$\ell^y = \ell^y_s < \ell^{\overline{V}_s}_s \leq \ell^{\overline{V}_s},$$

et donc $V(\mu s(1+\varepsilon)) \geq \mu s(1-\varepsilon)$. Pour conclure, il ne reste qu'à observer que, comme $S_s \leq \mu s(1+\varepsilon)$, on a nécessairement $\overline{V}_s \leq \mu s(1+\varepsilon)$, i.e. $\mu s \geq \overline{V}_s/(1+\varepsilon)$. □

Le corollaire 5.3 découle maintenant immédiatement de la proposition 1.1 et du lemme 5.5. Quant au corollaire 5.4, il se déduit de façon analogue du théorème 1.2, du lemme 5.5 et du fait que le résultat du test intégral est inchangé lorqu'on remplace $f(t)$ par $af(bt)$.

Remarque : Pour ce qui est du comportement de la limite supérieure de $\overline{V}$, on peut montrer que

$$\limsup_{t\to+\infty} \frac{\overline{V}_t}{\mu t} = 1.$$

Comme précédemment, l'argument repose sur le lemme 5.5; nous omettons les détails.

Remerciements : Nous remercions Jean-François Le Gall et Zhan Shi pour les remarques constructives qu'ils nous ont faites sur ce travail.

References

[1] R.F. Bass et P.S. Griffin : The most visited site of Brownian motion and simple random walk. *Z. Wahrscheinlichkeitstheorie verw. Gebiete* 70, 417-436. (1985)

[2] A.N. Borodin et P. Salminen : *Handbook of Brownian motion - Facts and formulae.* Probability and its applications, Birkhäuser. (1996)

[3] N. Eisenbaum : Un théorème de Ray-Knight lié au supremum des temps locaux browniens. *Probab. Theory Relat. Fields* 87, 79-95. (1990)

[4] C. Leuridan : Le point d'un fermé le plus visité par le mouvement brownien. *Ann. Probab.* 25, 953-996. (1997)

[5] J. Pitman et M. Yor : A decomposition of Bessel bridges. *Z. Wahrscheinlichkeitstheorie verw. Gebiete* 59, 425-457. (1982)

[6] S.I. Resnick : *Extreme values, regular variation, and point processes.* Applied Probability, vol. 4. Springer, Berlin (1987)

[7] D. Revuz et M. Yor : *Continuous martingales and Brownian motion*, 2nd edn. Springer, Berlin. (1994)

[8] L.C.G. Rogers et D. Williams : *Diffusions, Markov Processes, and Martingales* vol. 2 : Itô calculus. Wiley, New-York. (1987)

[9] F. Spitzer : *Principles of random walk.* Van Nostrand, Princeton. (1964)

Brownian motion, excursions, and matrix factors

Paul McGill[1]

Laboratoire de Probabilités, Université de Lyon I, 69622 Villeurbanne, France[2]

The Wiener-Hopf problem in analysis asks how one can factor a matrix function on the line as $\mathfrak{A} = \mathfrak{A}^+ * \mathfrak{A}^-$ where $*$ denotes convolution and $\mathfrak{A}^\pm$ are supported on $\mathbf{R}^\pm$ respectively. Existence is known [7], but a general algorithm seems to be out of reach: even the 2×2 case [5] presents an unexpected degree of complication.

The probabilistic counterpart (in dimension one) involves characterising a Lévy process Y in the form (Y^+, Y^-) where the laws $Y^\pm$ are supported on $\mathbf{R}^\pm$. These factors can be constructed by decomposing the sample path into its excursions from the maximum: Y^+ is then a Lévy process while Y^- takes its values in the space of paths. The probabilistic factorisation implies the (unique) factorisation of the generator $\mathcal{G} = \mathcal{G}^+ * \mathcal{G}^-$ — $Y^\pm$ are not unique — and allows us to describe the connection $Y^\pm \leftrightarrow \mathcal{G}^\pm$ in more detail. It is natural therefore to ask for an extension to higher dimensions: is there a probabilistic setting in which one can perform a sample path factorisation of certain matrices?

The difficulty is to know where to start. This note presents an example based on the idea in [10] and we recall the general problem: given a Markov process $X_t \in \mathcal{E}$ and a fluctuating additive functional V, there is a decomposition of the state space $\mathcal{E} = \mathcal{E}^+ \oplus \mathcal{E}^-$ determined by V increasing/decreasing on $\mathcal{E}^+/\mathcal{E}^-$. This leads to a probabilistic decomposition $X \to (X^+, X^-)$ and one studies the relationship $(X, V) \leftrightarrow (X^+, X^-)$; in terms of generators

$$(\mathcal{G}, V) \leftrightarrow \begin{pmatrix} \mathcal{G}_{++} & \mathcal{G}_{+-} \\ \mathcal{G}_{-+} & \mathcal{G}_{--} \end{pmatrix}$$

but see [10] for more explanation. The analogy here is with two-point Markov chains and excursions with 'interaction': $\mathcal{G}_{++}$ is the generator of X^+ in the interior of $\mathcal{E}^+$ while $\mathcal{G}_{-+}$ describes how X^+ returns to $\mathcal{E}^+$ after interacting with $\mathcal{E}^-$.

We treat only the simplest case where $X = B$ a real Brownian motion; the boundary set $\partial = \bar{\mathcal{E}}^+ \cap \bar{\mathcal{E}}^-$ then has a local time. As a first step towards calculating $\mathcal{G}_{-+}$ we derive a (vector) convolution equation for the entrance law of (B, V) into the right half plane. The equation formally resembles the first passage relation for a real Lévy process and is solved by Wiener-Hopf factorisation of a certain matrix $\mathfrak{S}$ where, and quite remarkably, the factors can be seen from a sample path decomposition. The proof echoes [4] — it uses excursions from maxima of V observed in

---

[1] Supported by the National Science Foundation and the Air Force Office of Scientific Research Grant No: F49620-92-J-0154 and the Army Research Office Grant No: DAAL03-92-G-0008.

[2] While visiting: Center for Stochastic Processes, University of North Carolina, Chapel Hill, NC 27599, USA.

the boundary local time — but the need to simultaneously track B means that the theory of homogeneous regenerative sets does not apply directly. Instead we employ Maisonneuve's theory [6] in multiple timescales, which we then patch together by viewing the 'boundary chain at the maximum' in its equilibrium distribution.

The matrix $\mathfrak{S}$ is quite special here: it is the generator of a two-dimensional Markov process and tri-diagonal to boot. Nevertheless, and even if the other manageable case, that of finite Markov chains [10], lies beyond our compass since there is nothing to play the crucial role of the boundary ∂, one suspects that the method should apply more generally.

1. Problem

We work with real Brownian motion B_t and a fixed $V_t = \int L_t^a m(da)$; L_t^a is a bicontinuous version of the B local time and the Hahn decomposition of the Radon measure $m = m^+ - m^-$ splits $\mathbf{R} = \mathcal{E}^+ \oplus \mathcal{E}^-$ with $\mathcal{E}^+$ defined as the closed support of m^+. We will assume throughout that the boundary $\partial = \mathcal{E}^+ \cap \bar{\mathcal{E}}^-$ is discrete — no limit points. Our results are not proved in full generality however; further restrictions, pertaining mainly to the set of maxima and used to simplify an already complicated proof, are stated at the beginning of sections 5 and 6.

To define the splitting of B induced by V we consider the increasing process $\bar{V}_t = \sup\{V_s : 0 < s \leq t\}$ and take $\tau_t^+ = \tau_t$ as its right continuous inverse. Then $X_t^+ = B_{\tau_t} \in \mathcal{E}^+$, but note that the process can jump since in general the intervals of constancy of $\bar{V}$ strictly contain $\{t : B_t \in \mathcal{E}^-\}$. A similar description of X^- provides the desired factorisation $(B, V) \leftrightarrow (X^+, X^-)$.

This external approach to defining (X^+, X^-) is quite useless. One should look instead for an intrinsic (internal) characterisation of each factor[1] which would amount to decomposing the generator $\mathcal{G}$ as above. As a first step in this direction we obtain an expression containing $\Pi^+(x, dy) = \mathbf{P}_x[B_\tau \in dy; \tau < \infty]$ with $\tau = \tau_0^+ = \inf\{t > 0 : V_t > 0\}$ the equalisation time. The method is to solve a vector equation of Wiener-Hopf type which we derive in the next section.

2. Equation for $\Pi^+(x, dy)$

We will derive an equation containing $\Pi^+(x, dy) = \mathbf{P}_x[B_\tau \in dy; \tau < \infty]$. Take T as the B hitting time of the boundary ∂ and let $L^\partial = \sum_{j \in \partial} L^j$ denote the boundary local time. Then, for each $j \in \partial$ we derive one scalar equation; our notation is that j_r is the closest boundary point to the right of j, j_l its closest point on the left.

Define functions $u(x, j) = \mathbf{E}_x\left[e^{-zV_T}; B_T = j\right]$ noting that they solve (eg.)

$$du_x = 2zudm \quad ; \quad u(j, j) = 1, \, u(j_r, j) = 0$$

[1] Think of B timechanged to stay above zero and its characterisation via Skorokhod's equation.

if $x \in (j, j_r)$. So by Itô one can write the martingale $\int_0^t e^{-\lambda L_s^\partial} e^{-zV_s} u_x(B_s, j) dB_s$ in the form

$$e^{-\lambda L_t^\partial} e^{-zV_t} u(B_t, j) + \lambda \int_0^t e^{-\lambda L_s^\partial} e^{-zV_s} dL_s^j - \tfrac{1}{2}\Delta u_x(j,j) \int_0^t e^{-\lambda L_s^\partial} e^{-zV_s} dL_s^j$$

$$- \tfrac{1}{2}\Delta u_x(j_r, j) \int_0^t e^{-\lambda L_s^\partial} e^{-zV_s} dL_s^{j_r} - \tfrac{1}{2}\Delta u_x(j_\ell, j) \int_0^t e^{-\lambda L_s^\partial} e^{-zV_s} dL_s^{j_\ell}$$

the process being uniformly bounded for z purely imaginary and $\lambda > 0$. If we stop it at first return to the boundary after time τ and take the expectation then

$$\mathbf{E}_x \left[e^{-\lambda L_\tau^\partial} e^{-zV_T \circ \theta_\tau}; B_T \circ \theta_\tau = j \right] = u(x,j) - \kappa_{jj}(z) \Psi_\lambda(z, j, x)$$

$$- \kappa_{jj_r}(z) \Psi_\lambda(z, j_r, x) - \kappa_{jj_\ell}(z) \Psi_\lambda(z, j_\ell, x) - \lambda \Psi_\lambda(z, j, x)$$

with $\Psi_\lambda(z, j, x) = \mathbf{E}_x \left[\int_0^\tau e^{-\lambda L_s^\partial} e^{-zV_s} dL_s^j \right]$. In vector form this reads

$$\mathbf{\Phi}_\lambda^+(z,x) = \mathbf{\Upsilon}^-(z,x) - \mathbf{K}(z)\mathbf{\Psi}_\lambda^-(z,x) - \lambda \mathbf{\Psi}_\lambda^-(z,x) \tag{2.1}$$

where $\mathbf{K}$, called the symbol, is a tri-diagonal matrix indexed by $\partial \times \partial$ and the superscripts $\pm$ are meant to suggest analyticity properties — the function $z \to \mathbf{\Phi}_\lambda^+(z,x)$ is bounded and analytic on the right half plane.

The solution of 2.1 is best explained analytically so we invert the Laplace transform, this being straightforward for all terms bar one, the symbol $\mathbf{K}(z)$ which we must examine in more detail. For the diagonal terms $\mathbf{K}_{jj}$ we denote by $\eta_t = \eta(j,t)$ the right continuous inverse of the local time L^j, whereupon $\kappa_{jj}(z) = -\frac{1}{2}\Delta u_x(j,j) = \int (1 - e^{-zx}) \nu_{jj}(dx)$ with ν_{jj} the Lévy measure of the excursion functional $t \to \int_{j_\ell}^{j_r} L(a, \eta_t) m(da)$ under the law of B_t conditioned to return to j before hitting the points j_r and j_ℓ — right and left excursions are therefore calculated by using the respective Bessel laws $\mathfrak{bes}_{j_r}(3)$ and $\mathfrak{bes}_{j_\ell}(3)$. This implies the existence of a Schwartz distribution defined on test functions by $\mathfrak{S}_{jj}(f) = \int [f(0) - f(x)] \nu_{jj}(dx)$, hence $\mathfrak{S}_{jj}(e^{-z\cdot}) = \mathbf{K}_{jj}(z)$. For the off-diagonal terms we have a similar description: $\kappa_{jj_r}(z) = \frac{1}{2}\Delta u_x(j_r, j) = \int (1 - e^{-zx}) \nu_{j_r j}(dx)$ where $\nu_{j_r j}$ is the Lévy measure of the excursion functional $\int_j^{j_r} L(a,T) m(da)$ computed under the law of B exiting j_r and conditioned to re-enter ∂ at j_r, in other words the excursion law of $\mathfrak{bes}_j(3)$ from j_r.

This means that we can interpret 2.1 as the Laplace transform of a vector convolution equation

$$\mathbf{\Pi}_\lambda^+(x, dt) \quad = \quad \mathbf{\Sigma}(x, dt) \quad - \quad \mathfrak{S} * \mathbf{r}_\lambda(x, dt) \quad - \quad \lambda \mathbf{r}_\lambda(x, dt) \tag{2.2}$$

where $\mathfrak{S}$ is a distributional matrix such that $\mathfrak{S}(e^{-z\cdot}) = \mathbf{K}(z)$ (since each $\mathfrak{S}_{jj}$ is the sum of a finite measure and a distribution with compact support $\mathfrak{S} * \mathbf{r}_\lambda$ is well-defined [9]). Note how the left side of 2.2 is supported on $[0, \infty)$ whereas the terms on the right — bar the distribution $\mathfrak{S}$ which is 'mixed' — are all supported on $(-\infty, 0]$. We therefore have an equation of Wiener-Hopf type for $\mathbf{\Pi}_\lambda^+(x, dt) = (\mathbf{E}_x[e^{-\lambda L_\tau^\partial} \mathbf{P}_{B_\tau}[V_T \in dt; B_T = j]])$.

3. Solution

The method for solving 2.2 is well-known [2] and depends on factoring the distributional matrix $\mathfrak{S} = \mathfrak{S}^+ * \mathfrak{S}^-$. Probabilistic proofs of the following results are postponed to sections 5 and 6 where the interpretation will make it plain as to why the various convolutions can be defined.

Lemma 3.1 The distribution $\mathfrak{S}$ has Wiener-Hopf factors $\mathfrak{S}^\pm$ where $(\mathfrak{S}^+)^{-1}$ is a matrix-valued positive σ-finite Radon measure on $(0,\infty)$ and $\mathfrak{S}^-$ is supported on $(-\infty, 0]$. Moreover:
(1) $(\mathfrak{S}^+)^{-1}_{ji} \leq (\mathfrak{S}^+)^{-1}_{ii}$.
(2) For K a fixed compact set, $\lim_{\lambda\downarrow 0} \lambda(\mathfrak{S}^+)^{-1} * \mathbf{r}_\lambda(K) = 0$.
(3) $\lim_{z\downarrow -\infty} z^{-1}\mathfrak{S}^-(e^{-z\cdot}) = 0$.

Remarks: (1) By definition the $\mathfrak{S}^\pm$ are unique modulo a distribution supported at the origin; 3.1 (3) restricts the choice even further.
(2) Although $\mathfrak{S}$ is tri-diagonal its factors are not and this poses a difficulty when ∂ is infinite. Strictly speaking, one should specify the solution space for 2.2 as perhaps an inductive limit space of vector measures on $\mathbf{R}$. But this would take us too far afield, particularly since our main concern here is with the factorisation method and not the solution of the equation. So convergence in 3.1, and elsewhere, will always be interpreted coordinatewise.

Recall the method for solving 2.2. Convolution on the left with $(\mathfrak{S}^+)^{-1}$ gives

$$(\mathfrak{S}^+)^{-1} * \Pi^+_\lambda \quad = \quad (\mathfrak{S}^+)^{-1} * \Sigma \quad - \quad \mathfrak{S}^- * \mathbf{r}_\lambda \quad - \quad \lambda(\mathfrak{S}^+)^{-1} * \mathbf{r}_\lambda$$

and the idea is to eliminate the middle term on the right by projection onto $\mathbf{R}^+$, denoted $\mathbf{P}^+$. We must therefore verify that the distribution does not charge the origin. In the case of $\mathbf{r}_\lambda$ this is straightforward, since by the definition of ∂ in the closed support of m^+ we have $\mathbf{P}_\partial[\tau > 0] = 0$. Moreover, since the singular support at zero of $\mathfrak{S}^-$ is necessarily a linear combination of a Dirac mass and its derivatives, and since 3.1 (3) excludes the latter, the distribution $\mathfrak{S}^- * \mathbf{r}_\lambda$ cannot charge zero. We therefore have $\mathbf{P}^+[\mathfrak{S}^- * \mathbf{r}_\lambda] = 0$, and taking $\lambda \downarrow 0$ and applying 3.1 (2) we deduce

$$\Pi^+_0(x, dt) = \mathfrak{S}^+ * \mathbf{P}^+\left[(\mathfrak{S}^+)^{-1} * \Sigma(x, .)\right](dt)$$

Remarks: (1) If ∂ is a singleton then the relation with the analytic problem is transparent: $\mathfrak{S}^+$ can be interpreted as the generator of V_t observed in the local time scale at its maximum. The complex variable approach, in the case of symmetric stable processes, is described in Ray's paper [8].
(2) We omit the details of how one recovers $\Pi^+(x, dy)$ from $\Pi^+_0(x, dt)$ since this involves inverting an integral transform and can lead to delicate uniqueness questions.
(3) In some simple cases 2.1 can be solved explicitly for $\Pi^+(x, dy)$ — but apparently not by using the factorisation described here. This is a mystery even for real processes.

4. Excursions

We prove 3.1 by decomposing the path at the maxima of V_t observed when L_t^∂ increases, this process being finite almost surely since ∂ is a discrete set. Start by writing σ^∂ for its right continuous inverse and define $Y_t = V(\sigma_t^\partial)$. Then the maximum $\bar{Y}_t = \sup\{0 < s \le t : Y_s\}$ defines a random set $\mathcal{M} = \{t : Y_t = \bar{Y}_t \text{ or } Y_{t-} = \bar{Y}_t\}$. We will factor our matrix by using excursions of Y from $\mathcal{M}$, but since the set is not homogeneous the method of [4] does not apply directly. Instead, we modify their argument by using excursions from the maximum as observed from a point on the boundary.

To make this precise we introduce $\mathcal{M}^j = \{t \in \mathcal{M} : B(\sigma_t^\partial) = j\}$. By the strong Markov property of B this is a closed regenerative set in the sense of Maisonneuve [6] and, being optional in the $\mathcal{B}(\sigma_t^\partial)$ filtration, it has an adapted local time $L^{\mathcal{M}^j}$ whose right continuous inverse we denote σ^j. Each gap $(\sigma_{t-}^j, \sigma_t^j)$ then defines an excursion with corresponding excursion measure denoted $\mathcal{Q}^j$. The basic formula of [6] says that if A is an additive functional of Y then

$$\sum_{0<s\le t} \mathcal{A}_s \circ \theta_{\sigma_{s-}^j} - \int_0^t \mathcal{Q}^j[\mathcal{A}_s]\, ds$$

is a martingale[1] — here $\mathcal{A}_t = \Delta A_{\sigma_t}$ is the change in the value of A over the excursion interval $(\sigma_{t-}^j, \sigma_t^j)$. By stochastic integration of the above martingale against a bounded predictable process h it follows that

$$\sum_{0<s\le t} h_s \mathcal{A}_s \circ \theta_{\sigma_{s-}^j} - \int_0^t h_s \mathcal{Q}^j[\mathcal{A}_s]\, ds$$

is also a martingale. We remark for later (and frequent) use that if h is càglàd then in particular h_{t+} will be defined and can be substituted for h_t in the integral on the right.

The excursion measures $\mathcal{Q}^j$ can be decomposed still further, according to whether the excursion from $\mathcal{M}^j$ straddles a point in $\mathcal{M}^k$ or not. If not, then we have an excursion interval common to $\mathcal{M}$ and $\mathcal{M}^j$ with excursion measure denoted $\mathcal{Q}^{jj}$. On the other hand, excursions from $\mathcal{M}^j$ which do straddle points in $\mathcal{M}$ have their excursion measure labelled $\mathcal{Q}^{jk}$ where k is the first point of maximum after leaving $\mathcal{M}^j$: for $k \ne j$ an excursion governed by $\mathcal{Q}^{jk}$ first re-enters $\mathcal{M}$ at a point of $\mathcal{M}^k$, but it does not die there, instead continuing on until it reaches a point of $\mathcal{M}^j$.

The advantage of working with $\mathcal{M}^j$, and not $\mathcal{M}$, is that the $\mathcal{M}^j$ are homogeneous regenerative sets [6]; the downside is that this involves the manipulation of multiple timescales $L_t^{\mathcal{M}^j}$, unique only up to constants, and it would be better if, somehow, we could patch all these together to obtain $L^{\mathcal{M}} = \sum_{j\in\partial} L^{\mathcal{M}^j}$. To do this canonically let us introduce $N_t = B_{\sigma_t^\partial} \in \partial$ and denote by $\sigma^{\mathcal{M}}$ the right-continuous inverse of $L^{\mathcal{M}}$. The process $N_{\sigma^{\mathcal{M}}}$ is a Markov chain which we call 'the boundary chain at the

[1] We emphasise that the original filtration has been timechanged twice — first do $t \to \sigma_t^\partial$ and then $t \to \sigma_t^j$.

maximum'. If $\mathcal{M}$ is recurrent, $N_{\sigma^{\mathcal{M}}}$ has an invariant measure $\{\pi_j : j \in \partial\}$ and we can normalise its exponential holding times by $\lambda_j = \pi_j$.

Notation: When convenient, we shall simplify by writing σ in place of $\sigma^{\mathcal{M}}$ or σ^j.

5. Existence

This is the main part of the paper. We prove existence of the factorisation under the assumption that the boundary is discrete, Y spends zero time in $\mathcal{M}$, and the boundary chain at the maximum is recurrent. From Rogozin's trichotomy [1] for Lévy processes, any individual $\mathcal{M}^j$ is either recurrent or transient almost surely; by the strong Markov property it follows that the $\mathcal{M}^j$ are either all unbounded or all transient. We prove existence in the former case only — if the maximum is transient, and $-Y$ satisfies the conditions indicated, then one can use the set of minima instead.

First we outline the method, starting with the problem of finding factors $\mathbf{K}^{\pm}$ such that $\mathbf{K}\mathbf{K}^{-}\mathbf{K}^{+} = \mathbf{I}$ where the tri-diagonal matrix $\mathbf{K}$ is defined by the martingale of section two

$$e^{-zV_t}e^{-\lambda L_t^{\partial}}u(B_t,j) + \kappa_{jj}(z)\int_0^t e^{-\lambda L_s^{\partial}}e^{-zV_s}dL_s^j + \kappa_{jj_r}(z)\int_0^t e^{-\lambda L_s^{\partial}}e^{-zV_s}dL_s^{j_r}$$

$$+\kappa_{jj_l}(z)\int_0^t e^{-\lambda L_s^{\partial}}e^{-zV_s}dL_s^{j_l} + \lambda\int_0^t e^{-\lambda L_s^{\partial}}e^{-zV_s}dL_s^j$$

For z purely imaginary and $\lambda > 0$ this is uniformly bounded. Starting at $B_0 = i \in \partial$ and applying martingale stopping as $t \uparrow$ then, since $L_t^{\partial} \uparrow$, boundedness of $u(x,j)$ gives

$$\kappa_{jj_l}(z)\mathbf{E}_i\left[\int_0^\infty e^{-\lambda L_s^{\partial}-zV_s}dL_s^{j_l}\right] + \kappa_{jj}(z)\mathbf{E}_i\left[\int_0^\infty e^{-\lambda L_s^{\partial}-zV_s}dL_s^{j}\right]$$

$$+\kappa_{jj_r}(z)\mathbf{E}_i\left[\int_0^\infty e^{-\lambda L_s^{\partial}-zV_s}dL_s^{j_r}\right] + \lambda\mathbf{E}_i\left[\int_0^\infty e^{-\lambda L_s^{\partial}-zV_s}dL_s^{j}\right] = \delta_{ji}$$

We do the factorisation from this. But first let us simplify our notation. Recall that when factorising real Lévy processes one does the calculations with $\lambda > 0$, taking $\lambda \downarrow 0$ at the end. In our case such reasoning leads to

$$\kappa_{jj_l}(z)\mathbf{E}_i\left[\int_0^\infty e^{-zV_s}dL_s^{j_l}\right] + \kappa_{jj}(z)\mathbf{E}_i\left[\int_0^\infty e^{-zV_s}dL_s^{j}\right]$$

$$+\kappa_{jj_r}(z)\mathbf{E}_i\left[\int_0^\infty e^{-zV_s}dL_s^{j_r}\right] = \delta_{ji}$$

where the expectations are interpreted as weak limits in λ. So the *claim* is that this doubly infinite family of equations indexed by $\partial \times \partial$ is precisely $\mathbf{K}\mathbf{K}^{-}\mathbf{K}^{+} = \mathbf{I}$ and that, moreover, one can determine the entries of $\mathbf{K}^{\pm}$ by decomposing the above integrals appropriately. This would give our desired factorisation $\mathfrak{G} = \mathfrak{G}^+ * \mathfrak{G}^-$ in the form

$$\mathrm{K}^-(z) = \left(\mathfrak{G}^-\right)^{-1} e^{-z\cdot} \quad ; \quad \mathrm{K}^+(z) = \left(\mathfrak{G}^+\right)^{-1} e^{-z\cdot}$$

using invertibility in the sense of distributions.

Of course to make this rigorous one should do the argument with $\lambda > 0$. The difficulty there is that keeping track of all the different timescales would present a notational nightmare and consequently, since λ does not figure in the final answer, we suppress all mention of it in our calculations with the *caveat* that $\lambda > 0$ is essential for justifying the various manipulations.

With this in mind we set out to identify the factor matrices $\mathrm{K}^{\pm}$. Recalling the notation $Y = V_{\sigma^\theta}$ and $N = B_{\sigma^\theta}$, we start from our conjecture in the form

$$\sum_k \mathrm{K}^-_{ik}\mathrm{K}^+_{kj} = \left(\mathrm{K}^-\mathrm{K}^+\right)_{ij} = \mathbf{E}_j\left[\int_0^\infty e^{-zV_s} dL^i_s\right] = \mathbf{E}_j\left[\int_0^\infty e^{-zY_s} 1_{(N_s=i)} ds\right] \quad (5.1)$$

and the idea of calculating with excursions from $\mathcal{M}$, the set of maxima of Y. Since Y spends no time in $\mathcal{M}$ we see that $\sigma = \sigma^j$, the right continuous inverse of $L^{\mathcal{M}^j}$, is a pure jump process. The integral therefore decomposes into its excursions from $\mathcal{M}^j$

$$\sum_{t>0} e^{-zY_{\sigma_t -}} \int_{\sigma_{t-}}^{\sigma_t} e^{-zY_u + zY_{\sigma_t -}} 1_{(N_u=i)} du = \sum_{t>0} e^{-zY_{\sigma_t -}} \left[\int_0^\zeta e^{-zY_u} 1_{(N_u=i)} du\right] \circ \theta_{\sigma_t -}$$

where $\zeta^j = \zeta$ is the excursion lifetime. Now apply the excursion theorem to see that

$$\sum_{0<s\le t} e^{-zY(\sigma_s -)} \int_0^\zeta e^{-zY_u} 1_{(N_u=i)} du \circ \theta_{\sigma_s -} - \int_0^t e^{-zY(\sigma_s)} ds Q^j \left[\int_0^\zeta e^{-zY_u} 1_{(N_u=i)} du\right]$$

is a martingale[1]. Taking the expectation therefore gives[2]

$$\mathbf{E}_j\left[\int_0^\infty e^{-zY_s} 1_{(N_s=i)} ds\right] = \mathbf{E}_j\left[\int_0^\infty e^{-zY(\sigma^j_s)} ds\right] Q^j \left[\int_0^\zeta e^{-zY_u} 1_{(N_u=i)} du\right]$$

which we propose to write as $\sum_k \kappa^-_{ik}/\kappa^+_{kj}$ by exploiting a suitable decomposition of Q^j— for convenience we take $\mathrm{K}^-_{ij} = \kappa^-_{ij}$ but $\mathrm{K}^+_{ij} = 1/\kappa^+_{ij}$.

The diagonal entries of K^+ have the most transparent definition: since Y_{σ^i} is a subordinator, with Laplace exponent κ^+_{ii} (say), we can take $\mathrm{K}^+_{ii} = 1/\kappa^+_{ii}$. For the other entries, recall that in the previous section we decomposed the excursion measures as $Q^i = \sum_{j\in\partial} Q^{ij}$; the measure Q^{ii} is supported on excursions from $\mathcal{M}^i$ which do not straddle any points in $\mathcal{M}$ and, since $Y_u \le 0$ throughout, it follows that $Q^{ii}[\int_0^\zeta e^{-zY_u} 1_{(N_u=i)} du]$ is the Laplace transform of a measure supported on $(-\infty, 0]$. The temptation then is to place these terms along the diagonal of K^-. But this is wrong. We show later that K^-_{ii} is more complicated and contains additional terms

---

[1] Thanks to our phantom λ.

[2] The price we pay for homogeneity is that Q^j is not necessarily supported on $(-\infty, 0]$.

coming from a path decomposition of the Q^{ij}. This observation — that 5.1 entails complicated cross-cancellations amongst the various excursion terms — prompted us to devise a notation for keeping track of the relevant components of the path. Our argument is best understood in the

2 × 2 case

Fix $i \neq j$ throughout. We start by decomposing $\mathbf{E}_i[\int_0^\infty e^{-zY_s} 1_{(N_s=i)} ds]$ using excursions from $\mathcal{M}^i$ in the timescale $\sigma_t^i = \sigma_t$. These split naturally into two kinds: a $\mathcal{Q}^i$ excursion either straddles a point of $\mathcal{M}^j$ or it doesn't. By the definition of κ_{ii}^+, the excursion theorem gives

$$(\mathbf{K}^-\mathbf{K}^+)_{ii} = \mathcal{Q}^{ii}\left[\int_0^\zeta e^{-zY_s} 1_{(N_s=i)} ds\right]/\kappa_{ii}^+ + \mathcal{Q}^{ij}\left[\int_0^\zeta e^{-zY_s} 1_{(N_s=i)} ds\right]/\kappa_{ii}^+$$

which we want in the form $\kappa_{ii}^-/\kappa_{ii}^+ + \kappa_{ij}^-/\kappa_{ji}^+$. The idea is to use path decomposition inside $\mathcal{Q}^{ij}[\int_0^\zeta e^{-zY_u} 1_{(N_u=k)} du]$ by noting that an excursion from $\mathcal{M}^i$ which straddles a point in $\mathcal{M}^j$ has three distinct components, each non-trivial:
$\mathfrak{a}$) The initial excursion until we arrive in $\mathcal{M}^j$.
$\mathfrak{b}$) Excursions from $\mathcal{M}^j$ such that, even if N visits i, Y cannot achieve a maximum there.
$\mathfrak{c}$) The final excursion from $\mathcal{M}^j$ back to $\mathcal{M}^i$.
We write all this as $\mathcal{Q}_k^{ij}[\mathfrak{a}+\mathfrak{b}+\mathfrak{c}]$ and remark immediately that $\mathcal{Q}_k^{ij}[\mathfrak{a}]$ is the LT of a measure supported on $(-\infty, 0]$, the same being true for

$$\mathcal{Q}_k^{i\to\mathcal{M}} = \mathcal{Q}^{ii}\left[\int_0^\zeta e^{-zY_u} 1_{(N_u=k)} du\right] + \mathcal{Q}_k^{ij}[\mathfrak{a}]$$

It is therefore legitimate to define the entries of $\mathbf{K}^-$ by

$$\mathbf{K}_{ij}^- = \kappa_{ij}^- = \mathcal{Q}_i^{j\to\mathcal{M}} \quad ; \quad \mathbf{K}_{ii}^- = \kappa_{ii}^- = \mathcal{Q}_i^{i\to\mathcal{M}}$$

thereby reducing the problem to defining the off-diagonal entries of $\mathbf{K}^+$ in a manner consistent with

$$(\mathbf{K}^-\mathbf{K}^+)_{ii} = \mathcal{Q}_i^{i\to\mathcal{M}}/\kappa_{ii}^+ + \mathcal{Q}_i^{ij}[\mathfrak{b}+\mathfrak{c}]/\kappa_{ii}^+$$

We start by writing $\mathcal{Q}_k^{ij}[\mathfrak{b}+\mathfrak{c}]$ in a more convenient form. First modify our notation so that $T = T_\mathcal{M}$ represents the first time Y enters $\mathcal{M}$ and recall that $N_{\sigma\mathcal{M}}$ is the 'boundary chain at the maximum' whose exponential holding times have parameters $\{\lambda_i : i \in \partial\}$. The passages $\mathcal{M}^i \to \mathcal{M}\setminus\mathcal{M}^i = \mathcal{M}^j$ are then Poisson processes of rate λ_i so that in obvious notation $\mathcal{Q}^{ij}[e^{-zY_T}] = \lambda_i \mathbf{E}^{ij}[e^{-zY_T}]$. Applying the excursion theorem from $\mathcal{M}^j$ (*sic*) now gives

$$\begin{aligned}\mathcal{Q}_k^{ij}[\mathfrak{b}] &= \lambda_i \mathbf{E}^{ij}\left[e^{-zY_T}\right]\mathbf{E}^{jj}\left[\int_0^\xi e^{-zY(\sigma_t)}dt\right]\mathcal{Q}^{jj}\left[\int_0^\zeta e^{-zY_u} 1_{(N_u=k)} du\right] \\ &= \lambda_i \lambda_j^{-1} \mathbf{E}^{ij}\left[e^{-zY_T}\right]\mathbf{E}^{jj}\left[e^{-zY_{\sigma_\xi}}\right]\mathcal{Q}^{jj}\left[\int_0^\zeta e^{-zY_u} 1_{(N_u=k)} du\right]\end{aligned}$$

where $\xi = \xi_j = L_\zeta^{\mathcal{M}^j} \sim \exp(\lambda_j)$ is the holding time at j. The final part of the Q^{ij} excursion is dealt with similarly: we use the strong Markov property to expand

$$\begin{aligned} \mathcal{Q}_k^{ij}[\mathfrak{c}] &= \lambda_i \mathbf{E}^{ij}\left[e^{-zY_T}\right] \mathbf{E}^{jj}\left[e^{-zY_{\sigma_\xi}}\right] \mathbf{E}^{ji}\left[\int_0^T e^{-zY_u} 1_{(N_u=k)} du\right] \\ &= \lambda_i \lambda_j^{-1} \mathbf{E}^{ij}\left[e^{-zY_T}\right] \mathbf{E}^{jj}\left[e^{-zY_{\sigma_\xi}}\right] \mathcal{Q}_k^{ji}[\mathfrak{a}] \end{aligned}$$

The shorthand[1]

$$e^{ij} = \lambda_i \lambda_j^{-1} \mathbf{E}^{ij}\left[e^{-zY_T}\right] \mathbf{E}^{jj}\left[e^{-zY_{\sigma_\xi}}\right] \quad ; \quad E_k^{ij} = \mathbf{E}^{ij}\left[\int_0^T e^{-zY_u} 1_{(N_u=k)} du\right]$$

now lets us write $\mathcal{Q}_k^{ij}[\mathfrak{b}+\mathfrak{c}] = e^{ij} Q_k^{j \to \mathcal{M}}$ which, in light of the desired decomposition of $(\kappa^- \kappa^+)_{ii}$, forces

$$\mathrm{K}_{ji}^+ = 1/\kappa_{ji}^+ = e^{ij}/\kappa_{ii}^+$$

We now see that we have a proof of 5.1 and so the proof of factorisation for the 2×2 case is complete.

The next step is to look at the

<u>3 × 3 case</u>

Here the main difference is that a Q^{ij} excursion may visit $\mathcal{M}^k$ for $k \neq i, j$. Taking our cue from the above, we define $\mathrm{K}_{ii}^+ = 1/\kappa_{ii}^+ = \mathbf{E}[\int_0^\infty e^{-zY(\sigma_t^i)} dt]$, and denoting $Q_k^{ij} = \mathcal{Q}^{ij}[\int_0^\zeta e^{-zY_t} 1_{(N_t=k)} dt]$, we propose

$$\mathrm{K}_{ij}^- = \kappa_{ij}^- = Q_i^{j \to \mathcal{M}} = Q_i^{jj} + \sum_{k \neq j} \mathcal{Q}_i^{jk}[\mathfrak{a}] = Q_i^{jj} + \lambda_j \sum_{k \neq j} p_{jk} E_i^{jk}$$

The problem is now to determine the off-diagonal entries in κ^+. For this we will decompose $\mathcal{Q}_k^{ij}[\mathfrak{b}+\mathfrak{c}] = Q_k^{ij} - \mathcal{Q}_k^{ij}[\mathfrak{a}]$ by tracking the boundary chain $N_{\sigma \mathcal{M}}$ whose transition matrix and holding times we denote respectively by p_{ij} and $\xi_j \sim \exp(\lambda_j)$. Also, taking i, j, j' all distinct, we introduce the notation $\mathfrak{f}_{ij} = 1 - p_{jj'} p_{j'j} e^{jj'} e^{j'j}$ with e^{ij} defined as before. Now to calculate. By definition a $\mathcal{Q}^{ij}$ excursion first visits $\mathcal{M}^j$ and the journey gives rise to a factor $\lambda_i p_{ij} \mathbf{E}^{ij}\left[e^{-zY_T}\right]$. The contribution from its initial sojourn in $\mathcal{M}^j$, which lasts for time ξ_j, is

$$\lambda_i p_{ij} \mathbf{E}^{ij}\left[e^{-zY_T}\right] \lambda_j^{-1} \mathbf{E}^{jj}\left[e^{-zY_{\sigma_\xi}}\right] Q_k^{jj} = p_{ij} e^{ij} Q_k^{jj}$$

with a factor $p_{ij}\lambda_j e^{ij}$ prefixing what remains — either termination at $\mathcal{M}^i$ which gives $p_{ij}\lambda_j e^{ij} p_{ji} E_k^{ji}$, or else a passage to $\mathcal{M}^{j'}$ which adds on

$$p_{ij}\lambda_j e^{ij}\left[p_{jj'} E_k^{jj'} + \frac{p_{jj'} \mathcal{Q}_k^{ij'}[\mathfrak{b}+\mathfrak{c}]}{\lambda_i p_{ij'} \mathbf{E}^{ij'}\left[e^{-zY_T}\right]}\right]$$

[1] Rem: caps for negative, positive in lower case.

The result is

$$\mathcal{Q}_k^{ij}[\mathfrak{b}+\mathfrak{c}] = p_{ij}e^{ij}\left[Q_k^{j\to\mathcal{M}} + \frac{p_{jj'}\,e^{jj'}}{p_{ij'}\,e^{ij'}}\mathcal{Q}_k^{ij'}[\mathfrak{b}+\mathfrak{c}]\right]$$

Noting that i is fixed here, switching $j \leftrightarrow j'$ yields a 2×2 system of equations of the form

$$\rho^j = Q_k^{j\to\mathcal{M}} + p_{jj'}e^{jj'}\rho^{j'}$$

which we can solve for $\rho^j = \mathcal{Q}_k^{ij}[\mathfrak{b}+\mathfrak{c}]/(p_{ij}e^{ij})$ to obtain

$$\mathcal{Q}_k^{ij}[\mathfrak{b}+\mathfrak{c}] = \frac{p_{ij}e^{ij}}{\mathfrak{f}_{ij}}\left[Q_k^{j\to\mathcal{M}} + p_{jj'}e^{jj'}Q_k^{j'\to\mathcal{M}}\right]$$

The diagonal term $\mathbf{E}_i[\int_0^\infty e^{-zY_t}1_{(N_t=i)}dt]$ therefore decomposes like

$$\sum_j \mathcal{Q}^{ij}\left[\int_0^\zeta e^{-zY_t}1_{(N_t=i)}dt\right]/\kappa_{ii}^+ = Q_i^{i\to\mathcal{M}}/\kappa_{ii}^+ + \sum_{j\neq i}\mathcal{Q}_i^{ij}[\mathfrak{b}+\mathfrak{c}]/\kappa_{ii}^+$$

$$= Q_i^{i\to\mathcal{M}}/\kappa_{ii}^+ + \sum_{j\neq i}p_{ij}e^{ij}\left[Q_i^{j\to\mathcal{M}} + p_{jj'}e^{jj'}Q_i^{j'\to\mathcal{M}}\right]/\mathfrak{f}_{ij}\kappa_{ii}^+ = \sum_j Q_i^{j\to\mathcal{M}}/\kappa_{ji}^+$$

which means we ought to take

$$\mathbf{K}_{ji}^+ = 1/\kappa_{ji}^+ = \frac{p_{ij}e^{ij} + p_{ij'}p_{j'j}e^{ij'}e^{j'j}}{\mathfrak{f}_{ij}\kappa_{ii}^+} \qquad (i\neq j)$$

This completes the proof of the 3×3 case.

Discrete Boundary

The boundary chain at the maximum is recurrent and so it has an invariant measure $\{\pi_i : i\in\partial\}$; as explained in §4 we normalise the holding times by $\lambda_i = \pi_i$. The proof here follows the plan for the 3×3 case: the diagonal terms of κ^+ are defined as above, likewise the entries of κ^-, and we look at the decomposition of $\mathcal{Q}_k^{ij}[\mathfrak{b}+\mathfrak{c}]$ to help discover the rest of κ^+. If we define $\rho^j = \mathcal{Q}_k^{ij}[\mathfrak{b}+\mathfrak{c}]/(p_{ij}e^{ij})$ then arguing as before gives a system of equations

$$\rho^j = Q_k^{j\to\mathcal{M}} + \sum_{j'\neq i,j} p_{jj'}e^{jj'}\rho^{j'}$$

which we write in vector form as $\rho = \mathbf{Q} + {}_i\mathfrak{P}\rho$ (i is a taboo point). The solution is therefore $\rho = \sum_{n=0}^\infty {}_i\mathfrak{P}^n\mathbf{Q}$: for $z\geq 0$ the series converges since (with notation from [3]) the entries of the sum matrix are dominated by $\sum_{n=0}^\infty p_{jk}^{(n)}\lambda_j\lambda_k^{-1} = \lambda_i\lambda_j^{-1}{}_i\pi_{ij}$ where ${}_i\pi_{ij}$ is a multiple of the invariant measure. This gives us

$$\mathcal{Q}_k^{ij}[\mathfrak{b}+\mathfrak{c}] = p_{ij}e^{ij}\left[\sum_{n=0}^\infty {}_i\mathfrak{P}^n\mathbf{Q}\right]_j$$

which must square with

$$\mathbf{E}_i\left[\int_0^\infty e^{-zY_t}1_{(N_t=i)}dt\right] = Q_i^{i\to\mathcal{M}}/\kappa_{ii}^+ + \sum_k Q_i^{ik}[\mathfrak{b}+\mathfrak{c}]/\kappa_{ii}^+$$

$$= Q_i^{i\to\mathcal{M}}/\kappa_{ii}^+ + \sum_k p_{ik}e^{ik}\left[\sum_{n=0}^\infty {}_i\mathfrak{P}^n\mathrm{Q}\right]_k /\kappa_{ii}^+ = \sum_j \kappa_{ij}^- \kappa_{ji}^+ = \sum_j Q_i^{j\to\mathcal{M}}\kappa_{ji}^+$$

Comparing coefficients of $Q_i^{j\to\mathcal{M}}$ we find

$$\kappa_{ji}^+ = \frac{1}{\kappa_{ii}^+}\sum_k p_{ik}e^{ik}\sum_{n=0}^\infty {}_i\mathfrak{P}_{kj}^n \qquad (j\neq i)$$

Since our decomposition satisfies 5.1, the proof of factorisation in the general discrete case is now complete.

The proof of 3.1 (1) follows, since for $z\geq 0$ we have $\kappa_{ji}^+/\kappa_{ii}^+ \leq \sum_k p_{ik}p_{kj}\lambda_k^{-1}\lambda_{j\,i}\pi_{ik} = \sum_k p_{ik}p_{kj} < 1$ using the normalisation $\lambda_j = \pi_j$ and the formula ${}_i\pi_{ik} = \pi_k/\pi_i$ of [3] 11.24. An immediate consequence is that the $\mathfrak{S}_{ij}^{-1}$ are positive σ-finite measures — they are dominated by $1/\kappa_{ii}$ which is the LT of a σ-finite Radon measure on the line (the potential of the Lévy process Y_{σ^i}). We have therefore proved the existence part of 3.1.

Remarks: (1) If ∂ is a singleton then the above probabilistic argument can be deduced from [4] but note that they deal with the process killed at an independent exponential time; this corresponds to factoring $\lambda+\kappa$, a task more difficult than factoring the symbol κ alone.
(2) We obtain an analytic interpretation for our factorisation by noting that κ is the LT of the generator of the Markov process $(V_{\sigma^\partial}, B_{\sigma^\partial})$.
(3) The set $\mathcal{M}$ is not homogeneous since the excursions of $Y = V_{\sigma^\partial}$ are only conditionally independent given the boundary process $N_{\sigma^\mathcal{M}}$. Nevertheless, one can use the exit system of $\mathcal{M}$ [6] to see (cf. after 5.1) that

$$\sum_{0<s\leq t} e^{-\lambda\sigma_{s-}^\mathcal{M} - zY(\sigma_{s-}^\mathcal{M})}\int_0^\zeta e^{-\lambda u - zY_u}1_{(N_u=i)}du\circ\theta_{\sigma_{s-}^\mathcal{M}}$$

$$-\int_0^t e^{-\lambda\sigma_s^\mathcal{M} - zY(\sigma_s^\mathcal{M})}Q^\mathcal{M}\left[\int_0^\zeta e^{-\lambda u - zY_u}1_{(N_u=i)}du; N_{\sigma_{s-}^\mathcal{M}}\right]ds$$

is a uniformly integrable martingale; $\mathbf{E}[\int_0^\infty e^{-\lambda L_t^\partial - zV_t}1_{(B_t=i)}dt]$ then takes the form

$$\mathbf{E}\left[\int_0^\infty e^{-\lambda\sigma_t^\mathcal{M} - zY(\sigma_t^\mathcal{M})}Q^\mathcal{M}\left[\int_0^{\zeta^\mathcal{M}} e^{-\lambda u - zY_u}1_{(N_u=i)}du; N_{\sigma_{t-}^\mathcal{M}}\right]dt\right]$$

and our factorisation is obtained by conditionally decomposing the expectation according to the values of $N_{\sigma_{t-}^\mathcal{M}}$.

6. Estimates

We turn now to the last step in justifying the solution of 2.1, which is the proof of 3.1 (2)-(3) — we dealt with 3.1 (1) at the end of §5. Our running assumption is that the maximum is recurrent and so $N_{\sigma\mathcal{M}}$ has an invariant measure $\{\pi_i : i \in \partial\}$. For the proof of the last part of 3.1 (3) we will assume that ∂ is finite.

In the previous section we obtained factors of the symbol matrix in the form $\kappa\kappa^-\kappa^+ = 1$ where the off-diagonal elements of $\kappa^\pm(z) = (\mathfrak{S}^\pm)^{-1}(e^{-z\cdot})$ are

$$(\mathfrak{S}^-)^{-1}_{ij}(e^{-z\cdot}) = Q_i^{j\to\mathcal{M}} \quad ; \quad (\mathfrak{S}^+)^{-1}_{ji}(e^{-z\cdot}) = \frac{1}{\kappa^+_{ii}}\sum_k p_{ik}e^{ik}\sum_{n=0}^{\infty} {}_i\mathfrak{P}^n_{kj}$$

From [9] we know that convolution is well-defined for

a) bounded measures,

b) distributions supported on the same half-line,

c) two distributions if one of them has compact support.

But to justify all the steps in the solution of 2.2 we also need the following: for $\mu = \{\mu_i\}$ a vector measure supported on $(-\infty, 0]$, such that $\sum_i \mu_i(\mathbf{R}^-) < \infty$, the formula $\sum_i (\mathfrak{S}^+)^{-1}_{ji} * \mu_i$ defines a Radon measure (we used $\mu = \Sigma, \mathbf{r}_\lambda$). To show this it suffices, by 3.1 (1), to prove that for any compact K containing zero

$$\sum_i (\mathfrak{S}^+)^{-1}_{ii} * \mu_i(K) = \sum_i \int_{-\infty}^0 \mu_i(dy)\mathbf{E}_i\left[\int_0^\infty 1_{(Y_t\in K-y)}dL_t^{\mathcal{M}^i}\right] < \infty$$

However, by the strong Markov property at first entry into $K - y$ and translation invariance, we can replace $K - y$ by K. Using $L^{\mathcal{M}^i} \le L^{\mathcal{M}}$ now gives the bound

$$\sum_i \mu_i(\mathbf{R}^-)\mathbf{E}_i\left[\int_0^\infty 1_{(Y_t\in K)}dL_t^{\mathcal{M}}\right] = \sum_i \mu_i(\mathbf{R}^-)\mathbf{E}_{\mu^\partial}\left[\int_0^\infty 1_{(Y_t\in K)}dL_t^{\mathcal{M}}\right]$$

with $\mu^\partial\{i\} = \mu_i(\mathbf{R}^-)/\sum_i \mu_i(\mathbf{R}^-)$ as initial probability. The expectation is finite since the Markov process $(Y_{\sigma\mathcal{M}}, N_{\sigma\mathcal{M}})$ is transient, and the result is proved.

The same reasoning gives a proof of 3.1 (2), where for fixed compact set K we need $\lim_{\downarrow 0} \lambda \sum_i (\mathfrak{S}^+)^{-1}_{ji} * r^i_\lambda(K) = 0$. Taking $\lambda \le 1$ we bound by

$$\sum_i r^i_\lambda(\mathbf{R}^-)\mathbf{E}_i\left[\int_0^\infty 1_{(Y_t\in K)}dL_t^{\mathcal{M}}\right] \le \sum_i r^i_\lambda(\mathbf{R}^-)\mathbf{E}_{\mu_1}\left[\int_0^\infty 1_{(Y_t\in K)}dL_t^{\mathcal{M}}\right]$$

with $\mu_1\{i\} = r^i_1(\mathbf{R}^-)/\sum_i r^i_\lambda(\mathbf{R}^-)$ and, since the expectation is bounded, it suffices to show $\lim_{\lambda\downarrow 0} \lambda \sum_i r^i_\lambda(\mathbf{R}^-)$. Recalling $r^i_\lambda(dx) = \mathbf{E}[\int_0^\tau e^{-\lambda L_t^\partial} 1_{(Y_t\in dx)}dL_t^i]$ we get

$$\lambda\sum_i r^i_\lambda(\mathbf{R}^-) = \sum_i \lambda\mathbf{E}\left[\int_0^\tau e^{-\lambda L_t^\partial}dL_t^i\right] = \lambda\mathbf{E}\left[\int_0^\tau e^{-\lambda L_t^\partial}dL_t^\partial\right] = \mathbf{E}\left[1 - e^{-\lambda L_\tau^\partial}\right]$$

This converges to zero by dominated convergence and $\tau < \infty$ a.s.

Turning now to the proof of 3.1 (3), we will write $(\mathfrak{S}^-)^{-1} = \mathfrak{D} - \mathfrak{O}$ where $\mathfrak{D}_{ii} = (\mathfrak{S}^-)^{-1}_{ii}$ is diagonal. The proof exploits the expansion $\mathfrak{S}^- = \mathfrak{D}^{-1}\sum_{n\geq 0}\mathfrak{D}^{-n}\mathfrak{O}^n$; we show:

d) $\lim_{z\downarrow -\infty} z^{-1}\mathfrak{D}^{-1}(e^{-z\cdot}) = 0$ pointwise.

e) $\sum_{n\geq 0}\mathfrak{D}^{-n}\mathfrak{O}^n(e^{z\cdot})$ converges in the supremum norm when $\Re z < 0$.

Let us start with d) where it suffices to see that $\lim_{z\downarrow -\infty} zQ_i^{ii} = \infty$ for each $i \in \partial$. Consider the Lévy process $\bar{Y}$ obtained by deleting from $Y = V_{\sigma^\bullet}$ all excursions from $\mathcal{M}^i$ which enter $\mathcal{M}\setminus\mathcal{M}^i$. By a duality argument (meaning here decomposition at the minimum) $1/Q_i^{ii}$ appears as the Laplace exponent of the positive factor $\bar{Y}^+$, and the result follows.

For the proof of e) it suffices to see that the entries of $\mathfrak{D}^{-1}\mathfrak{O}$ are uniformly strictly less than one. But these have the form $Q_i^{j\to\mathcal{M}}/Q_i^{i\to\mathcal{M}}$ where the only contribution comes when N visits i; we write T_i for the time of first passage. By the strong Markov property $Q_i^{j\to\mathcal{M}}$ decomposes as

$$Q_i^{j\to\mathcal{M}}\left[e^{-zY_{T_i}}\mathbf{E}_i\left[\int_0^\infty e^{-z\tilde{Y}_u}1_{(\tilde{N}_u=i,\tilde{Y}_u\leq -Y_{T_i})}du\right];T_i<\zeta\right]$$

with $(Y,N)\sim(\tilde{Y},\tilde{N})$ independent. Evaluating the expectation by using the excursion theorem for semi-regenerative sets (cf. remark at the end of §5) we have

$$\mathbf{E}_i\left[\int_0^\infty e^{-z\tilde{Y}(\sigma_u^i)}1_{(\tilde{Y}(\sigma_u^i)\leq -Y_{T_i})}du\right]Q_i^{i\to\mathcal{M}}$$

which gives the ratio $Q_i^{j\to\mathcal{M}}/Q_i^{i\to\mathcal{M}}$ in the form

$$Q_i^{j\to\mathcal{M}}\left[e^{-zY_{T_i}}\mathbf{E}_i\left[\int_0^\infty e^{-z\tilde{Y}(\sigma_u^i)}1_{(\tilde{Y}(\sigma_u^i)\leq -Y_{T_i})}du\right];T_i<\zeta\right]$$

As $z\downarrow -\infty$ this converges to zero and so we obtain a proof of 3.1(3) when ∂ is finite.

REFERENCES

1. **Bingham N.** *Fluctuation theory in continuous time. Adv. Appl. Prob.* 7 (1975) 705-766.

2. **Dieudonne J.** *Eléments d'analyse, Tome 7.* Gauthier-Villars, Paris 1978.

3. **Feller W.** *An Introduction to Probability Theory and its Applications, Vol. 1. 3rd Ed.*, Wiley, New York 1968.

4. **Greenwood P. and Pitman J.** *Fluctuation identities for Lévy processes and splitting at the maximum. Adv. Appl. Prob.*, 12 (1980) 893-902.

5. **Jones D.S.** *Wiener-Hopf splitting of a* 2×2 *matrix. Proc. Roy. Soc. Lond. A.* 434 (1991) 419-433.

6. **Maisonneuve B.** *Systèmes régénératifs. Astérisque No. 15*, Société Mathématique de France, Paris 1974.

7. **Muskelishvili N.I.** ***Singular integral equations.*** **Noordhoff, Groningen 1953.**

8. **Ray D.B.** ***Stable processes with absorbing barrier. Trans. Amer. Math. Soc.*** **89 (1958) 16-24.**

9. **Schwartz L.** ***Théorie des distributions.*** **Hermann, Paris 1966.**

10. **Williams D.** ***Some aspects of Wiener-Hopf factorization. Phil. Trans. R. Soc. Lond. A.*** **335 (1991) 593-608.**

Sur les temps de coupure des marches aléatoires réfléchies

Sandrine Lagaize
UMR 6628-MAPMO
Université d'Orléans et CNRS
B.P. 6759
45067 Orléans Cedex 2
FRANCE

Abstract : The paths of bilateral reflected simple random walks in dimension $d \geq 5$ have a.s. infinitely many cut times.

L'étude des points et temps de coupure du mouvement brownien en diverses dimensions est ancienne ([2]) et de récents travaux de Burdzy ([1]) et Lawler ([4]) ont provoqué un regain d'intérêt pour ces questions. Dans son livre ([3]), Lawler s'intéresse aux marches aléatoires simples. S'il apparaît que la récurrence empêche l'existence de temps de coupure en dimension un et deux, la situation est moins claire en dimension trois et quatre ([5]), et il est plus simple de considérer des marches bilatères, c'est-à-dire indexées par l'ensemble des entiers relatifs. Dans ce cadre, Lawler montre que, presque sûrement, il n'existe pas de temps de coupure en dimension $d \leq 4$, et qu'il y en a une infinité si $d \geq 5$. C'est ce résultat que nous nous proposons d'établir pour la marche aléatoire bilatère réfléchie sur un hyperplan. La ligne générale de la démonstration de Lawler nous servira de guide; cependant la réflexion va introduire un certain nombre de difficultés spécifiques.

Soit $(X_n, n \in \mathbb{Z})$ une suite de variables aléatoires indépendantes, identiquement distribuées dans $\mathbb{Z}^d$ de loi définie par

$$P(X_n = e) = \frac{1}{2d}$$

pour tout $n \in \mathbb{Z}$ et tout e dans $\mathbb{Z}^d$ vérifiant $|e| = 1$. On définit la *marche aléatoire simple bilatère* issue de $x \in \mathbb{Z}^d$ en posant

$$\begin{aligned} S_0 &= x \\ S_n &= S_{n-1} + X_n \quad \text{pour tout } n \in \mathbb{Z}. \end{aligned}$$

Pour tout $y = (y^1, \cdots, y^d) \in \mathbb{Z}^d$, on note

$$\begin{aligned} y^{|d|} &= (y^1, \cdots, y^{d-1}, |y^d|) \\ \widetilde{y} &= (y^1, \cdots, y^{d-1}, -y^d). \end{aligned}$$

La *marche aléatoire simple bilatère réfléchie* W associée à $S = (S_n)$ est définie par

$$W_n = S_n^{|d|} \quad \text{pour tout } n \in \mathbb{Z}.$$

Pour tout intervalle I d'entiers relatifs, on pose

$$SI = \{S_j \ ; \ j \in I\} \quad \text{et} \quad WI = \{W_j \ ; \ j \in I\}$$

et on dit que $j \in \mathbb{Z}$ est un *temps de coupure* de S, resp. W, si

$$S]-\infty, j] \cap S]j, +\infty] = \emptyset, \quad \text{resp.} \quad W]-\infty, j] \cap W]j, +\infty] = \emptyset.$$

Comme un temps de coupure de W est nécessairement un temps de coupure de S, le résultat de Lawler entraîne immédiatement que si $d \leq 4$, p.s., il n'existe aucun temps de coupure de W. La réflexion provoquant des intersections supplémentaires, il n'est pas a priori évident, quand $d \geq 5$, que le nombre de temps de coupure de W reste p.s infini. C'est ce que nous allons établir.

Théorème : *Si $d \geq 5$, pour tout $x \in (\mathbb{Z}^d)^+ := \mathbb{Z}^{d-1} \times \mathbb{N}$, le nombre de temps de coupure de la marche aléatoire simple bilatère réfléchie issue de x est p.s. infini.*

<u>1re étape</u> : Choisissons x quelconque dans $(\mathbb{Z}^d)^+$. Notons P^x la probabilité pour laquelle $S_0 = x$ et posons

$$\begin{aligned} g(x) &= P^x(W]-\infty, 0] \cap W]0, +\infty[= \emptyset) \\ R &= \sum_{i=0}^{\infty} \sum_{j=0}^{\infty} \mathbb{1}(W_{-i} = \dot{W}_j). \end{aligned}$$

D'après le théorème local de la limite centrale ([3], p.14),

$$E^x(R) \leq \sum_{i=0}^{\infty} \sum_{j=0}^{\infty} [P^x(S_{-i} = S_j) + P^x(S_{-i} = \tilde{S}_j)] \leq 2 \sum_{i=0}^{\infty} \sum_{j=0}^{\infty} P^0(S_{i+j} = 0) < \infty.$$

On remarquera que si $x^d = y^d$ alors $g(x) = g(y)$.
Commençons par vérifier que g n'est pas identiquement nulle. D'après le résultat ci-dessus, P^x-p.s. il n'existe qu'un nombre fini de couples (k, l) dans $\mathbb{N}^2$ tels que $W_{-k} = W_l$, et

$$\sum_{i=0}^{\infty} \sum_{j=0}^{\infty} P^x(\{W_{-i} = W_j\} \cap \{W]-\infty, -i] \cap W]j, +\infty[= \emptyset\}) \geq 1.$$

D'autre part,

$$\sum_{i=0}^{\infty}\sum_{j=0}^{\infty} P^x(\{W_{-i} = W_j\} \cap \{W]-\infty,-i] \cap W]j,+\infty[=\emptyset\})$$

$$= \sum_{y\in(\mathbb{Z}^d)^+}\sum_{i=0}^{\infty}\sum_{j=0}^{\infty} P^x(W_{-i} = W_j = y)P^y(W]-\infty,0] \cap W]0,+\infty[=\emptyset).$$

On en déduit qu'il existe y_0 tel que $g(y_0) > 0$.
2e étape : On définit la suite bilatère des temps aléatoires suivants

$$\begin{aligned} T_0 &= 0 \\ T_k &= \inf\{j > T_{k-1} \ ; \ S_j^d = x^d\} \quad \text{pour tout } k \geq 1 \\ T_k &= \sup\{j < T_{k+1} \ ; \ S_j^d = x^d\} \quad \text{pour tout } k \leq -1. \end{aligned}$$

Pour $n \geq 1$ et $i \geq 1$, on note

$$\begin{aligned} V_{i,n} &= \{W]-\infty, T_{(2i-1)n}] \cap W]T_{(2i-1)n}, +\infty[=\emptyset\} \\ W_{i,n} &= \{W[T_{(2i-2)n}, T_{(2i-1)n}] \cap W]T_{(2i-1)n}, T_{2in}] = \emptyset\} \\ \text{et} \quad g_n(x) &= P^x(W[T_{-n}, 0] \cap W]0, T_n] = \emptyset\}. \end{aligned}$$

La récurrence de la marche (S_n^d) montre que $P^x(T_k < \infty) = 1$. Pour tout $k \geq 1$, posons

$$W'_n = W_{T_k+n} - W_{T_k} + x \quad n \in \mathbb{Z}.$$

Il est facile de vérifier que pour P^x, (W_n) et (W'_n) ont même loi, et par conséquent, pour tout $n \geq 1$ et pour tout $i \geq 1$,

$$\begin{aligned} P^x(W_{i,n}) &= g_n(x) \\ P^x(V_{i,n}) &= g(x). \end{aligned}$$

On vérifie facilement aussi que les $(W_{i,n} \ ; \ i \geq 1)$ sont indépendants pour P^x.
3e étape : Nous sommes maintenant en mesure, en reprenant la démonstration de Lawler, de prouver le résultat proposé lorsque la marche est issue d'un point $x \in (\mathbb{Z}^d)^+$ vérifiant $g(x) > 0$. Notons X le nombre de temps de coupure positifs de $W]-\infty,+\infty[$. Pour tout $k \geq 1$, tout $m \geq 1$ et tout $n \geq 1$,

$$\begin{aligned} P^x(X \geq k) &\geq P^x(\sum_{i=1}^{m} \mathbf{1}(V_{i,n}) \geq k) \\ &\geq P^x(\sum_{i=1}^{m} \mathbf{1}(W_{i,n}) \geq k) - P^x(\sum_{i=1}^{m} \mathbf{1}(W_{i,n} \backslash V_{i,n}) \geq 1) \\ &\geq P^x(\sum_{i=1}^{m} \mathbf{1}(W_{i,n}) \geq k) - \sum_{i=1}^{m}(P^x(W_{i,n}) - P^x(V_{i,n})). \end{aligned}$$

D'après l'étape précédente, $\sum_{i=1}^{m} \mathbb{1}(W_{i,n})$ suit une loi binomiale de paramètres m et $g_n(x)$ avec $g_n(x) \geq g(x) > 0$. Il en résulte que pour tout $\varepsilon > 0$, il existe $m \geq 1$ tel que pour tout $n \geq 1$,

$$P^x(\sum_{i=1}^{m} \mathbb{1}(W_{i,n}) \geq k) \geq 1 - \varepsilon.$$

D'autre part, $\sum_{i=1}^{m}(P^x(W_{i,n}) - P^x(V_{i,n})) = m(g_n(x) - g(x))$, et comme $g_n(x) \to g(x)$, on peut choisir n assez grand pour que $P^x(\sum_{i=1}^{m} \mathbb{1}(V_{i,n}) \geq k) \geq 1 - 2\varepsilon$, et on en déduit que

$$P^x(X = +\infty) = 1.$$

Un raisonnement analogue montrerait que la marche réfléchie possède aussi une infinité de temps de coupure négatifs.

4e étape: Soit maintenant x quelconque dans $(\mathbb{Z}^d)^+$. Notons A l'événement "W admet une infinité de temps de coupure ". En sommant sur $m \geq 0$ et sur l'ensemble des chemins possibles C de longueur m qui mènent de x à un premier point x_m vérifiant $x_m^d = y_0^d$, on a

$$\begin{aligned} P^x(A) &= \sum P^x(A \cap (S_0, \cdots, S_m) = C) \\ &= \sum P^{x_m}(A \cap (S_{-m}, \cdots, S_0) = C) \\ &= \sum P^{x_m}((S_{-m}, \cdots, S_0) = C) \\ &= 1. \end{aligned}$$

□

Références

[1] Burdzy, K., (1989). Cut points on Brownian paths. Ann. Proba. **17**, 1012-1036.

[2] Dvoretsky, A., Erdös, P., Kakutani, S., (1950). Double points of paths of Brownian motions in n-space. Acta. Sci. Math. Szeged. **12**, 75-81.

[3] Lawler, G., (1991). *Intersections of Random Walks*. Birkhäuser, Boston.

[4] Lawler, G., (1996). Hausdorff dimension of cut points for Brownian motion. Electronical Journal of Probability **1**, paper n° 2.

[5] Lawler, G., (1996). Cut times for simple random walk. Electronical Journal of Probability **1**, paper n° 13.

NOTE TO CONTRIBUTORS

Contributors to the Séminaire are reminded that their articles should be formatted for the Springer Lecture Notes series.

The dimensions of the printed part of a page without running heads should be:

15.3 cm × 24.2 cm if the font size is 12 pt (or 10 pt magnified 120%),

12.2 cm × 19.3 cm if the font size is 10 pt.

Page numbers and running heads are not needed. Author(s)' address(es) should be indicated, either below the title or at the end of the paper.

Packages of TEX macros are available from the Springer-Verlag site

`http://www.springer.de/author/tex/help-tex.html`

Lecture Notes in Mathematics

For information about Vols. 1–1494 please contact your bookseller or Springer-Verlag

Vol. 1495: J. M. Bony, G. Grubb, L. Hörmander, H. Komatsu, J. Sjöstrand, Microlocal Analysis and Applications. Montecatini Terme, 1989. Editors: L. Cattabriga, L. Rodino. VII, 349 pages. 1991.

Vol. 1496: C. Foias, B. Francis, J. W. Helton, H. Kwakernaak, J. B. Pearson, H_∞-Control Theory. Como, 1990. Editors: E. Mosca, L. Pandolfi. VII, 336 pages. 1991.

Vol. 1497: G. T. Herman, A. K. Louis, F. Natterer (Eds.), Mathematical Methods in Tomography. Proceedings 1990. X, 268 pages. 1991.

Vol. 1498: R. Lang, Spectral Theory of Random Schrödinger Operators. X, 125 pages. 1991.

Vol. 1499: K. Taira, Boundary Value Problems and Markov Processes. IX, 132 pages. 1991.

Vol. 1500: J.-P. Serre, Lie Algebras and Lie Groups. VII, 168 pages. 1992.

Vol. 1501: A. De Masi, E. Presutti, Mathematical Methods for Hydrodynamic Limits. IX, 196 pages. 1991.

Vol. 1502: C. Simpson, Asymptotic Behavior of Monodromy. V, 139 pages. 1991.

Vol. 1503: S. Shokranian, The Selberg-Arthur Trace Formula (Lectures by J. Arthur). VII, 97 pages. 1991.

Vol. 1504: J. Cheeger, M. Gromov, C. Okonek, P. Pansu, Geometric Topology: Recent Developments. Editors: P. de Bartolomeis, F. Tricerri. VII, 197 pages. 1991.

Vol. 1505: K. Kajitani, T. Nishitani, The Hyperbolic Cauchy Problem. VII, 168 pages. 1991.

Vol. 1506: A. Buium, Differential Algebraic Groups of Finite Dimension. XV, 145 pages. 1992.

Vol. 1507: K. Hulek, T. Peternell, M. Schneider, F.-O. Schreyer (Eds.), Complex Algebraic Varieties. Proceedings, 1990. VII, 179 pages. 1992.

Vol. 1508: M. Vuorinen (Ed.), Quasiconformal Space Mappings. A Collection of Surveys 1960-1990. IX, 148 pages. 1992.

Vol. 1509: J. Aguadé, M. Castellet, F. R. Cohen (Eds.), Algebraic Topology - Homotopy and Group Cohomology. Proceedings, 1990. X, 330 pages. 1992.

Vol. 1510: P. P. Kulish (Ed.), Quantum Groups. Proceedings, 1990. XII, 398 pages. 1992.

Vol. 1511: B. S. Yadav, D. Singh (Eds.), Functional Analysis and Operator Theory. Proceedings, 1990. VIII, 223 pages. 1992.

Vol. 1512: L. M. Adleman, M.-D. A. Huang, Primality Testing and Abelian Varieties Over Finite Fields. VII, 142 pages. 1992.

Vol. 1513: L. S. Block, W. A. Coppel, Dynamics in One Dimension. VIII, 249 pages. 1992.

Vol. 1514: U. Krengel, K. Richter, V. Warstat (Eds.), Ergodic Theory and Related Topics III, Proceedings, 1990. VIII, 236 pages. 1992.

Vol. 1515: E. Ballico, F. Catanese, C. Ciliberto (Eds.), Classification of Irregular Varieties. Proceedings, 1990. VII, 149 pages. 1992.

Vol. 1516: R. A. Lorentz, Multivariate Birkhoff Interpolation. IX, 192 pages. 1992.

Vol. 1517: K. Keimel, W. Roth, Ordered Cones and Approximation. VI, 134 pages. 1992.

Vol. 1518: H. Stichtenoth, M. A. Tsfasman (Eds.), Coding Theory and Algebraic Geometry. Proceedings, 1991. VIII, 223 pages. 1992.

Vol. 1519: M. W. Short, The Primitive Soluble Permutation Groups of Degree less than 256. IX, 145 pages. 1992.

Vol. 1520: Yu. G. Borisovich, Yu. E. Gliklikh (Eds.), Global Analysis – Studies and Applications V. VII, 284 pages. 1992.

Vol. 1521: S. Busenberg, B. Forte, H. K. Kuiken, Mathematical Modelling of Industrial Process. Bari, 1990. Editors: V. Capasso, A. Fasano. VII, 162 pages. 1992.

Vol. 1522: J.-M. Delort, F. B. I. Transformation. VII, 101 pages. 1992.

Vol. 1523: W. Xue, Rings with Morita Duality. X, 168 pages. 1992.

Vol. 1524: M. Coste, L. Mahé, M.-F. Roy (Eds.), Real Algebraic Geometry. Proceedings, 1991. VIII, 418 pages. 1992.

Vol. 1525: C. Casacuberta, M. Castellet (Eds.), Mathematical Research Today and Tomorrow. VII, 112 pages. 1992.

Vol. 1526: J. Azéma, P. A. Meyer, M. Yor (Eds.), Séminaire de Probabilités XXVI. X, 633 pages. 1992.

Vol. 1527: M. I. Freidlin, J.-F. Le Gall, Ecole d'Eté de Probabilités de Saint-Flour XX – 1990. Editor: P. L. Hennequin. VIII, 244 pages. 1992.

Vol. 1528: G. Isac, Complementarity Problems. VI, 297 pages. 1992.

Vol. 1529: J. van Neerven, The Adjoint of a Semigroup of Linear Operators. X, 195 pages. 1992.

Vol. 1530: J. G. Heywood, K. Masuda, R. Rautmann, S. A. Solonnikov (Eds.), The Navier-Stokes Equations II – Theory and Numerical Methods. IX, 322 pages. 1992.

Vol. 1531: M. Stoer, Design of Survivable Networks. IV, 206 pages. 1992.

Vol. 1532: J. F. Colombeau, Multiplication of Distributions. X, 184 pages. 1992.

Vol. 1533: P. Jipsen, H. Rose, Varieties of Lattices. X, 162 pages. 1992.

Vol. 1534: C. Greither, Cyclic Galois Extensions of Commutative Rings. X, 145 pages. 1992.

Vol. 1535: A. B. Evans, Orthomorphism Graphs of Groups. VIII, 114 pages. 1992.

Vol. 1536: M. K. Kwong, A. Zettl, Norm Inequalities for Derivatives and Differences. VII, 150 pages. 1992.

Vol. 1537: P. Fitzpatrick, M. Martelli, J. Mawhin, R. Nussbaum, Topological Methods for Ordinary Differential Equations. Montecatini Terme, 1991. Editors: M. Furi, P. Zecca. VII, 218 pages. 1993.

Vol. 1538: P.-A. Meyer, Quantum Probability for Probabilists. X, 287 pages. 1993.

Vol. 1539: M. Coornaert, A. Papadopoulos, Symbolic Dynamics and Hyperbolic Groups. VIII, 138 pages. 1993.

Vol. 1540: H. Komatsu (Ed.), Functional Analysis and Related Topics, 1991. Proceedings. XXI, 413 pages. 1993.

Vol. 1541: D. A. Dawson, B. Maisonneuve, J. Spencer, Ecole d' Eté de Probabilités de Saint-Flour XXI - 1991. Editor: P. L. Hennequin. VIII, 356 pages. 1993.

Vol. 1542: J.Fröhlich, Th.Kerler, Quantum Groups, Quantum Categories and Quantum Field Theory. VII, 431 pages. 1993.

Vol. 1543: A. L. Dontchev, T. Zolezzi, Well-Posed Optimization Problems. XII, 421 pages. 1993.

Vol. 1544: M.Schürmann, White Noise on Bialgebras. VII, 146 pages. 1993.

Vol. 1545: J. Morgan, K. O'Grady, Differential Topology of Complex Surfaces. VIII, 224 pages. 1993.

Vol. 1546: V. V. Kalashnikov, V. M. Zolotarev (Eds.), Stability Problems for Stochastic Models. Proceedings, 1991. VIII. 229 pages. 1993.

Vol. 1547: P. Harmand, D. Werner, W. Werner, M-ideals in Banach Spaces and Banach Algebras. VIII, 387 pages. 1993.

Vol. 1548: T. Urabe, Dynkin Graphs and Quadrilateral Singularities. VI, 233 pages. 1993.

Vol. 1549: G. Vainikko, Multidimensional Weakly Singular Integral Equations. XI, 159 pages. 1993.

Vol. 1550: A. A. Gonchar, E. B. Saff (Eds.), Methods of Approximation Theory in Complex Analysis and Mathematical Physics IV, 222 pages, 1993.

Vol. 1551: L. Arkeryd, P. L. Lions, P.A. Markowich, S.R. S. Varadhan. Nonequilibrium Problems in Many-Particle Systems. Montecatini, 1992. Editors: C. Cercignani, M. Pulvirenti. VII, 158 pages 1993.

Vol. 1552: J. Hilgert, K.-H. Neeb, Lie Semigroups and their Applications. XII, 315 pages. 1993.

Vol. 1553: J.-L- Colliot-Thélène, J. Kato, P. Vojta. Arithmetic Algebraic Geometry. Trento, 1991. Editor: E. Ballico. VII, 223 pages. 1993.

Vol. 1554: A. K. Lenstra, H. W. Lenstra, Jr. (Eds.), The Development of the Number Field Sieve. VIII, 131 pages. 1993.

Vol. 1555: O. Liess, Conical Refraction and Higher Microlocalization. X, 389 pages. 1993.

Vol. 1556: S. B. Kuksin, Nearly Integrable Infinite-Dimensional Hamiltonian Systems. XXVII, 101 pages. 1993.

Vol. 1557: J. Azéma, P. A. Meyer, M. Yor (Eds.), Séminaire de Probabilités XXVII. VI, 327 pages. 1993.

Vol. 1558: T. J. Bridges, J. E. Furter, Singularity Theory and Equivariant Symplectic Maps. VI, 226 pages. 1993.

Vol. 1559: V. G. Sprindžuk, Classical Diophantine Equations. XII, 228 pages. 1993.

Vol. 1560: T. Bartsch, Topological Methods for Variational Problems with Symmetries. X, 152 pages. 1993.

Vol. 1561: I. S. Molchanov, Limit Theorems for Unions of Random Closed Sets. X, 157 pages. 1993.

Vol. 1562: G. Harder, Eisensteinkohomologie und die Konstruktion gemischter Motive. XX, 184 pages. 1993.

Vol. 1563: E. Fabes, M. Fukushima, L. Gross, C. Kenig, M. Röckner, D. W. Stroock, Dirichlet Forms. Varenna, 1992. Editors: G. Dell'Antonio, U. Mosco. VII, 245 pages. 1993.

Vol. 1564: J. Jorgenson, S. Lang, Basic Analysis of Regularized Series and Products. IX, 122 pages. 1993.

Vol. 1565: L. Boutet de Monvel, C. De Concini, C. Procesi, P. Schapira, M. Vergne. D-modules, Representation Theory, and Quantum Groups. Venezia, 1992. Editors: G. Zampieri, A. D'Agnolo. VII, 217 pages. 1993.

Vol. 1566: B. Edixhoven, J.-H. Evertse (Eds.), Diophantine Approximation and Abelian Varieties. XIII, 127 pages. 1993.

Vol. 1567: R. L. Dobrushin, S. Kusuoka, Statistical Mechanics and Fractals. VII, 98 pages. 1993.

Vol. 1568: F. Weisz, Martingale Hardy Spaces and their Application in Fourier Analysis. VIII, 217 pages. 1994.

Vol. 1569: V. Totik, Weighted Approximation with Varying Weight. VI, 117 pages. 1994.

Vol. 1570: R. deLaubenfels, Existence Families, Functional Calculi and Evolution Equations. XV, 234 pages. 1994.

Vol. 1571: S. Yu. Pilyugin. The Space of Dynamical Systems with the C^0-Topology. X, 188 pages. 1994.

Vol. 1572: L. Göttsche, Hilbert Schemes of Zero-Dimensional Subschemes of Smooth Varieties. IX, 196 pages. 1994.

Vol. 1573: V. P. Havin, N. K. Nikolski (Eds.), Linear and Complex Analysis – Problem Book 3 – Part I. XXII, 489 pages. 1994.

Vol. 1574: V. P. Havin, N. K. Nikolski (Eds.), Linear and Complex Analysis – Problem Book 3 – Part II. XXII, 507 pages. 1994.

Vol. 1575: M. Mitrea, Clifford Wavelets, Singular Integrals, and Hardy Spaces. XI, 116 pages. 1994.

Vol. 1576: K. Kitahara, Spaces of Approximating Functions with Haar-Like Conditions. X, 110 pages. 1994.

Vol. 1577: N. Obata, White Noise Calculus and Fock Space. X, 183 pages. 1994.

Vol. 1578: J. Bernstein, V. Lunts, Equivariant Sheaves and Functors. V, 139 pages. 1994.

Vol. 1579: N. Kazamaki, Continuous Exponential Martingales and *BMO*. VII, 91 pages. 1994.

Vol. 1580: M. Milman, Extrapolation and Optimal Decompositions with Applications to Analysis. XI, 161 pages. 1994.

Vol. 1581: D. Bakry, R. D. Gill, S. A. Molchanov, Lectures on Probability Theory. Editor: P. Bernard. VIII, 420 pages. 1994.

Vol. 1582: W. Balser, From Divergent Power Series to Analytic Functions. X, 108 pages. 1994.

Vol. 1583: J. Azéma, P. A. Meyer, M. Yor (Eds.), Séminaire de Probabilités XXVIII. VI, 334 pages. 1994.

Vol. 1584: M. Brokate, N. Kenmochi, I. Müller, J. F. Rodriguez, C. Verdi, Phase Transitions and Hysteresis. Montecatini Terme, 1993. Editor: A. Visintin. VII, 291 pages. 1994.

Vol. 1585: G. Frey (Ed.), On Artin's Conjecture for Odd 2-dimensional Representations. VIII, 148 pages. 1994.

Vol. 1586: R. Nillsen, Difference Spaces and Invariant Linear Forms. XII, 186 pages. 1994.

Vol. 1587: N. Xi, Representations of Affine Hecke Algebras. VIII, 137 pages. 1994.

Vol. 1588: C. Scheiderer, Real and Étale Cohomology. XXIV, 273 pages. 1994.

Vol. 1589: J. Bellissard, M. Degli Esposti, G. Forni, S. Graffi, S. Isola, J. N. Mather, Transition to Chaos in Classical and Quantum Mechanics. Montecatini Terme, 1991. Editor: S. Graffi. VII, 192 pages. 1994.

Vol. 1590: P. M. Soardi, Potential Theory on Infinite Networks. VIII, 187 pages. 1994.

Vol. 1591: M. Abate, G. Patrizio, Finsler Metrics – A Global Approach. IX, 180 pages. 1994.

Vol. 1592: K. W. Breitung, Asymptotic Approximations for Probability Integrals. IX, 146 pages. 1994.

Vol. 1593: J. Jorgenson & S. Lang, D. Goldfeld, Explicit Formulas for Regularized Products and Series. VIII, 154 pages. 1994.

Vol. 1594: M. Green, J. Murre, C. Voisin, Algebraic Cycles and Hodge Theory. Torino, 1993. Editors: A. Albano, F. Bardelli. VII, 275 pages. 1994.

Vol. 1595: R.D.M. Accola, Topics in the Theory of Riemann Surfaces. IX, 105 pages. 1994.

Vol. 1596: L. Heindorf, L. B. Shapiro, Nearly Projective Boolean Algebras. X, 202 pages. 1994.

Vol. 1597: B. Herzog, Kodaira-Spencer Maps in Local Algebra. XVII, 176 pages. 1994.

Vol. 1598: J. Berndt, F. Tricerri, L. Vanhecke, Generalized Heisenberg Groups and Damek-Ricci Harmonic Spaces. VIII, 125 pages. 1995.

Vol. 1599: K. Johannson, Topology and Combinatorics of 3-Manifolds. XVIII, 446 pages. 1995.

Vol. 1600: W. Narkiewicz, Polynomial Mappings. VII, 130 pages. 1995.

Vol. 1601: A. Pott, Finite Geometry and Character Theory. VII, 181 pages. 1995.

Vol. 1602: J. Winkelmann, The Classification of Three-dimensional Homogeneous Complex Manifolds. XI, 230 pages. 1995.

Vol. 1603: V. Ene, Real Functions – Current Topics. XIII, 310 pages. 1995.

Vol. 1604: A. Huber, Mixed Motives and their Realization in Derived Categories. XV, 207 pages. 1995.

Vol. 1605: L. B. Wahlbin, Superconvergence in Galerkin Finite Element Methods. XI, 166 pages. 1995.

Vol. 1606: P.-D. Liu, M. Qian, Smooth Ergodic Theory of Random Dynamical Systems. XI, 221 pages. 1995.

Vol. 1607: G. Schwarz, Hodge Decomposition – A Method for Solving Boundary Value Problems. VII, 155 pages. 1995.

Vol. 1608: P. Biane, R. Durrett, Lectures on Probability Theory. Editor: P. Bernard. VII, 210 pages. 1995.

Vol. 1609: L. Arnold, C. Jones, K. Mischaikow, G. Raugel, Dynamical Systems. Montecatini Terme, 1994. Editor: R. Johnson. VIII, 329 pages. 1995.

Vol. 1610: A. S. Üstünel, An Introduction to Analysis on Wiener Space. X, 95 pages. 1995.

Vol. 1611: N. Knarr, Translation Planes. VI, 112 pages. 1995.

Vol. 1612: W. Kühnel, Tight Polyhedral Submanifolds and Tight Triangulations. VII, 122 pages. 1995.

Vol. 1613: J. Azéma, M. Emery, P. A. Meyer, M. Yor (Eds.), Séminaire de Probabilités XXIX. VI, 326 pages. 1995.

Vol. 1614: A. Koshelev, Regularity Problem for Quasilinear Elliptic and Parabolic Systems. XXI, 255 pages. 1995.

Vol. 1615: D. B. Massey, Le Cycles and Hypersurface Singularities. XI, 131 pages. 1995.

Vol. 1616: I. Moerdijk, Classifying Spaces and Classifying Topoi. VII, 94 pages. 1995.

Vol. 1617: V. Yurinsky, Sums and Gaussian Vectors. XI, 305 pages. 1995.

Vol. 1618: G. Pisier, Similarity Problems and Completely Bounded Maps. VII, 156 pages. 1996.

Vol. 1619: E. Landvogt, A Compactification of the Bruhat-Tits Building. VII, 152 pages. 1996.

Vol. 1620: R. Donagi, B. Dubrovin, E. Frenkel, E. Previato, Integrable Systems and Quantum Groups. Montecatini Terme, 1993. Editors:M. Francaviglia, S. Greco. VIII, 488 pages. 1996.

Vol. 1621: H. Bass, M. V. Otero-Espinar, D. N. Rockmore, C. P. L. Tresser. Cyclic Renormalization and Auto-morphism Groups of Rooted Trees. XXI, 136 pages. 1996.

Vol. 1622: E. D. Farjoun, Cellular Spaces, Null Spaces and Homotopy Localization. XIV, 199 pages. 1996.

Vol. 1623: H.P. Yap, Total Colourings of Graphs. VIII, 131 pages. 1996.

Vol. 1624: V. Brinzanescu, Holomorphic Vector Bundles over Compact Complex Surfaces. X, 170 pages. 1996.

Vol.1625: S. Lang, Topics in Cohomology of Groups. VII, 226 pages. 1996.

Vol. 1626: J. Azéma, M. Emery, M. Yor (Eds.), Séminaire de Probabilités XXX. VIII, 382 pages. 1996.

Vol. 1627: C. Graham, Th. G. Kurtz, S. Méléard, Ph. E. Protter, M. Pulvirenti, D. Talay, Probabilistic Models for Nonlinear Partial Differential Equations. Montecatini Terme, 1995. Editors: D. Talay, L. Tubaro. X, 301 pages. 1996.

Vol. 1628: P.-H. Zieschang, An Algebraic Approach to Association Schemes. XII, 189 pages. 1996.

Vol. 1629: J. D. Moore, Lectures on Seiberg-Witten Invariants. VII, 105 pages. 1996.

Vol. 1630: D. Neuenschwander, Probabilities on the Heisenberg Group: Limit Theorems and Brownian Motion. VIII, 139 pages. 1996.

Vol. 1631: K. Nishioka, Mahler Functions and Transcendence. VIII, 185 pages. 1996.

Vol. 1658: A. Pumarino, J. A. Rodrıguez, Coexistence and Persistence of Strange Attractors. VIII, 195 pages. 1997.

Vol. 1659: V. Kozlov, V. Maz'ya, Theory of a Higher-Order Sturm-Liouville Equation. XI, 140 pages. 1997.

Vol. 1660: M. Bardi, M. G. Crandall, L. C. Evans, H. M. Soner, P. E. Souganidis, Viscosity Solutions and Applications. Montecatini Terme, 1995. Editors: I. Capuzzo Dolcetta, P. L. Lions. IX, 259 pages. 1997.

Vol. 1661: A. Tralle, J. Oprea, Symplectic Manifolds with no Kähler Structure. VIII, 207 pages. 1997.

Vol. 1662: J. W. Rutter, Spaces of Homotopy Self-Equivalences – A Survey. IX, 170 pages. 1997.

Vol. 1632: A. Kushkuley, Z. Balanov, Geometric Methods in Degree Theory for Equivariant Maps. VII, 136 pages. 1996.

Vol.1633: H. Aikawa, M. Essén, Potential Theory – Selected Topics. IX, 200 pages.1996.

Vol. 1634: J. Xu, Flat Covers of Modules. IX, 161 pages. 1996.

Vol. 1635: E. Hebey, Sobolev Spaces on Riemannian Manifolds. X, 116 pages. 1996.

Vol. 1636: M. A. Marshall, Spaces of Orderings and Abstract Real Spectra. VI, 190 pages. 1996.

Vol. 1637: B. Hunt, The Geometry of some special Arithmetic Quotients. XIII, 332 pages. 1996.

Vol. 1638: P. Vanhaecke, Integrable Systems in the realm of Algebraic Geometry. VIII, 218 pages. 1996.

Vol. 1639: K. Dekimpe, Almost-Bieberbach Groups: Affine and Polynomial Structures. X, 259 pages. 1996.

Vol. 1640: G. Boillat, C. M. Dafermos, P. D. Lax, T. P. Liu, Recent Mathematical Methods in Nonlinear Wave Propagation. Montecatini Terme, 1994. Editor: T. Ruggeri. VII, 142 pages. 1996.

Vol. 1641: P. Abramenko, Twin Buildings and Applications to S-Arithmetic Groups. IX, 123 pages. 1996.

Vol. 1642: M. Puschnigg, Asymptotic Cyclic Cohomology. XXII, 138 pages. 1996.

Vol. 1643: J. Richter-Gebert, Realization Spaces of Polytopes. XI, 187 pages. 1996.

Vol. 1644: A. Adler, S. Ramanan, Moduli of Abelian Varieties. VI, 196 pages. 1996.

Vol. 1645: H. W. Broer, G. B. Huitema, M. B. Sevryuk, Quasi-Periodic Motions in Families of Dynamical Systems. XI, 195 pages. 1996.

Vol. 1646: J.-P. Demailly, T. Peternell, G. Tian, A. N. Tyurin, Transcendental Methods in Algebraic Geometry. Cetraro, 1994. Editors: F. Catanese, C. Ciliberto. VII, 257 pages. 1996.

Vol. 1647: D. Dias, P. Le Barz, Configuration Spaces over Hilbert Schemes and Applications. VII. 143 pages. 1996.

Vol. 1648: R. Dobrushin, P. Groeneboom, M. Ledoux, Lectures on Probability Theory and Statistics. Editor: P. Bernard. VIII, 300 pages. 1996.

Vol. 1649: S. Kumar, G. Laumon, U. Stuhler, Vector Bundles on Curves – New Directions. Cetraro, 1995. Editor: M. S. Narasimhan. VII, 193 pages. 1997.

Vol. 1650: J. Wildeshaus, Realizations of Polylogarithms. XI, 343 pages. 1997.

Vol. 1651: M. Drmota, R. F. Tichy, Sequences, Discrepancies and Applications. XIII, 503 pages. 1997.

Vol. 1652: S. Todorcevic, Topics in Topology. VIII, 153 pages. 1997.

Vol. 1653: R. Benedetti, C. Petronio, Branched Standard Spines of 3-manifolds. VIII, 132 pages. 1997.

Vol. 1654: R. W. Ghrist, P. J. Holmes, M. C. Sullivan, Knots and Links in Three-Dimensional Flows. X, 208 pages. 1997.

Vol. 1655: J. Azéma, M. Emery, M. Yor (Eds.), Séminaire de Probabilités XXXI. VIII, 329 pages. 1997.

Vol. 1656: B. Biais, T. Björk, J. Cvitanic, N. El Karoui, E. Jouini, J. C. Rochet, Financial Mathematics. Bressanone, 1996. Editor: W. J. Runggaldier. VII, 316 pages. 1997.

Vol. 1657: H. Reimann, The semi-simple zeta function of quaternionic Shimura varieties. IX, 143 pages. 1997.

Vol. 1663: Y. E. Karpeshina; Perturbation Theory for the Schrödinger Operator with a Periodic Potential. VII, 352 pages. 1997.

Vol. 1664: M. Väth, Ideal Spaces. V, 146 pages. 1997.

Vol. 1665: E. Giné, G. R. Grimmett, L. Saloff-Coste, Lectures on Probability Theory and Statistics 1996. Editor: P. Bernard. X, 424 pages, 1997.

Vol. 1666: M. van der Put, M. F. Singer, Galois Theory of Difference Equations. VII, 179 pages. 1997.

Vol. 1667: J. M. F. Castillo, M. González, Three-space Problems in Banach Space Theory. XII, 267 pages. 1997.

Vol. 1668: D. B. Dix, Large-Time Behavior of Solutions of Linear Dispersive Equations. XIV, 203 pages. 1997.

Vol. 1669: U. Kaiser, Link Theory in Manifolds. XIV, 167 pages. 1997.

Vol. 1670: J. W. Neuberger, Sobolev Gradients and Differential Equations. VIII, 150 pages. 1997.

Vol. 1671: S. Bouc, Green Functors and *G*-sets. VII, 342 pages. 1997.

Vol. 1673: F. D. Grosshans, Algebraic Homogeneous Spaces and Invariant Theory. VI. 148 pages. 1997.

Vol. 1674: G. Klaas, C. R. Leedham-Green, W. Plesken, Linear Pro-*p*-Groups of Finite Width. VIII, 115 pages. 1997.

Vol. 1675: J. E. Yukich, Probability Theory of Classical Euclidean Optimization Problems. X, 152 pages. 1998.

Vol. 1676: P. Cembranos, J. Mendoza, Banach Spaces of Vector-Valued Functions. VIII, 118 pages. 1997.

Vol. 1677: N. Proskurin. Cubic Metaplectic Forms and Theta Functions. VIII, 196 pages. 1998.

Vol. 1678: O. Krupková, The Geometry of Ordinary Variational Equations. X, 251 pages. 1997.

Vol. 1679: K.-G. Grosse-Erdmann, The Blocking Technique. Weighted Mean Operators and Hardy's Inequality. IX, 114 pages. 1998.

Vol. 1680: K.-Z. Li, F. Oort, Moduli of Supersingular Abdelian Varieties. V, 116 pages. 1998.

Vol. 1681: G. J. Wirsching, The Dynamical System Generated by the 3n+1 Function. VII, 158 pages. 1998.

Vol. 1682: H.-D. Alber. Materials with Memory. X, 166 pages. 1998.

Vol. 1683: A. Pomp, The Boundary-Domain Integral Method for Elliptic Systems. XVI. 163 pages. 1998.

Vol. 1684: C. A. Berenstein, P. F. Ebenfelt, S. G. Gindikin, S. Helgason, A. E. Tumanov, Integral Geometry, Radon Transforms and Complex Analysis. Firenze, 1996. Editors: E. Casadio Tarabusi. M. A. Picardello, G. Zampieri. VII, 160 pages. 1998.

Vol. 1685: S. König, A. Zimmermann. Derived Equivalences for Group Rings. X, 146 pages. 1998.

Vol. 1686: J. Azéma, M. Émery, M. Ledoux, M. Yor (Eds.), Séminaire de Probabilités XXXII. VI, 440 pages. 1998.

General Remarks

Lecture Notes are printed by photo-offset from the master-copy delivered in camera-ready form by the authors. For this purpose Springer-Verlag provides technical instructions for the preparation of manuscripts.

Careful preparation of manuscripts will help keep production time short and ensure a satisfactory appearance of the finished book. The actual production of a Lecture Notes volume normally takes approximately 8 weeks.

Authors receive 50 free copies of their book. No royalty is paid on Lecture Notes volumes.

Authors are entitled to purchase further copies of their book and other Springer mathematics books for their personal use, at a discount of 33,3 % directly from Springer-Verlag.

Commitment to publish is made by letter of intent rather than by signing a formal contract. Springer-Verlag secures the copyright for each volume.

Addresses:

Professor A. Dold
Mathematisches Institut
Universität Heidelberg
Im Neuenheimer Feld 288
D-69120 Heidelberg, Germany

Professor F. Takens
Mathematisch Instituut
Rijksuniversiteit Groningen
Postbus 800
NL-9700 AV Groningen
The Netherlands

Professor Bernard Teissier
École Normale Supérieure
45, rue d'Ulm
F-7500 Paris, France

Springer-Verlag, Mathematics Editorial
Tiergartenstr. 17
D-69121 Heidelberg, Germany
Tel.: *49 (6221) 487-410

